Cyclodextrins and Their Complexes

Edited by
Helena Dodziuk

Related Titles

Elias, H.-G.

Macromolecules

Volume 1: Chemical Structures and Syntheses

2005

ISBN 3-527-31172-6

Munk, P., Aminabhavi, T. M.

Introduction to Macromolecular Science

2002

ISBN 0-471-41716-5

Bladon, C.

Pharmaceutical Chemistry

Therapeutic Aspects of Biomacromolecules

2002

ISBN 0-471-49636-7

Cyclodextrins and Their Complexes

Chemistry, Analytical Methods, Applications

Edited by Helena Dodziuk

WILEY-VCH

WILEY-VCH Verlag GmbH & Co. KGaA

The Editor

Prof. Dr. Helena Dodziuk
Inst. of Physical Chemistry
Polish Academy of Sciences
Kasprzaka 44
01-224 Warsaw
Poland
and
Graduate College ICM-IWR
Warsaw (Poland) – Heidelberg (Germany)

Library of Congress Card No.: applied for
British Library Cataloguing-in-Publication Data: A catalogue record for this book is available from the British Library.

Bibliographic information published by Die Deutsche Bibliothek
Die Deutsche Bibliothek lists this publication in the Deutsche Nationalbibliografie; detailed bibliographic data is available in the Internet at ⟨http://dnb.ddb.de⟩.

Printed in the Federal Republic of Germany.
Printed on acid-free paper.

Typesetting Asco Typesetters, Hong Kong
Printing betz-druck GmbH, Darmstadt
Binding Schäffer GmbH, Grünstadt
Cover Design Grafik-Design Schulz, Fußgönheim

ISBN-13: 978-3-527-31280-1
ISBN-10: 3-527-31280-3

Contents

Cyclodextrins and Their Complexes. Edited by Helena Dodziuk

ISBN: 3-527-31280-3

Preface

Cyclodextrins and their complexes constitute a fascinating subject that, due to its complexity, lacked an extended coverage in a monograph for long time. We are happy to bridge this gap with this book. The field has grown immensely, thus discussing or even mentioning many valuable works was not possible in the limited volume. However, we hope to have given an extensive description of all important aspects of research and applications involving cyclodextrins.

The creation of this book would not be possible without the assistance of several people deserving acknowledgement. They have helped by précising its content, choosing contributors, reviewing chapters, and in other ways. I am especially indebted to Prof. H. Ueda, Prof. R. Bilewicz and Prof. K. Harata for their involvement in the project. The critical remarks on specific chapters by Prof. A. Temeriusz, Prof. M. Kańska, Prof. J. Frelek, Prof. A. Holas, Prof. M. Jaszuński, Dr. R. Nowakowski and Dr. K. Chmurski are gratefully acknowledged. I would like also to thank Wacker, GmbH for a generous gift of cyclodextrins that helped me to carry out cyclodextrin experimental studies, and Oxford University that has enabled my literature research at Kebble College, leading to my deeper understanding of supramolecular complexes.

Warsaw, January 2006 *Helena Dodziuk*

Cyclodextrins and Their Complexes. Edited by Helena Dodziuk

ISBN: 3-527-31280-3

List of Contributors

Hidetoshi Arima
Graduate School of Pharmaceutical Sciences
Kumamoto University
5-1, Oe-honmachi
Kumamoto 862-0973
Japan

Monika Asztemborska
Institute of Physical Chemistry
Polish Academy of Sciences
01-224 Warsaw
Kasprzaka 44/52
Poland

Anna Bielejewska
Institute of Physical Chemistry
Polish Academy of Sciences
Kasprzaka 44/52
01-224 Warsaw
Poland

Renata Bilewicz
Department of Chemistry
University of Warsaw
Pasteura 1
02-093 Warsaw
Poland

Amélie Bochot
Laboratoire de Pharmacie galénique
UMR CNRS 8612
Centre d'Etudes Pharmaceutiques
Université Paris Sud
5 rue J.B. Clément
F-92290 Châtenay-Malabry
France

Udo H. Brinker
Institut für Organische Chemie der
Universität Wien
Währinger Straße 38
1090 Wien
Austria

Bezhan Chankvetadze
Molecular Recognition and Separation Science
Laboratory
School of Chemistry
Tbilisi State University, Chavchavadze Ave 1
380028 Tbilisi
Georgia

Kazimierz Chmurski
Department of Chemistry
University of Warsaw
Pasteura 1
02-093 Warsaw
Poland

Witold Danikiewicz
Institute of Organic Chemistry
Polish Academy of Sciences
01-224 Warsaw
Kasprzaka 44/52
Poland

Helena Dodziuk
Institute of Physical Chemistry
Polish Academy of Sciences
Kasprzaka 44/52
01-224 Warsaw
Poland
and
Graduate College ICM-IWR
Warsaw – Heidelberg

Dominique Duchêne
Laboratoire de Pharmacie galénique
UMR CNRS 8612
Centre d'Etudes Pharmaceutiques
Université Paris Sud
5 rue J.B. Clément
F-92290 Châtenay-Malabry

Andrzej Ejchart
Institute of Biochemistry and Biophysics
Polish Academy of Sciences
Pawińskiego 5A
02-106 Warszawa
Poland

Tomohiro Endo
Department of Physical Chemistry
Hoshi University
4-41, Ebara 2-chone
Shinagawa-ku
Tokyo 142-8501
Japan

Gottfried Grabner
Max F. Perutz Laboratories
Department of Chemistry
University of Vienna
Campus Vienna Biocenter 5
1030 Vienna
Austria

Akira Harada
Laboratory of Supramolecular Science
Department of Macromolecular Science
Graduate School of Science
Osaka University
Toyonaka
Osaka 560-0043
Japan

Kazuaki Harata
Biological Information Research Center
National Institute of Advanced Industrial
Science and Technology
Central 6, 1-1-1 Higashi
Tsukiba
Ibaraki 305-8566
Japan

Akihito Hashidzume
Laboratory of Supramolecular Science
Department of Macromolecular Science
Graduate School of Science
Osaka University
Toyonaka
Osaka 560-0043
Japan

Hitoshi Hashimoto
Bio Research Corporation of Yokohama
Yokohama Kanazawa High-Tech Center
Techno-Core
1-1-1 Fukuura
Kanazawa-Ku
Yokohama 236-0004
Japan

Kenjiro Hattori
Department of Nanochemistry,
Faculty of Engineering
Tokyo Polytechnic University
1583 Iiyama
Atsugi 243-0297
Japan

Fumitoshi Hirayama
Graduate School of Pharmaceutical Sciences
Kumamoto University
5-1, Oe-honmachi
Kumamoto 862-0973
Japan

Hiroshi Ikeda
Department of Bioengineering,
Graduate School of Bioscience and
Biotechnology
Tokyo Institute of Technology
4259-B-44 Nagatsuta-cho, Midori-ku
Yokohama 226-8501
Japan

Yoshihisa Inoue
Department of Applied Chemistry
Osaka University
2-1 Yamada-oka
Suita 565-0871
Japan

Makoto Komiyama
Research Center for Advanced Science and
Technology
The University of Tokyo
4-6-1 Komaba
Meguro-ku
Tokyo 153-8904
Japan

Wiktor Koźmiński
Department of Chemistry
Warsaw University
Pasteura 1
02-093 Warszawa
Poland

Daniel Krois
Institut für Organische Chemie der
Universität Wien
Währinger Straße 38
1090 Wien
Austria

Masashi Kunitake
Graduate School of Science and Technology
Department of Applied Chemistry and
Biochemistry
Kumamoto University
39-1 Kurokami 2-chome
Kumamoto 860-8555
Japan

Irene M. Mavridis
Institute of Physical Chemistry
N.C.S.R. "Demokritos"
P.O. Box 60228
153 10 Aghia Paraskevi
Athens
Greece

Masahiko Miyauchi
Laboratory of Supramolecular Science
Department of Macromolecular Science
Graduate School of Science
Osaka University
Toyonaka
Osaka 560-0043
Japan

Eric Monflier
Faculté des Sciences J. Perrin, Université
d'Artois
Rue Jean Souvraz/Sac postal 18
F-62307 Lens Cédex

Esmeralda Morillo
Instituto de Recursos Naturales y Agrobiología
de Sevilla
Avda. Reina Mercedes 10
Apdo 1052
41080-Sevilla
Spain

Akihiro Ohira
Graduate School of Materials Science
Nara Institute of Science and Technology
8916-5, Takayama, Ikoma
Nara 630-0101
Japan

Mikhail V. Rekharsky
Entropy Control Project
ICORP
Japan Science and Technology Agency
4-6-3 Kamishinden
Toyonaka 565-0085
Japan

Laury Trichard
Laboratoire de Pharmacie galénique
UMR CNRS 8612
Centre d'Etudes Pharmaceutiques
Université Paris Sud
5 rue J.B. Clément
F-92290 Châtenay-Malabry

Hauhisa Ueda
Department of Physical Chemistry
Hoshi University
4-41, Ebara 2-chone
Shinagawa-ku
Tokyo 142-8501
Japan

Kaneto Uekama
Graduate School of Pharmaceutical Sciences
Kumamoto University
5-1, Oe-honmachi
Kumamoto 862-0973
Japan

Konstantina Yannakopoulou
Institute of Physical Chemistry
N.C.S.R. "Demokritos"
P.O. Box 60228
153 10 Aghia Paraskevi
Athens
Greece

1
Molecules with Holes – Cyclodextrins

Helena Dodziuk

1.1
Introduction

Cyclodextrins, CyDs, are macrocyclic oligosugars most commonly composed of 6, 7, or 8 glucosidic units bearing the names α-**1**, β-**2** and γ-**3** CyD, respectively [1–3]. Other, usually smaller, molecules (called guests) can enter their cavity forming inclusion complexes with these hosts. α- and β-CyDs are believed to have been isolated for the first time at the end of nineteenth century [4] while their first complex seems to have been reported early in the twentieth century [5]. However, it took more than 50 years to establish and confirm the structure of CyDs. Today we take for granted the idea of inclusion complex formation by these macrocycles, but when it was suggested by Cramer in the 1940s the idea was not, to put it mildly, generally accepted. Cramer said later with considerable enjoyment [6]: "When I presented my results for the first time at a meeting in Lindau, Lake Konstanz, I met fierce opposition from some parts of the establishment. One of my older (and very important) colleagues even stated publicly and bluntly in the discussion that he would try to remove a young man with such crazy ideas from the academic scene. But there was also a good number of supporters, so I finally made it."

According to Stoddart [6], "Cyclodextrins are all-purpose molecular containers for organic, inorganic, organometallic, and metalloorganic compounds that may be neutral, cationic, anionic, or even radical." They are usually built of glucopyranoside units in the $^4C_1$ conformation (Fig. 1.1). In most cases these host molecules have an average structure of a truncated cone with a cavity lined with H3 and H5 protons and lone pairs of glycosidic oxygen atoms lying in a plane thus endowing the cavity with hydrophobic character, while the bases formed by the primary and secondary OH groups bestow a hydrophilic character (Fig. 1.2). The great significance of CyDs both in research and applications lies in their ability to selectively form inclusion complexes with other molecules, ions, or even radicals. This phenomenon bears the name *molecular recognition* while the selectivity in the formation of complexes with enantiomeric species as guests is called *chiral recognition*. Complex formation changes the properties of both host and guest, allowing one to monitor the process by several experimental techniques. On the basis of X-ray

Cyclodextrins and Their Complexes. Edited by Helena Dodziuk

ISBN: 3-527-31280-3

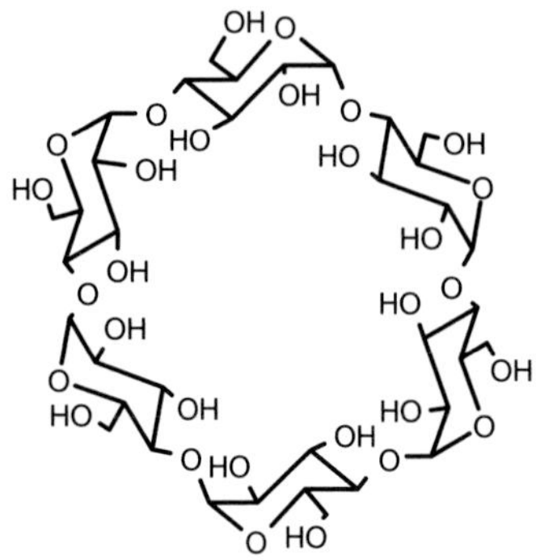

1

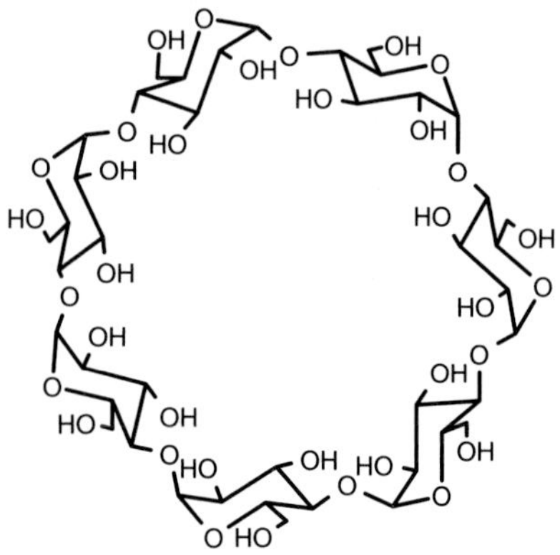

2

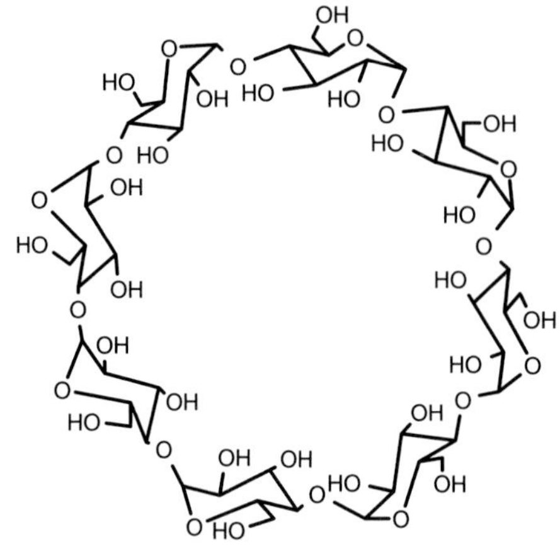

3

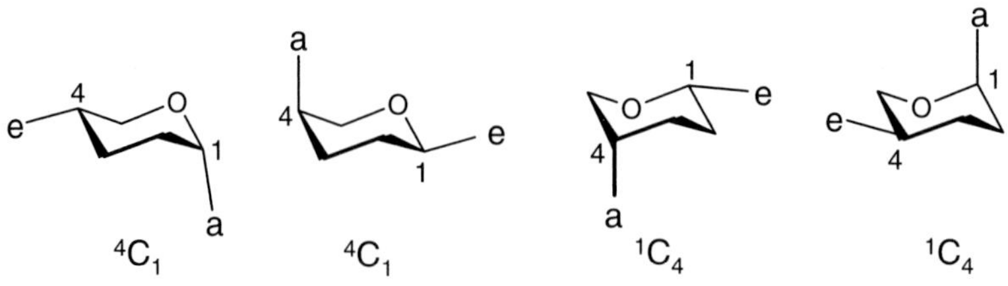

Fig. 1.1. Notation of conformations of the glucopyranoside ring.

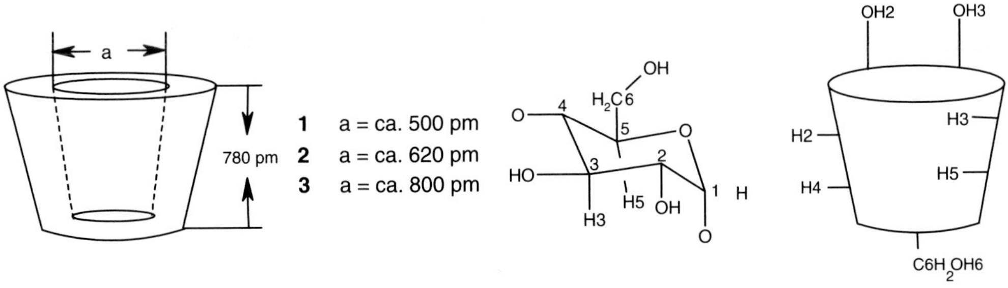

Fig. 1.2. Schematic view of the glucopyranoside ring with the atom numbering and native CyD sizes and the average orientation of the most important atoms and OH groups.

measurements, native CyDs **1–3** have for years been considered to possess a rigid, truncated-cone structure [7–9]. This view is inconsistent both with the CyDs' ability to selectively complex guests of various shapes and with several experimental and theoretical findings, discussed later in this chapter. These data reveal the amazing flexibility of the CyD macrocycles. The implications of the non-rigidity of CyDs for their complexing ability, and its influence on the results obtained using different experimental techniques, are also presented there. One of the most striking examples of this kind is provided by the different mode of entrance of the guest in the complex of nitrophenol **4** with permethylated α-CD **5** (R1 = R2 = R3 = Me) [10] in the solid state and in solution shown in Fig. 1.3.

The ability to predict recognition by CyDs would be of great practical value, especially for drug manufacturers. Consequently, several models of chiral recognition by CyDs have been proposed in the literature, neglecting the complexity of the complexation process involving very small energy differences between the complexes with enantiomeric species. The models critically reviewed later in this chapter are mostly based on very few experimental data and some of them contradict

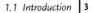

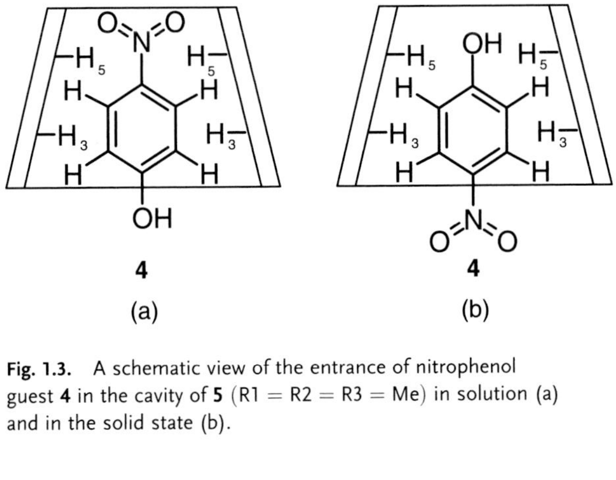

Fig. 1.3. A schematic view of the entrance of nitrophenol guest **4** in the cavity of **5** (R1 = R2 = R3 = Me) in solution (a) and in the solid state (b).

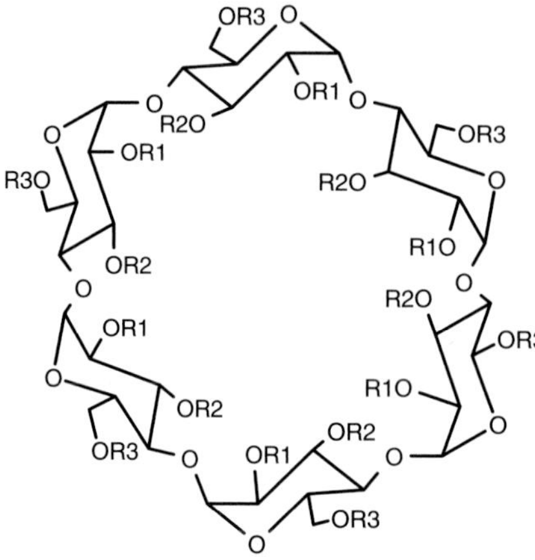

5

the basic properties of three-dimensional space. For instance, the most often used 3-point Dalgleish model [11] is incompatible with the basic requirement of three-dimensional space that at least four points (not lying in a plane) are necessary for an object to be chiral [12, 13].

Numerous CyD derivatives have been synthesized with the aim of improving their complexing properties and to make them suitable for various applications, in particular to increase the bioavailability of a drug complexed with a particular CyD derivative. By appropriate choice of host and guest one can achieve a very high selectivity. For instance, **6** is complexed much more strongly by the dimeric host **7** than is **8** [14]. Numerous CyD derivatives mono- or polysubstituted in positions 2, 3 and/or 6 by alkyl groups **5**, **9** as well as modifications of hydroxyl groups

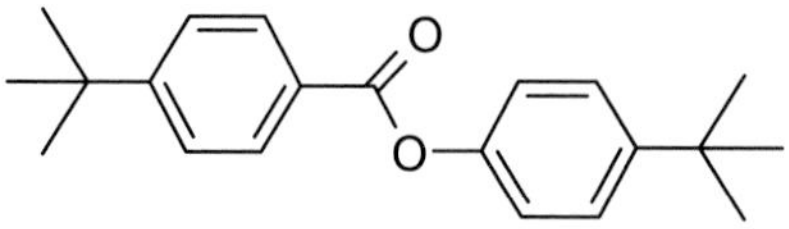

6

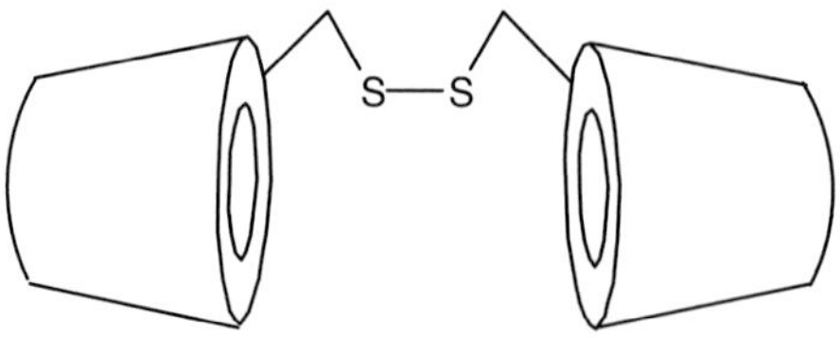

7

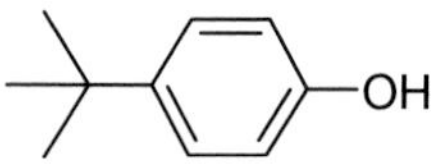

8

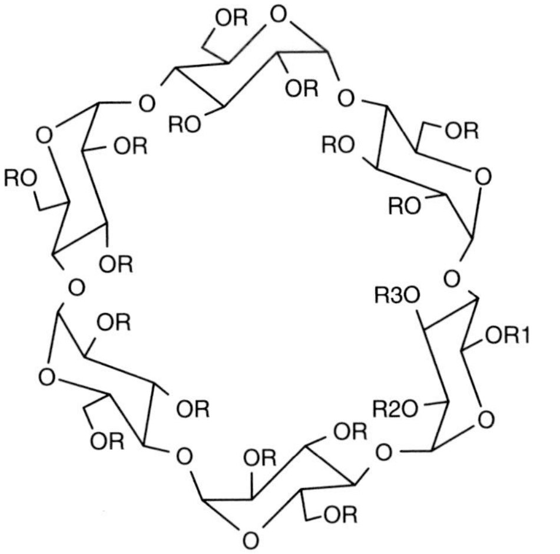

9

to sulfopropyl, carboxymethyl, tosyl, aldehyde, silyl, and many other groups have been obtained [15, 16]. The reactivity of CyD and the plethora of exciting CyD structures developed, among other reasons, to enhance and modify their complexing ability will be shown in Chapter 2. Fascinating CyD structures include, among others, **10** [17], **11** [18], amphiphilic **5** (R1 = R2 = OH, R3 = $CH_2S(CH_2)_3C_6F_{13}$ [19], capped **12** [20], peptide appended **13** [21] and 2:2 complex **14** formed by the CyD dimer with porphyrin and zinc ion [22]. On the other hand, obtaining dimers

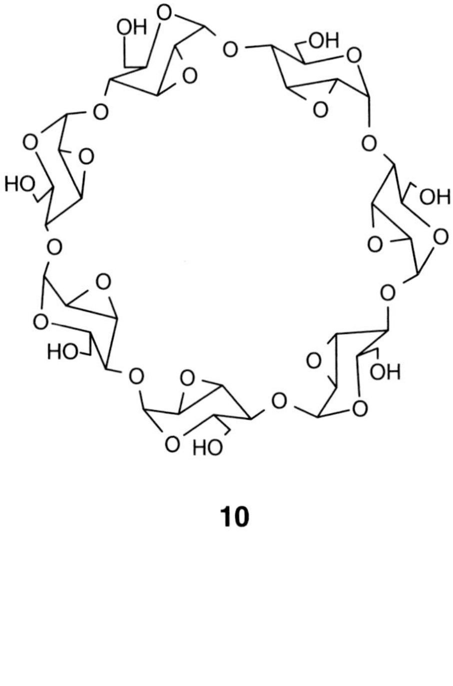

10

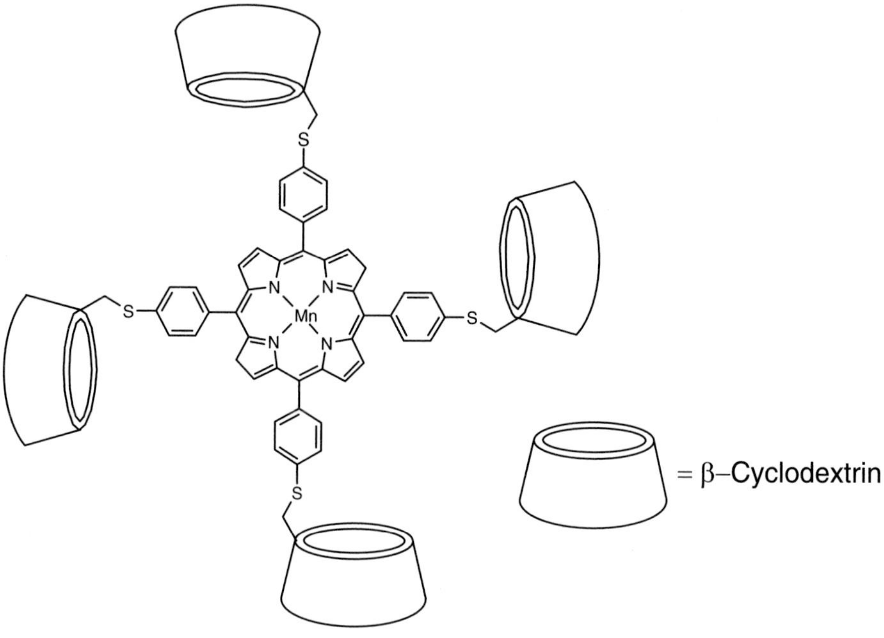

11

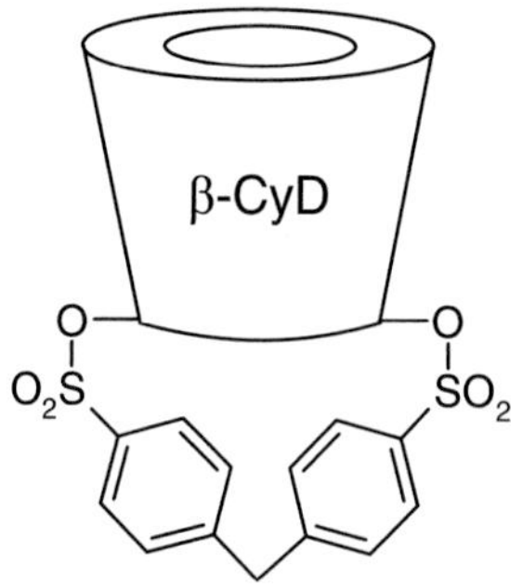

12

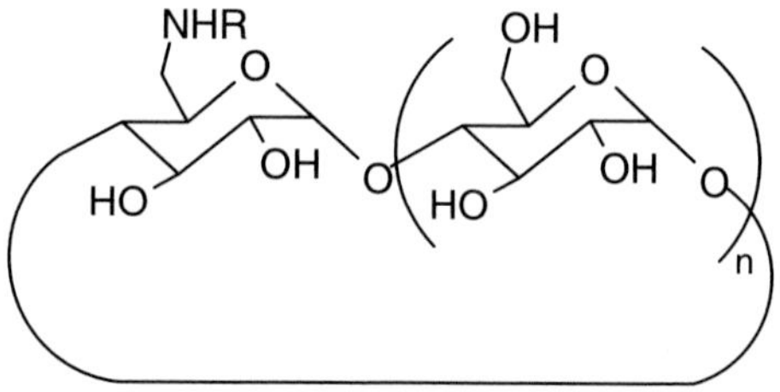

n = 6, 7 R = Glycyl-L-Phenylalanyl

13

of isomeric naphthalenic acid, using an appropriately substituted γ-CyD template to exert control of the reaction's stereochemistry, shows a very elegant method making use of the encapsulation of two naphthyl-involving substituents on different glucopyranoside rings (Fig. 1.4) [23].

Exciting CyDs involving oligomers and polymers both covalently bound and self-assembled will be presented in Chapter 3 while their SPM observations and some

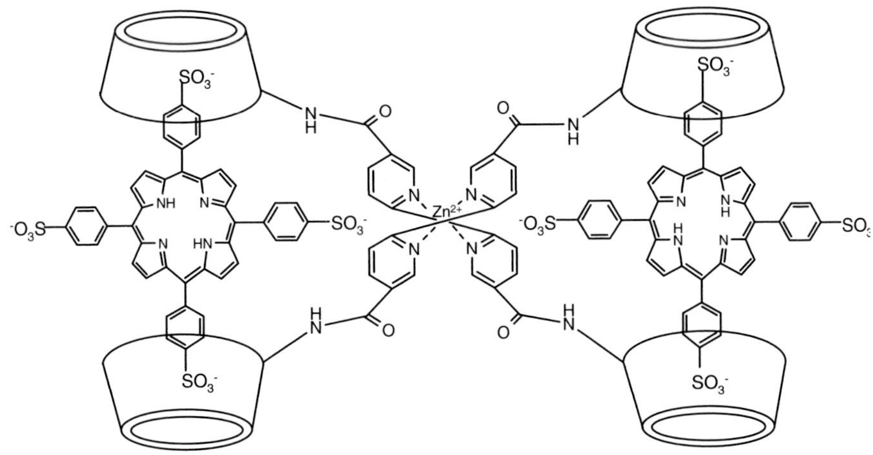

14

polymers having catenane or rotaxane structures (Fig. 1.5) will be discussed in Section 10.6 and Chapter 12.

CyD catalysis, discussed in Chapter 4, and the application of CyDs as enzyme models constitute a fascinating field. The influence of a specific CyD on the stability of the included molecule can have, like the two-headed god Janus, contrasting consequences. Mostly, luckily for pharmaceutical applications, complexation with CyD usually stabilizes the guest. (However, it can also catalyze its decomposition as is the case with aspirin **16** [24] preventing its application in the form of a CyD complex.) CyDs are known to catalyze numerous reactions. Notably, the catalytic activity is usually not high, but (a) it can reach high values in a few cases as evidenced by the ca. 1.3 million-fold acceleration of the acylation of β-CyD **2** by bound *p*-nitrophenyl ferrocinnamate **17** [25] and (b) CyDs may impose limitations on the reaction's regioselectivity. Chlorination of anisole in the absence or presence of α-CyD illustrates this point [26] (Fig. 1.6) since it is known to produce only *para*-substituted isomers in the presence of α-CyDs while both *meta*- and *para*-isomers are obtained in its absence [27]. It should be stressed that, although the catalytic activity of CyD can achieve high levels [25, 28], these oligosaccharides are much more effective in inducing stereo- or regioselectivity than in genuine catalytic action. Another example of regioselectivity induced by γ-CyD was presented earlier in Fig. 1.4.

An exciting field of considerable importance related to CyD catalysis is their use as enzyme models by testing the reactivity of appropriately substituted CyDs [29–32]. For instance, to mimic the cleavage of RNA followed by cyclization of phosphate ester with subsequent hydrolysis using imidazole groups of Histidine-12 and Histidine-119 of ribonuclease A, isomeric diimidazole-substituted at C6 positions β-CyDs **18–20** were synthesized [29] and checked for their influence on the

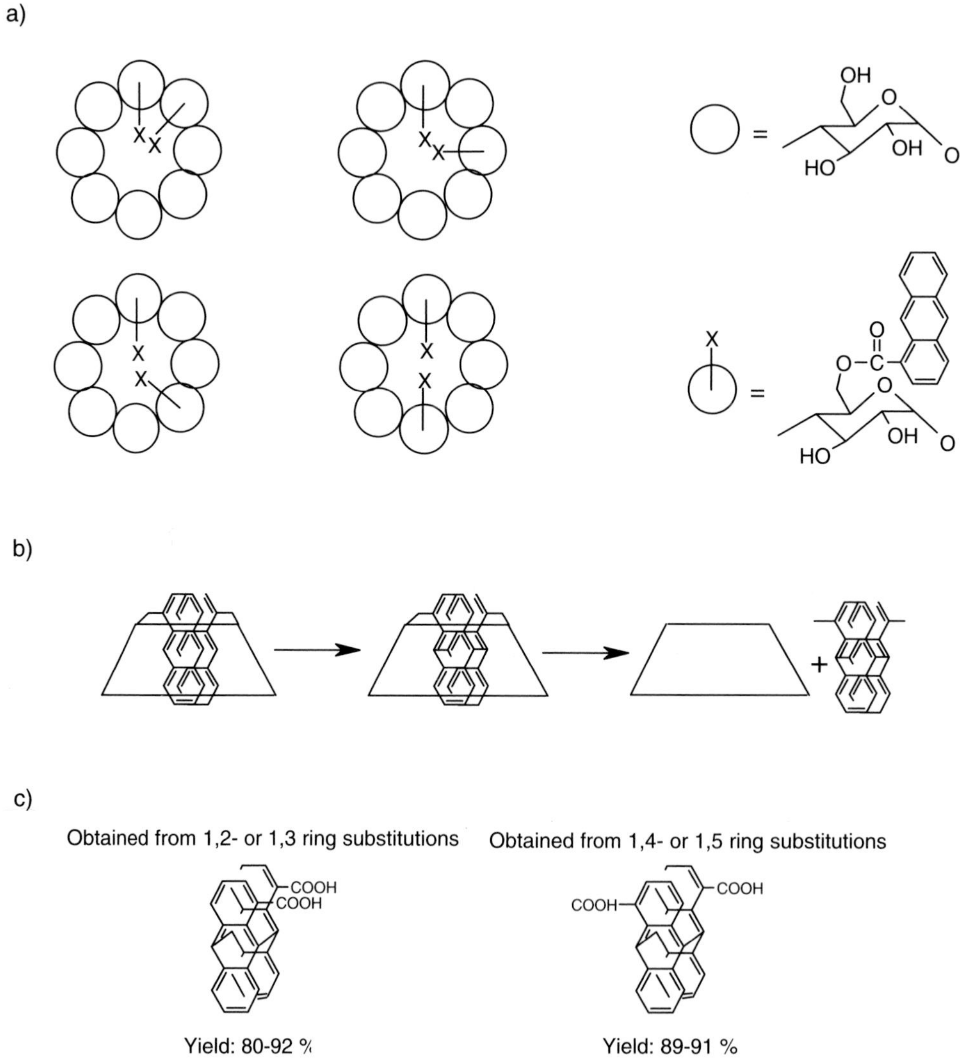

Fig. 1.4. Regioselectivity of the reaction of the included guest: schematic views of (a) disubstituted γ-CyDs, (b) the course of the reaction, (c) the product and yield of the reaction.

cleavage reaction. Interestingly, contrary to the classic mechanism proposed for this reaction, the close to linear arrangement of imidazole **21** groups in **20** did not lead to the most efficient catalysis, causing the abandonment of the mechanism commonly accepted in textbooks. The catalytic action of nuclease, ligase, phosphatase, and phosphorylase was also analyzed using more complicated CyD derivatives [33].

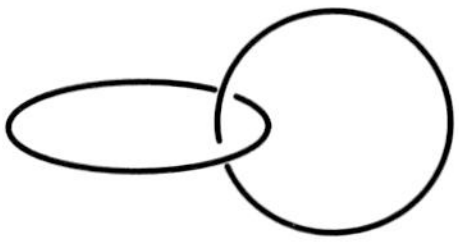

15a

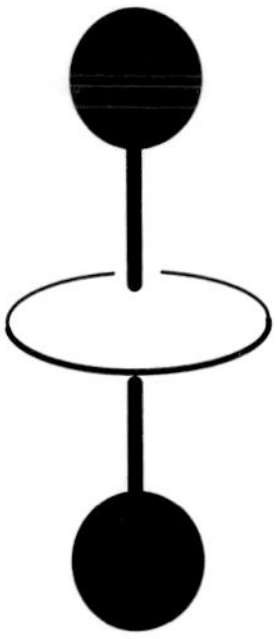

15b

(a) (b)

Fig. 1.5. Schematic catenane (a) and rotaxane (b) structures.

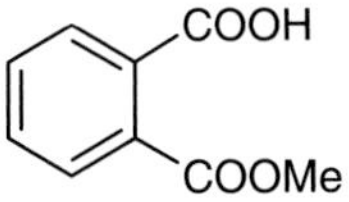

16

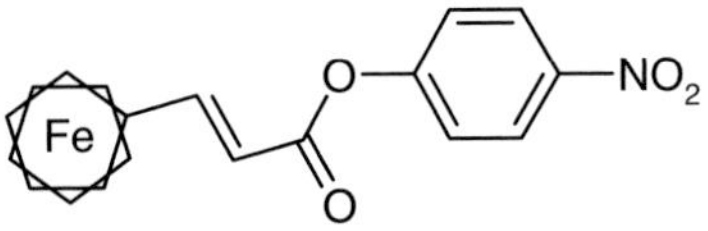

17

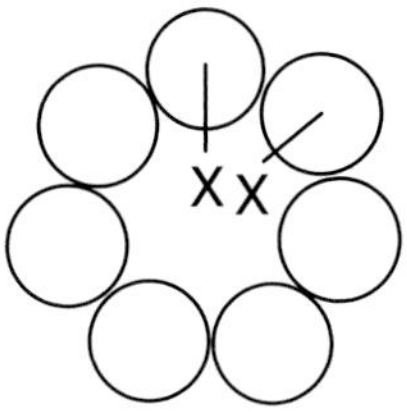

18

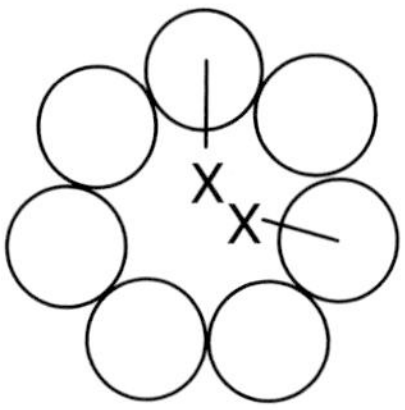

19

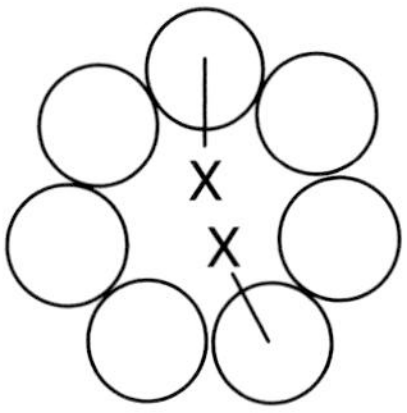

20

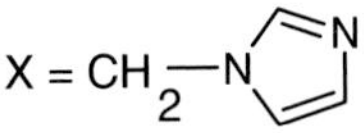

21

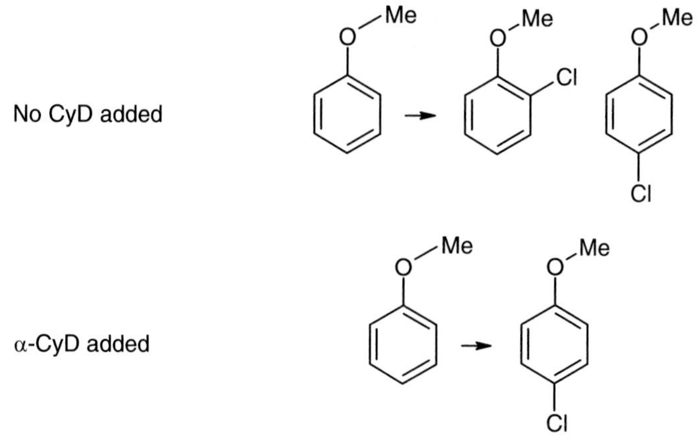

Fig. 1.6. Influence of α-CyD on anisole chlorination.

As mentioned before, CyDs and their complexes elicit a vivid interest as systems that allowing us to study the factors that drive the selective complexation known as molecular recognition. Particularly great interest is focused on the differentiation between enantiomers of guest molecules by their complexation with CyDs, i.e. on chiral recognition. This will be of great importance in future CyD applications, in particular in the pharmaceutical industry, since most drugs and the active sites in which they operate are chiral. As a consequence (remember a small child trying to put his foot into the other shoe?), enantiomers of several drugs have been found to exhibit different pharmacological activities [34]. The differences may go so far as to result in harming instead of healing. (The old thalidomide **22** tragedy, when pregnant women taking this drug in the racemate form later gave birth to babies with crippled extremities, was sometimes interpreted in terms of the teratogenic activity of its second enantiomer [35]. However, the recent revival of interest in thalidomide drug for various illnesses [36–38] should be acknowledged.

Chromatography is one of the most important methods for direct studies of molecular and chiral recognition by CyDs. Today it has split into several branches, e.g. gas chromatography, GC, high-performance liquid chromatography, HPLC, and capillary electrophoresis and other electromigration techniques, that enable us not only to detect the recognition but also to estimate the complex stoichiometry and formation constant and, consequently, the enthalpies and entropies of complex for-

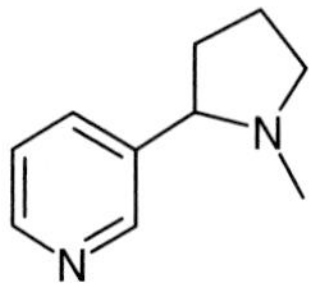

22

mation. The amazing sensitivity of CyDs to the shapes of guest molecules or ions may be illustrated by the big difference among retention times of the complex of 1,8-dimethylnaphthalene **23h** with **2** on the one hand and those of other isomers **23a–23g** on the other, determined by gas–liquid chromatography (discussed in Chapter 4 in more detail) [39]. Such a big difference is most probably caused by a difference in the stoichiometry of these complexes. Namely, the latter complexes are of 1:1 stoichiometry while in the former one guest molecule is mostly embedded in a capsule formed by two host CyDs [40]. The striking change in elution order with temperature rise indicates the importance of the entropy factor and of CyD flexibility on the complex stability [41]. Similarly, the dependence of the elution order of the enantiomers of phenothiazines **24** complexed with γ-CyD **3** on the buffer used shows the complexity of the complexation process [42]. Molecular and chiral recognition by CyDs, as studied mainly by HPLC and GC, will be presented in Chapter 5 together with the application of this method for studying complex stoichiometries and stability constants while a wide range of chromatographic methods used for enantioseparations will be discussed in Chapter 6.

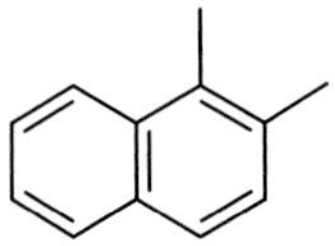
23a

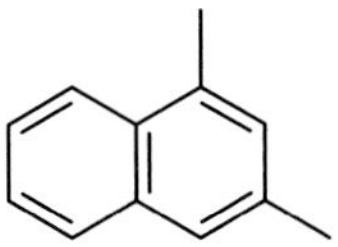
23b

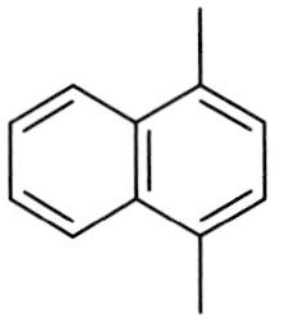
23c

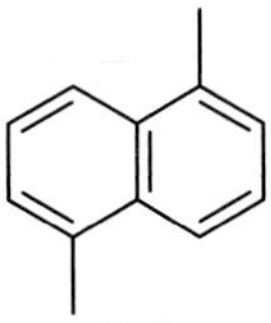
23d

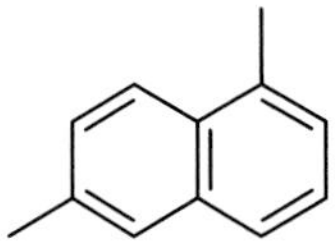
23e

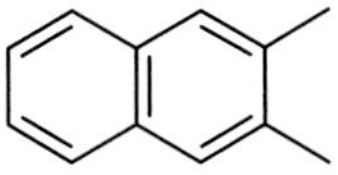
23f

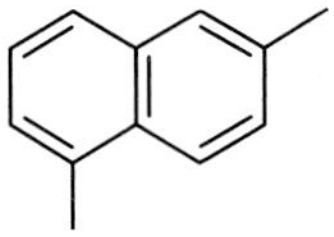
23g

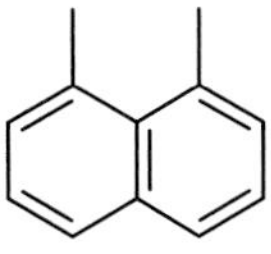
23h

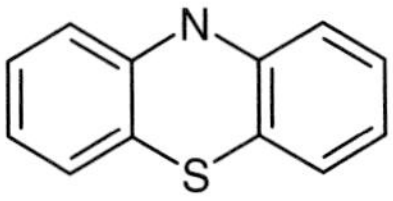

24

In most cases, CyD structures are elucidated on the basis of X-ray studies which will be presented in Chapter 7 together with the results of a few, but very interesting, neutron diffraction investigations. They include systems of O2H $\rightarrow$ O3 and O3H $\rightarrow$ O2 hydrogen bonds in β-CyD **2** rapidly interchanging at room temperature (Fig. 1.7) [43]. Freezing the process and accurately determining the positions of hydrogen atoms using neutron diffraction [9] allowed the determination of the

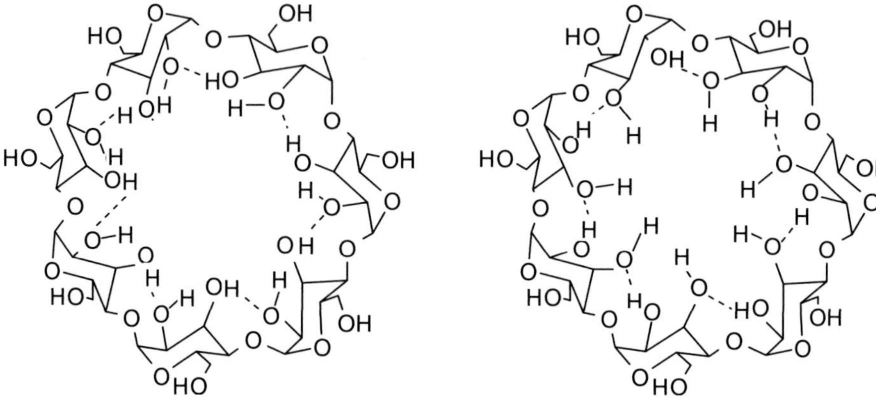

Fig. 1.7. Rapidly interconverting circular systems of hydrogen bonds in β-CyD 2 (flip-flop, HO2 → O3 simultaneously and reversibly changing to O3H → O2).

circular systems of the bonds shown in the figure. The mechanism of simultaneous change of directions of hydrogen bonds in **2** is called "flip-flop". Other fascinating examples of X-ray determined CyD structures are provided among others by [3] catenane-type CyDs **25** [44], the sixteenfold deprotonated γ-CyD dimer with Pb ions **26** [45], and the complex of an alkali ion buried inside a capsule formed by two crown ethers, in turn inserted in another capsule built of two γ-CyD molecules, the whole system resembling a Russian doll [46, 47].

The forces driving complexation by CyDs cannot be understood without a knowledge of the energy differences and barriers involved in the complexation. The calorimetric measurements, involving isothermal titration calorimetry and differential scanning calorimetry are discussed in Chapter 8. They give the most accurate thermodynamic data characterizing the complexes. In particular, these data provide further examples of the amazing enthalpy–entropy compensation that is not limited to CyD complexes [48]. Exciting studies of isotope effects on complex formation are also discussed there.

X-ray and neutron diffraction studies yield precise information on the CyD structure in the solid state. On the other hand, in addition to the information on the structure and dynamics of complexes in the solid state, NMR spectra allow elucidation of the structure in solution, which is of particular importance since most CyD applications take place in this state. (Even if we take a drug as a CyD complex in the form of a pill it dissolves in the stomach before acting.) NMR studies can give not only unequivocal proof of the complex formation in form of, usually small, chemical shifts but also, by studying the NOE effect [49], they can show how the organic guest molecule enters the host cavity in the solid state and in solution. The spectra are also sensitive to the dynamics of the complex and so they provide

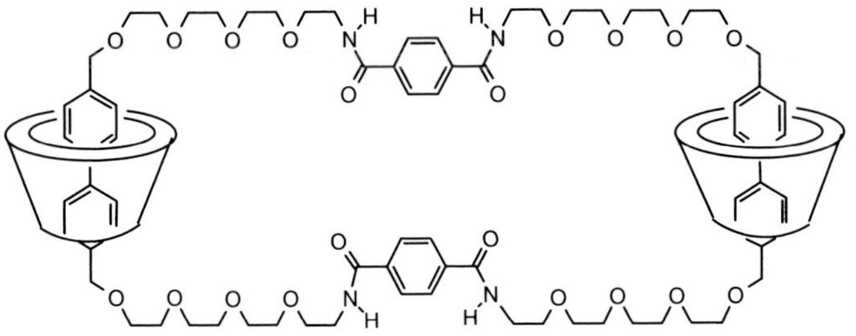

25a

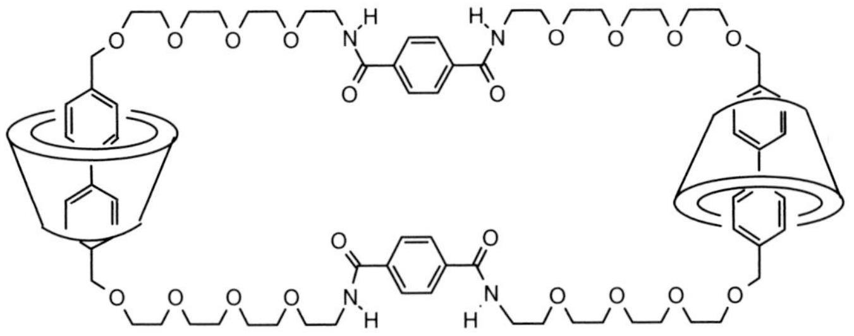

25b

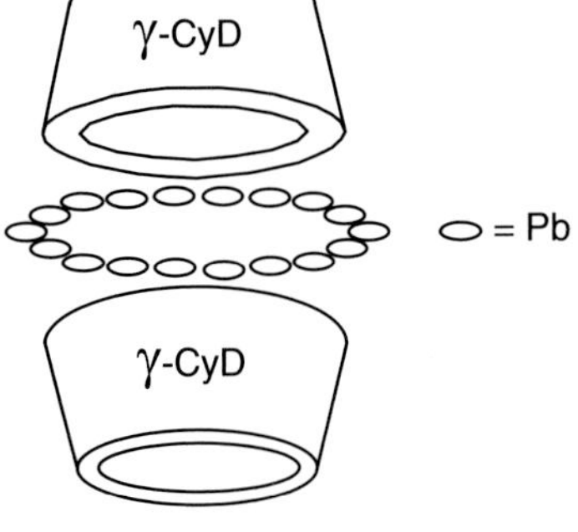

26

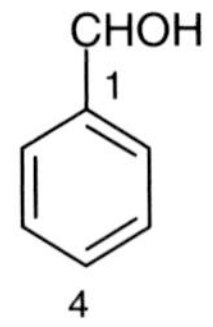

27

information on the complex's nonrigidity showing the host and/or guest movement even in the solid state where, owing to positional and time averaging, X-ray results point to a single rigid structure. This is the case for the complex of benzyl alcohol **27** with **2** for which ^{2}H NMR spectra indicate a rapid flip of the aromatic ring around the C1C4 axis [50]. In addition to information on the complex's

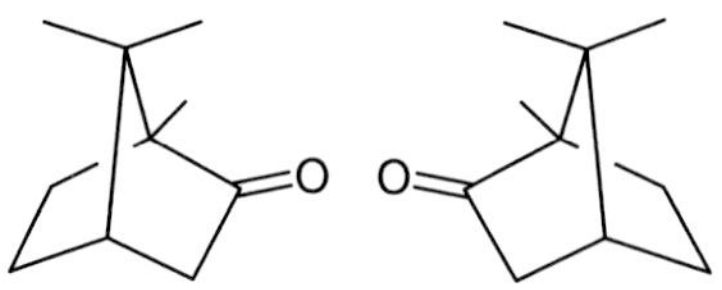

28

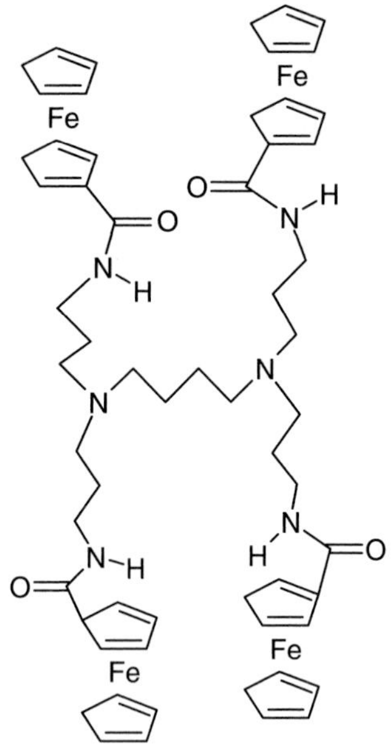

29

stoichiometry and stability constants, in favorable cases the study of relaxation rates in ^{1}H NMR spectra can show the orientation of the guest in the host cavity in solution, which no other technique can give, as shown for the complexes of camphor enantiomers **28** with **1** [51]. The kind of information on CyDs and their complexes that can be provided by NMR studies is discussed in Chapter 9 while Chapter 10 is devoted to the application of other physicochemical methods (UV, circular dichroism, mass spectroscopy, electrochemistry, AFM and STM, etc.) to the elucidation of the structure of CyDs and their complexes. Some of these methods are usually less sensitive to complex formation involving CyDs, but the effect can be considerable in specific cases and is of importance for applications in sensors and other devices. Although CyDs themselves do not have electroactive groups, electrochemical studies of their complexes form the basis of their future applications. Dendrimers with electroactive end groups like **29** [52] forming multiple CyD complexes are, probably, one of the most interesting examples in this area. Of course, mass spectra are most frequently used to prove the synthesis of a CyD derivative but, as shown in this chapter, they can be a source of valuable information on CyD complexes. New, rapidly developing AFM and STM techniques allowing the study of CyD aggregates on surfaces will also be presented there. They provide information on a single molecular aggregate or superstructures formed by them. In particular, rotaxane-type structures **15b** (discussed in detail in Chapter 12) can be observed with atomic resolution by the latter method.

The possibility of predicting the molecular and chiral recognition ability of CyDs would be of great value, in particular for the pharmaceutical industry. The need for reliable theoretical treatment of CyD complexes is also reflected in several chapters in this book. Numerous studies applying quantum mechanics [53], molecular mechanics [54], and molecular dynamics [55] have been published by researchers fascinated by beautiful computer models and the ease of carrying out the calculations. The complexity of the complexation process and its consequences for the nonrigidity of CyDs, as well as the limited accuracy of the calculations, are neglected in most of these studies. Modeling of CyDs and their complexes and the dependence of the results of calculations on the assumed model and its parameterization are critically reviewed in Chapter 11.

The exciting catenanes, like **25** [44], and rotaxane molecular necklace **30** of 1:n stoichiometry incorporating $n = 20$–22 α-CyD macrocycles [56], respectively, falling into the realm of topological chemistry [57] will be shown in Chapter 12. These systems, also discussed in Chapters 10.6 and 16, form the basis of exciting applications. Large CyDs such as the 12-membered **31**, which differ dramatically from native CyDs in properties and, most probably, in complexing ability, will be discussed in Chapter 13. In particular, contrary to the structure of **1–3**, large CyDs do not have truncated-cone average structures with glycosidic oxygen atoms lying approximately in a single plane, but some of them are known to be twisted allowing for formation of hydrogen bonds between OH groups of distant glycosidic units [58].

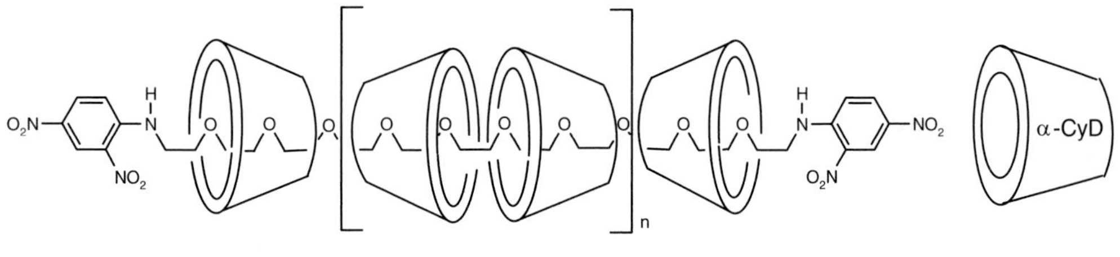

30

Chiral recognition by CyDs is of primary importance for the pharmaceutical industry since the second enantiomer of a drug, usually present as 50% impurity as the result of chemical synthesis, can be harmful. Therefore, an effective preparative separation of enantiomers is one of the important goals of applied CyD research since at present it has not reached the industrial scale. Today the main CyD application in the pharmaceutical industry is their use as drug carriers, since CyD containers in most cases stabilize and solubilize the included drugs (see, how-

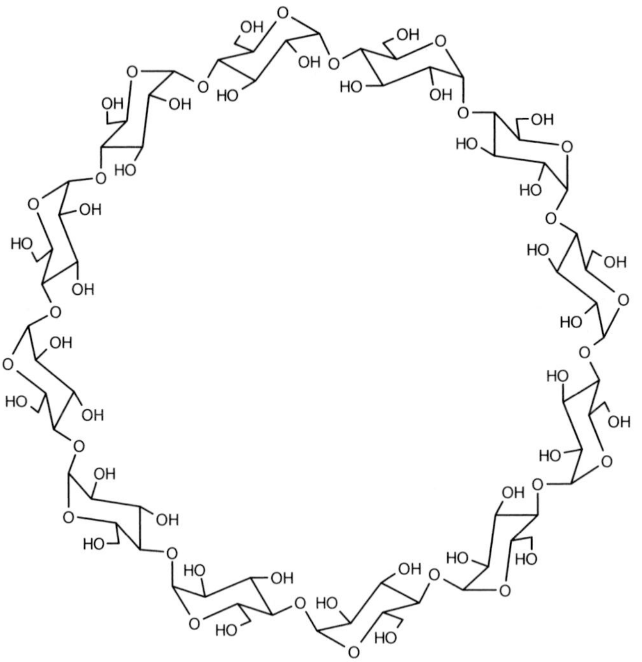

31

ever, the aspirin case mentioned above). Moreover, the slow release of a drug from the complex results in its higher and more uniform content in the organism, allowing less frequent administration of the drug. Interestingly, CyD applications in drug delivery were considerably delayed by the erroneous determination of their toxicity at an early stage of development [59]. Today we know that they are not harmful in most cases by oral, parenteral, nasal, or skin administration [60]. CyD applications in the form of inclusion complexes in the pharmaceutical industry in general are presented in Chapter 14 with a detailed discussion of the ways in which various CyD types (hydrophilic, hydrophobic, or ionizable ones) affect the bioavailability of drugs by influencing their solubility and the rate of release from the CyD complex. A small section on site-specific drug delivery is also included. The even better therapeutic effect of drugs in the form of emulsions, microparticles, nanoparticles, and higher aggregates is given in Chapter 15.

CyD applications are by no means restricted to the pharmaceutical industry. Several examples, mainly prospective ones, are scattered throughout this book. CyDs are used to remove unpleasant tastes, odors, or other undesirable components in the food industry, in agrochemistry, cosmetics, dying, cleaning, and in many other areas. To name just a few examples of numerous CyDs applications: grapefruit juice loses its unpleasant taste when its bitter component naringine is removed by complexation with β-CyD [61]; similarly, removal of phenylalanine and tyrosine makes the food harmless for those suffering from phenylketonuria [62]. Garlic retains all its health benefits but lacks its annoying odor when applied in the form of a CyD complex. Similarly, other flavor components (such as apple, citrus fruits, and plums) and spices (cinnamon, ginger, horseradish, menthol, etc.) are marketed as β-CyD complexes characterized by high stability exhibited towards heating during industrial food processing [63–66]. The problems encountered by the use of CyDs for stabilization of mixtures has to be mentioned here since the components involved can be released at different rates, changing the taste or smell of the product. The creation and higher stability of foams used in both the food and cosmetic industries is also favored by the complexation. Interestingly, CyD applications should not only improve the properties of a marketed product but, consistently with modern trends, they are also aiming at creating new needs by producing new types of products with unheard of properties. A long-lifetime fragrance-releasing room-decorating paint is one example of such applications [67]. A few examples of numerous CyD applications in industries other than pharmaceuticals are briefly presented in Chapter 16, which concludes with the presentation of their thought-challenging prospective applications in molecular devices and machines.

The cyclodextrin field is rapidly expanding. According to information from the Cyclolab website, 3.9 articles per day on CyDs were published on average in 1995. This figure increased to 4.4 in 2004 (http://www.cyclolab.hu/literature_0.htm) and is still growing. The field is not only very large but also highly diversified. This is probably the reason why no general modern comprehensive monograph on CyDs has been available [1–3, 16, 68–71] for so long. We would like to bridge this gap, providing a survey of the whole exciting area with numerous references to help novices to enter this domain on the one hand and to give researchers in a specific field a broader insight into the whole area on the other. We apologize for not being able to mention, even briefly, numerous valuable works on CyDs but we hope to have given the picture of the whole CyDs field, illustrated by representative examples of the respective research and applications.

1.2 Cyclodextrin Properties

In the standard way, native CyDs **1–3** are obtained by enzymatic degradation of starch. First obtained in minute amounts and very expensive, in particular α- and γ-CyD, they now cost less than $10/kg, making their large-scale industrial use feasible [72]. IUPAC names of these macrocycles are cumbersome; 5,10,15,20,25,30,35-heptakis-(hydroxymethyl)-2,4,7,9,12,14,17,19,22,24,27,29,32,34-tetradecaoxaocta-cyclo[$31.2.2.2.^{3,6}.2^{8,11}.2^{13,16}.2^{18,21}.2^{23,26}.2^{28,31}$]nonatetracontane-36,37,38,39,40, 41,42,43,44,45,46,47,48,49-tetradecol for β-CyD makes the use of trivial names necessary. Lichtenthaler and Immel proposed a general system of naming macrocyclic oligosaccharides [73] but it has not been generally accepted and used. The chemical, physical, and biological properties of CyDs, and in particular their toxicity by various type of administration, are summarized in Ref. [60] while their stability when

Table 1.1. Some properties of native CyDs [60][a]

	α-CyD 1	β-CyD 2	2γ-CyD 3
Number of glucose units	6	7	8
Molecular weight	972	1134	1296
Approximate inner cavity diameter (pm)	500	620	800
Approximate outer diameter (pm)	1460	1540	1750
Approximate volume of cavity (10^6 pm^3)	174	262	427
$[\alpha]_D$ at 25 °C	150 ± 0.5	162.5 ± 0.5	177.4 ± 0.5
Solubility in water (room temp., g/100 mL)	14.5	1.85	23.2
Surface tension (MN/m)	71	71	71
Melting temperature range (°C)	255–260	255–265	240–245
Crystal water content (wt.%)*	10.2	13–15	8–18
Water molecules in cavity	6	11	17

[a] Some data on these and larger CyDs are also given in Tables 9.1 and 13.1.

treated by various enzymes is presented in Ref. [74]. Some of their properties are given in Table 1.1.

As mentioned earlier, native CyDs are usually obtained by biochemical processes [72]. However, a 21-step synthesis of **1** with 0.3% total yield [75] and that of **3** [76] with 0.02% yield are worth mentioning. The macrocycle built of only five glucopyranose rings **32** [77] (thought to be too strained to exist on the basis of model calculations [78]) and probably thousands of CyD derivatives have been synthesized [15, 16]. Several exciting CyDs have been presented earlier. Here CyDs having glucopyranoside ring(s) in $^1C_4$ [79–81] or *skew* conformations [82], those incorporating other than glucopyranoside units [83] and large CyDs having from 9 to

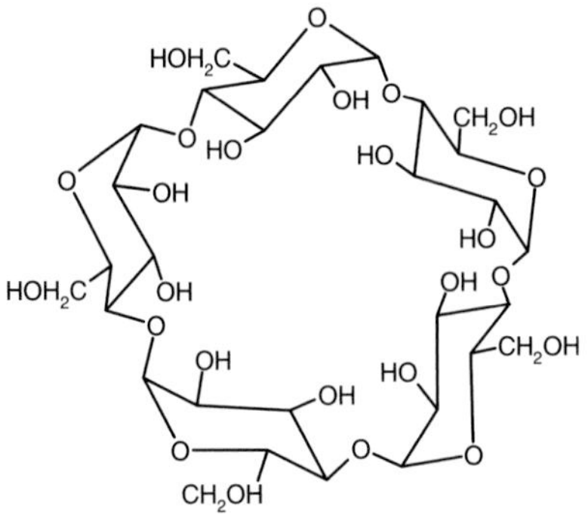

32

more than 100 monosaccharide rings (discussed in Chapter 13) should be mentioned [58, 84, 85]. Interestingly, three unusual CyD derivatives have been found in nature [86].

As discussed in Chapter 7, native CyDs can form complexes with differing amounts of water. CyDs are seldom really empty. Even if they do not contain another guest there is usually at least one solvent molecule in their cavity. Almost all their applications involve inclusion complex formation when one or more molecules are at least partly immersed in the CyD cavity. The complexes can be obtained both in solution (sometimes requiring heating or the use of a cosolvent) or in the solid state, e.g. by cogrinding or milling [87]. In solution, they exist in a rapidly exchanging equilibrium of the free CyD host and guest. As mentioned before, CyD complexes of different stoichiometry are known. In addition to 1:1 complexes like **4@5** (R1 = R2 = R3 = Me) [10], the 1:2 ones like those of camphor enantiomers **28** embedded in a capsule formed by two α-CD **1** [51] and that of C_{60} buried in two γ-CyD molecules **3** in a similar way [88], 2:2 complex **14** [22] or even rotaxane molecular necklace **29** of 1:n stoichiometry involving $n = 20$–22 α-CyD macrocycles [56] are known. In spite of the years that have passed from the publication of the Szejtli review [89], his statement "The 'driving force' of complexation, despite the many papers dedicated to this problem, is not yet fully understood." is still valid. The complexation process in a, mostly water, solvent is considered to involve a release of water molecule(s) from the relatively hydrophobic CyDs' cavity, removal of the polar hydration shell of the apolar guest molecule, entry of the guest into the empty CyD cavity where it is stabilized mainly by weak but numerous van der Waals attractive interactions, restoration of the structure of water around the exposed part of the guest, and integration of this with the hydration shell of host macrocycle. Thus, a change in both enthalpic and entropic contributions occurs in complex formation that depends on the host- and guest-induced fit [90], the solvent used, and numerous other factors. On the other hand, in the solid state the magnitude of the crystal forces is comparable with the forces keeping the complex together. Thus, as exemplified by **4@5** (R1 = R2 = R3 = Me) [10], they can influence the complex's structure and dynamics which is considerable even in the solid state. To summarize: CyD complexes are very difficult to study since (1) for poorly soluble species the complexation process can be much more effective for impurities (present in minute amount in the solution) than for the guest under investigation; (2) the process depends heavily on the experimental conditions (pH, cosolvent, temperature, etc.); (3) the complexes can involve species of different stoichiometries, e.g. dimethylnaphthalenes **23a–g** and **23h** [40], or an additional solvent molecule can enter the cavity as a second guest resulting in ternary complex formation [91, 92] (interestingly, complexes of even higher stoichiometry, involving two CyDs, one pyrene, and two cyclohexanol guest molecules [92] are known); (4) the experimental results for a CyD complex can depend on the technique used since the CyD complexes are held together by weak forces (one example of this kind was shown in Fig. 1.3 [10]); (5) Reliable theoretical studies for CyD complexes are extremely difficult to obtain since, as discussed in detail in Chapter 11, these large systems are characterized by energy surfaces exhibiting numerous very shallow local minima separated by low energy barriers [93, 94]. The size of the system and its n-fold degeneracy ($n = 6, 7, 8$ for α-, β-, or γ-CyD, respectively) also make it difficult to compare X-ray geometrical parameters for, for example, complexes with different guests. As discussed in detail in Section 1.4, one can either attempt to analyze the values of all internal parameters (e.g. of 126 bond lengths in α-CyD, etc.) in the series, which gives a very nontransparent picture, or compare the values of, for instance, the C2O2 bond length averaged over all saccharide rings, losing a lot of information by such an approach.

1.3
Cyclodextrin Nonrigidity [94, 95]

On the basis of X-ray studies [7–9], for tens of years CyDs structure was thought to resemble a rigid truncated cone of the high C_6, C_7, or C_8 symmetry for α-, β-, or γ-CyD, respectively, with a planar ring of glycosidic oxygen atoms [8, 96, 97]. Numerous experimental and theoretical data are incompatible with the concept of rigid CyDs. First of all, a general analysis of these systems shows that there is no physical reason for the rigidity, since the macrocycles are built of relatively rigid glucopyranose rings connected by ether C–O bonds, characterized by a low barrier to internal rotation of ca. 1 kcal mol^{-1} [98]. This reasoning was supported by model molecular mechanics calculations on α-CyD [93] showing that (a) the usually depicted structure with planar rings formed by glycosidic oxygen atoms does not correspond to the energy minimum and (b) the energy hypersurface exhibits several energy minima separated by low barriers. With regard to CyD complexes, they are held together by weak intermolecular interactions which somewhat limit the macrocycle's mobility but cannot endow the macrocycle with considerable rigidity.

It should be emphasized that a rigid structure for CyDs is also incompatible with the ease of formation of inclusion complexes of various shapes, since the latter implies an effective fitting of the host and guest to each other [90]. Most experimental proofs of the nonrigidity of CyDs come from NMR studies not only in solution but even in the solid state. If CyDs were not flexible then the spectra of complexes with aromatic guests in solution should exhibit several signals for, e.g. H3 CyD protons on different glucopyranose rings pointing into the cavity (Fig. 1.2). This is not the case [99]. Moreover, NMR studies in the solid state show that the rings included in the CyD cavity can exhibit a rapid flip around the C1C4 axis. One example is provided by **27@2** for which ^{2}H NMR spectra are incompatible with the rigid structure [50]; similar evidence has been obtained on the basis of ^{13}C NMR spectra [100, 101]. The rapid inversion of *cis*-decalin **33** in the complex with **2** at room temperature frozen at 233 K, observed in both ^{1}H and ^{13}C NMR spectra [102, 103], is also incompatible with CyD rigidity.

The very fast internal movement of native CyDs and of most of their derivatives, leading to the observation of averaged structures by most experimental techniques, is frequently overlooked. In addition to the temperature-dependent process of self-inclusion of substituent(s) [104–107], we were able to find only two studies of substituted CyDs in which movement of the macrocycles was at least partly frozen [104, 108]. Some other experimental results proving CyD flexibility using NMR

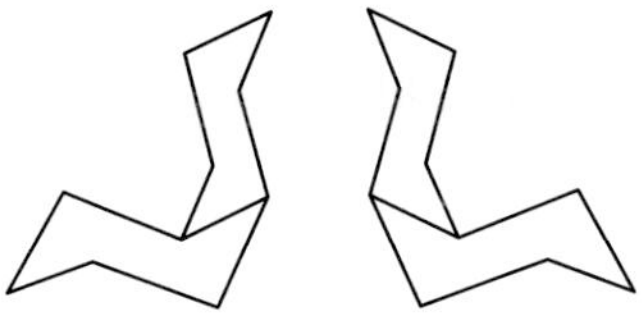

33

and/or other methods [109–114] (discussed in more detail in Ref. [95] have also been reported. Raman studies of H/D and/or D/H exchange rates and those of $H_2{}^{17}O$ in the solid CyD hydrates also indirectly proved the nonrigidity of CyDs [115, 116]. Interestingly, in spite of the time and space averaging characteristic of X-ray and neutron diffraction, arguments in favor of the nonrigidity of native CyDs have been also recently reported on the basis of these techniques. For instance, neutron diffraction study of powder and single-crystal samples of β-CyD crystallized from D_2O shows 11 lattice D_2O molecules occupying 16 positions while deuterium nuclear resonance spectra of the same species exhibited only a single exchange-averaged doublet, proving nonrigidity by showing that D_2O molecules freely move between these sites in a time less than 10^{-6} s [117]. Similarly, the positional disorder of 6 water molecules included in the cavity of β-CyD crystal containing 11 H_2O molecules per oligosaccharide and the orientational disorder (flip-flop, Fig. 1.7) exhibited by the hydroxyl groups also point to CyD nonrigidity [118].

On the basis of X-ray analysis, methylated or acetylated CyDs are sometimes considered to be more flexible than the native ones (see Chapter 7). Such a conclusion seems unfounded since X-ray diffraction can yield straightforwardly only the structure of the macrocyclic ring averaged over time and space, not its mobility. As a matter of fact, native CyDs are more flexible, and thus more difficult to freeze, than permethylated ones. This fact is frequently overlooked since the above mentioned averaging is not taken into account. As shown above, NMR spectra in solution and in the solid state are much more sensitive to CyD flexibility and clearly prove their nonrigidity.

The shape and flexibility of CyDs larger than γ-CyD ($n > 8$) are briefly discussed in Section 13.4. Of course, they must be influenced by the macrocycle ring size and should be different for those with $n = 9$ or 10 from the giant ones with $n = 50$ or 100. It is interesting to note that for $n = 26$ the macrocyclic ring is twisted, with hydrogen bonds linking the hydroxyl groups on opposite sides of the ring [119]. Some interesting observations on the considerable flexibility of large CyDs are also given in Section 13.4.

1.4 Models of Chiral Recognition by Cyclodextrins

As saccharides, CyDs are chiral and exhibit *chiral recognition*, that is they form diastereomeric complexes, usually of different stability, with enantiomeric species.

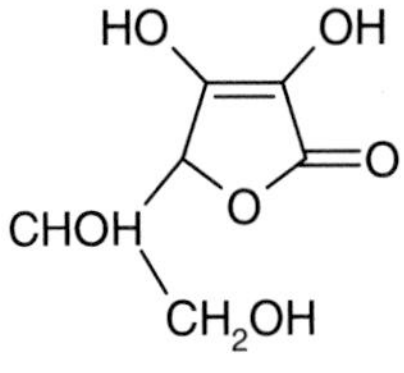

34

The latter observation is of importance in particular for separation science and the pharmaceutical industry owing to the usually different pharmacological activity of enantiomers of a chiral compound [34] as evidenced, for instance, by the action of (+)- and (−)-ascorbic acid **34** in humans [120, 121]. The second enantiomer of a chiral drug, usually present as 50% impurity, can even be harmful. Therefore, the administration of drugs in the form of one enantiomer that exhibits the desired pharmacological activity is of vital importance for the pharmaceutical industry. The possibility of differentiating between natural enantiopure fragrances and their synthetic racemic counterparts is also of value for the cosmetic industry [122]. Therefore, the ability to predict which enantiomer forms a stronger complex with a particular CyD and to estimate the energy difference between these diastereomeric complexes is of particular significance for the pharmaceutical and some branches of the cosmetic industry as well as for separation science. However, these energy differences are usually very small. As stated in Chapter 6, a difference as small as 10 cal is sufficient to achieve a satisfactory chromatographic resolution today. The impossibility of formulating any model of chiral recognition allowing one to make the predictions so desirable for the commercial applications is a consequence of these small energy differences combined with the flexibility of CyDs and the complexity of complexation process. The situation was probably most precisely formulated long ago by Pirkle and Pochapsky [123] in their review on chiral recognition, although it was not devoted to CyD research. Their statements are long but so precise that they should be cited explicitly:

> Owing to the nature of chromatographic processes, relatively small values of $\Delta\Delta G$ suffice to afford observable chromatographic separations. A value of 50 small calories (note that today the limit is 10 not 50, HD) affords a separation of 1.09, easily observable on a high-efficiency HPLC system. There is justifiable skepticism concerning the validity of any mechanism purporting to explain such small energy differences, despite a strong tendency among workers in the field to advance chiral recognition rationales, even when comparatively few data are available upon which to base such a rationale.... Typically, chromatographic separation of enantiomers involves solution interactions between CSP (i.e. chiral stationary phase, HD) and analyte for which free energies are small with respect to kT. This implies that the molecules are relatively free to tumble with respect to each other and exert relatively little mutual conformational control.... It is es-

sentially to recognize that it is the *weighted time average* of all possible solution interactions that is important for determining retention and enantioselectivity.

These statements determine the author's whole attitude to models of chiral recognition, presented below, based both on experimental data and/or on calculations.

The following models used for the prediction of chiral recognition by CyDs have been proposed and/or applied to explain the recognition:

1. Dalgleish three-points model [11].
   This most frequently invoked model of chiral recognition proposed that three pairs of point-to-point interactions between a solute and a selector can explain enantioselectivity of the chiral stationary phase, CSP. Such an approach is incompatible with the properties of three-dimensional (3D) space and the foundations of molecular chirality since, as pointed out by Prelog and Helmchen [12] and Dodziuk and Mirowicz [13], three points do not form a chiral figure in 3D space. Interestingly [123], a chiral stationary phase developed by Baczuk and coworkers [124] using a three-points recognition model based on *l*-arginine to separate the *l*-enantiomer of the antiparkinson's drug DOPA (dihydroxyphenylalanine, Fig. 1.8) [124] separated the DOPA enantiomers but, contrary to the Dalgleish model, the *d*-enantiomer was found to be mostly retained on the *l*-CSP, indicating that the mechanism of action was not that originally proposed. It is noteworthy that, in spite of its incorrectness, the three-points model of chiral recognition has been and is still used to rationalize experimental results on chiral recognition by CyDs [125–128]. Bentley [129] commented on the model stating that four-points interactions are necessary. However, he claimed that with a drug (or substrate) approaching the receptor (or enzyme) surface from only one direction [an assumption that should be proved in a general case, HD] from the exterior but not the interior of the protein three-point attachment (interactions between six centers) suffices. In addition, it should be stressed that in the case of CyDs the host–guest interactions cannot usually be described as point-

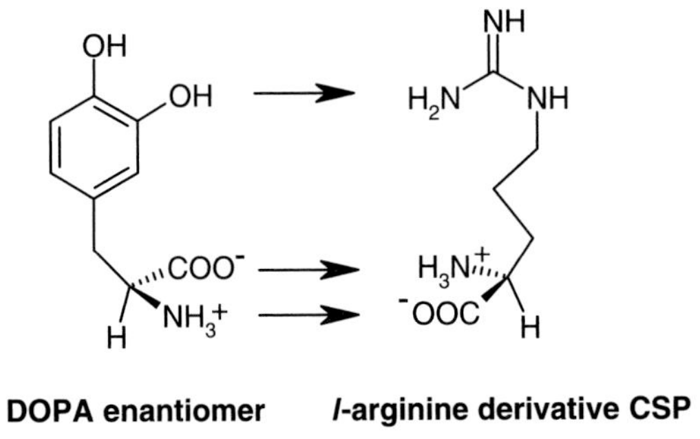

Fig. 1.8. The model of three pair-wise interactions on which the development of Baczuk's CSP was based.

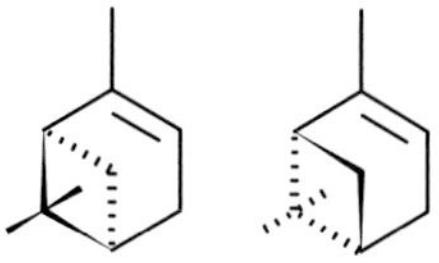

35

to-point ones (e.g. for pinene **35** complexes with **1**). Moreover, the Dalgleish description is sometimes inappropriately used. For instance, in Ref. [126] the whole aromatic ring is treated as a point.

2. A model relating the lower symmetry and greater flexibility of permethylated α-**5** (R1 = R2 = R3 = Me), β- and γ-CyDs, with their supposedly better chiral recognition has been proposed [130, 131]. It should be stressed that (i) as discussed in the previous section, X-ray analysis can provide data on the average molecular symmetry, not on the flexibility and (ii) the model has been founded on very few established X-ray structures of the complexes involving both enantiomers, and does not find support in chromatographic studies. The studies involving the latter method show that, in general, permethylated hosts are not better chiral selectors than native parent CyDs **1–3** [132, 133]. The problems with the description of geometrical parameters characterizing the structures of CyD complexes with enantiomeric guests are worth mentioning here since they are not always realized:
   (a) Different numbers of crystalline water are often observed for such complexes. For instance, the complexes of *R*- and *S*-mandelic acid **36** with permethylated α-CyD **5f** were found in the form of di- and trihydrate, respectively [131].
   (b) The difficulty of analyzing geometrical parameters in such big systems have been mentioned earlier. For them one can either compare all numerous bond lengths, bond angles, etc. losing transparency or one can compare one type of parameter in two (or more) structures, e.g. the distance between glycosidic oxygen atoms on neighboring rings $O4_n$–$O4_{n+1}$ in the complexes of fenoprofen **37** with β-CyD [134]. The data collected in Table 1.2 (obscured

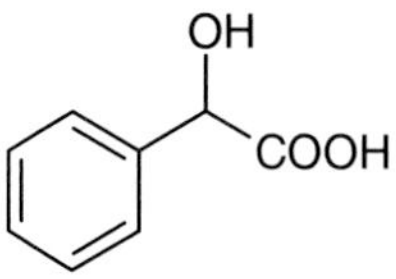

36

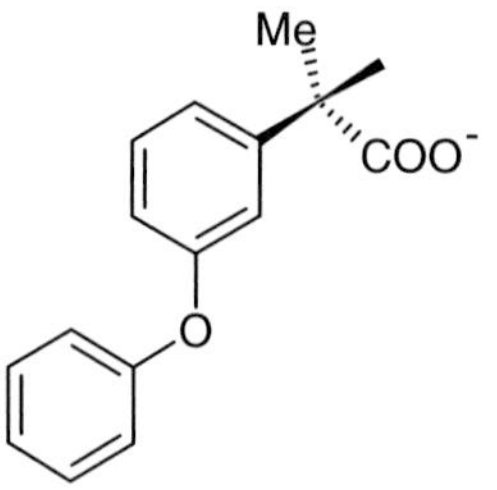

37

Tab. 1.2. $O4_n$–$O4_{n+1}$ distances in pm for the complexes of two (*R*)- and two (*S*)-isomers of **37** with **2** [134]

$O4_n$–$O4_{n+1}$	(*R*)-37@CyD1	(*R*)-37@CyD2	(*S*)-37@CyD1	(*S*)-37@CyD2
$O4_1$–$O4_2$	430 (2)	433 (2)	425 (1)	430 (2)
$O4_2$–$O4_3$	442 (1)	444 (1)	459 (3)	456 (3)
$O4_3$–$O4_4$	439 (3)	446 (3)	426 (1)	426 (2)
$O4_4$–$O4_5$	435 (1)	433 (2)	433 (2)	428 (2)
$O4_5$–$O4_6$	432 (2)	434 (2)	435 (2)	445 (2)
$O4_6$–$O4_7$	435 (2)	438 (2)	453 (2)	447 (3)
$O4_7$–$O4_1$	446 (2)	446 (3)	424 (2)	426 (2)
Mean value	437 (2)	438 (2)	436 (2)	437 (2)

by the existence of two guest molecules in the complexes involving a β-CyD capsule in the crystal) visualize the problem showing that mean values are practically the same while individual ones differ considerably.

3. The requirements for efficient chiral recognition formulated in 1987 [135] demanded inclusion complex formation with a tight fit of the included guest in the host cavity (implying the formation of a strong complex) and the formation of a strong interaction of the hydroxyl groups at the CyD cavity entrance with a guest stereogenic center as the conditions. The first requirement of inclusion complex formation was later lifted. It seems that, in general, the tight fit implying the formation of strong complexes can be unfavorable for enantiodifferentiation, as evidenced by the case of α-pinene **35** the enantiomers of which form stronger complexes with β-CyD **2** than with α-CyD **1** although they are discriminated by the latter host [136].
4. As concerns the formation of hydrogen bonds between the host and guest as a condition of effective chiral recognition, numerous examples of chiral hydrocarbons effectively discriminated by CyDs [102, 137, 136] clearly show that this condition does not hold. Interestingly, by studying chiral recognition of several terpenes by CyDs Sybilska and coworkers [138] showed that for this specific group of compounds the condition for efficient separation could be the formation of 1:2 complexes of these guests with two CyD host molecules forming a head-to head, head-to-tail or tail-to-tail capsule (Fig. 1.9) in which a guest enantiomer is located.
5. Computational models. Easy access to computers and user-friendly program packages as well as the beauty of molecular models have resulted in a plethora of theoretical papers aiming at a rationalization of chiral recognition by CyDs. Computational studies of CyDs and their complexes, and in particular those referring to chiral recognition, are described in Chapter 11 in some detail. Here it should suffice to say that such calculations are mostly treated as operations on a

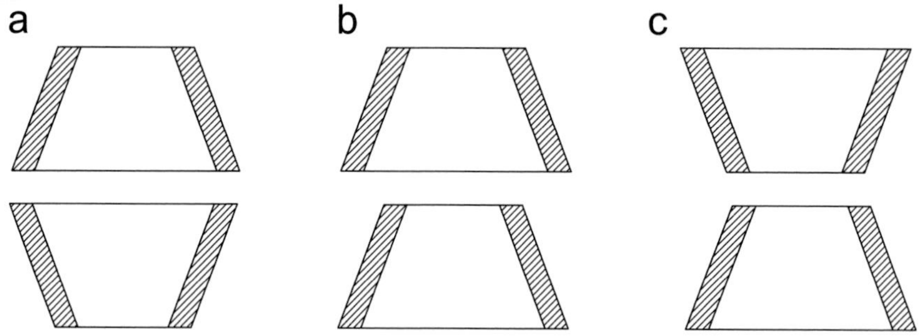

Fig. 1.9. Head-to-head (a), head-to-tail (b) and tail-to-tail (c) orientations in a CyD dimer.

black box. Therefore, their reliability is practically never checked in the situation when the results obtained do depend on the assumed parameterization and their accuracy is much lower than the value of the energy difference for the complexes involving enantiomeric guests they aim to reproduce.

References

1 Szejtli, J. *Comprehensive Supramolecular Chemistry*; Pergamon: Oxford, 1996; Vol. 3.

2 Bender, M. L.; Momiyama, M. *Cyclodextrin Chemistry*; Springer: Berlin, 1978.

3 *Chem. Rev.* 1998, *98*.

4 Villiers, A. *Comp. Rend.* 1896, *112*, 536.

5 Schardinger, F. *Zentralbl. Bacteriol. Parasitenkd., Infektionskrank. Hyg. Abt.2 Naturwiss: Mikrobiol. Landwirtschaft Technol. Umweltschutzes* 1911, *29*, 188.

6 Stoddart, J. F. *Carbohydr. Res.* 1989, *192*, xii–xv.

7 Saenger, W. *Angew. Chem. Int. Ed. Engl.* 1980, *19*.

8 Harata, K. In *Comprehensive Supramolecular Chemistry*; Szejtli, J., Ed.; Pergamon: Oxford, 1996; Vol. 3, p 279.

9 Zabel, W.; Saenger, W.; Mason, S. A. *J. Am. Chem. Soc.* 1986, *108*, 3664–3673.

10 Inoue, Y.; Takahashi, Y.; Chujo, R. *Carbohydr. Res.* 1985, *144*, c9–c11.

11 Dalgliesh, C. E. *J. Chem. Soc.* 1952, 3940–3942.

12 Prelog, V.; Helmchen, G. *Angew. Chem.* 1982, *21*, 567.

13 Dodziuk, H.; Mirowicz, M. *Tetrahedron Asymm.* 1990, *1*, 171.

14 Breslow, R.; Halfon, S.; Zhang, B. *Tetrahedron* 1995, *51*, 377–388.

15 Jicsinszky, L.; Hashimoto, H.; Fenyvesi, E.; Ueno, A. In *Comprehensive Supramolecular Chemistry*; Szejtli, J., Ed.; Pergamon: Oxford, 1996; Vol. 3, p 57.

16 Easton, C. J. *Modified Cyclodextrins Scaffolds and Templates for Supramolecular Chemistry*; Imperial College Press: London, 1999.

17 Khan, A. R.; Barton, L.; D'Souza, V. T. *J. Org. Chem.* 1996, *61*, 8301–8303.

18 Mori, M.; Ito, Y.; Ogawa, T. *Tetrahedron Lett.* 1990, *31*, 3029–3030.

19 Peroche, S.; Parrot-Lopez, H. *Tetrahedron Lett.* 2003, *44*, 241–245.

20 Tabushi, K.; Shimokawa, K.; Fujita, K. *Tetrahedron Lett.* 1977, *18*, 1527.

21 Djeddaini-Pilard, F.; Azaroual-Bellanger, N.; Gosnat, M.; Vernet, D.; Perly, B. *J. Chem. Soc. Perkin Trans. 2* 1995, 723.

22 Venema, F.; Rowan, A. E.; Nolte, R. J. M. *J. Am. Chem. Soc.* 1996, *118*, 257–258.

23 Ueno, A.; Moriwaki, F.; Iwama, Y.; Suzuki, I.; Osa, T.; Ohta, T.; Nozoe, S. *J. Am. Chem. Soc.* 1991, *113*, 7034.

24 Tee, O. S.; Takasaki, B. K. *Can. J. Chem.* 1985, *63*, 3540.

25 Czarniecki, M. F.; Breslow, R. *J. Am. Chem. Soc.* 1978, *100*, 7771–7772.

26 Breslow, R. In *Inclusion compounds*; Atwood, J. L., Davies, J. E. D., MacNicol, D. D., Eds.; Academic Press: London, 1984.

27 Breslow, R.; Campbell, P. *J. Am. Chem. Soc.* 1969, *91*, 3084.

28 Trainor, G. L.; Breslow, R. *J. Am. Chem. Soc.* 1981, *103*, 154–158.

29 Breslow, R.; Dong, S. D. *Chem. Rev.* 1998, *98*, 1997–2011.

30 Komiyama, M. In *Comprehensive Supramolecular Chemistry*; J, S., Ed.; Pergamon: Oxford, 1996; Vol. 3, pp 401–421.

31 Murakami, Y.; Kikuchi, J.-I.; Hisaeda, Y.; Hayashida, O. *Chem. Rev.* 1996, *96*, 721–758.

32 Dodziuk, H. *Introduction to Supramolecular Chemistry*, Kluwer: Dordrecht, 2003, pp 104–105, 152–155.

33 Han, M. J.; Yoo, K. S.; Chang, J. Y.; Ha, T.-K. *Angew. Chem. Int. Ed. Engl.* 2000, *39*, 347.

34 Leffingwell, J. C. *Leffingwell Reports* 2003, *3*, 1–27.

35 Cayen, M. N. *Chirality* 1991, *3*, 94.

36 Dispenzieri, A. *New Engl. J. Med.* 2005, *352*, 2546.

37 Gullestad, L.; Semb, A. G.; Fjeld, J. G.; Holt, E.; Gundersen, T.; Breivilk, K.; Folling, M.; Hodt, A.; Skardal, R.; Andreassen, A.; Kjekshus, J.; Ueland, T.; Kjekshus, E.; Yndestad, A.; Froland, S. S.; Aukrust, P. *Circulation* 2004, *110*, 3349 Suppl. S.

38 Bulic, S. O.; Fassihi, H.; Mellerio, J. E.; McGrath, J. A.; Atherton, D. J. *Brit. J. Dermatology* 2005, *152*, 1332.

39 Sybilska, D.; Asztemborska, M.; Bielejewska, A.; Kowalczyk, J.; Dodziuk, H.; Duszczyk, K.; Lamparczyk, H.; Zarzycki, P. *Chromatogr.* 1993, *35*, 637.

40 Dodziuk, H.; Sitkowski, J.; Stefaniak, L.; Sybilska, D.; Jurczak, J.; Chmurski, K. *Pol. J. Chem.* 1997, *71*, 757.

41 Morin, N.; Cornet, S.; Guinchard, C.; Rouland, J. C.; Guillaume, Y. C. *J. Liq. Chromatog. Relat. Technol.* 2000, *23*, 727–739.

42 Lin, C. E.; Chen, K. H. *Electrophoresis* 2003, *24*, 3139–3146.

43 Betzel, C.; Saenger, W.; Hingerty, B. E.; Brown, G. M. *J. Am. Chem. Soc.* 1984, *106*, 7545.

44 Armsbach, D.; Ashton, P. R.; Ballardini, R.; Balzani, V.; Godi, A.; Moore, C. P.; Prodi, L.; Spencer, N.; Stoddart, J. F.; Tolley, M. S.; Wear, T. J.; Williams, D. J. *Chem. Eur. J.* 1995, *1*, 33.

45 Kluefters, P.; Schumacher, J. *Angew. Chem. Int. Ed. Engl.* 1994, *33*, 1863.

46 Voegtle, F.; Mueller, W. M. *Angew. Chem.* 1979, *91*, 676.

47 Kamitori, S.; Hirotsu, K.; Higuchi, T. *J. Am. Chem. Soc.* 1987, *109*, 2409–2414.

48 Rekharsky, V. M.; Inoue, Y. *Chem. Rev.* 1998, *98*, 1857–1917.

49 Neuhaus, D.; Williamson, M. P. *The Nuclear Overhauser Effect in Structural and Conformational Analysis*; VCH, 1989.

50 Usha, M. G.; Wittebord, R. J. *J. Am. Chem. Soc.* 1992, *114*, 1541–1548.

51 Anczewski, W.; Dodziuk, H.; Ejchart, A. *Chirality* 2003, *15*, 654–659.

52 Castro, R.; Cuadrado, I.; Alonso, B.; Casado, C. M.; Moran, M.; Kaifer, A. E. *J. Am. Chem. Soc.* 1997, *119*, 5760–5761.

53 Piela, L. *Ideas of Quantum Chemistry*; Elsevier: New York, 2006.

54 Allinger, N. L.; Burkert, U. *Molecular Mechanics*; American Chemical Society Monograph, 1982.

55 van Gunsteren, W. F.; Berendsen, H. J. C. *Angew. Chem. Int. Ed. Engl.* 1990, *29*, 992–1023.

56 Harada, A.; Kamachi, J. *Nature* 1992, *356*, 325.

57 Dodziuk, H. In *Introduction to Supramolecular Chemistry*; Kluwer: Dordrecht, 2003, pp 27–35, 275–284.

58 Gessler, K.; Uson, I.; Takaha, T.; Krauss, N.; Smith, S. M.; Okada, S.; Sheldrick, G. M.; Saenger, W. *Proc. Natl. Acad. Sci. USA.* 1999, *96*, 4246–4251.

59 Szejtli, J. In *Comprehensive Supramolecular Chemistry*; J, S., Ed.; Pergamon: Oxford, 1996; Vol. 3, p 1.

60 Szejtli, J. In *Comprehensive Supramolecular Chemistry*; Szejtli, J., Ed.; Pergamon: Oxford, 1996; Vol. 3, p 5.

61 Toda, J.; Misaki, M.; Konno, A.; Wada, T.; Yasumatsu, K. In *The International Flavor Conference*, 1985.

62 Spect, M.; Rothe, M.; Szente, L.; Szejtli, J.: Ger. Pat. (DDR), 1981.

63 Hashimoto, H.; Hara, K.; Kuwahara, N.; Mikuni, K.; Kainuma, K.; Kobayashi, S.: *Jpn Kokai*, 1985.

64 Kobayashi, M.; Yamashita, K.; Matukura, K.; Okumura, J.: *Jpn Kokai*, 1981.

65 Kijima, I.: *Jpn Kokai*, 1981.

66 Kominato, Y.; Nishimura, S.: *Jpn Kokai*, 1985.

67 Fukunaga, T.; (kunjudu Koho K. K.) In *Chem. Abstr., 1978, 90, 12181*: Jpn Kokai, 1978.

68 Duchene, D. *New Trends in Cyclodextrins and Derivatives*; Edition de Sante: Paris, 1991.

69 Duchene, D. *Cyclodextrins and Their Industrial Uses*; Edition de Sante: Paris, 1987.

70 Szejtli, J. *Cyclodextrin Technology*; Kluwer: Dordrecht, 1988.

71 Issue, s. *Chem. Rev.* 1989, *98*, 1741.

72 Schmid, G. In *Comprehensive Supramolecular Chemistry*; J, S., Ed.; Pergamon: Oxford, 1996; Vol. 3, p 41.

73 Lichtenthaler, F. V.; Immel, S. *Tetrahedron Asymm.* 1994, *5*, 2045.

74 Schmid, G. In *Comprehensive Supramolecular Chemistry*; J, S., Ed.; Pergamon: Oxford, 1996; Vol. 3, p 615.

75 Takahashi, Y.; Ogawa, T. *Carbohydr. Res.* 1987, *164*, 277.

76 Takahashi, Y.; Ogawa, T. *Carbohydr. Res.* 1987, *169*, 127.

77 Nakagawa, T.; Koi, U.; Kashiwa, M.; Watanabe, J. *Tetrahedr. Lett.* 1994, *35*, 1921.

78 Sundararajan, P. R.; Rao, V. S. R. *Carbohydr. Res.* 1970, *13*, 351.

79 Gadelle, A.; Defaye, J. *Angew. Chem., Int. Ed. Eng.* 1991, *30*, 78.

80 Ashton, P. R.; Ellwood, P.; Staton, I.; Stoddart, J. F. *Angew. Chem., Int. Ed. Eng.* 1991, *30*, 80.

81 Caira, M. R.; Griffith, V. J.; Nassimbeni, L. R.; van Oudtshoorn, B. *J. Chem. Soc. Perkin Trans. 2* 1994, 2071.

82 Caira, M. R.; Griffith, V. J.; Nassimbeni, L. R.; van Oudtshoorn, B. *Supramol. Chem.* 1996, *7*, 119.

83 Yang, C.; Yuan, D. Q.; Nogami, Y.; Fujita, K. *Tetrahedron Lett.* 2003, *44*, 4641.

84 Ueda, H. *J. Incl. Phenom. Macro. Chem.* 2002, *44*, 53.

85 Takaha, T.; Yanase, M.; Takata, S.; Okada, S.; Smith, S. M. *J. Biol. Chem.* 1996, *271*, 2902.

86 Entzeroth, M.; Moore, R. E.; Niemczura, W. P.; Patterson, M. L. G.; Schoolery, J. N. *J. Org. Chem.* 1986, *51*, 5307.

87 Szente, L. In *Comprehensive Supramolecular Chemistry*; J, S., Ed.; Pergamon: Oxford, 1996; Vol. 3, p 243.

88 Andersson; Nilsson, K.; Sundahl, M.; Westman, G.; Wennerström. *Chem. Commun.* 1992, 604.

89 Szejtli, J. In *Comprehensive Supramolecular Chemistry*; J, S., Ed.; Pergamon: Oxford, 1996; Vol. 3, p 189.

90 Koshland, D. E. *Angew. Chem. Int. Ed. Engl.* 1994, *33*, 2475.

91 Hamai, S. *J. Phys. Chem.* 1989, *93*, 2074.

92 Munoz de la Pena, A.; Ndou, T. T.; Zung, J. B.; Greene, K. L.; Live, D. H.; Warner, I. M. *J. Am. Chem. Soc.* 1991, *113*, 1572.

93 Dodziuk, H.; Nowinski, K. *J. Mol. Struct. (THEOCHEM)* 1994, *304*, 61.

94 Dodziuk, H. In *Modern Conformational Analysis, Elucidating Novel*

Exciting Molecular Structures; Wiley-VCH: New York, 1995, p 219.
95 DODZIUK, H. *J. Mol. Struct.* 2002, *614*, 33.
96 LI, S.; PURDY, W. C. *Chem. Rev.* 1992, *92*, 1457.
97 MARQUES, H. M. C. *Rev. Port. Farm.* 1994, *44*, 77.
98 ELIEL, E. L.; ALLINGER, N. L.; ANGYAL, S. J.; MORRISON, G. A. *Conformational Analysis*; Interscience: New York, 1965.
99 INOUE, Y. *Annu. Rep. NMR Spectrosc.* 1993, *27*, 59.
100 INOUE, Y.; OKUDA, T.; CHUJO, R. *Carbohydr. Res.*, 1985, *141*, 179.
101 YAMAMOTO, Y.; ONDA, M.; TAKAHASHI, Y.; INOUE, Y.; CHUJO, R. *Carbohydr. Res.*, 1988, *182*, 41.
102 DODZIUK, H.; SITKOWSKI, J.; STEFANIAK, L.; JURCZAK, J.; SYBILSKA, D. *Chem. Commun.* 1992, 207.
103 DODZIUK, H.; SITKOWSKI, J.; STEFANIAK, L.; JURCZAK, J.; SYBILSKA, D. *Supramol. Chem.* 1993, *3*, 79.
104 JULLIEN, J.; CANCEILL, J.; LACOMBE, L.; LEHN, J.-M. *J. Chem. Soc., Perkin Trans. 2* 1994, 989.
105 MACALPINE, S. R.; GARCIA-GARRIBAY, M., JACS 1996 118 2750; *J. Am. Chem. Soc.* 1996, *118*, 2750.
106 MACALPINE, S. R.; GARCIA-GARRIBAY, M. A. *J. Org. Chem.* 1996, 8307.
107 DODZIUK, H.; CHMURSKI, K.; JURCZAK, J.; KOZMINSKI, W.; LUKIN, O.; SITKOWSKI, J.; STEFANIAK, L. 2000, *519*, 33.
108 ELLWOOD, P.; SPENCER, C. M.; SPENCER, N.; STODDART, J. F.; ZARZYCKI, R. *J. Inc. Phenom. Macro.* 1992, *12*, 121–150.
109 CORRADINI, R.; DOSSENA, A.; MARCHELLI, R.; PANAGIA, A.; SARTOR, G.; SAVIANO, M.; LOMBARDI, A.; PAVONE, V. *Chem. Eur. J.* 1996, *2*, 373.
110 BANGAL, P. R.; CHAKRAVORTI, S.; MUSTAFA, G. *J. Photochem. Photobiol. A-Chem.* 1998, *113*, 35.
111 GRABNER, G.; RECHTHALER, K.; MAYER, B.; KOHLER, G.; ROTKIEWICZ, K. *J. Phys. Chem. A* 2000, *104*, 1365.
112 DU, X. Z.; ZHANG, Y.; HUANG, X. Z.; LI, V. Q.; JIANG, Y. B.; CHEN. G. Z. *Spectrochim. Acta A* 1996, *52*, 1541.
113 BRIGHT, F. V.; CAMENA, G. C.; HUANG, J. *J. Am. Chem. Soc.* 1990, *112*, 1343.
114 BELL, A. F.; HECHT, L.; BARRON, L. D. *Chem. Eur. J.* 1997, *3*, 1292.
115 STEINER, T.; MOREIRA, A. M.; TEXEIRA-DIAS, J. J. C.; MÜLLER, J.; SAENGER, W. *Angew. Chem. Int. Ed. Engl.* 1995, *34*, 1452.
116 AMADO, A. M.; RIBEIROCALRO, P. J. A. *J. Chem. Soc., Faraday Trans* 1997, *93*, 2387.
117 STEINER, T.; MASON, S. A.; SAENGER, W. *J. Am. Chem. Soc.* 1991, *113*, 5676.
118 STEINER, T.; SAENGER, W.; LECHNER, R. E. *Mol. Phys.* 1991, *72*, 1211.
119 GESSLER, K.; USON, I.; TAKAHA, T.; KRAUSS, N.; SMITH, S. M.; OKADA, S.; SHELDRICK, G. M.; SAENGER, W. *Proc. Natl. Acad. Sci. USA* 1999, *96*, 4246.
120 PARK, C. H. *Cancer Res.* 1985, *45*, 3969.
121 FRIEDMAN, P. A.; ZEIDEL, M. L. *Nat. Med.* 1999, *5*, 620.
122 KRAFT, P.; FRATER, G. *Chirality* 2001, *13*, 388.
123 PIRKLE, W. H.; POCHAPSKY, T. C. *Chem. Rev.* 1989, *89*, 347.
124 BACZUK, R. J.; LANDRAM, G. K.; DUBOIS, R. J.; DEHM, H. C. *J. Chromat.* 1971, *60*, 351.
125 KANO, K. *J. Phys. Org. Chem.* 1997, *10*, 286.
126 CAMILLERI, P.; EDWARDS, A. J.; RZEPA, H. S.; GREEN, S. M. *Chem. Commun.* 1992, 1122.
127 AHN, S.; RAMIREZ, J.; GRIGOREAN, G.; LEBRILLA, C. B. *J. Am. Soc. Mass Spectrosc.* 2001, *12*, 278.
128 LEBRILLA, C. B. *Acc. Chem. Res.* 2000, *34*, 653.
129 BENTLEY, R. *Chem. Brit.* 1994, *30*, 191.
130 HARATA, K. In *Minutes, 5th International Symposium on Cyclodextrins*; Paris, 1991.
131 HARATA, K.; UEKAMA, K.; OTAGIRI, M.; HIRAYAMA, F. *Bull. Chem. Soc. Jpn.* 1987, *60*, 497.
132 BIELEJEWSKA, A.; NOWAKOWSKI, R.; DUSZCZYK, K.; SYBILSKA, D. *J. Chromatogr. A* 1999, *840*, 159–170.
133 SYBILSKA, D.; BIELEJEWSKA, A.;

Nowakowski, R.; Duszczyk, K. *J. Chromatogr.* 1992, *625*, 349–352.

134 Hamilton, J. A.; Chen, L. *J. Am. Chem. Soc.* 1988, *110*, 4379–4391.

135 Armstrong, D. W.; Yang, X.; Han, S. M.; Menges, R. A. *Anal. Chem.* 1987, 2594.

136 Asztemborska, M.; Nowakowski, R.; Sybilska, D. *J. Chromat.* 2000, *902*, 381.

137 Dodziuk, H.; Sitkowski, J.; Stefaniak, L.; Sybilska, D. *Pol. J. Chem.* 1996, *70*, 1361.

138 Asztemborska, M.; Sybilska, D.; Nowakowski, R.; Perez, G. *J. Chromatogr. A* 2003, *1010*, 233.

2
Modification Reactions of Cyclodextrins and the Chemistry of Modified Cyclodextrins

Kenjiro Hattori and Hiroshi Ikeda

2.1
Scope of This Chapter

The classic chemistry of native CyDs is now a kind of completed area, described in a book [1] and several reviews that summarize state-of-the-art syntheses of these macrocycles [2–4], while syntheses of several novel CyD structures or their analogues are given in the literature [5–8]. Studies of CyDs as enzyme models, involving syntheses of numerous selectively substituted derivatives, have been reviewed [9–11]. Numerous journals and proceedings can also be mentioned here [12, 13].

The first part of this chapter describes methods for modifying CyDs in which the number and exact positions of substituents have been ascertained and where pure compounds with unambiguous structures have been obtained.

The second half of the chapter focuses on the chemistry involved for the modified CyDs, such as CyD oligomers, charged CyDs, CyD chemosensors, CyD analogues, coupling of CyDs with other kinds of host and with peptides and saccharides, while CyD polymers and catenanes and rotaxanes of nontrivial topology will be discussed in Chapters 3 and 12, respectively. Improvements in the binding, selectivity, and control by photons, metal ions, or pH have made possible the use of modified CyDs as chemosensors, artificial enzymes, carriers, and reaction fields, and also as scaffolds for highly preorganizing functional units. Applications of such CyDs in the pharmaceutical industry are presented in Chapters 14 and 15 while their present and prospective uses in other domains are discussed in Chapter 16.

2.2
Modification Reactions of Cyclodextrins

When highly water-soluble CyD derivatives are investigated for application in drug formulations, random modifications of hydroxyl groups to hydroxypropyl, sulfopropyl, carboxymethyl, or silyl groups can be easily achieved [14]. On the other hand, if a supramolecular behavior is to be investigated using CyD derivatives, the

Cyclodextrins and Their Complexes. Edited by Helena Dodziuk

ISBN: 3-527-31280-3

compound needs to be pure and well characterized. The number and the exact position of substituents have to be established before mechanistic information about the supramolecular behavior can be reliably derived. This section provides the systematic description of modification methods that is a prerequisite for any detailed discussion of CyD functionalizations. Two predominant factors have to be considered for modification reactions involving CyD; the nucleophilicity of the hydroxyl groups of the C2-, C3- and C6-positions and the ability of CyDs to form complexes with the reagents used. Since hydroxyl groups are nucleophilic in nature, the initial reaction is an electrophilic attack on these positions. Among three kinds of hydroxyl groups, primary side C6–OH groups are most basic and most nucleophilic, C2–OH groups are most acidic, and C3–OH groups are the most inaccessible. Therefore electrophilic reagents will initially attack C6–OH groups. More-reactive reagents will react with C6–OH and also with C2–OH and C3–OH. The size of the CyD cavity also has an effect on the strength and the orientation of the intermediate complex and affects the product structure. For example, in alkaline aqueous solution, tosyl chloride reacts with α-CyD to give C2-tosyl α-CyD, whereas with β-CyD, it gives C6-tosyl β-CyD [15].

Conversion of CyDs to polymeric materials is of great value for the practical use of CyD in the field of pharmaceutical and environmental applications such as drug delivery, slow release of guest molecules, and removal of pollutants, and are described in many papers and patents but are omitted here.

2.2.1 Modification Reactions at the Primary Side

2.2.1.1 Mono-modification at the C6-Position

6-Tosyl β-CyDs are important precursors for a variety of modified CyDs because a nucleophile can attack the electrophilic carbon at the 6-position to produce a corresponding functionality. The reaction of β-CyD with tosyl chloride in aqueous alkaline medium gives mono-6-tosylated β-CyD in fairly good yield [15].

Modifications of C6 of CyD other than by tosylation are very rare. However, a single-step quantitative-yield synthesis of CyD monoaldehydes has been published. The CyD was dissolved in an organic solvent, Dess–Martin perodinane was added, and the mixture was stirred for 1 h at room temperature. Addition of acetone and cooling allowed isolation of the crude product by filtration [16]. Another synthesis was performed by using IBX (1-hydroxy-1,2-benziodoxol-3(1*H*)-one 1-oxide) as oxidant in DMSO. Mono-oxidation of β-CyD was performed along with its incorporation into chitosan by a reductive coupling reaction [17]. A direct azidation of CyDs with sodium azide in the presence of triphenylphosphine–carbon tetrabromide has also been reported [18].

Starting from 6-monotosyl CyD, many derivatives can be obtained, including aldehyde [19], iodide, chloride, azide, amino-, and alkyldiamino-CyDs. Dimeric β-CyD receptors were synthesized from mono-6-iodo-CyD with a dithiol core, and they have potential as selective receptors of α-helical peptides [20]. Fullerene derivatives bearing α-, β- and γ-CyD units were synthesized by the reaction of C_{60} and peracetylated CyD 6-azides [21]. From 6-monoazide-β-peracetylated CyD or

β-CyD → 6-*mono*-tosyl CyD [15] → 6-*mono*-X-CyD
β-CyD → 6-*mono*-N_3-CyD [18]
β-CyD → 6-*mono*-CHO-CyD [16,17]

X = -I [20],-NH_3 [21,22]
-SR [23,28]
-CHO [19,24]
-NH_2 [29–35,37,38]
-NHR [25–27,36,39–44]

Scheme 2.1. Mono-modification at the C6-position of *β*-CyD.

heptakis-(6-azide)-*β*-CyD, in a polymer-assisted "one-pot" phosphine imide reaction using polymer-bonded triphenylphosphine was performed and the dimer and the isocyanate were obtained [22].

Mono-6-mercapt-*β*-CyD has been synthesized from mono-tosyl *β*-CyD. This compound was then combined with tri- or tetravalent carbosilane bromide in liquid ammonia to obtain three or four CyD derivatives as a new core substance for the construction of a variety of functional materials [23]. Mono-6-modified-*β*-CyDs bearing N-attached aminobenzoic acids can be prepared from mono-6-formyl-*β*-CyD derived from mono-6-tosyl-*β*-CyD [24]. A new fluorescence probe for a selective, rapid, and simple analytical procedure for the determination of nucleic acids, mono[6-*N*(4-carboxy-phenyl)]-*β*-CyD, was synthesized using C6-monotosyl CyD [25]. Starburst polyamidoamine dendrimer conjugates with *α*-, *β*-, and *γ*-CyDs were synthesized from monotosyl-*α*- and *β*-CyDs and mononaphthalenesulfonyl-*γ*-CyD as transfection-efficient nonviral vectors [26]. The enatiomeric separation of *α*-hydroxy acids and carboxylic acids was performed using 6-*N*-histamino-*β*-CD [27]. Trifluoroethylthio-*β*-CyDs were prepared by way of C6-monotosyl CyD with the hope that the fluorine reporter group by NMR (1H and 19F) may assist in complexation of the drug carriers as well as solubility in water [28]. Heptakis(6-*O*-tert-butyldimethylsilyl)-*β*-CyD involved the permethylation of the secondary hydroxyl groups with MeI and NaH, disylation of the primary hydroxyls, monotosylation of O-6, nucleophilic replacement of the tosyl group with azide anion, and reduction of the azide group to an amino group [29].

A series of CyD trimers and dimers were prepared and examined as binders for appropriate trimeric and dimeric amino acid amides [30]. The synthesis of chitosan-bearing *β*-CyD by the reaction of succinyl chitosan with a small amount of mono-C6-amino-CyD has been improved by the use of an alternative condensing agent DMT-MM (4-(4,6-dimethoxy-1,3,5-triazin-2-yl)-4-methylmorpholinium chloride) [31]. *β*-CyD was converted to C6-heptakis(*tert*-butyldimethylsilyl) CyD. The C2–OH groups, which are more acidic than the C3–OH groups, were selectively deprotonated with NaH and reacted with 1-azido-3-tosyloxypropane. The azide was reduced to the amine. Coupling of the monofunctionalized CyD with deactivated esters produced CyD dimers [32]. Cationically charged C6-monoamino-*β*-CyD was used as a chiral selector in electrokinetic chromatography for chiral separation [33]. Targeted drug delivery systems have been built from *β*-CyDs by monoconjugation of C6-amino-*β*-CyD with mannosyl-coated dendritic branches following an interactive thiourea-forming convergent strategy [34]. C6-Mono-glucose-branched CyDs were synthesized from amino-β-CyD and arbutin. The glucose residues acted as reactive glycosyl acceptors for the transglycosylation by the enzyme to afford sialo-complex type oligosaccharide-branched CyDs [35]. Mono-6-alkylamino-β-CyDs are prepared by way of 6-tosyl CyD. Precise procedures have been described [36]. Lactose-appended CyDs have been synthesized and threaded onto hydrophobic polymers in aqueous solution to form dynamic multivalent lactosides for binding to lectins. The lactose-bearing α- and β-CyDs were obtained by coupling lactosyl propionic acid derivatives with C6-monoamino CyDs using HBTU-BF_4 to activate the carboxylic acid [37]. The synthesis of monosubstituted β-CyDs with β-D-glucose, β-D-galactose, α-D-mannose, β-L- and β-D-fucose linked to the macrocycle via a C9 spacer chain is based on the highly efficient coupling of the NCS sugar derivatives to monomethyl nonanedicarboxylate to generate a stable amide linkage [38]. Bovine pancreatic trypsin was chemically modified using mono-C6-amino-β-CyD and other C6-mono-alkylene-diamino-β-CyDs [31] with EDC as coupling agent. The catalytic and thermal stability properties of trypsin were improved by the attachment of CyD residues [39]. Branched β-CyDs substituted with mannosyl mimetic derivatives have been synthesized in which self-inclusion in water does not preclude the inclusion of other guests [40]. C6 and C3 mono-positively charged CyDs were prepared to allow systematic studies of the interaction with anionic guests [41]. Novel indolyl-containing β-CyDs were synthesized by the condensation of indol-3-ylbutylic acid with the oligo(aminoethylamino)-β-CyDs of N,N′-dicyclohexylcarbodiimide (DCC) [42]. C6-Mono-substituted α-CyD and β-CyD with monoazacoronand were derived from the acylation of the 6-aminohexylamino-CyDs [43]. The technique of coupling saccharide antennae onto CyDs allows mobility for the biologically active galactose head group, and allows recognition by the lectin. A chemical synthesis of β-CyD derivatives using spacer arms of 3, 4, 5, 6, and 9 carbons has been reported (Scheme 2.2) [44].

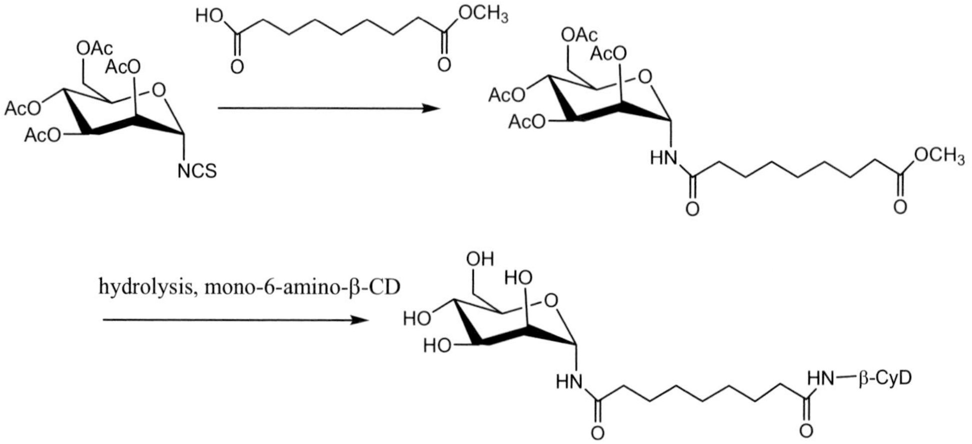

Scheme 2.2. Synthesis of mannose-branched β-CyD with a spacer arm of nine carbons [44].

2.2.1.2 **Per-modification at the C6-Position**

Fulton and Stoddart [45] proposed neoglycoconjugates based on CyDs and calixarenes. Molecules displaying multiple carbohydrate ligands may have the ability to interact with lectines with much stronger affinities than their monovalent counterparts. Research in the field of multivalent CyDs to date has largely focused on the synthesis of persubstituted CyDs (Scheme 2.3), in which each glucosyl residue of the CyD torus has an appended carbohydrate ligand [45].

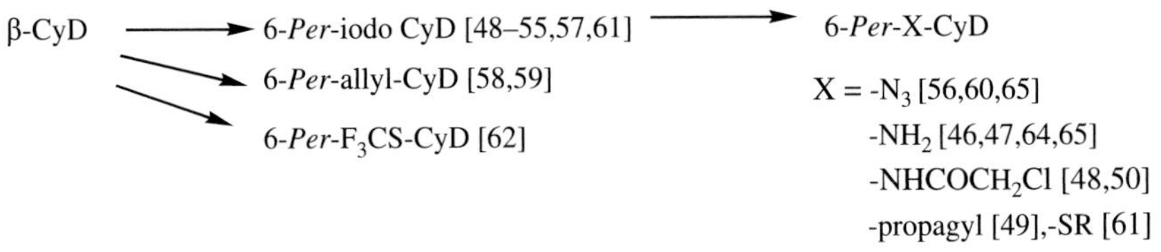

Scheme 2.3. Per-modification at the C6-position of *β*-CyD [37].

β-D-Glycopyranosyl isothiocyanates in the glucopyranose, cellobiose, and lactose series react readily with per-6-amino-*β*-CyD to afford the fully substituted glycoclusters in good yields [46]. Reaction of per-6-amino-*β*-CD with chloroacetic anhydride affords the heptachloro-*β*-CD derivative qualitatively. When the compound is treated with the isothiouronium-*β*-D-glucoside, the glycocluster was obtained [47]. The synthesis of CyD-based cluster *N*-glycoside involves conversion of the chloroacetyl *N*-glycosides into their isothiouronium derivatives, then attachment of the *N*-glycosides onto heptakis(6-iodo) and heptakis(6-chloroacetamide) *β*-CyD with cesium carbonate in DMF [48]. Galactosyl nitrile oxide derivatives, when reacted with the per-6-propargyl-*β*-CD derivatives, afford the cluster compound in 79% yield [49]. The thioacetate of *N*-acetylneuraminic acid was coupled in a one-pot reaction to heptakisC6-(chloroacetamide)-*β*-CyDs, yielding multivalent sialosides [50]. Glycoside-CyDs were synthesized by glycosylation of the oxyethylene arm and the attachment of the *O*-glycosides onto the heptakis(6-iodo)-*β*-CyD derivatives using cesium carbonate [51].

Synthesis of three first-order dendrimers based on a *β*-CyD core containing 1-thio-*β*-D-glucose, 1-thio-*β*-mannose, and 1-thio-*β*-rhamnose residues was performed involving preparation of a thiolated bis-branched glycoside building block and its attachment onto heptakis(6-iodo)-*β*-CyD [52]. A series of cluster galactosides constructed on *β*-CyD scaffolds containing seven 1-thio-*β*-lactoses of *β*-lactosylamine bound to the macrocyclic core through different spacer arms have been synthesized [53].

The coupling reaction of heptakis6-deoxy-6-iodo-*β*-CyD with unprotected sodium thiolates derived from 3-(3-thioacetyl propionamido)propyl glycosides proceeded smoothly, giving novel perglycosylated CyDs with Gal, GlcNAc, lactose, and LacNAc in excellent (78–88%) yields [54, 55]. A new intermediate with seven azido groups on the primary rim and fourteen palmitoyl chains on the secondary one was obtained in 40% yield, and a one-pot coupling reaction between the intermediate and an excess of the amino-terminal unprotected glucosamine derivative was carried out [56]. Persubstituted macrocycles were prepared from heptakis6-iodo-β-CyD using the unprotected sodium thiolate of 3-(3-thioacetyl propionamido) propyl glycosides (galactose, lactose, and *N*-acetyllactosamine). The hepta-antennated β-CyDs produced were found to be potent inhibitors of protein–carbohydrate recognition using galactoside-specific plant and, notably, mammalian lectins as well as immunoglobulin G [57].

Reaction of a CyD with sodium hydride and allylbromide afforded (32%) the CyD substituted on its primary face with seven allyl ether functions. The synthesis of a β-CyD derivative persubstituted on its secondary face with allyl functions was accomplished with sodium hydride and allylbromide (yield 26%). Reaction of β-CyD with allylbromide in the presence of BaO and $Ba(OH)_2/8H_2O$ gave a β-CyD derivative (17%), persubstituted with seven allyl ether functions on each face of the CyD. Reaction of per-2,6-diallyl-β-CyD with β-D-thioglucose or β-D-thiolactose with UV light gave the desired protected glucose and lactose cluster compounds (70%) [58, 59].

Treating the hepta-iodo-β-CyD derivative with NaN_3 in DMF gave, in 96% yield, the hepta-azide, which was methylated using a large excess of MeI and NaH to afford the permethylated derivative quantitatively. Treatment with Ph_3P followed by aqueous NH_3 furnished the hepta-amino-β-CyD. The disaccharide-containing ligand was coupled, affording the persubstituted product (79%) [60].

Amphiphilic heptakis(6-alkylthio)-β-CyD derivatives were synthesized and their monolayer behavior on a water surface was studied on the basis of surface–molecular area (π–A) isotherms [61]. β-CyD functionalized at the 6-position with trifluoromethylthio groups have been obtained from the native β-CyD [62]. Heptakis(2,3-dihexanoyl)-β-CyD, heptakis(6-hexanamido)-β-CyD, and heptakis(6-myristamido)-β-CyD have been synthesized and nanocapsules prepared by nanoprecipitation [63]. Heptakis(6-*O*-amphiphilic)-β-CyDs with substituents of varying chain lengths (C_6 and C_{14}) and bond types (ester or amide) have been studied [64].

A novel chiral stationary phase was prepared by immobilization of heptakis(6-azido-6-deoxy-2,3-di-*O*-phenylcarbamoylated)-β-CyD onto the surface of amino-functionalized silica gel via multiple urea linkages [65].

2.2.1.3 **Multi-modification at the C6-Position**

Bis-A,D-galactose-branched-CyDs with different spacer arm lengths between two terminal galactose residues were found to have the optimum length for association with PNA lectin, and drugs. The dual association of these compounds was quantitatively evaluated by SPR and compared to the other oligosaccharide-branched CyDs [66]. Novel branched β-CyDs bearing β-D-galactose residues, 6A,6D-di-*O*-(β-D-galactosyl)-β-CyD, 6A,6D-di-*O*-(4′-*O*-β-D-lactosyl)-β-CyD, 6A,6D-di-*O*-(4′-*O*-β-D-galactosyl-β-lactosyl)-β-CyD, have been synthesized using the trichloroacetimide method [67].

6A-dansyl-6X-tosyl-modified (X = B or G, C or F, and D or E, see Scheme 2.4) β-CyDs and γ-CyDs were synthesized to investigate their chemo-sensor potential [68].

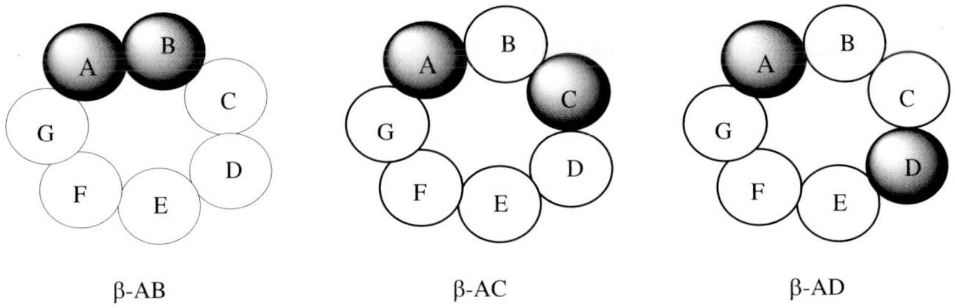

Scheme 2.4. Positional isomers of di-modified β-CyD [66–68, 71, 72, 74, 77].

Quadruply γ-CyD-linked cofacial porphyrin can be synthesized by the one-pot reaction of pyrrole and acetylated γ-CyD 6A, 6E-bis-benzaldehyde [69]. A series of molecular sugar bowls composed of a γ-CyD wall and a trehalose floor was synthesized by reacting 6A,6X-dimercapto-γ-CyD and 6,6′-diiodotrehalose (X = C, D, E) [70]. On treatment with imidazole, 6A,6B-mesitylenedisufonyl-capped β-CyD demonstrated a reactivity about 10 times higher at the 6B position than at the 6A position, leading to the selective formation of hetero-disubstituted CyDs [71]. β-CyD is transannularly disulfonylated at the 6A- and 6B-positions, and then converted to the corresponding 6A,6B-diiodide and 6A,6B-dithiol. Cross-coupling of the latter two species yields a single head-to-head-coupled β-CyD dimer with two sulfur linkers at adjacent 6-methylene carbons [72]. α-CyD reacts with dibenzofuran-2,8-disulfonyl chloride leading to the selective formation of 6A,6C-transannularly disulfonylated α-CyD (18%) [73].

Two isomeric compounds were synthesized carrying a pyridoxamine on 6A of β-CyD and an imidazole unit on 6B of the neighboring glucose residue. Each one stereoselectively transaminates phenylpyruvic acid to produce phenylalanine, with opposite stereochemical preferences [74]. A new capillary gas chromatography stationary phase, monokis(2,6-di-*O*-benzyl-3-*O*-propyl-(3′)-hexakis(2,6-di-*O*-benzyl-3-*O*-methyl)-β-CyD bonded to polysiloxane through a single spacer of 3-*O*-propyl, has been synthesized [75]. CyD dimer derivatives were synthesized as novel GC stationary phases by linking two single CyD derivatives with a difunctional spacer of the primary side of CyD [76]. Substituted CyDs carrying methyl groups on the primary rim undergo highly regioselective de-*O*-methylation in the presence of benzyl groups, giving access to A,D- or A,B-di-6-*O*-demethylated derivatives [77]. β-CyD was coupled to chitosan by the intermediate of its monochlorotriazinyl derivative [78]. Hydroxybutenyl CyD ethers have been prepared by a base-catalyzed reaction of 3,4-epoxy-1-butene with CyDs [79].

By reacting with a rigid disulfonyl chloride, β-CyD 6-tosylate is readily disulfonated to give 6A,6C,6E-trisulfonylated β-CyDs in good yields [80]. Novel β-CyD-based terpyridine derivatives have been prepared by the coupling of terpyridine with mono-hydroxy permethylated β-CyD. The CyD dimer was synthesized by reaction of terpyridine-dicarbonitrile with an excess of 6-tosyl-β-CyD [81]. The synthesis of the chloroacetylate derivatives of α-, β-, and γ-CyDs has been reported [82].

2.2.2
Modification Reactions at the Secondary Side

2.2.2.1 Mono-modification at the C2-Position

Regioselective mono-2-sulfonation of γ-CyD was conveniently achieved using a combination of sulfonyl imidazole and molecular sieves (36%). The reactions do not require strict anhydrous or basic conditions, or specific sulfonyl groups [83]. 2-Toluylsulfonyl-β-CyD was obtained (42%) by reaction of *p*-tosyl-(1*H*)-1,2,4-triazole with NaH-deprotonated β-CyD and converted into the mono-2,3-manno-β-CyD [84]. Mono-2-benzylated eicosa-methyl-β-CyD was prepared in a one-pot reaction from β-CyD (33%) and converted to mono-2OH-free eicosa-methyl-β-CyD. The benzyl ether group was easily restored to the hydroxyl group by Pd-catalyzed hydrogenation. This key compound could be transformed to the acetate, dimethylcarbamate, sulfonate, and dithiocarbamate [85]. β-CyD was partially substituted mainly on the C2 hydroxyl group of glycosyl units by 2-octen-1-ylsuccinic anhydride, via its oxyanion prepared with NaH in DMF [86]. Mono(2-phenylseleno)-β-CyD, which showed high enatioselectivity towards L/D-amino acids, was obtained by adding sodium borohydride was added to diphenyldiselenide and 2-tosyl-β-CyD [87]. β-CyD dimers linked at their primary and secondary faces by spacers of varying lengths were synthesized as carrier systems. Starting with mono-2-tosyl-β-CyD, alkylation by diaminoalkanes followed by treatment with *N*-hydroxysuccinimide esters produced CyD dimers [88]. A β-CyD-2-chitosan conjugate was synthesized by reacting mono C2 tosyl-β-CyD with chitosan [89]. Na_2S was found to attack the C2 of β-CyD mannoepoxide more readily than various other nucleophiles, and with the subsequent formation of CyD-2-S^- only the abnormal ring opening was observed [91].

2.2.2.2 Mono-modification at Any One of the C2-, C3-, or C6-Positions

Three types of reaction of 2,3-anhydro-β-CyDs, namely nucleophilic ring-opening, reduction to 2-enopyranose, and reduction to 3-deoxypyranose, were investigated to regio- and stereoselectively functionalize the secondary face of β-CyD. Upon treatment with various nucleophiles, both 2,3-mannoepoxy and 2,3-alloepoxy-β-CyDs undergo a nucleophilic ring-opening reaction generating 2- and 3-modified CyD derivatives (Scheme 2.5). The 3-position is more easily accessible than the 2-position. Thiourea also reacts with the CyD epoxides, generating thiirane and olefin species instead of any ring-opening products. By ameliorating the reaction conditions, CyD olefin, diene, and triene are prepared in moderate to good yields [91].

A method for selective monoalkylation of the C2- or C3-mono-hydroxy groups of

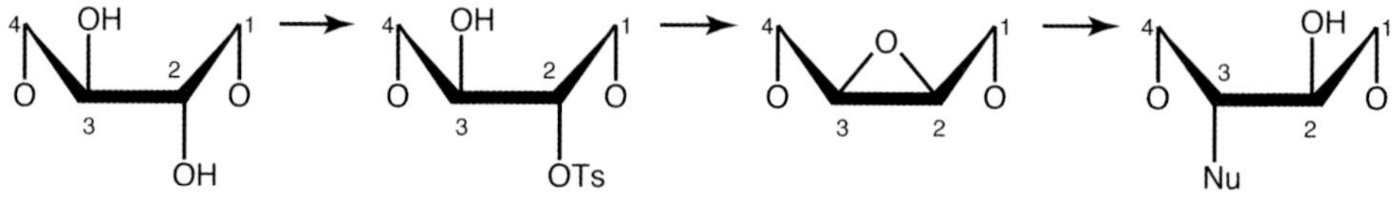

Scheme 2.5. Nucleophilic ring-opening of 2,3-mannoepoxy to 3-deoxypyranose [92].

β-CyD and exploitation of the method to synthesize β-CyD dimers that are ether linked in a 2–2′ or 3–3′ fashion at their secondary faces has been proposed [92]. The carboxyl group of predonisolone-21-hemisuccinate was mainly connected to the C2 and C3 hydroxyl groups of α-, β-, and γ-CyDs using carbonyldiimidazole as a coupling agent [93].

Heptakis-6-(tert-butyldimethylsilyl)-β-CyD was treated with excess *t*-butylsilyl chloride (TBSCl) resulting in a single octasilyl derivative at the more acidic C2 position. Permethylation under strongly basic conditions (MeI, NaH, THF), led to migration of the 2-*O*-silyl group to the 3-*O* position. Treatment of the products with tetrabutylammonium fluoride gave the versatile monofunctionalized β-CyDs in high yields [94].

Mono-altro-β-CyD was selectively sulfonylated by 2-naphthalenesulfonyl chloride at the 2A-OH of the altrose residue. By using 1-naphthalenesulfonyl chloride as reactant, the 3G-OH of the neighboring glucose became available for selective sulfonylation [95]. Secondary imidazole-appended β-CyD, with a nondistorted cavity built of glucoside units synthesized from the novel intermediate 3-amino-β-CyD, exhibits much greater catalytic activity in ester hydrolysis than its isomer with the distorted cavity obtained from altroside units [96].

A β-CyD derivative modified with *p*-xylylenediamine at the 3-position, mono-3-[4-(aminomethyl)benzylamino]-β-CyD, was prepared by the reaction of β-CyD-2,3-manno-epoxide with *p*-xylylenediamine [97]. The three isomeric mono-2-, 3-, or 6-hydroxy permethylated β-CyDs are good precursors for a wide variety of monofunctionalized "permethyl" β-CyDs. As protecting groups, the benzyloxy group was used for C2, and the *t*-butyldimethylsilyl group for C6. Mono-C3 hydroxy CyD was obtained by partial methylation of 2,6-dimethyl CyD [98].

The unambiguous regioselective synthesis of chiral polysiloxane-containing CyDs as chiral stationary phases, with the mono-octamethylene spacer in either the O2, O3, or O6 position was performed and the products applied to enantioselective GC separations [99]. Subtle differences in the chemistry of the hydroxyl groups at the 2-, 3-, and 6-positions of CyDs can be exploited to direct an electrophilic reagent to the desired site. Selective monoalkylation at the primary side of α-CyDs involves the reaction of α-CyD with 4-methylamino-3-nitrobenzyl chloride in 2,6-lutidine. Monoalkylation at the 2-position of β-CyD is accomplished by the reaction of β-CyD with 1′-bromo-4-methylamino-3-nitroacetophenone [100].

Novel branched β-CyDs with β-D-galactose residues were chemically synthesized using the trichloroacetimidate method. Tetraacetyl galactosyl trichloroacetimidate was obtained with trichloacetonitrile in the presence of 1,8-diazabicyclo[5.4.0]undec-7-ene (DBU) [101].

2.2.2.3 **Per-modification at the C3-, C2-, or C6-Position**

Per(3-deoxy)-γ-cyclomannin was synthesized starting from γ-CyD. Per(6-silyl)-per(2,3-mannoepoxy)-γ-CyD was prepared using the same procedure as for the β-CyD derivative. Reduction of the epoxide was performed using $LiAlH_4$ [102].

Photoaddition of the thiol 2,3,4,6-tetra-*O*-acetyl-β-D-1-thioglucopyranose to the allyl ether functions of per-2-allyl-, per-6-allyl-, and per-2,6-diallyl-β-CyD derivatives provides a remarkably simple and efficient way of attaching glucopyranose units onto the secondary face, as well as the primary face, of β-CyD (70%) [58, 59]. Addition of ethylene carbonate to heptakis(6-alkylthio-6-deoxy)-β-CyDs occurs exclusively at the C2 position and results in a degree of substitution of 8–22 ethylene glycol units per CyD [103]. Trimethylsilylation of α- and β-CyD with *N*-(trimethylsilyl)acetamide gave the per-2,6-*O*-trimethylsilyl derivatives with high yield and selectivity [104].

2.2.3 Per-modification Reactions at All Three Positions

A simple versatile method has been reported for the preparation of chemically uniform per-2,3,6-, per-2,3-, and per-2,6-hydroxyethylated and carboxymethylated β-CyD derivatives via (i) introduction of allyl groups into the desired positions and alkylation of the remaining free hydroxylic groups; (ii) dihydroxylation of the allylic double bonds; (iii) diol cleavage and reduction to hydroxyethyl derivatve in "one pot"; and (iv) tetramethylpiperidyl-1-oxy- (TEMPO-) mediated oxidation to the carbomethyl derivative [105]. The copper (I)-mediated living radical polymerization of methyl methacrylate (MMA) was carried out with a CyD-core-based initiator with 21 independent discrete initiation sites of heptakis[2,3,6-tri-*O*-(2-bromo-2-methylpropionyl)]-β-CyD [106]. Pertrimethylsilylated α-, β-, and γ-CyDs were prepared with trimethylsilylimidazole [107]. High pressures are required to dissolve peracetylated β-CyD in liquid and supercritical carbon dioxide [108]. Heptakis(2,6-di-*O*-methyl-3-*O*-acetyl)-β-CyD was prepared by acetylation at the 3-position of DM-β-CyD to investigate the inclusion and pharmaceutical properties [109]. The 2,3-di-*O*-methyl-, 2,3-di-*O*-ethyl-, and 2,3-di-*O*-acetyl-derivatives of 6-*t*-hexyldimethylsilyl-β- and γ-CyDs were synthesized to evaluate their ability to recognize a series of racemates [110]. Two pairs of new chiral selectors, derived from octakis (6-*O*-*t*-butyldimethylsilyl)- and octakis (6-*O*-hexyldimethylsilyl)-γ-CyD, were synthesized with inverse substitution patterns at the secondary positions. 2-*O*-Methyl-3-*O*-acetyl- and 2-*O*-acetyl-3-*O*-methyl derivatives are suggested as useful models to further investigate the effect of substituents at C2 and C3 on the efficiency of enantioseparations using GC ([111], and see Chapter 5). Naphthylcarbamate-substituted (3-(2-*O*-β-CyD)-2-hydroxypropoxy)-propylsilyl-appended silica gel has been synthesized and shows excellent selectivity for the separation of disubstituted benzenes and enantiomers of chiral aromatic compounds [112]. A method of preparing heptakis(2,6-di-*O*-alkyl)-β-CyDs, heptakis(2-*O*-alkyl)-β-CyDs, and heptakis(6-*O*-alkyl)-β-CyDs has been developed involving treatment of β-CyD with sodium or barium hydroxide and various alkyl halides [113].

2.2.4 Enzymatic Modification Reactions of Cyclodextrins

Transgalactosylated derivatives of CyDs and glucosyl and maltosyl CyDs were synthesized using α-galactosidases from coffee beans and *Mortierella vinacea*. Coffee-bean α-galactosidase transferred a galactosyl residue not only to side chains of glucosyl-CyDs and maltosyl-CyDs but also directly to CyD rings. *M. vinacea* α-galactosidase transferred a galactosyl residue only to side chains of maltosyl-CyDs [114]. Mannosylated derivatives of CyDs, mannosyl-α-, β-, and γ-CyD were respectively synthesized from a mixture of mannose and α-, β-, and γ-CyD by the reverse reaction of α-mannosidase from Jack bean [115]. The novel heterogeneous branched CyDs having 6-*O*-α-D-galactosyl, 6-*O*-α-D-mannosyl, and 6-*O*-α-D-glucosyl groups were prepared as described in an earlier paper [116, 117]. Galactose CyDs, synthesized by an enzymatic method using β-galactosidase, have a specific interaction with hepatocytes and may be useful as a targeting carrier to hepatocytes [118]. Glucuronylglucosyl-β-CyD, prepared by oxidation of maltosyl-β-CyD with *Pseudogluconobacter saccharoketogenes*, showed lower hemolytic activity and negligible cytotoxity as well as greater affinity for the basic drugs [119].

2.2.5 Construction Reactions for Cyclodextrin Ring Formation

The reconstruction of the fully characterized new asymmetric cyclodextrins involves a three-step procedure: ring opening of fully methylated CyDs, elongation of the chain with correctly modified monosaccharide derivatives, and finally macrocyclization to obtain the desired compounds. This strategy has been applied to achieve the synthesis, by insertion of glucopyranose derivatives, of a series of asymmetric γ-CyDs directly useful in chiral GC [120]. The α- and γ-CyD analogues, which possess a hexa-2,5-diyne-1,6-dioxy unit, were synthesized by intramolecular coupling of bis-*O*-propargylated maltohexaoside or the analogous maltooctaoside [121]. All the glycosilation for chain elongation and cyclization of saccharides was carried out after tethering the donor to the acceptor by a phthaloyl bridge to give the desired saccharides in good yields [122].

2.3 Chemistry of Modified Cyclodextrins

Various modified CyDs have been prepared to improve their binding affinities and selectivities controlled by external factors such as photons, metal ions, or pH. Such CyDs can act as chemosensors, artificial enzymes, drug carriers, or reaction fields. CyDs have been used as scaffolds for highly preorganizing functional units. There are already many reviews on modified CyDs [1–6, 8, 9, 123–125], so here we summarize only developments reported in the last seven years.

2.3.1 Cyclodextrin Dimers and Trimers

CyD dimers and trimers have been developed to improve the binding ability for specific applications. Although the dimer linked with a short alkyl chain **1** showed higher affinity for ditopic guest molecules like *p*-toluidino-6-naphthalene sulfonate

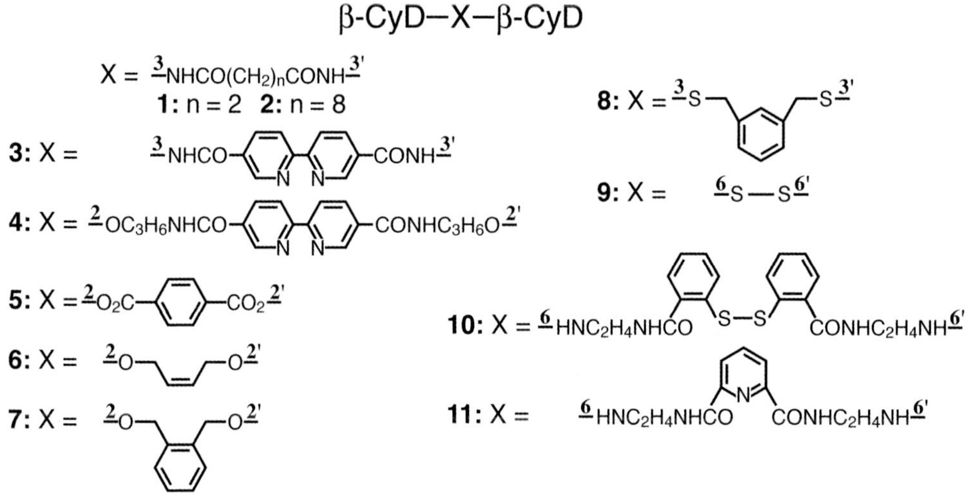

derivatives and porphyrins with cooperative binding, the self-inclusion of the linkers in the case of the dimer **2** led to lower binding affinities for the guests [126, 127].

The rigid 2,2′-bipyridyl group connecting the two CyD moieties in **3** resulted in a higher affinity for porphyrin guests since self-inclusion did not occur [127]. An additional alkyl chain between the 2,2′-bipyridyl group and the *β*-CyD moiety **4** caused self-inclusion of the 2,2′-bipyridyl group by one of the CyD moieties [32]. When all hydroxyl groups of the CyD were methylated, an azobenzene linker was also self-included by inversion of the CyD ring [128]. Other bonds including –S–S– [129, 130], –S– [130], –Se–Se– [131, 132], –Te–Te– [133], and –PO– [134] have also been used for making CyD dimers.

Some CyD dimers have been prepared for selective binding of peptides [135]. CyD dimers linked at the secondary hydroxyl side made by reaction of the 2,3-manno-epoxide with a thiol **8** show only a weak binding ability for peptides, because one of the glucose units of the CyD undergoes a conformational change, causing an indentation in the CyD cavity. Linkage of two CyDs on their secondary faces by direct alkylation or acylation of the C2 hydroxyl groups **5–7** caused a strong binding ability for peptides with sequence selectivity by cooperative chelate binding. For example, **5** shows good selectivity for binding peptides with sequences Phe–D–Pro–X–Phe–D–Pro, and to a lesser extent Phe–D–Pro–X–Val–D–Ala, but does not bind Trp–X–Trp, Phe–X–Phe, or Trp–Trp. By contrast, **9** with a short linker on the primary face binds Trp–Trp well.

A CyD dimer **10** with a functional tether was synthesized for residue- and sequence-selective binding of nonaromatic dipeptides, binding Gly–Leu significantly better than Leu–Gly [136]. This remarkable difference in the binding between Leu–Gly and Gly–Leu can be accounted for in terms of the attractive/repulsive interaction between the protonated amino group in the tether ($-NH_2^+-$)

and the relevant charged group (CO_2^- or NH_3^+) remaining in the amino acid residues. Dimer **11** with a pyridine-2,6-dicarboxamide linker showed a pH-dependent binding affinity for oligopeptides resulting in a stronger association with oligopeptides in neutral media because of electrostatic interactions between the diamidopyridine group and the carboxylate of the peptides [137]. **11** afforded length-selectivity up to 4.7 for the Gly–Gly/Gly–Gly–Gly pair at pH 2.0 and sequence-selectivity up to 4.2 for the Gly–Leu/Leu–Gly pair at pH 7.2. Self-inclusion of the pyridine moiety occurred in an acidic medium, leading to a weaker binding affinity.

The doubly bridged dimer and trimer were prepared [138, 139]. The large binding constant of 27 000 M^{-1} of **15** by **13** may be due to the statistical enhancement of binding by three CyD units in the host and the entropic advantage of two-points fixation of the CyDs at the A,D-positions. In contrast, smaller association constants of **12** and **14** of 5200 and 8600 M^{-1} respectively have been determined for the same guest. The entropic advantage is largely cancelled by the extra energy for opening the stacked form of the linker of **12** during the association processes.

Irradiation with light of **16**, with two CyD cavities linked through their secondary sides by a photochromic dithienylethene unit [140–142], switches these dimers between a relatively flexible (open) and a rigid (closed) form. Dimer **16** bound tetrakissulfonatophenyl porphyrin (TSPP) 35 times more strongly in the open form **16b** than in the closed form **16a**, owing to the loss of cooperativity between the CyD cavities in the closed form.

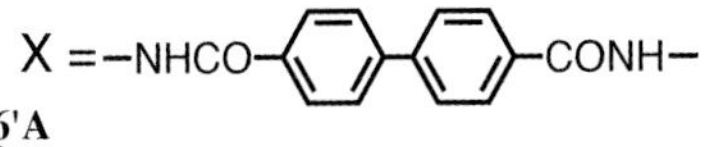

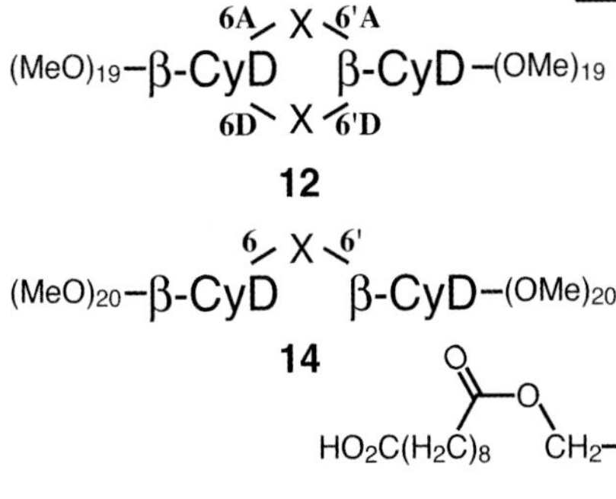

12

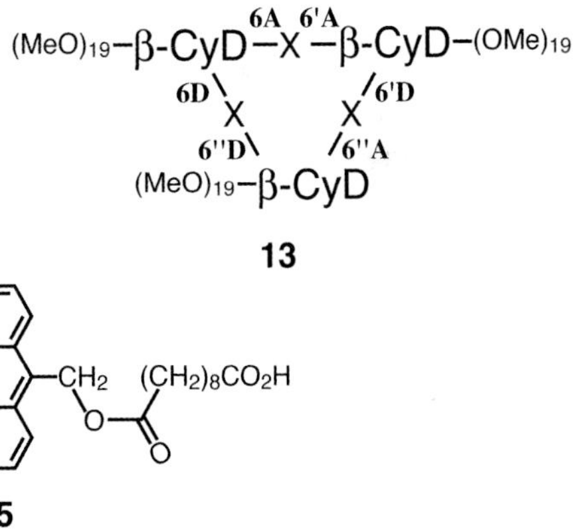

13

14

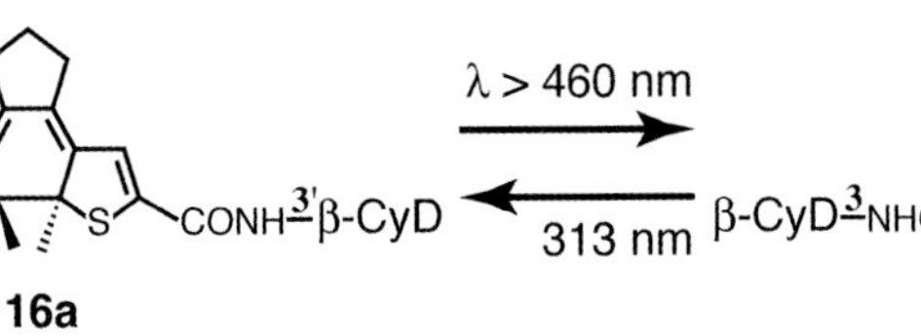

15

16a **16b**

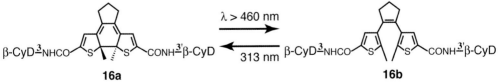

CyD dimers **17** and **18** can be used for transporting saccharides through a liquid membrane. The transport rate of D-ribose and methyl D-galactopyranoside through a chloroform liquid membrane was over 15 times faster with dimer **17a** as the transporter than with monomer **17b** [143, 144]. CyD trimers **19** showed selective and strong binding to trimeric amino acid amide **21** ($K_b = 3.5 \times 10^6$ M^{-1}) in spite of weak binding to **20** ($K_b = 650$ M^{-1}) [30].

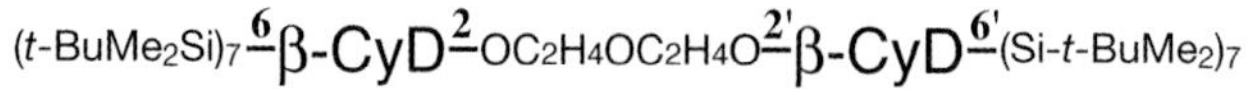

$$(t\text{-BuMe}_2\text{Si})_7\text{-}^{6}\text{β-CyD-}^{2}\text{OC}_2\text{H}_4\text{OC}_2\text{H}_4\text{O-}^{2'}\text{β-CyD-}^{6'}(\text{Si-}t\text{-BuMe}_2)_7$$

17a

$$(t\text{-BuMe}_2\text{Si})_7\text{-}^{6}\text{β-CyD}$$

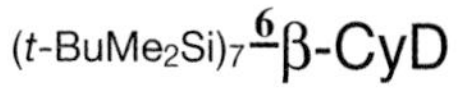

17b

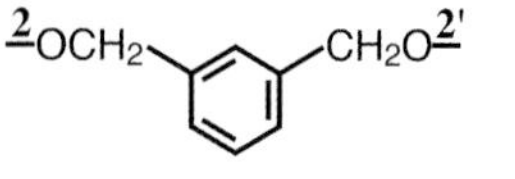

18

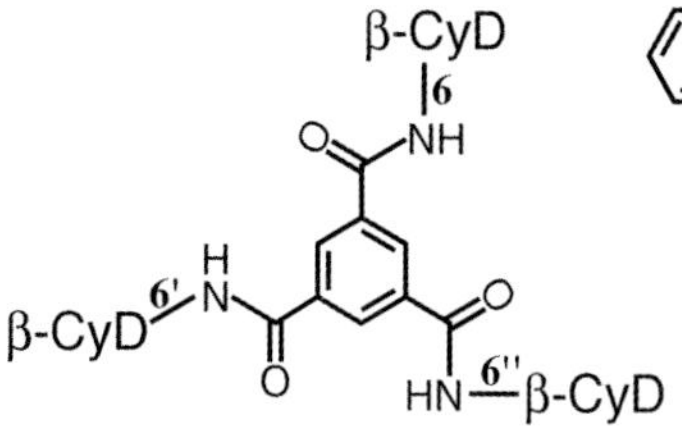

19 **20** **21**

A novel trimer receptor molecule **22** consisting of two α-CyDs and one β-CyD was synthesized to create an effective reaction field by an optimum arrangement between a catalyst **23** and a substrate **24** [145]. This led to a large increase in the rate of the hydrolytic reaction because the catalyst **23** was selectively accommodated by the β-CyD and the carbonyl group of the substrate, which was accommodated by two α-CyDs, was close to the catalyst.

$$\text{α-CyD-}^{6'}\text{NHCOCH}_2\text{S-}^{6A}\text{β-CyD-}^{6D}\text{SCH}_2\text{CONH-}^{6''}\text{α-CyD}$$

22

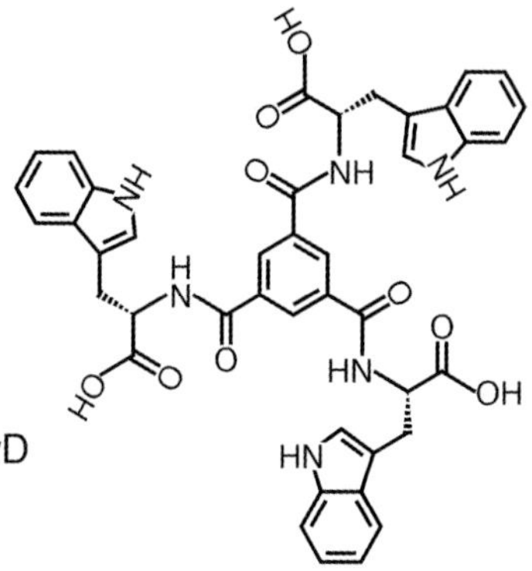

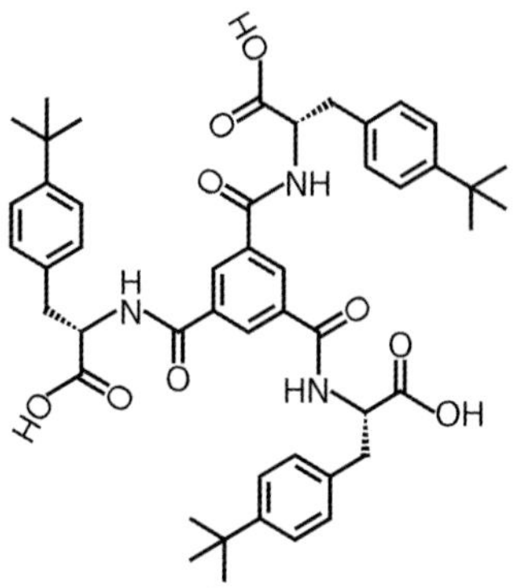

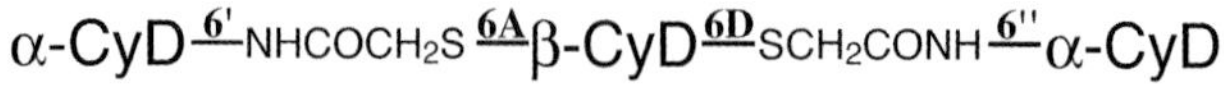

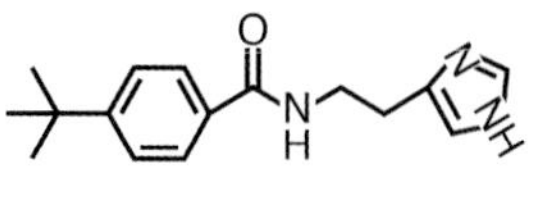

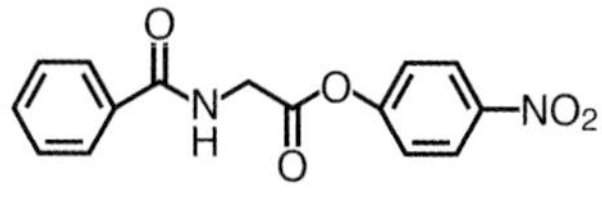

2.3.2
Charged Cyclodextrins

Because one of the main driving forces for the formation of inclusion complexes by the CyD molecule in water is hydrophobic interaction, a more hydrophobic guest is apt to be accommodated in the CyD cavity and any hydrophilic functional groups on the guests tend to reduce the binding affinity. On the other hand, modification by charged functional units can improve the binding affinity of CyDs for oppositely charged guests.

Heptakis-6-amino-6-deoxy-β-CyD **25** is protonated at neutral pH [146]. The binding constant of **25** for phosphate **26** was three times larger than that of the native β-CyD, whereas the binding constant of **25** for phenol **27** was only 2% of that of the native β-CyD.

Heptakis-6-(2-hydroxyethylamino)-6-deoxy-β-CyD **28** can be used as an inhibitor of phosphate ester hydrolysis catalyzed by λ-protein phosphatase and acid phosphatase, because **28** can bind to phosphotyrosine **29** and other aryl phosphate esters with significant affinity [147].

Guanidinium-appended β-CyD was also prepared for the complexation of phosphates (e.g. **32**) [148, 149]. A,D-bisammonium-appended or A,D-bisguanidinium-appended β-CyD (**30** and **31**, respectively) displayed significantly enhanced binding of phosphotyrosine, whereas monoammonium-appended β-CyD showed low affinity and monoguanidinium-appended β-CyD showed no detectable binding [149].

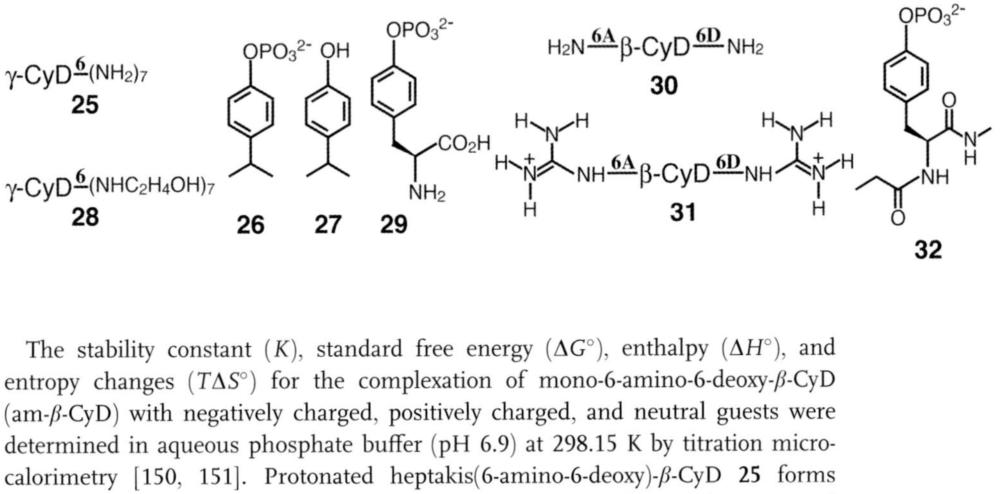

25 **28** **26** **27** **29** **30** **31** **32**

The stability constant (K), standard free energy (ΔG°), enthalpy (ΔH°), and entropy changes ($T\Delta S^\circ$) for the complexation of mono-6-amino-6-deoxy-β-CyD (am-β-CyD) with negatively charged, positively charged, and neutral guests were determined in aqueous phosphate buffer (pH 6.9) at 298.15 K by titration microcalorimetry [150, 151]. Protonated heptakis(6-amino-6-deoxy)-β-CyD **25** forms stronger complexes with the (*S*)-enantiomers of *N*-acetylated Trp, Phe, Leu, and Val in their anionic forms than with the (*R*)-enantiomers. Monoaminated am-β-CyD also recognized the chirality of the amino acids while the native CyDs such as α- and β-CyDs did not.

Orientation or motion of a guest in the CyD cavity can be controlled by electrostatic interaction [152, 153]. The *p*-nitrophenolate ion **35** was strongly bound to the mono- and dipyridinio derivatives of α-CyD **33** in an aqueous 0.1-M Na_2CO_3 solution, and the molecular rotation of **35** was significantly retarded within the cavities of the pyridinio derivatives of α-CyD by electrostatic and van der Waals interactions, whereas **36** could rotate in the CyD cavity of **33**, because the C–C bond between the carboxylate group and the benzene ring of **36** can freely rotate, even when the carboxylate group is trapped by the pyridinio groups of **33**. $6^A,6^D$-Bispyridinio-appended γ-CyD **34** effectively enhanced the excimer fluorescence of 2-naphthylacetate [154] owing to formation of the 1:2 complex between the positively charged γ-CyD and 2-naphthylacetate favored by the electrostatic interaction between the host and the guest.

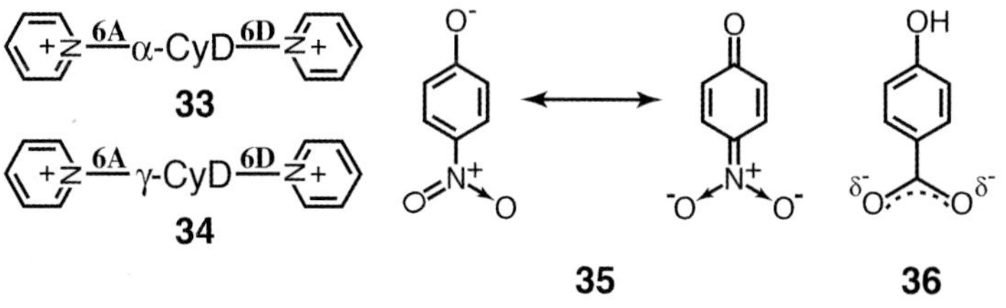

Encapsulation of neuromuscular blockers by an exogenous host molecule such as CyD would promote dissociation of neuromuscular blockers from their site of action, and result in the reversal of neuromuscular blockage. Negatively charged CyD **37**, which has a high affinity for rocuronium **38**, a steroidal muscle relaxant, reversed the neuromuscular blocking effect of rocuronium bromide *in vitro* (mouse hemi-diaphragm) and *in vivo* (anaesthetized monkeys) [155, 156].

Carboxylate-modified or ammonium-modified CyDs show pH-dependent inclusion phenomena. Therefore it is expected that these charged hosts can be used as pH-triggered guest release systems [157]. For example, 1-pyrenesulfonate was included in the cavity of heptakis(2,3-di-*O*-carboxymethyl)-β-CyD **39** with a binding constant of 2300 M^{-1} at pD = 2.0, while no complexation occurred at pD > 6.0.

If a negatively charged functional unit is removed from charged CyD by an enzyme reaction, its binding affinity for a positively charged guest is weakened and the enzyme-controlled drug delivery system (DDS) system can be prepared [158]. 6-Phosphorylated β-CyD **40** formed an inclusion complex with a positively charged molecule (e.g. an antineoplastic agent, berenil) up to 100-fold more effectively than the native β-CyD by phosphate–cation salt bridges. Cleaving the phosphate group in the phosphatase-catalyzed hydrolysis induces release of the guest from the cavity.

Per-amino-CyDs form various self-assemblies by hydrogen bonding in water [159–161]. Cationic amphiphilic β-CyDs **41**, **42**, substituted with hydrophobic *n*-alkylthio chains at the primary hydroxyl side and hydrophilic ω-amino-oligo-(ethylene glycol) units at the secondary side, formed bilayer vesicles with a diameter of 30–35 nm (when alkyl = hexadecyl) or nanoparticles with a diameter of ca. 120 nm (when alkyl = hexyl) in water [159]. Sulfated amphiphilic CyDs **43–45**

$\gamma\text{-CyD}\underline{^{6}}(SC_2H_4CO_2Na)_7$
37

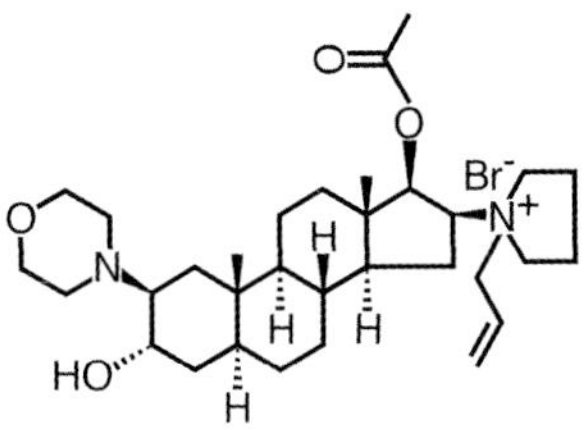

Rocuronium (**38**)

$\beta\text{-CyD}\underline{^{2,3}}(OCH_2CO_2Na)_{14}$
39

$\beta\text{-CyD}\underline{^{6}}OPO_3^{2-}$
40

$(RS)_7\underline{^{6}}\beta\text{-CyD}\underline{^{2}}(OC_2H_4OC_2H_4OC_2H_4NH_2)_7$
41: R = C_6H_{13}
42: R = $C_{16}H_{33}$

$(Me_3N^+\ {}^-O_3SO)_n\underline{^{6}}\text{CyD}\underline{^{2}}(OCOC_{15}H_{31})_n$
43: n = 6 (α-CD)
44: n = 7 (β-CD)
45: n = 8 (γ-CD)

$(H_3N^+)_6\underline{^{6}}\alpha\text{-CyD}\,^{\underline{2}(OCH_2CO_2^-)_6}_{\underline{3}(OMe)_6}$
46

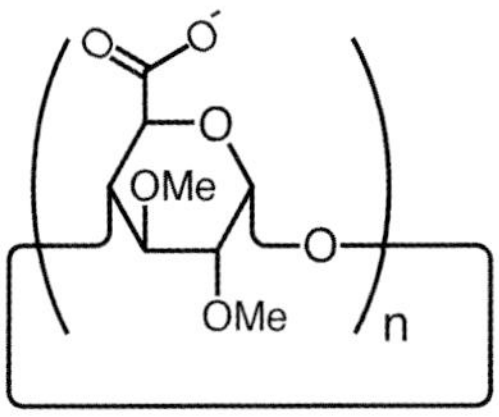

47: n = 6
48: n = 7

formed stable monolayers at an air–water interface and a mixture of compound **44** with dipalmitoyl phosphatidylcholine and cholesterol formed specific erythrocyte-like liposomes with a controlled encapsulation capacity for fluorescent-labeled compounds [160]. Hexakis(6-amino-2-*O*-carboxymethyl-6-deoxy-3-*O*-methyl)-α-CyD **46** formed helical nanotubes by intermolecular hydrogen bonds between neighboring CyDs [161]. The formation of hydrogen bonds between carboxyl units leads to the production of very strong dimers of per(2,3-di-*O*-methyl)-per(5-carboxy-5-dehydroxymethyl)-CyD **47**, **48** in 1,2-dichloroethane and in chloroform, as evidenced by molecular masses determined by vapor-phase osmometry (VPO) [162].

2.3.3
Chemosensors Using Modified Cyclodextrins

Studies involving CyD derivatives bearing bridging caps, either purely organic caps or caps composed of one or more metal centers, which could be used as sensors, have been reviewed [7]. In particular, numerous dye-modified CyDs were prepared as chemosensors for highly selective and sensitive molecular recognition [163]. For instance, dansyl-modified CyD dimers and trimers with larger cavities formed strong inclusion complexes with bile acids [164, 165].

The 2-naphthylamine-appended *β*-CyD-calix[4]arene couple **49** showed sensitivity for analytes such as steroids and terpenes with different selectivity from the native CyD. On the other hand, the dansyl-appended *β*-CyD-calix[4]arene couple **50** did not show any change in fluorescence intensity upon the addition of guests because of the strong inclusion of the dansyl group into the CyD cavity [166].

Monensin is an antibacterial hydrophobic agent that is expected to act as a hydrophobic cap responsive to sodium ion, if monensin is introduced to the primary face of the CyD. A monensin-capped dansyl-modified *β*-CyD **51** was prepared as

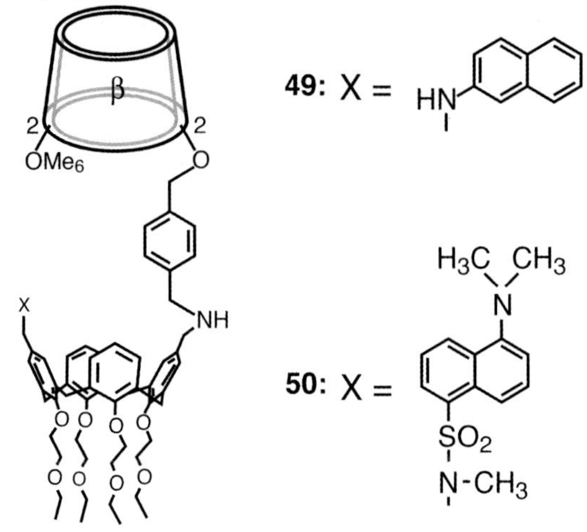

a fluorescent chemosensor for molecule recognition and the presence of sodium cation in a solution of **51** enhanced its binding ability and increased the guest-dependent fluorescence variation [167].

Biotin-appended dansyl-modified CyDs **52**, **53** were prepared to examine the effects of avidin binding on guest-inclusion phenomena and fluorescence properties [168]. The fluorescence intensities of these CyDs were enhanced more than three times by avidin. The avidin forces the dansyl moiety into the CyD cavity, so that the dansyl moiety is then excluded to the hydrophobic environment of the avidin rather than to the bulk water associated with guest binding. The binding is stronger in the presence of avidin as a hydrophobic cap.

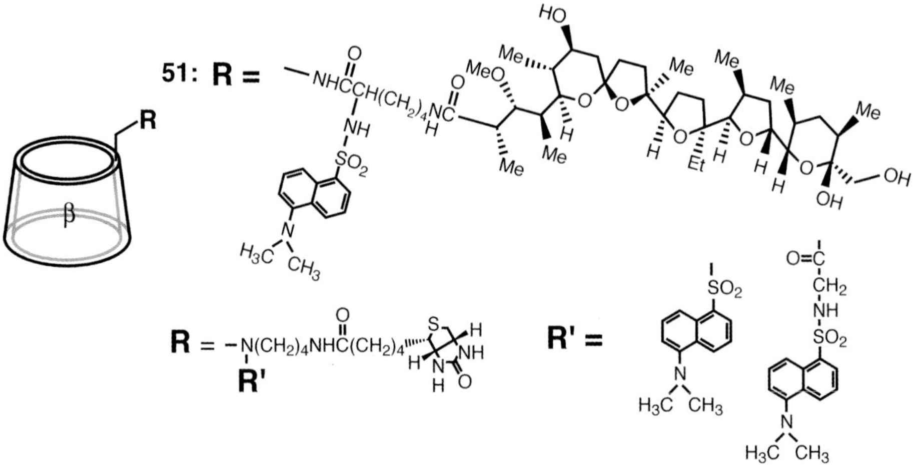

52 **53**

Introducing multiple functional groups on CyD to improve the sensitivity and selectivity of a CyD-based chemosensor is no easy task, because there are so many possible positional isomers of multiply-functionalized CyD. If the rigid α-helix scaffold of peptides was used, two different photoreactive moieties could easily be placed at both the primary and secondary hydroxyl CyD bases [169–171]. For example, pyrene (electron donor) and nitrobenzene (electron acceptor) groups were inserted at both sides of CyD bases on the peptide scaffold **54**, **55** [169]. In the absence of guests effective quenching occurred between included nitrobenzene and pytene, both closely located in the CyD cavity. The fluorescence intensity dramatically increases with increasing concentration of a guest (e.g. lithocholic acid), because the nitrobenzene moiety is excluded from the CyD cavity. Pyrene (donor) and coumarin (acceptor) were introduced at one side **56** or both sides **57** of CyD on the α-helix scaffold of a peptide to make a fluorescence resonance energy transfer (FRET) system for a chemosensor [170, 171]. Similarly, the CyD molecule and the dansyl moiety can also be deployed on the rigid α-helix scaffold of peptides in various directions or positions and exhibit a selective response for complexation of steroids **58–61** [172].

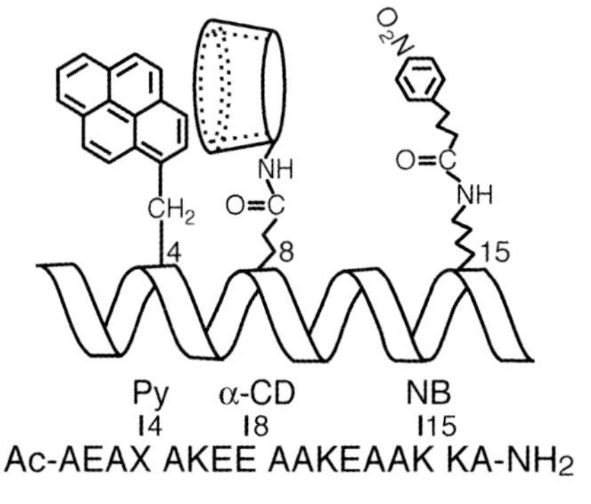

54

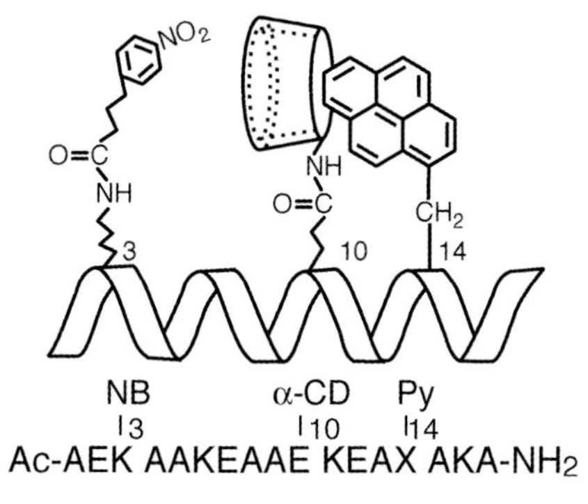

55

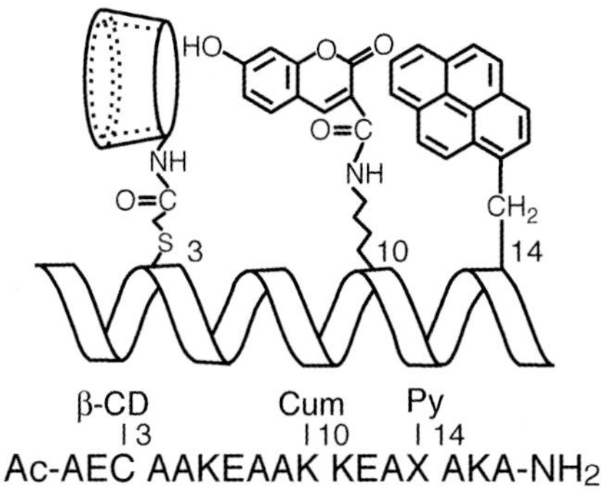

56

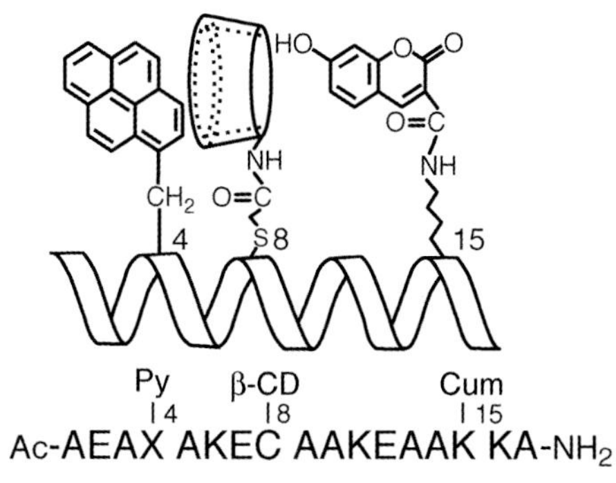

57

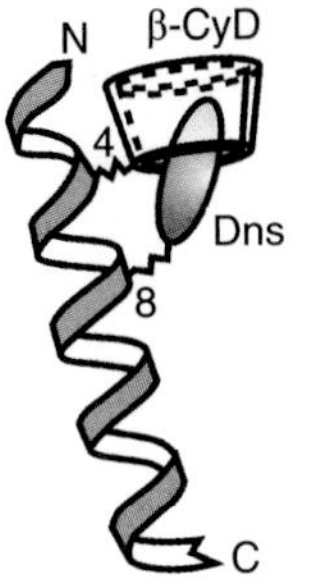

58

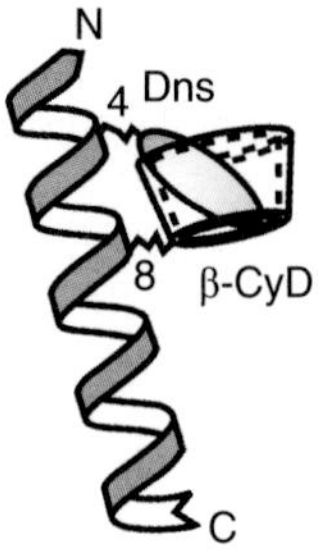

59

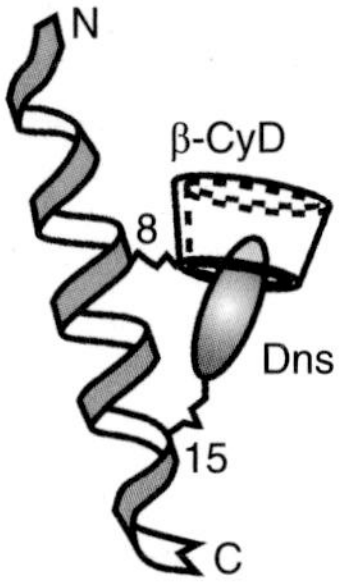

60

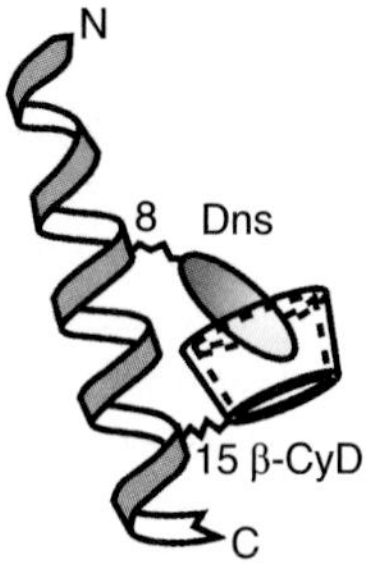

61

2.3.4
Cyclodextrin Analogues

CyD analogues have been prepared by insertion of noncarbohydrate species into the CyD skeleton to improve molecular recognition abilities [6, 120–122].

The CyD analogue with an isophthaloyl spacer **62** shows a binding ability for *p*-nitrophenol almost the same as that of permethylated *β*-CyD, whereas the binding constant of the CyD analogue with a 2,6-pyridinedicarbonyl spacer **63** is only one thirteenth of that of permethylated *β*-CyD [173].

Insertion of two triazole rings into the cyclomannoside ring **64** does not alter the binding ability of *β*-CyD [174].

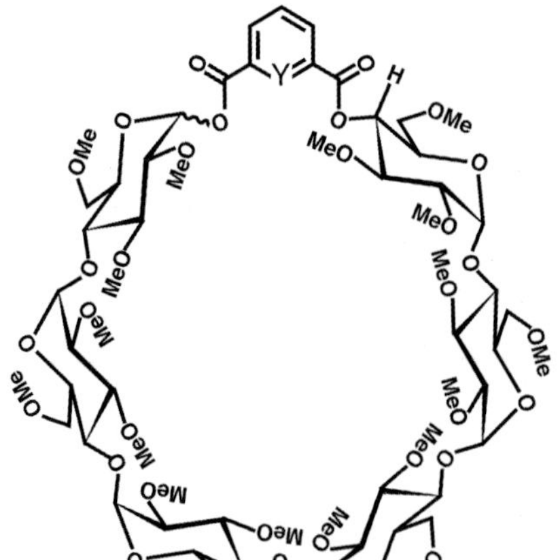

62: Y = CH
63: Y = N

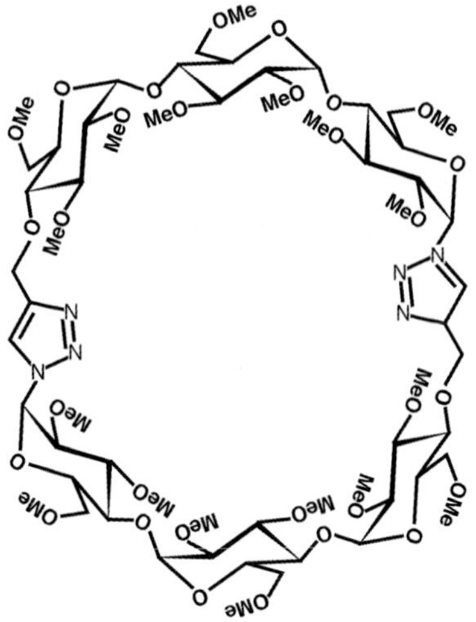

64

CyD analogues **65**, **66** incorporating one β-(1,4)-glucosidic bond were synthesized in three steps from permethylated α- and β-CyD [175]. The binding abilities of these CyD analogues were lower than those of permethylated α- and β-CyDs but the conversion of one α-(1,4)-glucosidic bond of permethylated α- and β-CyDs into a β-(1,4)-glucosidic bond caused a reversal of the original inclusion selectivity. These CyD analogues showed *m*-selectivity for nitrobenzoate, whereas permethylated α- and β-CyD showed *p*-selectivity.

Glucopyranose units of CyDs were converted to other pyranose units. A CyD analogue with six 3,6-anhydroglucose units and a 6-amino-glucopyranose unit was synthesized (**67**) [176]. This analogue had binding affinity for metal ions with Cs^+ selectivity.

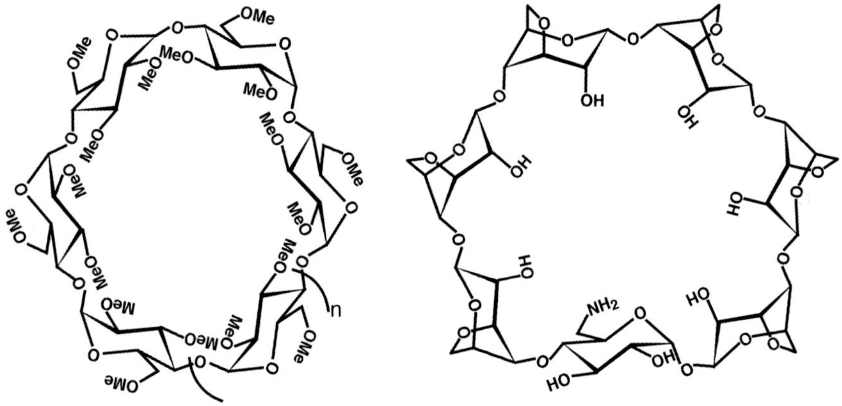

65: n = 1
66: n = 2

67

A series of cyclic oligosaccharides of even numbers of sugar units, 6–10, built up from (1 → 4)-linked alternating D- and L-pyranosidic units in a polycondensation-cycloglycosylation reaction has been described. In the solid state, X-ray crystal structure analysis of the cyclic oligosaccharide containing the alternating 1,4-linked α-L-rhamnopyranosyl and α-D-mannopyranosyl residues revealed a nanotube with a diameter of ca. 1 nm [177].

2.3.5 Cyclodextrins Conjugated with Other Kinds of Hosts

CyDs were coupled with calixarene or crown ether to improve their binding affinity. β-CyD functionalized with diaza-18-crown-6 at the primary face **68** showed 7–10-fold greater affinities for aromatic ammonium ions [178] and the introduction of a benzo-18-crown-6 moiety at the secondary-hydroxy side **69** improved chiral recognition for tryptophan (Trp), whereas its introduction at the primary-hydroxy side **70** was not effective for the Trp recognition [179].

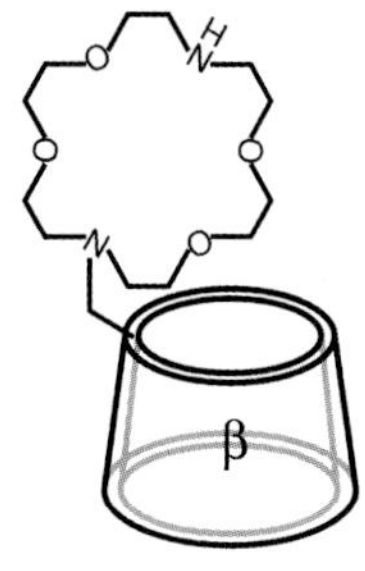

68

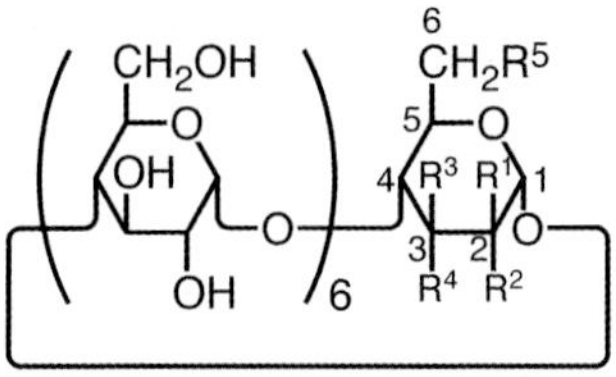

	R1	R2	R3	R4	R5
69	OH	H	H	18Cr6	OH
70	H	OH	OH	H	18Cr6

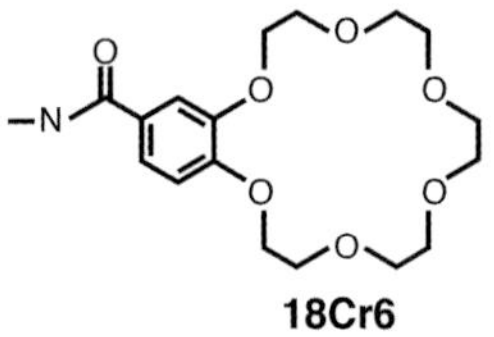

18Cr6

Crown ether-capped β-CyD **71** showed a binding affinity for 8-anilino-1-naphthalenesulfonate (ANS) 88 times higher than the native β-CyD, but the affinity for Acridine Red and Rhodamine B was lower than that of the native β-CyD [180].

A β-CyD-calix[4]arene dyad **72** and a calixarene-bridged bis(β-CyD) **73** showed enhanced binding ability with some fluorescence dyes compared to the native β-CyD [181].

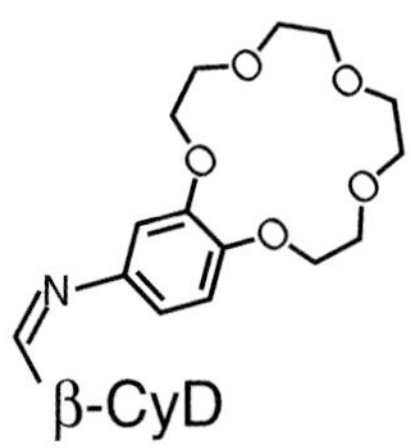

71

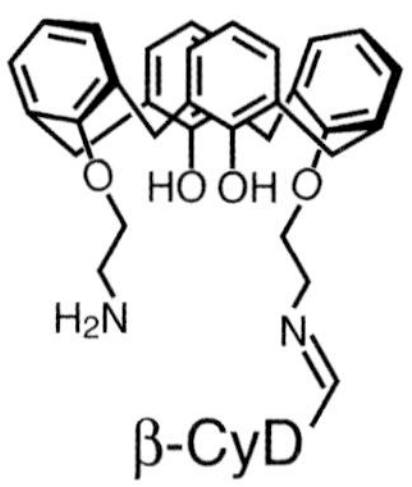

72

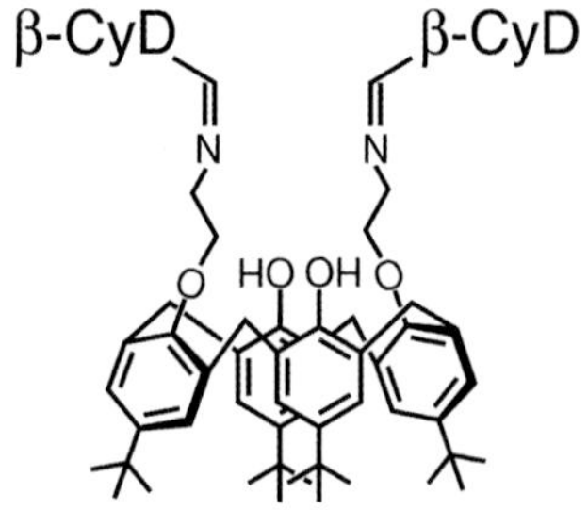

73

The CyD framework can be used as a scaffold to construct polyvalent receptors [182]. Seven dibenzo[24]crown-8 **74** or seven benzometaphenylene[25]crown-8 **75**

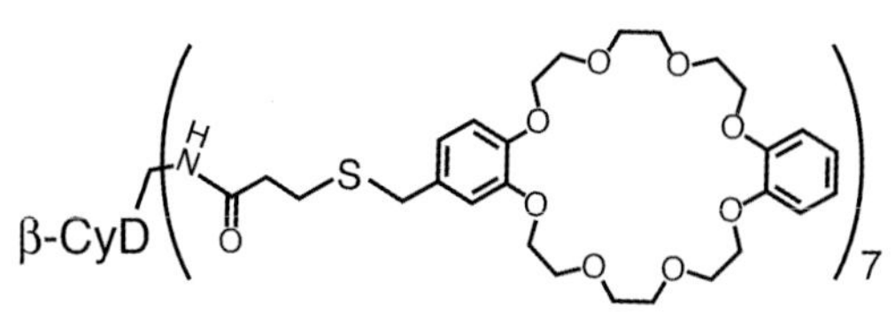

74

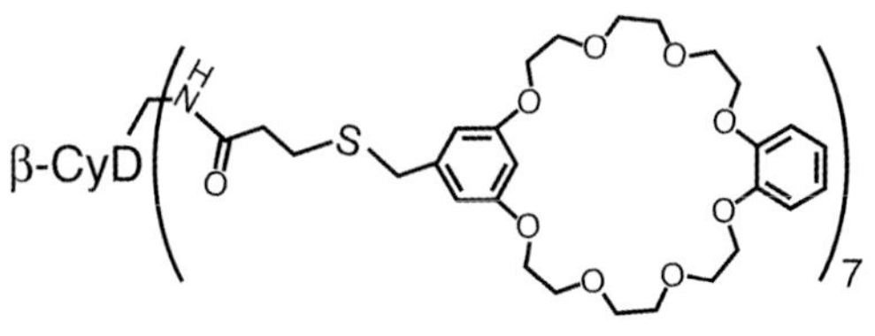

75

moieties appended to β-CyD increased their polyvalent interactions for a divalent bisammonium dication by approximately two orders of magnitude.

2.3.6
Cyclodextrin–Peptide Conjugates

When a site-directed drug carrier system is constructed using the CyD framework, it is necessary to introduce affinity for protein or cell into the CyD framework. CyD–peptide conjugates are promising compounds for targeting host/guest complexes or covalently bound cytotoxic drugs to specific tumor cells for receptor-mediated internalization, but peptide with the relatively large CyD carrier is liable to impair recognition processes at a molecular level. The C-terminal tetrapeptide amide of gastrointestinal hormone was linked to mono-(6-succinylamino-6-deoxy)-β-CyD **76** or heptakis-(6-succinylamino-6-deoxy)-β-CyD **77** to analyze the effect of the bulky cyclic carbohydrate moiety on recognition of the peptides by the G-protein-coupled CCK-B/gastrin receptor and on their signal transduction potencies [183]. The receptor binding affinities clearly show that the β-CyD moiety causes a significant decrease in the affinity of the peptide conjugates relative to the unconjugated peptide ligands. With the four-carbon succinyl spacer and particularly with the additional tripeptide spacer in the heptapeptide/β-CyD conjugate **78**, full recognition by the receptor was obtained with binding affinities identical to those of the unconjugated tetrapeptide and with potency comparable to that of full agonists.

A site-directed drug carrier system **79** was prepared by conjugation of β-CyD with an *endo*-epoxysuccinyl peptide inhibitor MeO–Gly–Gly–Leu–(2*S*,3*S*)–*t*EPS–Leu–Pro–OH as an appropriate address sequence for cathepsin B [184]. Several tumor cell lines are known to secrete cathepsin B and/or to contain membrane-associated cathepsin B, which is thought to be involved in invasion and metastasis. This peptide selectively inhibits cathepsin B among the cysteine proteinases by the specific salt-bridge interaction of the C-terminal carboxylate function with histidine 110 and 111 of the occluding loop and spans the whole substrate binding groove at the S subsite with the propeptide portion (46–48).

β-CyD$\overset{6}{—}$$(NHCOC_2H_4CO—R)_n$

76: n = 1, R = Trp-Nle-Asp-Phe-NH_2
77: n = 7, R = Trp-Nle-Asp-Phe-NH_2
78: n = 1, R = Ala-Tyr-Gly-Trp-Nle-Asp-Phe-NH_2

β-CyD$\overset{6}{—}$$NHCOC_5H_{10}NH—R$

79: R = Gly-Gly-Leu-(2S,3S)-*t*EPS-Leu-Pro-OH

The CyD-peptide conjugate **79** quantitatively inhibited the endogenous cathepsin B activity of lysates of MCF-7 breast cancer cells at nanomolar concentrations identical to those determined for the isolated enzyme. Even at the concentrations required for full inhibition of extracellular and/or membrane-bound cathepsin B the conjugate was not membrane-permeant. Thus, the synthesized conjugate **79** exhibits all the characteristics necessary to act as a site-selective drug carrier system.

β- or *γ*-CyDs were linked with a peptidic moiety consisting of the neuropeptide Substance P (SP) or one of its shorter derivatives, Substance P_{4-11} **80–82** [185]. These conjugates were largely protected against enzymatic degradation in plasma, exhibiting a half-life close to 6 h compared with less than 10 min for SP. This probably results from the steric hindrance introduced by coupling one or two CyDs to the SP backbone.

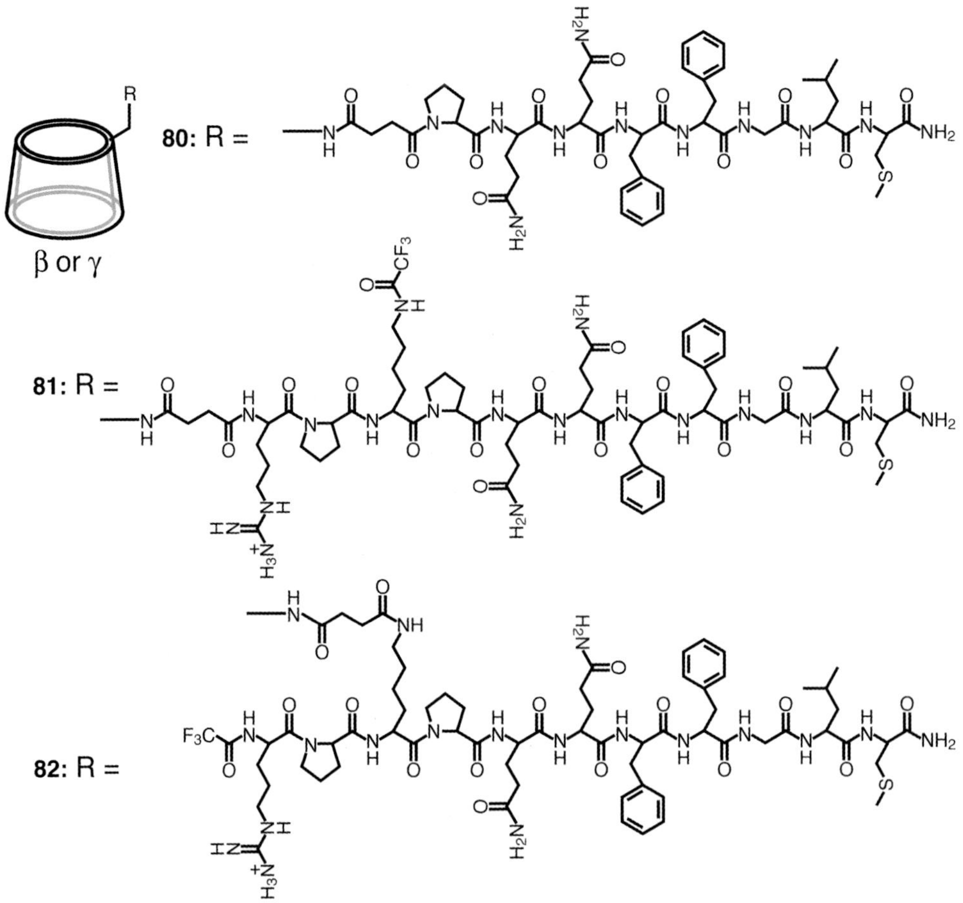

First-order dendrimers based on a *β*-CyD core containing fourteen Val, Phe, and Val–Phe were synthesized **83–86** [186]. This tetradecapeptidyl CyD could form a 2:1 host/guest inclusion complex with adamantanecarboxylic acid, although the native *β*-CyD makes a 1:1 host/guest complex with this guest.

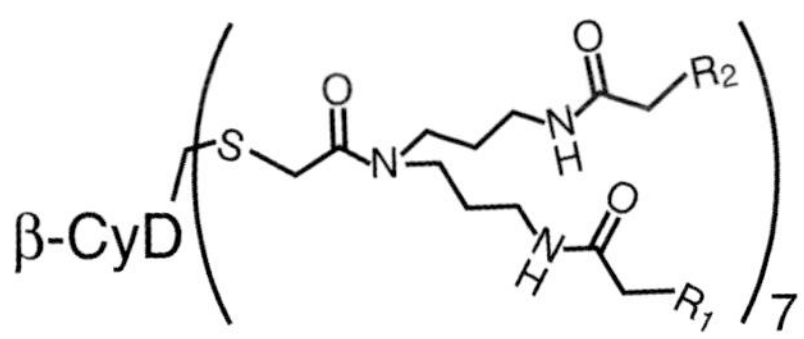

83: $R_1 = R_2 =$ MeO-Val-$COCH_2S$-
84: $R_1 = R_2 =$ MeO-Phe-$COCH_2S$-
85: $R_1 = R_2 =$ MeO-Val-Phe-$COCH_2S$-
86: $R_1 =$ MeO-Val-$COCH_2S$- ; $R_2 =$ MeO-Phe-$COCH_2S$-

CyD-peptide conjugates are also used in the construction of chemosensors (see above) or artificial enzymes.

2.3.7
Cyclodextrin–Saccharide Conjugates

Protein–carbohydrate interactions play an essential role in biological communication. On cell surfaces, protein–carbohydrate interactions are involved in the mediation of important biological processes such as embryogenesis, cancer metastasis, inflammation, and bacterial and viral infection. Although the intrinsic binding ability of monovalent protein carbohydrate interactions is low (K_b values of the order of 10^3 to 10^6 M^{-1}), the presence of carbohydrate epitopes in multivalent arrays usually results in highly specific and effective ligands, which could be used as drug carriers or inhibitors of undesired carbohydrate–receptor associations [45, 187]. Persubstituted CyD-based glycoclusters can be used as sensors to delineate topological differences between the two dimeric prototype proteins. Several CyD-saccharide conjugates have been presented earlier in this chapter.

Sugar density and topology are emerging as crucial factors modulating lectin binding, and many kinds of saccharide-modified CyD have been examined. A nearly 400-fold increase in inhibitory potency of each galactose moiety in the heptavalent form relative to free lactose was detected [57]. These perglycosylated *β*-CyDs had potency to discriminate between two prototypes and between prototype and chimera-type galectins.

The glycosidic moieties were scaffolded onto the *β*-CyD core by direct binding of the anomeric position to the primary position of *β*-CyD or through a short spacer [47]. *β*-CyD derivatives with longer spacer arms between the CyD core and the haptenic sugar moieties **91–94** showed improved inhibitory properties in comparison to derivatives with shorter spacer arms **87–90**. Perglycosylated *β*-CyD showed higher binding ability to lectin than the corresponding monosaccharide and also showed the ability of cross-linking reagents to precipitate lectins.

Targeted drug delivery systems for the anticancer drug Taxotère were built from *β*-CyD by monoconjugation with mannosyl-coated dendritic branches. Binding inhibition of horseradish peroxidase-labeled Concanavalin A to yeast mannan by mannosylated *β*-CyDs was examined. The IC_{50} values of **95**, **96**, **97**, **98**, and **99** were 800, 780, 91, 110, and 8 μM, respectively. **97** solubilized Taxotère similarly to monobranched CyDs [34]. The solubility of Taxotère in water can by increased 1000-fold by **97**.

Dendritic *β*-CyD derivatives bearing multivalent mannosyl ligands **99**, **100** were

β-CyD$\overset{6}{—}$(SR)$_7$ R =

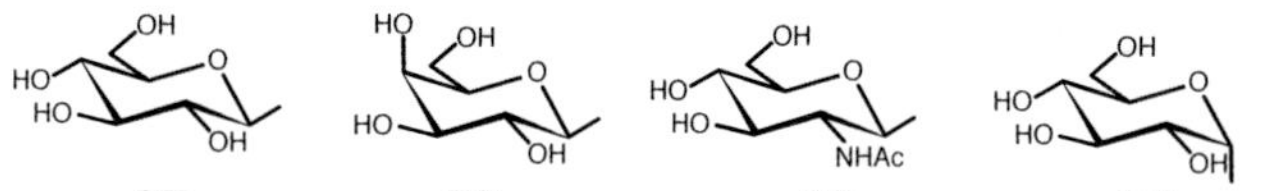

87 **88** **89** **90**

β-CyD$\overset{6}{—}$(NHCOCH$_2$SR')$_7$ R' =

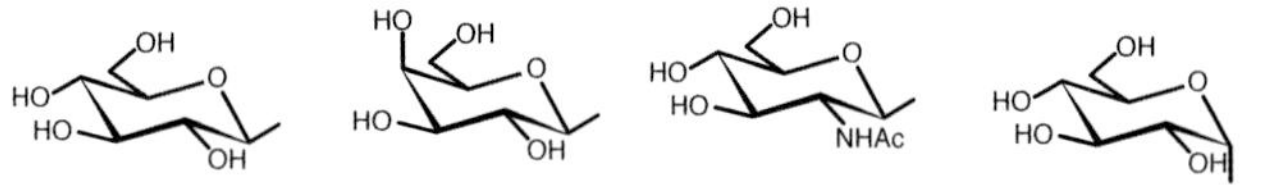

91 **92** **93** **94**

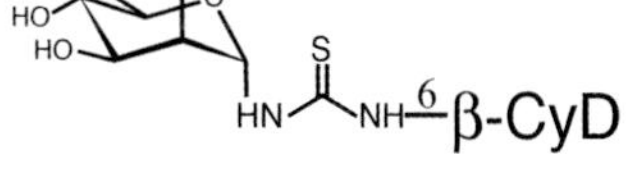

95

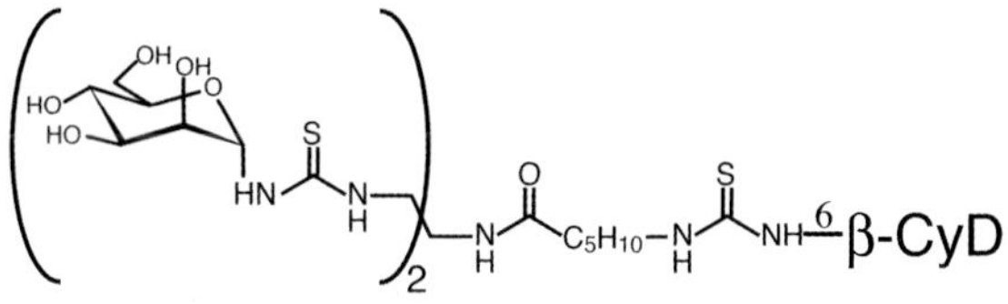

96

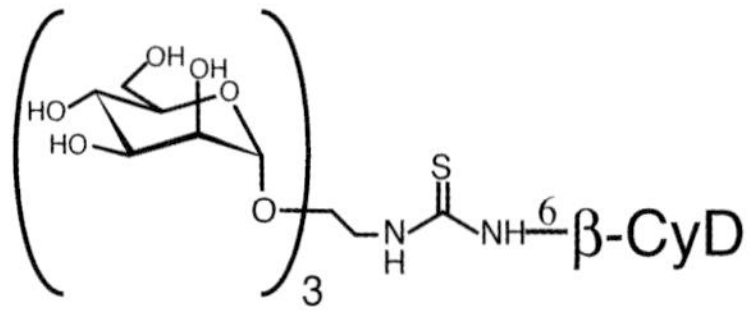

97

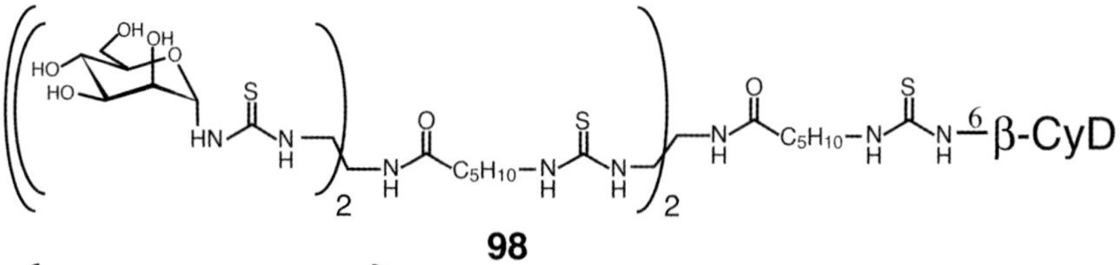

98

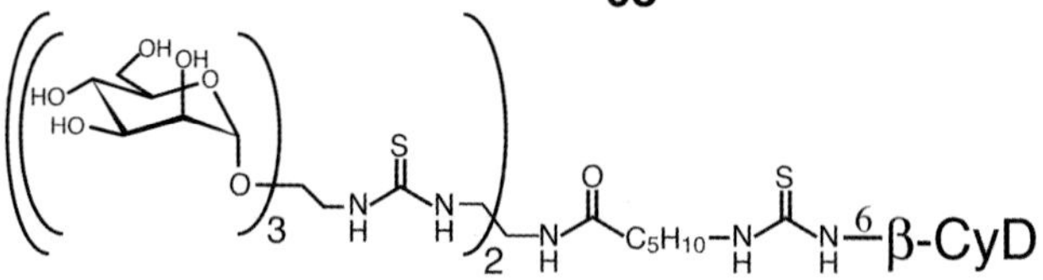

99

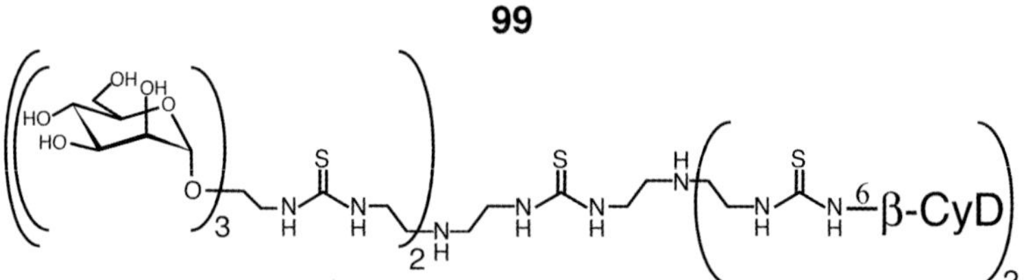

100

also tested as carriers for Taxotère [188]. The association constant of the dimeric host **100** for Taxotère is 1.5×10^5 M^{-1}, which is 37.5-fold larger than that of the monomeric host **99**.

2.3.8
Metallocavitands Using Modified Cyclodextrins

CyDs were used as scaffolds for making metallocavitands [7]. Two phosphine units can be highly preorganized on the CyD rim and a metal center can be rigidly held above the entrance of the opened CyD cavity [189–193]. These metallocavitands are expected to lead to the development of new metal catalysts with enhanced selectivities and activities (CyD catalysis is presented in Chapter 4), in particular because of the protection against undesired side reactions provided by substrate inclusion into the CyD cavity and by stabilization of unusual coordination modes.

A C_2-symmetrical diphosphine based on α-CyD **101** was synthesized and several C_2-symmetric metal complexes, **102–106**, were prepared [189], in which the orientation of the phosphorus lone pairs in the diphosphine is controlled by steric interactions between the CyD matrix and the closely appended phosphanyl groups. Treatment of **101** with [Au(thf)(SC_4H_8)]BF_4 led to a high yield of the unusually air-stable C_2-symmetric complex **103**, which remained intact even when subjected to chromatography on silica gel. **101** with [$PtCl_2$(PhCN)$_2$] led to the exclusive formation of *trans-P,P*-chelate complex **104a**. Exposure of a refluxing ethanolic solution of **104a** to sunlight for a week produced a 65:35 equilibrium mixture of **104a** and *cis* chelate **104b**. The reaction between the ligand **101** and [$PdCl_2$(PhCN)$_2$] (*cis* and *trans* isomers) afforded a rapidly interconverting 80:20 equilibrium mixture of *trans*-chelate **105a** and *cis*-chelate **105b**. The reaction of [{Rh(CO)$_2$Cl}$_2$] in dichloromethane with **101** quantitatively yielded complex **106**, which catalyzed the conversion of oct-1-ene into the corresponding aldehydes in H_2O/MeOH. Selectivity of linear to branched aldehydes (*n*:*i*) is of the order of 70:30.

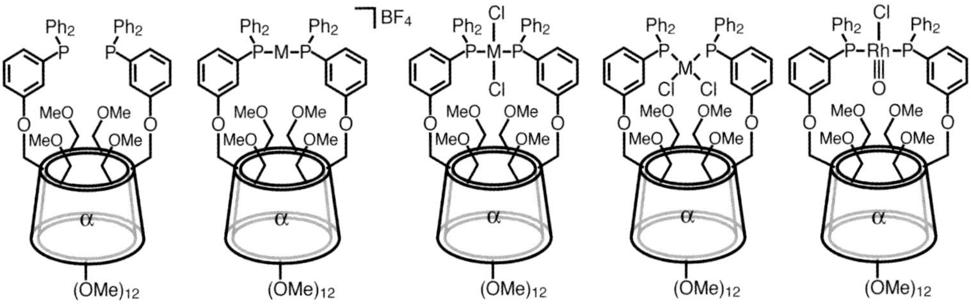

101 **102:** M = Ag **103:** M = Au **104a:** M = Pt **105a:** M = Pd **104b:** M = Pt **105b:** M = Pd **106**

The reaction of ligand **107**, which has two C(5)-linked CH_2PPh_2 units at the A and D positions of α-CyD, with one equivalent of $AgBF_4$ in MeCN led to quantitative formation of the complex [Ag(**107**)(CH_3CN)$_2$]BF_4 (**108**), which is only stable in the presence of a large excess of MeCN (>15 equiv.) [190]. The coordination of MeCN molecules is stabilized by the CyD cavity. The addition of benzonitrile (ca. 8 equiv.) to a solution of **108** in a $CHCl_3$/MeCN mixture (**108**/MeCN = 1:15) resulted in the quantitative formation of complex **109**, in which a single benzonitrile molecule, accommodated in the CyD cavity, is coordinated to the silver ion.

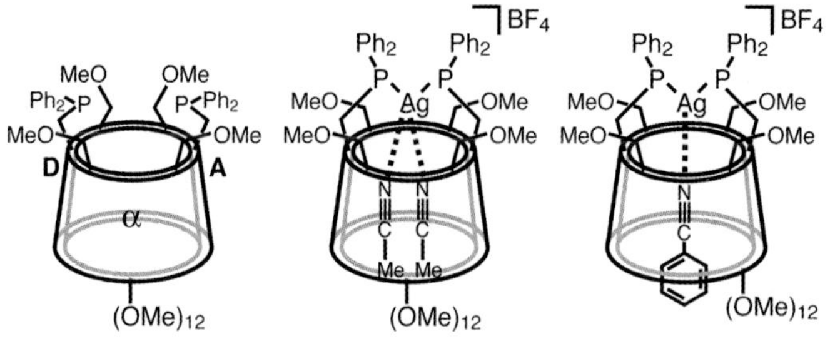

107 **108** **109**

The binding properties of two α-CyDs, each containing two C(5)-linked CH_2PPh_2 units, **107** (A,D-substituted) and **110** (A,C-substituted) were studied [191–193]. The reaction of the symmetrical ligand **107** with [$PdCl_2(PhCN)_2$] and [$PtCl_2(PhCN)_2$] afforded C_2-symmetrical complexes **111** and **112**, respectively, while the unsymmetrical ligand **110** produced complexes **113** and **114**. The P–Pt vectors are slightly bent towards the cavity. The ligand possessed a certain degree of flexibility and allowed the stabilization of a trigonal silver complex in which the bite angle dropped

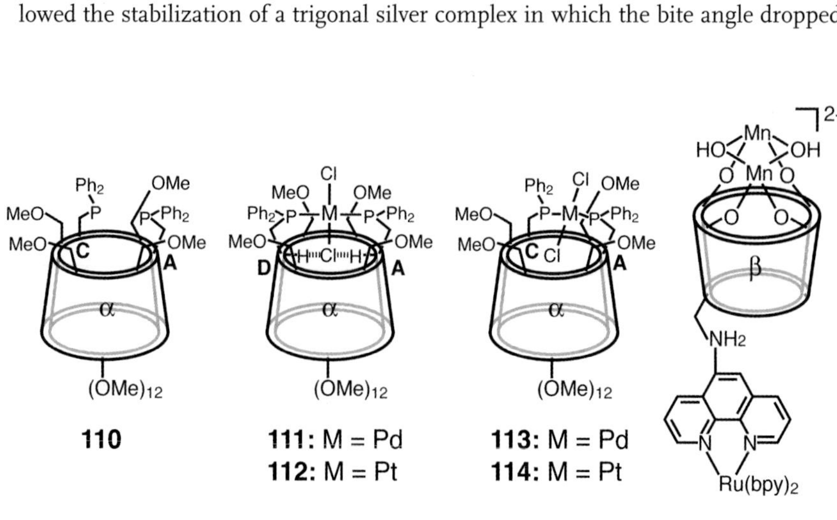

110

111: M = Pd
112: M = Pt

113: M = Pd
114: M = Pt

115

to 143°. Furthermore, the particular structures of **107** and **110**, characterized by the presence of P(III) units lying close to the cavity entrance, lead upon complexation to complexes in which the first coordination sphere is partly entrapped in the CyD. Thus, when treated with metal chlorides, both ligands systematically produce complexes in which the M–Cl unit is maintained inside the CyD through weak Cl · · · H-5 interactions.

The photoinduced electron-transfer system **115** was constructed using a donor–acceptor complex comprising a photosensitizing ruthenium polypyridyl moiety covalently linked to a β-CyD unit bearing a hydroxy-bridged manganese(III) dimer as a model system for elucidating aspects of photosystem II, in which photosynthetic water oxidation is carried out [194]. The guest binding to this system is expected to allow direct control over the strength of electronic coupling between donor and acceptor.

References

1 Easton CJ, Lincoln SF, *Modified Cyclodextrins: Scaffolds and Templates for Supramolecular Chemistry*, Imperial College Press, London, **1999**; pp 43–100.
2 Khan AR, Forgo P, Stine KJ, D'Souza VT, *Chem. Rev.* **1998**, *98*, 1977–1996.
3 Takahashi K, *Chem. Rev.* **1998**, *98*, 2013–2034.
4 Jicsinszky L, Hashimoto H, Fenivesi É, Ueno A, in *Comprehensive Supramolecular Chemistry*, Vol. 3, Lehn JM, Atwood JL, Davies JE, MacNicol DD, Vögtle F, Eds, Szejtli J. Osa T, Vol. Eds., Pergamon: Oxford, **1996**; pp 57–188.
5 Nepogodiev SA, Stoddart JF, *Chem. Rev.* **1998**, *98*, 1959–1976.
6 Gattuso G, Nepogodiev SA, Stoddart JF, *Chem. Rev.* **1998**, *98*, 1919–1958.
7 Engeldinger E, Armspach D, Matt D, *Chem. Rev.* **2003**, *103*, 4147–4173.
8 Ashton PR, *Chem. Eur. J.* **1997**, *3*, 1299–1314.
9 Komiyama M, Shigekawa H, in *Comprehensive Supramolecular Chemistry*, Vol. 3, Lehn JM, Atwood JL, Davies JE, MacNicol DD, Vögtle F, Eds, Szejtli J, Osa T, Vol. Eds., Pergamon: Oxford, **1996**; pp 401–421.
10 Breslow R, Dong SD, *Chem. Rev.* **1998**, *98*, 1997–2011.
11 Dodziuk H, in *Introduction to Supramolecular Chemistry*, Kluwer Academic, Dordrecht, **2002**, Chapter 5.3.5.
12 For instance, Hattori K, Kano K, Eds., *Special Issue of 2nd Asian Cyclodextrin Conference* in *J. Inclusion Phenom. Macrocycl. Chem.* **2004**, *50*, 1–127.
13 Duchene D, Fattal E, Eds, *Proceedings for the 12th International Cyclodextrin Symposium*, **2004**, May 16–19, Montpellier, France.
14 Thompson DO, Nagai T, Ueda H, Usuda M, Endo T, *Pharm. Tech. Jpn* **2002**, *18*, 63.
15 Takahashi K, Hattori K, Toda F, *Tetrahedron Lett.* **1984**, *25*, 3331–1334.
16 Cornwell MJ, Huff JB, Bieniarz C, *Tetrahedron Lett.* **1995**, *36*, 8371.
17 Jimenez V, Belmar J, Alderete JB, *J. Inclusion Phenom.* **2003**, *47*, 71–75.
18 Jimenez-Blanco JL, Garcia-Fernandez JM, Gadelle A, Defaye J, *Carbohydr. Res.* **1997**, *303*, 367–372.
19 Martin KA, Czarnik AW, *Tetrahedron Lett.* **1994**, *35*, 6781–6782.
20 Wilson D, Perlson L, Breslow R, *Bioorg. Med. Chem.* **2003**, *11*, 2649–2653.
21 Yuan DQ, Koga K, Kourogi Y, Fujita K, *Tetrahedron Lett.* **2001**, *42*, 6727–6729.

22 Porwanski S, Kryczka B, Marsura A, *Tetrahedron Lett.* **2002**, *43*, 8441–8443.
23 Matsuoka K, Terabatake M, Saito Y, Hagihara C, Terunuma D, Kuzuhara H, *Bull. Chem. Soc. Jpn.* **1998**, *71*, 2709–2713.
24 Eliadou K, Giastas P, Yannakopoulou K, Mavridis IM, *J. Org. Chem.* **2003**, *68*, 8550–8557.
25 Gao F, Shang YJ, Zhang L, She SK, Wang L, *Anal. Lett.* **2004**, *37*, 1285–1295.
26 Arima H, Kihara F, Hirayama F, Uekama K, *Bioconjugate Chem.* **2001**, *12*, 476–484.
27 Galaverna G, Corradini R, Dossena A, Marchelli R, *Electrophoresis* **1999**, *20*, 2619–2629.
28 Diakur J, Zuo Z, Wiebe LI, *J. Carbohydr Chem.* **1999**, *18*, 209–223.
29 Carofiglio T, Cordioli M, Fornasier R, Jicsinsky L, Tonellato U, *Carbohydr. Res.* **2004**, *339*, 1361–1366.
30 Leung DK, Atkins JH, Breslow R, *Tetrahedron Lett.* **2001**, *42*, 6255–6258.
31 Aoki N, Arai R, Hattori K, *J. Inclusion Phenom.* **2004**, *50*, 115–120.
32 Nelissen HFM, Feiters MC, Nolte RJM, *J. Org. Chem.* **2002**, *67*, 5901–5906.
33 Garcia-Ruiz C, Crego AL, Marina ML, *Electrophoresis* **2003**, *24*, 2657–2664.
34 Baussanne I, Benito JM, Mellet CO, Garcia-Fernandez JM, Law H, Defaye J, *Chem. Commun.* **2000**, 1489–1490.
35 Yamanoi T, Yoshida N, Oda Y, Akaike E, Tsumida M, Kobayashi N, Osumi K, Yamamoto K, Fujita K, Takahashi K, Hattori K, *Bioorg. Med. Lett.* **2005**, *15*, 1009–1013.
36 Petter RC, Salek JS, Sikorski CT, Kumaravel G, Lin FT, *J. Am. Chem. Soc.* **1990**, *112*, 3860–3868.
37 Nelson A, Stoddart JF, *Org. Lett.* **2003**, *5*, 3783–3786.
38 Parrot-Lopez H, Leray E, Coleman AW, *Supramol. Chem.* **1993**, *3*, 37–42.
39 Fernandez M, Fragoso A, Cao R, Villalonga R, *J. Mol. Catalysis B: Enzymatic* **2003**, *21*, 133–141.
40 Yockot D, Moreau V, Demailly G, Djedaïni-Pilard F, *Org. Biomol. Chem.* **2003**, *1*, 1810–1818.
41 Alvarez-Parrilla E, Cabrer PR, De La Rosa LA, Al-Soufi W, Meijide F, Tato JV, *Supramol. Chem.* **2003**, *15*, 207–211.
42 Liu Y, You CC, He S, Chen GS, Zhao YL, *J. Chem. Soc., Perkin Trans. 2* **2002**, 463–469.
43 Lock JS, May BL, Clements P, Lincoln SF, Easton CJ, *Org. Biomol. Chem.* **2004**, 1381–1386.
44 Kassab R, Felix C, Parrot-Lopez H, Bonaly R, *Tetrahedron Lett.* **1997**, *38*, 7555–7558.
45 Fulton DA, Stoddart JF, *Bioconjugate Chem.* **2001**, *12*, 655–672.
46 Mellet CO, Benito JM, Garcia-Fernandes JM, Law H, Chmurski K, Defaye J, O'Sullivan ML, Caro H, *Chem. Eur. J.* **1998**, *4*, 2523–2531.
47 Garcia-Lopez JJ, Hernandes-Mateo F, Isac-Garcia J, Kim JM, Roy R, Santoyo-Gonzalez F, Vargas-Berenguel A, *J. Org. Chem.* **1999**, *64*, 522–531.
48 Garcia-Lopez JJ, Santoyo-Gonzale F, Vargas-Berenguel A, Gimerenz-Martinez JJ, *Chem. Eur. J.* **1999**, *5*, 1775–1784.
49 Calvo-Balderas F, Isac-Garcia J, Hernandes-Mateo F, Perez-Balderas F, Calvo-Astin JA, Sacchez-Vaquero E, Santoyo-Gonzarez F, *Org. Lett.* **2000**, *2*, 2499–2502.
50 Roy R, Hernandes-Mateo F, Santoyo-Gonzalez F, *J. Org. Chem.* **2000**, *65*, 8743–8746.
51 Garcia-Barrientos A, Garcia-Lopez JJ, Isac-Garcia J, Ortega-Caballero F, Uriel C, Vagas-Berenguel A, Santoyo-Gonzalez F, *Synthesis* **2001**, 1057–1064.
52 Ortega-Caballero F, Gimerenz-Martinez JJ, Garcia-Fuentes L, Ortiz-Salmeron E, Vargas-Berenguel A, *J. Org. Chem.* **2001**, *66*, 7786–7795.
53 Vargas-Berenguel A, Ortega-Caballero F, Santoyo-Gonzalez F, Garcia-Lopez JJ, Gimenez-Martinez JJ, Garcia-Fuentes L, Ortiz-

Salmeron E, *Chem. Eur. J.* **2002**, *8*, 812–827.
54 Furuike T, Aiba S, *Chem. Lett.* **1999**, 69–70.
55 Furuike T, Aiba S, Nishimura S, *Tetrahedron* **2000**, *56*, 9909–9915.
56 Sallas F, Niikura K, Nishimura S, *Chem. Commun.* **2004**, 596–597.
57 Andre S, Kaltner H, Furuike T, Nishimura S, Gabius HJ, *Bioconjugate Chem.* **2004**, *15*, 87–98.
58 Fulton DA, Stoddart JF, *J. Org. Chem.* **2001**, *66*, 8309–8319.
59 Fulton DA, Stoddart JF, *Org. Lett.* **2000**, *2*, 1113–1116.
60 Fulton DA, Pease AR, Stoddart JF, *Israel J. Chem.* **2000**, *40*, 325–333.
61 Kobayashi K, Kajikawa K, Sasabr H, Knoll W, *Thin Solid Films* **1999**, *349*, 244–249.
62 Granger CE, Felix CP, Parrot-Lopez HP, Lamglois BR, *Tetrahedron Lett.* **2000**, *41*, 9257–9260.
63 Ringard-Lefebvre C, Bochot A, Memisoglu E, Charon D, Duchene D, Baszkin A, *Colloids Surfaces B: Biointerfaces* **2002**, *25*, 109–117.
64 Memisoglu E, Bochot A, Sen M, Charon D, Duchene D, Hincal AA, *J. Pharm. Sci.* **2002**, *91*, 1214–1224.
65 Chen L, Zhan LF, Ching CB, Ng SC, *J. Chromatog. A* **2002**, *950*, 65–74.
66 Abe H, Kenmoku A, Yamaguchi N, Hattori K, *J. Inclusion Phenom. Macrocycl. Chem.* **2002**, *44*, 39–47.
67 Ikuta A, Mizuta N, Kitahata S, Murata T, Usui T, Koizumi K, Tanimoto T, *Chem. Pharm. Bull.* **2004**, *52*, 51–56.
68 Narita M, Hamada F, *J. Chem. Soc., Perkin Trans. 2* **2000**, 823–832.
69 Chen WH, Yan JM, Tagashira Y, Yamaguchi M, Fujita K, *Tetrahedron Lett.* **1999**, *40*, 891–894.
70 Koga K, Ishida K, Yamada T, Yuan DQ, Fujita K, *Tetrahedron Lett.* **1999**, *40*, 923–926.
71 Yuan DQ, Yamada T, Fujita K, *Chem. Commun.* **2001**, *24*, 2706–2707.
72 Yuan DQ, Immel S, Koga K, Yamaguchi M, Fujita K, *Chem. Eur. J.* **2003**, *9*, 3501–3506.
73 Koga K, Yuan DQ, Fujita K, *Tetrahedron Lett.* **2000**, *41*, 6855–6857.
74 Fasella E, Dong SD, Breslow R, *Bioorg. Med. Chem.* **1999**, *7*, 709–714.
75 Shi XY, Fu RN, Gu LJ, *J. Beijin Inst. Tech.* **2002**, *11*, 285–289.
76 Shi XY, Zhang YQ, Han JH, Fu RN, *Chromatographia* **2000**, *52*, 200–204.
77 Wang W, Pearce AJ, Zhang Y, Sinay P, *Tetrahedron: Asymmetry* **2001**, *12*, 517–523.
78 Martel B, Devassine M, Morcellent M, Crini G, Weltrowski M, Bourdonneau M, *J. Polymer Sci. A* **2001**, *39*, 169–176.
79 Buchanan CM, Alderson SR, Cleven CD, Dixon DW, Ivanyi R, Lambert JL, Lowman DW, Offerman RJ, Szejtli J, Szente L, *Carbohydr. Res.* **2002**, *337*, 493–507.
80 Atsumi M, Izumida M, Yuan DQ, Fujita K, *Tetrahedron Lett.* **2000**, *41*, 8117–8120.
81 Yin XH, *Chin. Chem. Lett.* **2003**, *14*, 445–447.
82 Carpov A, Mocanu G, Vizitiu D, *Angew. Makromol. Chem.* **1998**, *256*, 75–79.
83 Teranishi K, Tanabe S, Hisamatsu M, Yamada T, *Biosci. Biotechnol. Biochem.* **1998**, *62*, 1249–1252.
84 Law H, Baussanne I, Garcia-Fernandes JM, Defaye J, *Carbohydr. Res.* **2003**, *338*, 451–453.
85 Suzuki M, Nozoe Y, *Carbohydr. Res.* **2002**, *337*, 2393–2397.
86 Choi JK, Girek T, Shin DH, Lim ST, *Carbohydr. Polym.* **2002**, *49*, 289–296.
87 Liu Y, You CC, Zhang HY, Zhao YL, *Eur. J. Org. Chem.* **2003**, 1415–1422.
88 Ruebner A, Kirsch D, Andrees S, Decker W, Roeder B, Spengler B, Kaufmann R, Moser JG, *J. Inclusion Phenom.* **1997**, *27*, 69–84.
89 Chen S, Wang Y, *J. Appl. Polym. Sci.* **2001**, *82*, 2414–2421.
90 Yan J, Watanabe R, Yamaguchi M, Yuan DQ, Fujita K, *Tetrahedron Lett.* **1999**, *40*, 1513–1514.
91 Yuan DQ, Tahara T, Chen WH, Okabe Y, Yang C, Yagi Y, Nogami T, Fukudome M, Fujita K, *J. Org. Chem.* **2003**, *68*, 9456–9466.
92 Chiu SH, Myles DC, Garrel RL, Stoddart JF, *J. Org. Chem.* **2000**, *65*, 2792–2796.

93 Yano H, Hirayama F, Arima H, Uekama K, *J. Pharm. Sci.* **2001**, *90*, 493–503.

94 Chiu SH, Myles DC, *J. Org. Chem.* **1999**, *64*, 332–333.

95 Fukudome M, Oiwane K, Mori T, Yuan DQ, Fujita K, *Tetrahedron Lett.* **2004**, *45*, 3383–3386.

96 Chen WH, Hayashi S, Tahara T, Nogami Y, Koga T, Yamaguchi M, Fujita K, *Chem. Pharm. Bull.* **1999**, *47*, 588–589.

97 Park KK, Kim YS, Lee SY, Song HE, Park JW, *J. Chem. Soc., Perkin Trans. 2*, **2001**, 2114–2118.

98 Cousin H, Cardinael P, Oulyadi H, Pannnecoucke X, Combert JC, *Tetrahedron: Asymmetry* **2001**, *12*, 81–88.

99 Cousin H, Trapp O, Peulon-Agasee V, Pannecoucke X, Banspach L, Trapp G, Jiang Z, Combret JC, Schurig V, *Eur. J. Org. Chem.* **2003**, 3273–3287.

100 Tian S, Zhu H, Forgo P, D'Souza VT, *J. Org. Chem.* **2000**, *65*, 2624–2630.

101 Ikuta A, Mizuta N, Kitahata S, Murata T, Usui T, Koizumi K, Tanimoto T, *Chem. Pharm. Bull.* **2004**, *52*, 51–56.

102 Yang C, Yuan DQ, Nogami Y, Fujita K, *Tetrahedron Lett.* **2003**, *44*, 4641–4644.

103 Mazzaglia A, Donohue R, Ravoo BJ, Darcy R, *Eur. J. Org. Chem.* **2001**, 1715–1721.

104 Bukowska M, Maciejewski M, Prejzner J, *Carbohydr. Res.* **1998**, *308*, 275–279.

105 Kraus T, Budesinsky M, Zavada J, *J. Org. Chem.* **2001**, *66*, 4595–4600.

106 Ohno K, Wong B, Haddleton DM, *J. Polymer Sci., A, Polymer Chem.* **2001**, *39*, 2206–2214.

107 Harabagiu V, Simionescu BC, Pinteala M, Merrienne C, Mahuteau J, Guegan P, Cheradame H, *Carbohydr. Polym.* **2004**, *56*, 301–311.

108 Potluti VK, Xu J, Enick R, Beckman E, Hamilton AD, *Org. Lett.* **2002**, *4*, 2333–2335.

109 Hirayama F, Mieda S, Miyamoto Y, Arima H, Uekama K, *J. Pharm. Sci.* **1999**, *88*, 970–975.

110 Bicchi C, Cravotto G, D'Amato A, Rubiolo P, Galli A, Galli M, *J. Microcolumn Sep.* **1999**, *11*, 487–500.

111 Cravotto G, Palmisano G, Panza L, Tagliapiertra S, *J. Carbohydr. Chem.* **2000**, *19*, 1235–1245.

112 Gong Y, Lee HK, *J. Sep. Sci.* **2003**, *26*, 515–520.

113 Bansal PS, Francis CL, Hart NK, Henderson SA, Oakenfull D, Robertson AD, Simpson GW, *Aust. J. Chem.* **1998**, *51*, 915–923.

114 Hara K, Fujita K, Kuwahara N, Tanimoto T, Hashimoto H, Koizumi K, Kitahata S, *Biosci. Biotech. Biochem.* **1994**, *58*, 652–659.

115 Hamayasu K, Hara K, Fujita K, Kondo Y, Hashimoto H, Tanimoto T, Koizumi K, Nakano H, Kitahata S, *Biosci. Biotech. Biochem.* **1997**, *61*, 825–829.

116 Okada Y, Matsuda K, Hara K, Hamayasu K, Hashimoto H, Koizumi K, *Chem. Pharm. Bull.* **1999**, *47*, 1564–1568.

117 Koizumi K, Okada Y, Kubota Y, Utamura T, *Chem. Pharm. Bull.* **1987**, *35*, 3413–3418.

118 Shinoda T, Maeda A, Kagatani S, Konno Y, Sonobe T, Fukui M, Hashimoto H, Hara K, Fujita K, *Internat. J. Pharm.* **1998**, *167*, 147–154.

119 Tavornvipas S, Arima H, Hirayama F, Uekama K, Ishiguro T, Oka M, Hamayasu K, Hashimoto H, *J. Inclusion Phenom.* **2002**, *44*, 391–394.

120 Bourgeaux E, Combret JC, *Terahedron: Asymmetry* **2000**, *11*, 4189–4205.

121 Hoffmann B, Bernet B, Vasella A, *Helvetica Chim. Acta* **2002**, *85*, 265–287.

122 Wakao M, Fukase K, Kusumoto S, *J. Org. Chem.* **2002**, *67*, 8182–8190.

123 Bender ML, Komiyama M, *Cyclodextrin Chemistry*, Springer-Verlag: Berlin, **1978**.

124 Szejtli S, *Cyclodextrin Technology*, Kluwer Academic: Dordrecht, **1988**.

125 Wenz G, *Angew. Chem., Int. Ed. Engl.* **1994**, *33*, 803–822.

126 Birlirakis N, Henry B, Berthault P, Venema F, Nolte RJM, *Tetrahedron* **1998**, *54*, 3513–3522.

127 Venema F, Nelissen HFM, Berthault P, Birlirakis N, Rowan AE, Feiters MC, Nolte RJM, *Chem. Eur. J.* **1998**, *4*, 2237–2250.

128 Yamada T, Fukuhara G, Kaneda T, *Chem. Lett.* **2003**, *32*, 534–535.

129 Fukudome M, Okabe Y, Yuan DQ, Fujita K, *Chem. Commun.* **1999**, 1045–1046.

130 Yamamura H, Yamada S, Kohno K, Okuda N, Araki S, Kobayashi K, Katakai R, Kano K, Kawai M, *J. Chem. Soc., Perkin Trans. 1* **1999**, 2943–2948.

131 Liu Y, You CC, Chen Y, Wada T, Inoue Y, *J. Org. Chem.* **1999**, *64*, 7781–7787.

132 Liu Y, Li L, Zhang HY, Song Y, *J. Org. Chem.* **2003**, *68*, 527–536.

133 Ren X, Xue Y, Liu J, Zhang K, Zheng J, Luo G, Guo C, Mu Y, Shen J, *Chem. Bio. Chem.* **2002**, 356–363.

134 Liu Y, Li L, Li XY, Zhang HY, Wada T, Inoue Y, *J. Org. Chem.* **2003**, *68*, 3646–3657.

135 Breslow R, Yang Z, Ching R, Trojandt G, Odobel F, *J. Am. Chem. Soc.* **1998**, *120*, 3536–3537.

136 Liu Y, Yang YW, Song Y, Zhang HY, Ding F, Wada T, Inoue Y, *Chem. Bio. Chem.* **2004**, *5*, 868–871.

137 Liu Y, Chen GS, Chen Y, Ding F, Liu T, Zhao YL, *Bioconjugate Chem.* **2004**, *15*, 300–306.

138 Sasaki K, Nagasaka M, Kuroda Y, *Chem. Commun.* **2001**, 2630–2631.

139 Lecourt T, Mallet JM, Sinaÿ P, *Tetrahedron Lett.* **2002**, *43*, 5533–5536.

140 Mulder A, Jukovic A, Huskens J, Reinhoudt DN, *Org. Biomol. Chem.* **2004**, 1748–1755.

141 Mulder A, Jukovic A, Lucas LN, van Esch J, Feringa BL, Huskens J, Reinhoudt DN, *Chem. Commun.* **2002**, 2734–2735.

142 Mulder A, Jukovic A, van Leeuwen FWB, Kooijman H, Spek AL, Huskens J, Reinhoudt DN, *Chem. Eur. J.* **2004**, *10*, 1114–1123.

143 Ikeda H, Matsuhisa A, Ueno A, *Chem. Eur. J.* **2003**, *9*, 4907–4910.

144 Kida T, Ohe T, Higashimoto H, Harada H, Nakatsuji Y, Ikeda I, Akashi M, *Chem. Lett.* **2004**, *33*, 258–259.

145 Nakajima H, Sakabe Y, Ikeda H, Ueno A, *Bioorg. Med. Chem. Lett.* **2004**, *14*, 1783–1786.

146 Vizitiu D, Thatcher GRJ, *J. Org. Chem.* **1999**, *64*, 6235–6238.

147 Ghosh M, Sanders TC, Zhang R, Seto CT, *Org. Lett.* **1999**, *1*, 1945–1948.

148 Cotner ES, Smith PJ, *J. Org. Chem.* **1998**, *63*, 1737–1739.

149 Hauser SL, Johanson EW, Green HP, Smith PJ, *Org. Lett.* **2000**, *2*, 3575–3578.

150 Rekharsky MV, Inoue Y, *J. Am. Chem. Soc.* **2002**, *124*, 813–826.

151 Rekharsky M, Yamamura H, Kawai M, Inoue Y, *J. Am. Chem. Soc.* **2001**, *123*, 5360–5361.

152 Ohtsuki H, Ahmed J, Nagata T, Yamamoto T, Matsui Y, *Bull. Chem. Soc. Jpn.* **2003**, *76*, 1131–1138.

153 Ahmed J, Nagata T, Imaoka S, Matsui Y, Yamamoto T, *Chem. Lett.* **2000**, *29*, 960–961.

154 Ikeda H, Iidaka Y, Ueno A, *Org. Lett.* **2003**, *5*, 1625–1627.

155 Bom A, Bradley M, Cameron K, Clark JK, van Egmond J, Feilden H, MacLean EJ, Muir AW, Palin R, Rees DC, Zhang MQ, *Angew. Chem., Int. Ed.* **2002**, *41*, 265–270.

156 Cameron KS, Clark JK, Cooper A, Fielding L, Palin R, Rutherford SJ, Zhang MQ, *Org. Lett.* **2002**, *4*, 3403–3406.

157 Kano K, Horiki Y, Mabuchi T, Kitagishi H, *Chem. Lett.* **2004**, *33*, 1086–1087.

158 Cho A, Lara KLO, Yatsimirsky AK, Eliseev AV, *Org. Lett.* **2000**, *2*, 1741–1743.

159 Donohue R, Mazzaglia A, Ravoo BJ, Darcy R, *Chem. Commun.* **2002**, 2864–2865.

160 Sukegawa T, Furuike T, Niikura K, Yamagishi A, Monde K, Nishimura S, *Chem. Commun.* **2002**, 430–431.

161 Kraus T, Budesínsky M, Císarová I, Závada J, *Angew. Chem., Int. Ed.* **2002**, *41*, 1715–1717.

162 Kraus T, Budesínsky M, Závada J, *Eur. J. Org. Chem.* **2000**, 3133–3137.
163 Ueno A, Ikeda H, in *Molecular and Supramolecular Photochemistry 8*, Ramamurthy V, Schanze KS, Eds., Marcel Dekker, New York, **2001**, pp 461–503.
164 de Jong MR, Engbersen JFJ, Huskens J, Reinhoudt DN, *Chem. Eur. J.* **2000**, *6*, 4034–4040.
165 Kikuchi T, Narita M, Hamada F, *Tetrahedron* **2001**, *57*, 9317–9324.
166 Bügler J, Engbersen JFJ, Reinhoudt DN, *J. Org. Chem.* **1998**, *63*, 5339–5344.
167 Ueno A, Ikeda A, Ikeda H, Ikeda T, Toda F, *J. Org. Chem.* **1999**, *64*, 382–387.
168 Ikunaga T, Ikeda H, Ueno A, *Chem. Eur. J.* **1999**, *5*, 2698–2704.
169 Hossain MA, Hamasaki K, Takahashi K, Mihara H, Ueno A, *J. Am. Chem. Soc.* **2001**, *123*, 7435–7436.
170 Hossain MA, Mihara H, Ueno A, *Bioorg. Med. Chem. Lett.* **2003**, *13*, 4305–4308.
171 Hossain MA, Mihara H, Ueno A, *J. Am. Chem. Soc.* **2003**, *125*, 11178–11179.
172 Matsumura S, Sakamoto S, Ueno A, Mihara H, *Chem. Eur. J.* **2000**, *6*, 1781–1788.
173 Kida T, Michinobu T, Zhang W, Nakatsuji Y, Ikeda I, *Chem. Commun.* **2002**, 1596–1597.
174 Bodine KD, Gin DY, Gin MS, *J. Am. Chem. Soc.* **2004**, *126*, 1638–1639.
175 Kida T, Kikuzawa A, Nakatsuji Y, Akashi M, *Chem. Commun.* **2003**, 3020–3021.
176 Yamamura H, Yotsuya T, Usami S, Iwasa A, Ono S, Tanabe Y, Iida D, Katsuhara T, Kano K, Uchida T, Araki S, Kawai M, *J. Chem. Soc., Perkin Trans 1* **1998**, 1299–1303.
177 Nepogodiev SA, Gattuso G, Stoddart JF, *J. Inclusion Phenom. Mol. Recog. Chem.* **1996**, *25*, 47–52.
178 Park JW, Lee SY, Park KK, *Chem. Lett.* **2000**, *29*, 594–595.
179 Suzuki I, Obata K, Anzai J, Ikeda H, Ueno A, *J. Chem. Soc., Perkin Trans 2* **2000**, 1705–1710.
180 Liu Y, Yang YW, Li L, Chen Y, *Org. Biomol. Chem.* **2004**, 1542–1548.
181 Liu Y, Chen Y, Li L, Huang G, You CC, Zhang HY, Wada T, Inoue Y, *J. Org. Chem.* **2001**, *66*, 7209–7215.
182 Fulton DA, Cantrill SJ, Stoddart JF, *J. Org. Chem.* **2002**, *67*, 7968–7981.
183 Schaschke N, Fiori S, Weyher E, Escrieut C, Fourmy D, Muller G, Moroder L, *J. Am. Chem. Soc.* **1998**, *120*, 7030–7038.
184 Schaschke N, Assfalg-Machleidt I, Machleidt W, Laßleben T, Sommerhoff CP, Moroder L, *Bioorg. Med. Chem. Lett.* **2000**, *10*, 677–680.
185 Péan C, Créminon C, Wijkhuisen A, Grassi J, Guenot P, Jéhan P, Dalbiez JP, Perly B, Djedaïni-Pilard F, *J. Chem. Soc., Perkin Trans. 2* **2000**, 853–863.
186 Muhanna AMA, Ortiz-Salmerón E, García-Fuentes L, Giménez-Martínez JJ, Vargas-Berenguel A, *Tetrahedron Lett.* **2003**, *44*, 6125–6128.
187 Mellet CO, Defaye J, Garcia-Fernández JM, *Chem. Eur. J.* **2002**, *8*, 1982–1990.
188 Benito JM, Gómez-García M, Mellet CO, Baussanne I, Defaye J, Garcia-Fernández JM, *J. Am. Chem. Soc.* **2004**, *126*, 10355–10363.
189 Armspach D, Matt D, *Chem. Commun.* **1999**, 1073–1074.
190 Engeldinger E, Armspach D, Matt D, *Angew. Chem., Int. Ed.* **2001**, *40*, 2526–2529.
191 Engeldinger E, Armspach D, Matt D, Jones PG, Welter R, *Angew. Chem., Int. Ed.* **2002**, *41*, 2593–2596.
192 Engeldinger E, Armspach D, Matt D, Jones PG, *Chem. Eur. J.* **2003**, *9*, 3091–3105.
193 Engeldinger E, Poorters L, Armspach D, Matt D, Toupet L, *Chem. Commun.* **2004**, 634–635.
194 Hoof NV, Keyes TE, Forster RJ, McNally A, Russell NR, *Chem. Commun.* **2001**, 1156–1157.

3
Polymers Involving Cyclodextrin Moieties

Akira Harada, Akihito Hashidzume, and Masahiko Miyauchi

Polymers involving CyD moieties can be either regular polymers or supramolecular ones. The latter ones are novel and promising future applications while the former, more traditional ones have already found numerous applications. Since supramolecular oligomers and polymers seem to be more interesting they will be presented first. It should be stressed that most these systems have the rotaxane structure, which together with the structure of catenanes involving CyDs are discussed in Chapter 14.

3.1
Supramolecular Polymers Formed by Cyclodextrin Derivatives

3.1.1
Introduction

Supramolecular polymers are ubiquitous in nature, especially in biological systems. Microtubules, microfilaments, and flagella are helical supramolecular polymers formed by proteins. In recent years, much attention has been focused on supramolecular polymers formed by synthetic molecules, because of their unique structures and properties (Fig. 3.1) [1, 2]. Supramolecular oligomers formed by hydrogen bonds were reported for the first time by Lehn et al. (Fig. 3.2) [1, 3–5]. Supramolecular polymers formed by four hydrogen bonds of bifunctional ureidopyrimidinone derivatives were reported by Meijer et al. [6–8]. Rebeck et al. [9–12] reported a supramolecular polymer formed by utilizing hydrogen bonding between bifunctionalized calixarene derivatives. Rehahn et al. [13] reported supramolecular polymers formed by porphyrin dimers with metal coordination. However, there are few reports on the formation of supramolecular polymers by guest–host interactions in aqueous solutions, although the formation of supramolecular polymers by these interactions is important in biological systems.

Cyclodextrins and Their Complexes. Edited by Helena Dodziuk

ISBN: 3-527-31280-3

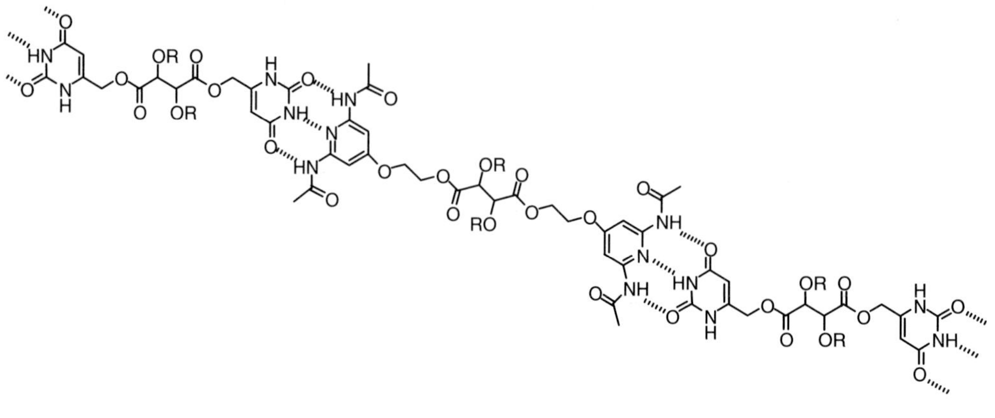

Fig. 3.1. Structure of Lehn's supramolecular polymer.

3.1.2
Preparation of Mono-substituted Cyclodextrins

Although the native CDs themselves are of interest as host molecules, their modification changes the size, shape, and physical properties of the CD ring. Therefore, great effort has been put into the preparation of CDs covalently attached to guest molecules. When a guest group is covalently attached to a cyclic host molecule, it could form intramolecular [14–23] or intermolecular complexes [24–47] in aqueous solutions by guest–host interactions, depending on the flexibility and length of the substitution part. Many modified CDs have been prepared and their supramolecular structure characterized in aqueous solutions. Most of them formed intramolecular complexes. Ueno et al. [14–16] reported that chromophor-modified CDs can act as chemosensors by a location change of the substitution part from inside to outside the CD cavities with inclusion of another guest molecule (Fig. 3.3).

Benzoyl-modified β-CD did not form supramolecular polymers [27]. This result

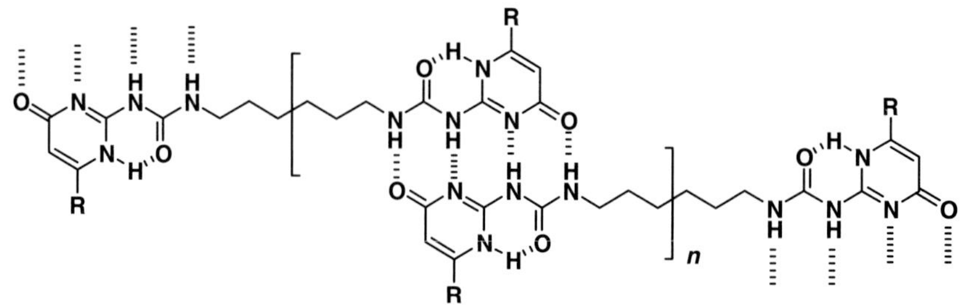

Fig. 3.2. Structure of Meijer's supramolecular polymer.

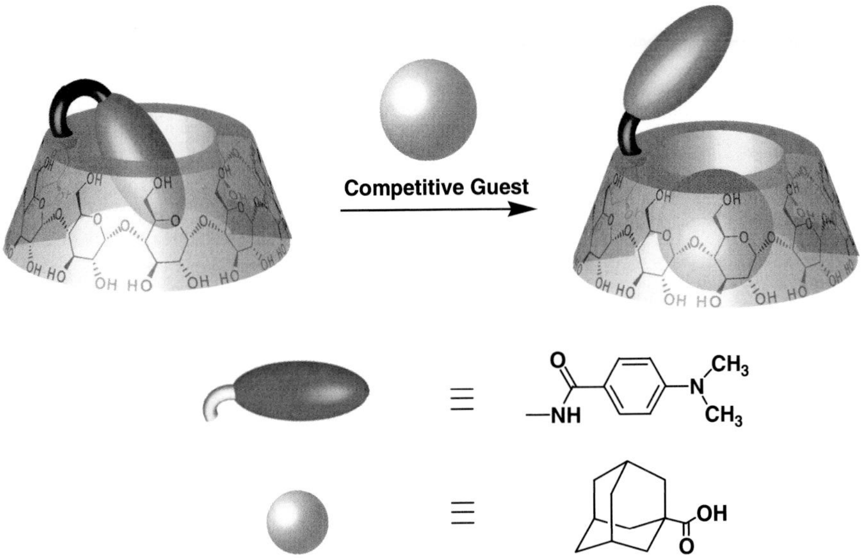

Fig. 3.3. Schematic representation of a fluorescent chemosensor involving an intramolecular complex bearing a chromophor.

suggests that some spacer groups are required for efficient formation of intermolecular complexes. Therefore, Harada et al. [30, 31, 41–47] have been studying the supramolecular structures of modified CDs bearing a hydrocinnamoyl or cinnamoyl group in aqueous solution (Fig. 3.4).

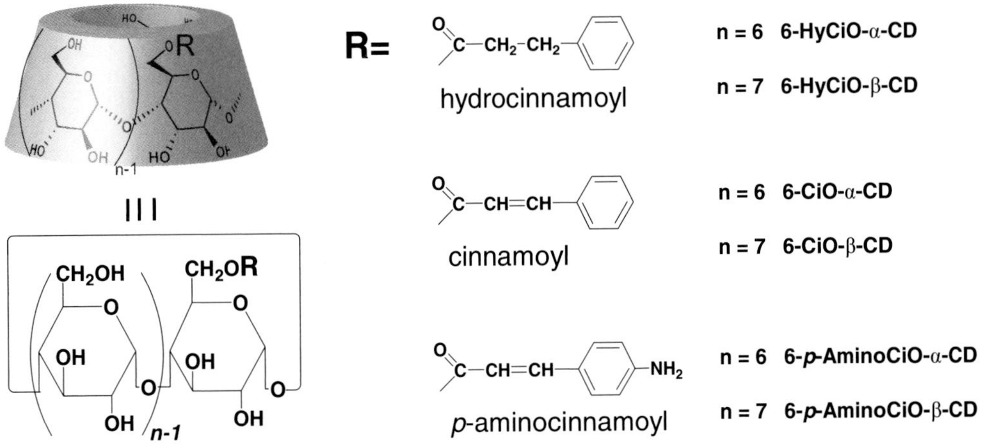

Fig. 3.4. Structures of hydrocinnamoyl- and cinnamoyl-modified CDs.

3.1.3
Formation of Inclusion Complexes [30, 31]

Harada et al. found that the ^{1}H NMR spectra of 6-hydroxycinnnamoyl β-CD (6-HyCiO-β-CD) showed no concentration dependence in D_2O solutions. Moreover, the spectra are very similar to those of the 1:1 mixture of β-CD and an ethylcinnamate (a model compound), suggesting that they formed intramolecular complexes. The 2D ROESY NMR spectrum shows that there are correlation peaks between the aromatic protons and inner protons of β-CD, indicating that guest parts are included in its cavities. All the signals of β-CD in the ^{1}H NMR spectrum of 6-HyCiO-β-CD are assigned by measuring various 2D NMR spectra (COSY, TOCSY, ROESY, HMQC) (Fig. 3.5). Some C(3) and C(5) protons in the cavity shifted to upper field. When the glucose unit with a guest part is named the A ring, the protons of the glucose D and E rings showed large shifts, indicating that the phenyl ring is in-

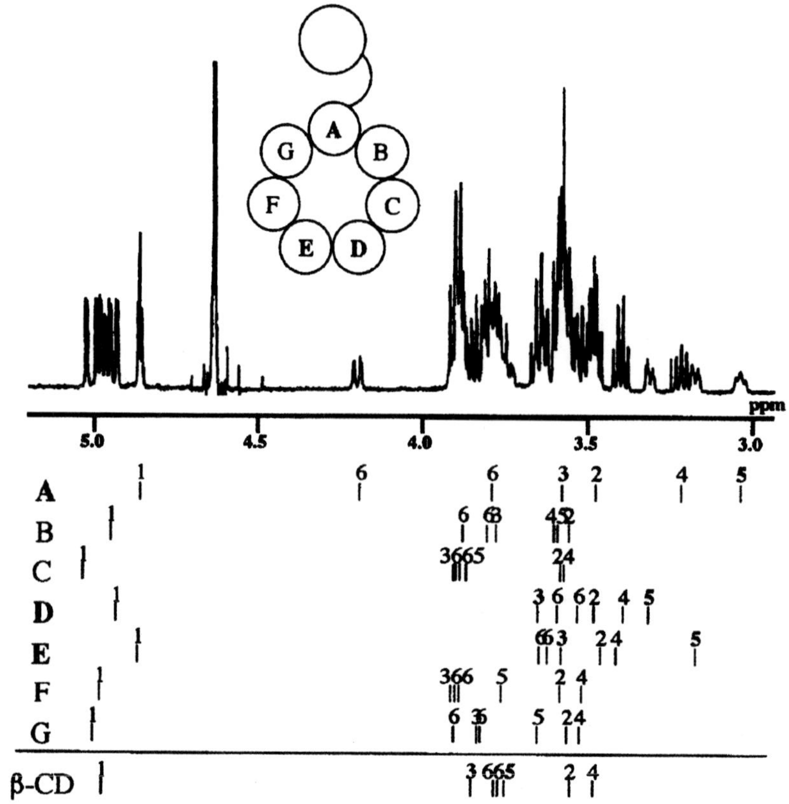

Fig. 3.5. 600-MHz NMR spectrum of 6-HyCiO-β-CD.

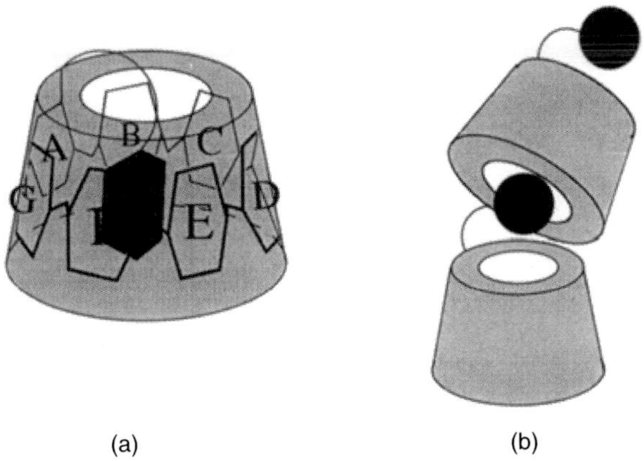

Fig. 3.6. Proposed structures of 6-HyCiO-β-CD (a) and 6-HyCiO-α-CD (b).

cluded in the cavity sandwiched by ring A and rings D and E. A proposed structure of 6-HyCiO-β-CD in water is depicted in Fig. 3.6 (a).

3.1.4
Polymer Formation by Intermolecular Interactions

Kaifer et al. [29] reported electrochemically controlled intermolecular dimeric complexation of viologen-modified β-CDs in aqueous solutions (Fig. 3.7). Harada et al. [30] have found that the ^{1}H NMR signal of the phenyl part of the 6-hydroxycinnnamoyl α-CD (6-HyCiO-α-CD) gave slight peak-shifts as the concentration increased. They found that 6-HyCiO-α-CD formed a weak intermolecular complex in aqueous solutions (Fig. 3.6 (b)). Kaneda et al. [32, 33] previously reported the formation of a cyclic dimer formed by azobenzene-derivative-substituted permethylated α-CDs in equilibrium (Fig. 3.8). Liu et al. [34, 35] reported on the dimerization

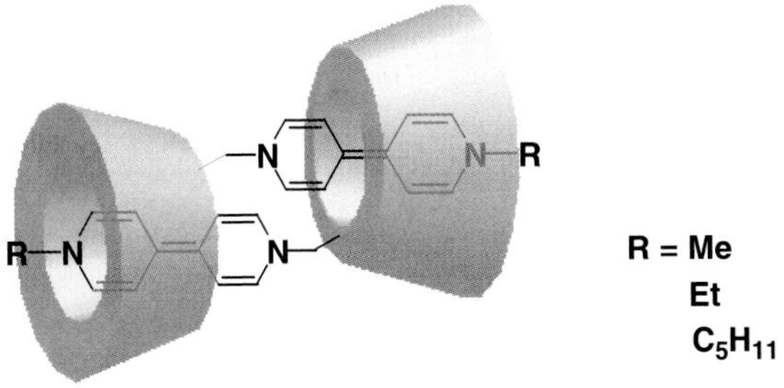

Fig. 3.7. Structures of Kaifer's intermolecular dimeric complexes formed by viologen-substituted β-CDs [29].

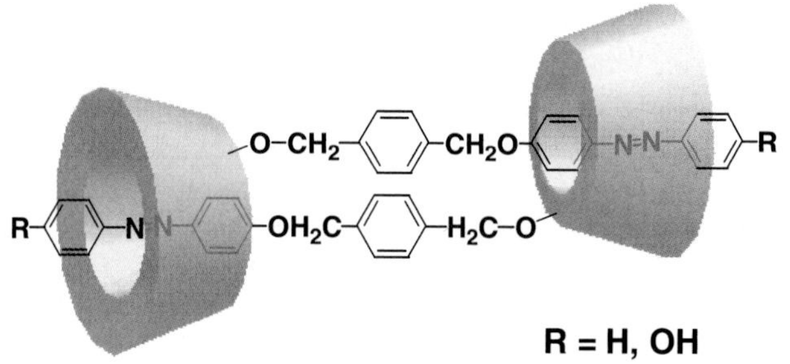

Fig. 3.8. Structures of Kaneda's intermolecular dimeric complexes and tetramer formed by azobenzene-derivative-substituted permethylated α-CDs [31].

of *p*-hydroxybenzoyl-substituted β-CD and *p*-hydroxybenzoyl-substituted β-CD in aqueous solution, respectively (Fig. 3.9).

6-HyCiO-β-CD was found to form intramolecular complexes and 6-HyCiO-α-CD was found to form weak intermolecular complexes. These results indicate that a hydrocinnamoyl group is too flexible to form intermolecular complexes. As a result, Harada et al. have attempted to use a more rigid spacer such as a cinnamoyl derivative with a double bond.

3.1.5
Supramolecular Dimers

Harada et al. [44] have found that 6-aminocinnamoyl β-CD (6-aminoCiO-β-CD) is sparingly soluble in water, although most 6-modified CDs are soluble. Figure 3.10 shows the structures of 6-aminoCiO-β-CD in the crystal. 6-AminoCiO-β-CD was found to form a dimer tightly stacked by intermolecular hydrogen bonding to form a head-to-head channel-type structure.

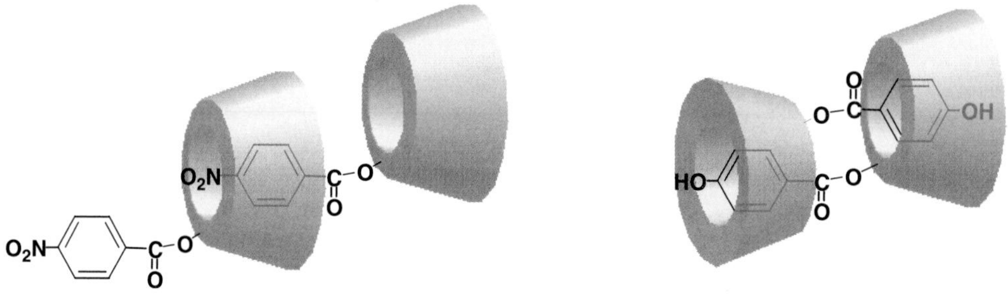

Fig. 3.9. Structures of Liu's intermolecular dimeric complexes formed by *p*-nitrobenzoyl-substituted β-CDs and *p*-hydroxybenzoyl-substituted β-CDs [36, 37].

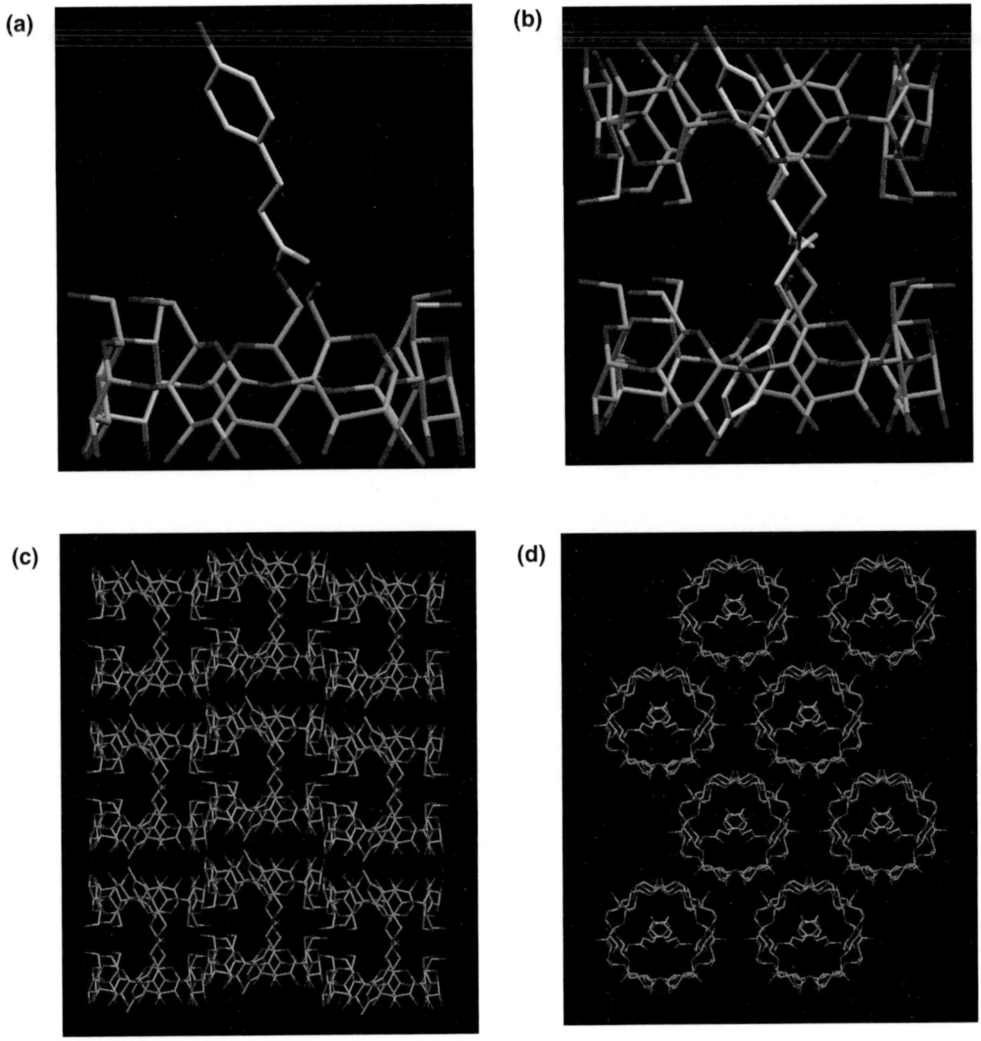

Fig. 3.10. Structures of 6-aminoCiO-β-CD in the crystal: (a) monomeric, (b) dimeric structures. Packing structures of 6-aminoCiO-β-CD (c) and (d).

3.1.6
Formation of Supramolecular Trimers

Harada et al. [30, 31, 41] also have studied the supramolecular structure of 6-cinnamoyl α-CD in aqueous solution. The ^{1}H NMR and 2D ROESY NMR spectra of 6-cinnamoyl α-CD in various concentrations in D_2O showed the formation of

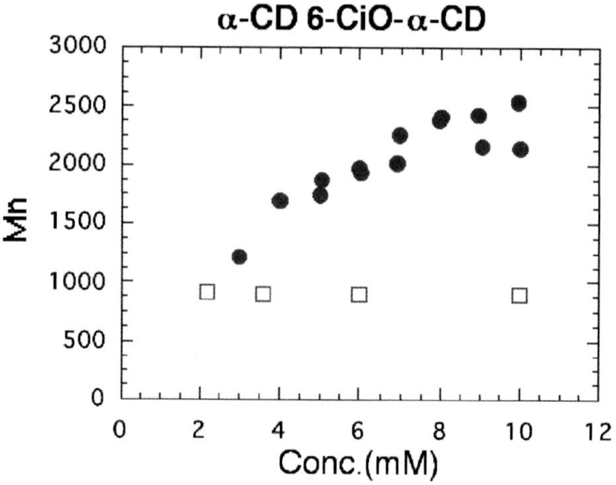

Fig. 3.11. Concentration dependence of the number-average molecular weight of α-CD (□) and 6-CiO-α-CD (●) in aqueous solutions measured by VPO at 40 °C.

intermolecular complexes. Figure 3.11 shows the results of vapor pressure osmometry (VPO) measurements of the native α-CD and 6-CiO-α-CD at various concentrations. Although α-CD is independent of the concentrations, 6-CiO-α-CD shows the concentration dependence. The molecular weight increased with an increase in the concentrations and reached saturation at about 3000 Da. 6-CiO-α-CD was found to form a trimer in aqueous solutions.

3.1.7
Cyclic Daisy Chain [31, 42]

When supramolecular polymers are treated with bulky stopper groups, they may form poly[2]rotaxane "daisy chains" [45–53]. Harada et al. [31] treated 6-*p*-aminoCiO-α-CD (40 mM) with 2M excess 2,4,6-trinitrobenzenesulfonic acid sodium salt (TNBS) as bulky stoppers in aqueous solutions. The resulting precipitate was found to be mainly a cyclic trimer by ^{1}H NMR and TOF mass spectra. After purification of the crude product, the 2D ROESY spectrum of the cyclic trimer shows cross-peaks between phenyl protons close to an amino group and secondary hydroxyl groups (O(2)H). A trinitrophenyl group is found at the secondary hydroxyl group side. A proposed structure of a cyclic trimer (cyclic daisy chain) is shown in Fig. 3.12. Kaneda et al. [38] reported the preparation of cyclic di[2]rotaxane fashion constructed tail-to-tail by azobenzene derivatives of permethylated α-CDs and showed its computer-generated supramolecular structures (Fig. 3.13). Easton et al. [39] also reported the preparation of cyclic di[2]rotaxane constructed by stilbene-appended α-CDs in tail-to-tail fashion (Fig. 3.14). Kaneda et al. [40]

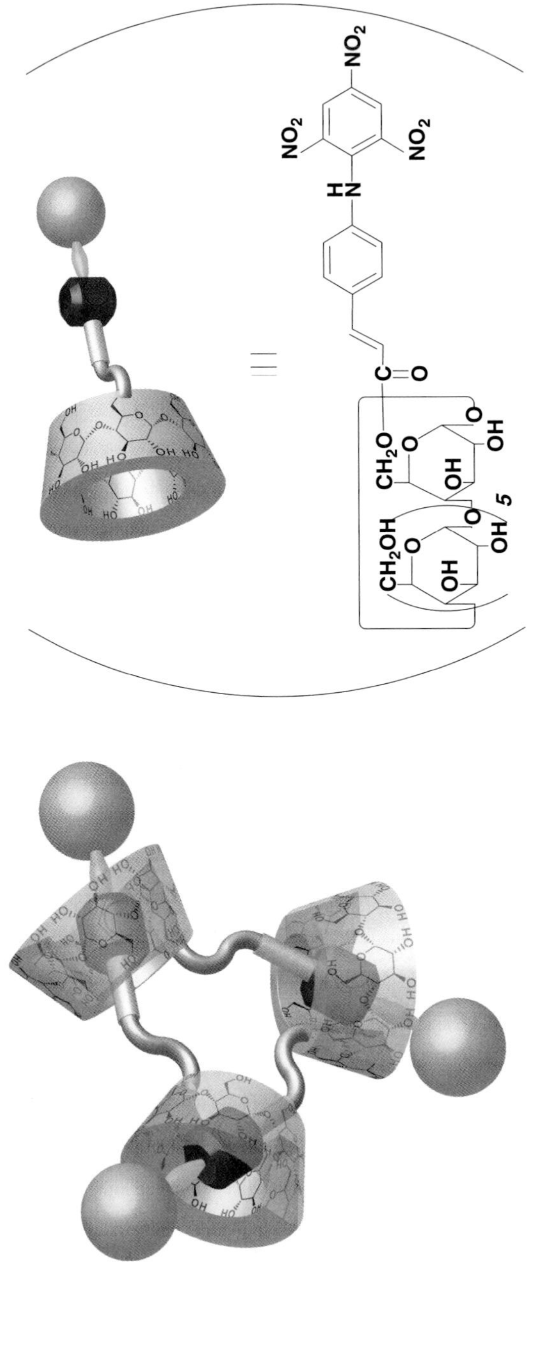

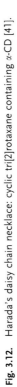

Fig. 3.12. Harada's daisy chain necklace: cyclic tri[2]rotaxane containing α-CD [41].

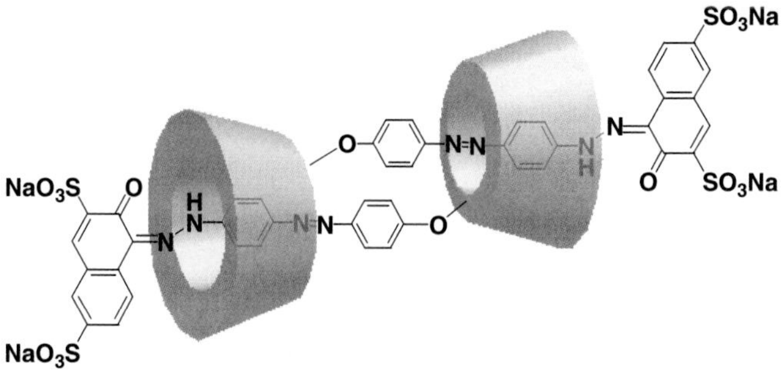

Fig. 3.13. Structure of Kaneda's cyclic di[2]rotaxane (a) and cyclic tetra[2]rotaxane (b) formed by permethylated α-CDs with an azobenzene derivative [31, 32].

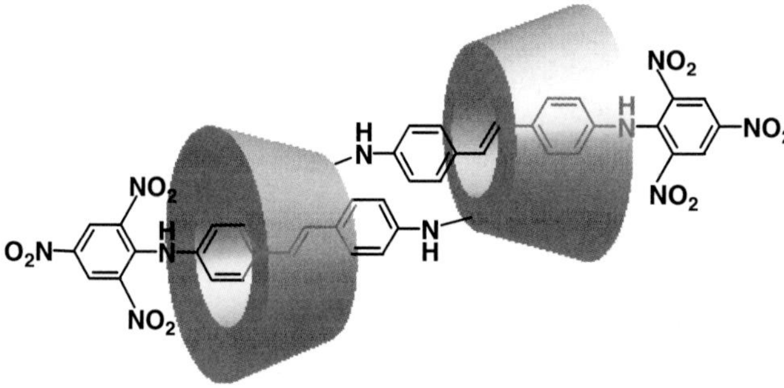

Fig. 3.14. Structure of Easton's cyclic di[2]rotaxane formed by stilbene-appended α-CDs [34].

also prepared cyclic tetra[2]rotaxane from azobenzene-modified permethylated α-CDs (Fig. 3.15).

3.1.8
Supramolecular Polymers [45]

As mentioned above, Harada et al. found that an α-CD derivative which has a cinnamoyl group as a guest moiety on the 6-position of the CD formed intermolecular complexes, giving rise to oligomeric a supramolecular structure in aqueous solutions [30, 31, 43]. An α-CD derivative with a cinnamamide group on its 6-position (6-CiNH-α-CD) (Fig. 3.16) was also prepared by Harada et al. and found to form a dimer in aqueous solutions (Fig. 3.17) [43]. In order to obtain linear supramolecular polymers, the formation of cyclic oligomers must be avoided. The same workers have prepared an α-CD derivative with a cinnamamide group on its 3-position (3-CiNH-α-CD). The molecular weight observed by VPO measurements in aqueous solution drastically increased with an increase in the concentrations (Fig. 3.18).

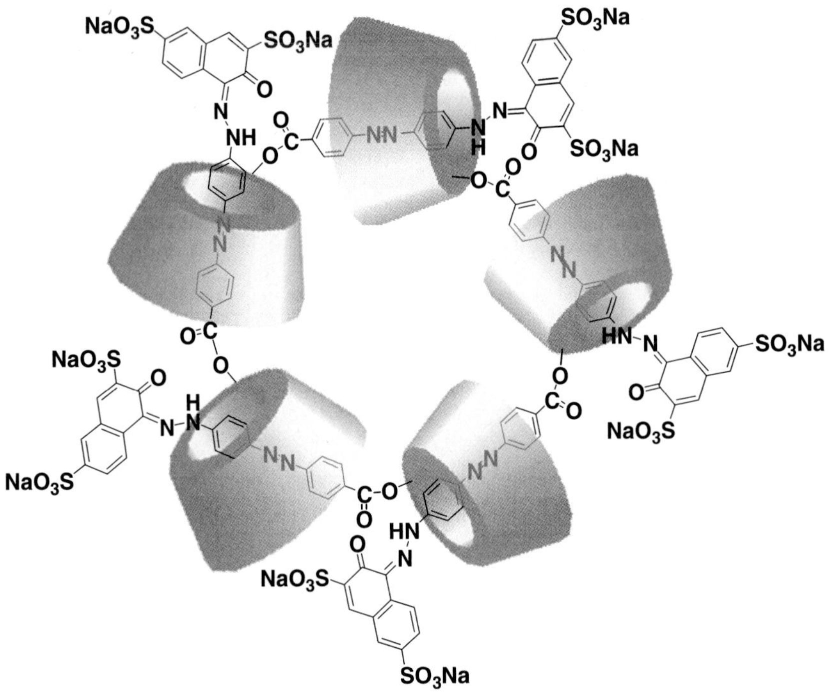

Fig. 3.15. Structure of Kaneda's cyclic tetra[2]rotaxane formed by permethylated α-CDs with an azobenzene derivative [32].

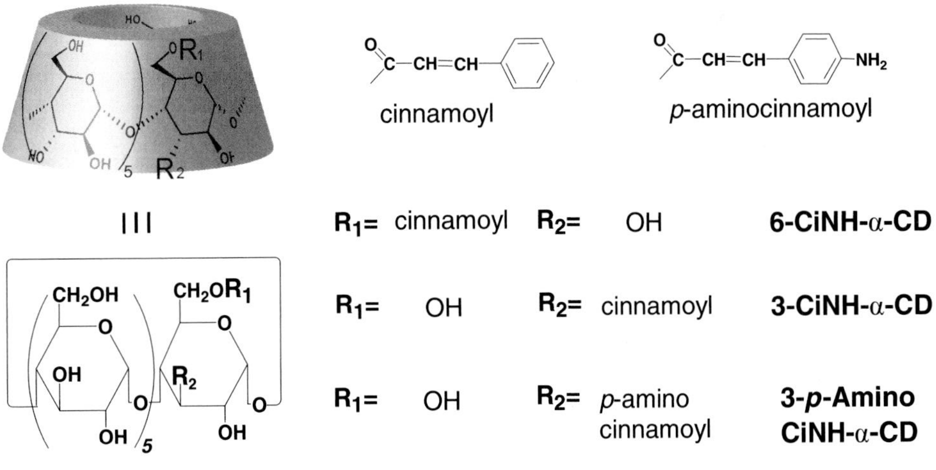

Fig. 3.16. Structures of cinnamamide-modified CDs.

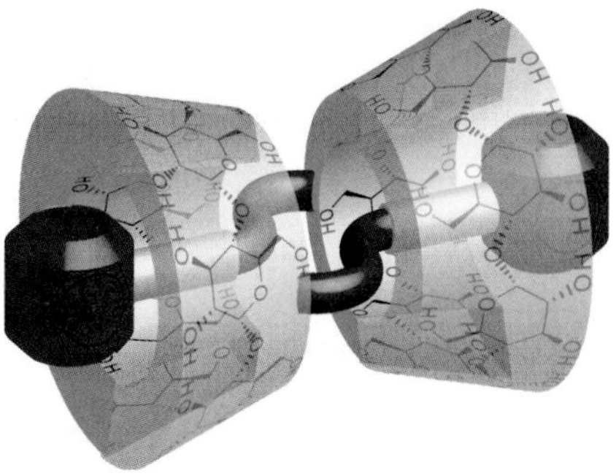

Fig. 3.17. The structure of supramolecular dimers constructed by 6-CiNH-α-CD in aqueous solution [53].

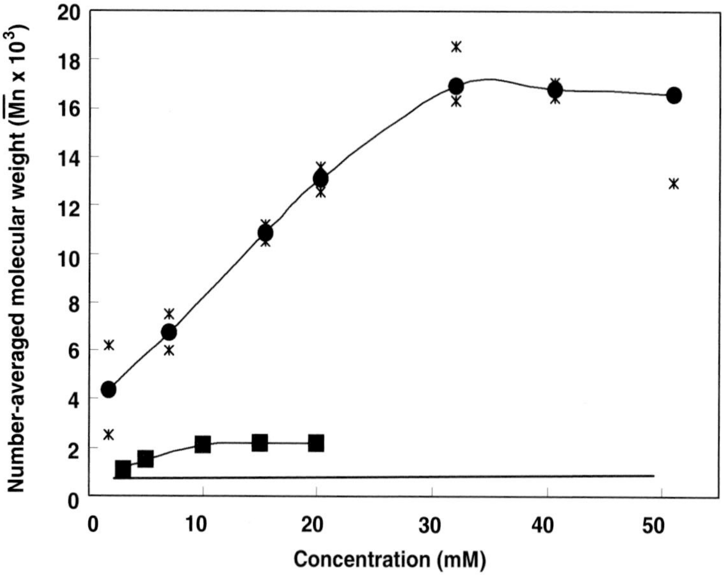

Fig. 3.18. The effect of concentration on the molecular weight of 3-CiNH-α-CD (●), 6-CiNH-α-CD (■), and α-CD (▲).

The molecular weight reached saturation at about 18 000 Da. 3-CiNH-α-CD was found to form supramolecular polymers in aqueous solutions (Fig. 3.19).

3.1.9
Poly[2]rotaxanes (Daisy Chain)

3-*p*-AminoCiNH-α-CD was also found to form supramolecular polymers as well as 3-CiNH-α-CD. After treatment of 3-*p*-aminoCiNH-α-CD (40 mM) with 3M excess

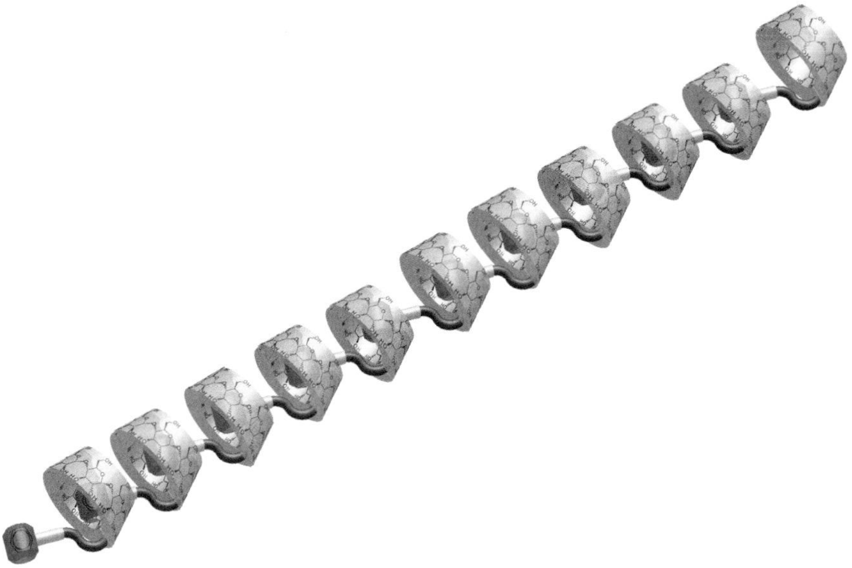

Fig. 3.19. Schematic representation of the supramolecular polymer constructed by 3-CiNH-α-CD in aqueous solution.

TNBS in water, the aqueous solution was poured into acetone. Figure 3.20 shows the MALDI/TOF mass spectra (the method is discussed in Section 10.2) of the resulting products. These signals can be clearly assigned as poly[2]rotaxane. The molecular weight of poly[2]rotaxane was found to be up to 18 000 Da (13 mer). The intervals of signals correspond to the molecular weight of 3-*p*-(trinitrobenzene)-aminoCiNH-α-CD ($M_w = 1328.1$ Da). Figure 3.21 shows a schematic representation of a poly[2]rotaxane based on a supramolecular polymer formed by 3-*p*-aminoCiNH-α-CD.

3.1.10
Helical Supramolecular Polymers [45, 46]

Harada et al. have also prepared an α-CD with a *p*-*tert*-butoxy-aminocinnamoylamino group at the 3-position (3-*p*-^{t}BocCiNH-α-CD) (Fig. 3.22). This compound was found to form intermolecular complexes that gave rise to stable supramolecular polymers up to 15 mer even in lower concentrations, as shown by NMR, VPO, and ESI MS measurements (Fig. 3.23). The circular dichroism spectra of 3-*p*-^{t}BocCiNH-α-CD gave rise to the splitting positive and negative Cotton bands of ^{1}La transition moments of the substituent part, indicating that substitution/substitution interactions exist among the adjacent monomers of the supramolecular polymer (Fig. 3.24). Using model compounds, the substitution part of 3-*p*-

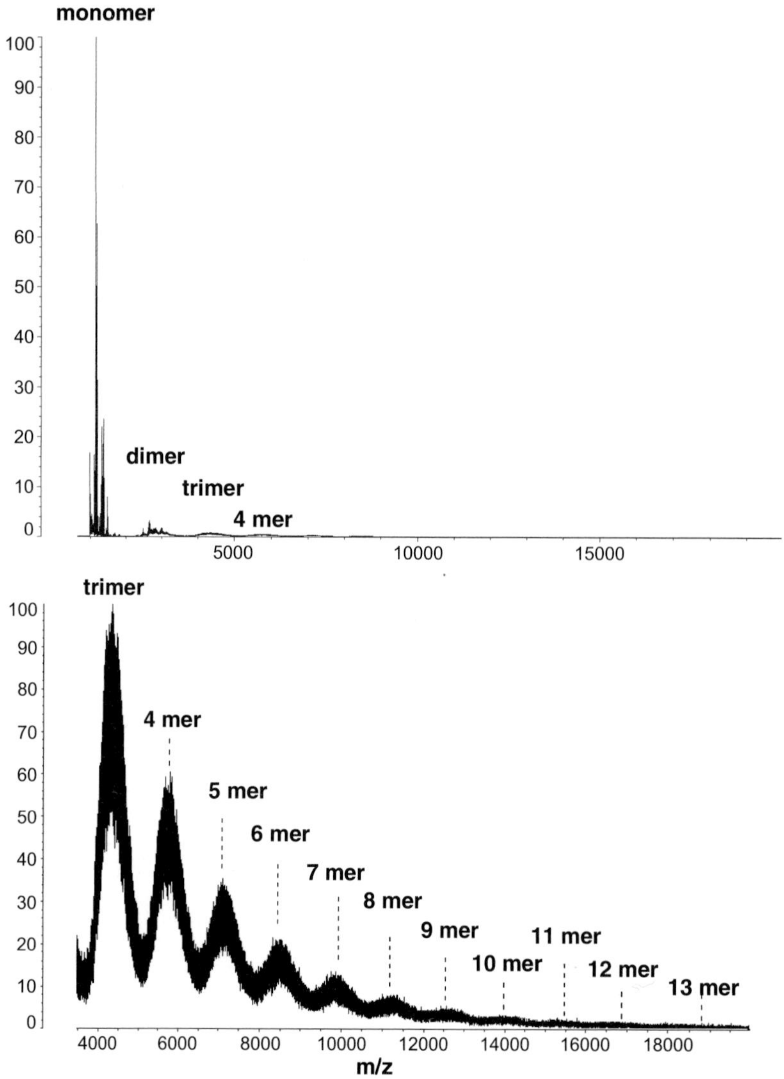

Fig. 3.20. MALDI/TOF mass spectra of poly[2]rotaxane.

^{t}BocCiNH-α-CD was found to be incorporated into the CD cavity of an adjacent monomer in a slantwise state. The substitutions were found to exist as a left handed-*anti*-configuration. The formation of a helical supramolecular polymer with some cooperativity is shown by the increasing intensities of Cotton effect peaks with increase in the concentrations. Fig. 3.25 shows an STM image of the

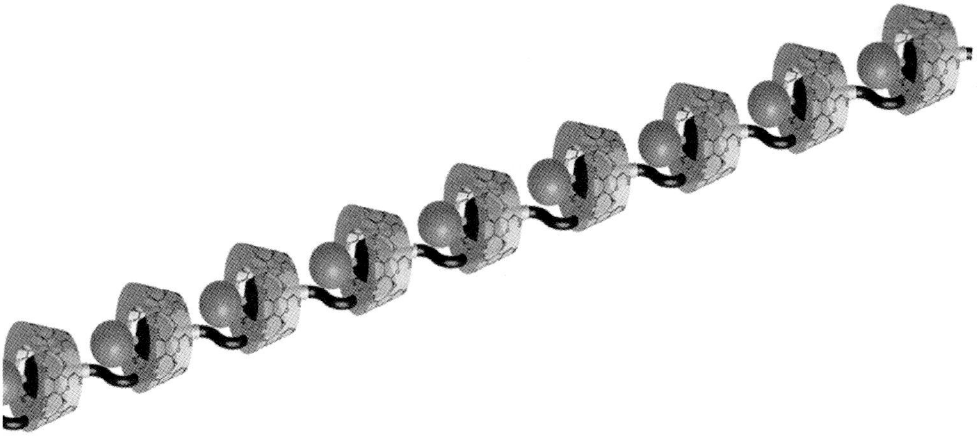

Fig. 3.21. Schematic representation of the supramolecular polymer constructed by 3-CiNH-α-CD in aqueous solution.

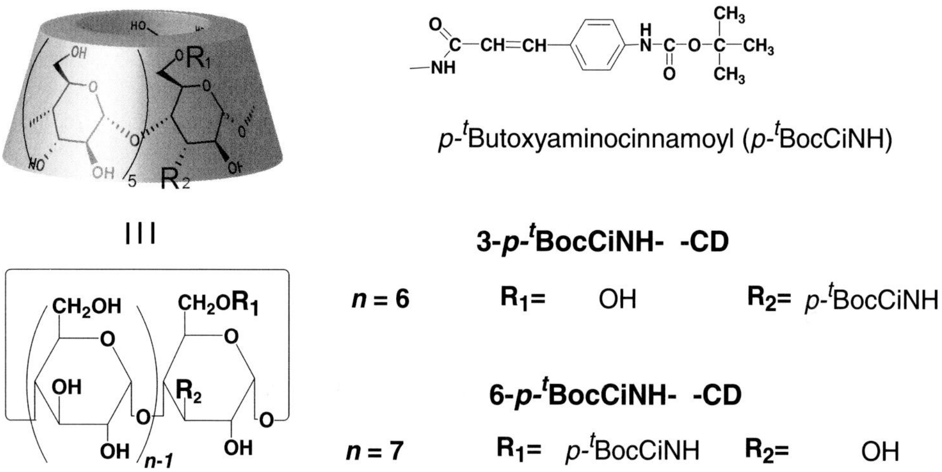

Fig. 3.22. Structures of *tert*-butoxycarboxyaminocinnamoyl-substituted CDs.

helical supramolecular polymer formed by 3-*p*-ᵗBocCiNH-α-CD on a MoS_2 substrate. A proposed structure for the supramolecular polymer formed by 3-*p*-ᵗBocCiNH-α-CD is shown in Fig. 3.25. These results remind us that microtubules, microfilaments, and flagella are helical supramolecular polymers formed by guest–host cooperative interactions in the biological system.

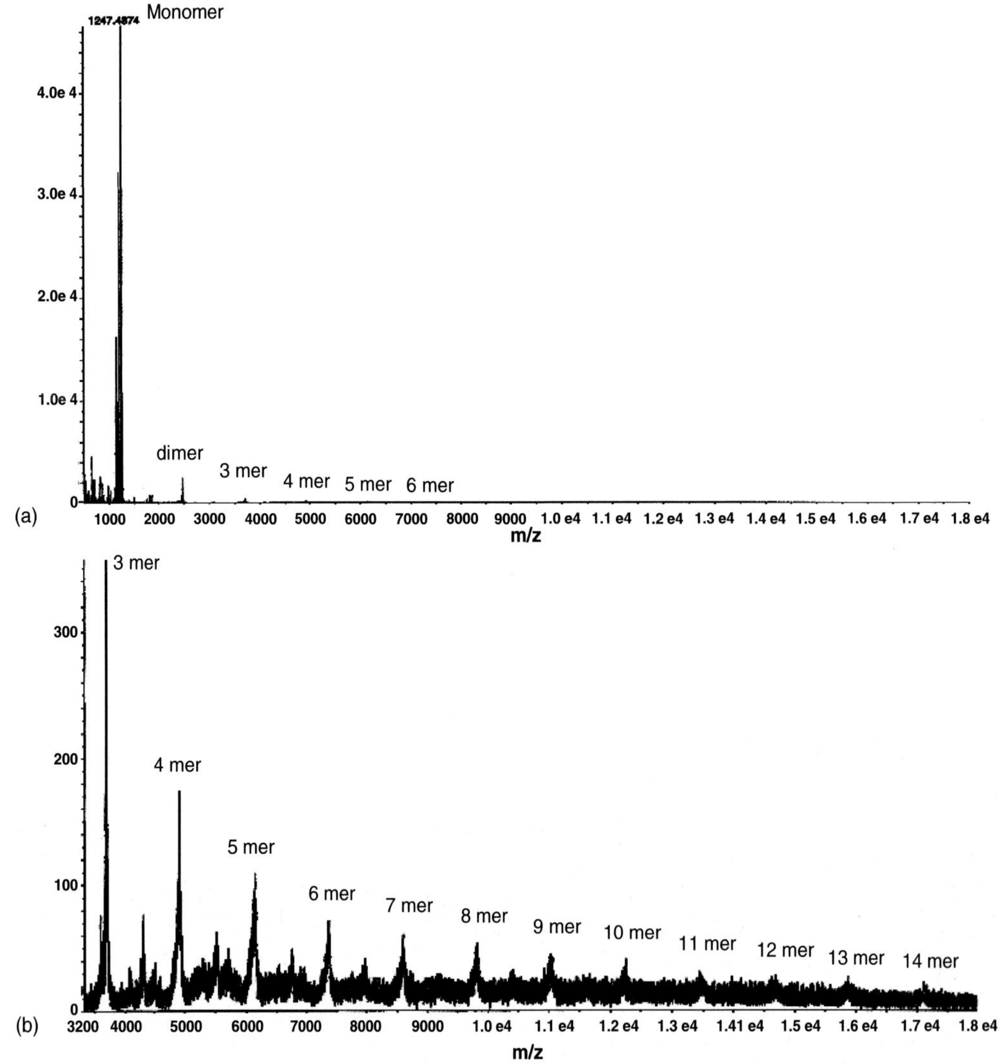

Fig. 3.23. ESI mass spectra of 3-*p*-^{t}BocCiNH-α-CD with $AcONH_4$ in aqueous solution in the range of 0–18 000 (a) and 3200–18 000 (b).

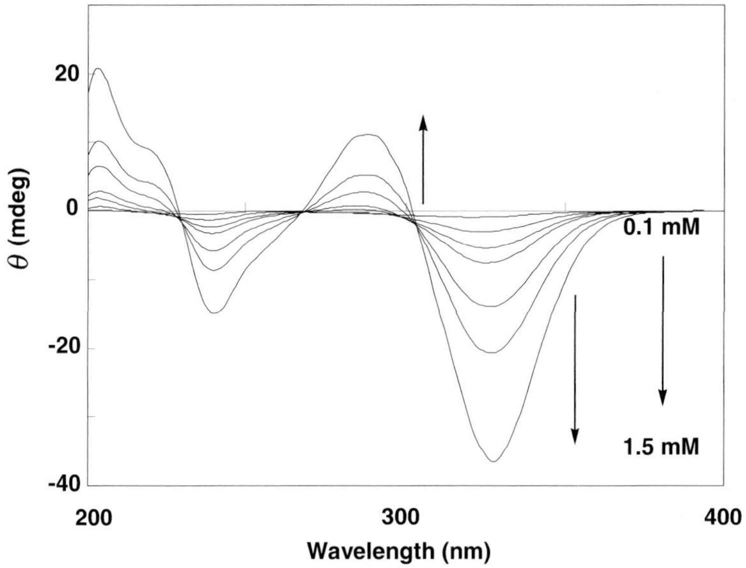

Fig. 3.24. Circular dichroism spectra of 3-*p*-^{t}BocCiNH-α-CD in aqueous solution.

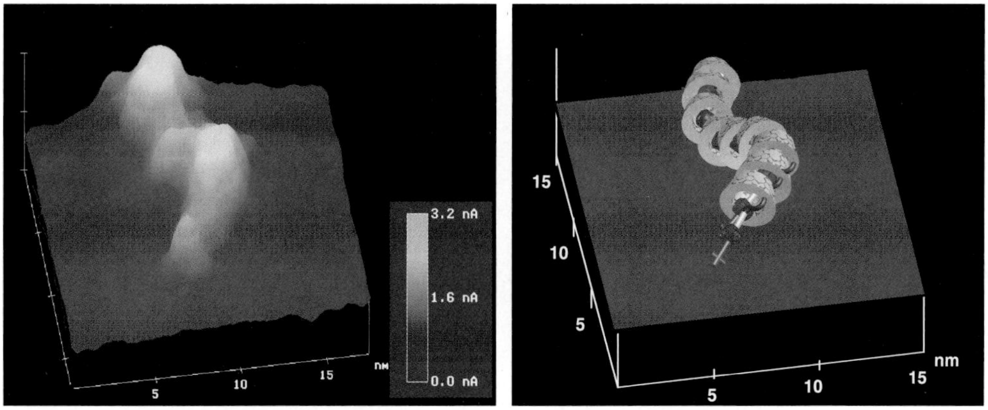

Fig. 3.25. STM image of 3-*p*-^{t}BocCiNH-α-CD from concentrated aqueous solution on a MoS_2 substrate (a) and its schematic structure (b).

Fig. 3.26. Proposed structure for the supramolecular polymer formed by 3-*p*-^{t}BocCiNH-α-CD in aqueous solution [45, 46].

3.1.11
Alternating α-, β-Cyclodextrin Supramolecular Polymers [47]

As mentioned above, 3-*p*-^{t}BocCiNH-α-CD forms helical supramolecular polymers in aqueous solutions. In contrast, 6-*p*-^{t}BocCiNH-β-CD was synthesized and found to form an intramolecular complex using NMR measurements. When adamantanecarboxylic acid (1-AdCA) was added as a competitive guest to an aqueous solution of 6-*p*-^{t}BocCiNH-β-CD, NMR spectra showed that the ^{t}Boc-cinnamamide part in its own β-CD cavity was expelled into water. The 6-*p*-^{t}BocCiNH-β-CD underwent a conformational change to form an inclusion complex with AdCx. Moreover, with an addition of an excess amount of α-CD, the cinnamoyl part was found to be included in α-CD to form a hetero dimer (Fig. 3.27).

Harada et al. also synthesized α-CD carrying an AdCx group (3-AdHexNH-α-CD) (Fig. 3.28). 3-AdHexNH-α-CD did not form a supramolecular structure in water. As mentioned above, 6-*p*-^{t}BocCiNH-β-CD undergoes a conformational change to give inclusion complexes with AdCx. Figure 3.29 shows the ^{1}H NMR spectra of 1:1 mixtures of 6-*p*-^{t}BocCiNH-β-CD and 3-AdHexNH-α-CD in D_2O solutions at various concentrations. The peaks of the adamantane group shifted, as the concentrations increased, indicating that the Ad group was included in a β-CD cavity. Moreover, the peaks of the cinnamoyl group also shifted, indicating that the cinnamoyl group was included in an α-CD cavity. Since the Ad group has specific interactions with β-CD, the Ad group was found to be displaced from the ^{t}Boc phenyl group of the β-CD derivative from the CD cavity. Accordingly, the phenyl ring was exposed to water. Then the phenyl group was included in an α-CD cavity of 3-AdHexNH-α-CD, because the phenyl ring can be easily included in an α-CD cavity (see the preceding section). As a result, 3-AdHexNH-α-CD and 6-*p*-^{t}BocCiNH-β-CD was found to form intermolecular complexes with an alternating order, αβ–αβ–αβ. The molecular weight of the intermolecular complexes was found to be more than 10 000 Da (>10 mM), indicating that the formation of a heterodimer leads to a supramolecular polymer having an alternating α-, β-CD units by means of a conformational change (Fig. 3.30).

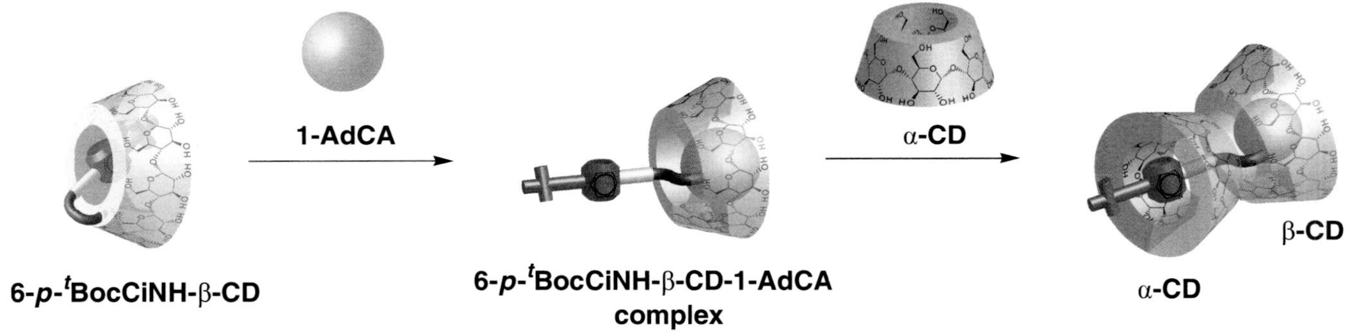

Fig. 3.27. Structural changes of 6-*p*-^{t}BocCiNH-α-CD with 1-AdCA and α-CD.

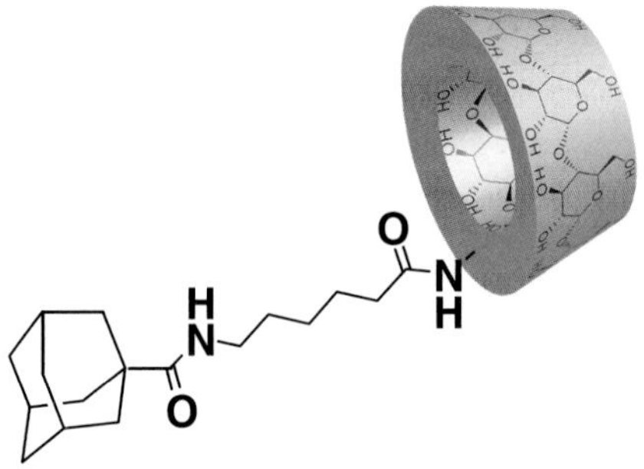

Fig. 3.28. The structure of α-CD carrying an AdCx group (3-AdHexNH-α-CD).

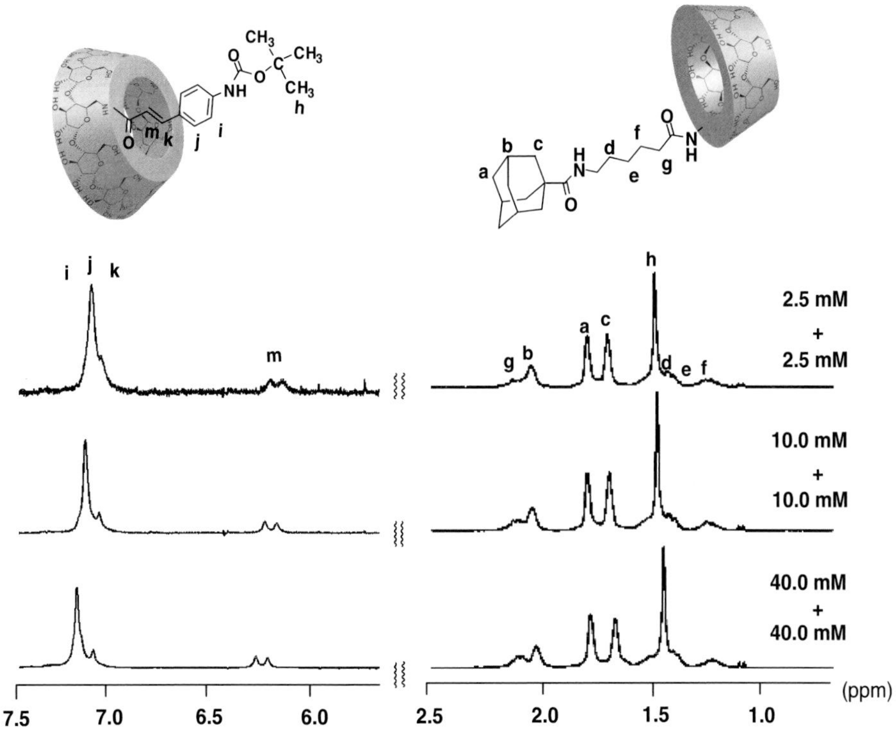

Fig. 3.29. ^{1}H NMR spectra of 3-AdHexNH-α-CD + 6-*p*-*t*BocCiNH-β-CD in a 1:1 ratio at various concentrations in D_2O.

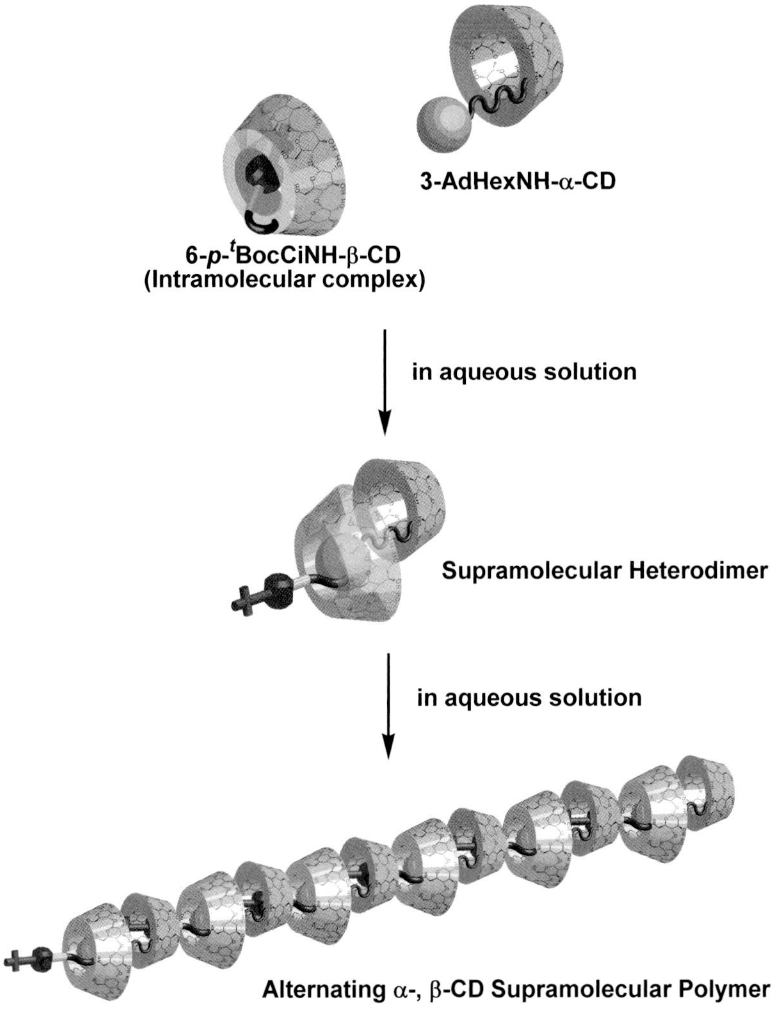

Fig. 3.30. Formation of 3-AdHexNH-α-CD + 6-*p*-^{t}BocCiNH-*β*-CD heterodimers and the corresponding supramolecular polymer.

3.1.12
Supramolecular [2]rotaxane Polymer [43]

CDs have been extensively used as rotors of rotaxanes and catenes [57–69]. If a [2]rotaxane containing a CD at one end of the axle and a guest at the other end is available, intermolecular complexes will be formed to give supramolecular poly-

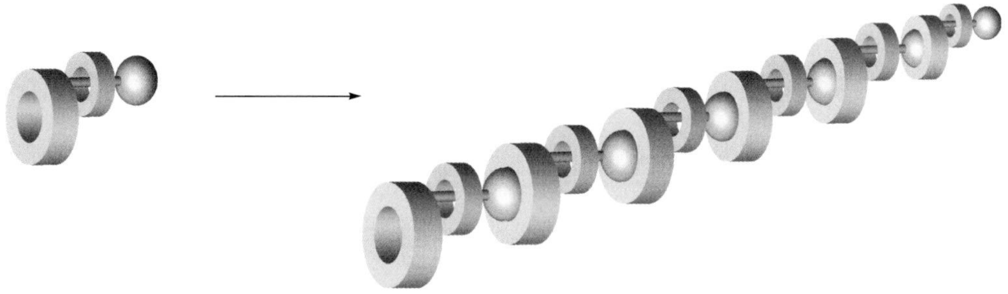

Fig. 3.31. Formation of a supramolecular [2]rotaxane polymer.

mers, because a rotaxane structure is rigid enough to prevent intramolecular complexation (Chapter 3.1). Harada et al. previously found that 6-aminocinnamoyl-α-CD (6-aminoCiO-α-CD) formed supramolecular oligomers in aqueous solutions [31]. When the supramolecular polymers were stabilized by attaching a bulky substituent (a trinitrophenyl group), a cyclic tri[2]rotaxane was obtained. In contrast, 6-aminoCiO-β-CD was sparingly soluble in water owing to the hydrogen-bonded stacking of its dimer in the crystal (Fig. 3.11) [44]. When 1-AdCx was added to an aqueous suspension of 6-aminoCiO-β-CD, the CD was solubilized to give 6-aminoCiO-β-CD–1-AdCx complexes (Scheme 3.1). Moreover, with addition of α-CD to the aqueous solution, the cinnamoyl group was found to be included in an α-CD cavity. After treatment of the complexes with TNBS and extraction of 1-AdCA, β-CD capped [2]rotaxane was obtained. α-CD and β-CD were found to face in tail-to-tail fashion by 2D ROESY NMR measurements. The ^{1}H NMR spectra of [2]rotaxane at various concentrations in D_2O shows the peak-shifts of the TNBS group as the concentration increased, indicating that β-CD binds the trinitrophenyl group to give intermolecular complexes. The molecular weight of the intermolecular complexes was found to be more than 10 000 Da (>10 mM) by VPO measurements, indicating that a β-CD stopper binds a trinitrophenyl group stopper of another [2]rotaxane to give supramolecular [2]rotaxane polymers (Fig. 3.31). Fig. 3.32 shows the STM image of the supramolecular [2]rotaxane polymer concentrated aqueous solution on HOPG.

3.1.13
Conclusion

Many reports on the formation of supramolecular polymer in solution have been mentioned in this chapter. In particular, the interest in cyclodextrin-based supra-

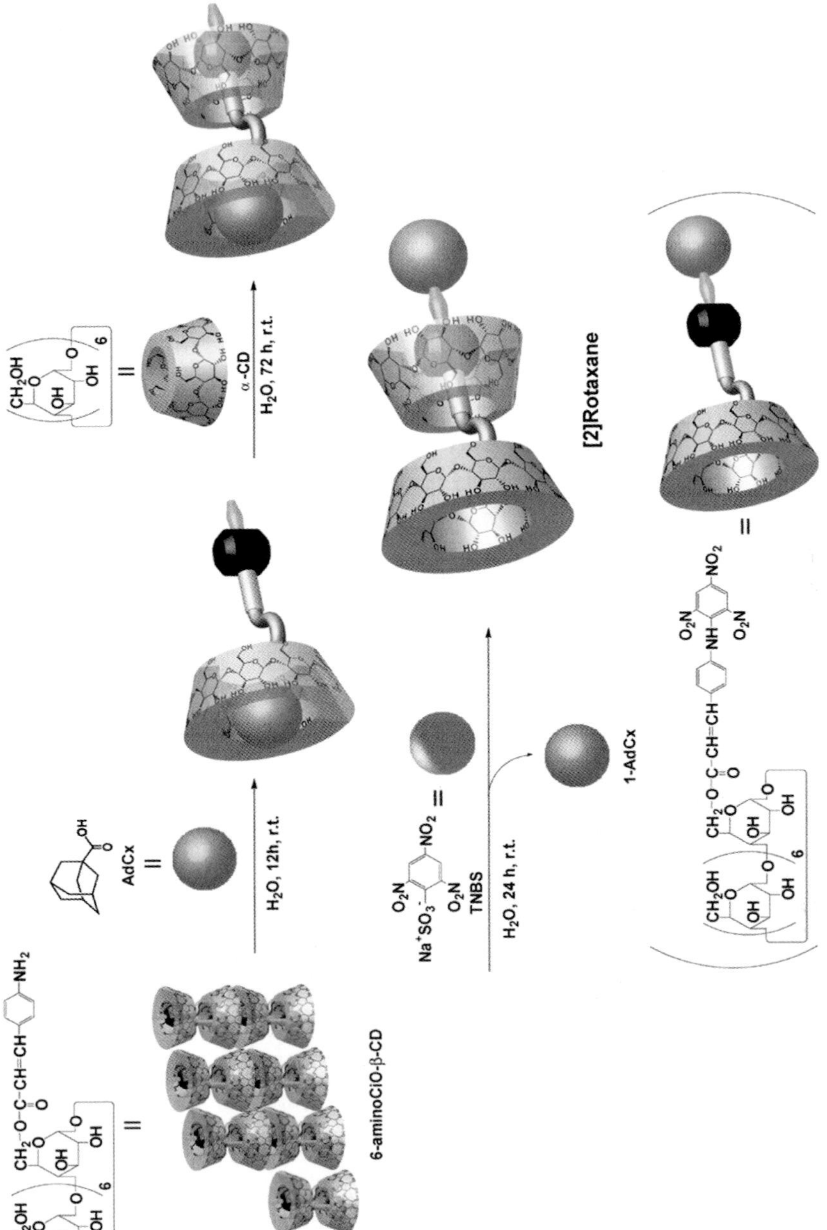

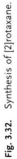

Fig. 3.32. Synthesis of [2]rotaxane.

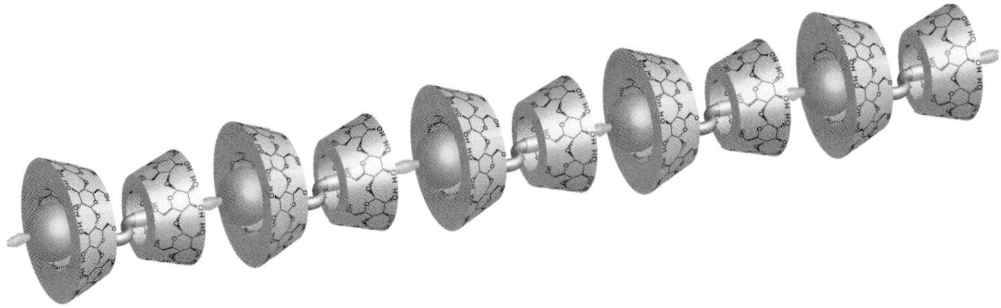

Fig. 3.33. A proposed structure of a supramolecular [2]rotaxane polymer.

molecular polymers using guest–host interactions in aqueous solutions have been increasing year by year as shown in recent papers. The construction of supramolecular polymers often reminds us of the supramolecular structures such as microtubule, microfilaments, and flagella that occur in nature. Moreover, such supramolecular polymers could be constructed not only in order to mimic the supramolecular structure in a biological system, but also to give rise to new functions beyond the scope of synthetic covalent polymers.

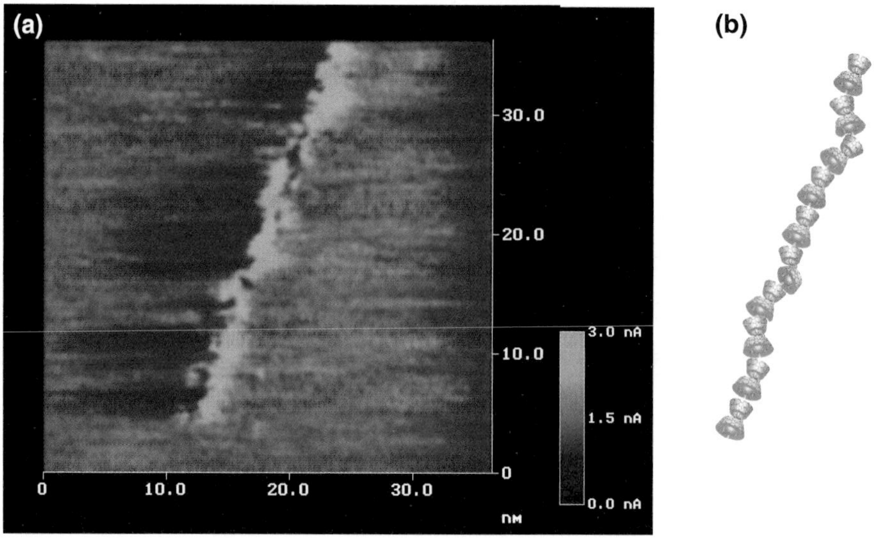

Fig. 3.34. STM image of the supramolecular [2]rotaxane polymer (a) and its schematic structure (b).

3.2 Supramolecular Complexes of Polymers Bearing Cyclodextrin Moieties with Guest Molecules

Polymers bearing CD moieties have been synthesized by cross-linking of CDs using epichlorohydrin [69, 70], by coupling reaction between modified or native CDs and polymers [71–75], and by homo- and copolymerization of monomers bearing CD moieties [76–78].

Polymers bearing CD moieties form supramolecular complexes with guest molecules. We investigated the interaction of β-CD polymers, poly(acryloyl-β-CD) and poly(*N*-acrylyl-6-aminocaproyl-β-CD), with fluorescent guest molecules [79, 80]. When guest molecules, such as 2-*p*-toluidinylnaphthalene-6-sulfonate, 1-anilino-naphthalene-8-sulfonate, or 5-dimethylaminonaphthalenesulfonylphenylalanine, formed 2:1 inclusion complexes of β-CD moieties, cooperative binding was observed. Werner et al. [81, 82] and Hollas et al. [83] also reported cooperative binding for epichlorohydrin-linked CD polymers with 2-(*N*-methylanilino)naphthalene-6-sulfonic acid or with pyrene, and for poly(allylamine)-modified with β-CD moieties with pyrene, respectively.

Weickenmeier and Wenz [84] investigated the inclusion complex formation of poly[(*N*-vinyl-2-pyrrolidone)-co-(maleic acid)] modified with β-CD moieties with 1-adamantanamine and with 1-adamantanecarboxylic acid by isothermal titration calorimetry. They reported 1:1 inclusion complexes between β-CD moieties and adamantane derivatives with relatively large binding constants (ca. $3\text{–}5 \times 10^3\ M^{-1}$).

Amiel et al. [85, 86] investigated the formation of ternary complexes of a water soluble epichlorohydrin-linked β-CD polymer, cationic surfactant dodecyltrimethylammonium bromide (DTAB), and a polyanion sodium salt dextran sulfate (NaDxS) by viscometry and small angle neutron scattering. They proposed a structure of polyion complexes between NaDxS and the supramolecular polycation formed from β-CD polymer and DTAB. More recently, Amiel et al. [87] have characterized in detail the supramolecular polycation.

Yashima et al. [88] investigated the interaction of poly(phenylacetylene) modified with β-CD moieties with guest molecules, including 1-adamantanol, (−)-borneol, and (*R*)- and (*S*)-1-phenylethylamine. They observed a color change upon complexation because of a change in the helicity pitch arising from helicity inversion.

Supramolecular complex formation of polymers bearing CD moieties have been investigated for a broad range of applications, including catalytic systems [89–94], separation or recognition systems [95–105], and controlled release or drug delivery systems [106–110].

References

1 Lehn J-M (1995) Supramolecular Chemistry; VCH, Weinheim.

2 Ciferri A (2000) Supramolecular Polymers, Marcel Dekker, New York.

3 Lehn J-M (1995) Adv Mater 5: 254.
4 Gulik-Krzywicki T, Fouquey C, Lehn J-M (1993) Proc Natl Acad Sci USA 90: 163.
5 Lehn J-M (1993) Makromol Chem Macromol Symp 69: 1.
6 Sijbesma R P, Beijer F H, Brunsveld L, Folmer B J B, Hirschberg J H K K, Lange R F M, Lowe J K L, Meijer E W (1997) Science 278: 1601.
7 Hirschberg J H K K, Brunsveld L, Ramzi A, Vekemans J A J M, Sijbesma R P, Meijer E W (2000) Nature 407: 167.
8 Brunsveld L, Folmer B J B, Meijer E W, Sijbesma R P (2001) Chem Rev 101: 4071.
9 Castellano R K, Rudkevich D M, Rebek J Jr (1997) Proc Natl Acad Sci USA 94: 7132.
10 Castellano R K, Rebek J Jr (1998) J Am Chem Soc 120: 3657.
11 Castellano R K, Clark R, Craig S L, Nuckolls C, Rebek J Jr (2000) Proc Natl Acad Sci USA 97: 12418.
12 Castellano R K, Nuckolls C, Eichhorn S H, Wood M R, Lovinger A J, Rebek J Jr (1999) Angew Chem Int Ed 38: 2603.
13 Knapp R, Schott A, Rehahn M (1996) Macromolecules 29: 478.
14 Hamasaki K, Ikeda H, Nakamura A, Ueno A, Toda F, Suzuki I, Osa T (1993) J Am Chem Soc 115: 5035.
15 Ueno A, Minato S, Suzuki I, Fukushima M, Ohkubo M, Osa T, Hamada F, Murai K (1990) Chem Lett 605.
16 Wang Y, Ikeda T, Ikeda H, Ueno A, Toda F (1994) Bull Chem Soc Jpn 67: 1598.
17 Corradini R, Dossena A, Marchelli R, Panagia A, Sartor G, Saviano M, Lombardi A, Pavone V (1996) Chem Eur J 2: 373.
18 Corradini R, Dossena A, Galaverna G, Marchelli R, Panagia A, Sartor G (1997) J Org Chem 62: 6283.
19 Takahashi K, Imohori K, Kitsuka M (2001) Polym J 33: 242.
20 Inoue Y, Wada T, Sugahara N, Yamamoto K, Kimura K, Tong L-H, Gao X-M, Hou Z-J, Liu Y (2000) J Org Chem 65: 8041.
21 Yamada T, Fukuhara G, Kaneda T (2003) Chem Lett 32: 534.
22 Fukuhara G, Fujimoto T, Kaneda T (2003) Chem Lett 32: 536.
23 McAipine S R, Garcia Garibay M A (1996) J Am Chem Soc 118: 2750.
24 Zanotti-Gerosa A, Solari E, Giannini L, Floriani C, Chiesi-Villa A, Rizzoli C (1996) Chem Commun 119.
25 Yamaguchi N, Nagvekar D S, Gibson H W (1998) Angew Chem Int Ed 37: 2361.
26 Bugler J, Sommerdik N A J M, Visser A J W G, van Hoek A, Nolte R J M, Engbersen J F J, Reinhoudt D N (1999) J Am Chem Soc 121: 28.
27 Tong L-H, Hou Z-J, Inoue Y, Tai A (1992) J Chem Soc Perkin Trans 2: 1253.
28 McAipine S R, Garcia-Garibay M A (1998) J Am Chem Soc 120: 4269.
29 Mirzoian A, Kaifer A E (1999) Chem Commun 1603.
30 Harada A, Miyauchi M, Hoshino T (2003) J Polym Sci Part A Polymer Chemistry 41: 3519.
31 Hoshino T, Miyauchi M, Kawaguchi Y, Yamaguchi H, Harada A (2000) J Am Chem Soc 122: 9876.
32 Fujimoto T, Uejima Y, Imaki H, Kawarabayashi N, Jung J H, Sakata Y, Kaneda T (2000) Chem Lett 564.
33 Fujimoto T, Sakata Y, Kaneda T (2000) Chem Lett 764.
34 Liu Y, You C-C, Zhang M, Weng L-H, Wada T, Inoue Y (2000) Org Lett 2: 2761.
35 Liu Y, Fan Z, Zhang H-Y, Yang Y-W, Liu S-X, Wu X, Wada T, Inoue Y (2003) J Org Chem 68: 8345.
36 Park J W, Song H E, Lee S Y (2002) J Phys Chem B 106: 5177.
37 Gao X M, Zhang Y L, Tong L H, Ye Y H, Ma X Y, Liu W S, Inoue Y (2001) J Inclusion Phenom Macrocyclic Chem 39: 77.
38 Fujimoto T, Sakata Y, Kaneda T (2000) Chem Commun 2143.
39 Onagi H, Easton C J, Lincoln S F (2001) Org Lett 3: 1041.
40 Kaneda T, Yamada T, Fujimoto T, Sakata Y (2001) Chem Lett 1264.

41 Harada A, Kawaguchi Y, Hoshino T (2001) J Inclusion Phenom. Macromol Chem 41: 115.
42 Harada A (2001) Acc Chem Res 34: 456.
43 Miyauchi M, Kawaguchi Y, Harada A (2004) J Inclusion Phenom. Macromol Chem 50: 57.
44 Miyauchi M, Hoshino T, Yamaguchi H, Kamitori S, Harada A (2005) J Am Chem Soc 127: 2034.
45 Miyauchi M, Harada A (2005) Chem Lett 34: 104.
46 Miyauchi M, Takashima Y, Yamaguchi H, Harada A (2005) J Am Chem Soc 127: 2984.
47 Miyauchi M, Harada A (2005) J Am Chem Soc 126: 11418.
48 Sauvage J-P (1999) Molecular Catenanes, Rotaxanes and Knots, Dietrich Buchecker C (Ed), Wiley-VCH, Weinheim, Germany.
49 Amabilino D B, Stoddart J F (1995) Chem Rev 95: 2725.
50 Balzani V, Credi A, Raymo F M, Stoddart J F (2000) Angew Chem Int Ed 39: 3348.
51 Ashton P R, Baxter I, Cantrill S J, Fyfe M C T, Glink P T, Stoddart J F, White A J P, Williams D J (1998) Angew Chem Int Ed 37: 1294.
52 Ashton P R, Parsons I W, Raymo F M, Stoddart J F, White A J P, Williams D J, Wolf R (1998) Angew Chem Int Ed 37: 1913.
53 Rowan S J, Cantrill S J, Stoddart J F, White A J P, Williams D J (2000) Org Lett 2: 759.
54 Cantrill S J, Youn G J, Stoddart J F (2001) J Org Chem 66: 6857.
55 Jimenez M C, Dietrich-Buchecker C, Sauvage J-P, De Cian A (2000) Angew Chem Int Ed 39: 1295.
56 Jimenez M C, Dietrich-Buchecker C, Sauvage J-P, De Cian A (2000) Angew Chem Int Ed 39: 3284.
57 Ogino H, Ohta K (1984) Inorg Chem 23: 3312.
58 Manka J S, Lawrence D S (1990) J Am Chem Soc 112: 2440.
59 Rao T V S, Lawrence D S (1990) J Am Chem Soc 112: 3614.
60 Isnin R, Kaifer A E (1991) J Am Chem Soc 113: 8188–8190.
61 Wylie R S, Macartney D H (1992) J Am Chem Soc 114: 3136.
62 Wenz G, von der Bey E, Schmidt L (1992) Angew Chem Int Ed Engl 31: 783.
63 Harada A (1997) Adv Polym Sci 133: 141.
64 Harada A, Li J, Kamachi M (1992) Nature 356: 325.
65 Harada A, Li J, Kamachi M (1993) Macromolecules 26: 5698.
66 Harada A, Li J, Kamachi M (1993) Nature 364: 516.
67 Harada A, Li J, Kamachi M (1994) Nature 370: 126.
68 Harada A, Li J, Kamachi M (1994) J Am Chem Soc 116: 3192.
68a Kawaguchi Y, Harada A (2000) Org Lett 2: 1353.
69 Wiedenhof N, Lammers JNJJ, van Panthaleon van Eck CL (1969) Staerke 21:119.
70 Renard E, Deratani A, Volet G, Sebille B (1997) Eur Polym J 33:49.
71 Deratani A, Poepping B, Muller G (1995) Macromol Chem Phys 196:343.
72 Martel B, Leckchiri Y, Pollet A, Morcellet M (1995) Eur Polym J 31:1083.
73 Crini G, Torri G, Guerrini M, Martel B, Lekchiri Y, Morcellet M (1997) Eur Polym J 33:1143.
74 Ruebner A, Statton GL, James MR (2000) Macromol Chem Phys 201:1185.
75 Renard E, Volet G, Amiel C (2005) Polym Int 54:594.
76 Harada A, Furue M, Nozakura S-i (1976) Macromolecules 9:701.
77 Liu Y-Y, Fan X-D (2002) Polymer 43:4997.
78 Liu Y-Y, Fan X-D (2003) J Appl Polym Sci 89:361.
79 Harada A, Furue M, Nozakura S-i (1977) Macromolecules 10:676.
80 Harada A, Furue M, Nozakura S (1981) Polym J 13:777.
81 Werner TC, Warner IM (1994) J Inclusion Phenom Mol Recognit Chem 18:385.
82 Werner TC, Iannacone JL, Amoo MN (1996) J Inclusion Phenom Mol Recognit Chem 25:77.
83 Hollas M, Chung M-A, Adams J

(1998) Polym Prepr (Am Chem Soc, Div Polym Chem) 39(2):786.
84 Weickenmeier M, Wenz G (1996) Macromol Rapid Commun 17:731.
85 Galant C, Amiel C, Wintgens V, Sébille B, Auvray L (2002) Langmuir 18:9687.
86 Galant C, Amiel C, Auvray L (2004) J Phys Chem B 108:19218.
87 Galant C, Wintgens V, Amiel C, Auvray L (2005) Macromolecules 38:5243.
88 Yashima E, Maeda K, Sato O (2001) J Am Chem Soc 123:8159.
89 Harada A, Furue M, Nozakura S-i (1976) Macromolecules 9:705.
90 Suh J, Lee SH, Zoh KD (1992) J Am Chem Soc 114:7916.
91 Martel B, Morcellet M (1995) Eur Polym J 31:1089.
92 Suh J, Hah SS, Lee SH (1997) Bioorganic Chemistry 25:63.
93 Suh J, Kwon WJ (1998) Bioorganic Chemistry 26:103.
94 Zhu M, Huang X-M, Shen H-X (1999) J Inclusion Phenom Macrocycl Chem 33:243.
95 Harada A, Furue M, Nozakura S (1978) J Polym Sci, Polym Chem Ed 16:189.
96 Crini G, Torri G, Lekchiri Y, Martel B, Janus L, Morcellet M (1995) Chromatographia 41:424.
97 Sasaki H, Ohara M (1998) Chem Pharm Bull 46:1924.
98 David C, Millot MC, Sebille B (2001) J Chromatogr B 753:93.
99 Hishiya T, Shibata M, Kakazu M, Asanuma H, Komiyama M (1999) Macromolecules 32:2265.
100 Asanuma H, Akiyama T, Kajiya K, Hishiya T, Komiyama M (2001) Anal Chim Acta 435:25.
101 Akiyama T, Hishiya T, Asanuma H, Komiyama M (2001) J Inclusion Phenom Macrocycl Chem 41:149.
102 Hishiya T, Asanuma H, Komiyama M (2003) Polym J 35:440.
103 Hishiya T, Akiyama T, Asanuma H, Komiyama M (2003) J Inclusion Phenom Macrocycl Chem 44:365.
104 Asanuma H, Hishiya T, Komiyama M (2004) J Inclusion Phenom Macrocycl Chem 50:51.
105 Tsai H-A, Syu M-J (2005) Biomaterials 26:2759.
106 Pun SH, Bellocq NC, Liu A, Jensen G, Machemer T, Quijano E, Schluep T, Wen S, Engler H, Heidel J, Davis ME (2004) Bioconjugate Chem 15:831.
107 Li J, Xiao H, Li J, Zhong Y (2004) Int J Pharm 278:329.
108 Liu Y-Y, Fan X-D, Hu H, Tang Z-H (2004) Macromol Biosci 4:729.
109 Zhang J-T, Huang S-W, Liu J, Zhuo R-X (2005) Macromol Biosci 5:192.
110 Zhang J-T, Huang S-W, Gao F-Z, Zhuo R-X (2005) Colloid Polym Sci 283:461.

4
Cyclodextrin Catalysis

Makoto Komiyama and Eric Monflier

4.1
Introduction

As is well known, CyDs show catalytic activities on various reactions [1]. One of the pioneering works in CyD catalysis was accomplished by Bender and co-workers on the hydrolysis of phenyl acetates [2, 3]. On the basis of detailed and systematic studies, they elegantly showed that these CyD-induced reactions are associated with significant substrate specificity. For the hydrolysis of *meta*-substituted phenyl acetates ("specific substrates"), CyDs achieved remarkable acceleration. For *para*- and *ortho*-substituted phenyl acetates ("non-specific substrates"), however, their rate-effects were marginal. These results attracted much interest from many researchers in the world, who started working on CyDs. As a result, notable specific catalyses by CyDs (both substrate specific and/or product specific) were found in many kinds of reactions. CyDs are now being widely accepted as superb models of enzymes. Furthermore, the scope of catalytic reactions by CyDs has been greatly extended by attaching various catalytic residues to CyDs. In the very early stage of all these studies, the primary goal of the studies was to use CyDs as simple models of naturally occurring enzymes and gain a better understanding of the origins of the superiority of the enzymes. However, the aims gradually shifted towards the development of useful catalysts for practical applications (e.g. selective syntheses of target products). Undoubtedly, CyDs are also highly appropriate for these purposes.

All the reactions catalyzed by CyDs (and by their derivatives) proceed via their complexes with substrates, in which the chemical transformation takes place. This reaction scheme is exactly parallel to that employed by naturally occurring enzymes, and both high specificity and large reaction rates are primarily associated with this reaction scheme. Catalyses by CyDs are divided into three categories: (1) "covalent catalysis" in which a covalent intermediate is first formed from CyD and substrate and this intermediate is converted to the final products in the following step, (2) "general acid–base catalysis" by OH groups, and (3) "non-covalent catalysis" in which CyDs participate in the reactions only in a noncovalent fashion without even proton-transfer processes. The number of papers on catalysis by CyDs has

Cyclodextrins and Their Complexes. Edited by Helena Dodziuk

ISBN: 3-527-31280-3

been exponentially increasing, and it is impossible to cover all of them in the present limited space. Thus, we here put our emphasis on the basic principles of catalyses so that the readers can draw an appropriate picture for the reactions they are really interested in. Several examples of catalysis involving CyDs are presented in Chapter 2, one example of electrocatalysis is briefly discussed in Section 10.5 while a detailed discussion of the role of CyD catalysis in drugs is given in Section 14.3. In addition, a role for CyD catalysis in the degradation of pesticides is presented in Section 16.3.

4.2 Covalent Catalysis

In this catalysis, the hydroxy groups of CyDs directly react with the reagents and form a covalent intermediate. The catalysis occurs intramolecularly in the CyD–substrate complexes and thus can be extremely efficient. Furthermore, the chemical transformation is selective (with respect to substrates, products, and/or stereochemistry), mainly because CyDs provide their cavities as chemically and sterically specific reaction fields. The number of covalent catalyses by CyDs is rather limited, mainly because the reactivity of these hydroxyl groups is not very versatile. From the viewpoint of the history of CyD chemistry, however, this type of catalysis is undoubtedly essential. The hydrolysis of phenyl acetates reported by Bender et al. is a typical example, and opened the way to the fruitful CyD chemistry of today.[2,3] The first step of these reactions is the nucleophilic attack of the secondary OH group of CyD (as an alkoxide ion) towards the carbonyl carbon of the substrate which is accommodated in the CyD cavity. The phenol (leaving group) is removed from the carbonyl carbon, and acetyl-CyD is formed. In the following step, this covalent intermediate is hydrolyzed to the final products. Interestingly, *meta*-substituted phenyl acetates are hydrolyzed much faster than the corresponding *ortho*- and *para*-substituted phenyl acetates. This substrate-specificity comes from the difference in the mutual position of the secondary OH group and the carbonyl carbon of the substrate [4]. In the inclusion complexes of the *meta*-substituted phenyl acetates, the nucleophilic center is placed near the electrophilic center. Thus, only a small structural change (slight movement of the substrate in the cavity) is necessary to proceed from the initial state to the transition state, and thus the reaction is very rapid. For the *para*-compounds, however, the distance between the nucleophilic center and the electrophilic center is so large that the reaction process accompanies a significant structural change. The reaction is thus inefficient. This interpretation has been confirmed by ^{1}H- and ^{13}C-NMR spectroscopy [5]. In order to shed light on whether 2′-OH or 3′-OH of CyD (the alkoxide) is really the nucleophile in these reactions, these two types of hydroxy groups were regioselectively converted to an –SH group and the modified CyD was used for the cleavage of *m*-nitrophenyl acetate (one of the "specific substrates"). It was found that the 3-SH is much more effective than the 2-SH in promoting the acyl-transfer [6]. These results seem to favor nucleophilic attack by the alkoxide of 3′-OH in CyD catalysis, although the situation still remains unclear.

Another important pioneering work in CyD chemistry was reported by Breslow and Campbell [7] on the chlorination of anisole with HOCl. In the presence of α-CyD, the chlorination occurs exclusively at the *para*-position of anisole. This reaction also takes advantage of covalent catalysis. Here, HOCl first reacts with the secondary OH group of CyD, and the chlorine atom is selectively transferred to the *para*-position of anisole. Since the anisole penetrates into the cavity with the methoxy group first, the *para*-position of anisole is located near the secondary OH group (and thus near the CyD–OCl group). Apparently, the product selectivity comes from a "proximity effect", as is often observed in enzymatic reactions.

4.3 General Acid–Base Catalysis by OH Groups

The secondary OH groups of CyDs are more acidic ($pK_a = 12$) than normal aliphatic OH groups, and are eminent general acid catalysts. Because of this low pK_a, they are easily dissociated under physiological conditions to $CyD{-}O^-$ which acts as a general base catalyst. In some reactions, the primary OH groups of CyD can also be general acid catalysts. General acid–base catalysis by these OH groups can be efficient and specific, because the catalysis can occur in the inclusion complexes where these OH groups and substrates are placed in appropriate positions by the steric restraints. Cooperation by several OH groups is also possible.

Hydrolysis of alkyl esters by CyD proceeds via general base catalysis by $CyD{-}O^-$ [8], which is in contrast with nucleophilic catalysis by CyD (direct attack by $CyD{-}O^-$ to the substrate) in the hydrolysis of aryl esters (see above). The leaving groups of alkyl esters (alkoxide ions) are too poor to be removed from the tetrahedral intermediate formed by the nucleophilic attack by $CyD{-}O^-$. Thus, as an alternative way, $CyD{-}O^-$ activates the water in solutions by general base catalysis and this activates the attacks of water towards the substrate.

General acid catalysis by CyD is apparent in the hydrolysis of RNA [9, 10]. This reaction involves intramolecular attack by 2′-OH of ribose (as alkoxide) towards the phosphate, which provides the 2′,3′-cyclic monophosphate of the ribonucleotide as an intermediate. In the second step, either the P–O(2′) linkage or the P–O(3′) linkage in this intermediate is hydrolyzed, producing the RNA fragments bearing the terminal phosphates at either the 3′-position or the 2′-position (Fig. 4.1). In non-enzymatic hydrolysis, these two pathways A and B proceed at almost the same rate. In the RNA hydrolysis catalyzed by ribonuclease, however, only the P–O(2′) linkage is hydrolyzed and the P–O(3′) linkage is kept intact. This regioselective reaction is successfully mimicked by CyDs: either the P–O(2′) or P–O(3′) bond of the intermediate is selectively hydrolyzed, depending on the kind of CyD used. In the presence of α-CyD, the P–O(2′) bond is preferentially cleaved, giving the corresponding 3′-phosphate in high selectivity. This scission manner is the same as that exhibited by ribonuclease. The hydroxyl groups of α-CyD are essential

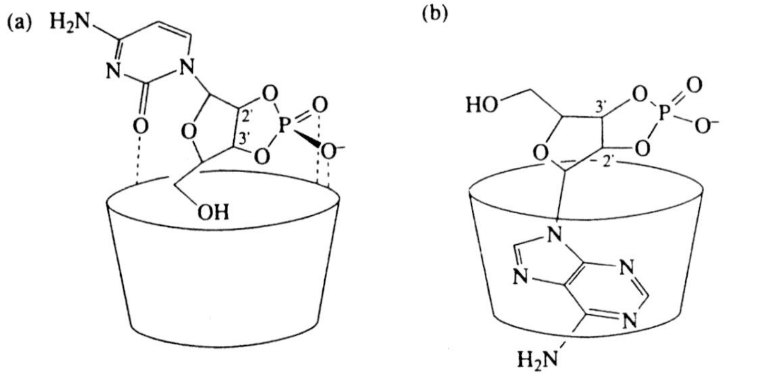

Fig. 4.1. Complexes of α-CyD (a) and β-CyD (b) for regioselective hydrolysis of RNA.

for this catalysis, so that hexa-2,6-dimethyl-α-CyD is totally inactive. In contrast, β- and γ-CyDs enhance P–O(3′) cleavage, producing the RNA fragments bearing the phosphate at the 2′-position. The hydroxyl groups of CyDs are again essential for the regioselective reaction. This remarkable dependence of regioselectivity on the kind of CyD comes from the difference in the structure of the complex between CyD and the 2′,3′-cyclic monophosphate (see Fig. 4.1). In the of α-CyD complex, the phosphate residue of the 2′,3′-cyclic monophosphate forms hydrogen bonds with the secondary hydroxyl group of α-CyD, and the oxygen atom at the C2 carbon of the nucleobase (in this case, cytosine) forms another hydrogen bond with the secondary hydroxyl group of the farthest glucose (Fig. 4.1(a)). The five-membered cyclic ring of the monophosphate is standing parallel to the longitudinal axis of the α-CyD cavity. As a result, the chemical environments of the P–O(2′) and the P–O(3′) bonds, which are otherwise almost identical, are largely differentiated. Scission of the P–O(3′) bond, which is located near the apolar cavity of α-CyD, is highly suppressed, since the formation of an alkoxide intermediate for the P–O scission is energetically unfavorable in an apolar environment. On the other hand, the P–O(2′) bond exists in a relatively polar medium and thus is preferentially cleaved by nucleophilic attack by hydroxide ion. On the other hand, β-CyD (or γ-CyD) and the cyclic monophosphate form conventional inclusion complexes in which the nucleobases are accommodated in the cavity (Fig. 4.2(b)). Thus, the P–O(3′) bond is placed far from the cavity and preferentially cleaved by hydroxide ion, in the same way as described above for the selective P–O(2′) cleavage by α-CyD.

4.4 Noncovalent Catalysis

In many other catalyses by CyDs, these cyclic oligonucleotides do not directly react with substrates. Direct proton-transfer between them does not occur either. Instead, CyDs merely offer their cavity as a hydrophobic and constrained reaction field. When substrate molecules are accommodated in the cavity, they take specific orientation there and thus two otherwise equivalent positions in these substrates can be clearly differentiated. Under these conditions, two reagents for bimolecular reactions are placed near each other with a specific mutual orientation. This situation is of course favorable for prompt and selective reactions. Other reactions are accelerated simply because the transition state is stabilized in the apolar cavity of CyD. Furthermore, the substrates and/or the products are protected from undesired side reactions (e.g. decomposition by other reagents). Therefore, notable specificity (regio-, stereo-, and enantio-selectivity), together with relatively high yields, are satisfactorily accomplished in CyD-catalyzed reactions.

4.4.1
Regulation of the Mutual Conformation of Reactants

By using CyD in alkaline solutions, both formyl and carboxy groups are selectively introduced to the *para*-position of phenol [4, 12]. Both products (*para*-hydroxybenzaldehyde and *para*-hydroxybenzoic acid) are important in industry for the production of various fine chemicals. When phenol is treated with $CHCl_3$ and sodium hydroxide in the presence of α-CyD, for example, *para*-hydroxybenzaldehyde is formed with almost 100% selectivity [11]. β-CyD is also effective for this *para*-selective reaction. The yield is satisfactorily high. In the absence of CyD, however, the product is a 1:1 mixture of *ortho*- and *para*-hydroxybenzaldehyde and the yields are very low. Apparently, the attack of dichlorocarbene (the active species formed from $CHCl_3$ and sodium hydroxide) to the *para*–position of phenol is promoted by CyD, and its attack to the *ortho*–position is suppressed. The reaction mechanism is depicted in Fig. 4.2. The key step is the formation of a ternary complex from CyD, phenol, and $CHCl_3$. Note that the phenol penetrates the CyD cavity with the *para*-carbon atom first, since this side is more apolar (the phenoxy group exists as an anion under the reaction conditions and thus this side is more polar). The $CHCl_3$ is accommodated in the cavity so that it is placed near the *para*-carbon of phenol. When $CHCl_3$ is transformed to dichlorocarbene by the action of hydroxide ion, this active species immediately

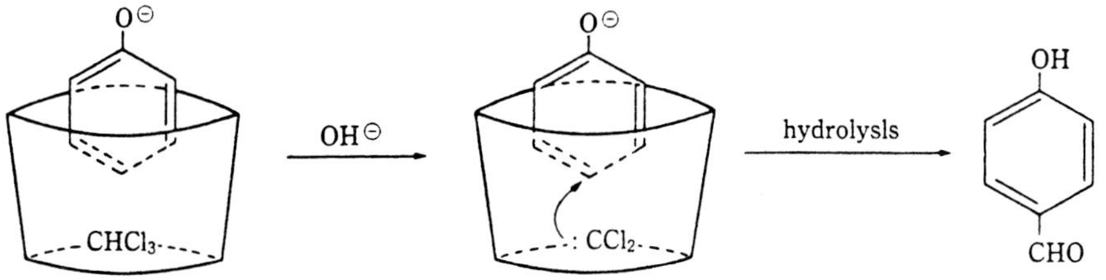

Fig. 4.2. Mechanism of *para*-selective formylation of phenol using CyDs as catalysts.

attacks the *para*-carbon atom of the phenol, resulting in the desired *para*-selective reaction. Furthermore, the product is protected from undesired side-reactions. Recently, the fate of carbene in the CyD cavity, as well as the orientation of its precursor in the cavity, has been analyzed in detail by spectroscopic and crystallographic means [13].

In a similar manner, carboxylation of phenol by CCl_4 (instead of $CHCl_3$) proceeds with high *para*-selectivity [12]. Copper powder is used as cocatalyst here. In this reaction, the active species is trichloromethyl cation, which is formed from CCl_4 on the surface of copper powder. Since this active species is positively charged and CyD is negatively charged (because of the dissociation of the OH groups), an inclusion complex between them is easily formed. Furthermore, the phenol forms an inclusion complex with the CyD. Exactly as described above for the *para*-selective formylation, the trichloromethyl cation is located near the *para*–carbon atom of the phenol in the resultant ternary complex, and thus preferentially attacks this carbon atom for the *para*-selective reaction. When HCHO is used in place of $CHCl_3$ or CCl_4, *p*-hydroxylmethylphenol is obtained (this product is easily converted by oxidation to *para*-hydroxybenzaldehyde and *para*-hydroxybenzoic acid). Although CyDs also promote *para*-selectivity in this reaction, the regioselectivity accomplished by unmodified CyDs is not so remarkable as is observed in the formylation and the carboxylation described above. This is primarily because HCHO is rather polar and is not readily included in the CyD cavity. In order to solve this problem, hydroxypropyl-β-CyD, in place of unmodified β-CyD, is quite useful and provides much better results. Here, the hydroxypropyl groups form hydrogen bonds with HCHO and force it into the cavity. As the result, nearly 100% *para*-selectivity is successfully accomplished [14].

CyDs also affect the stereochemistry of the products of Diels–Alder reactions [15, 16]. When ethyl hydrogen fumarate and cyclopentadiene are reacted in water (without CyDs), both the *endo*-adduct and the *exo*-adduct are formed in almost a 1:1 ratio. On the addition of β-CyD (0.015 M) to the solution, the formation of the *endo*-adduct is notably promoted (the *endo:exo* ratio = 2.2) [16]. In the cycloaddition between diethyl malonate and cyclopentadiene, β-CyD also promotes the formation of the *endo*-adduct, increasing the *endo:exo* ratio from 48.5 to 112. It has been proposed that the geometry of the transition-state for the formation of the *endo*-adduct is more favorably accommodated in the cavity of β-CyD. Consistently, the addition reactions are retarded by α-CyD which has a smaller cavity. This cavity is too small to accommodate both the diene and the dienophile simultaneously (and also the reaction transition state) and can bind only one of them. Thus the reactants are separated by α-CyD molecules and accordingly the formation of adducts is suppressed. In the 1,3-dipolar cycloaddition of 4-*tert*-butylbenzonitrile oxide in the presence of 6^A-acrylamido-6^A-deoxy-β-CyD in aqueous solution, 4-substituted isoxazoline is preferentially formed [17]. In its absence, however, the formation of the 5-substituted isomer is predominates. The dipole–dipole interaction between the acrylamido substituent and the guest in the cavity governs the regioselectivity of the reaction.

4.4.2
Regulation of Photoreactions

CyDs effectively regulate photoreactions producing the desired products in high selectivity and high yield. In the conventional photochemical Fries rearrangement of phenyl acetate in water, both 2- and 4-hydroxyacetophenone are formed. Here, the acetyl radical generated by the scission of the ester bond attacks the *para*- and *ortho*-positions almost equally (the ratio of 4-hydroxyacetophenone to 2-hydroxyacetophenone is close to 1). In the presence of β-CyD, however, the reaction at the *ortho*-position is greatly suppressed and thus the *para*-reaction is dominant (the *para*:*ortho* ratio = 6.2) [18]. Similar *para*-selectivity by CyDs was also reported in the photo-Fries rearrangement of other phenyl esters and anilides. In addition to the increase in *para*-selectivity, the formation of phenol, which results from the escape of acyl radical from the solvent cage, is strikingly suppressed by CyD, resulting in a notable increase in the yields of the *para*-product.

In all the reactions described above, CyDs promote *para*-selectivity. In some reactions, however, CyDs catalyze the formation of *ortho*-products. When anisole and pentyl chloroacetate in the presence of α-CyD (3 mM), *ortho*-alkylated anisole is efficiently and selectively formed (80% selectivity) [18a]. The selective reactions are ascribed to a cage effect and geometric control to the CyD cavity.

When anthracene-2-sulfonate is irradiated with light in the absence of β-CyD, four [2+2] photo-adducts are formed (Fig. 4.3 (a)). Upon the irradiation in its presence, however, only one isomer (**1**) is dominantly formed (see Fig. 4.3(a)) [19]. Anthracene-2-sulfonate forms a 2:2 complex with β-CyD, in which the mutual orientation of two anthracene derivative molecules are regulated by two β-CyD molecules (Fig. 4.3(b)). Accordingly, isomer **1** is selectively formed by the photoreaction. Consistently, little product-selectivity is observed when γ-CyD is used in place of β-CyD. The cavity of γ-CyD is large enough to accommodate two anthracene-2-sulfonate molecules and in this complex, they move rather freely relative to each other (Fig. 4.3(c)).

Upon UV irradiation in the presence of β-CyD, (*E*)-4,4′-bis(dimethylammoniomethyl)stilbene is transformed to its (*Z*)-isomer. In the presence of γ-CyD, however, only the [2 + 2] cycloaddition products are formed. Apparently, this kind of primary photoreaction is governed by the size of cavity of the CyD. This result is attributable to the difference in stoichiometry of the inclusion complex (1:1 vs. 2:1) [20].

4.4.3
Use of the CyD Cavity as a Specific Reaction Field

The cavity of CyD is apolar, and thus any reactions, which proceed rapidly in apolar media, should be accelerated simply by a "microsolvent effect". A classic example of this effect of a CyD is decarboxylation of anions of activated acids (e.g. α-cyano

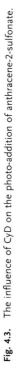

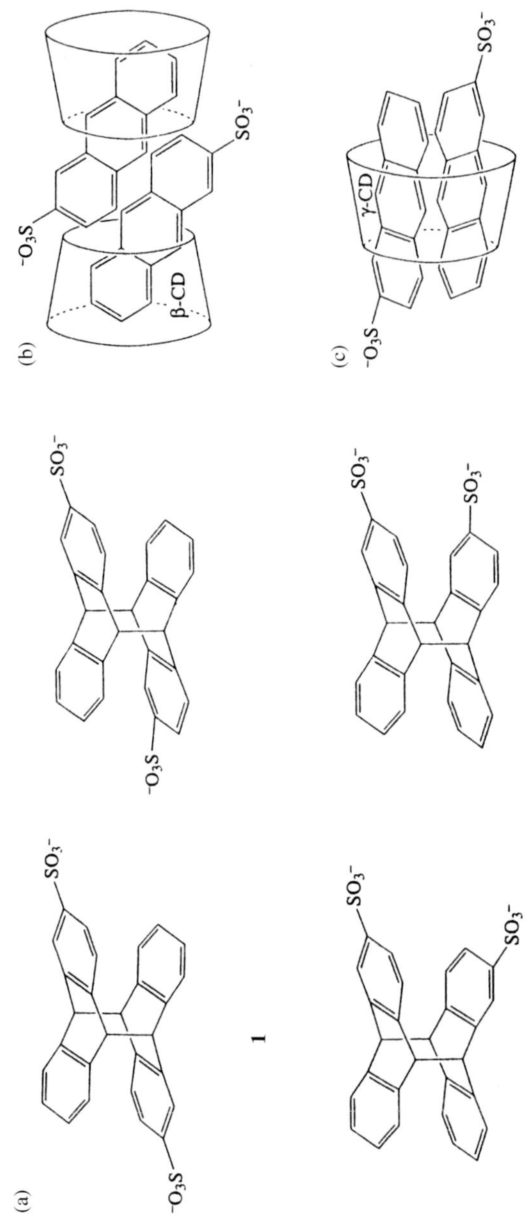

Fig. 4.3. The influence of CyD on the photo-addition of anthracene-2-sulfonate.

and β-keto acids) [21]. These reactions are extremely solvent-dependent, proceeding much faster in solvents of lower dielectric constant (the rate-determining step is heterolytic cleavage of the carbon-carbon bond adjacent to the carboxylate group and thus is efficient in apolar solvents). As expected, the acceleration is due to the decrease in activation enthalpy.

The cavity of CyDs is surrounded by glycosidic oxygen atoms and their lone pairs are directed into the interior of the cavity. Thus, the surface of the internal wall is abundant in electrons and has a Lewis-basic character. In some aromatic electrophilic substitution reactions, the cavity serves as an electronically negative environment and accelerates the reactions. A coupling between substituted phenyldiazonium and phenol is one of the examples (for the reaction of *p*-methoxyphenyldiazonium a six-fold rate enhancement is accomplished) [22]. It has been proposed that a reaction intermediate with a positive charge on the phenyl ring is stabilized by the electrostatic interactions, although mechanistic details are not yet clear. The trapping of CyD in the rotaxane by a positively charged stopper, in which these caps interlock the CyD molecules and avoid their slipping off the molecular necklace, is probably associated with this kind of interaction between the CyD cavity and the positive charges [23].

4.5 Catalysis by Chemically Modified CyD

Various advanced enzyme models have been envisioned by chemical modification of CyDs according to detailed molecular design. Technical developments in the precise chemical modification of CyDs have made an essential contribution to this remarkable progress. In these chemically modified CyDs, the catalytic residues are located in precise positions so that the catalysis is effective and selective. In order to achieve remarkable catalyses, multiple molecular recognition by CyDs in both the initial state and the transition state is necessary. One of the greatest advantages of these artificial enzymes is freedom of molecular design. One can simply choose an appropriate catalyst for the target transformation, and attach this catalyst to the appropriate position of a CyD.

For example, a superb mimic of the charge-relay system in serine proteases has been prepared by attaching both carboxylate and imidazole to α-, β-, and γ-CyDs [24]. The hydroxy group, the last component of the charge-relay system, is provided by the CyDs. The activity (kinetic parameters) of the β-CyD-based artificial enzyme for ester hydrolysis is close to that of α-chymotrypsin. These artificial enzymes show acylation, deacylation, and turnover, as is observed in the reactions of chymotrypsin. The substrate-specificity is dependent on the kind of CyD used, since it is primarily governed by the substrate-binding process. In phenyl ester hydrolysis, α- and β-CyD-based artificial enzymes are better than the γ-CyD-based artificial enzyme. For the hydrolysis of tryptophan ethyl ester, however, the γ-CyD-based artificial enzyme is the best. In another serine protease model, tripeptide (Ser–His–Asp) is directly introduced to the primary hydroxyl side of β-CyD [25]. This

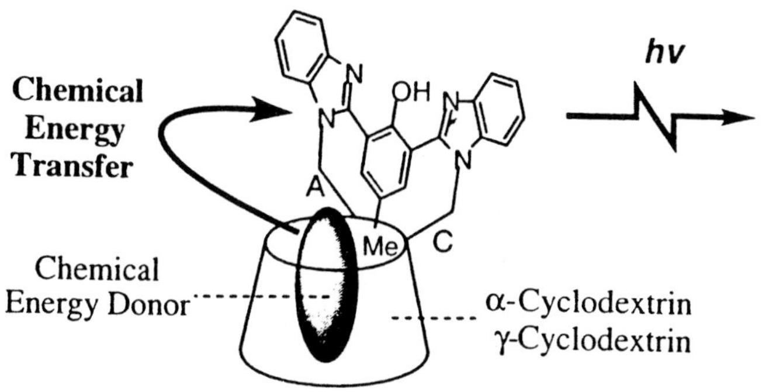

Fig. 4.4. Efficient conversion of chemical energy to photo-energy using a CyD.

compound hydrolyzes *p*-nitrophenyl esters with various alkyl group lengths. For the hydrolysis of *p*-nitrophenyl hexanoate, it is 3.4-fold more active than the tripeptide itself. However, the product is amide which is inconsistent with enzymatic reactions.

α-CyD (or γ-CyD) bearing bisbenzimidazolyl-*p*-cresol (fluorophore) at the primary side efficiently harvests chemical energy from the oxidation reaction of bis(2,4,6-trichlorophenyl)oxalate. The efficiency of light emission is two orders of magnitude greater than that from a typical dye, fluorescein [26]. When bis(2,4,6-trichlorophenyl)oxalate reacts with hydrogen peroxide and is converted to a high-energy intermediate, its chemical energy is efficiently used to elevate the fluorophore to the excited state, since both are included in the same cavity and located in close proximity (Fig. 4.4). A model for the chemiluminescent reaction of luminol has also been reported [27].

In principle, appropriate artificial enzymes for any target reaction should be obtained when the corresponding catalyst is covalently bound to CyDs. The validity of this idea has been substantiated in various examples, making this a quite important field. However, there are many review articles that deal with these subjects, and thus they are only briefly described here to avoid unnecessary repetition [28].

4.6
Phase-transfer Catalysis

As is well known, the solubility of apolar guest compounds in water is (in general) increased when they forms inclusion complexes with CyD. Thus, CyDs are potent phase-transfer catalysts [29]. Applications to organometallic catalysis are especially attractive from practical viewpoints [30]. As shown in Fig. 4.5, CyDs form inclusion complexes with hydrophobic substrates (S) at the liquid/liquid interface and transfer them into the aqueous phase where they contact the water-soluble organometallic catalyst. After the reaction, the product (P) is released in the organic phase and the transfer cycle can go on.

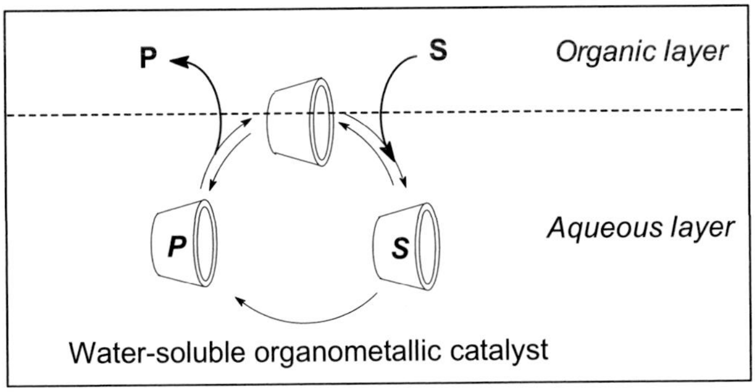

Fig. 4.5. Reaction scheme for phase-transfer catalysis by CyDs in organometallic catalysis.

Unmodified CyDs are effective as inverse phase-transfer catalysts for the deoxygenation of allylic alcohols [31], epoxydation [32], oxidation [33], or hydrosilylation [34] of olefins, and reduction of α,β-unsaturated acids [35], α-keto esters [36], conjugated dienes [37], or aryl alkyl ketones [38]. In many reactions (e.g. Wacker oxidation [39], hydrogenation of aldehydes [40], Suzuki cross-coupling reactions [41], cleavage of allylic substrates [42], hydroformylation [43], and hydrocarboxylation [44] of olefins), randomly *O*-methylated β-CyD is a better catalyst than an unmodified CyD. The outstanding effects of this modified β-CyD was attributed to both its small surface activity and the presence of a deeper hydrophobic cavity that properly accommodates the substrate. Substrate-selective reactions were also accomplished in other systems [45]. For instance, a 97:3 product ratio was observed for the palladium-catalyzed cleavage of a 50:50 mixture of *N*-hexadecyl *O*-allyl urethane and *N*,*N*-dioctyl *O*-allyl urethane. Finally, more sophisticated approaches involving covalent attachment of the phosphane ligand to CyD through a spacer have also been performed to combine molecular recognition, phase-transfer properties, and aqueous organometallic catalysis [46, 32d]. The rhodium complexes of diphosphane-modified β-CyD successfully showed catalyses for hydrogenation and hydroformylation in a biphasic medium.

4.7 Conclusion

Catalyses by CyDs are characterized by high selectivity and large reaction rate. The scope of catalytic reactions is wide and can be further extended by modifying CyDs according to detailed molecular design. All these properties are making CyDs one of the most promising enzyme models. By assembling CyDs in an ordered fashion, still better catalysts (and better enzyme models) should be obtained. In order to prepare these ordered assemblies, the molecular imprinting technique is very useful as recently reported [47, 48].

References

1 (a) Komiyama, M. "Cyclodextrin as Enzyme Model", in *Comprehenseve Supramolecular Chemistry* Vol 3, Ed by Szejtli, J.; Osa, T., Chapter 12, Pergamon (1996). (b) Osa, T.; Suzuki, I. "Reactivity of Included Guest", in *Comprehensive Supramolecular Chemistry* Vol 3, Ed by Szejtli, J. and Osa, T., Chapter 11, Pergamon (1996).

2 VanEtten, R. L.; Sebastian, J. F.; Clowes, G. A.; Bender, M. L. *J. Am. Chem. Soc.*, 1967, *89*, 3242.

3 VanEtten, R. L.; Clowes, G. A.; Sebastian, J. F.; Bender, M. L. *J. Am. Chem. Soc.*, 1967, *89*, 3253.

4 Komiyama, M.; Bender, M. L. *J. Am. Chem. Soc.*, 1978, *100*, 2259.

5 Komiyama, M.; Hirai, H. *Chem. Lett.*, 1980, *1467*, 1471.

6 Fukudome, M.; Okabe, Y.; Yuan, D.-Q.; Fujita, K. *Chem. Commun.*, 1999, 1045.

7 Breslow, R.; Campbell, P. *J. Am. Chem. Soc.*, 1969, *91*, 3084.

8 Komiyama, M.; Hirai, H. *Chem. Lett.*, 1980, *9*, 1251.

9 Komiyama, M. *J. Am. Chem. Soc.*, 1989, *111*, 3046.

10 Komiyama, M.; Takeshige, Y. *J. Org. Chem.*, 1989, *54*, 4936.

11 Komiyama, M.; Hirai, H. *J. Am. Chem. Soc.*, 1983, *105*, 2018.

12 Komiyama, M.; Hirai, H. *J. Am. Chem. Soc.*, 1984, *106*, 174.

13 (a) Bobek, M. M.; Giester, G.; Kählig, H.; Brinker, U. H. *Tetrahedron Lett.*, **2000**, *41*, 5663. (b) Bobek, M. M.; Krois, D.; Brinker, U. H. *Org. Lett.*, **2000**, *2*, 1999. (c) Mieusset, J.-L.; Krois, D.; Pacar, M.; Brecker, L.; Giester, G.; Brinker, U. H. *Org. Lett.*, **2004**, *6*, 1967.

14 Komiyama, M. *J. Chem. Soc., Chem. Commun.*, 1988, 651.

15 Rideout, D. C.; Breslow, R. *J. Am. Chem. Soc.*, 1980, *102*, 7816.

16 Schneider, H.-J.; Sangwan, N. K. *Angew. Chem., Int. Ed. Engl.*, 1987, *26*, 896.

17 Meyer, A. G.; Easton, C. J.; Lincoln, S. F.; Simpson, G. W. *Chem. Commun.*, 1997, 1517.

18 Ohara, M.; Watanabe, K. *Angew. Chem., Int. Ed. Engl.*, 1975, *14*, 820.

18a Bantu, N. R.; Kupfer, R.; Brinker, U. H. *Tetrahedron Lett.*, **1994**, *35*, 5117.

19 (a) Tamaki, T.; Kokubu, T. *J. Inclusion Phenom.*, 1984, *2*, 845. (b) Tamaki, T.; Kokubu, T.; Ichimura, K. *Tetrahedron*, 1987, *43*, 1485.

20 Herrmann, W.; Wehrle, S.; Wenz, G. *Chem. Commun.*, 1997, 1709.

21 Straub, T. S.; Bender, M. L. *J. Am. Chem. Soc.*, 1972, *94*, 8875.

22 Ye, H.; Rong, D.; D'Souza, V. T. *Tetrahedron Lett.*, 1991, *32*, 5231.

23 Kawaguchi, Y.; Harada, A. *J. Am. Chem. Soc.*, 2000, *122*, 3797.

24 D'Souza, V. T.; Bender, M. L. *Acc. Chem. Res.*, 1987, *20*, 146.

25 Ekberg, B. E.; Andersson, L. I.; Mosbach, K. *Carbohydr. Res.*, 1989, *192*, 111.

26 Yuan, D.-Q.; Kishikawa, N.; Yang, C.; Koga, K.; Kuroda, N.; Fujita, K. *Chem. Commun.*, 2003, 416.

27 Yuan, D.-Q.; Lu, J.; Atsumi, M.; Izuka, A.; Kai, M.; Fujita, K. *Chem. Commun.*, 2002, 730.

28 For example, Komiyama, M.; Shigekawa, H. in *Comprehensive Supramolecular Chemistry*, Vol. 3, Ed by Szejtli, J.; Osa, T. Chap. 11, Elsevier Science (1996).

29 Trifonou, A. Z.; Nikiforov, T. T. *J. Mol. Catal.*, 1984, *24*, 15.

30 CyDs were previously used for phase-transfer catalysis by metal ions: (a) Harada, A.; Hu, Y.; Takahashi, S. *Chem. Lett.*, 1986, 2083. (b) Hu, Y.; Uno, M.; Harada, A.; Takahashi, S. *Chem. Lett.*, 1990, 797.

31 Lee, J. T.; Alper, H. *Tetrahedron Lett.*, 1990, *31*, 4101.

32 Ganeshpure, P. A.; Satish, S. *J. Chem. Soc., Chem. Commun.*, 1988, 981.

33 (a) Zahalka, H. A.; Januszkiewicz, K.; Alper, H. *J. Mol. Catal.*, 1986, *35*, 249. (b) Harada, A.; Hu, Y.; Takahashi, S. *Chem. Lett.*, 1986, 2083.

(c) Karakhanov, E. A.; Filippova, T. Y.; Martynova, S. A.; Maximov, A. L.; Predeina, V. V.; Topchieva, I. N. *Cat. Today*, 1998, *44*, 189. (d) Karakhanov, E.; Maximov, A.; Kirillov, A. *J. Mol. Catal. A: Chem.*, 2000, *157*, 25.

34 Lewis, L. N.; Sumpter, C. A.; Stein, J. *J. Inorg. Organomet. Polymers*, 1996, *6*, 123.

35 (a) Lee, J. T.; Alper, H. *Tetrahedron Lett.*, 1990, *31*, 1941. (b) Arzoumanian, H.; Nuel, D.; *C. R. Acad. Sci. Paris*, 1999, *Série IIc*, 289.

36 Pinel, C.; Gendreau-Diaz, N.; Bréhéret, A.; Lemaire, M. *J. Mol. Catal. A.: Chem.*, 1996, *112*, L157.

37 Lee, J. T.; Alper, H. *J. Org. Chem.*, 1990, *55*, 1854.

38 Zahalka, H.; Alper, H. *Organometallics*, 1986, *5*, 1909.

39 Monflier, E.; Tilloy, S.; Blouet, E.; Barbaux, Y.; Mortreux, A. *J. Mol. Catal. A: Chem.*, 1996, *109*, 27.

40 Tilloy, S.; Bricout, H.; Monflier, E. *Green Chem.*, 2002, *4*, 188.

41 Hapiot, F.; Lyskawa, J.; Tilloy, S.; Bricout, H.; Monflier, E. *Adv. Synth. Catal.*, 2004, *346*, 83.

42 Widehem, R.; Lacroix, T.; Bricout, H.; Monflier, E. *Synlett*, 2000, *5*, 722.

43 (a) Monflier, E.; Fremy, G.; Castanet, Y.; Mortreux, A. *Angew. Chem. Int. Ed. Engl.*, 1995, *34*, 2269. (b) Dessoudeix, M.; Urrutigoïty, M.; Kalck, P. *Eur. J. Inorg. Chem.*, 2001, 1797.

44 Tilloy, S.; Bertoux, F.; Mortreux, A.; Monflier, E. *Catal. Today*, 1999, *48*, 245.

45 (a) Cabou, J.; Bricout, H.; Hapiot, F.; Monflier, E. *Catal. Commun.*, 2004, *5*, 265. (b) Torque, C.; Bricout, H.; Hapiot, F.; Monflier, E. *Tetrahedron*, 2004, *60*, 6487.

46 Reetz, M. T.; Waldvogel, S. R. *Angew. Chem., Int. Ed. Engl.*, 1997, *36*, 865.

47 Komiyama, M.; Takeuchi, T.; Mukawa, T.; Asanuma, H. *Molecular Imprinting – From Fundamentals to Applications*, Wiley-VCH, Weinheim (2003).

48 Asanuma, H.; Hishiya, T.; Komiyama, M. *Adv. Mater.*, 2000, *12*, 1019.

5
Chromatographic Studies of Molecular and Chiral Recognition

Monika Asztemborska and Anna Bielejewska

5.1
Introduction

Chromatography on the basis of CyDs is applied today primarily for the separation of enantiomers [1–4] occurring mainly as the result of complexation of a chiral guest-molecule by an enantiomerically pure host CyD. However, the complexation abilities of these hosts may also be utilized for molecular recognition and separation of closely related compounds, including geometrical and structural isomers [5–8]. In this chapter we shall present some examples of molecular and chiral recognition by CyDs and applications of chromatographic techniques for characterization of CyDs complexes, such as the determination of stoichiometries and stability constants while enantioseparations involving CyDs are discussed in the next chapter. Both native and modified CyDs are used in chromatography. Some CyD derivatives developed especially for this purpose are presented in Chapter 2.

Two basic chromatographic techniques, high-performance liquid chromatography (HPLC) and gas chromatography (GC), are presented here. In the latter, CyDs are applied as a component of the stationary phase. They can work in two main modes of gas chromatography: gas–solid and gas–liquid. In gas–solid chromatography, β-CyD polyurethane resins have successfully been applied to the separation of xylene isomers and pyridine derivatives [9]. α-, β- and γ-CyDs and their methylated derivatives deposited on Chromosorb have been used to separate isomeric trimethylbenzenes, *cis*- and *trans*-3-methyl-2-pentenes, and *ortho*-, *para*-, and *meta*-xylenes, cymenes, and diethylbenzenes [10, 11]. α- and β-CyDs and their acetylated derivatives chemically bonded to silica gel have been used as a gas–solid support in packed columns for the separation of light hydrocarbons and inorganic gases [12, 13]. The same compounds were separated using α- and β-CyD chemically bonded to fused silica capillary columns [14]. Using β-CyD as the solid stationary phase in a classical packed column, the heats of adsorption of selected organic compounds on β-CyD were determined [15].

In gas–liquid chromatography with formamide solutions of α- and β-CyD as stationary phases it was found that α-CyD exhibits great selectivity towards enantiomers of monoterpenes [16, 17] while β-CyD selectively recognizes constitutional

Cyclodextrins and Their Complexes. Edited by Helena Dodziuk

ISBN: 3-527-31280-3

isomers of *ortho*-, *para*-, and *meta*-xylenes, *para* and *meta*-dialkylbenzenes, and diastereomers of *cis*- and *trans*-decalin, and *cis*- and *trans*-1,2-dimethylocyclohexanes [18–21].

Interestingly, capillary columns with permethylated-β-CyD were successfully used for the separation of isotopomers [22].

The 12 tobacco alkaloids and nicotine metabolites **1–12** were very well separated on a β-CyD column in liquid chromatography [23]. The very close similarity of some of these compounds, the similarity of their basicity, and the similar molecular size throughout the entire set makes the analytical challenge significant. β- or γ-

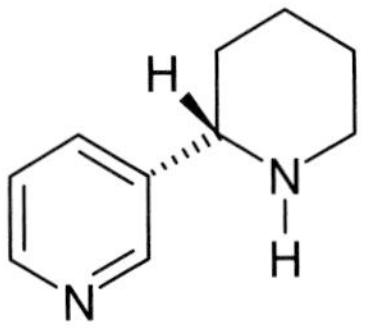

(***S***)-(-)-anabasine **1**

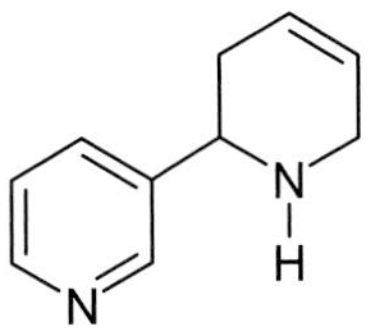

(***R***,***S***)-anatabine **2**

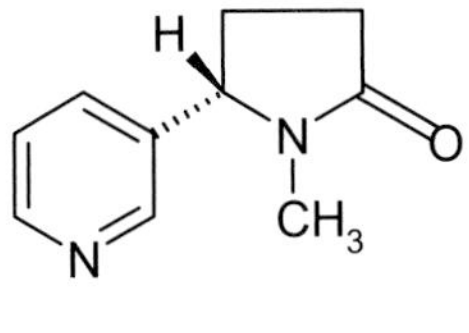

(***S***)-(-)-cotinine **3**

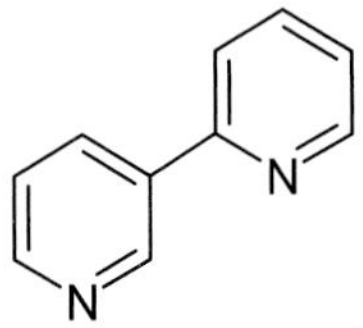

2,3′-dipyridyl **4**

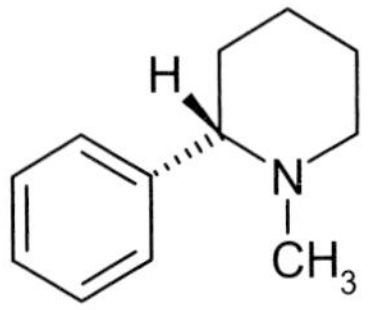

(***S***)-(-)-***N′***-methylanabasine **5**

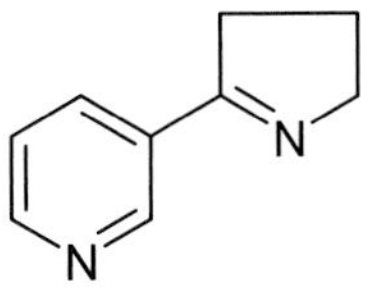

Myosmine **6**

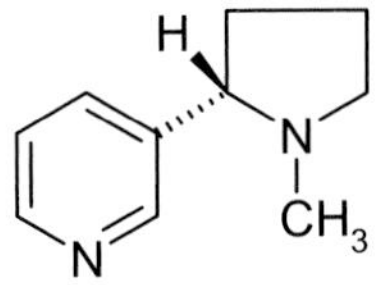

(***S***)-(-)-nicotine **7**

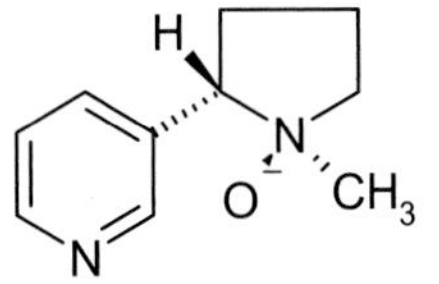

(1***R***,2***S***)-antinicitine-*N′*-oxide **8**

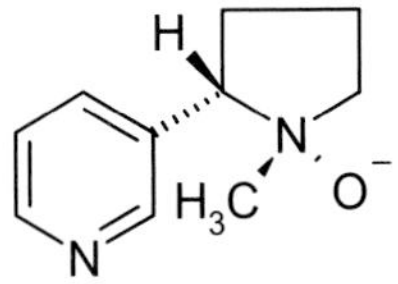

(1***S***,2***S***)-syn-nicitine-*N′*-oxide **9**

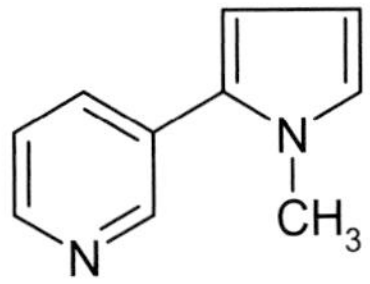

Nicotyrine **10**

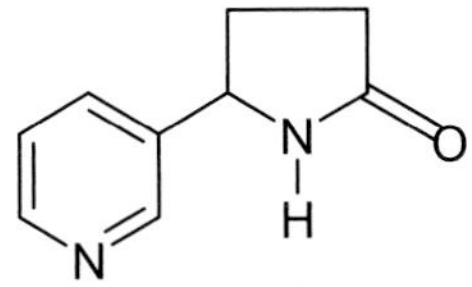

(***R***,***S***)-norcotinine **11**

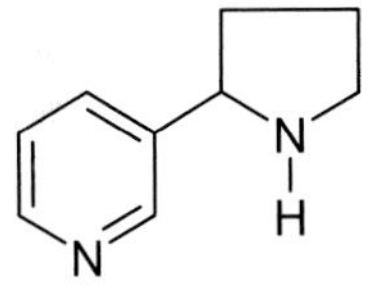

(***R***,***S***)-nornicotine **12**

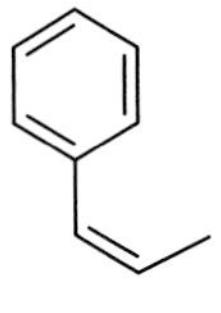

cis-anethole **13**

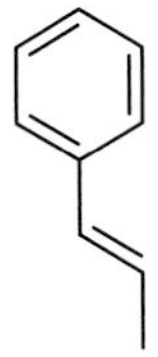

trans-anethole **14**

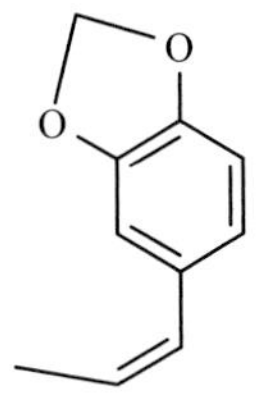

cis-isosafrole **15**

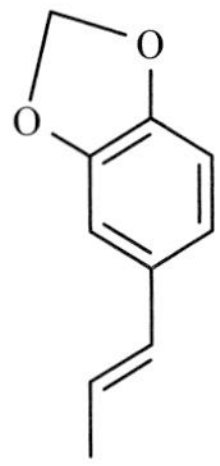

trans-isosafrole **16**

CyDs added to the mobile phase in liquid chromatography were successfully used to separate the *cis*- and *trans*-isomers of isoeugenole and isosafrole **13–16** on Li-Chrosorb RP 18 stationary phase [24].

A wide variety of native and derivatized CyDs are available for use as mobile phase additives in HPLC. In liquid chromatography to study the interaction of CyDs with guest molecules, the information about cyclodextrin adsorption on the stationary phase is very substantial. It should be noted that the separation ability of bonded CyDs and CyDs added to the mobile phase in HPLC is not always the same (see below). To assess the adsorption of CyDs on the stationary phase, the chromatographic properties of native and permethylated CyDs applied in RP18 and porous graphitic carbon (PGC) column have been studied [25–29]. On the RP18 column the adsorption of β-CyD is much stronger than that of α- or γ-CyDs. On the PGC, the order of elution is α- < β- < γ-CyD, in accordance with the increase of their molecular weight, the retention increasing with the increasing CyD size. It seems that for the RP18 column the solubility of CyDs in the eluent plays a greater role in the adsorption process, and consequently in separation. The solubility of β-CyD in water is much lower than those of α- and γ-CyD [30, 31]. The methyl derivatives of α-, β-, and γ-CyD are much more strongly adsorbed on the hydrophobic stationary phase than their native counterparts [29]. Elution of the methyl derivatives requires higher concentrations of organic additives in the aqueous mobile phase solutions for the RP18 column. On the PGC column, it was impossible to investigate the retention behavior of permethylated CyD in an ethanol–water solution even using a high concentration of ethanol. For the PGC column the permethylated CyDs were eluted in a dioxane/2-propanol mixture (20:80 v/v). The elution order differences for the two columns prove that the mechanism of separation is different for each of them, although both RP-18 and PGC columns act in reverse chromatographic mode.

As discussed in Chapter 1 in Section 1.4, contrary to Harata's opinion [32, 33], methylated CyDs are in general not better selectors than the native ones [34, 35].

5.2
Determination of the Stoichiometry and Stability of the Complexes

Determination of the stability constants and stoichiometry of CyD inclusion complexes is important for understanding the mechanism of complexation.

As mentioned before, chromatographic methods that are very sensitive to structure, size, shape, and dynamics of the analytes have been used to characterize CyD complexes, in addition to X-ray, spectrophotometric, NMR, calorimetric, and other methods presented in Chapters 6–10.

The stability constants of some CyD complexes estimated by HPLC and GC are collected in Table 5.1. It is clear that the stability constants depend on the structure of the guest compound, on the kind of CyD, and the solvent and additives applied. Very interesting information was obtained from HPLC systems with the CyD added to the mobile phase. The change in retention factors of each compound

Tab. 5.1. Stability constants of inclusion complexes of selected compounds with CyDs.

Compound	CyD/solvent	Stoichi-ometry	Stability constant		Technique	Ref.
			K_1	K_2		
Estriol	β-CyD/MeOH-H_2O	1:1	300	–	HPLC	49
	β-CyD/MeCN-H_2O		152			
	γ-CyD/MeOH-H_2O		523			
	γ-CyD/MeCN-H_2O		313			
Estradiol	β-CyD/MeOH-H_2O	1:1	465	–	HPLC	49
	β-CyD/MeCN-H_2O		191			
	γ-CyD/MeOH-H_2O		882			
	γ-CyD/MeCN-H_2O		384			
Ethinyloestradiol	β-CyD/MeOH-H_2O	1:1	364	–	HPLC	49
	β-CyD/MeCN-H_2O		125			
	γ-CyD/MeOH-H_2O		910			
	γ-CyD/MeCN-H_2O		414			
Estrone	β-CyD/MeOH-H_2O	1:1	267	–	HPLC	49
	β-CyD/MeCN-H_2O		92			
	γ-CyD/MeOH-H_2O		352			
	γ-CyD/MeCN-H_2O		182			
Mandelic acid (MA)	α-CyD/MeOH-H_2O	1:1	25 ± 4	–	HPLC	43
	β-CyD/MeOH-H_2O		55 ± 5			
	γ-CyD/MeOH-H_2O		35 ± 8			
Methyl ester of MA	α-CyD/MeOH-H_2O	1:1	23 ± 1	–	HPLC	43
	β-CyD/MeOH-H_2O		42 ± 3			
	γ-CyD/MeOH-H_2O		31 ± 7			
Ethyl ester of MA	α-CyD/MeOH-H_2O	1:1	34 ± 4	–	HPLC	43
	β-CyD/MeOH-H_2O		58 ± 3			
	γ-CyD/MeOH-H_2O		43 ± 4			
n-Propyl ester of MA	α-CyD/MeOH-H_2O	1:1	32 ± 2	–	HPLC	43
	β-CyD/MeOH-H_2O		115 ± 4			
	γ-CyD/MeOH-H_2O		66 ± 9			
n-Butyl ester of MA	α-CyD/MeOH-H_2O	1:1	64 ± 3	–	HPLC	43
	β-CyD/MeOH-H_2O		116 ± 6			
	γ-CyD/MeOH-H_2O		95 ± 18			
(+)-Camphor	α-CyD/20:80 MeOH:H_2O	1:2	147 ± 44	417 ± 170	HPLC	53
	α-CyD/35:65 MeOH:H_2O		39 ± 5	277 ± 45		
	α-CyD/35:65 EtOH:H_2O		7 ± 2.5	200 ± 87		

Tab. 5.1 *(continued)*

Compound	CyD/solvent	Stoichi-ometry	Stability constant		Technique	Ref.
			K_1	K_2		
(−)-Camphor	α-CyD/20:80 $MeOH:H_2O$	1:2	170 ± 24	194 ± 46	HPLC	53
	α-CyD/35:65 $MeOH:H_2O$		31 ± 3	189 ± 27		
	α-CyD/35:65 $EtOH:H_2O$		5.8 ± 1.8	126 ± 50		
(+)-Carvone	α-CyD/glycerol (70 °C)	1:2	9	5	GC	60
(−)-Carvone	α-CyD/glycerol (70 °C)	1:2	9	3	GC	60
(−)-Isopinocampheol	α-CyD/glycerol (70 °C)	1:2	7	8	GC	60
(+)-Isopinocampheol	α-CyD/glycerol (70 °C)	1:2	6	7	GC	60
(+/−)-Menthol	α-CyD/glycerol (70 °C)	1:1	3	–	GC	59
(+/−)-Neomenthol	α-CyD/glycerol (70 °C)	1:1	1	–	GC	59
(+/−)-Pulegone	α-CyD/glycerol (70 °C)	1:1	5	–	GC	59
1,3-DMN[a]	β-CyD/glycerol (80 °C)	1:1	10.0	–	GC	59
1,4-DMN	β-CyD/glycerol (80 °C)	1:1	10.0	–	GC	59
1,7-DMN	β-CyD/glycerol (80 °C)	1:1	8.0	–	GC	59
1,8-DMN	β-CyD/glycerol (80 °C)	1:2	5.5	333.0	GC	59
cis-Anethole	α-CyD/glycerol (80 °C)	1:1	6.0	–	GC	58
trans-Anethole	α-CyD/glycerol (80 °C)	1:2	1.3	196.0	GC	58
cis-Isosafrole	α-CyD/glycerol (80 °C)	1:1	11.0	–	GC	58

Tab. 5.1 *(continued)*

Compound	CyD/solvent	Stoichi-ometry	Stability constant		Technique	Ref.
			K_1	K_2		
trans-Isosafrole	α-CyD/glycerol (80 °C)	1:2	2.0	116.0	GC	58
cis-Decalin	α-CD/formamide	1:1	128.0	–	GC	58
trans-Decalin	α-CD/formamide	1:1	45.0	–	GC	58
(+)-Fenchone	α-CyD/glycerol (60 °C)	1:2	34	10	GC	60
	β-CyD/glycerol (60 °C)		12	44		
(−)-Fenchone	α-CyD/glycerol (60 °C)	1:2	13	11	GC	60
	β-CyD/glycerol (60 °C)		10	57		
(+)-Isomenthone	α-CyD/glycerol (60 °C)	1:2	6.4	9.0	GC	60
	β-CyD/glycerol (60 °C)		6.9	15.0		
(−)-Isomenthone	α-CyD/glycerol (60 °C)	1:2	7.4	11.0	GC	60
	β-CyD/glycerol (60 °C)		7.4	13.0		
(+)-Menthone	α-CyD/glycerol (60 °C)	1:2	3.8	8.9	GC	60
	β-CyD/glycerol (60 °C)		8.6	–		
(−)-Menthone	α-CyD/glycerol (60 °C)	1:2	4.4	6.1	GC	60
	β-CyD/glycerol (60 °C)		8.6	–		

[a] DMN: dimethylnaphthalene

was monitored as the concentration of CyD changed [36–42]. Stability constant and stoichiometry could be determined from these data, assuming that the complexes and CyDs are not adsorbed on the stationary phase. It was found that two α-CyD molecules bind prostaglandin B1 (Fig. 5.1), prostaglandin B2, 4,4′-biphenol and *p*-nitroaniline. (The applications of CyD in prostaglandin drug formulations

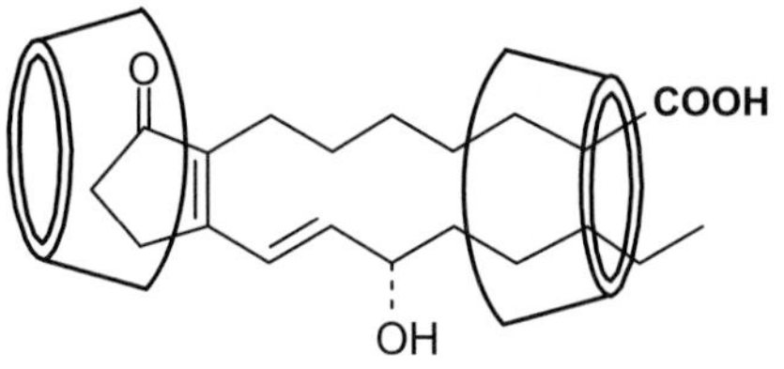

Fig. 5.1. Schematic representation of the 1:2 complex of prostaglandin B1 with α-CyDs.

are discussed in Chapter 14). Conversely *o*-nitroaniline, and *m*-nitroaniline form 1:1 complexes. It is interesting to note that very closely related compounds such as isomeric nitroanilines can exhibit different binding behaviors [38].

The chromatographic behavior of mandelic acid and its esters was studied in reversed-phase HPLC (RP-HPLC) with α-, β-, γ-, and permethylated-β-CyD as additive to the mobile phase. It was found that native CyDs do not recognize enantiomers of esters although they form relatively stable 1:1 complexes with them. Estimated stability constants for these guests are presented in Table 5.1 [43].

The inclusion of imidazole derivatives (bifonazole, clotrimazole, econazole, sulconazole, miconazole, oxiconazole) with β-CyD and hydroxypropyl-β-CyD (HP-β-CyD) was investigated over a wide range of column temperatures in RP-HPLC [44, 45]. Interestingly, at 20 °C the inclusion of imidazole derivatives in the HP-β-CD cavity was from 5 to 16 times weaker than in the β-CyD one. At 5 °C the six imidazole derivatives had nil stability constant with HP-β-CyD, whereas the stability constants with β-CyD were in the range 21.5–185 M^{-1}.

Chromatographic, NMR, and molecular modeling studies [46] indicate that benzene-*sym*-tris-*N,N,N*-carbonyltriglycylglycine *N'*-1-adamantylamide **17** forms the strongest complex with β-CyD encapsulating the terminal adamantyl groups. The analogous complex with γ-CyD is weaker with a deeper penetration of the guest into the host cavity while the complex with α-CyD is very weak. Chromato-

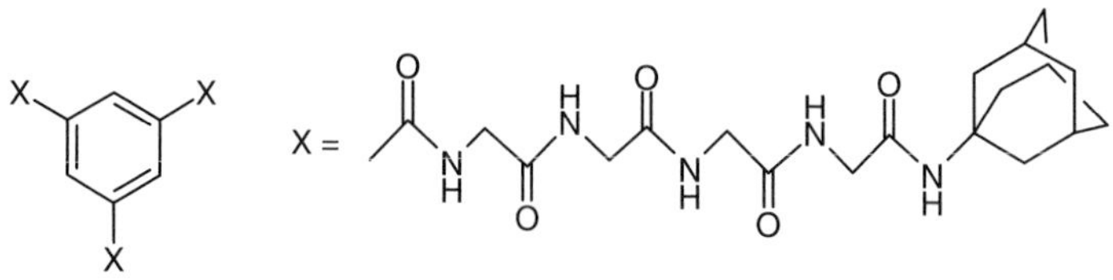

17

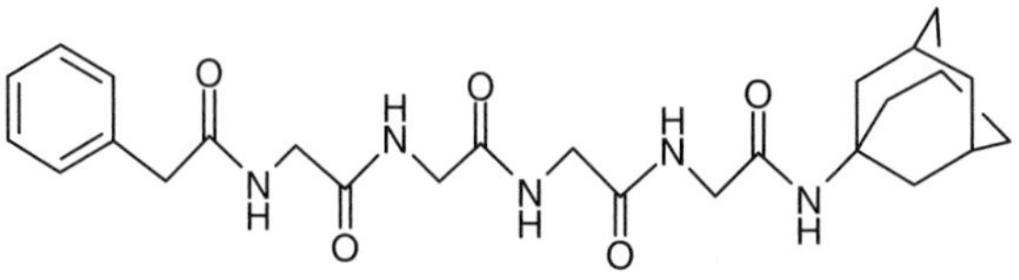

18

graphic measurements for the complexes with β- and γ-CyD with the dendrimer revealed interesting differences in the stoichiometry of the dendrimer complexes in mixed solvents chosen because of the low solubility of the dendrimer under study and its CyD complexes. Namely, in a 60:40 (v/v) ethanol:water mixture the latter complex was undoubtedly of 1:1 stoichiometry while for the former the most plausible stoichiometry was 1:3. The stability constants of the multiple dendrimer complexes with β-CyD (in a 20:80 (v/v) ethanol:water mixture) of 1:1, 1:2 and 1:3 stoichiometry were estimated to be equal to ca 4×10^2 M^{-1}, 1×10^2 M^{-1} and 25×10^3 M^{-1}, respectively, while only a 1:1 complex with a binding constant of ca 6×10^2 M^{-1} was found for the dendrimer complex with the bigger γ-CyD (in a 60:40 (v/v) methanol:water mixture). The considerably higher value of K_3 compared to K_1 and K_2 seems to indicate that in that case complex formation with dendrimer end groups is cooperative.

Chromatographic measurements for compound **18** mimicking one dendrimer branch yielded a 1:1 complex with γ-CyD with a binding constant of 1×10^2 M^{-1} while it was shown to complex at both adamantyl and benzene ends with β-CyD with constants of 20×10^2 M^{-1} and 2×10^2 M^{-1}, respectively.

Reversed-phase liquid chromatography and gas–liquid chromatography have been applied to studies on the complexation of cyclodextrin with isomeric dimethylnaphthalenes (DMN) **19–22**. Both β- and γ-CyD have been found to form inclusion complexes with them. With the exception of the complex of 1,8-DMN **22** with β-CyD, all other complexes under investigation are of weak stability. The stability constant of the former complex is always about 1.5–2.0 orders of magnitude greater than the corresponding value for the other isomers [47]. For example, at 40 °C in 40% of ethanol in water solvent, the stability constants of the complexes of β-CyD with 1,8-DMN **22** and 1,3-DMN **19** are equal to ca. 500 M^{-1} and ca. 8 M^{-1}, respectively.

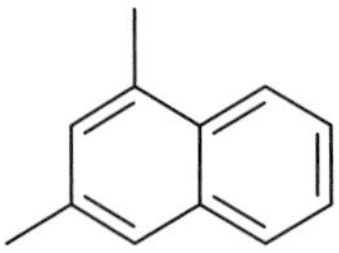

1,3-DMN **19**

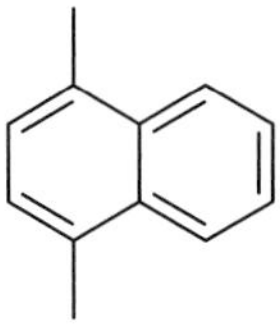

1,4-DMN **20**

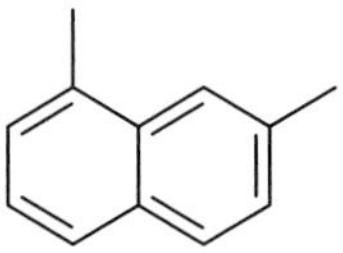

1,7-DMN **21**

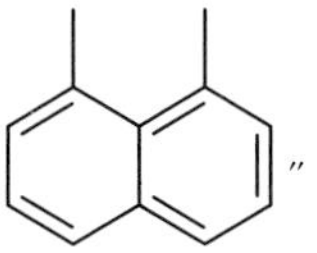

1,8-DMN **22**

Hummel and Dreyer developed a method for determining the stability constant from chromatographic data [48]. A modification of the method was used to calculate the apparent stability constants of steroid-CyD inclusion complexes [49–51]. In this method, varying amounts of CyD are injected on the column equilibrated with the guest compound present in the mobile phase. A negative peak obtained on the chromatogram corresponds to the amount of guest consumed to produce the complex. The area of the positive peak is a measure of the concentration of the complex formed. Using this method Sadlej-Sosnowska [49] determined the association constants of four steroids (estriol, estradiol, ethinyloestradiol, and estrone) with β- and γ-CyDs in methanol–water (45:55 v/v) and acetonitrile–water (30:70 v/v) as a function of temperature (see Table 5.1). The thermodynamic data obtained in this paper are presented in Table 5.2 and discussed in Section 5.3. The complexes of the four steroids with γ-CyD are stronger than those with β-CyD. In addition, the values of the stability constants are greater in the methanol solution than in the acetonitrile one. There is evidence that the nature of the solvent may influence or control the structure of the CyD complexes [52]. The influence of organic modifiers on complex formation is very well reflected in the chromatographic study. Some interesting results have been obtained in RP-HPLC systems for camphor and α-pinene enantiomers. The observed results clearly demonstrate that in a chromatographic system an achiral solvent may promote or inhibit chiral recognition of the solute [53]. It is also noteworthy that the separation abilities of bonded CyDs and CyDs added to the mobile phase in HPLC are not always the same [35, 40, 54].

The static head-space gas chromatography method (in which samples are injected as vapor from above the water solution of CyD-guests) was used to determine stoichiometry and stability constants of aqueous solutions of α-CyD with benzene and alkylbenzenes [55]. 1:1 and 1:2 stability constants obtained by this method were in a reasonable agreement with the values determined by other methods. For less volatile compounds from the alkanediol group a competitive method was applied together with head-space GC [56]. The solubility method based on increase in the guest solubility in water as a result of complex formation by a CyD connected with head-space chromatography allowed the determination of stability constants of monoterpenoids with 2-HP-β-CyD [57]. Head-space chromatography was also used to determine the stability constants of crystalline inclusion complexes of permethylated-β-CyD with selected hydrocarbons [58].

Gas–liquid chromatography with α- and β-CyD dissolved in glycerol or formamide as stationary phases was used to establish a relationship between selectivity towards isomers and the stoichiometry of CyD complexes [59, 60]. The model tested compounds were: *cis*/*trans* anetholes **13**, **14**, *cis*/*trans* isosafroles **15**, **16**, *cis*/*trans* decalins as diastereomers and 1,3-, 1,4-, 1,7-, and 1,8-DMNs **19–22** (briefly discussed earlier) and α- and β-pinenes as constitutional isomers as well as selected hydrocarbons, ketones, and alcohols of monoterpenoids as enantiomers. Experimental retention data were used to confirm a simple theoretical model that allows differentiation between the formation of 1:1 (G CyD) complexes and 1:2 (G CyD_2) complexes. (In line with the convention adopted in chromatography, the complexes are often denoted as of 1:2 stoichiometry although the mixture of rapidly interconverting free host and guest and 1:1 and 1:2 complexes is present in solution.) Examples of stability constants obtained for such complexes are presented in Table 5.1. It has been found that remarkable selectivity towards constitutional isomers and diastereomers may appear both for 1:1 stoichiometry (β-CD complexes of decalins and of α- and β-pinenes) and for 1:2 stoichiometry (α-CD complexes with (+/−)-α-pinenes and (+/−)-camphenes). Occasionally, selectivity arises from different compositions, when one isomer forms a 1:1 stoichiometry complex while

Tab. 5.2. Thermodynamic parameters of complexation of selected compounds by CyDs.

Compound	CD/solvent	Thermodynamic parameters			Technique	Ref.
		$-\Delta H$ [kJ mol^{-1}]	$-T\Delta S$ [kJ mol^{-1}]	$-\Delta G$ [kJ mol^{-1}]		
(+)-Camphor	α-CyD/35:65 MeOH:H_2O	60	38.2	22.6	HPLC	53
	α-CyD/35:65 EtOH:H_2O	26.4	8.7	17.7		
(−)-Camphor	α-CyD/35:65 MeOH:H_2O	54	32.9	21.1	HPLC	53
	α-CyD/35:65 EtOH:H_2O	19.9	1.8	16.1		
Estriol	β-CyD/MeOH-H_2O	33.1	18.3	14.8	HPLC	49
	β-CyD/MeCN-H_2O	23.8	10.9	12.9		
	γ-CyD/MeOH-H_2O	26.8	10.7	16.1		
	γ-CyD/MeCN-H_2O	18.8	4.3	14.5		
Estradiol	β-CyD/MeOH-H_2O	37.2	21.2	16.0	HPLC	49
	β-CyD/MeCN-H_2O	25.1	11.6	13.5		
	γ-CyD/MeOH-H_2O	26.4	9.1	17.3		
	γ-CyD/MeCN-H_2O	17.6	2.5	15.1		
Ethinyloestradiol	β-CyD/MeOH-H_2O	35.6	19.9	15.7	HPLC	49
	β-CyD/MeCN-H_2O	21.8	9.2	12.6		
	γ-CyD/MeOH-H_2O	31.0	13.0	18.0		
	γ-CyD/MeCN-H_2O	17.6	2.3	15.2		
Estrone	β-CyD/MeOH-H_2O	33.5	18.8	14.7	HPLC	49
	β-CyD/MeCN-H_2O	21.8	9.4	12.3		
	γ-CyD/MeOH-H_2O	27.2	12.0	15.2		
	γ-CyD/MeCN-H_2O	17.6	4.2	13.4		
(+)-Fenchone	α-CyD/glycerol	42.2	26	16	GC	61
	β-CyD/glycerol	32.4	15.1	17.3		
(−)-Fenchone	α-CyD/glycerol	30.3	16	14	GC	61
	β-CyD/glycerol	33.2	15.7	17.5		
(+)-Isomenthone	α-CyD/glycerol	18.8	8	11	GC	61
	β-CyD/glycerol	18.0	5.1	12.9		
(−)-Isomenthone	α-CyD/glycerol	24.9	13	12	GC	61
	β-CyD/glycerol	15.5	2.9	12.6		
(+)-Menthone	α-CyD/glycerol	22.0	12	10	GC	61
	β-CyD/glycerol	9.8	4	6		
(−)-Menthone	α-CyD/glycerol	19.6	11	9	GC	61
	β-CyD/glycerol	9.8	4	6		

another forms 1:2 complexes (dimethylnaphthalenes, *cis/trans* anetholes **13**, **14** and *cis/trans* isosafroles **15**, **16**). In contrast, for the compounds studied, enantioselectivity is observed only for the complexes of 1:2 stoichiometry, involving two CyDs (for example, fenchone, carvone, α-pinene, β-pinene, isomenthone, and isopinocampheol) while 1:1 stoichiometry does not lead to chiral recognition (pulegone, menthol, neomenthol, terpinen-4-ol, α-terpineol).

As discussed in the next section, the magnitude of the stability constant does not determine the enantioselectivity. For example the enantiomers of fenchone and isomenthone form more stable complexes with β- than with α-CyD while the observed enantioselectivity is much higher for α- than for β-CyD (see Table 5.1) [61].

Numerous chromatographic studies have also been carried out to investigate possible applications of CyDs as multifunctional drug carriers, discussed in more detail in Chapters 14 and 15. For example the retention of psoralen derivatives was investigated on a C-18 column with HP-β-cyclodextrin as mobile phase additive [62]. Assuming 1:1 stoichiometry, in a 52:48 (v/v) methanol water mixture and temperature −5 °C, the corresponding stability constants were 30, 75, and 40 M^{-1} for the complexes of 8-methoxypsoralen, 5-methoxypsoralen and trimethylpsoralen with HP-β-CyD, respectively.

5.3 Thermodynamics

In spite of very diverse successful practical applications, the mechanism of complexation and the relationship between structure and selectivity are still at best only partly solved and remain open for discussion. Thermodynamic studies could supply some valuable information facilitating an understanding of the physicochemical basis of the complexation processes. The GC modified with CyDs is one of a variety of experimental methods employed in the determination of thermodynamic quantities for the formation of CyD inclusion complexes (see Chapters 8–10). The thermodynamic parameters for separation of the enantiomers were determined for various derivatives of CyDs dissolved in various stationary phases [63–65] or as a "liquid derivatized" form [66]. Interesting observations were made by Armstrong et al. [66]. The authors postulated two different retention mechanisms. The first involved classical formation of the inclusion complex with high thermodynamic values of ΔH, $\Delta\Delta H$, and $\Delta\Delta S$ and a relatively low column capacity and the second loose, probably external, multiple association with the CyD characterized by lower ΔH, $\Delta\Delta H$, and $\Delta\Delta S$ values. The thermodynamic parameters of complexation processes obtained from liquid and gas chromatography measurements are collected in Table 5.2. It is clear from those data that for all the compounds presented the complexation processes are enthalpy-driven since in all cases ΔH is more negative than $T\Delta S$.

The thermodynamic parameters for estriol, estradiol, ethinyloestradiol, and estrone with β- and γ-CyD measured by Sadlej-Sosnowska [49] in methanol–water and acetonitrile–water mixtures indicate that the values of enthalpy changes due to complex formation with β- and γ-CyD are more negative in methanol than in acetonitrile solution. The values of the stability constants are greater in the first solution than in the second one. These facts are interpreted as being due to the more positive enthalpy change connected with the removal of the cosolvent molecules from the CyD cavity, and thus the different competition of methanol and acetonitrile with the steroid for binding with CyD.

References

1 T. Ward, D.W. Armstrong in M. Zief, L.J. Crane (Editors), Chromatographic Chiral Separation, Marcel Dekker, New York, 1988, Ch. 5, 131.
2 S.M. Han, D.W. Armstrong in A.M. Krstulović (Editor), Chiral Separation by HPLC, John Wiley & Sons, New York, 1989, Ch. 10, p. 208.
3 V. Schurig, H.-P. Novotny, Angew. Chem. Int. Ed. Engl., 29 (1990) 939.
4 W.A. König, Gas Chromatographic Enantiomer Separation with Modified Cyclodextrins, Hüthig, Heidelberg, 1992.
5 D.W. Armstrong, A. Alak, K. Bui, W. Demond, T. Ward, T.E. Richl, W.L. Hinze, J. Incl. Phenom. 2 (1984) 533.
6 K. Shimada, T. Oe, C. Kanno, T. Nambara, Anal. Sci. 4 (1988) 377.
7 K. Shimada, T. Masue, K. Takani, T. Nambara, J. Liq. Chromatogr. 11 (1988) 1475.
8 M. Gazdag, G. Szepesi, L. Huszar, J. Chromatogr. 371 (1986) 227.
9 Y. Mizobuchi, M. Tanaka, T. Shono, J. Chromatogr. 194 (1980) 153.
10 J. Mraz, L. Feltl, E. Smolkova-Keulemansova, J. Chromatogr. 286 (1984) 17.
11 E. Smolkova-Keulemansova, E. Neumannova, L. Feltl, J. Chromatogr. 365 (1986) 279.
12 G.L. Reid, III, C.A. Monge, M.T. Wall, D.W. Armstrong, J. Chromatogr. 633 (1993) 135.
13 G.L. Reid, III, M.T. Wall, D.W. Armstrong, J. Chromatogr. 633 (1993) 143.
14 G.L. Reid, III, D.W. Armstrong, J. Microcolumn Separation 6 (1994) 151.
15 D. Sun, J. Chen, W. Lu, X. Zheng, J. Chromatogr. 864 (1999) 293.
16 T. Kościelski, D. Sybilska, J. Jurczak, J. Chromatogr. 280 (1983) 131.
17 T. Kościelski, D. Sybilska, J. Jurczak, J. Chromatogr. 364 (1986) 299.
18 D. Sybilska, T. Kościelski, J. Chromatogr. 261 (1983) 357.
19 T. Kościelski, D. Sybilska, J. Chromatogr. 349 (1985) 3.
20 T. Kościelski, D. Sybilska, J. Lipkowski, A. Miediokritskaja, J. Chromatogr. 351 (1986) 512.
21 D. Sybilska, J. Jurczak, Carbohydrate Research 192 (1989) 243.
22 A.R. Andrews, Z. Wu, A. Zlatkis, Chromatographia 34 (1992) 457.
23 D.W. Armstrong, G.L. Bertrand, K.D. Ward, T.J. Ward, Anal. Chem. 62 (1990) 332.
24 A. Bielejewska, K. Duszczyk, D. Sybilska, "Resolution of some selected steroisomers in RP-HPLC system via cyclodextrin complexation" in Proceedings Natural Product Analysis 1998, 75.
25 R.A. Bielejewska, M. Koźbiał, R. Nowakowski, K. Duszczyk, D. Sybilska, Anal. Chim. Acta 300 (1995) 2011.
26 R. Nowakowski, P.J.P. Cardot, A.W. Coleman, E. Villard, G. Guiochon, Anal. Chem. 67 (1995) 259.
27 A.K. Chatjigakis, Ph.J.P. Cardot, A.W. Coleman, H. Parrot-Lopez, Chromatographia 36 (1993) 174.
28 I. Clarot, D. Clédat, L. Boulkanz, E. Assidjo, T. Chianéa, P. Cardot, J. Chromatogr. Sci. 38 (2000) 38.
29 A. Kwaterczak, A. Bielejewska, Anal. Chim. Acta 537 (2005) 41.
30 K. Fromming, J. Szejtli, Cyclo-

dextrins in Pharmacy. Kluwer Academic, Dordrecht (1994).
31 J. Szejtli, Cyclodextrins and their inclusion complexes. Akademiai Kiado, Budapest (1982).
32 K. Harata, K. Uekama, M. Otagiri, F. Hirayama, Bull. Chem. Soc. Jpn. 60 (1987) 497.
33 K. Harata, In Minutes, 5th International Symposium on Cyclodextrins: Paris, 1991.
34 D. Sybilska, A. Bielejewska, R. Nowakowski, K. Duszczyk, J. Chromatogr. 625 (1992) 349.
35 A. Bielejewska, R. Nowakowski, K. Duszczyk, D. Sybilska, J. Chromatogr. A 840 (1999) 159.
36 K. Uekama, F. Hirayama, S. Nasau, N. Matsuo, T. Irie, Chem. Pharm. Bull 26 (1978) 3477.
37 K. Fujimura, T. Ueda, M. Kitagawa, H. Takayranagi, T. Ando, Anal. Chem. 58 (1986) 2668.
38 D.W. Armstrong, F. Nome, L.A. Spino, T.D. Golden, J. Am. Chem. Soc. 108 (1986) 1418.
39 C. Horwath, W. Melander, J. Melander, A. Nahum, J. Chromatogr. 186 (1980) 1416.
40 D. Sybilska, J. Żukowski in A.M. Krstulovic (Editor), Chiral Separation, J. Wiley & Sons, New York 1989, Ch. 7, p. 147.
41 R. Nowakowski, A. Bielejewska, K. Duszczyk, D. Sybilska, J. Chromatogr. A 782 (1997) 1.
42 R.M. Mohseni, R.J. Hurtubise, J. Chromatogr. 499 (1990) 395.
43 A. Bielejewska, B. Łukasik, K. Duszczyk, D. Sybilska, Chem. Anal. (Warsaw) 47 (2002) 419.
44 N. Morin, Y.C. Guillaume, E. Peyrin, J.C. Rouland, J. Chromatogr. A 808 (1998) 51.
45 N. Morin, S. Cornet, C. Guinchard, J.C. Rouland, Y.C. Guillaume, J. Liq. Chromatogr. Relat. Technol. 23 (2000) 727.
46 H. Dodziuk, O.M. Demchuk, A. Bielejewska, W. Koźmiński, G. Dolgonos, Supramolecular Chem. 16 (2004) 287.
47 D. Sybilska, M. Asztemborska, A. Bielejewska, J. Kowalczyk, H. Dodziuk, K. Duszczyk, H. Lamparczyk, P. Zarzycki, Chromatographia 35 (9, 12) (1993) 637.
48 J.P. Hummel, W.J. Dreyer, Biochem. Biophs. Acta 63 (1962) 352.
49 N. Sadlej-Sosnowska, J. Chromatogr. A 728 (1–2) (1996) 89–95.
50 N. Sadlej-Sosnowska, J. Inclusion Phenom. Mol. Recognition Chem. 27 (1997) 31.
51 K.G. Flood, E.R. Reynolds, N.H. Snow, J. Chromatogr. 913 (2001) 261.
52 K.A. Connors, Chem. Rev. 97 (1997) 1325.
53 A. Bielejewska, K. Duszczyk, D. Sybilska, J. Chromatogr. A 931 (2001) 81–93.
54 J. Zukowski, M. Pawlowska, HRC 16 (1993) 505.
55 Y. Saito, K. Yoshihara, I. Tanemura, H. Ueda, T. Sato, Chem. Pharm. Bull. 45 (1997) 1711.
56 K. Yoshihara, I. Tanemura, Y. Saito, H. Ueda, T. Sato, Chem. Pharm. Bull. 45 (1997) 2076.
57 I. Tanemura, Y. Saito, H. Ueda, T. Sato, Chem. Pharm. Bull. 46 (1998) 540.
58 P. Cardinaël, V. Peulon-Agasse, G. Coquerel, Y. Combret, J.C. Combret, J. Sep. Sci. 24 (2001) 109.
59 M. Asztemborska, R. Nowakowski, D. Sybilska, J. Chromatogr. A 902 (2000) 381.
60 M. Asztemborska, D. Sybilska, R. Nowakowski, J. Chromatogr. A 1010 (2003) 233.
61 M. Skórka, M. Asztemborska, J. Żukowski, J. Chromatogr. A 1078 (2005) 136.
62 L. Ismaili, C. Andre, L. Nicod, J.L. Mozer, J. Millet, B. Refouvelet, S. Makki, J.F. Robert, A. Xicluna, Y.C. Guillaume, J. Liq. Chromatogr. Relat. Technol. 26 (6) (2003) 871.
63 M. Jung, D. Schmalzing, V. Schurig, J. Chromatogr. 552 (1991) 43.
64 V. Schurig, M. Juza, J. Chromatogr. A 757 (1997) 119.
65 V. Schurig, R. Schmidt, J. Chromatogr. A 1000 (2003) 311.
66 A. Berthod, W. Li, D.W. Armstrong, Anal. Chem. 64 (1992) 873.

6
The Application of Cyclodextrins for Enantioseparations

Bezhan Chankvetadze

6.1
Introduction

Cyclodextrins (CyDs) are among the most remarkable macrocyclic molecules, with significant theoretical and practical impacts in chemistry. CyDs are produced on a multi-tonne scale and widely used in chemical, pharmaceutical, food, and other technologies as enzyme mimics, enantioselective catalysts, drug carriers, and odor and taste-masking compounds [1]. Another important field of application of cyclodextrins is separation science [2]. At present, CyDs are used as chiral selectors in gas chromatography (GC), high-performance liquid chromatography (HPLC), supercritical fluid chromatography (SFC), capillary electrophoresis (CE), capillary electrochromatography (CEC) and most recently in lab-on-chip enantioseparations [3]. It must be noted that among currently applied chiral selectors only CyDs are effectively used in all enantioseparation techniques [2]. The discovery of CyDs as probably the most universal chiral selectors contributed enormously to the maturation of instrumental enantioseparation techniques. *Vice versa*, it can be noted that during the last few years enantioseparation techniques have advanced cyclodextrin chemistry because recent developments, especially in CE, allow a better understanding of inclusion complex formation and the chiral recognition mechanisms by CyDs [4].

In what follows, some properties of cyclodextrins are mentioned that have a major impact on their widespread application in enantioseparation techniques.

Schardinger made an assumption regarding the cyclic structure of two types of dextrins produced by "thermophilic" bacteria when digesting starch 100 years ago [5]. Thirty-five years after this publication, Freudenberg and Mayer-Delius confirmed the assumption made by Schardinger and noted that the Schardinger dextrins are cyclic oligosaccharides composed solely of D-glucosyl residues bonded by α-(1,4)-glucosidic linkages [6]. The macrocyclic structure of CyDs is one of the major contributors to their ability to form intermolecular complexes. This property of CyDs was already known to Schardinger [7]. Freudenberg et al. were the first to assume that these complexes are of the inclusion type [8]. The first direct evidence for molecular inclusion by CyDs in the solid state was provided by Hybl et al. using

Cyclodextrins and Their Complexes. Edited by Helena Dodziuk

ISBN: 3-527-31280-3

X-ray crystallography [9]. Demarco and Thakkar demonstrated inclusion complex formation by CyDs in solution using ^{1}H NMR spectroscopy [10]. Another remarkable property of CyDs, i.e. their chiral recognition ability by complex formation, was discovered in the 1950s by Cramer [11]. Most applications of CyDs are based primarily on two of the aforementioned properties, namely complex formation and chiral recognition ability. Together with these two important properties some other properties of these molecules must be mentioned, which are of key importance either in all or in selected separation techniques. Thus, for example, the presence of free hydroxyl groups on the outer surface of CyDs allows a variety of derivatives (ionic or nonionic, hydrophilic or lipophilic, etc.) to be synthesized, as well as for linkage of CyDs to various surfaces. This property is of major importance for CyD applications in all kinds of enantioseparation techniques. In addition, hydroxyl groups are responsible for the solubility of CyDs in aqueous media that is of key importance for their application as chiral selectors in CE. The transparency of CyDs in the UV-Vis range is also an advantage in their application in CE. Their availability with various cavity dimensions and the low cost of CyDs are also significant. The nontoxic character of CyDs together with their applicability in pharmaceutical, food, and cosmetic sciences is also important for their laboratory and environmentally friendly analytical applications. Some special properties of CyDs relevant to one or other applications are addressed in the following sections.

6.2 Gas Chromatography

Gas chromatography (GC) represents historically the first instrumental technique used for enantioseparations. The first GC enantioseparation was reported in 1966 by Gil-Av et al. [12]. High efficiency, speed, and sensitivity are advantages of this technique. GC can be performed with both packed and capillary columns. In addition, quite sophisticated hyphenated and multidimensional versions are available as well as various special, universal, and high-sensitivity detectors.

Chiral GC is successfully used for the precise determination of the enantiomeric composition of volatile natural compounds, authenticity control of essential oils, flavors, fragrances, some chiral research chemicals, auxiliaries, a few volatile drugs, drug metabolites, pesticides, fungicides, herbicides, pheromones, etc. [13].

Cyclodextrins have been used in noninstrumental separation techniques for obtaining enantiomerically enriched samples by several groups [14–16].

Cyclodextrins were introduced as chiral selectors in GC in the 1980s and rapidly became established as the most powerful and versatile materials for GC enantioseparations. The first GC enantioseparation employing CyD as a chiral selector was reported by Sybilska's group in 1983 [17]. A packed column was used in this study with a rather short lifetime and rather low peak efficiency. Later Venema and Tolsma reported enantioseparation with a capillary column which was modified by heptakis(2,3,6-tri-*O*-methyl)-β-CyD (TM-β-CyD) [18]. This kind of capillary column could be employed at temperatures that exceeded the melting point of CyD. Schurig and co-workers provided the solution for the problem of the high melting point of CyDs by dissolving CyDs in moderately polar polysiloxanes [19]. The combination of the enantio- and chemoselectivities of CyDs with the chemoselectivity of polysiloxanes allowed chiral GC columns with great enantiomer-resolving potential to be prepared. Many commercially available chiral GC columns (ChirasilDex series) are now prepared based on this technology [13]. König and co-workers found that per-*n*-pentylated CyDs are liquid at room temperature [20]. Commercially available chiral GC columns of the Lipodex series are prepared based on per-*n*-pentylated CyDs. Capillary columns prepared by coating their inner surface with rather polar derivatives of CyDs were introduced by Armstrong and co-workers as very useful chiral GC columns [21]. A significant contribution to the development of new CyD-based chiral columns for GC was made by Mosandl [22] and Bicchi and co-workers [23].

The number of chiral GC separations summarized in appropriate databases (for instance, Chirbase GC) currently exceeds 20 000. CyD-based columns are used in most GC separations. Thus, a comprehensive overview of CyD-based GC enantioseparations would be beyond the scope of this short section, but a few important trends from the author's personal viewpoint are mentioned below and exemplified with separations taken from recent literature.

As shown in Fig. 6.1 very rapid enantioseparations can be performed with a short CyD-based chiral column in GC. The enantiomers of the chiral inhalation

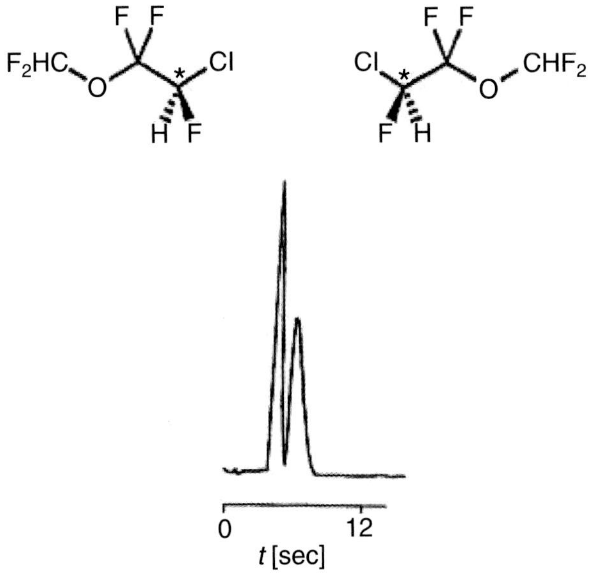

Fig. 6.1. Enantioseparation of enflurane with very short analysis time. (Reproduced from Ref. [24] with permission.)

anesthetic agent enflurane were resolved to the baseline in just a few seconds with a short Chirasil-γ-Dex column [24].

Interesting results on mechanistic studies, calculation of thermodynamic characteristics of enantioseparation and enantiomerization processes, assignment of absolute configuration, enantiomer-labeling methodology, etc. based on CyDs as chiral GC selectors are summarized in Ref. [13]. As absolutely correctly mentioned in this excellent and authoritative overview, the multimodal recognition process, which may involve *inter alia* hydrogen bonding, dispersion forces, dipole–dipole interactions, and electrostatic and hydrophobic interactions make it extremely difficult to intuitively predict or compute the outcome of the CyD–analyte interactions responsible for enantioseparations in GC. The author of Ref. [13] considers the enantioseparation with per-*n*-pentylated amylose as an indication that inclusion complex formation may not be necessary for the enantioselective recognitions responsible for GC separations. We share the idea that inclusion complex formation may not be a prerequisite for enantioselective recognition of chiral analytes by CyDs but the example with per-*n*-pentylated amylose may not be an unambiguous indication of this. This is due the fact that amylose and its derivatives may form helical structures mimicking the cavity of CyDs where the analytes may be included.

Based on the fact that owing to the high resolving power of GC, Gibbs free energy differences as low as 0.02 kcal mol^{-1} corresponding to a separation factor of only 1.02 is sufficient for enantioseparations, the author of Ref. [13] correctly considers correlations between molecular modeling calculations and experimental chiral GC results useless. This does not mean, however, that these methods should not be used in combination. We support the idea of applying chiral GC and other high-efficiency separation techniques for examining the results of molecular modeling calculations. This will help to further advance and refine the calculation techniques (see, however Chapter 11 on the CyD modelling in which the opposite opinion is expressed.)

As an example of interesting applications of chiral GC, a few preparative scale separations must be mentioned. The large separation factors observed for the chiral inhalation anesthetic agents enflurane, isoflurane, and desflurane allowed preparative-scale separation of the enantiomers required for biomedical studies [25]. Later, one of these compounds, enflurane, was also resolved on a preparative scale with the first enantioselective GC simulating moving bed (SMB) technique [26]. There is no problem of removal of a mobile phase in preparative scale GC separations, which is the advantage of this technique compared to liquid-phase enantioseparation techniques.

Prerequisites for the use of GC, such as volatility and thermal stability of the analytes, restrict exclusive application of this technique. Thus, this very powerful technique for enantioselective analysis of many natural compounds (essential oils, terpenoids, food components, etc.) may fail for most pharmaceutical and biomedical applications, where the samples are polar and nonvolatile or too thermolabile.

6.3
High-performance Liquid Chromatography and Related Techniques

CyDs have been used as major chiral mobile phase additives (CMPAs) for enantioseparations in HPLC. The first application of β-CyD as a CMPA in combination with an achiral reversed-phase material for HPLC enantioseparations was reported by Sybilska and co-workers in 1982 [27]. These authors could achieve partial resolution of the enantiomers of mandelic acid and derivatives. The CMPA method played an important role in HPLC enantioseparations before the development of effective chiral stationary phases (CSPs) but is now rarely used. The major disadvantage of this technique, together with difficulties associated with the isolation of resolved enantiomers, is the rather large consumption of chiral selector.

The first successful CSP for HPLC enantioseparations containing CyDs were developed by Armstrong and co-workers [28, 29]. Initially, CyD-based CPSs were developed for reversed-phase chromatography while inclusion complex formation was considered as a prerequisite for enantiorecognition. The latter is commonly favored in aqueous medium. Reversed-phase chromatography offers certain advantages from the viewpoint of bioanalysis, because biofluids are more compatible with it. Moreover, biomaterials can be directly injected without any special sample treatment on some kinds of CyD-based chiral columns [30, 31]. Together with native CyDs, the derivatives of CyDs also show interesting enantiomer-resolving abilities in HPLC. Derivatization is very useful when CyDs are used as chiral mobile phase additives because many derivatives of CyDs are characterized by better solubility in aqueous solution compared to native CyDs. This is particularly true for β-CyD and its derivatives.

Some delay in the development of CyD-based CSPs for normal-phase HPLC separations was caused by the suspicion that the more nonpolar component of the mobile phase would occupy the CyD cavity thereby blocking inclusion complex formation with concomitant loss of chiral recognition. To circumvent this problem, several derivatized CyD-based CSPs for HPLC were synthesized in the 1990s by Armstrong and co-workers [32–33] and Stalcup [34]. The derivatization of CyDs was performed in order to introduce into the structure new sites for π–π, dipole–dipole, or steric interactions that are unavailable with the native CyDs [34]. One of the interesting derivatives of β-CyD is naphthylethylcarbamoylated β-CyD synthesized by Armstrong and co-workers [34]. Incorporation of a naphthylethylcarbamate (NEC) moiety onto the CyD introduces additional stereogenic centers. It has been shown that for some analytes the NEC substituent plays a critical role in the enantioseparation, and the elution order of enantiomers was defined by the configuration of the substituent. Together with its theoretical interest this finding is of certain practical importance and it may allow a desirable adjustment of the enantiomer migration order, which represents a particular problem for the natural chiral selectors commonly available in a single stereochemical configuration similar to cyclodextrins. NEC-substituted β-CyD appears to be a useful CSP for normal, as well as for reversed-phase conditions. The enantiomer-resolving abilities of this material in different modes appear to be complementary to each other. Thus, for example, the enantiomers of many analytes not resolvable with β-CyD or with NEC-β-CyD under normal-phase conditions could be resolved under reversed-phase conditions. CyD-based CSPs can be also successfully used in combination with recently evolving so-called polar organic mobile phases [34]. The aforementioned NEC-β-CyD may be the first effective multimodal CSP for liquid chromatography.

CyD-based columns for HPLC enantioseparations are well established and several tens of these columns are commercially available at present. Although highly suitable for analytical-scale enantioseparations these columns hardly compete with polysaccharide and macrocyclic antibiotic-based CSPs for preparative-scale enantioseparations. More research is needed in this area.

Counter-current chromatography (CCC) and centrifugal partition chromatography (CPC) although among the liquid-phase separation techniques, are a long way from HPLC in their separation principle, potential, and characteristics. These techniques are characterized by low efficiency (in the range of 1000) and are more suitable for preparative-scale separations. Owing to the low separation efficiency, a chiral selector used in this technique needs to exhibit high enantioselectivity. Not many chiral selectors meet this requirement, which makes progress in chiral CCC and CPC rather slow [35]. Breinholt et al. reported the application of sulfated β-CyD for enantioseparation of 7-demethylormeloxifene using CCC [36]. Not many other examples occur in the literature on the use of CyDs in CCC and related techniques.

6.4 Super (Sub)-critical Fluid Chromatography

Supercritical (SFC), sub-critical (SubFC), and enhanced fluidity chromatography are commonly used terms to describe the use of mobile phases operating near or above the critical parameters. Chiral SFC started in 1985 [37] and is currently performed in packed columns, open tubular capillary columns, and packed capillary columns [38]. Operating conditions in SFC are typically mild, affording long column lifetime and highly reproducible separations. The enhanced fluidity of the mobile phase in SFC may allow higher plate numbers than HPLC. One important technical advantage over HPLC, especially for preparative-scale separations is that there is no problem with removal of the mobile phase in SFC. For the latter, preparative SFC instruments as well as packed column supercritical fluid simulated moving bed chromatography (SF-SMB) instruments are available.

The columns used in SFC may in principle be the same as in HPLC. Some kinds of detectors impractical for HPLC (for example, flame-ionization) can be used in SFC. Application of Fourier-transform infrared spectrometry and evaporative light scattering detectors may also appear easier in SFC than in HPLC.

Although CyD-based CSPs do not dominate in SFC enantioseparations with packed columns, there are some applications reported also with this kind of material [38].

The first chiral separation with open-tubular columns in SFC was published by Roder et al. in 1987 [39]. Schurig and co-workers [40] linked permethylated β-CyD via an octamethylene spacer to polydimethylsiloxane forming a chiral polymer Chirasil-Dex. The polymer was immobilized on the inner surface of fused-silica capillaries and the capillaries were used for so-called unified chromatography including GC, LC, SFC, and capillary electrochromatography (CEC).

SFC offers some advantages over HPLC for enantioseparations both on the analytical scale (wider choice of available detectors, higher peak efficiency) and on the preparative scale (easy removal of the mobile phase). However, the technique has not yet become a serious competitor to HPLC for either analytical or preparative scale enantioseparations.

6.5 Capillary Electrophoresis

Capillary electrophoresis (CE) is one of the most promising microanalytical separation techniques for enantiomers. Chiral CE which has had 20 years of development [41] offers some interesting advantages warranting further expansion of this technique in the field of analytical enantioseparation.

As already mentioned, CE is a microanalytical separation technique and possesses advantages such as low sample and reagent consumption, environmental friendliness, low costs, etc. Together with these technical advantages, which are common to all microanalytical techniques, CE also offers some unique advantages owing to the original sample migration and separation mechanisms in this technique.

At the start it is important to note some special properties of electrokinetic migration. One rather early recognized feature of electrokinetic flow is its plug-like profile. This profile is responsible for the higher plate numbers observed in CE compared to pressure-driven techniques. High peak efficiency is of crucial importance also for chiral CE, because it allows the observation of enantioseparations in those cases where other techniques, commonly with lower peak efficiencies, do not provide an indication of chiral recognition. The thermodynamic selectivity of enantioselective selector–selectand interactions in the range 1.01 is sufficient for observation of baseline-resolved peaks in CE, whereas this number is above 1.02 in GC and even higher in other instrumental enantioseparation techniques [42, 43].

The second, less-addressed advantage of electrokinetic migration is that it allows counter flows in the separation chamber [43]. Many special features of chiral CE compared to pressure-driven techniques are associated with this property of electrokinetically driven flow. The most significant differences between pressure-driven and electrokinetically driven separations can be briefly summarized as follows:

1. It is feasible in chiral CE but not in chromatographic techniques that the selectivity of an enantioseparation exceeds the thermodynamic selectivity of chiral recognition and approaches an infinitely high value [42–44].

2. It is possible in chiral CE to adjust the enantiomer migration order without changing the affinity pattern of the analyte enantiomers towards a chiral selector. This is impossible in chromatographic techniques unless using a chiral selector as a CMPA [42].
3. The most striking difference between these two techniques seems to be the fact that CE allows, in principle, the separation of enantiomers in the case when the equilibrium constants of both enantiomers with the chiral selector are equal [42, 43, 45].

CyDs are among the most powerful and widely used chiral selectors for CE [42, 43, 46]. The very first applications of CyDs as chiral selectors in this technique were reported in the papers by Smolkova-Keulemansova et al. [47], Fanali [48], Karger and co-workers [49] and Terabe [50]. Interestingly, these authors used CyDs in different modes of CE. Smolkova-Keulemansova's group used CyDs as chiral selectors in capillary isotachophoresis (CITP), Fanali in so-called free zone capillary electrophoresis (CZE), Karger's group in capillary gel electrophoresis (CGE) and Terabe in capillary electrokinetic chromatography (CEKC). It must be noted that although the techniques were different in all these studies, the principles of enantioseparation were the same in all the techniques. This means that the mechanism of enantioseparation was an enantioselective interaction of chiral analytes with CyDs (separation principle of CEKC) and not a separation principle of CITP, CZE, or CGE.

CyDs offer several advantages as chiral selectors for CE. The most important is that these macrocyclic molecules possess a quite universal chiral recognition ability for many different classes of organic compounds. In addition, CyDs are water soluble, transparent in the UV range, relatively inexpensive, and nontoxic. All of these contribute significantly to the status of CyDs as one of the most useful chiral selectors in CE. Free hydroxyl groups on the outer rim of CyDs offers various derivatization possibilities for introduction of nonionic and ionic groups into the structure of CyDs. Several CyD derivatives developed to be used in enantioseparations are presented in Chapter 2.

CyDs can be used alone as chiral selectors in CE or in combination with other achiral and chiral buffer additives. Native CyDs and their neutral and charged derivatives are currently available as chiral selectors for CE. Derivatized CyDs can be randomly substituted multicomponent mixtures or a selectively derivatized single-component CyD. Randomly substituted mixtures offer the advantages of being readily available and less expensive but are less suitable for mechanistic studies and method validation. Single-component mixtures are more expensive but better suited to mechanistic studies and method validation.

Chiral CE with CyDs as chiral selectors can be used to evaluate the enantiomeric purity of chiral chemicals, agrochemicals, food additives, pharmaceuticals, and drug formulations as well as for investigations of endogenous compounds and enantioselective drug metabolism and pharmacokinetics [51, 52]. A few selected examples of applications of chiral CE with CyDs as chiral selectors as well as of mechanistic studies on CyD-analyte interactions are mentioned below.

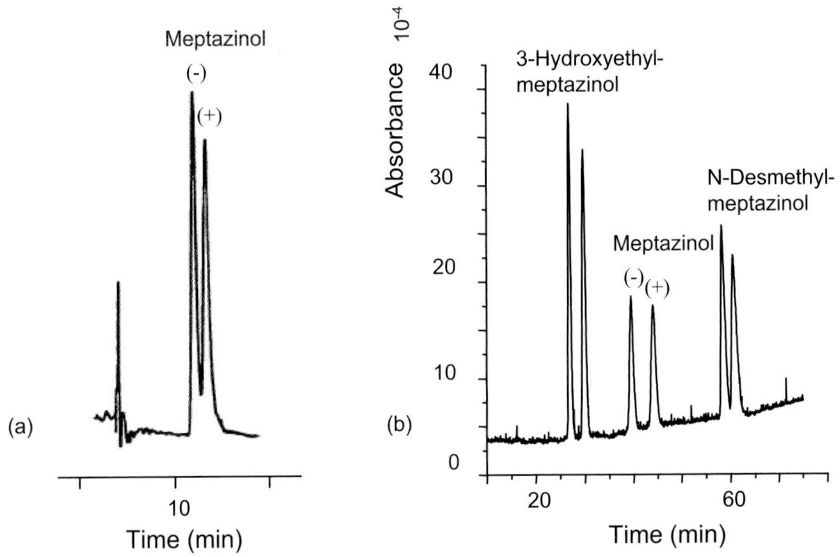

Fig. 6.2. Separation of meptazinol enantiomers by HPLC with a β-CyD-based column (a) and simultaneous separation of meptazinol and metabolites by CE with β-CyD as a chiral selector (b).

CE offers advantages such as the simultaneous detection of a parent chiral drug and its phase 1 and charged phase 2 metabolites [43]. Sometimes chiral selectors ineffective as components of HPLC chiral columns may appear effective for enantioseparations of the same compounds in CE. For example, the enantiomers of the chiral opioid drug meptazinol can scarcely be resolved using β-CyD-based HPLC column in HPLC and the enantiomers of its metabolites cannot be resolved. Application of the same β-CyD in CE allows simultaneous baseline enantioseparation of meptazinol and its metabolites (Fig. 6.2) [53]. Although CE has been considered as a separation method for charged analytes, owing to its unique separation principle chiral CE is also useful for neutral analytes. This is illustrated below with the example of thalidomide and its metabolites. Thalidomide (TD) is an active ingredient of the well-known chiral drug Contergan, which was introduced into the clinical practice in early the 1960s but soon withdrawn because of its severe teratogenic effects. Since the late 1990s there has been revived interest in this compound due to some unique pharmacological effects it shows against leprosies and some kinds of cancer. TD was recommended for clinical use again in 1998 [54]. The reason for the teratogenicity of TD is not yet known, and the simultaneous enantioselective determination of TD, which is a neutral compound, and its (also neutral) phase 1 metabolite was a challenging task. A combination of neutral and charged CyDs allowed the simultaneous enantioseparation mentioned above, as shown in Fig. 6.3 [55].

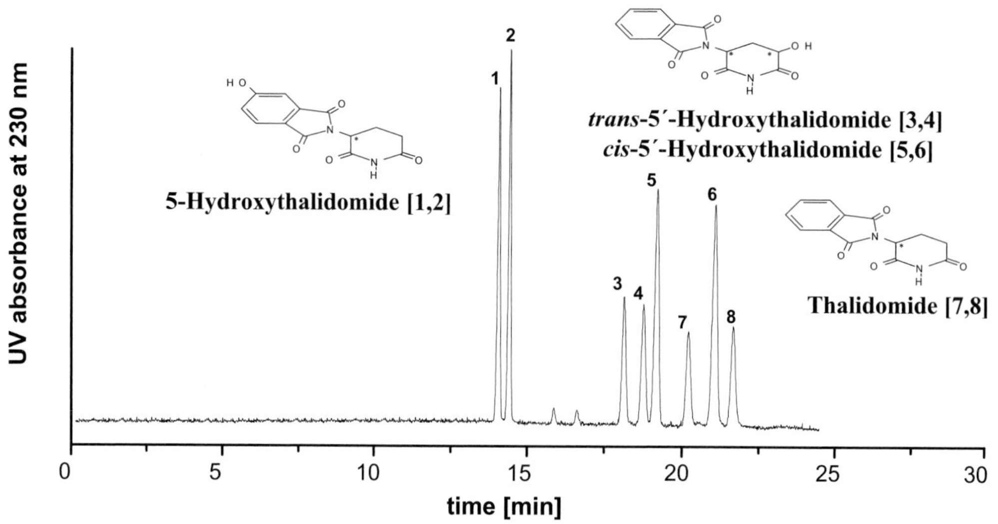

Fig. 6.3. Simultaneous enantioseparation of thalidomide and its metabolites by CE. (Reproduced from Ref. [55] with permission.)

Together with the aforementioned applications, chiral CE also offers interesting advantages in comparison with other instrumental techniques for better understanding of enantioselective analyte–CyD binding mechanisms. The major advantages of CE compared to chromatographic techniques from the viewpoint of molecular recognition science are the following [56]:

1. CE allows very fast screening of analyte–CyD interactions in order to find the most interesting pairs from the huge number of chiral analytes and numerous CyDs. There is no other instrumental separation or nonseparation technique that can compete with CE from this point of view.
2. The high peak efficiency in CE allows one to observe (enantio)selective effects in selector–selectand interactions which are invisible by other (separation) techniques.
3. A small thermodynamic selectivity of recognition can be transformed into a high separation factor in CE.
4. CE is more flexible than chromatographic techniques from the viewpoint of the adjustment of the (enantio)separation factor.

The reversal of the affinity pattern of enantiomers towards a chiral selector is the most dramatic change that may take place due to any chemical or structural modification of a chiral selector. The screening of the affinity pattern of a wide range of chiral analytes towards CyD-type hosts using CE indicated that the affinity pattern

Table 6.1. Enantiomer affinity pattern of selected chiral analytes towards native CyDs having different cavity size (Reproduced with permission from Ref. 56).

Analyte	Chiral selector and the first migrating enantiomer		
	α-CyD	β-CyD	γ-CyD
Aminoglutethimide	(*R*)	(*S*)	(*R*)
Ephedrine	(−)	(+)	No separation
Ketamine	(*S*)	(*R*)	(*R*)
Ketoprofen	(*S*)	(*R*)	(*R*)
Mefloquine	(−)	(−)	(−)
Metharaminol	(+)	(+)	(+)
Norephedrine	(+)	(−)	No separation
Promethazine	(−)	(−)	(−)
Tetramisole	(*S*)	(*R*)	(*R*)
AlaPheOMe	(*R*,*R*)	(*S*,*S*)	(*S*,*S*)
Ala-Tyr	(*R*,*R*)	(*S*,*S*)	(*S*,*S*)
Asp-PheOMe	(*R*,*R*)	(*S*,*S*)	(*S*,*S*)

may change depending on the type and position of the substituent on the CyD rim and even depending on the cavity size of the CyD. Some recent examples are shown in Tables 6.1–6.3 [56].

In general, for a given pair of an analyte and a CyD, one may expect that the

Table 6.2. Enantiomer affinity pattern of selected chiral analytes towards native and selectively methylated CyDs. (Reproduced with permission from Ref. 56).

Analyte	Chiral selector and the first migrating enantiomer		
	β-CyD	DM-β-CyD	TM-β-CyD
Aminoglutethimide	(*S*)	(*S*)	(*R*)
Brompheniramine	(−)	No separation	(+)
Chlorpheniramine	(−)	(+)	(+)
Dimethindene	(*S*)	(*R*)	(*R*)
Ephedrine	(+)	(−)	(−)
Ketoprofen	(*R*)	No separation	(*S*)
Metharaminol	(+)	(−)	(+)
Tetramisole	(*R*)	(*R*)	(*S*)
Verapamil	(−)	(−)	(+)

Table 6.3. Enantiomer affinity pattern of some chiral analytes towards β-CyD, HDA-β-CyD and randomly acetylated β-CyD. (Reproduced with permission from Ref. 56).

Chiral analyte	β-CyD	HDA-β-CyD	Acetylated β-CyD
Aminoglutethimide	(−)	No separation	(+)
Clenbuterol	(−)	(+)	(−)
DNS-phenylalanine	L	D	not studied
Ephedrine	(+)	(−)	(+)
Mefloquine	(−)	(+)	(−)
Metaraminol	(+)	(+)	(−)
Tetramisole	(+)	(−)	(+)

enantioseparation power in CE will be higher than the enantiorecognition ability in NMR spectrometry. In particular cases, however, NMR spectroscopy may provide an indication for chiral recognition for those selector–selectand pairs that have been considered to be unsuccessful based on the CE experiment alone. For instance, native β-CyD has been suggested to be unsuitable as a chiral selector for the enantioseparation of the cationic form of the chiral cholinergic drug aminoglutethimide (AGT) in contrast to α- and γ-CyD which allow baseline enantioseparations of AGT [57, 58]. In contrast to the CE data, NMR studies indicated the most pronounced interactions between AGT and β-CyD among the three native CyDs (Fig. 6.4) [59]. Together with the signal splitting pattern due to the nonequivalence of the complexation-induced chemical shifts (CIS) of AGT protons, interesting effects were also observed for CyD protons. In particular, the resonance signals due to the H-5 protons, which are located inside the cavity close to the narrower primary ring of the CyD, were strongly shifted upfield in the case of β-CyD whereas only moderate effects were observed in the case of α- and γ-CyD. These data were also supported by electrospray ionization mass spectrometry (ESI/MS) studies on the comparative affinity of AGT enantiomers towards these CyDs [59]. Careful optimization of the separation in CE also allowed the resolution of the enantiomers of AGT with β-CyD. The migration times in the presence of β-CyD were the longest, which also indicates the strongest interaction with this CyD. In addition, the enantiomer migration order with β-CyD was opposite to that with the two other native CyDs (Fig. 6.5) [59]. Thus, in this particular case NMR and ESI/MS studies allowed optimization of the enantioseparation in CE. The combination of these techniques revealed an example of the opposite affinity of the AGT enantiomers towards native CyDs. Examples of an affinity reversal of enantiomers depending solely on the size of the CyD cavity are rather rare (some examples are shown in Tables 6.1–6.3) [56].

The structural reasons for the above-mentioned opposite affinity of the enantiomers of AGT towards native β- and γ-CyD are reported in Ref. [59]. The nuclear Overhauser effect (NOE) data shown in Fig. 6.6 allow us to deduce the structure of

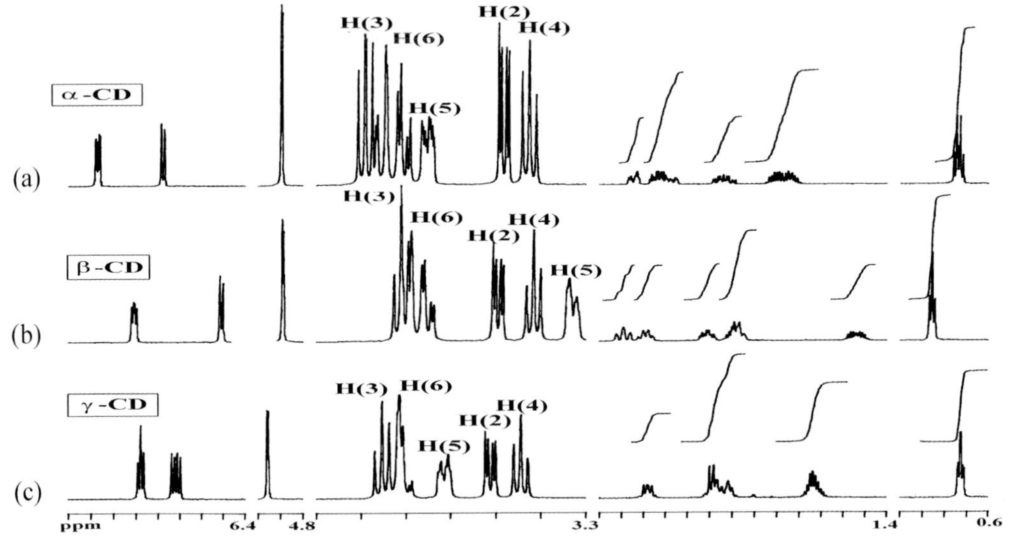

Fig. 6.4. The 600-MhZ ^{1}H NMR spectra of (±)-AGT in the presence of two equivalents of α-CyD (a), β-CyD (b), and γ-CyD (c). (Reproduced with permission from Ref. [59].)

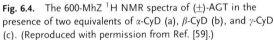

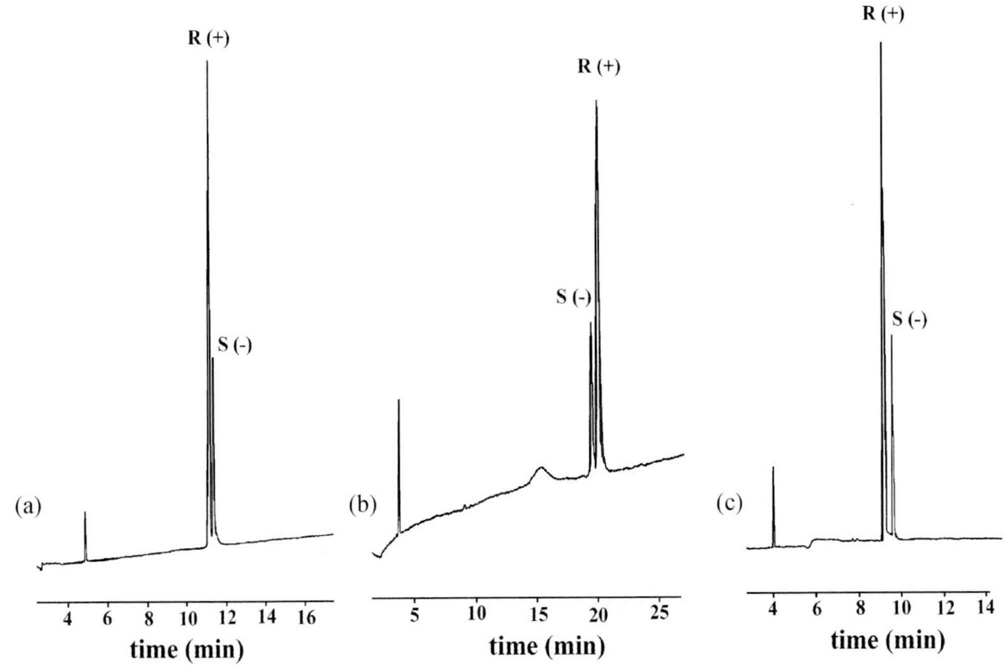

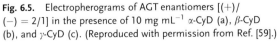

Fig. 6.5. Electropherograms of AGT enantiomers [(+)/(−) = 2/1] in the presence of 10 mg mL^{-1} α-CyD (a), β-CyD (b), and γ-CyD (c). (Reproduced with permission from Ref. [59].)

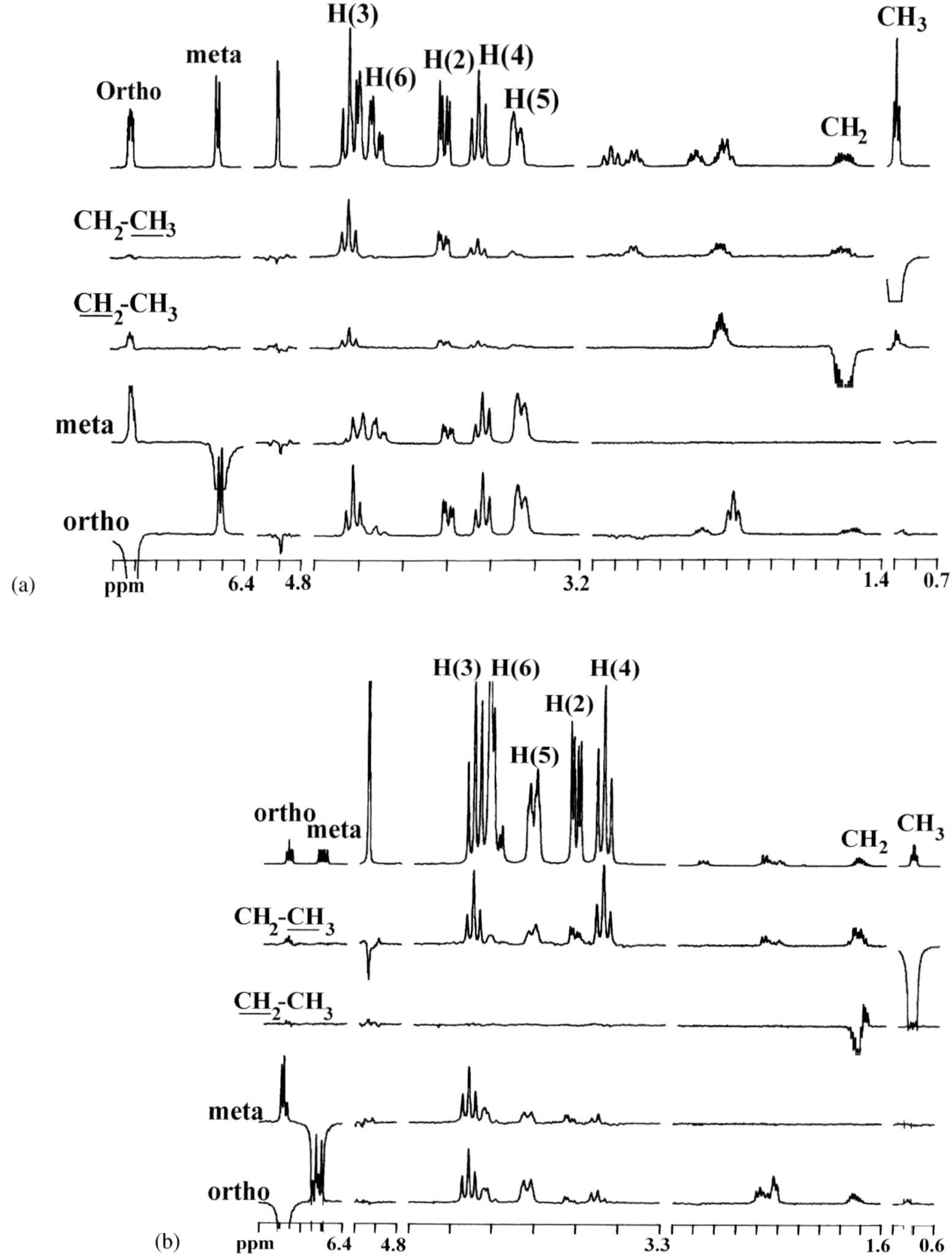

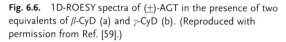

Fig. 6.6. 1D-ROESY spectra of (±)-AGT in the presence of two equivalents of β-CyD (a) and γ-CyD (b). (Reproduced with permission from Ref. [59].)

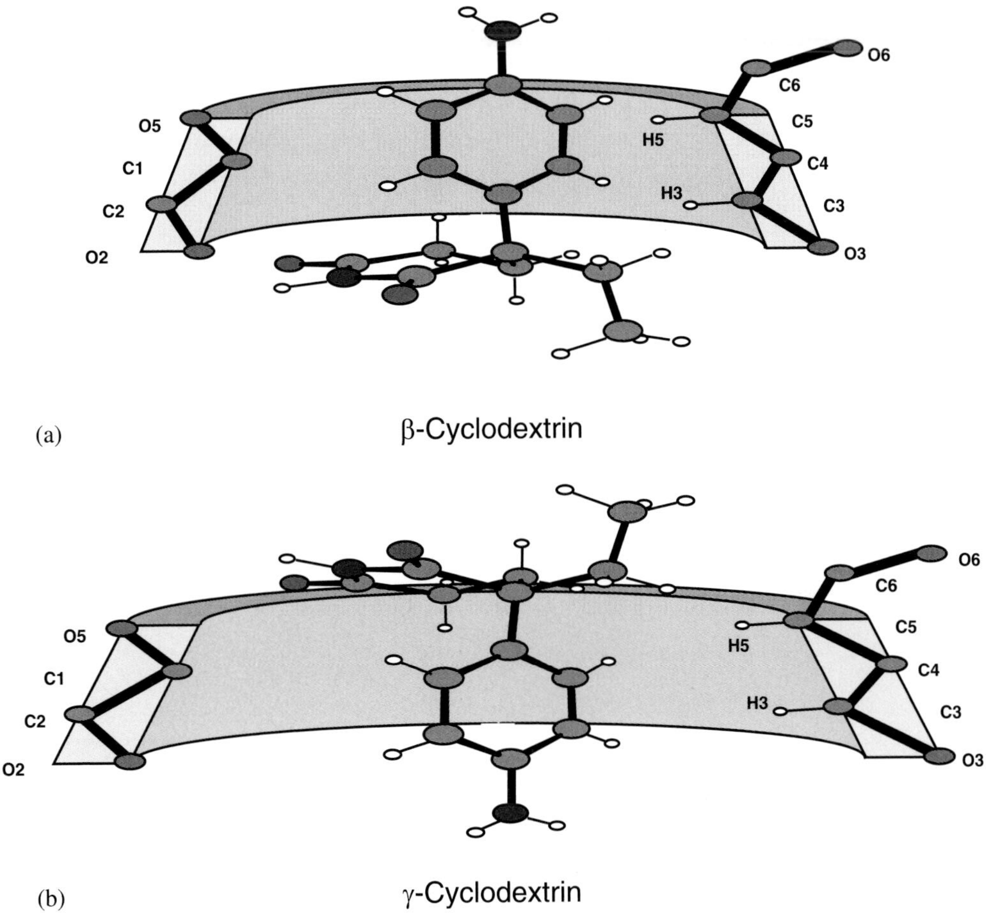

Fig. 6.7. Structure of AGT complexes with β-CyD (a) and γ-CyD (b). (Reproduced with permission from Ref. [59].)

the complexes depicted in Fig. 6.7. Thus, by selective saturation of the aromatic protons in the *ortho* position of ($\pm$)-AGT equally strong intermolecular NOE-effects were observed for both H-3 and H-5 protons of β-CyD (Fig. 6.6a). However, by irradiation of the aromatic protons in the *meta* position only a minor effect was observed for the H-3 protons of β-CyD and the NOE-effect appeared instead at the H-6 protons. These data support a deep inclusion of the *p*-aminophenyl moiety of AGT into the cavity of β-CyD entering it from the wider secondary side (Fig. 6.7a). The deep inclusion of the aromatic moiety of the AGT molecule into the cavity of β-CyD on the secondary side is supported also by a significant NOE effect observed between the H-3 protons of CyD and the ethyl moiety of AGT. Rather strong "NOE-like" effects observed on the external CyD protons in this experiment make it questionable whether the structure represented in Fig. 6.7a is the only possible structural element of this complex or if the alternative structures are also present.

In contrast to the AGT/β-CyD complex, the NOE effect decreased for the H-5 protons and remained almost unchanged for H-3 protons when irradiating the protons in the *meta* position instead of the protons in the *ortho* position of the aromatic ring of AGT in the ($\pm$)-AGT/γ-CyD complex (Fig. 6.6b). These data support a complex formation from the narrower primary side of γ-CyD with the amino group ahead (Fig. 6.7b). The glutarimide ring is apparently less involved in the complex formation in this case. However, the involvement of the methyl group in complex formation by a still unknown mechanism cannot be completely excluded. Thus, as shown with this example, combined application of CE and a ROESY experiment may indicate the principal differences in the structure of analyte-CyD complexes in solution.

Another example when ROESY experiments may provide information about distinct differences between the structures of analyte–CyD complexes is the complex of chiral β_2-sympathomimetic agent clenbuterol (CL) with β-CyD and heptakis (2,3-di-*O*-acetyl)-β-CyD (HAD-β-CyD). Again it was noted from the CE experiments that the enantiomers of CL exhibit an opposite affinity pattern towards these two CyDs (Fig. 6.8) [60].

A 1D ROESY spectrum of the ($\pm$)-CL complex with β-CyD is shown in Fig. 6.9a. Upon saturation of the aromatic protons of CL a significant response was observed for the H-3 protons and a rather weak but measurable response for the H-5 protons of β-CyD. Both the H-3 and H-5 protons are located inside the cavity of β-

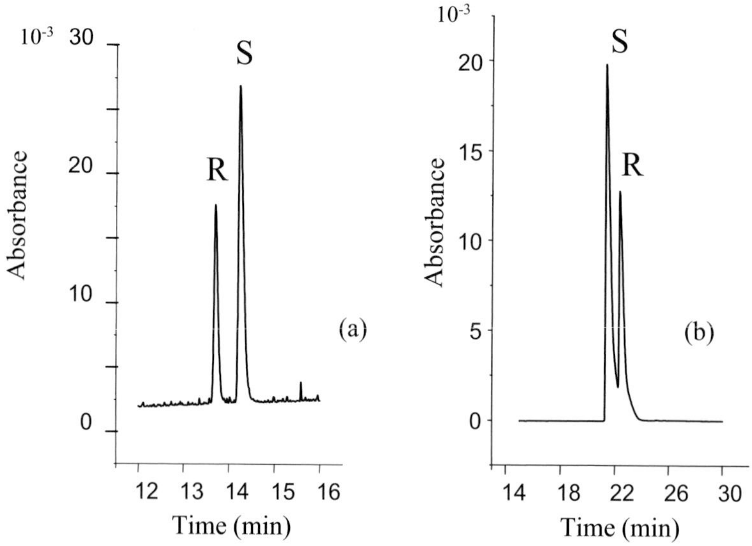

Fig. 6.8. Electropherograms of CL enantiomers [(*R*)/(*S*) = 2/1] in the presence of 18 mg mL^{-1} β-CyD (a) and 12 mg mL^{-1} HAD-β-CyD (b). (Reproduced with permission from Ref. [60].)

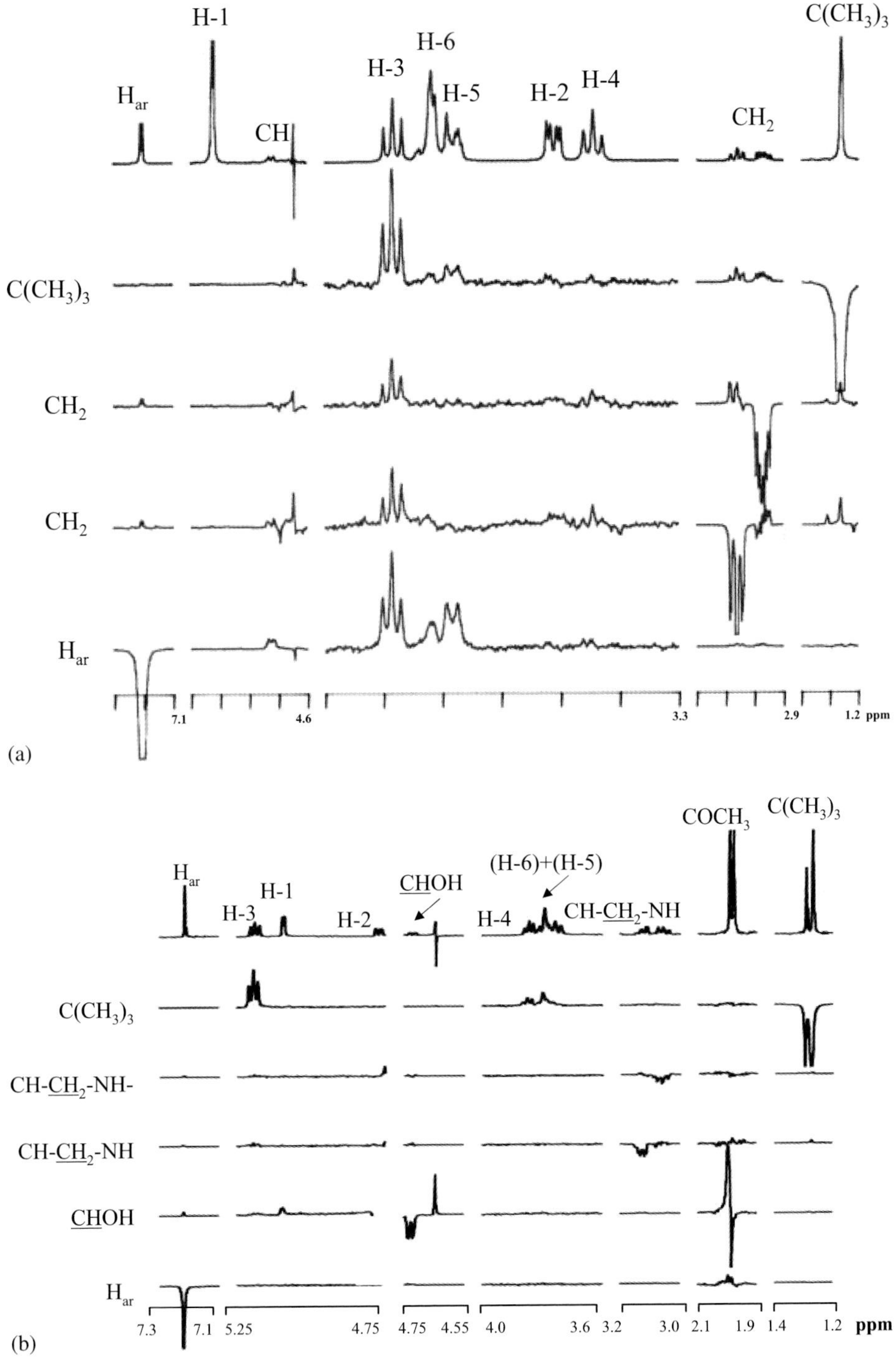

Fig. 6.9. 1D ROESY spectra of CL [(*R*)/(*S*) = 2/1] in the presence of one equivalent of *β*-CyD (a) and of HAD-*β*-CyD. (b) (Reproduced with permission from Ref. [60].)

CyD close to the secondary and primary CyD rims, respectively. Thus, these data clearly indicate that the CL molecule forms an inclusion complex with β-CyD, entering the cavity from the secondary wider side with the substituted phenyl moiety ahead. An additional indication for the analyte approaching the cavity of β-CyD from the secondary side is the following: when the protons of CH_2 and $C(CH_3)$ groups of CL were saturated, the NOE-response was observed on H-3 protons which are located inside the cavity of β-CyD close to the secondary rim. At the same time almost no effect was observed on the H-5 and H-6 protons, which are located close to the primary narrower rim of β-CyD. Thus, based on the 1D ROESY spectrum shown in Fig. 6.9a the structure depicted in Fig. 6.10a can be proposed for the (±)-CL/β-CyD complex. A rather strong NOE effect was observed between the *tert*-butyl moiety of CL and the H-3 protons of β-CyD. The structure shown in Fig. 6.10a allows some spatial proximity between these groups. However, at present it is impossible to conclude whether this proximity or the formation of a complex with an alternative structure (*tert*-butyl moiety inside the cavity of β-CyD) takes place for a significant population. The structure represented in Fig. 6.10a was unambiguously supported by a 1D ROESY experiment in which the protons were saturated in the β-CyD molecule and the response was observed on the protons of CL [60]. Thus, the response was found for the protons of CL only when the H-3 and H-5 protons of β-CyD were saturated. In addition, upon saturation of the H-3 protons both the aromatic and alkyl parts of CL responded equally, whereas only a rather weak effect was observed for the aromatic protons of CL and no effect at all on the alkyl part upon saturation of H-5 protons of β-CyD [60].

The information regarding the structure of the (*R*,*S*)-CL complexes with HDA-β-CyD was obtained using ROESY experiments, which were performed with three different pulse sequence for the complex between CyDs and a 3:1 (w/w) mixture of (*S*)- and (*R*)-CL.

The 1D ROESY spectrum of the (*R*,*S*)-CL/HDA-β-CyD complex (Fig. 6.9b) was

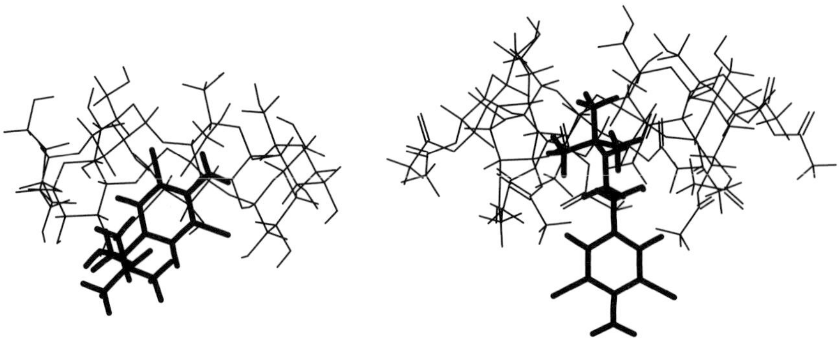

(a) **(b)**

Fig. 6.10. Structure of CL complexes with β-CyD (a) and HAD-β-CyD (b). (Reproduced with permission from Ref. [60].)

significantly different from that of the (*R,S*)-CL/*β*-CyD complex. Thus, upon irradiation of the aromatic protons in CL, no response was observed for the H-3 and H-5 protons located inside the cavity of HDA-*β*-CyD and a measurable effect was observed for the protons of the acetyl groups of HDA-*β*-CyD (Fig. 6.9b) [60]. When the protons of the *tert*-butyl moiety were irradiated, a very strong effect was observed for the H-3 protons located inside the cavity of HDA-*β*-CyD. These data indicate that CL enters the cavity of HDA-*β*-CyD also from the secondary wider side (similar to *β*-CyD) but in contrast to *β*-CyD not with the phenyl moiety but with the *tert*-butyl moiety ahead. The most likely structure of the (*R,S*)-CL/HDA-*β*-CyD complex based on these data is shown in Fig. 6.10 (b) [60].

This structure was clearly supported by the data obtained by irradiation of the protons of HDA-*β*-CyD and observing the response for the protons of CL [60]. Thus, upon irradiation of the H-3 protons of HDA-*β*-CyD an intermolecular NOE was observed only for the protons of the *tert*-butyl moiety of CL. In combination with the NOE response observed for the aromatic protons of CL upon irradiation of the acetyl group of HDA-*β*-CyD these data indicate that the *tert*-butyl moiety is included in the cavity and the phenyl moiety is located outside the cavity close to the secondary rim of HDA-*β*-CyD. Thus, the ROESY experiment shows a significant difference between the structures of the CL complexes with *β*-CyD and HDA-*β*-CyD. 1D and 2D transversal ROESY (T-ROESY) experiments confirmed that the effect observed in the 1D ROESY spectra were solely of intermolecular origin and that there was no significant contribution due to intramolecular TOCSY (total correlation spectroscopy) magnetization transfer. Thus, all ROESY experiments clearly indicated that CL forms intermolecular inclusion complexes with *β*-CyD and HDA-*β*-CyD. The CL molecule is included in the cavity of both CyDs from the secondary wider rim. The most distinct difference between the two complexes is that the phenyl moiety of CL is most likely included in the cavity of *β*-CyD whereas the *tert*-butyl moiety is included in the cavity of HDA-*β*-CyD.

The opposite enantiomer affinity pattern of the antihistaminic drug brompheniramine (BrPh) towards *β*-CyD and TM-*β*-CyD provoked us to study the structure of the corresponding diastereomeric complexes in solution using 1D-ROESY [61]. For the complexes of (+)-BrPh with both CyDs, unambiguous confirmation was obtained indicating the inclusion of the 4-bromophenyl moiety of the analyte in the cavity of the CyD. In addition, in the case of the (+)-BrPh complex with *β*-CyD, a weak but positive NOE effect was also observed for the protons of the maleate counter anion when saturating the CyD protons H-3 and H-5 located inside the cavity. This observation may indicate the simultaneous inclusion of the 4-bromophenyl moiety and maleate counter anion in the cavity of *β*-CyD but this contradicts simple geometric considerations and the assumption that the stoichiometry of the complex is 1:1.

X-Ray crystallographic study performed on the single crystals obtained from a 1:1 aqueous solution of (+)-BrPh maleate and *β*-CyD (Fig. 6.11) provides a plausible explanation for the above-mentioned contradiction. In particular, as shown in Fig. 6.11 (+)-BrPh forms with *β*-CyD, at least in the solid state, not a 1:1 complex

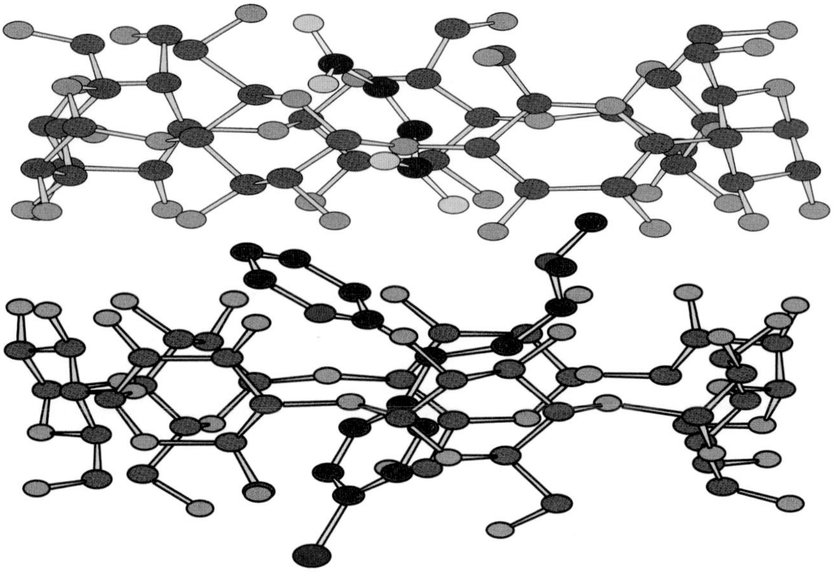

Fig. 6.11. Structure of the (+)-BrPh maleate β-CyD complex in the solid state determined by X-ray crystallography. (Reproduced with permission from Ref. [61].)

but a complex with 1:2 stoichiometry. In this complex the (+)-BrPh maleate is sandwiched between two molecules of β-CyD. The 4-bromophenyl moiety of (+)-BrPh enters the cavity of one of the β-CyD molecules whereas the cavity of another β-CyD molecule is occupied by the maleate counteranion [61].

One of the interesting questions of CyD chemistry is whether inclusion complexation represents a prerequisite for chiral recognition and, if not, which part of the CyD, external or internal, provides a more favorable environment for enantioselective recognition? The synthesis of highly crowded heptakis-(2-*O*-methyl-3,6-di-*O*-sulfo)-β-CyD (HMdiSu-β-CyD) with 14 bulky sulfate substituents on both primary and secondary CyD rims can provide insights to this problem [62] since the bulky substituents on both sides of the cavity entrance may hinder inclusion complex formation between chiral analytes and HMdiSu-β-CyD. In one study, 27 cationic chiral analytes were resolved in CE using native β-CyD and HMdiSu-β-CyD [63]. For 12 of 16 chiral analytes resolved with both chiral selectors the enantiomer migration order was opposite. Analysis of the structures of analyte–CyD complexes in solution indicated that in contrast to mainly inclusion-type complexation between chiral analytes and β-CyD, external complexes are formed between the chiral analytes and HMdiSu-β-CyD [63].

As can be seen from Fig. 6.12, the enantiomers of AGT enantioselectively bind to HMdiSu-β-CyD and are baseline resolved with this chiral selector in CE. In addition, the significant CIS differences were observed for the protons of AGT enan-

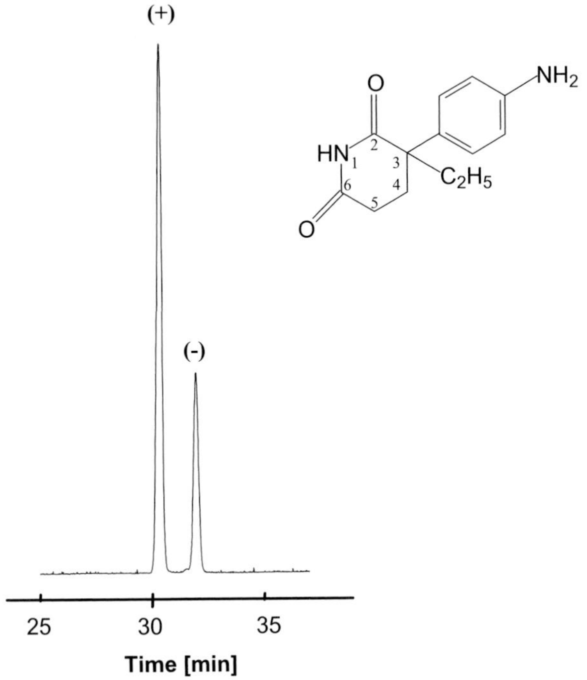

Fig. 6.12. Enantioseparation of AGT with 20 mg mL^{-1} HMdiSu-β-CyD. (Reproduced with permission from Ref. [63].)

tiomers in the ^1H NMR spectrum of the complex (Fig. 6.13). As the 1D T-ROESY spectra shown in Fig. 6.13 indicate, AGT most likely does not form an inclusion complex with HMdiSu-β-CyD. No intermolecular NOE was observed on any of the HMdiSu-β-CyD protons upon irradiation of analyte protons (Fig. 6.13). Some exception represents the OCH_3 protons of HMdiSu-β-CyD which are located on the secondary rim of the CyD cavity (Fig. 6.13). The weak NOE was observed on the protons of OCH_3 group upon saturation of almost all protons of the analyte except those of the CH_2-CH_3 group (Fig. 6.13).

Thus, inclusion complex formation between CyDs and their chiral guests does not seem to be a prerequisite for chiral recognition. CyDs are able to form quite strong external complexes enantioselectively with some chiral guest molecules. However, chiral recognition in the inclusion type complex, at least in this particular case, appears to be somewhat superior [63].

As shown by the aforementioned studies, the structural reasons responsible for the affinity reversal between the enantiomers and CyDs may vary from analyte to analyte and from CyD to CyD. Most likely there is no universal structural or chemical reason for affinity reversal from the side of either the chiral analyte or the CyD. This means that even significant differences observed in the structures of analyte–

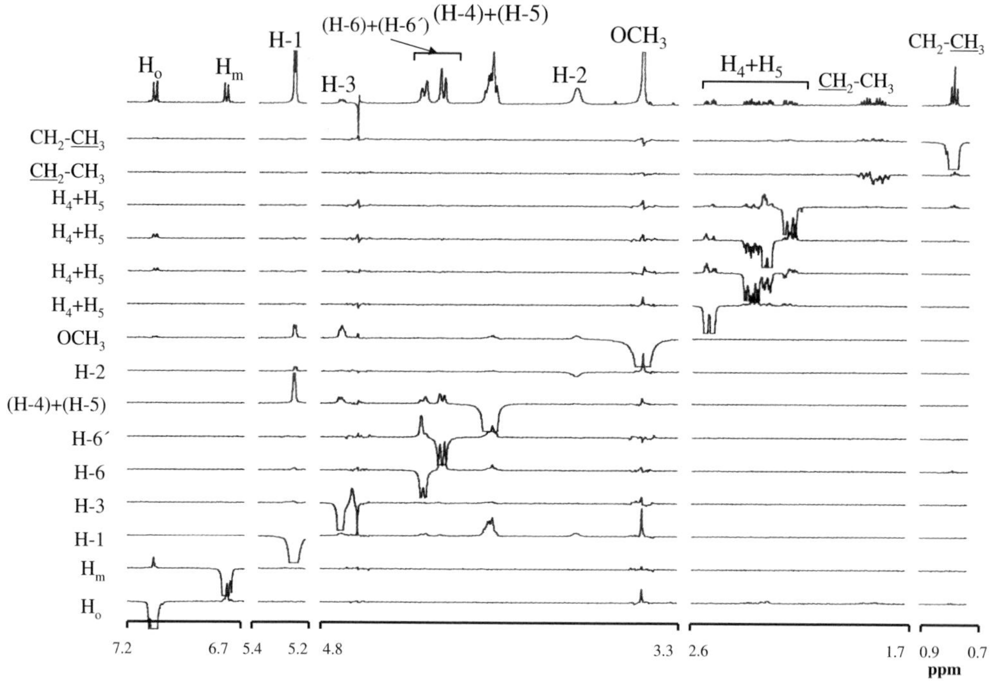

Fig. 6.13. 1D T-ROESY spectrum of AGT/HMdiSu-β-CyD complex. (Reproduced with permission from Ref. [63].)

CyD complexes may not be unambiguously considered to be the reason for affinity reversal. The most likely solution of this dilemma appears to be development of techniques for energy calculations based on the structure, dynamics, and statistical weight of given complexes. The contributions of individual intermolecular forces must be assigned and calculated accurately in total energy terms. As discussed in Chapter 11, the methods of molecular modeling and molecular mechanics calculations available at present do not completely meet this challenge.

CyDs are rather rigid molecules of medium size and therefore suitable for molecular modeling calculations (the opposite opinions on the rigidity as well as on CyDs modelling are expressed in Chapters 13–15). In addition, many CyDs have been well studied by alternative techniques of structure elucidation. Among these, X-ray crystallographic and NMR spectral data are of special interest.

A thermodynamic term describing chiral recognition is determined by the difference between the formation free energies of the transient diastereomeric complexes between the enantiomers and a chiral selector. Therefore, exact calculation of the absolute energy values is not necessarily required in molecular modeling studies related to enantioseparations. This simplifies the calculations (the opposite opinion on this topic is expressed in Chapter 11). On the other hand, due to the extremely high efficiency of CE the technique allows the observation of enantioseparations even for those selector–selectand pairs where the difference between the free energies of formation of the diastereomeric complexes is extremely small. The precise calculation of very small energy differences remains a challenging task even for the most sophisticated energy minimization techniques. Additional care must be taken in order to maximally approach a model of the real separation conditions. Thus, for instance, molecular modeling calculations are often performed in a vacuum without taking into account the effect of the medium. However, the aqueous medium commonly used in CE dramatically affects the hydrophobic and hydrogen-bonding interactions. Moreover, the ionic strength of the buffer plays a decisive role in the electrostatic intermolecular interactions. Another important point is the correct selection of the starting and the boundary conditions for energy minimization. Incorrectly defined conditions may totally confuse the calculations. For instance, when performing molecular modeling calculations for the complex between TM-β-CyD and (+)-BrPh in a neutral form the energy values indicated that complex formation with the alkyl amino moiety included in the cavity of TM-β-CyD would be energetically favorable. The structure with the alkyl amino moiety included in the cavity was also observed by an X-ray experiment performed on the single crystals obtained from the mixture of an aqueous suspension of deprotonated (+)-BrPh as a free base and TM-β-CyD (Fig. 6.14) [60]. These results are contradictory to the structure derived from the 1D ROESY experiment in solution. The intermolecular NOE effects clearly indicated the inclusion of the 4-bromophenyl moiety in the cavity of TM-β-CyD as shown in Fig. 6.15 [60]. Taking into consideration that the (+)-BrPh maleate, i.e. the protonated form of (+)-BrPh molecule, was used for the 1D-ROESY studies in solution, force-field calculations were performed again for interactions of a single positively charged (+)-BrPh with TM-β-CyD. The energy values obtained in this case clearly indicate that the complex formation with the 4-bromophenyl moiety of the (+)-BrPh molecule included in the cavity of TM-β-CyD is energetically favorable, which is in agreement with the structure observed using 1D-ROESY studies in solution (Fig. 6.15) [60].

In the author's opinion the most useful application of molecular modeling and molecular mechanics calculations to enantioselective analyte–CyD interactions would be a computation of individual intermolecular forces based on the structure, dynamics, and population of the complexes determined by instrumental techniques. Researchers working on molecular modeling of enantioselective CyD–chiral analyte interactions may use CE as a very powerful experimental technique for evaluating the reliability of their calculations. This may significantly contribute to further refinement of calculation techniques.

6.6 Capillary Electrochromatography

Capillary electrochromatography (CEC) represents a hybrid of capillary electrophoresis and HPLC. This technique relies on the migration principle of CE and ap-

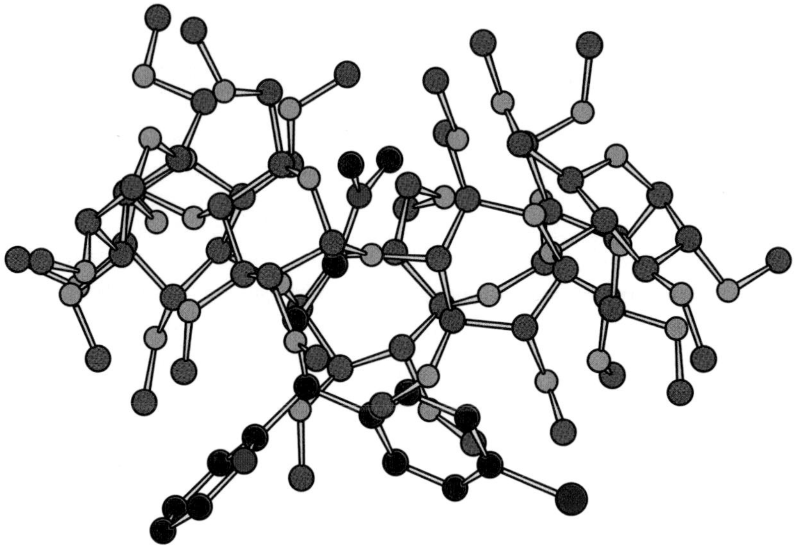

Fig. 6.14. Structure of (+)-BrPh maleate TM-β-CyD complex in the solid state determined by X-ray crystallography. (Reproduced with permission from Ref. [60].)

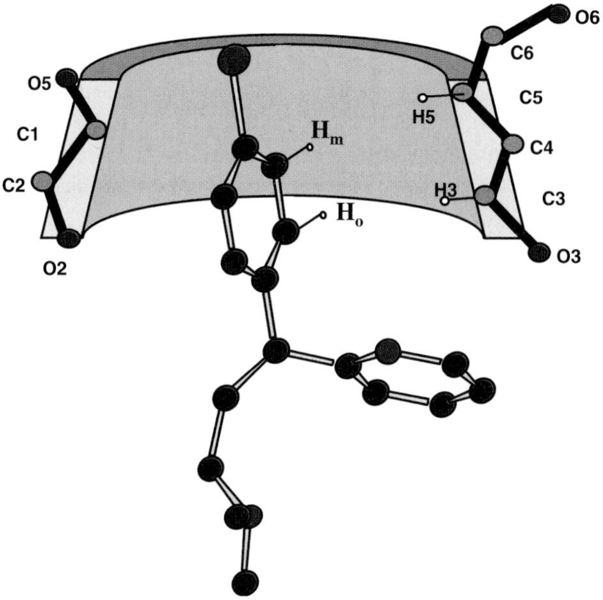

Fig. 6.15. Structure of (+)-BrPh maleate β-CyD complex in solution derived from a 1D ROESY experiment. (Reproduced with permission from Ref. [60].)

plies the separation principle of HPLC. For some separations of charged analytes both separation principles may operate simultaneously in CEC, although this statement does not apply to enantiomers. The enantiomers cannot be differentiated based on the separation principle of CE. Therefore, only the HPLC enantioseparation principle operates in chiral CEC whereas sample migration occurs as mentioned above through the electrokinetic mechanism.

At present CSPs based on polysaccharide derivatives, chiral ion exchangers, and macrocyclic antibiotics are most commonly used for CEC enantioseparations, but CyD-based CSPs played an important role in establishing of this technique in both open tubular [64, 65] and packed capillary format [65].

In 1992 Mayer and Schurig showed for the first time the possibility of enantioseparations in open tubular capillaries modified with a permethylated CyD derivative. This technique was used later with different chiral selectors but did not mature to become the method of choice for CEC enantioseparations, most likely due to the following conflict inherent in this technique. For a separation which occurs on the mobile phase/stationary phase interface and not in the bulk solution, retention of the analyte on the capillary wall (chiral stationary phase) is necessary for achieving a separation. On the other hand, any retentive analyte–capillary wall interactions are associated with a drastic decrease in peak efficiency in capillary electromigration techniques. However, this study stimulated research in both capillary enantioseparation techniques and in the use of CyD-based CSPs for CEC enantioseparations.

In principle, it is not difficult to find ways of increasing retention on the walls even of the capillaries having a diameter of 50 μm. This can be achieved by increasing the film thickness of a chiral selector [66] or increasing a capillary inner surface/volume ratio by etching the surface [67, 68] or by other special treatments of the capillary inner wall. However, this does not seem to be a solution for the problem because of the above mentioned conflict. The problem can be solved only by using a CSP with the highest possible enantiomer recognition ability. This may allow successful enantioseparations with the smallest possible capacity factors (k').

Cyclodextrin derivatives have been successfully used also for enantioseparations in packed capillary CEC. In this technique CyDs are used covalently immobilized to silica gel [69], to particulate silica sintered as a monolith [70], or to a monolithic organic matrix [71, 72]. In addition, CyDs have been used as dynamic modifiers of the stationary phase [73], as well as chiral mobile phase additives in combination with an achiral stationary phase [74, 75].

6.7 Lab-on-a-chip Technologies

Lab-on-a-chip separation of enantiomers have so far only been reported in the electrophoretic mode but not yet in the chromatographic or electrochromatographic modes. This technique represents a further miniaturization of capillary techniques and offers the advantages of ultra-short separation times, high throughput, and miniaturization of instruments. High throughput is particularly required with the rapid development of combinatorial synthetic technologies, screening of chiral catalysts, etc.

Enantiomeric separations on micromachined electrophoretic devices (MED) have been reported and about 10 research papers [76–87] and one review paper [88] summarize the developments in this field. In all enantioseparations performed using MED, CyDs have been used as chiral selectors except in one case [85] in which a chiral crown ether, (−)-(18-crown-6)-tetracarboxylic acid (18C6H4), was used as an effective chiral selector for resolving gemifloxacin in sodium-containing media.

In the very first enantioseparation using MED, Hutt et al. reported the enantioseparation of amino acids extracted from the Murchison meteorite using γ-CyD as a chiral selector [77]. Charged chiral analytes can be used with neutral native or derivatized CyDs as is common in chiral CE. However, fluorescent derivatization might be required for some analytes in order to enhance a detection sensitivity which represents a serious challenge in separations using MED. The derivatization may turn the analytes to neutral compounds which can be resolved with charged CyDs or with a combination of neutral (or charged) CyDs and micelles. Another argument for using charged CyDs as chiral selectors in MED-based enantioseparations is their commonly higher enantiomer-resolving ability as illustrated in the studies by Belder and co-workers. These authors could perform enantioseparations in a few seconds [87]. One additional advantage of charged selectors in MED-based enantioseparations is the directed mobility of these chiral selectors. The practical use of this advantage has still to be explored.

6.8 Future Trends

Owing to some unique properties, cyclodextrins have been successfully established as very useful chiral selectors in all instrumental analytical enantioseparation techniques. Further studies are required for a better understanding of the fine mechanisms of enantioselective recognition of these wonderful macrocyclic molecules as well as for evaluation of their potential for preparative scale enantioseparations.

References

1 J. Szejtli, *J. Mat. Chem.* **1997**, 7, 575–587.
2 B. Chankvetadze (Ed), *Chiral Separations*, Elsevier, Amsterdam, **2001**, 489 pp.
3 L.D. Hutt, D.P. Glavin, J.L. Bada, R.A. Mathies, *Anal. Chem.* **1999**, 71, 4000–4007.
4 B. Chankvetadze, *Rev. Chem. Soc.* **2004**, 33, 337–347.
5 F. Schardinger, *Z. Unters. Nahr. Genussm.* **1903**, 6, 865–880.
6 K. Freudenberg and M. Mayer-Delius, *Chem. Ber.* **1938**, 71, 1596–1600.

7 F. Schardinger, *Zentralbl. Bakteriol.*, Abt. II, **1911**, 29, 188–197.

8 K. Freudenberg, E. Schaaf, G. Dumpert and T. Ploetz, *Naturwissenschaften*, **1939**, 27, 850–853.

9 A. Hybl, R.E. Rundle and D.E. Williams, *J. Am. Chem. Soc.*, **1965**, 87, 2779–2788.

10 P.V. Demarco and A.L. Thakkar, *Chem. Commun.*, **1970**, 2–4.

11 F. Cramer, *Angew. Chem.* **1952**, 64, 136.

12 E. Gil-Av, B. Feibush, R. Charles-Sigler, *Tetrahedron Lett.* **1966**, 1009.

13 V. Schurig, *J. Chromatogr. A* **2001**, 906, 275–299.

14 F. Cramer, *Rev. Pure Appl. Chem.* **1955**, 143–164.

15 F. Cramer, W. Dietsche, *Chem. Ber.* **1959**, 92, 378–384.

16 M. Mikolajczyk, J. Drabowicz, F. Cramer, *J. Chem. Soc. Chem. Commun.* **1971**, 317–318.

17 T. Koscielski, D. Sybilska, J. Jurczak, *J. Chromatogr.* **1983**, 261, 357–362.

18 A. Venema, P.J.A. Tolsma, *J. High Resolut. Chromatogr. Chromatogr. Commun.* **1989**, 12, 32–34.

19 V. Schurig, H.-P. Novotny, *J. Chromatogr.* **1988**, 441, 155–163.

20 W. König, *Enantioselective Gas Chromatography with Modified Cyclodextrins*, Huthig, Heidelberg, **1992**.

21 D.W. Armstrong, W.Y. Li, J. Pitha, *Anal. Chem.* **1990**, 62, 214–217.

22 A. Mosandl, Food Rev. Int. 11, **1995**, 597–664.

23 C. Bicchi, G. Artuffo, A.D. Amato, A. Galli, M. Galli, Chirality 4, **1992**, 125–131.

24 V. Schurig, H. Grosenick, M. Juza, Recl. Trav. Chim. Pays-Bas 114, **1995**, 211–219.

25 M. Juza, E. Braun, V. Schurig, *J. Chromatogr. A* 769, **1997**, 119–127.

26 M. Juza, O. Di Giovanni, G. Biressi, V. Schurig, M. Mazotti, M. Morbidelli, *J. Chromatogr. A* 813, **1998**, 333–347.

27 J. Debowski, D. Sybilska, J. Jurczak, *J. Chromatogr.* **1982**, 237, 303–307.

28 D.W. Armstrong, US Patent 4539399, **1985**.

29 T.J. Ward, D.W. Armstrong, *J. Liq. Chromatogr.* **1986**, 9, 407–418.

30 J. Haginaka, J. Wakai, *Anal. Chem.* **1990**, 62, 997–1000.

31 A. Stalcup, K.L. Williams, *J. Liq. Chromatogr.* **1992**, 15, 29–37.

32 D.W. Armstrong, A.M. Stalcup, M.L. Hilton, J.D. Duncan, J.R. Jr. Foulkner, S.-C. Chang, *Anal. Chem.* **1990**, 62, 1612–1615.

33 D.W. Armstrong, M.L. Hilton, L. Coffin, *LC-GC* **1992**, 9, 647–652.

34 A.M. Stalcup, in: G. Subramanian, *A Practical Approach to Chiral Separations by Liquid Chromatography*, VCH, **1994**, Chapter 5, 95–114.

35 A.P. Foucault, *J. Chromatogr. A* **2001**, 906, 365–378.

36 J. Breinholt, S.V. Lehmann, A.R. Varming, *Chirality* **1999**, 11, 768–771.

37 P.A. Mourier, E. Eliot, M.H. Caude, R.H. Rosset, A.G. Tambute, *Anal. Chem.* **1985**, 57, 2819–2823.

38 G. Terfloth, *J. Chromatogr. A* **2001**, 906, 301–307.

39 W. Roder, F.J. Ruffing, G. Schomburg, W.H. Pirkle, *J. High Resolut. Chromatogr. Chromatogr. Commun.* **1987**, 10, 665–667.

40 V. Schurig, M. Jung, S. Mayer, M. Fluck, S. Negura, H. Jakubetz, Angew. Chem. **1994**, 106, 2265–2267.

41 E. Gassmann, J.E. Kuo, R.N. Zare, Science **1985**, 230, 813–815.

42 B. Chankvetadze, *J. Chromatogr. A* **1997**, 792, 269–295.

43 B. Chankvetadze, G. Blaschke, *J. Chromatogr. A* **2001**, 906, 309–363.

44 B. Chankvetadze, N. Burjanadze, D. Bergenthal, G. Blaschke, *Electrophoresis* **1999**, 20, 2680–2685.

45 B. Chankvetadze, W. Lindner, G.K.E. Scriba, *Anal. Chem.* **2004**, 76, 4256–4260.

46 B. Chankvetadze, *Capillary Electrophoresis in Chiral Analysis*, J. Wiley & Sons, Chichester, UK, **1997**, 555 pp.

47 J. Snopek, I. Jelinek, E. Smolkova-Keulemansova, *J. Chromatogr.* **1989**, 438, 211–218.

48 S. Fanali, *J. Chromatogr.* **1989**, 474, 441–446.
49 A. Guttman, A. Paulus, A.S. Cohen, N. Grinberg, B.L. Karger, *J. Chromatogr.* **1988**, 448, 41–53.
50 S. Terabe, *Trends Anal. Chem.* **1989**, 8, 129–134.
51 G. Scriba, *J. Pharm. Biomed. Anal.* **2002**, 26, 373–399.
52 G. Blaschke, B. Chankvetadze, *J. Chromatogr. A* **2000**, 875, 3–25.
53 K. Lomsadze, B. Chankvetadze, Poster presentation on HPCE **2004**, Salzburg, Austria, **2004**.
54 K.S. Bauer, S.C. Dixon, W.D. Figgi, *Biochem. Pharmacol.* **1998**, 55, 1827–1834.
55 M. Meyring, B. Chankvetadze, G. Blaschke, *Electrophoresis* **1999**, 20, 2425–2431.
56 B. Chankvetadze, *Electrophoresis* **2002**, 23, 4022–4035.
57 V.C. Anigbogu, C.L. Copper, M.J. Sepaniak, *J. Chromatogr. A* **1995**, 705, 343–349.
58 E. Francotte, S. Cherkauoi, M. Faupel, *Chirality* **1993**, 5, 516–526.
59 B. Chankvetadze, M. Fillet, N. Burjanadze, D. Bergenthal, K. Bergander, H. Luftmann, J. Crommen, G. Blaschke, *Enantiomer* **2000**, 5, 313–322.
60 B. Chankvetadze, K. Lomsadze, N. Burjanadze, J. Breitkreutz, G. Pintore, M. Chessa, K. Bergander, G. Blaschke, *Electrophoresis* **2003**, 24, 1083–1091.
61 B. Chankvetadze, N. Burjanadze, G. Pintore, D. Bergenthal, K. Bergander, C. Mühlenbrock, J. Breitkreutz, G. Blaschke, *J. Chromatogr. A* **2000**, 875, 471–484.
62 K.D. Maynard, G. Vigh, *Carbohydr. Res.* **2000**, 328, 277–285.
63 B. Chankvetadze, N. Burjanadze, K.D. Maynard, K. Bergander, D. Bergenthal, G. Blaschke, *Electrophoresis* **2002**, 23, 3027–3034.
64 S. Mayer, V. Schurig, *J. High Resolut. Chromatogr.* **1992**, 15, 129–131.
65 V. Schurig, D. Wistuba, *Electrophoresis* **1999**, 20, 2313–2328.
66 E. Francotte, M. Jung, *Chromatographia* **1996**, 42, 521–527.
67 J.J. Pesek, M.T. Matyska, *J. Chromatogr. A* **1996**, 736, 255–264.
68 J.J. Pesek, M.T. Matyska, *J. Chromatogr. A* **1996**, 736, 313–320.
69 D. Witsuba, H. Czesla, M. Roeder, V. Schurig, *J. Chromatogr. A* **1998**, 815, 183–188.
70 D. Wistuba, V. Schurig, *Electrophoresis* **2000**, 21, 3152–3159.
71 T. Koide, K. Ueno, *Anal. Sci.* **1998**, 14, 1021–1023.
72 T. Koide, K. Ueno, *J. Chromatogr. A* **2000**, 898, 177–187.
73 M. Zhang, Z. El Rassi, *Electrophoresis* **2000**, 21, 3135–3140.
74 F. Lelievre, C.- Yan, R.N. Zare, *J. Chromatogr. A* **1996**, 723, 145–156.
75 W. Wei, G. Luo, R. Xiang, *J. Microcol. Sep.* **1999**, 11, 263–269.
76 M.T. Reetz, *Angew. Chem. Int. Ed.* **2001**, 40, 284–310.
77 L.D. Hutt, D.P. Glavin, J.L. Bada, R.A. Mathies, *Anal. Chem.* **1999**, 71, 4000–4006.
78 I. Rodriguez, L.J. Jin, S.F.Y. Li, *Electrophoresis* **2000**, 21, 211–219.
79 S.N. Wallenborg, I.S. Lurie, D.W. Arnold, C.G. Bailey, *Electrophoresis* **2000**, 21, 3257–3263.
80 M.T. Reetz, K.M. Kuhling, A. Deege, H. Hinrichs, D. Belder, *Angew. Chem. Int. Ed.* **2000**, 39, 3891–3893.
81 H. Wang, Z.P. Dai, L. Wang, J.L. Bai, B.C. Lin, *Chinese J. Anal. Chem.* **2002**, 30, 665–669.
82 D. Belder, A. Deege, M. Maass, M. Ludwig, *Electrophoresis* **2002**, 23, 2355–3261.
83 M.A. Schwarz, P.C. Hauser, *J. Chromatogr. A*, **2001**, 928, 225–232.
84 E. Olvecka, M. Maser, D. Kaniansky, M. Johnk, B. Stanislawski, *Electrophoresis* **2001**, 22, 3347–3353.
85 S.I. Cho, K.-N. Lee, Y.-K. Kim, J. Jang, D.S. Chung, *Electrophoresis* **2002**, 23, 972–977.
86 M. Ludwig, D. Belder, *Electrophoresis* **2003**, 24, 2481–4286.
87 N. Piehl, M. Ludwig, D. Belder, *Electrophoresis* **2004**, 25, 3848–3852.
88 D. Belder, M. Ludwig, *Electrophoresis* **2003**, 24, 2422–2430.

7
Crystallographic Study of Cyclodextrins and Their Inclusion Complexes

Kazuaki Harata

7.1
Introduction

From the discovery of CyDs at the end of 19th century, it was more than 70 years before X-ray analysis explored the three-dimensional structure of CyDs, a macrocyclic oligosaccharide consisting of α-1,4 linked D-glucose units. In the early stage of CyD study, the cyclic structure of CyDs was expected from various chemical analyses; however, without a knowledge of the three-dimensional structure, it was difficult to fully understand the unique characteristics of CyDs capable of forming inclusion complexes with a variety of guest compounds. X-ray structures provided the direct evidence of the inclusion of the guest molecule in the CyD cavity.

The development of the structural study of CyDs in the last century has been summarized in several reviews [1–7]. In 1942, the crystallographic method was first applied to CyDs to determine molecular weight [8]. James et al. proposed a packing structure [9] for several crystalline complexes of α-CyD with small guest molecules in 1959. Preliminary crystallographic data of β-CyD complexes with benzene derivatives were reported in 1968 [10]. The crystal structure first solved by X-ray analysis in 1965 was that of the α-CyD complex with potassium acetate [11]. Structures of two other well-known CyDs, β-CyD [12] and γ-CyD [13, 14], were determined in 1976 and 1980, respectively. An up-to-date list of crystal structures determined by X-ray and/or neutron diffraction is collected in the Cambridge Crystallographic Database (http://www.ccdc.cam.ac.uk/).

7.2
Native CyDs

CyDs consisting of 6 to 31 glucose units have been separated from a reaction product of cyclodextrin glycosyltransferase, a CyD-producing enzyme that degrades soluble starch or amylose to produce cyclic oligosaccharides. (So called large CyDs with more than eight glucose units are discussed in Chapter 13.) Three-dimensional structures of CyDs have been determined for α-, β-, γ-, δ-, ε-, and

Cyclodextrins and Their Complexes. Edited by Helena Dodziuk

ISBN: 3-527-31280-3

ι-CyDs which consist of 6, 7, 8, 9, 10, and 14 glucose units, respectively, by X-ray and neutron diffraction analyses [15–24]. The enzymatic degradation of starch by potato D-enzyme, a kind of α-amylase, produces CyDs having more than 17 glucose units. Among these large CyDs, the crystal structure of cyclomaltohexaicosaose consisting of 26 glucose units has been determined [25].

7.2.1
Structure of the Glucose Unit

Figure 7.1 shows the structure of CyDs consisting of 6–9 glucose units. Some average parameters describing the macrocyclic conformation of native CyDs are summarized in Table 7.1. The pyranose ring of each glucose unit in native CyDs is relatively rigid and assumes the $^4C_1$ chair conformation. Some structural characteristics of CyDs are illustrated in Fig. 7.2. Primary hydroxyl groups have rotational

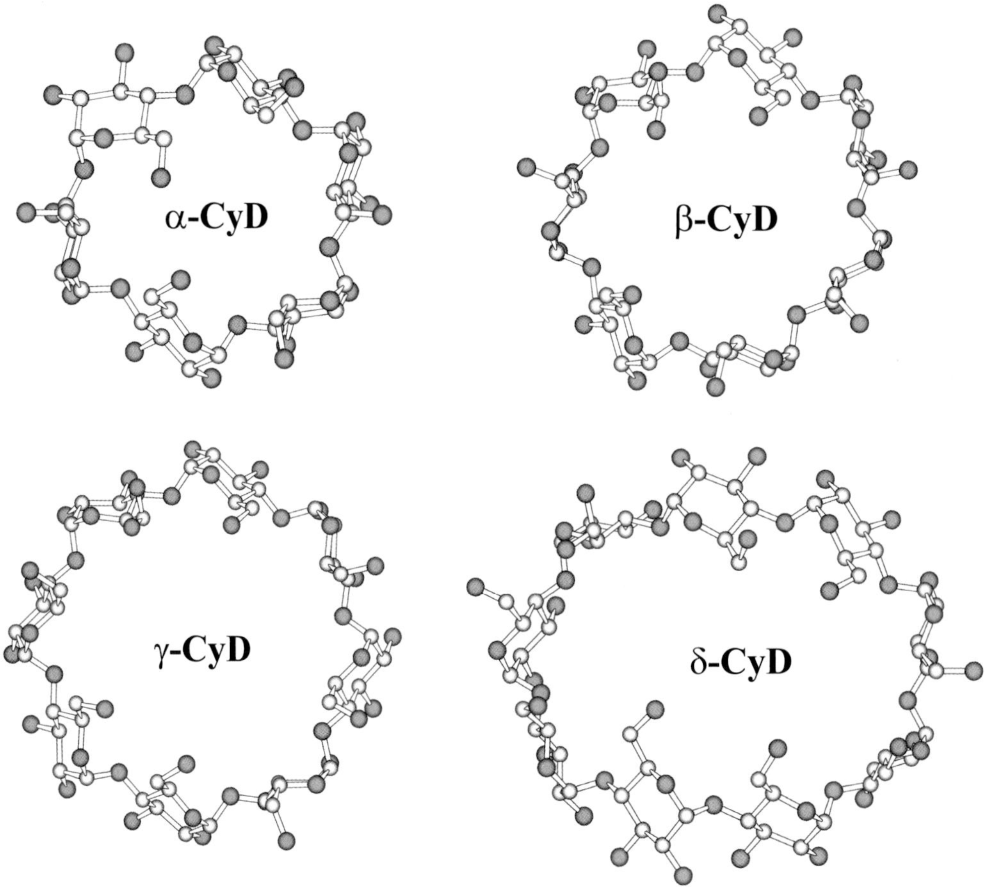

Fig. 7.1. Doughnut-shaped macrocyclic structures of α-, β-, γ-, and δ-CyD.

Tab. 7.1. Geometrical data of CyD macrocycles[a]

	α-CyD	β-CyD	γ-CyD	δ-CyD	ε-CyD	ι-CyD
Radius of the O-4 polygon (Å)	4.2(0.1)	5.0(0.2)	5.9(0.1)	6.4(0.7)	6.9(0.4)	8.4(1.0)
O-4–O-4′ distance (Å)	4.2(0.1)	4.3(0.1)	4.5(0.1)	4.5(0.1)	4.5(0.1)	4.5(0.1)
O-2–O-3′ distance (Å)	3.0(0.1)	2.9(0.1)	2.8(0.1)	2.9(0.2)	3.2(0.2)	2.9(0.2)
Planarity of the O-4 polygon (Å)[b]	0.10	0.16	0.11	0.80		
Glycosidic O-4 angle (°)	119(1)	118(1)	117(1)	116(2)	117(1)	118(1)
Tilt angle (°)[c]	13(10)	14(10)	19(9)	25(20)		

[a] Some data in this table are also given in Tables 1.1 and 13.1.
[b] Root-mean-square deviation of O-4 atoms from their least-squares plane.
[c] Average angle between the O-4 plane and the plane through C-1, C-4, O-4, and O-4′ of each glucose unit.

flexibility around the C5–C6 bond. However, their conformation is confined to two types, *gauche–gauche* and *gauche–trans*. In the former type, the C6–O6 bond is *gauche* to both the C5–O5 and C4–C5 bonds while in the other type the C6–O6 bond is *gauche* to the C5–O5 bond and *trans* to the C4–C5 bond. Secondary hy-

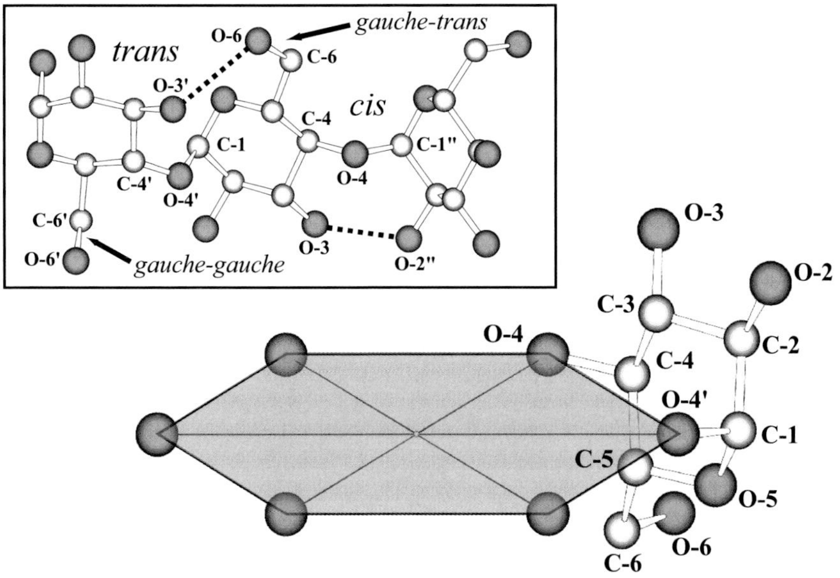

Fig. 7.2. Illustrations showing characteristics of the CyD structure. The plane composed of glycosidic O4 atoms is shaded. Intramolecular hydrogen bonds are denoted by dotted lines.

droxyl groups are circularly aligned like the rim of a torus and hydrogen bonds are formed between an O2H hydroxyl group and the O3H hydroxyl group of an adjacent glucose unit.

The nonrigidity of the CyD ring, discussed in detail in Chapter 13, is mostly ascribed to the rotational flexibility of each glucose unit around the glycosidic linkage and to small changes in the endocyclic torsion angles of the pyranose ring. In the latter, the change of each torsion angle is too small to detect, but the accumulation over six bonds in the pyranose ring, which is referred as the "torson-angle index", clearly reveals the change [26]. The torsion-angle index is negatively correlated with the distance between the two glycosidic O4 atoms bonded to C1 and C4 of each glucose unit. The average distance between these O4 atoms is smallest for α-CyD and increases with increasing number of glucose units, indicating that the glucose unit in α-CyD is forced to be slightly bent by the cyclization. The effect of cyclization is also observed in the average bond angle of O4 atoms, which is decreased from 119° to 116° in the order α-, β-, γ-, and δ-CyD.

7.2.2
Average Structural Characteristics of the Macrocyclic Ring

CyDs consisting of 6–9 glucose units exhibit a doughnut-shaped structure [15–21] (Fig. 7.1). These CyDs are approximately in the shape of a truncated polygonal cone. The macrocycle of CyDs is distorted from its ideal structure with *n*-fold symmetry to relieve the steric hindrance between glucose units and distortion of the glycosidic linkage. Upon complex formation, CyDs change their macrocyclic structure and adjust the structure of the cavity to accommodate the guest molecule. The shape of the CyD ring is well characterized by using a polygon consisting of glycosidic O4 atoms which are nearly coplanar (Fig. 7.2 and Table 7.1). The radius of the polygon, that is the average of the distance from the center of the polygon to each O4 atom, is a good measure to estimate the size of the macrocycle. The radius varies in the range from 4.2 Å in α-CyD to 6.4 Å in δ-CyD.

The α-1,4 linkage is responsible for the conformational flexibility of the macrocycle. The pyranose ring is almost perpendicular to the O4 plane but the average over all the glucose units indicates that the pyranose ring is generally tilted with its primary hydroxyl side towards the inside of the macrocyclic ring. The rotational movement of glucose units around the glycosidic linkage distorts the macrocyclic ring by increasing the degree of inclination of each glucose unit against the plane through the glycosidic O4 atoms. The tilt of each glucose unit is evaluated by the "tilt-angle", which is defined as the angle measured between the plane of the O4 polygon and the plane through C1, C4, O4, and O4′ of each glucose unit. The tilt-angle is largely in the range from 10° to 20°. Such flexibility of glucose units is still restrained by intramolecular hydrogen bonds of secondary hydroxyl groups formed between adjacent glucose units. When these hydrogen bonds are disrupted, glucose units gain high flexibility around the glycosidic linkage as we see in the structures of permethylated CyDs and peracetylated CyDs. (See, however, another opinion expressed in Section 1.3.)

α-CyD crystallizes from water in three forms, hexahydrate (Form I and Form II) and 7.57 hydrate (Form III). The flexible macrocyclic ring is demonstrated by the distinct conformational changes of α-CyD observed among these crystals. The α-CyD ring in the Form I crystal including two water molecules is less symmetrical than that in the form III crystal which contains 2.57 water molecules in the cavity of the round macrocyclic ring. β-CyD crystallizes in two forms, undecahydrate and dodecahydrate, in the crystals of which the structure of β-CyD is essentially identical but the change occurs in the arrangement of water molecules in the cavity. γ-CyD has the most symmetrical structure among the four CyDs. In contrast, δ-CyD is elliptically distorted and the ring deviates considerably from the planar structure.

In CyDs consisting of more than nine glucose units, the macrocyclic ring is no longer doughnut shaped, as shown in Fig. 7.3. Because of the strain imposed on the glycosidic linkage, the molecule cannot retain the round structure; instead, the elliptical ring is bent into a shape like a saddle [22–24] in CyDs composed of 10 and 14 glucose units, ε- and ι-CyD, respectively. At two diametrically opposed positions in these CyDs, the glucose unit of the non-reducing side of the glycosidic linkage is rotated about 180° around the C4–O4 bond. Therefore, these two adja-

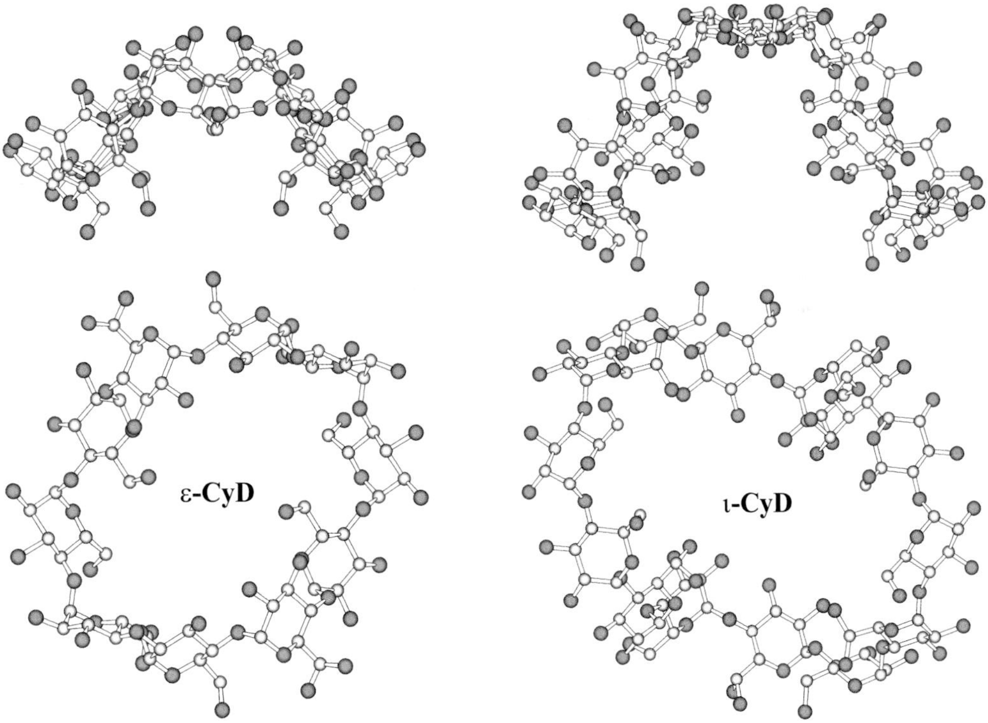

Fig. 7.3. Side and top views of the structures of ε- and ι-CyD consisting of 10 and 14 glucose units, respectively.

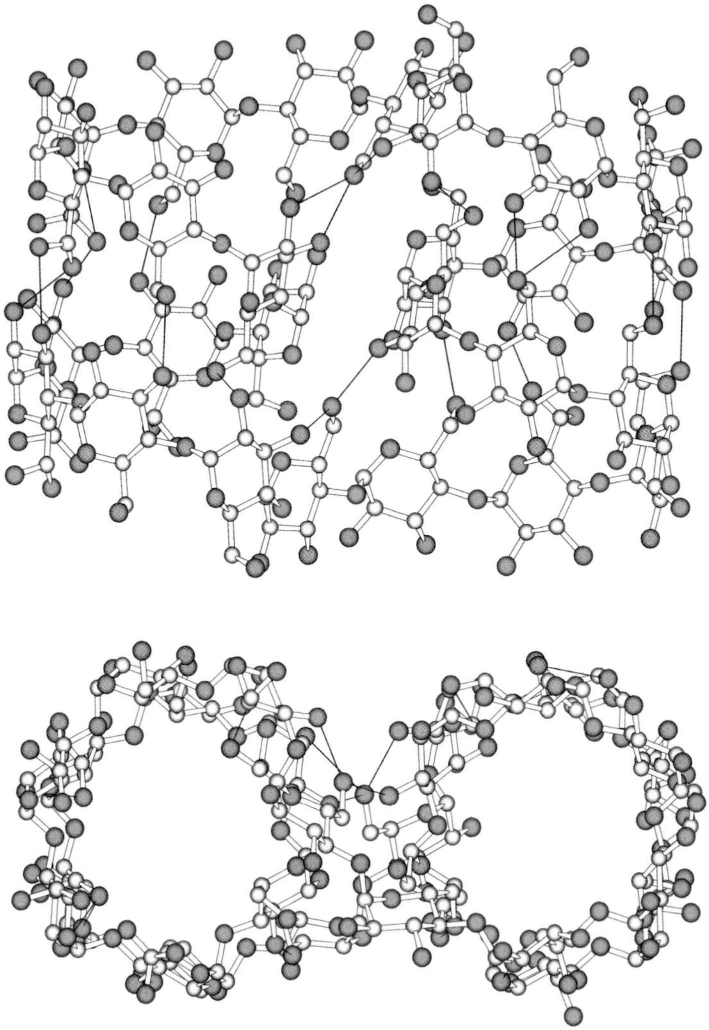

Fig. 7.4. Side and top views of the structure of cyclomaltohexaicosaose consisting of 26 glucose units. The molecule consists of two left-handed helical structures related by a two-fold symmetry.

cent glucose units are *trans* in respect of the glycosidic linkage. The structure of a CyD consisting of 26 glucose units, the largest one whose structure has been determined by X-ray analysis, is not circular but consists of two left-handed amylose-like helixes [25] (Fig. 7.4). The molecule has two-fold symmetry and the two asymmetric parts are connected by glucose units with the *trans* arrangement.

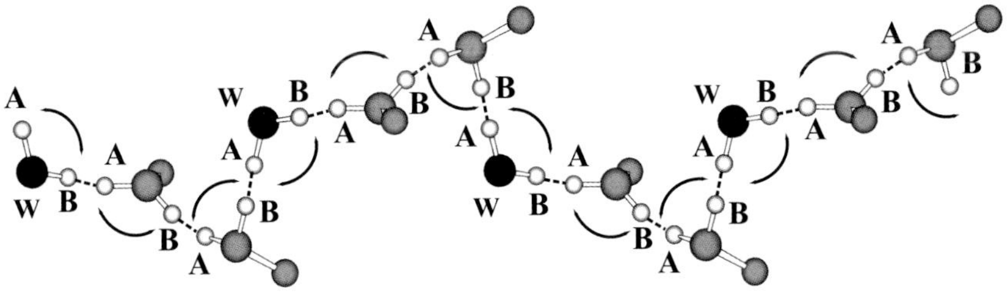

Fig. 7.5. A linear hydrogen-bond chain containing water molecules shown by filled circles. Hydrogen atoms cannot simultaneously occupy positions A and B.

7.2.3 Hydrogen Bonds

In the crystalline state, CyDs form both intra- and intermolecular hydrogen bonds, which stabilize not only the molecular conformation but also the crystal packing. In CyDs consisting of 6 to 9 glucose units [15–21], two adjacent glucose units are in a *cis* arrangement and secondary hydroxyl groups, either O2H or O3H, form hydrogen bonds. On the other hand, the *trans* arrangement observed in larger CyDs allows the formation of a hydrogen bond between an O3H hydroxyl group and an O6H hydroxyl group of an adjacent glucose unit (Fig. 7.2) [22–25]. These hydrogen bonds are incorporated in a hydrogen-bond network that stabilizes the crystal structure by a cooperative effect. In crystals of α-CyD hydrate, the O–H direction in the hydrogen bond network is not random but the bonds point in the same direction [16, 17]. On the other hand, in β-CyD crystals, the direction of the O–H bond can change cooperatively through the network, shown as the flip-flop hydrogen bond disorder [27] shown in Fig. 1.7. Neutron diffraction studies have revealed that the hydrogen atoms of the secondary hydroxyl groups of β-CyD are not statistically disordered and at room temperature the hydrogen bond was observed as the average of O2H–O3 and O2–HO3 hydrogen bonds while at low temperature the former direction prevails [28–33]. Figure 7.5 shows the flip-flop disorder in a hydrogen-bond chain involving water molecules. The direction of the O–H bond is cooperatively changed and such a flip-flop network is considered to be entropically more favored than a network with ordered hydrogen bonds.

7.3 Modified CyDs

7.3.1 Methylated CyDs

The macrocyclic structure of β-CyD is not affected by methylation of the O2H hydroxyl groups because the intramolecular O3H–O2 hydrogen bonds retain the

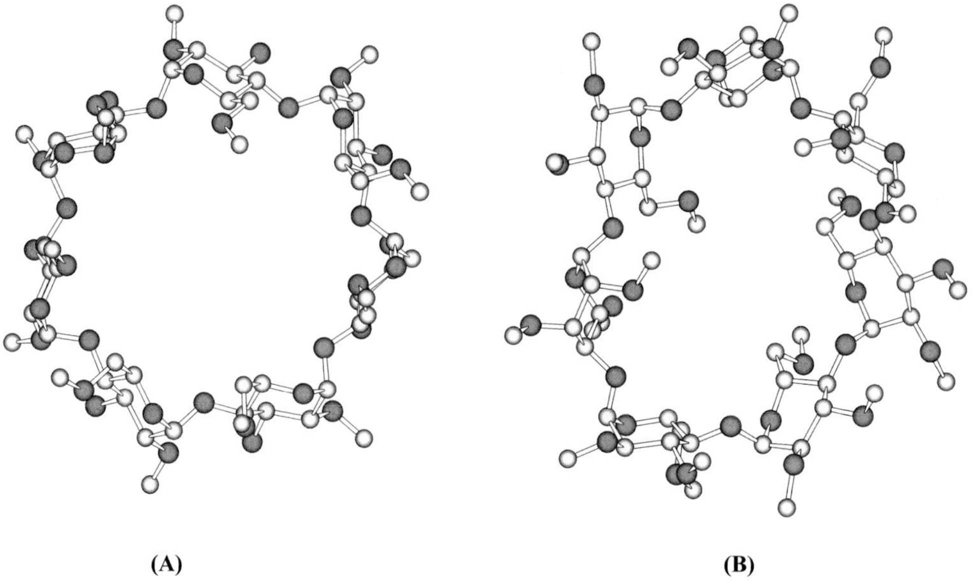

Fig. 7.6. Structure of heptakis(2,6-di-*O*-methyl)-β-CyD (A) and heptakis(2,3,6-tri-*O*-methyl)-β-CyD (B).

round shape of the molecule and all methyl groups point away from the center of the macrocycle [34]. Further introduction of methyl groups to O6H hydroxyl groups increases the depth of the cavity by about 1 Å compared to the cavity of native CyDs, although the diameter of the cavity does not change. Since the solubility of methylated CyDs is negatively correlated to the temperature, that is they are less soluble at higher temperature, methylated CyDs and their complexes have been crystallized mostly at temperatures from 30 °C to 50 °C. Structures of methylated β-CyD are shown in Fig. 7.6.

The crystal packing and hydration of hexakis(2,6-di-*O*-methyl)-α-CyD differ between the crystals obtained at room temperature (monohydrate) and 90 °C (anhydrate) [35]. The structure of α-CyD in the both crystals is almost the same but the α-CyD cavity of the room temperature crystal includes one water molecule. Heptakis(2,6-di-*O*-methyl)-β-CyD crystallizes in different forms at room temperature, low temperature, and high temperature [37–40]. The β-CyD cavity in the room-temperature crystal contains one water molecule [38] while the high-temperature crystal is anhydrous [37]. The low-temperature crystal contains about 15 water molecules which are all located outside the CyD cavity [39, 40]. A β-CyD derivative in which all O2H and O6H hydroxyl groups are ethylated is sparingly soluble in water and was crystallized from *n*-hexane or methanol [41]. The ethyl groups attached to both O2 and O6 point away from the macrocyclic ring and do not affect the round structure of the *β*-CyD ring.

The methylation of all hydroxyl groups increases the depth of the cavity by about 1.5 Å but such methylated CyDs do not retain the round shape because of steric hindrance and disruption of the intramolecular hydrogen bonds of secondary hydroxyl groups between adjacent glucose units. Anhydrous crystals of hexakis(2,3,6-tri-*O*-methyl)-*α*-CyD were obtained at 40 °C and also at room temperature in the presence of NaCl [42]. Two 2,3,6-tri-*O*-methyglucose units at diagonal position are sharply inclined with their O6 side towards the inside of the macrocycle. Heptakis(2,3,6-tri-*O*-methyl)-*β*-CyD was crystallized as monohydrate at 50 °C. The macrocyclic ring is elliptically distorted and one 2,3,6-tri-*O*-methylglucose unit is in the $^{1}C_4$ chair conformation [43]. The axial O2CH$_3$ methoxyl group points to the center of the cavity (Fig. 7.6B), while a water molecule is located outside the CyD ring. Two types of octakis(2,3,6-tri-*O*-methyl)-*γ*-CyD crystals were obtained at room temperature: 4.5 hydrate [44] and 19.3 hydrate containing four CyD molecules in the asymmetric unit [45]. The macrocyclic ring is strongly asymmetrical and two 2,3,6-tri-*O*-methylglucose units are *trans* in relation with the glycosidic linkage. One water molecule is located at the center of the cavity and the vacant space accommodates 6-*O* methoxyl groups.

7.3.2 Acylated CyDs

Acylated CyDs are insoluble in water and therefore crystallized from organic solvents. Peracetylation induces steric hindrance and blocks the formation of intramolecular hydrogen bonds and the CyD ring is markedly distorted from the round structure to relieve the steric hindrance caused by acetyl groups. The cavity of hexakis(2,3,6-tri-*O*-acetyl)-*α*-CyD crystallized from aqueous 2-propanol looks like a rectangular box because two 2,3,6-tri-*O*-acetylglucose units are tilted with the O2, O3 side towards the inside of the macrocycle as shown in Fig. 7.7A. A water molecule located at the center of the cavity links these 2,3,6-tri-*O*-acetylglucose units by hydrogen bonds with acetyl groups attached to O3 to stabilize the distorted macrocyclic structure [46]. Heptakis(2,3,6-tri-*O*-acetyl)-*β*-CyD [47] crystallized from methanol is elliptically distorted and one 2,3,6-tri-*O*-acetylglucose unit is in the $^{3,0}B$ boat conformation (Fig. 7.7B). The molecular cavity accommodates the 2-*O* acetyl group of this glucose unit, the 6-*O* acetyl group of the opposite glucose unit, and also acetyl groups of an adjacent molecule. The macrocyclic structure of heptakis(2,3,6-tri-*O*-propanoyl)-*β*-CyD is very similar to the structure of acetylated *β*-CyD, and the $^{3,0}B$ boat conformation is also observed in one 2,3,6-tri-*O*-propanoylglucose unit [47]. On the other hand, heptakis(2,3,6-tri-*O*-butanoyl)-*β*-CyD crystallized from hexane is less elliptical and all 2,3,6-tri-*O*-butanoylglucose units are in the $^{4}C_1$ chair conformation [47]. A butanoyl group attached to O6 is inserted into the molecular cavity and this self-inclusion makes the macrocyclic ring more round than those of the other two acylated *β*-CyDs.

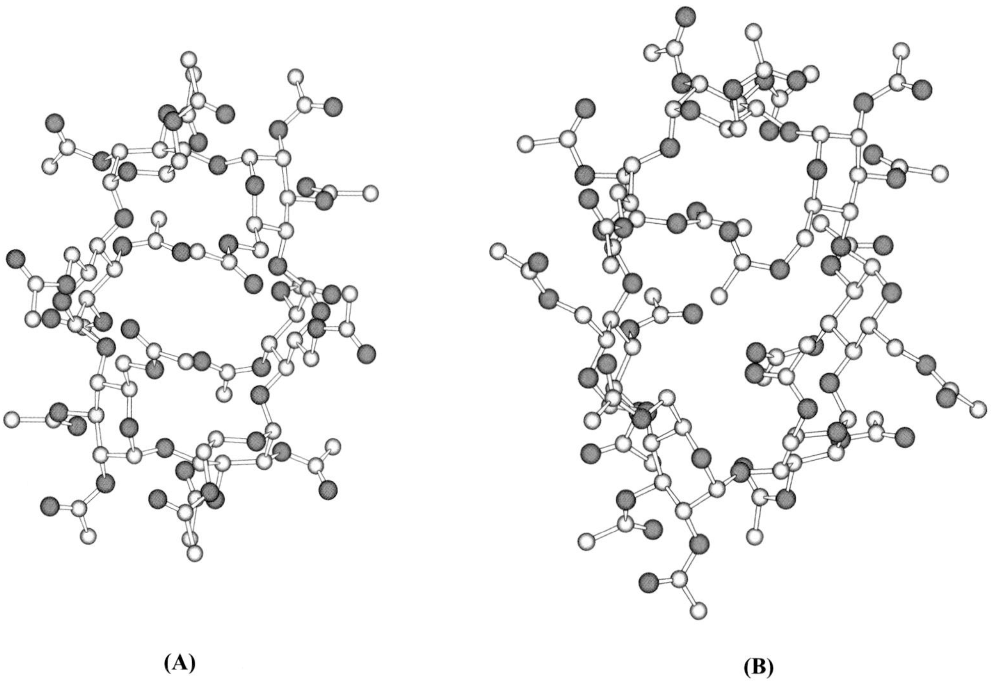

Fig. 7.7. Structure of hexakis(2,3,6-tri-O-acetyl)-α-CyD (A) and heptakis(2,3,6-tri-O-acetyl)-β-CyD (B).

7.3.3
Monosubstituted CyDs

As described in Chapter 2, mono-substituted CyDs have been extensively studied for the purpose of introducing a functional group on the macrocyclic ring. Because of the difference in the reactivity of three hydroxyl groups, O2H, O3H, and O6H, a number of 6-*O*-mono-substituted CyDs have been synthesized and their X-ray structures reported. Some typical molecular arrangements of 2-*O* substituted and 6-*O* substituted CyDs are shown in Fig. 7.8. The introduction of a substituent group to 6-*O* or 2-*O* position does not cause a large change in the macrocyclic conformation of the CyD ring. A substituent group attached to O2 is stretched out from the secondary hydroxyl side and behaves like a guest which is inserted into the cavity of an adjacent molecule. In the crystal of 2-*O*-[(*S*)-2-hydroxypropyl]-β-CyD (Fig. 7.8A), the hydroxypropyl group is inserted into the cavity of the next β-CyD ring from the secondary hydroxyl side and linked to an O6H hydroxyl group by a water-mediated hydrogen bond bridge [49]. A similar packing of α-CyD ring is observed in the crystal of mono-2-*O*-allyl-α-CyD where the α-CyD ring including the allyl group of the adjacent molecule is closely stacked along the two-fold screw axis [48].

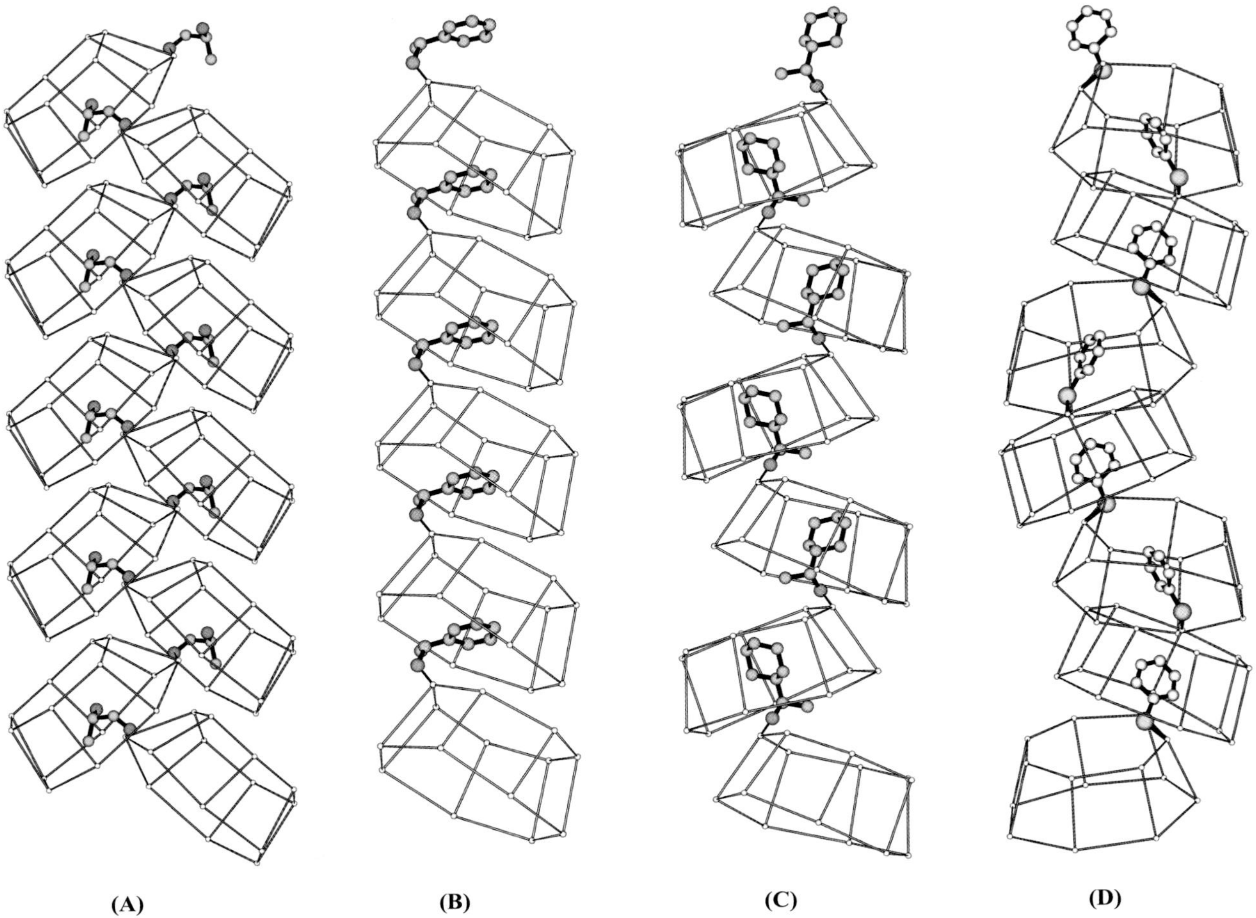

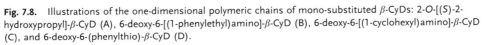

Fig. 7.8. Illustrations of the one-dimensional polymeric chains of mono-substituted β-CyDs: 2-*O*-[(*S*)-2-hydroxypropyl]-β-CyD (A), 6-deoxy-6-[(1-phenylethyl)amino]-β-CyD (B), 6-deoxy-6-[(1-cyclohexyl)amino]-β-CyD (C), and 6-deoxy-6-(phenylthio)-β-CyD (D).

A substituent group bonded to 6-*O* or 6-*C* points away from the macrocycle and is mostly inserted into the cavity of an adjacent CyD molecule from the secondary hydroxyl side. Because of the rotational flexibility of the C-6–X-6 (X = O, N, or S) bond, the substituent group gains more conformational flexibility than those attached to the secondary hydroxyl side. A helically extended polymeric chain is formed in the crystal of β-CyD derivatives with various substituent groups. The inclusion of the substituent group adjusts the relative orientation of the β-CyD ring and, as a result, the pitch of the helical arrangement changes according to the shape, size, and flexibility of the substituent group. For a small substituent group, the primary hydroxyl side of the β-CyD ring is in close contact with the secondary hydroxyl side of the next molecule to push the substituent group into the cavity. For example, a hydroxylpropyl group is deeply inserted into the β-CyD ring related by two-fold screw symmetry and forms a hydrogen bond with a primary hydroxyl group [51].

The inclusion of a bulky and rigid substituent group imposes a restriction on the arrangement of the CyD ring because the orientation of the substituent group is sterically restricted and the next CyD ring is arranged to include it. In the crystal of β-CyD derivatives carrying a 2-hydroxyindane-1-ylamino group **1** or 1-phenylethylamino group **2**, the β-CyD ring is linearly stacked [50] (Fig. 7.8B) along a lattice axis. On the other hand, a cyclohexyl [50] or *tert*-butyl group [52], which are well fitted to the β-CyD cavity, are arranged on a two-fold screw axis in a manner similar to the helical arrangement of the hydroxyalkyl derivatives (Fig. 7.8C), while the 6-aminohexyl group penetrates through the cavity and the terminal amino group contacts a secondary hydroxyl group of the second molecule [53]. A similar helical arrangement was also observed in the crystals of phenylsulfinyl [54] **3** and *p*-carboxyphenylamino [55] **4** derivatives. In contrast, the β-CyD ring substituted with a phenylthio group is stacked along the four-fold screw axis [54] (Fig. 7.8D). In the crystal of 6-*O*-monoglucosyl-α-CyD, the glucosyl group appended by an α-1,6 linkage is so bulky that half of the substitutent group is included in the cavity of the adjacent molecule related by a two-fold screw symmetry [56].

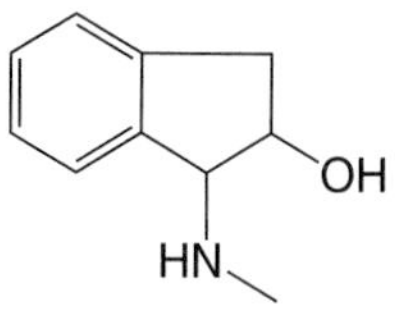

1

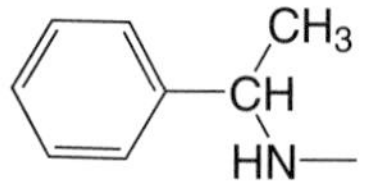

2

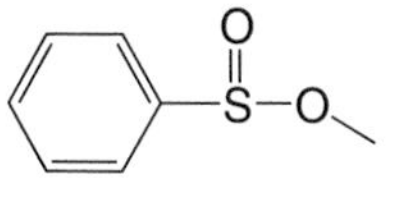

3

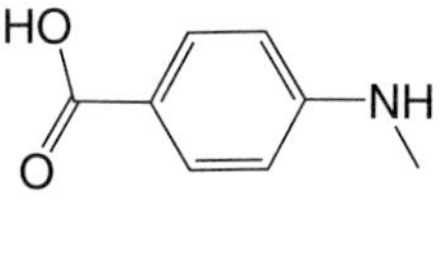

4

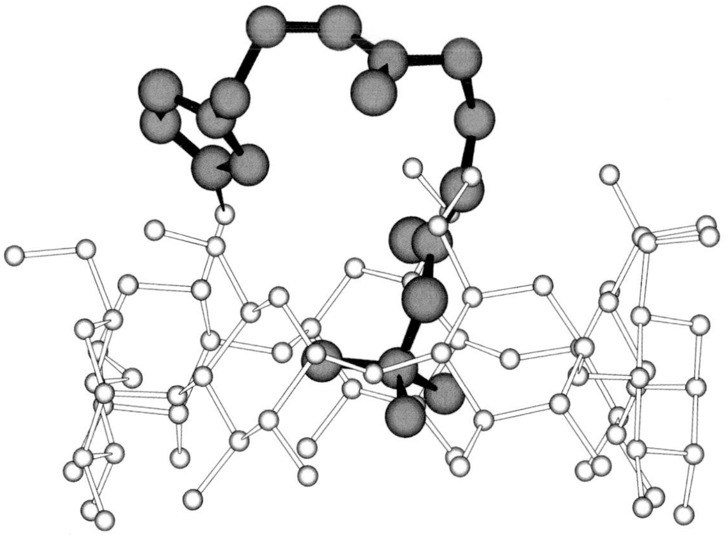

Fig. 7.9. Structure showing the self-inclusion of a butoxycarbonyl group of 6-deoxy-6-{4-[*N*-*tert*-butoxycarbonyl-*N*-(*N*′-ethyl)propanamide] imidazolyl-*β*-CyD.

In some 6-*O* mono-substituted derivatives, a substituent group attached to C-6 is bent and inserted into the CyD cavity, exhibiting a "sleeping-swan"-like structure as shown by 6-deoxy-6-{4-[*N*-tert-butoxycarbonyl-*N*-(*N*′-ethyl)propanamide]-imidazolyl}-*β*-CD **5** in Fig. 7.9. Imidazolyl derivatives with the imidazolyl ring directly bound to C-6 of *β*-CyD show self-inclusion of the terminal hydrophobic *iso*-propyl [58] and *tert*-butyl [59, 60] groups. The rigid imidazolyl group pointing to the inside of the *β*-CyD ring adjusts the orientation of the substituent group so that the terminal group can be inserted into the cavity. In contrast, no self-inclusion occurs when the amino group of imidazolylethylamino [61] **6** or boc-L-phenylalanylamino [62] **7** group is bound to C-6 of *β*-CyD. In the *β*-CyD derivative having a diphenylphosphinoethylenethio group **8**, one phenyl group is inserted into the cavity from the primary hydroxyl side while the other phenyl group is sandwiched between the *β*-CyD rings [57].

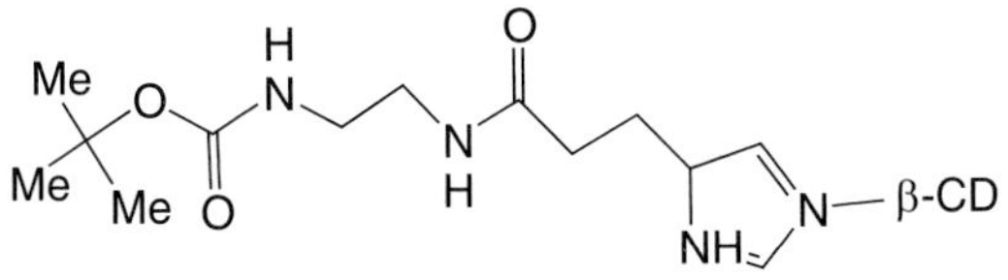

5

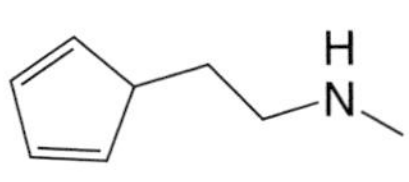

6

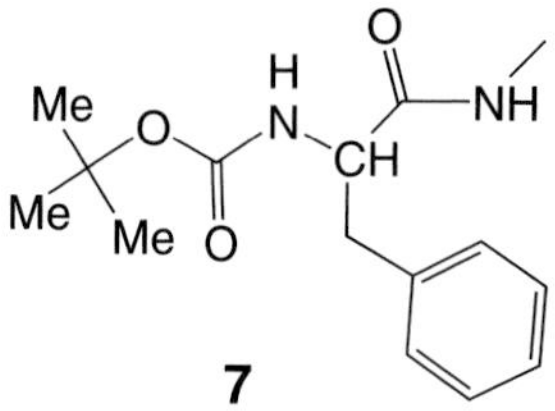

7

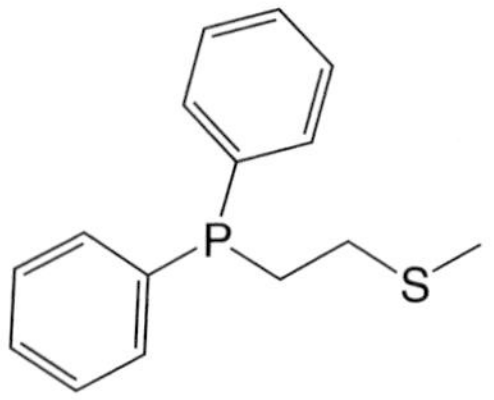

8

7.3.4
Disubstituted CyDs

The introduction of amino groups to the 6-*C* position of two vicinal glucose units of β-CyD does not affect the macrocyclic conformation [63]. In the metal-bound state, the two amino groups of $6^{I},6^{II}$-diamino-β-CyD are coordinated to platinum(II) in the *cis* form and the two chloride ions are also bound to Pt(II) to form a square plane [64]. The coordination of the amino groups to Pt(II) makes the inclination of these glucose units so large that the chloride ions are inserted into the β-CyD cavity. The β-CyD rings to which two 2-pyridylethylamino **9** groups are appended at the 6-*C* position of two vicinal glucose units are linearly stacked to form a column-like structure [65]. One pyridyl group is included at the primary hydroxyl end of the cavity and the other is inserted into the cavity of the next molecule from the secondary hydroxyl side. These pyridyl groups are stacked, with a distance of about 4 Å between the aromatic planes. A similar arrangement of the β-CyD ring was observed in the crystal of $6^{I},6^{II}$-disubstituted β-CyD with *p*-methylbenzylthio **10** and phenylthio groups [66]. However, the molecule is arranged on a two-fold screw axis, against which the β-CyD ring is so tilted and laterally shifted that the benzene ring cannot form the linear stack.

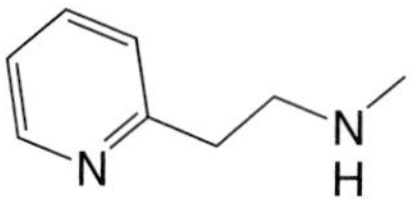

9

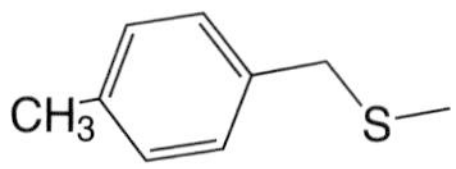

10

7.3.5
Capped CyDs

Capped CyDs (also discussed in Chapter 2) have been investigated for the purpose of increasing the hydrophobicity of the molecular cavity by closing its primary hydroxyl end. The capping is made by forming a bridge between the diagonally op-

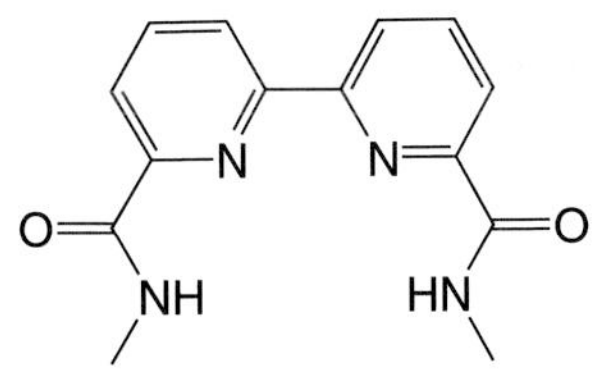

11

posed two glucose units at the primary hydroxyl side. Recently, crystal structures of permethylated α-CyD capped with a 2,2′-bipyridyl-6,6′-dicarboxyamino [67] group **11**, phenylene [68] group, or metal cations [69, 70] have been reported. Bridge formation at the O6 side imposes a restriction on the tilt of the glucose units and changes the macrocyclic structure. The coordination of 6-*N* amino groups of two diagonally positioned glucose units to Pt(II) causes a large tilt of these glucose units and distorts the macrocylic ring into a rectangular shape because the two glucose units are connected by a short bridge [69]. In contrast, the round macrocyclic ring is retained in the bipyridyl-capped α-CyD where the bridge is formed by a group with a suitable size for linking the diagonal positions (Fig. 7.10).

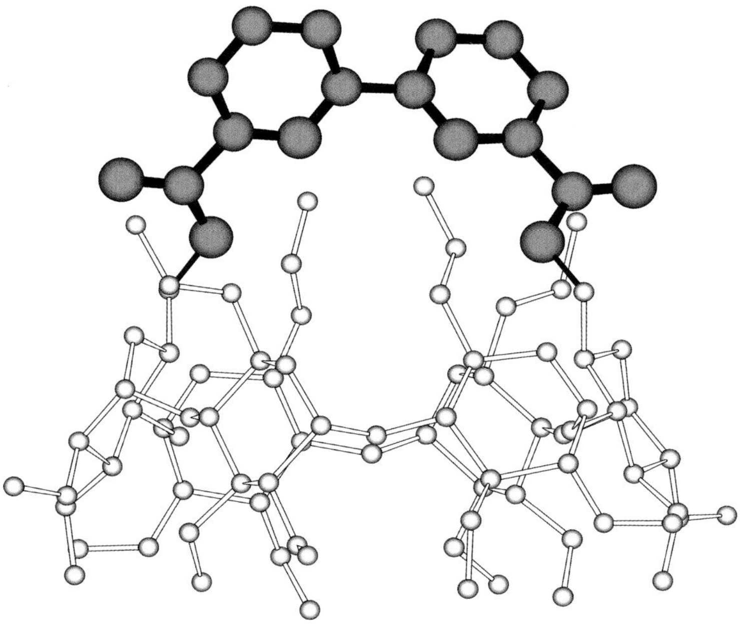

Fig. 7.10. Structure of 2,2′-bipyridyl-capped permethylated α-CyD.

7.3.6
CyDs with Nonglucose Units

Modification of the secondary hydroxyl side via 2,3-epoxide formation produces CyDs with nonglucose units. The aminolysis of mono-2,3-epoxy-α-CyD produces an α-CyD derivative with altrosamine. This molecule is elliptically distorted because of the strain imposed on the glycosidic linkage of the altrosamine with the $^{3,O}B$ boat conformation [71]. A similarly distorted macrocyclic structure was observed in mono[3-(2-imidazolylthio)]-altro-β-CyD which contains an altrose unit with the $^{1}C_{4}$ conformation [72]. On the other hand, all glucose units are converted to altrose in α-cycloaltrin which is prepared by the hydrolysis of 2,3-anhydro-α-cyclomannin, a round macrocycle that can include ethanol [74] and 1-propanol [75]. α-Cycloaltrin shows a symmetrical structure in which altrose units with the $^{1}C_{4}$ and $^{4}C_{1}$ chair conformations are alternately linked [73] as shown in Fig. 7.11. Dehydration between O2-H and O3H of the vicinal glucose unit produces an inter-residue anhydro-α-CyD in which two glucose units have no rotational freedom around the glycosidic linkage [76]. The intra-glucose dehydration of γ-CyD forms a highly asymmetrical macrocyclic compound with the saddle-like shape, octakis-(3,6-anhydro)-γ-CyD, because of the disruption of intramolecular hydrogen bonds between adjacent glucose units with the $^{1}C_{4}$ chair conformation [77].

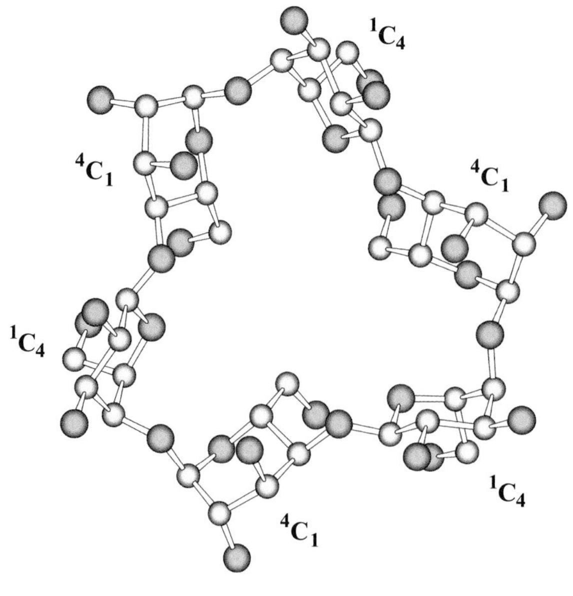

Fig. 7.11. Structure of α-cycloaltrin.

7.4 Crystalline Inclusion Complexes

7.4.1 Host–Guest Interactions

The intermolecular cavity of CyDs includes water molecules which are surrounded by a hydrophobic wall mainly composed of hydrogen atoms. Therefore, these water molecules cannot fully form hydrogen bonds and are energetically less stable than bulk water molecules. When a guest molecule, of suitable size and preferably hydrophobic character, is added to a CyD solution, water molecules in the CyD cavity are replaced by the guest molecule. Inclusion of the guest molecule induces a structural change in the CyD. As shown in the structure of δ-CyD and its complex, complex formation changes the elliptically distorted macrocyclic ring to a round structure with pseudo nine-fold symmetry [21, 78]. When CyDs form a complex with a guest molecule larger than the cavity space, the guest molecule is only partially included in the host cavity. In the crystal of such a complex, the guest molecule is in contact not only with the inner surface of the macrocyclic ring but also with adjacent CyD molecules and/or solvent molecules incorporated in the crystal.

Hydrogen bonds, van der Waals interaction, electrostatic interaction, etc., have been mentioned as attractive interactions to stabilize the structure of CyD complexes. Some short-range interactions observed in the crystalline state are illustrated in Fig. 7.12. In the crystal of the α-CyD complex with *p*-nitrophenol, the nitrophenyl group is well fitted to the α-CyD ring, which is rather elliptical in shape for the accommodation of the planar group [79]. The benzene ring is sandwiched by the two pyranose rings and two C–H bonds point to glycosidic oxygen atoms to form C–H–O hydrogen bonds. These hydrogen atoms of the benzene ring are also in van der Waals contact with hydrogen atoms bonded to C-3 and C-5 of α-CyD. The C–H bonds of methine groups, C-3H and C-5H, of the two pyranose rings facing the benzene ring are perpendicular to the benzene plane, indicating stabilization by C–H–π interaction. In addition to these short-range interactions, the crystal structure is also stabilized by long-range forces derived from electrostatic interaction, dipole–dipole interactions, etc.

In crystals of CyD complexes, the crystal packing is generally governed by the arrangement of CyD molecules because they dominate the intermolecular contact to form the crystal lattice. However, the packing mode is not unique for a particular CyD but varies according to the guest molecule. There are three packing types, cage type, channel type, and layer type, which are widely observed (Fig. 7.13). Guest molecules determine the selection of one of these packing modes, which provides a suitable cavity into which they can be stably accommodated.

The cage-type packing structure is frequently observed for relatively small guest molecules which can be enclosed in the host cavity. CyD molecules are arranged in a herring-bone fashion and both ends of the host cavity are closed by adjacent molecules to create an isolated "cage" (Fig. 7.13A). The channel-type structure is

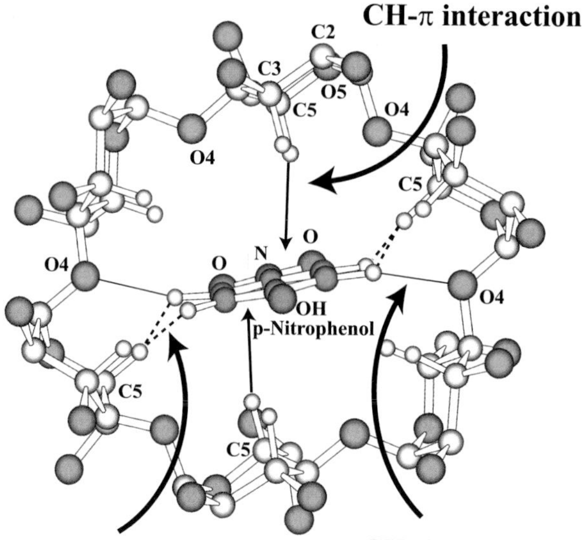

Fig. 7.12. Host–guest interaction observed in the α-CyD complex with *p*-nitrophenol. Hydrogen atoms of the phenylene ring are in van der Waals contact with hydrogen atoms of C-5H methine groups (dashed lines) and are also hydrogen bonded to O4 atoms of α-CyD (thin lines). The hydrogen atoms of C-5H methine groups of the two pyranose rings facing to the benzene ring are in the C–H–π contact as shown by arrows.

formed by linear stacking of CyD rings. This column-like structure has an infinite cylindrical channel that can accommodate a long molecule such as an alkyl chain or a linear polymer. There are two types of CyD arrangement, called "head-to-head" and "head-to-tail". The first structure is formed by the linear arrangement of head-to-head CyD dimers. In the dimer unit, the secondary hydroxyl side of two molecules is facing each other and connected by hydrogen bonds to create a barrel-like cavity (Fig. 7.13B). In the head-to-tail packing structure, CyD rings are linearly stacked and the primary hydroxyl side faces the secondary hydroxyl side of the next molecule by hydrogen bonds.

The layer-type packing structure has been sometimes observed when the guest molecule is so large that a part of the molecule cannot be accommodated within the CyD cavity. CyD rings are arranged in a plane to make a molecular layer and two adjacent layers are shifted with respect to each other by half a molecule, showing a brick-work pattern (Fig. 7.13C). Both ends of the cavity are open to an intermolecular space of the adjacent layers. A part of the guest molecule not included in the host cavity protrudes into the intermolecular space and is in contact with host molecules of the adjacent layer.

Guest molecules included in CyDs are noncovalently bound and have some degree of freedom for translational and rotational motion. In the crystal of the

(A)

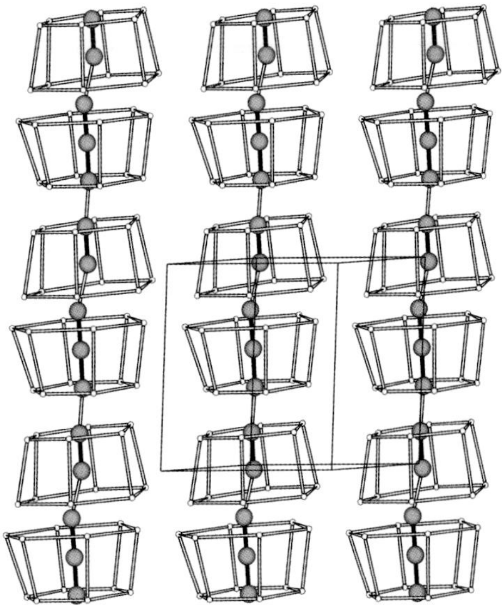

(B)

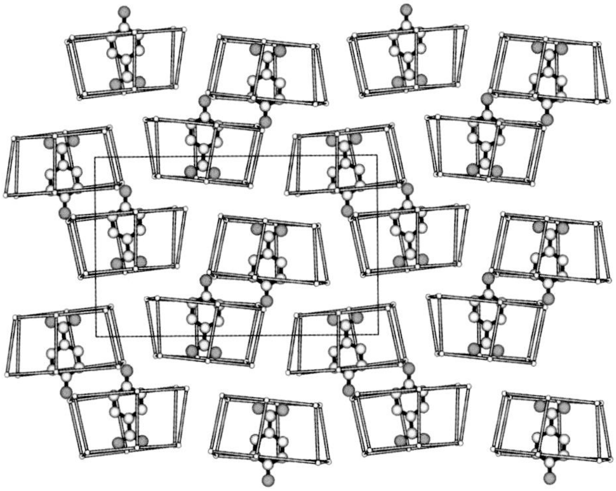

(C)

Fig. 7.13. Schematic drawing of crystal packings: cage-type observed in the β-CyD complex with hexamethylenetetramine (A), head-to-head channel-type observed in the α-CyD complex with iodine-iodide (B), and layer-type observed in the α-CyD complex with *p*-nitrophenol (C).

heptakis(2,6-di-*O*-methyl)-β-CyD complex with 2-naphthoic acid [80] **12**, the atomic motion of the guest molecule is mostly derived from rigid-body motion with an average amplitude of about 6° in rotation and 0.2 Å in translation [81]. A similar result has been derived from the rigid-body analysis of hexamethylenetetramine **13** included in β-CyD [82]. The rotational motion of this guest molecule is prominent around the axis through two amino groups that are fixed by hydrogen bonds with an adjacent host molecule [83]. The flexibility of the β-CyD ring is deduced from the rotational motion of pyranose rings with an average amplitude of 5.2° around the axis through glycosidic O4 atoms linking glucose units. The problem of CyD flexibility is discussed in detail in Chapter 3.

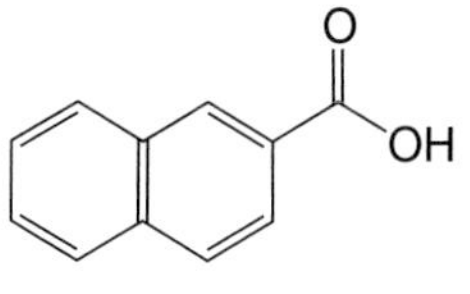

12

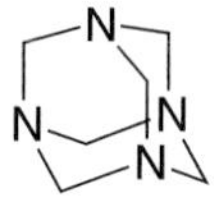

13

7.4.2
α-CyD Complexes

The radius of the α-CyD cavity is 4.2 Å as measured in the O4 hexagon (Table 7.1). This cavity size is most suitable for the inclusion of a benzene ring with its long axis parallel to the molecular axis of α-CyD if we consider van der Waals radii. Therefore, a guest molecule smaller than benzene can be fully included in the α-CyD cavity and many of these complexes crystallize in the cage-type packing structure as observed in the complexes with iodine [84], krypton [85], methanol [86], 1-propanol [87], 3-iodopropionic acid [88], nitromethane [85], and acetonitrile [86]. Figure 7.14 shows a typical cage-type packing in the crystal of the iodine complex. The α-CyD ring in hydrate crystals is distorted because of the disruption of some intermolecular hydrogen bonds between glucose units. The inclusion of a guest molecule makes the α-CyD ring more symmetrical with full hydrogen-bond formation between O2H and O3H of adjacent glucose unit. Krypton, a noble gas, partially occupies the α-CyD cavity and its content in the crystal is increased at high pressure [85]. The krypton atom is disordered and shares its position with water molecules; therefore, more water molecules are replaced by krypton at higher pressure. The hydroxyl group of methanol [86] and 1-propanol [87] is hydrogen bonded to a primary hydroxyl group of α-CyD, stabilizing the guest molecules which are too small to be fixed by van der Waals interactions with the inner surface of the cavity.

The channel-type structure is formed with a variety of guest molecules. Generally, anionic guests, such as, acetate [91], 1-propanesulfonate [92], benzenesulfonate [93], γ-aminobutyrate [94], and Methyl Orange [95] **14** are preferentially included in positively charged cylindrical channels. In contrast, metal cations are

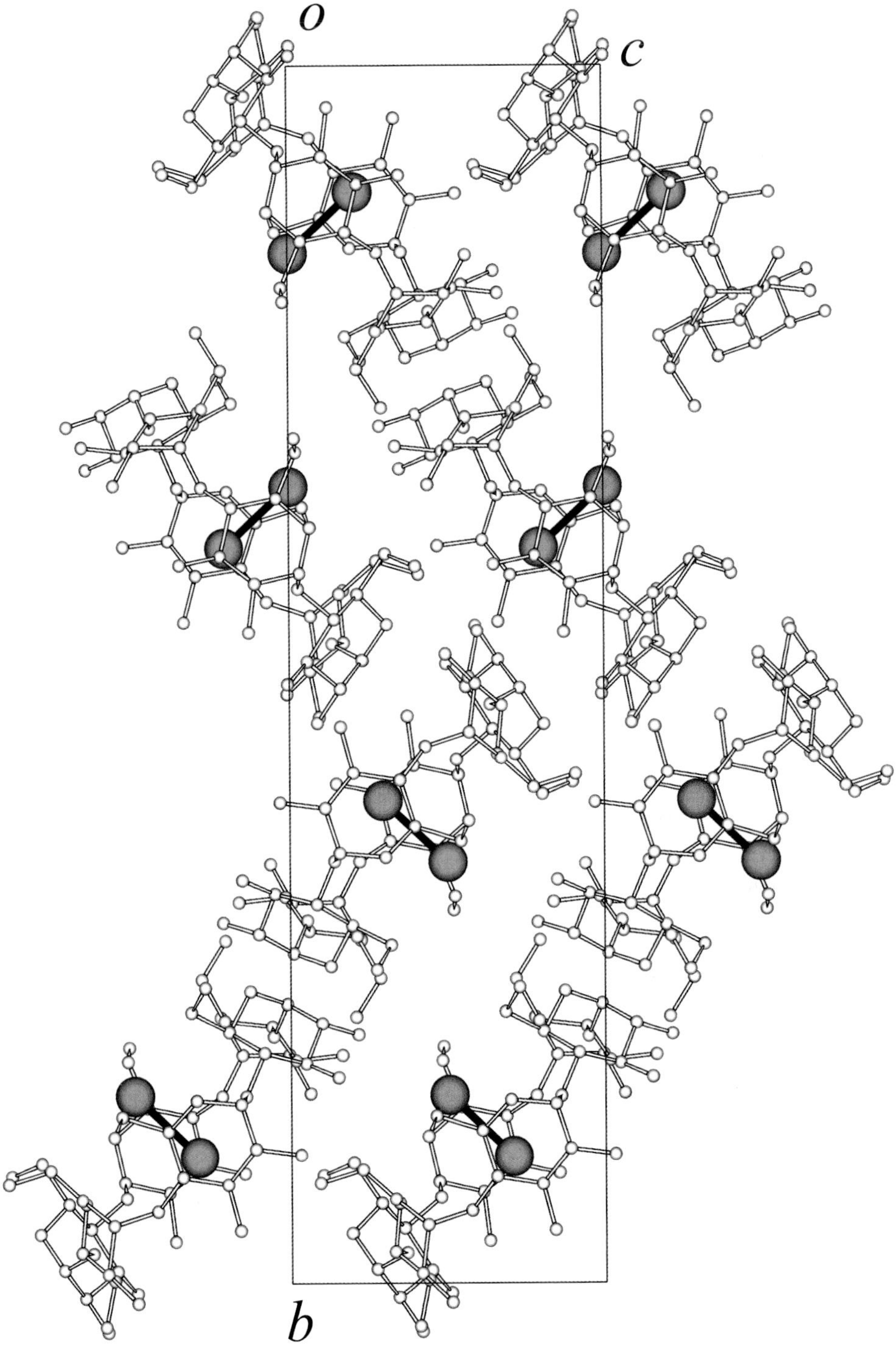

Fig. 7.14. Crystal structure of the α-CyD complex with iodine.

located outside the α-CyD ring and coordinated by hydroxyl groups and/or water molecules. In these complexes, α-CyD molecules are arranged in the head-to-tail mode (Fig. 7.15A). In the 1:2 complex with *m*-nitrophenol [96] or *m*-bromophenol [97], one guest molecule is included in the host channel and the other is located in the intermolecular space between α-CyD columns. A similar but not identical packing mode is observed in the complexes with the nonionic benzene derivatives benzaldehyde [98], 1-phenylethanol [99], hydroquinone [100], *p*-cresol [101], *m*-nitroaniline [102], and *o*- and *p*-fluorophenol [103]. In these crystals, the α-CyD ring is not perpendicular to the channel axis but a cylindrical column is formed by the stack of tilted α-CyD rings. Since the guest molecule is linearly aligned within the channel cavity, the size of the guest molecule is usually less than the depth of the α-CyD cavity that corresponds to the repetition unit. The Methyl Orange molecule is twice as large as a single host cavity and, as a result, the molecule penetrates one α-CyD ring and forms contacts with two other host molecules.

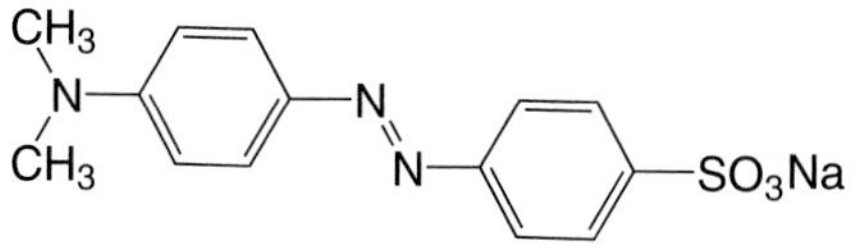

14

α-CyD complexes with iodine-iodide crystallize in five forms depending on the counter cations and two crystal structures have been determined by X-ray analysis [104]. In these crystals, the iodine-iodide anion is linearly aligned to form a polyiodide chain passing through the channel (Fig. 7.15B) formed by the head-to-head arrangement of α-CyD. A similar packing structure is formed in the complexes of the nonionic guests butyl isothiocyanate [105] and acetone [106]. The head-to-head dimer cavity includes isosorbide dinitrate [107] **15** and 4,4′-biphenyl dicarboxylic acid [108] **16** to form a complex with 2:1 stoichiometry. The latter complex forms a one-dimensional column-like structure while in the complex of isosorbide dinitrate the channel is in a zigzag shape because the α-CyD molecules in the dimer unit are laterally shifted away from each other to create a cavity fitted to the shape of the guest molecule.

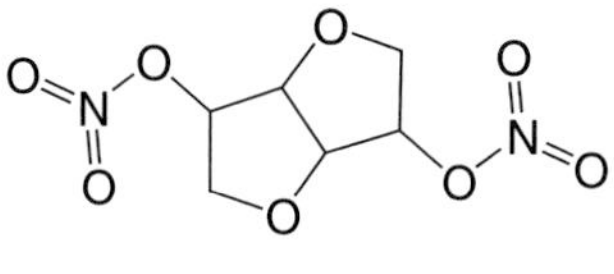

15

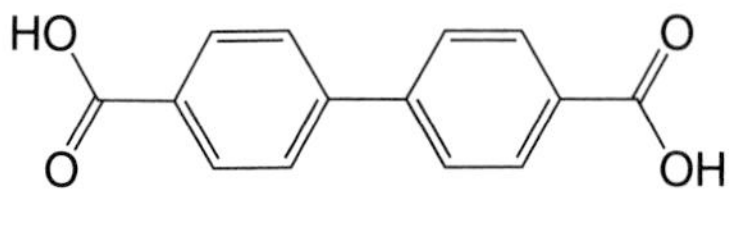

16

The layer-type packing structure is observed in the complexes with *para*-isomers of disubstituted benzenes, such as *p*-iodoaniline [109], *p*-iodophenol [110], *p*-

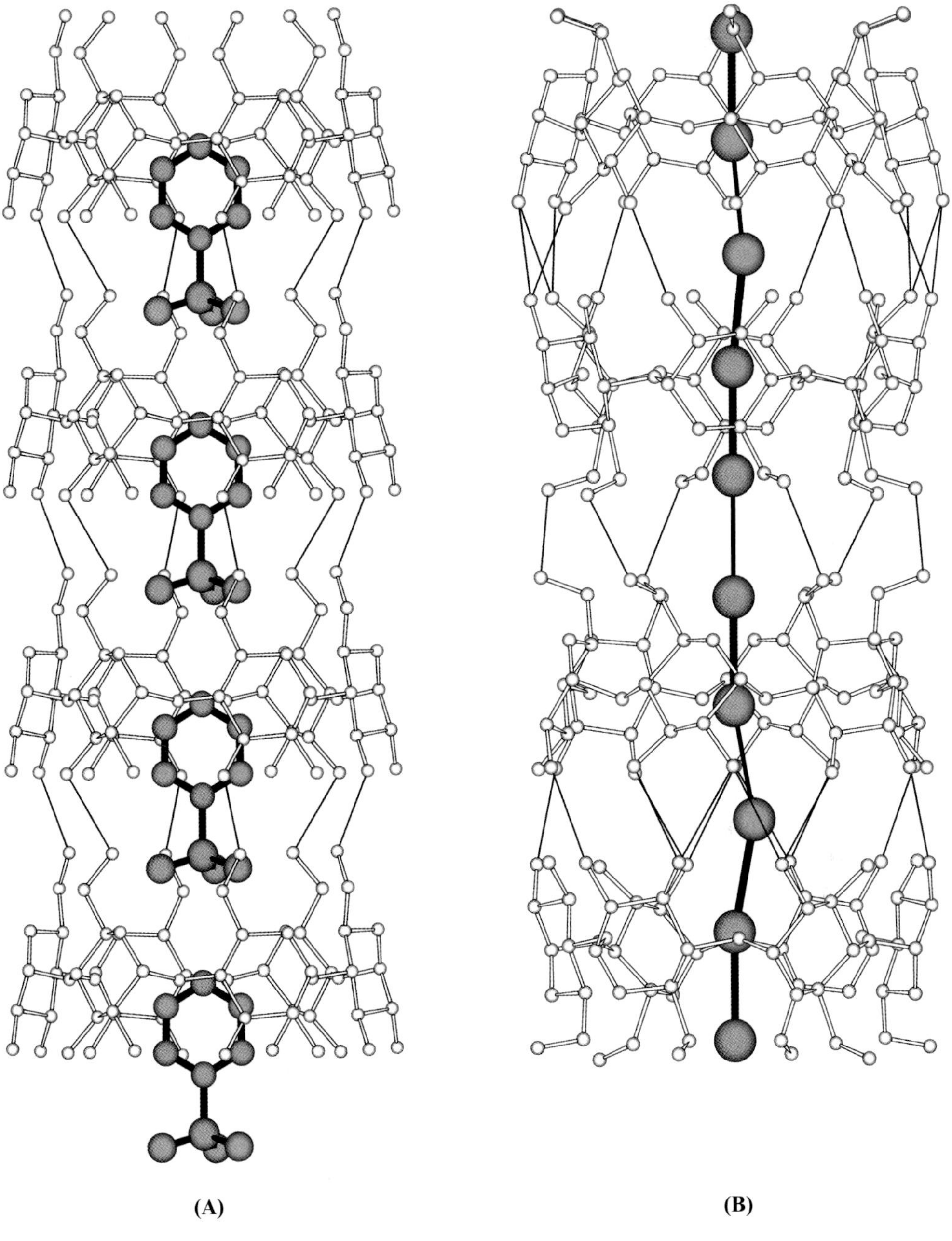

(A) (B)

Fig. 7.15. Channel-type packing structures observed in α-CyD complexes with benzenesulfonate (A) and iodine-iodide (B).

nitrophenol [79], *p*-chlorophenol [101], and *p*-bromophenol [97]. The benzene ring with the larger substituent group is included in the α-CyD cavity while a smaller amino or hydroxyl group protrudes from the secondary hydroxyl side of the host cavity and forms hydrogen bonds with adjacent α-CyD and/or water molecules. The α-CyD complexes with 2-pyrrolidone [111] and *N,N′*-dimethylformamide [111] also crystallize in the same packing structure although these guest molecules are small enough to be enclosed in the cavity. In the α-CyD crystal obtained from a DMSO/methanol solution, DMSO and methanol molecules are simultaneously included in the host cavity, but a part of the DMSO molecule is located outside the cavity [113]. In the complex with 1,12-diaminododecane, the alkyl chain of the guest molecule penetrates through two α-CyD molecules arranged to form a head-to-head dimer [114]. The primary hydroxyl ends of the dimer cavity are open to intermolecular space and each amino group protrudes from the primary hydroxyl side, forming hydrogen bonds with hydroxyl groups in the adjacent layer.

7.4.3 β-CyD Complexes

The cavity of β-CyD is so wide that it can accommodate the naphthalene ring axially. The mode of inclusion depends not only on the size and shape of the guest molecule but also on its physicochemical properties. Nonionic and/or hydrophobic small molecules tend to form the cage-type packing structure. Hydrogen iodide [115], methanol [115], ethanol [116], butane-1,4-diol [117], butyne-1,4-diol [118], ethylene glycol [119], glycerol [119], formic acid [120], acetic acid [120], and dimethylsulfoxide [121] are so small that they cannot fill the cavity and water molecules are co-included. The β-CyD cavity is fully occupied by squaric acid [122] (**17**), cyclohexane-1,4-diol [123], and hexamethylenetetramine [82]. The benzene ring of *m*-aminophenol [124], benzyl alcohol [125], and nicotinamide [126] is included with its long axis perpendicular to the molecular axis of β-CyD (Fig. 7.16A). 2,7-Dihydroxynaphthalene [127] is the largest guest so far observed in crystals with the cage-type packing structure. The naphthalene ring is accommodated in the β-CyD cavity but both the hydroxyl groups of the guest protrude outside the cavity (Fig. 7.16B).

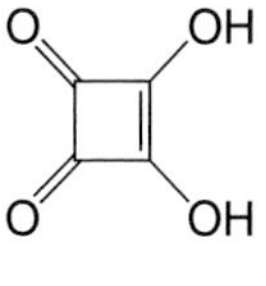

17

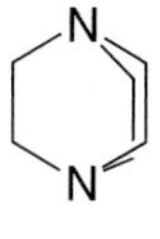

18

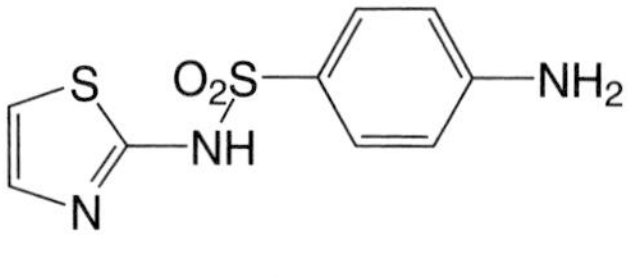

19

The layer-type structure is observed in β-CyD complexes with triethylenediamine [128] **18** and sulfathiazole [129] **19**. The β-CyD ring is arranged in a molecular layer stacked along the crystallographic two-fold axis. A part of the guest molecule

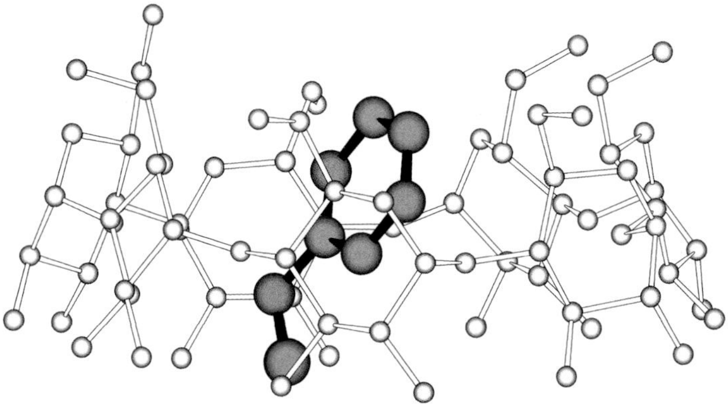

(A)

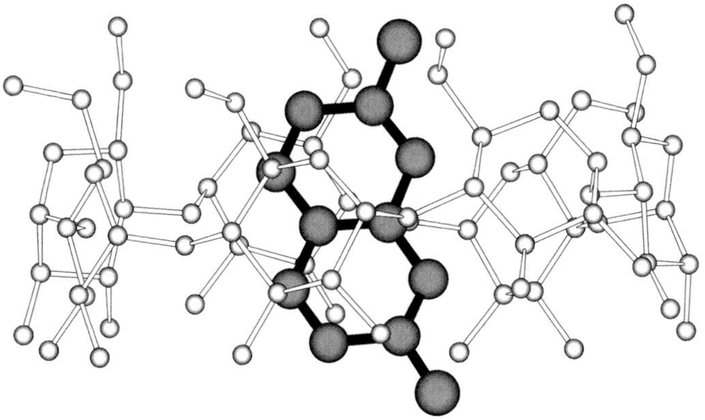

(B)

Fig. 7.16. Structures of the β-CyD complex with benzylalcohol (A) and 2,7-dihydroxynaphthalene (B).

protrudes from the secondary hydroxyl side into the intermolecular space between the two layers (Fig. 7.17). It is of interest that triethylenediamine, which is smaller than hexamethylenetetramine [82], does not form the cage-type packing structure.

β-CyD complexes with some drug guests (presented in Chapters 14 and 15) that are considerably larger than the host cavity form a head-to-tail channel-type structure. In complexes with piroxicam sodium [130] **20** and mefenamic acid [131] **21**, β-CyD rings are linearly stacked to form the column-like structure. The guest molecule is sandwiched between the two β-CyD rings into which a part of the guest molecule is inserted. Niflumic acid **22** forms a 1:2 complex in which two guest

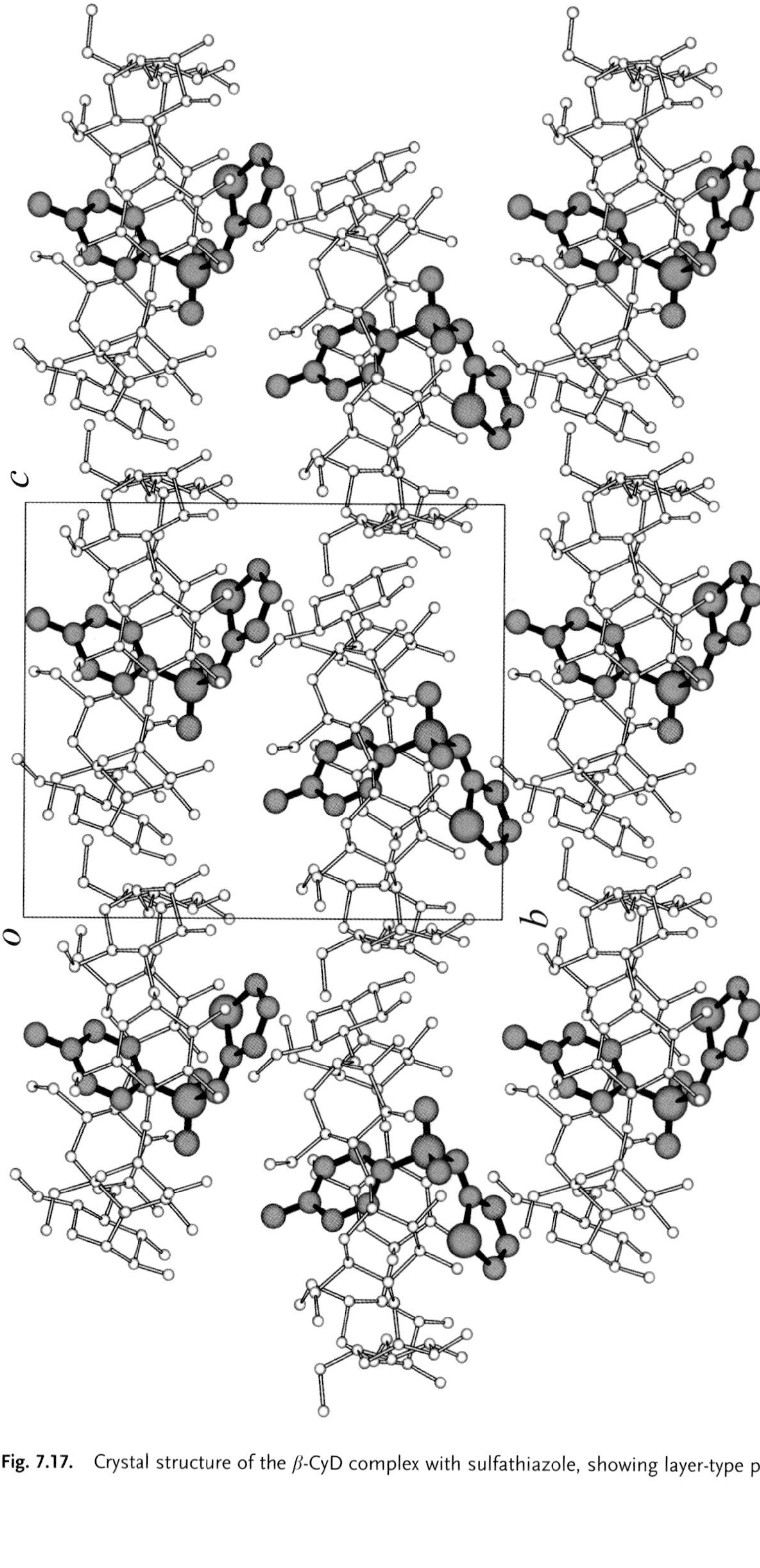

Fig. 7.17. Crystal structure of the β-CyD complex with sulfathiazole, showing layer-type packing.

molecules are inserted into the β-CyD cavity from both the primary hydroxyl and secondary hydroxyl sides [132]. On the other hand, the meclofenamate sodium complex **23** [133] is arranged on the two-fold screw axis to form a zigzag channel and the guest molecule is sandwiched between β-CyD molecules (Fig. 7.18A). A 1:1 complex of diclofenac sodium **24** is helically stacked along the six-fold screw axis [134] although the structure of the host–guest interaction is very similar to that of the meclofenamate sodium complex.

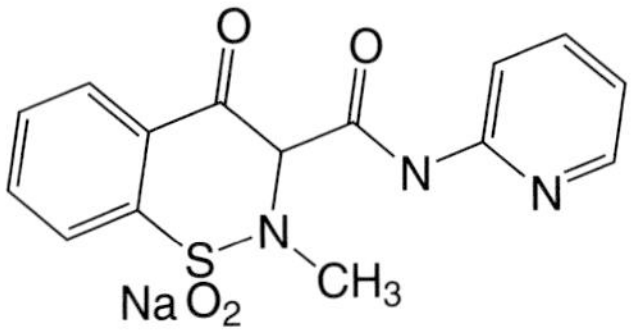

20

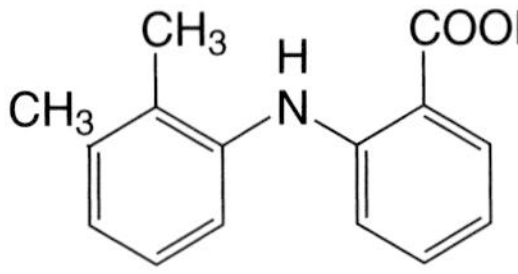

21

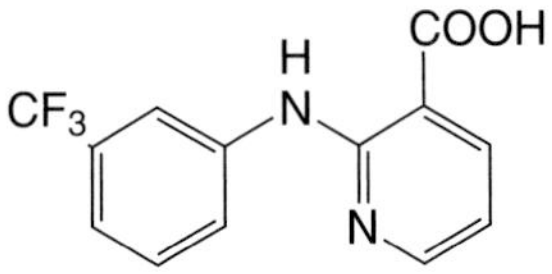

22

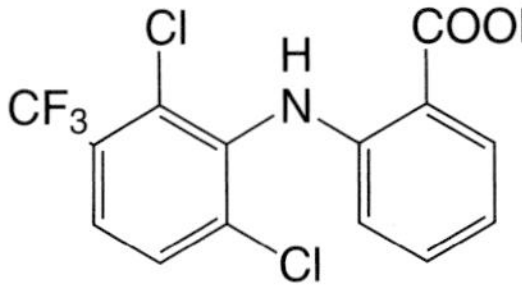

23

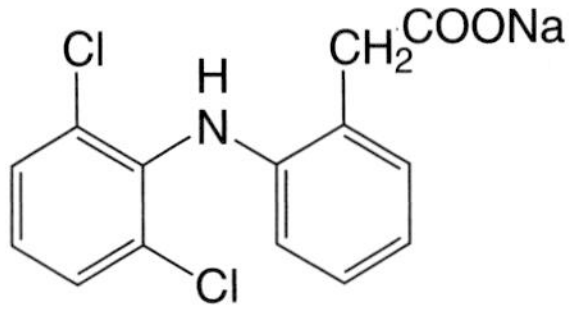

24

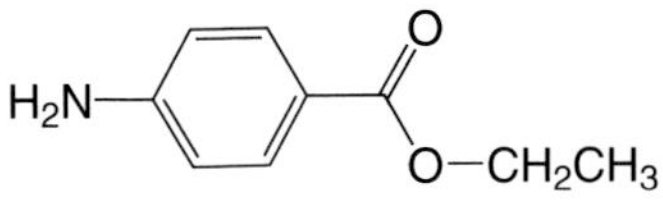

25

The dimeric structure of β-CyD is most frequently observed as a building block of crystal structures. There are two types of arrangement of the dimer unit. In the head-to-head channel-type structure, the dimer unit is arranged to form an infinite linear channel. The other type is the layer-type structure, in which dimer units are arranged to form a chess-board-like pattern. Both ends of the dimer cavity are open to the intermolecular space in the adjacent layer.

In the channel-type structure, the dimer units are linearly stacked or helically arranged on the crystallographic two-fold axis. In the former structure, the dimer unit coincides with the repetition unit. A variety of guest molecules, such as ethanol [135], *m*-iodobenzoic acid [136], benzocaine [137] **25**, iodine-iodide [138], 3,3-dimethylbutylamine [139], pyrene [140], *trans*-cinnamic acid [141], 4,7-dimethyl-

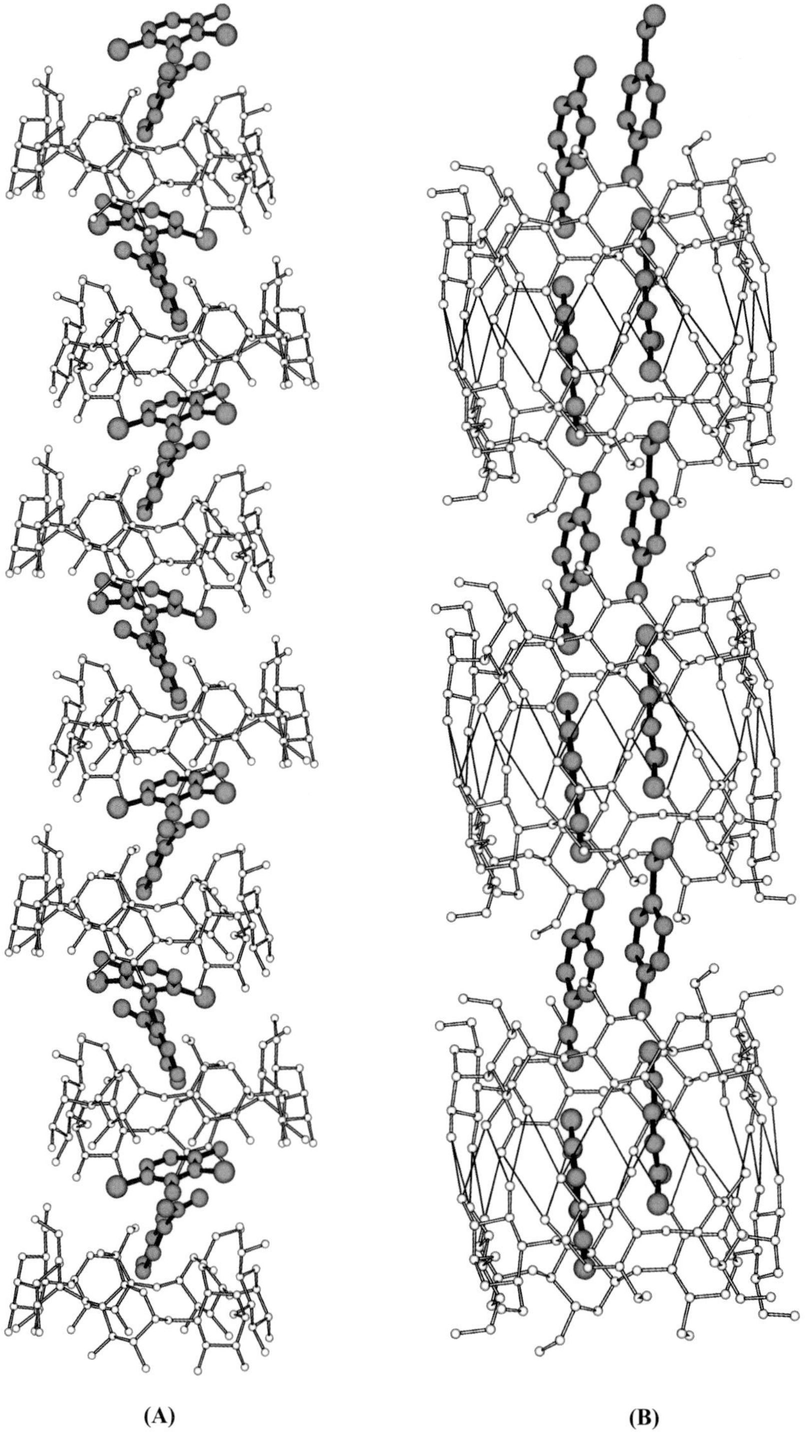

Fig. 7.18. A head-to-tail channel-type structure observed in the β-CyD complex with meclofenamate (A) and a head-to-head channel-type structure in the p-hydroxybenzaldehyde complex (B).

coumarin [142] **26**, acetaminophen [143] **27**, *β*-naphthyloxyacetic acid [144], and (*S*)-ibuprofen [145] **28** are included in the channel cavity. In these complexes, the dimer unit has a crystallographic two-fold symmetry but the arrangement of the guest molecules in the dimer cavity is sometimes inconsistent with this symmetry rule because they are bound by weak interactions and have no contact with the outside of the channel. A typical example is the complex with a linear polymer such as poly(trimethylene oxide) [146] or polyethylene glycol [147]. The repetition unit of these polymers coincides with the depth of the dimer cavity but has no two-fold symmetry. The iodine-iodide anion does not form the kind of linear polymeric chain observed in the *α*-CyD complex [138]. Two I_2 molecules are included in the dimer cavity and the I_3^- ion is sandwiched between two dimer units. On the other hand, in the ternary complex with pyrene and *n*-octanol both guest molecules are located on the two-fold axis. The pyrene molecule is sandwiched by *β*-CyD molecules at the center of the dimer cavity and the *n*-octanol molecule, located at the primary hydroxyl side, links two dimer units [140].

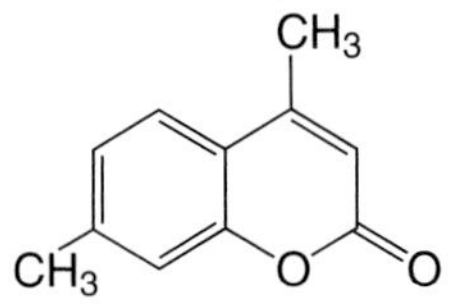

26

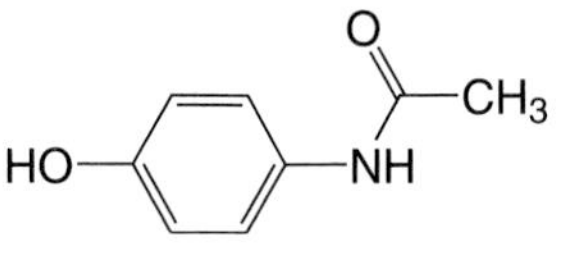

27

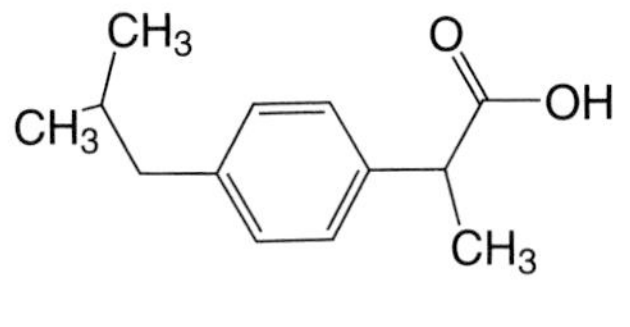

28

In the complexes with *p*-nitroacetanilide [148], *p*-bromoacetanilide [149], 4-*tert*-butyltoluene [150], benzoic acid [151], *p*-hydroxybenzaldehyde [152], 3,5-dimethylbenzoic acid [153], tridecanoic acid [154], (*Z*)-tetradec-7-enoic acid [154], and 1,2-bis(4-aminophenyl)ethane [160], the guest molecules are mostly ordered in a dimer cavity that has no crystallographic symmetry. The dimer cavity is so wide that two hydroxybenzaldehyde molecules are packed with their benzene planes facing each other (Fig. 7.18B). On the other hand, the *β*-CyD dimer is aligned along a two-fold screw axis in the complexes with *p*-iodoaniline [155], *p*-ethylaniline [155], carmofur [156] **29**, phenoprofen [157, 158] **30**, methyl-*p*-toluylsulfoxide [159], and 4,4′-diaminobiphenyl [160], which are included in a channel that is in the zigzag shape. These crystals are not necessarily isomorphous, since the guest molecule causes variation in the arrangement of the dimer unit.

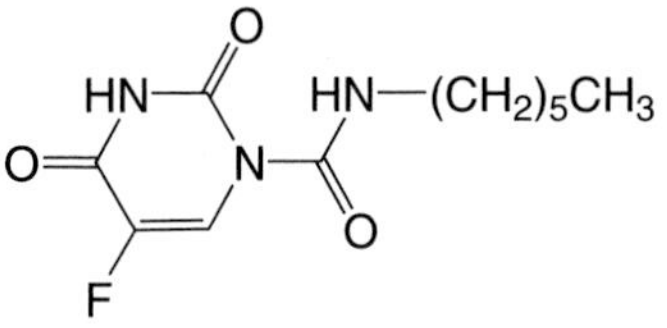

29

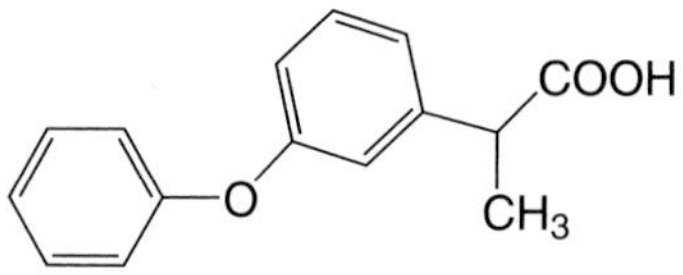

30

Layer-type packing of the β-CyD dimer is observed in the complexes with 1-propanol [161], *p*-iodophenol [161], ethylcinnamate [162], flurbiprofen [163, 164] (**31**), 1-adamantanecarboxylic acid [165], *tert*-butylbenzoic acid [166], hydroquinone [167], L-menthol [168], nonanoic acid [169], 1,2-dodecanedioic acid [170], 1,14-tetradecanedioic acid [171], 1,12-dodecanediol [172], (*Z*)-tetradec-7-en-1-al [173], 7-hydroxy-4-methylcoumarin [174], and phenylalanine derivatives [175–177]. The dimer cavity has a pseudo two-fold symmetry, and two guest molecules are included, mostly in a head-to-head arrangement. The cavity accommodates a bulky and hydrophobic moiety of the guest molecule but polar groups protrude and form hydrogen bonds with adjacent β-CyD/water molecules. In the β-CyD complex with 4-*tert*-butylbenzyl alcohol, the dimer unit is arranged almost parallel to the two-fold screw axis but does not form a continuous channel because of a lateral shift away from the screw axis [178].

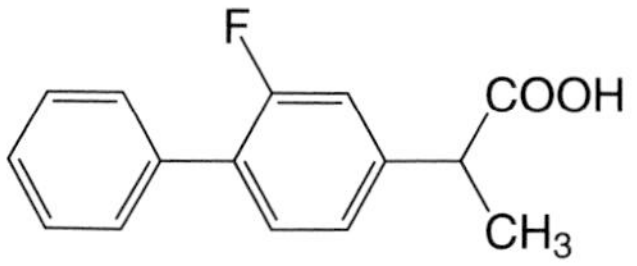

31

7.4.4 Complexes of Larger CyDs

γ-CyD crystallizes with the cage-type packing structure from water. However, crystals of inclusion complexes have so far only been found in the channel-type structure. In complexes with methanol [179] and 1-propanol [180], the γ-CyD molecule is on the four-fold axis and the repetition unit of the channel consists of three molecules, a head-to-head dimer with the third molecule attached in the head-to-tail mode. Methanol and 1-propanol molecules could not be located in the γ-CyD cavity because of statistical disorder or diffusive motion in the wide cavity. The dimer cavity accommodates a 2:1 complex of 12-crown-4 **32** with a metal cation [181, 182]. Thus, a complex of 2:2:1 stoichiometry is formed. This complex crystallizes

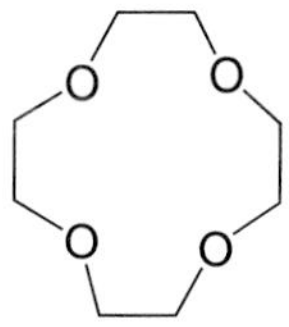

32

in the same packing structure as that observed in the methanol complex. In the complex with potassium 12-crown-4, the crown ring is in direct contact with the secondary hydroxyl side of γ-CyD while the potassium cation located at the center of the dimer cavity is sandwiched by two 12-crown-4 molecules (Fig. 7.19A). A metal-free 12-crown-4 molecule is included in the third γ-CyD molecule.

A crystal of the δ-CyD hydrate shows the cage-type packing structure [21]. In the solid state, δ-CyD forms complexes with several macrocyclic compounds, such as cyclononanone to cyclopentadecanone [183]. Currently only the structure of the cycloundecanone complex has been reported. The crystal shows a typical channel-type packing structure consisting of dimeric δ-CyD with the head-to-head mode [78]. Four cycloundecanone molecules are included in the dimer cavity (Fig. 7.19B). Two guest molecules are perpendicularly located at the center of the dimer cavity and each of the other two molecules is coaxially included at the primary hydroxyl end of the cavity.

Cyclomaltohexaicosaose forms a complex with iodine-iodide anion. Each of the two amylase-like short helixes includes a polymeric chain of I_3^- anions. The counter ion, Ba^{2+} or NH_4^+, is located outside the helices and hydroxyl groups, ethylene glycol, and PEG400 molecules are coordinated to the Ba^{2+} ion [184].

7.4.5
Complexes of Methylated CyDs

Hexakis(2,6-di-*O*-methyl)-α-CyD complexes with a small guest molecule, such as iodine and 1-propanol, crystallize with the cage-type packing structure [185]. Compared with the structure of the corresponding α-CyD complex, the guest molecules in the both complexes are shifted to the secondary hydroxyl side from the center of the cavity. 3-Iodopropionic acid [186], *m*-nitroaniline [187], and acetonitrile [188] are also fully accommodated in the host "cage". A 3-*O* acetylated host, hexakis(2,6-di-*O*-methyl-3-*O*-acetyl)-α-CyD, was crystallized from butylacetate [189]. In spite of the disruption of intramolecular hydrogen bonds, the host molecule is in a round shape because of the inclusion of butylacetate.

Heptakis(2,6-di-*O*-methyl)-β-CyD includes acetic acid [194], 2-adamantanol [190], *n*-butylacrylate [191], and isobornylacrylate [191] in its cavity. The complex with carmofur, which is larger than the host cavity, crystallizes in the layer-type packing structure [156]. Compared with the crystal structures of other methylated CyD

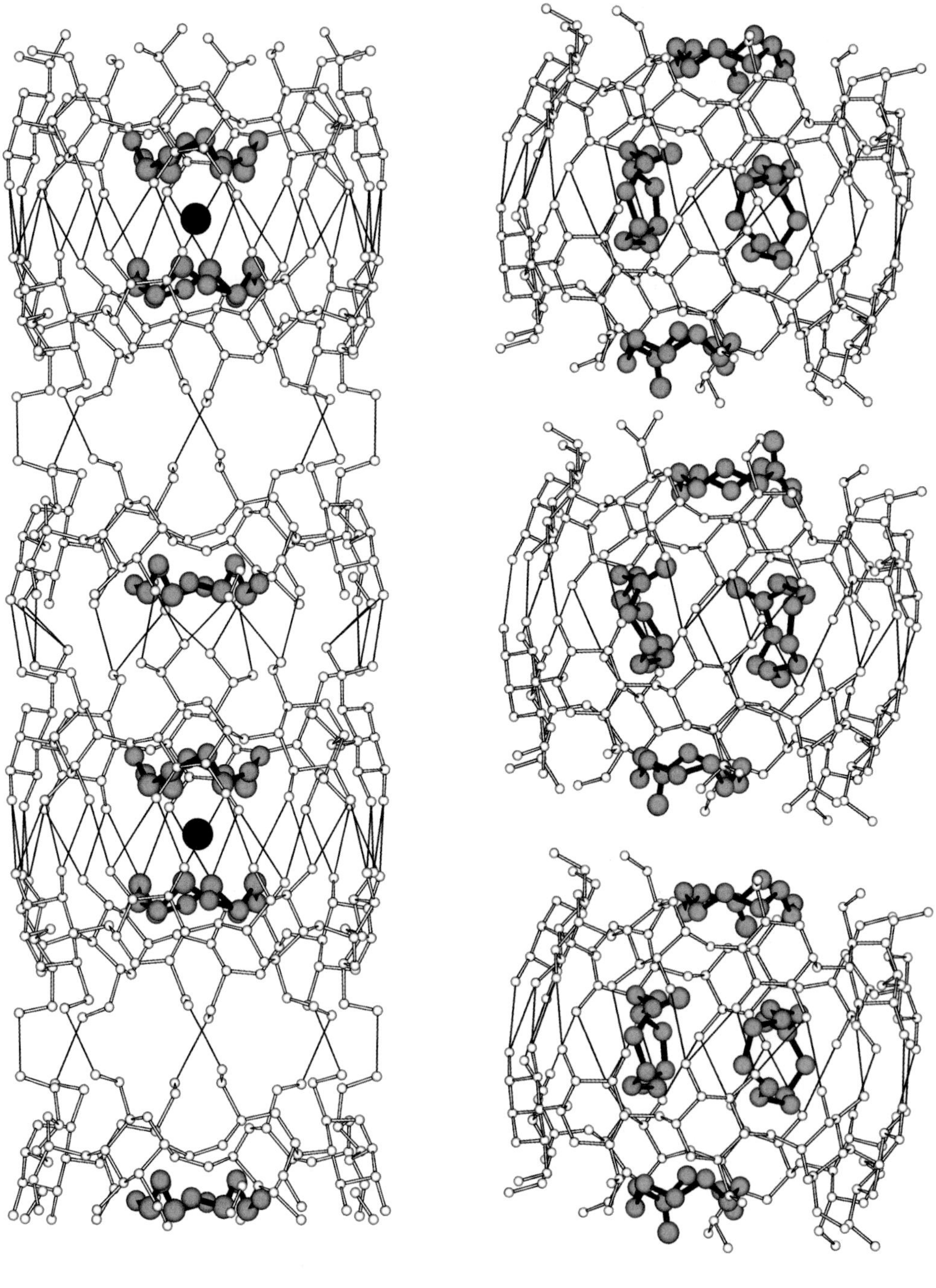

Fig. 7.19. Head-to-head channel-type structures observed in the γ-CyD complex with potassium (12-crown-4)$_2$ (A) and the δ-CyD complex with cycloundecanone (B).

complexes, a prominent difference has been observed in complexes with such aromatic guests as *p*-iodophenol [192], *p*-nitrophenol [192], 2,4-dichlorophenoxyacetic acid [193], and 2-naphthoic acid [80]. In these crystals, the guest molecules are excluded from the host cavity or only partially included at the O6 side as shown in Fig. 7.20. The host molecule is arranged along a two-fold screw axis and a 6-*O* methoxyl group is inserted into the cavity of the next molecule from the secondary hydroxyl side.

Permethylation of hydroxyl groups makes the O6 side narrower because of the sharp inclination of 2,3,6-tri-*O*-methylglucose units caused by the steric hindrance between methyl groups bonded to O3 and O2 of adjacent 2,3,6-tri-*O*-methylglucose unit. As the result, *m*-nitroaniline [187] and *p*-nitrophenol [197] are included with their hydroxyl or amino group located at the O6 side, that is, in the orientation upside down compared with the corresponding complex with the parent α-CyD (Fig. 7.21). A similar upside-down inclusion is observed in the benzaldehyde complex [196], but *p*-iodoaniline [195] is included in the same orientation as that observed in the α-CyD complex, although the molecule is shifted to the wider O2, O3 side of the cavity. These crystalline complexes with *para*-disubstituted benzenes show a channel-type packing structure. The *p*-nitrophenol and *p*-iodoaniline molecules are larger than the depth of the α-CyD cavity, but the permethylation makes the cavity so much deeper that these guest molecules are smaller than the repetition unit constructing the column-like structure [198]. A channel-type packing is also observed in the complex with D-mandelic acid [200] and D- and L-1-phenylethanol [199]. On the other hand, the complexes with L-mandelic acid [200] and *m*-nitroaniline [187] crystallize with layer-type packing.

Heptakis(2,3,6-tri-*O*-methyl)-*β*-CyD, which has an asymmetrical macrocyclic ring, forms crystalline complexes with a variety of guest molecules [202–212]. The complexes with indole-3-butyric acid [202] and 2,4-dichlorophenoxyacetic acid [202] crystallize in a typical head-to-tail channel-type packing and the guest molecules are almost fully included in the channel. A distorted channel-type structure is observed in the complexes with relatively large guest molecules, such as biphenyl derivatives, naphthalene derivatives, and compounds having a long alkyl chain, which cannot be accommodated in a single host cavity. The channel is in the zigzag shape because the distorted macrocyclic ring arranged on the two-fold screw axis is laterally shifted. The biphenyl moiety of flurbiprofen [203] and biphenylacetic acid [204] is inserted into the host cavity from the O2, O3 side and the carboxyl group protrudes from the slit-like crevice of the channel. The O6 side of the cavity is so narrow that the biphenyl group cannot penetrate the macrocyclic ring. A similar structure is observed in the complexes with naproxen [205] **33** and ibuprofen

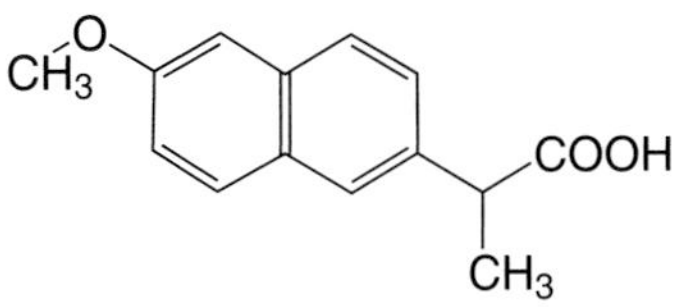

33

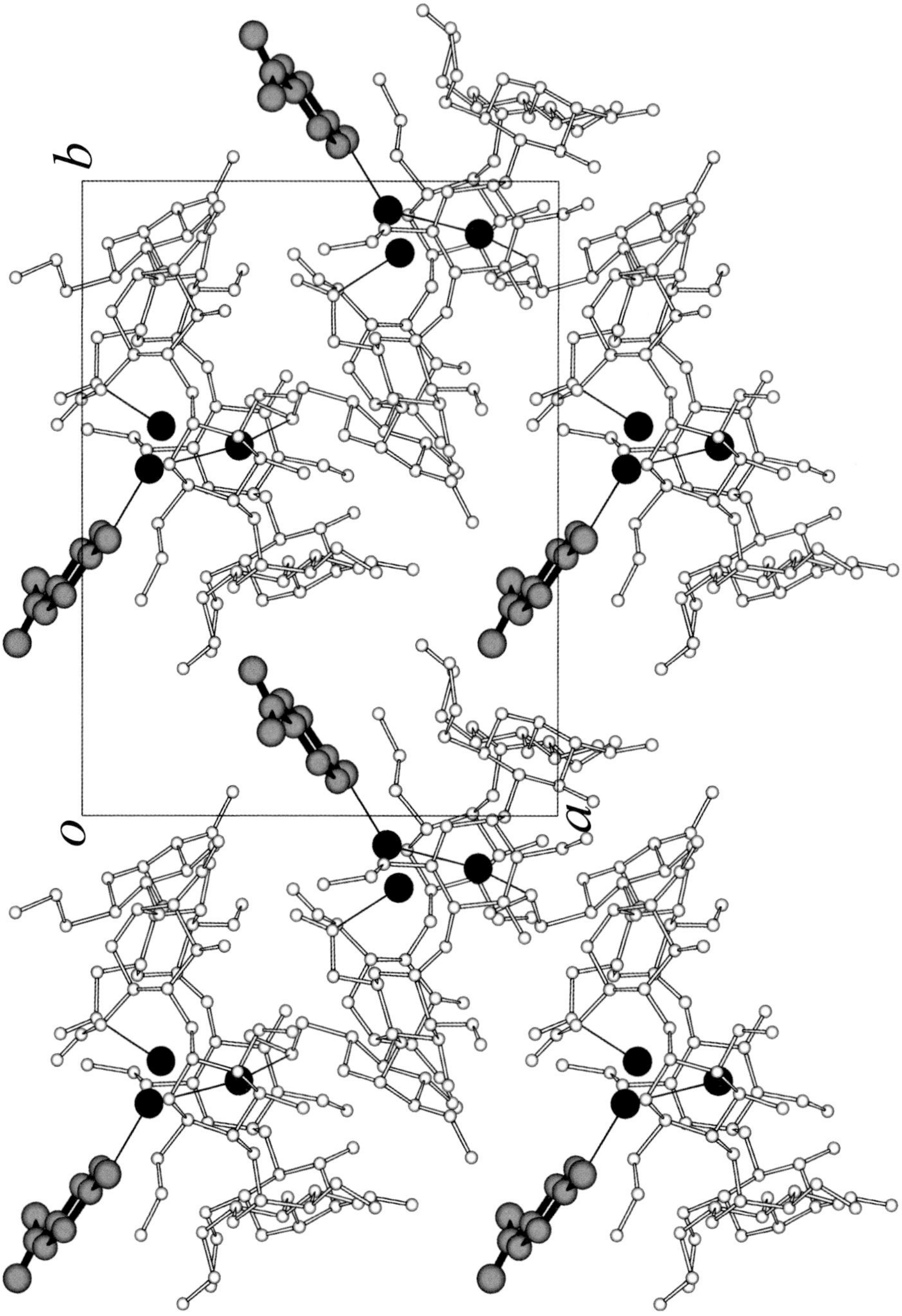

Fig. 7.20. Crystal structure of the heptakis(2,6-di-*O*-methyl)-β-CyD complex with *p*-nitrophenol. Water molecules are shown with filled circles and thin lines denote hydrogen bonds.

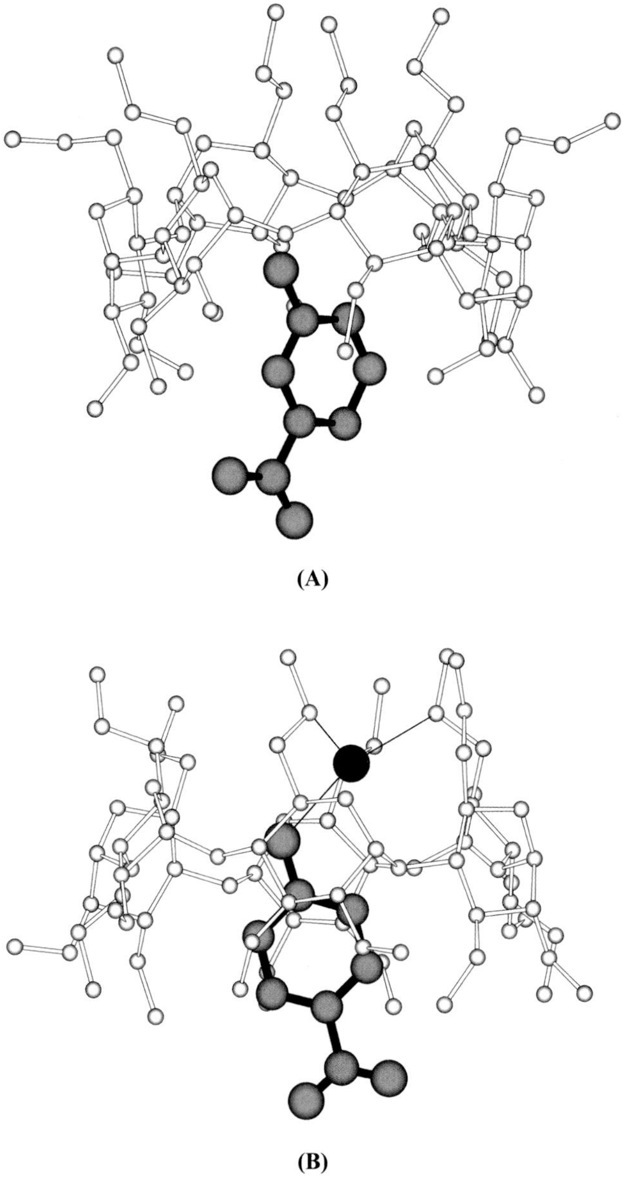

Fig. 7.21. Structures of hexakis(2,3,6-tri-*O*-methyl)-α-CyD complexes with *m*-nitroaniline (A) and *p*-nitrophenol (B).

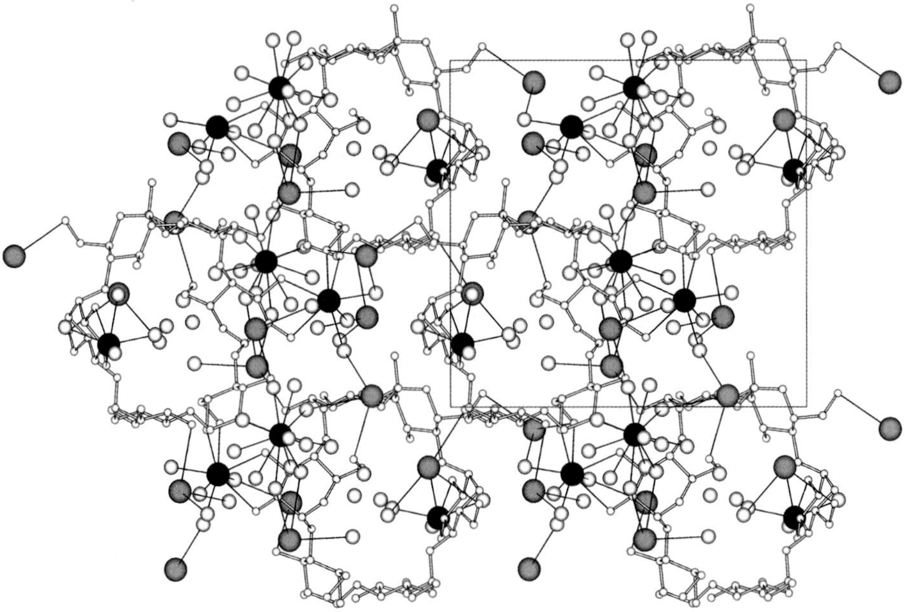

Fig. 7.22. Crystal structures of the calcium chloride complexes of α-CyD (A) and β-CyD (B). Calcium ions are shown by filled circles and chloride ions are shaded. Small circles represent water molecules.

[206]. Full accommodation of the guest molecule in the channel cavity is observed in the complex with ethyl laurate [207]. The dodecanoyl group penetrates through the host cavity and its terminal methyl group is inserted into the next CyD ring while the ethyl group protrudes from the O2, O3 side and is in contact with the O6 side of another symmetry-related host molecule. In the *m*-iodophenol [208] complex, one tri-*O*-methylglucose unit is in the ^{3,0}B boat conformation, which elliptically distorts the macrocyclic ring to create a cavity suitable for the planar guest molecule.

7.4.6
Complexes with Metals and Organometals

Usually metal cations are incorporated in the crystal as the counter ion of an anion included in the CyD cavity. Hydroxyl groups of CyD are sometimes directly coordinated to the metal cation. In the complexes of α-CyD with ionic guests, such

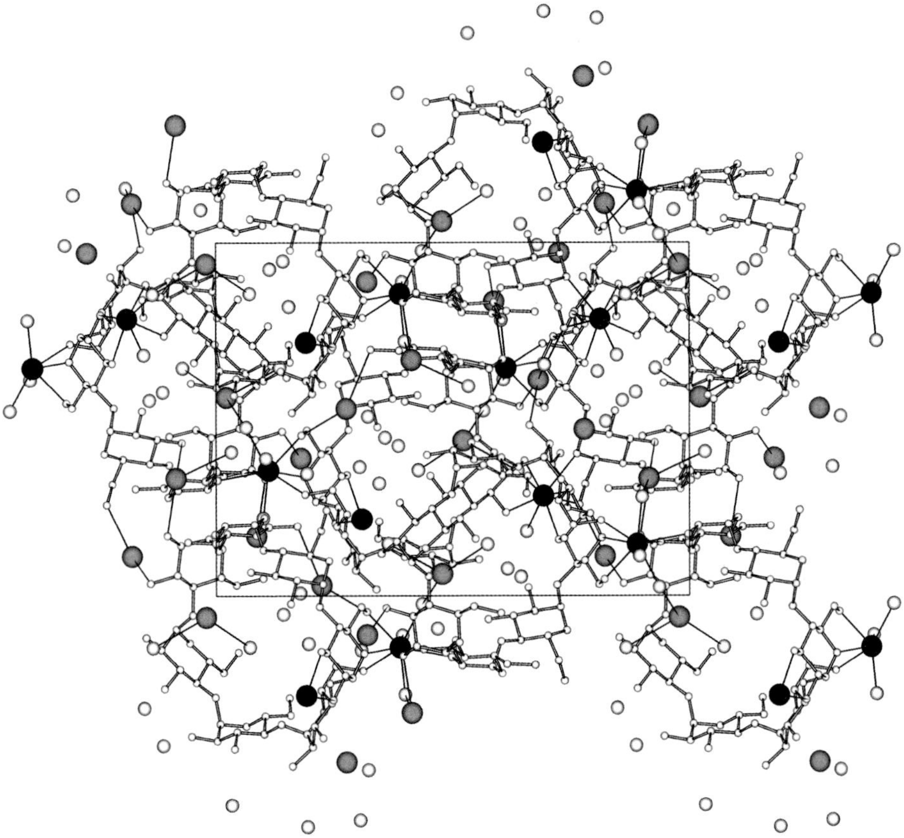

(B)

as Methyl Orange, sodium benzenesulfonate, and potassium acetate, the organic anions are included in the channel formed by the head-to-tail arrangement of α-CyDs while the sodium or potassium cation located outside the cavity is coordinated by hydroxyl groups and water molecules. A similar feature is observed in the iodine-iodide complex with the head-to-head channel-type structure where lithium or cadmium ion is separated from the iodide anion by the α-CyD ring.

α-CyD forms a 1:3 complex with calcium chloride [213]. Calcium ions are not included in the host cavity but directly bound to hydroxyl groups of α-CyD. Water molecules are also coordinated to the calcium ion (Fig. 7.22). β-CyD was co-crystallized with potassium hydroxide [214]. The crystal structure is isomorphous with β-CyD dodecahydrate and the potassium cation is incorporated in the intermolecular space by replacing a water molecule. In the 1:2 complex of β-CyD with calcium chloride [215], calcium and chloride ions are distributed outside the β-CyD ring, which is arranged in a cage-type like packing. On the other hand, in the 1:2 complex with magnesium chloride [216], the packing structure is composed of two alternately stacked layers, a β-CyD layer and an inorganic layer containing magnesium chloride and water. Six water molecules are coordinated to the magnesium ion, showing octahedral coordination.

Under alkaline conditions, sixteen Pb(II) ions link the secondary hydroxyl side of γ-CyD to form a metal-ion-mediated head-to-head dimer [217]. All the secondary hydroxyl groups are deprotonated and coordinated to bind Pb(II) ions forming a hexadecanuclear lead(II) alkoxide. Introduction of ionic substituents on CyDs enhances their metal-binding ability. Two amino groups introduced on the primary hydroxyl side of β-CyD can chelate a platinum ion [64]. Imidazole-appended β-CyD forms a ternary complex with a Cu(II) ion and L-tryptophanate [61]. The 6-amino and imidazolyl groups of the host molecule and the carboxyl and amino groups of L-tryptophanate are coordinated to the Cu(II) ion.

CyDs form crystalline complexes with organometals and coordination compounds. A metal ion surrounded by ligand molecules or groups has no direct contact with CyD, so the interaction is called "second sphere" coordination. A rhodium compound coordinated by 1,5-cyclooctadiene forms a crystalline complex with α-CyD [218]. The host cavity partly includes the 1,5-cyclooctadiene ligand and the structure is stabilized by hydrogen bonds between ammine ligands and secondary hydroxyl groups of α-CyD. The cyclobutene ring of carboplatin, cyclobutene-1,1-dicarboxylatodiamine platinum(II), is inserted into the α-CyD cavity from the secondary hydroxyl side and the two ammine ligands are hydrogen bonded to secondary hydroxyl groups [218]. In the α-CyD complex with aqua(*n*-butyl)cobaloxamine [219], the *n*-butyl group is inserted into the host cavity and the square plane made by the coordination of two dimethylglyoxime molecules is stacked on the secondary hydroxyl side (Fig. 7.23).

In the β-CyD complex with platinum phosphine [220], the trimethylphosphine ligand is bound at the primary hydroxyl side of the β-CyD cavity. On the other hand, alkyl ligands, *i*-butyl, *n*-butyl, *n*-hexyl, and cyclohexyl groups of alkyl(aqua)-cobaloximes are inserted into the cavity from the secondary hydroxyl side [221–224]. The channel cavity formed by β-CyD dimers includes diaqua(benzoate)-hydroxydioxouranium(VI). Two benzoate moieties are inserted into the dimer cavity and the U(VI) atom is located at the primary hydroxyl side without direct contact with the host molecule [225].

CyDs form crystalline complexes with metallocenes. Crystallographic study of several metallocene complexes of α-CyD and β-CyD has been reported [226–228]. These complexes mostly crystallize in a head-to-head channel-type packing structure. In the α-CyD complex with ferrocene, the metal ion is located at the center of the dimer cavity and the host α-CyD molecule has direct contacts only with the cyclopentadienyl ligands [226]. Some metallocenium complexes of α-CyD exhibit a similar structure [227, 228]. Anions such as PF_6^- are bound at the primary hydroxyl side (Fig. 7.24).

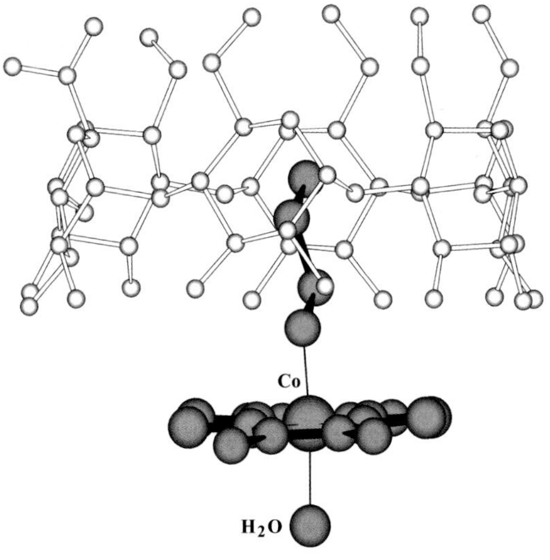

Fig. 7.23. Structure of the α-CyD complex with aqua(*n*-butyl)cobaloxime.

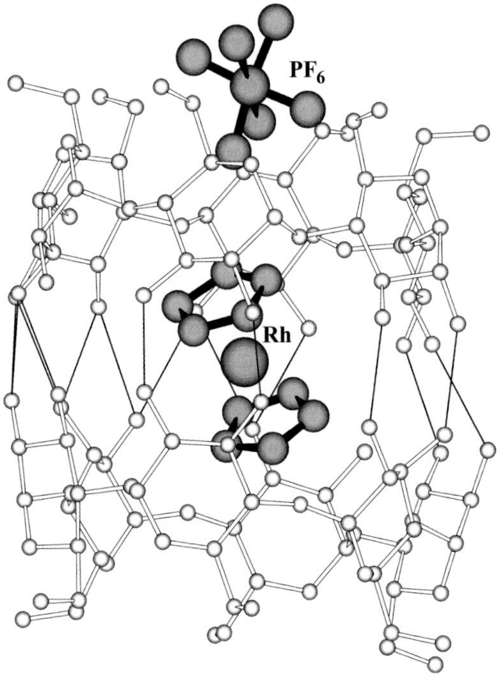

Fig. 7.24. Structure of the α-CyD complex with $Rh(cp)_2PF_6$.

7.5 Molecular Recognition in the Solid State

CyDs have a cylindrical cavity with a depth of about 8 Å and a diameter specific for each CyD. Therefore, CyDs include molecules or groups with a size and shape suitable to fit into the cavity. This leads to guest selectivity in complex formation. CyDs recognize not only the molecular size but also the molecular shape; that is, CyDs differentiate between geometrical isomers and selectively include the specific one that has a shape suitable for accommodation in the cavity. Since CyDs consist of optically active D-glucose units, they form a pair diastereoisomers with the racemic compound. Therefore, as discussed in Chapters 5 and 6, CyDs are potent reagents for chiral resolution.

7.5.1 Selectivity for Geometrical Isomers (see also Chapter 5)

A highly selective inclusion of geometrical isomers was observed in the CyD-catalyzed hydrolysis of substituted phenylesters. When disubstituted phenylesters are hydrolyzed in the presence of α-CyD, a distinct difference in the reaction rate is observed among *ortho*-, *meta*-, and *para*-isomers. The crystal structures of CyD complexes with several substituted benzenes have been determined and the difference in the catalytic reaction has been well explained on the basis of the geometry of inclusion [96]. *p*-Nitrophenol [79] is deeply inserted into the α-CyD cavity and the nitrophenyl group is tightly bound while only half of the nitrophenyl group of *m*-nitrophenol [96] is inserted from the secondary hydroxyl side. *o*-Nitrophenol forms no crystalline complex with α-CyD. Most of the α-CyD complexes with *para*-disubstituted benzenes form a layer-type packing structure [79, 97, 101, 109, 110] because the guest molecule is too long to be fully included in the cavity. *meta*-Disubstituted benzenes are included in a channel formed by the head-to-tail arrangement of α-CyD molecules [96, 97, 102]. Channel-type packing requires a suitable size of guest molecule that should coincide with the repetition unit of the column-like structure. α-CyD complexes with *o*- and *p*-disubstituted benzenes with small substituent groups, such as, *o*- and *p*-fluorophenol, also form the channel-type structure [103].

7.5.2 Chiral Recognition (see also Chapters 1, 5, 6)

Several attempts to separate optical isomers by co-precipitation with CyDs have suggested that CyDs are not good reagents for this purpose. CyDs sometimes co-crystallize with racemic compounds, as X-ray structures have demonstrated. In the α-CyD complex with racemic 1-phenylethanol, both (*R*)- and (*S*)-isomers occupy the same position with the disorder of the hydroxyl group in the host cavity [99]. The hydroxyl group of both isomers forms hydrogen bonds equally with secondary hydroxyl groups of α-CyD. Another example of a crystalline complex with a racemic compound is the β-CyD complex with flurbiprofen. In the crystal, two β-CyD molecules form a head-to-head dimer and a pair of (*R*)- and (*S*)-flurbiprofen is packed in the barrel-like cavity (Fig. 7.25A) [163]. In contrast, an excess of (*S*)-isomer was detected in the β-CyD complex with racemic fenoprofen [157]. β-CyD molecules form a same dimer structure in the crystal of the complex with each isomer, but the arrangement of guest molecules in the dimer cavity differs between the two crystals. Two (*R*)-isomer molecules are included in the head-to-head mode (Fig. 7.25B) while the head-to-tail arrangement is observed in the (*S*)-isomer complex (Fig. 7.25C) [158]. In β-CyD complexes with *N*-acetylphenylalanine, the L-isomer is disordered in the dimer cavity while two molecules of D-isomer are included in a head-to-head mode [177].

Permethylated CyDs have been used for crystallographic analysis of chiral recognition. The asymmetric macrocyclic ring was expected to recognize the chirality of the guest molecule (see, however, Section 1.4). The crystal structures of hexakis(2,3,6-tri-*O*-methyl)-α-CyD complexes with a series of chiral benzene derivatives have been determined. The complexes with (*R*)- and (*S*)-mandelic acid crystallize in different packing structures [200]. In the crystal of the (*R*)-isomer complex, which shows a channel-type packing structure, the guest molecule is deeply inserted into the host cavity and forms a hydrogen bond with O2 of the host molecule (Fig. 7.26A). On the other hand, the (*S*)-isomer complex forms a layer-type packing structure and the host molecule includes only half of the phenyl group (Fig. 7.26B). A similar structure is observed in the complexes with (*R*)- and (*S*)-2-phenylpropionic acid [201], which has no hydroxyl group to form a hydrogen bond with the host molecule. Comparison of the structure with the mandelic acid complex indicates that the host–guest hydrogen bond is not a determinant of the chiral recognition. Crystals of the complexes with (*R*)- and (*S*)-isomer of 1-phenylethanol [199] are isomorphous and both isomers are included in the same manner as occurs in the (*R*)-mandelic acid complex. In contrast, the both optical isomers of 2-phenylbutyric acid [201] form complexes with the same structure as that of (*S*)-mandelic acid.

Recently the optical resolution of 1-(*p*-bromophenyl)ethanol has been successfully achieved by repeated crystallization of the complex with heptakis(2,3,6-tri-*O*-methyl)-β-CyD [210]. In the complex of the (*R*)-isomer, the guest molecule is almost fully included in the host cavity (Fig. 7.27A). The bromophenyl group is axially inserted and the hydroxyl group forms hydrogen bonds with O2. A similar hydrogen bond is formed in the (*S*)-isomer complex. However, in this case the bromophenyl group is equatorially bound to the O2, O3 side and the bromine atom is located outside the host cavity (Fig. 7.27B). These two types of inclusion may be in equilibrium in solution, and the small difference in the binding energy could be amplified in the crystallization.

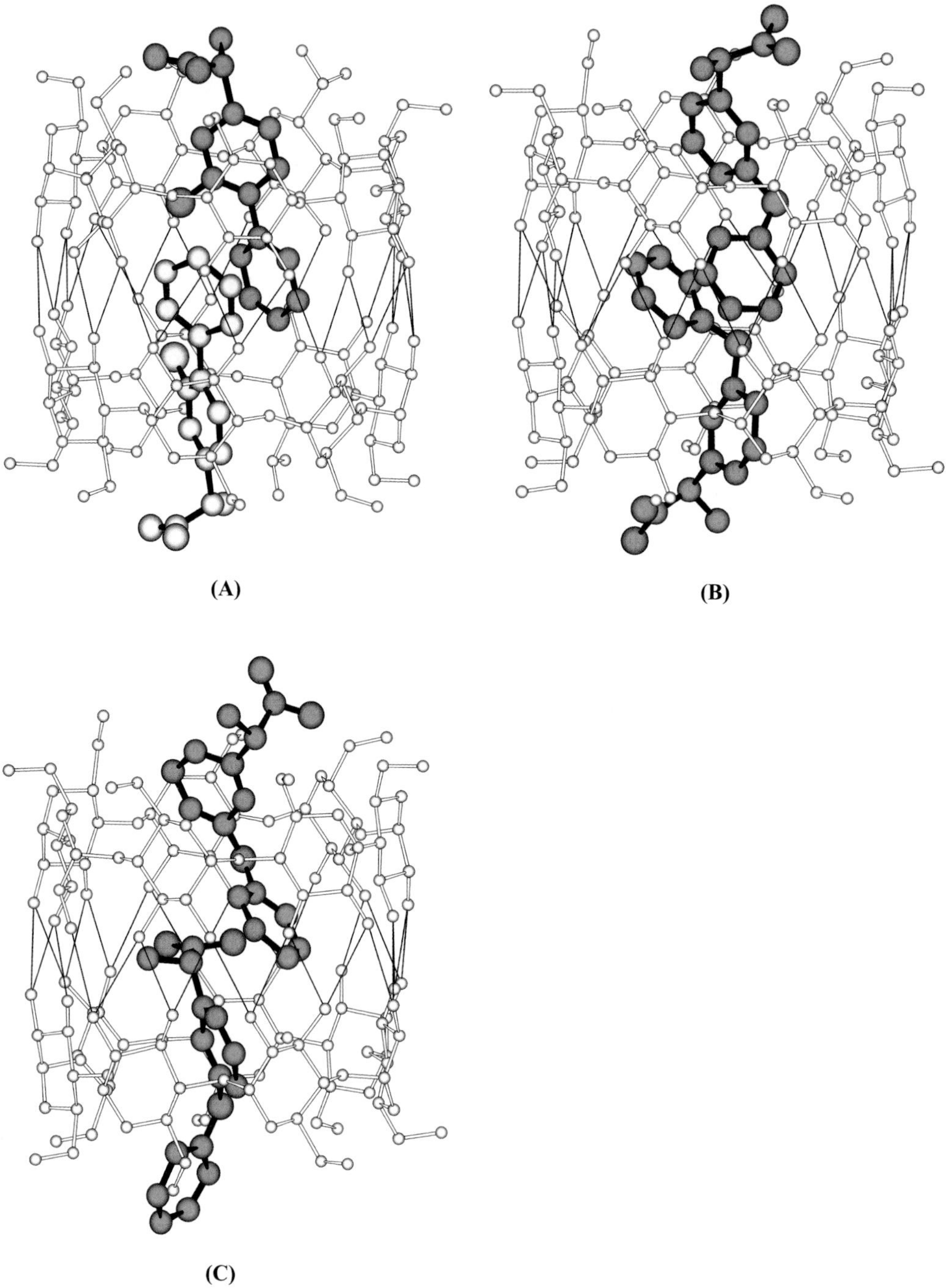

Fig. 7.25. β-CyD complexes with chiral guests: racemic flurbiprofen (A), (R)-fenoprofen (B), and (S)-fenoprofen (C).

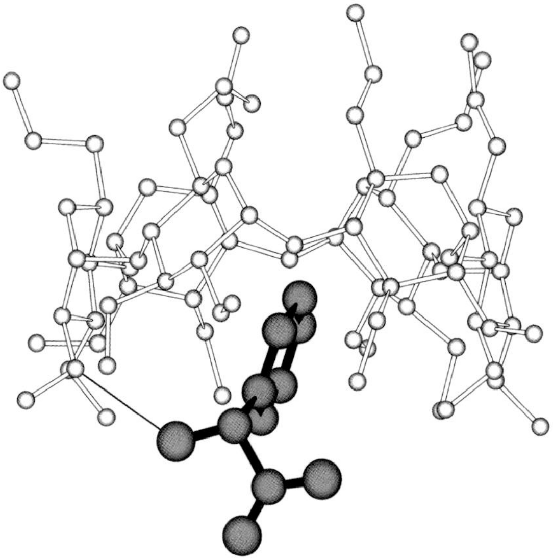

(A)

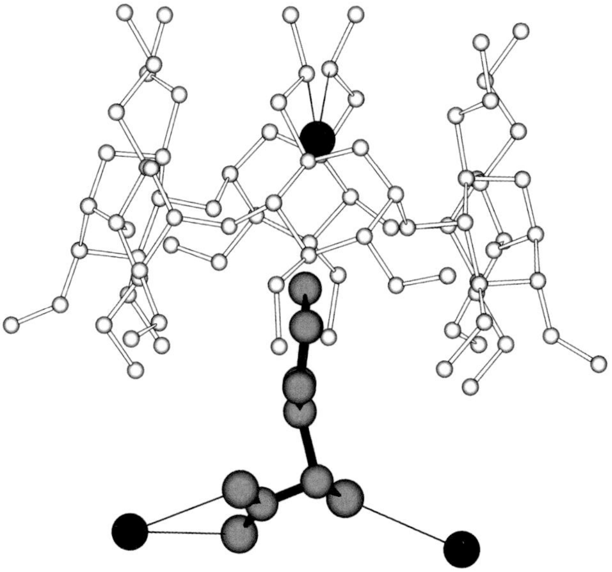

(B)

Fig. 7.26. Structure of hexakis(2,3,6-tri-*O*-methyl)-α-CyD complexes with (*R*)-mandelic acid (A) and (*S*)-mandelic acid (B). Water molecules are shown by filled circles. Thin lines denote hydrogen bonds.

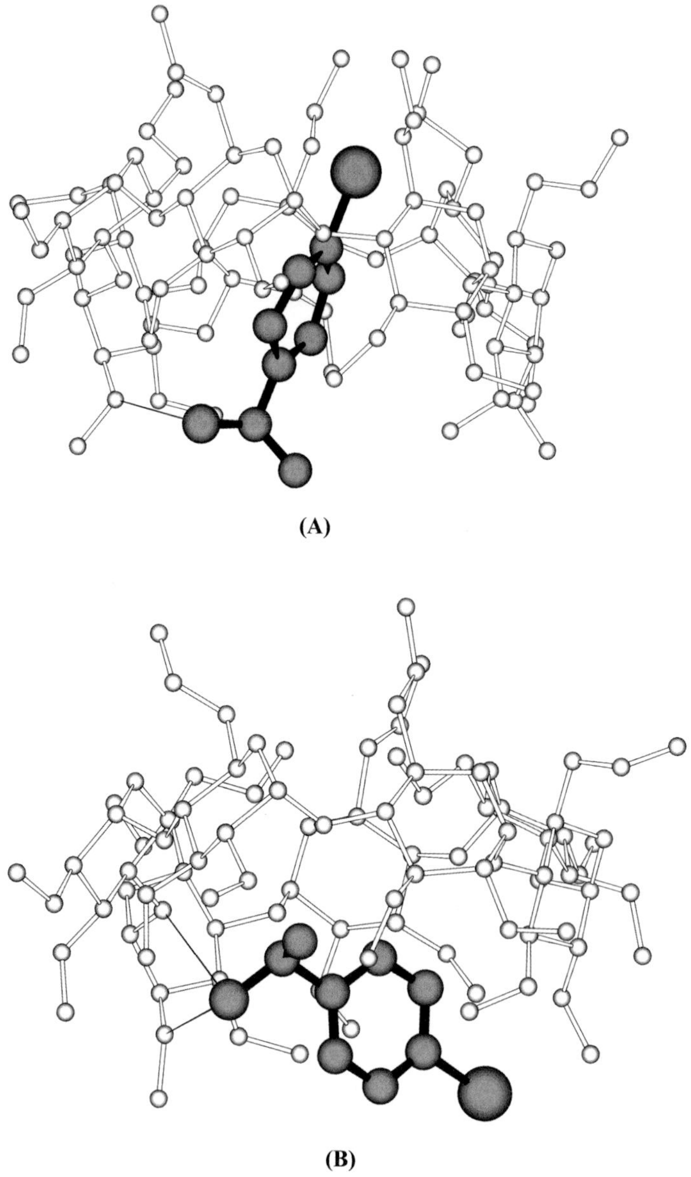

Fig. 7.27. Structure of heptakis(2,3,6-tri-*O*-methyl)-β-CyD complexes with (*R*)-1-(bromophenyl)ethanol (A) and (*S*)-1-(bromophenyl)ethanol (B). Hydrogen bonds are shown by thin lines.

7.6 Concluding Remarks

Throughout this chapter we have attempted to present an overview of the crystallographic study of CyDs and their inclusion complexes. In the last decade, the development of single-crystal X-ray analysis has much accelerated the structural study of CyD complexes and made it possible to determine the structure of large CyDs consisting of more than ten glucose units. Although the list of references included in this review is not complete, the important data discussed explore many of the geometric features of inclusion complexes of CyDs. To outline this aspect of the field, some representative structures in individual packing types are only briefly described and the author would recommend that readers refer to the original papers for a deeper understanding. We believe that the various types of crystal structure characterizing the host–guest complexes of CyDs and their derivatives will be useful not only for understanding the structural properties of CyD complexes but also in predicting the structures of new inclusion complexes of CyDs.

References

1 French, D. (1957). Adv. Carbohydr. Chem., 12, 189–260.
2 Saenger, W. (1980). Angew. Chem. Int. Ed. Engl., 19, 344–362.
3 Saenger, W. (1984). Inclusion compounds. Vol. 2, Atwood, J.L., Davies, J.E.D.D., and MacNicol, D.D. Eds., Academic Press (London), pp. 231–259.
4 Harata, K. (1991). Inclusion Compounds, Vol. 5, Atwood, J.L., Davies, J.E.D.D., and MacNicol, D.D. Eds., Oxford University Press (New York), pp. 311–344.
5 Harata, K. (1996). Comprehensive Supramolecular Chemistry, Vol. 3, Cyclodextrins. Azejtli, J. and Osa, T. Eds., Pergamon Press (UK), pp. 279–304.
6 Saenger, W., Jacob, J., Gessler, K., Steiner, T., Hoffman, D., Sanbe, H., Koizumi, K., Smith, M., and Takaha, T. (1998). Chem. Rev., 98, 1787–1802.
7 Harata, K. (1998). Chem. Rev., 98, 1803–1827.
8 French, D. and Rundle, R.E. (1942). J. Am. Chem. Soc., 64, 1651–1653.
9 James, W.J., French, D., and Rundle, R.E. (1959). Acta Crystallogr., 12, 385–389.
10 Hamilton, J.A., Steinrauf, L.K., and VanEtten, R.L., Acta Crystallogr. Sect. B, 24, 1560–1562.
11 Hybl, A., Rundle, R.E., and Williams, D.E. (1965). J. Am. Chem. Soc., 87, 2779–2788.
12 Hamilton, J.A., Sabesan, M.N., Steinrauf, L.K., and Gebbs, A., Biochem. Biophys. Res. Commun., 73, 659–664.
13 MacLennan, J.M. and Stezowski, J.J. (1980). Biochem. Biophys. Res. Commun., 92, 926–932.
14 Lindner, K. and Saenger, W. (1980). Biochem. Biophys. Res. Commun., 92, 933–938.
15 Manor, P.C. and Saenger, W. (1974). J. Am. Chem. Soc., 96, 3630–3639.
16 Chacko, K.K. and Saenger, W. (1981). J. Am. Chem. Soc., 103, 1708–1715.
17 Lindner, K. and Saenger, W. (1982). Acta Crystallogr. Sect. B, 38, 203–210.
18 Lindner, K. and Saenger, W. (1982). Carbohydr. Res., 99, 103–115.
19 MacLennan, J.M. and Stezowski, J.J.

(1980). Biochem. Biophys. Res. Commun., 92, 926–932.
20 Harata, K. (1987). Bull. Chem. Soc. Jpn., 60, 2763–2767.
21 Fujiwara, T., Tanaka, N., and Kobayashi, S. (1990). Chem. Lett., 739–742.
22 Endo, T., Nagase, H., Ueda, H., Kobayashi, S., and Shiro, M. (1999). Anal. Sci., 15, 613–614.
23 Jacob, J., Gebler, K., Hoffmann, D., Sanbe, H., Koizumi, K., Smith, S.M., Takaha, T., and Saenger, W. (1999). Carbohydr. Res., 322, 228–246.
24 Harata, K., Endo, T., Ueda, H., and Nagai, T. (1998). Supramol. Chem., 9, 143–150.
25 Gessler, K., Uson, I., Takaha, T., Krauss, N., Smith, S.M., Okada, S., Sheldrick, G.M., and Saenger, W. (1999). Proc. Natl. Acad. Sci. USA, 96, 4246–4251.
26 French, A.D. and Murphy, V.G. (1973). Carbohydr. Res., 27, 391–406.
27 Saenger, W., Betzel, C., Hingerty, B., and Brown, G.M. (1983). Angew. Chem. Int. Ed. Engl. 22, 883–884.
28 Klar, B., Hingerty, B., and Saenger, W. (1980). Acta Crystallogr. Sect. B, 36, 1154–1165.
29 Betzel, C., Saenger, W., Hingerty, B.E., and Brown, G.M. (1984). J. Am. Chem. Soc., 106, 7545–7557.
30 Zabel, V., Saenger, W., and Mason, S.A. (1986). J. Am. Chem. Soc., 108, 3664–3673.
31 Steiner, T., Mason, S.A., and Saenger, W. (1990). J. Am. Chem. Soc., 112, 6184–6190.
32 Steiner, T., Mason, S.A., and Saenger, W. (1991). J. Am. Chem. Soc., 113, 5676–5687.
33 Ding, J., Steiner, T., Zabel, V., Hingerty, B.E., Mason, S.A., and Saenger, W. (1991). J. Am. Chem. Soc., 113, 8081–8089.
34 Reibenspies, J.H., Maynard, D.K., Derecskei-Kovacs, A., and Vigh, G. (2000). Carbohydr. Res., 328, 217–227.
35 Harata, K. (1995). Supramol. Chem., 5, 231–236.
36 Steiner, T., Hirayama, F., and Saenger, W. (1996). Carbohydr. Res., 296, 69–82.
37 Steiner, T. and Saenger, W. (1995). Carbohydr. Res., 275, 73–82.
38 Aree, T., Saenger, W., Leibnitz, P., and Hoier, H. (1999). Carbohydr. Res., 315, 199–205.
39 Aree, T., Hoier, H., Schulz, B., Reck, G., and Saenger, W. (2000). Angew. Chem. Int. Ed. Engl., 39, 897–899.
40 Stezowski, J., Parker, W., Hilgenkamp, S., and Gdaniec, M. (2001). J. Am. Chem. Soc., 123, 3919–3926.
41 Harata, K., Hirayama, F., and Uekama, K. (2000). Carbohydr. Res., 329, 597–607.
42 Steiner, T., and Saenger, W. (1996). Carbohydr. Res., 282, 53–63.
43 Caira, M.R., Griffith, V.J., Nassimbeni, L.R., and Oudtshoorn, B. (1994). J. Chem. Soc., Perkin Trans. 2, 2071–2072.
44 Aree, T., Hoier, H., Schulz, B., Reck, G., and Saenger, W. (2000). Carbohydr. Res., 328, 399–407.
45 Aree, T., Uson, I., Schulz, B., Reck, G., Hoier, H., Sheldrick, G.M., and Saenger, W. (1999). J. Am. Chem. Soc., 121, 3321–3327.
46 Harata, K. (1998). Chem. Lett., 1998, 589–590.
47 Anibarro, M., Gessler, K., Uson, I., Scheldrick, G.M., Harata, K., Uekama, K., Hirayama, F., Abe, Y., and Saenger, W. (2001). J. Am. Chem. Soc., 123, 11854–11862.
48 Hanessian, S., Benalil, A., Simard, M., and Belanger-Gariepy, F. (1995). Tetrahedron, 51, 10149–10158.
49 Harata, K., Rao, C.T., Pitha, J., Fukunaga, K., and Uekama, K. (1991). Carbohydr. Res., 222, 37–45.
50 Harata, K., Takenaka, Y., and Yoshida, N. (2001). J. Chem. Soc., Perkin Trans. 2, 1667–1673.
51 Harata, K., Rao, C.T., and Pitha, J. (1993). Carbohydr. Res., 247, 83–98.
52 Hirotsu, K. and Higuchi, T. (1982). J. Org. Chem, 47, 1143–1144.
53 Mentzafos, D., Terzis, A., Coleman, A.W., and Rango, C. (1996). Carbohydr. Res., 282, 125–135.

54 Kamitori, S., Hirotsu, K., and Higuchi, T. (1987). J. Chem. Soc., Perkin Trans. 2, 7–14.

55 Eliadou, K., Giastas, P., Yannakopoulou, K., and Mavridis, I.M. (2003). J. Org. Chem., 68, 8550–8557.

56 Fujiwara, T., Tanaka, N., Hamada, K., and Kobayashi, S. (1989). Chem. Lett., 1131–1134.

57 Reetz, M.T., Rudolph, J., and Goddard, R. (2001). Can. J. Chem., 79, 1806–1811.

58 Blasio, B., Pavone, V., Nastri, F., Isernia, C., Saviano, M., Pedone, C., Cucinotta, V., Impellizzeri, G., Rizzarelli, E., and Vecchio, G. (1992). Proc. Natl. Acad. Sci. USA, 89, 7218–7221.

59 Blasio, B., Galdiero, S., Saviano, M., Simone, G., Benedetti, E., Pedone, C., Gibbons, W.A., Deschenaux, R., Rizzarelli, E., and Vecchio, G. (1996). Supramol. Chem., 7, 47–54.

60 Impellizzeri, G., Pappelardo, G., D'Alessandro, F., Rizzarelli, E., Saviano, M., Iacovino, R., Benedetti, E., and Pedone, C. (2000). Eur. J. Org. Chem., 1065–1076.

61 Bonomo, R.P., Blasio, B., Maccarrone, G., Pavone, V., Pedone, C., Rizzarelli, E., Saviano, M., and Vecchio, G. (1996). Inorg. Chem., 35, 4497–4504.

62 Selkti, M., Lopez, H.P., Navaza, J., Villain, F., and Rango, C. (1995). Supramol. Chem., 5, 255–266.

63 Rizzarelli, E., Pedotti, S., Vecchio, G., and Gibbons, W.A. (1996). Carbohydr. Res., 282, 41–52.

64 Cucinotta, V., Grasso, G., Pedotti, S., Rizzarelli, E., Vecchio, G., Blasio, B., Isernia, C., Saviano, M., and Pedone, C. (1996). Inorg. Chem., 35, 7535–7540.

65 Saviano, M., Benedetti, E., Blasio, B., Gavuzzo, E., Fierro, O., Pedone, C., Iacovino, R., Rizzarelli, E., and Vecchio, G. (2001). J. Chem. Soc., Perkin Trans. 2, 946–952.

66 Lichtenthaler, F.W., Lindner, H.J., Fujita, K., Yuan, D.Q., and Ren, Y. (2003). Acta Crystallogr. Sect. E, 59, 0408–0411.

67 Armspach, D., Matt, D., and Kyritsakas, N. (2001). Polyhedron, 20, 663–668.

68 Engeldinger, E., Armspach, D., Matt, D., Toupet, L., and Wesolek, M. (2002). C. R. Chimie, 5, 359–372.

69 Armspach, D. and Matt, D. (2001). Inorg. Chem., 40, 3505–3509.

70 Engeldinger, E., Armspach, D., Matt, D., Jones, P.G., and Welter, R. (2002). Angew. Chem. Int. Ed. Engl., 41, 2593–2596.

71 Harata, K., Nagano, Y., Ikeda, H., Ikeda, T., Ueno, A., and Toda, F. (1996). J. Chem. Soc. Chem. Commun., 1996, 2347–2348.

72 Lindner, H.J., Yuan, D.Q., Fujita, K., Kubo, K., and Lichtenthaler, F.W. (2003). Chem. Commun., 1730–1731.

73 Nogami, Y., Nasu, K., Koga, T., Ohta, K., Fujita, K., Immel, S., Lindner, H., Schmitt, G.E., and Lichtenthaler, F.W. (1997). Angew. Chem. Int. Ed. Engl., 36, 1899–1902.

74 Immel, S., Fujita, K., Lindner, H.J., Nogami, Y., and Lichtenthaler, F.W. (2000). Chem. Eur. J., 6, 2327–2333.

75 Immel, S., Lichtenthaler, F.W., Lindner, H.J., Fujita, K., Fukudome, M., and Nogami, Y. (2000). Tetrahedron: Asymmetry, 11, 27–36.

76 Immel, S., Fujita, K., Fukudome, M., and Bolte, M. (2001). Carbohydr. Res., 336, 297–308.

77 Yamamura, H., Masuda, H., Kawase, Y., Kawai, M., Butsugan, Y., and Einaga, H. (1996). Chem. Commun., 1069–1070.

78 Harata, K., Akasaka, H., Endo, T., Nagase, H., and Ueda, H. (2002). Chem. Commun., 2002, 1968–1969.

79 Harata, K. (1977). Bull. Chem. Soc. Jpn., 50, 1416–1424.

80 Harata, K. (1993). J. Chem. Soc., Chem. Commun., 1993, 546–547.

81 Harata, K. (1999). Chem. Commun., 1999, 191–192.

82 Harata, K. (1984). Bull. Chem. Soc. Jpn., 57, 2596–2599.

83 Harata, K. (2003). Carbohydr. Res., 338, 353.

84 McMullan, R.K., Saenger, W., Fayos, J., and Mootz, D. (1973). Carbohydr. Res., 31, 211–227.
85 Saenger, W. and Noltemeyer, M. (1976). Chem. Ber., 109, 503–517.
86 Hingerty, B. and Saenger, W. (1976). J. Am. Chem. Soc., 98, 3357–3365.
87 Saenger, W., McMullan, R.K., Fayos, J., and Mootz, D. (1974). Acta Crystallogr. Sect. B, 30, 2019–2028.
88 Harata, K., Uekama, K., Otagiri, M., and Hirayama, F. (1983). J. Chem. Soc. Jpn., 173–180.
89 Nakagawa, T., Immel, S., Lichtenthaler, F.W., and Lindner, H.J. (2000). Carbohydr. Res., 324, 141–146.
90 Aree, T., Jacob, J., Saenger, W., and Hoier, H. (1998). Carbohydr. Res., 307, 191–197.
91 Hybl, A., Rundle, R.E., and Williams, D.E. (1965). J. Am. Chem. Soc., 87, 2779–2788.
92 Harata, K. (1977). Bull. Chem Soc. Jpn., 50, 1259–1266.
93 Harata, K. (1976). Bull. Chem. Soc. Jpn., 49, 2066–2072.
94 Tokuoka, R., Abe, M., Matsumoto, K., Shirakawa, K., Fujiwara, T., and Tomita, K. (1981). Acta Crystallogr. Sect. B, 37, 445–447.
95 Harata, K. (1976). Bull. Chem. Soc. Jpn., 49, 1493–1501.
96 Harata, K., Uedaira, H., and Tanaka, J. (1978). Bull. Chem. Soc. Jpn., 51, 1627–1634.
97 Kamitori, S., Toyama, Y., Matsuzaka, O. (2001). Carbohydr. Res., 332, 235–240.
98 Harata, K., Uekama, K., Otagiri, M., Hirayama, F., and Ogino, H. (1981). Bull. Chem. Soc. Jpn., 54, 1954–1959.
99 Harata, K. (1982). Bull. Chem. Soc. Jpn., 55, 1367–1371.
100 Steiner, T. and Saenger, W. (1994). Carbohydr. Lett., 1, 143–150.
101 Muraoka, S., Matsuzaka, O., Kamitori, S., and Okuyama, K. (1999). Carbohydr. Res., 320, 261–266.
102 Harata, K. (1980). Bull. Chem. Soc. Jpn., 53, 2782–2786.
103 Shibakami, M. and Sekiya, A. (1992). J. Chem. Soc., Chem. Commun., 1742–1743.
104 Noltemeyer, M. and Saenger, W. (1980). J. Am. Chem. Soc., 102, 2710–2722.
105 Roselli, C.S., Perly, B., and Bas, G. (2001). J. Incl. Phenom. Mol. Recogn. Chem., 39, 333–337.
106 Nicolis, I., Villain, F., Coleman, A.W., and Rango, C. (1994). Supramol. Chem., 3, 251–259.
107 Harata, K. and Kawano, K. (2002). Carbohydr. Res., 337, 537–547.
108 Kamitori, S., Muraoka, S., Kondo, S., and Okuyama, K. (1998). Carbohydr. Res., 312, 177–181.
109 Harata, K. (1975). Bull. Chem. Soc. Jpn., 48, 2409–2413.
110 Harata, K. (1976). Carbohydr. Res., 48, 265–270.
111 Harata, K. (1979). Bull. Chem. Soc. Jpn., 52, 2451–2459.
112 Shibakami, M. and Sekiya, A. (1994). Carbohydr. Res., 260, 169–179.
113 Harata, K. (1978). Bull. Chem. Soc. Jpn., 51, 1644–1648.
114 Rontoyianni, A. and Mavridis, I.M. (1999). Supramol. Chem., 10, 213–218.
115 Lindner, K. and Saenger, W. (1982). Carbohydr. Res., 107, 7–16.
116 Tokuoka, R., Abe, M., Fujiwara, T., Tomita, K., and Saenger, W. (1980). Chem. Lett., 491–494.
117 Steiner, T., Koellner, G., and Saenger, W. (1992). Carbohydr. Res., 228, 321–332.
118 Steiner, T. and Saenger, W. (1995). J. Chem. Soc., Chem. Commun., 2087–2088.
119 Gebler, K., Steiner, T., Koellner, G., and Saenger, W. (1993). Carbohydr. Res., 249, 327–344.
120 Aree, T., Schulz, B., and Reck, G. (2003). J. Incl. Phenom. Macrocycl. Chem., 47, 39–45.
121 Aree, T. and Chaichit, N. (2002). Carbohydr. Res., 337, 2487–2494.
122 Crisma, M., Fornasier, R., and Marcuzzi, F. (2001). Carbohydr. Res., 333, 145–151.
123 Steiner, T. and Saenger, W. (1998). J. Chem. Soc., Perkin Trans., 2, 371–377.

124 Wang, J.L., Zhou, W.H., Ma, S.K., and Miao, F.M. (2002). Chinese J. Chem., 20, 358–361.

125 Harata, K., Uekama, K., Otagiri, M., Hirayama, F., and Ohtani, Y. (1985). Bull. Chem. Soc. Jpn., 58, 1234–1238.

126 Harata, K., Kawano, K., Fukunaga, K., and Ohtani, Y. (1983). Chem. Pharm. Bull., 31, 1428–1430.

127 Anibarro, M., Gebler, K., Uson, I., Sheldrick, G.M., and Saenger, W. (2001). Carbohydr. Res., 333, 251–256.

128 Harata, K. (1982). Bull. Chem. Soc. Jpn., 55, 2315–2320.

129 Caira, M., Griffith, V., Nassimbeni, L., and Oudtshoorn, B. (1994). J. Incl. Phenom. Mol. Recogn. Chem., 17, 187–201.

130 Chiesi-Villa, A., Rizzoli, C., Amari, G., Delcanale, M., Redenti, E., and Ventura, P. (1998). Supramol. Chem., 10, 111–119.

131 Pop, M.M., Goubitz, K., Borodi, G., Bogdan, M., Ridder, D.J.A., Peschar, R., and Schenk, H. (2002). Acta Crystallogr. Sect. B, 58, 1036–1043.

132 Bogdan, M., Caira, M.R., and Farcas, S.I. (2002). Supramol. Chem., 14, 427–435.

133 Caira, M., Griffith, V., and Nassimbeni, L. (1998). J. Incl. Phenom. Mol. Recogn. Chem, 32, 461–476.

134 Caire, M.R., Griffith, V.J., Nassimbeni, L.R., and Oudtshoorn, B. (1994). J. Chem. Soc., Chem. Commun., 1061–1062.

135 Aree, T. and Chaichit, N. (2003). Carbohydr. Res., 338, 1581–1589.

136 Hamilton, J.A., Sabesan, M.N., and Steinrauf, L.K. (1981). Carbohydr. Res., 89, 33–53.

137 Hamilton, J.A. and Sabesan, M.N. (1982). Carbohydr. Res., 102, 31–46.

138 Betzel, C., Hingerty, B., Noltemeyer, M., Weber, G., and Saenger, W. (1983) J. Incl. Phenom., 1, 181–191.

139 Mavridis, I.M. and Hadjoudis, E. (1991). Carbohydr. Res., 220, 11–21.

140 Udachin, K.A. and Ripmeester, J.A. (1998). J. Am. Chem. Soc., 120, 1080–1081.

141 Kokkinou, A., Makedonopoulou, S., and Mentzafos, D. (2000). Carbohydr. Res., 328, 135–140.

142 Brett, T.J., Alexander, J.M., and Stezowski, J.J. (2000). J. Chem. Soc., Perkin Trans., 2, 1095–1103.

143 Caira, M.R. and Dodds, D. (2000). J. Incl. Phenom. Maclocycl. Chem., 38, 75–84.

144 Kokkinou, A., Yannakopoulou, K., Mavridis, I.M., and Mentzafos, D. (2001). Carbohydr. Res., 332, 85–94.

145 Braga, S.S., Goncalves, I.S., Herdtweek, E., and Teixeira-Dias, J.J.C. (2003). New J. Chem., 27, 597–601.

146 Kamitori, S., Matsuzaka, O., Kondo, S., Muraoka, S., Okuyama, K., Noguchi, K., Okada, M., and Harada, A. (2000). Macromolecules, 33, 1500–1502.

147 Udachin, K.A., Wilson, L.D., and Ripmeester, J.A. (2000). J. Am. Chem. Soc., 122, 12375–12376.

148 Harding, M.M., MacLennan, J.M., and Paton, R.M. (1978). Nature, 274, 621–623.

149 Caira, M.R. and Dodds, D.R. (1999). J. Incl. Phenom. Mol. Recogn. Chem, 34, 19–29.

150 Mavridis, I.M. and Hadjoudis, E. (1992). Carbohydr. Res., 229, 1–15.

151 Aree, T. and Chaichit, N. (2003). Carbohydr. Res., 338, 439–446.

152 Braga, S.S., Aree, T., Imamura, K., Vertut, P., Palheiros, I.B., Saenger, W., and Teixeira-Dias, J.J.C. (2002). J. Incl. Phenom. Mol. Recogn. Chem, 43, 115–125.

153 Rontoyianni, A. and Mavridis, I.M. (1994). J. Incl. Phenom. Mol. Recogn. Chem, 18, 211–227.

154 Makedonopoulou, S., Papaioannou, J., Argyroglou, I., and Mavridis, I.M. (2000). J. Incl. Phenom. Mol. Recogn. Chem, 36, 191–215.

155 Tokuoka, R., Fujiwara, T., and Tomita, K. (1981). Acta Crystallogr. Sect. B, 37, 1158–1160.

156 Harata, K., Hirayama, F., Uekama, K., and Tsoucaris, G. (1988). Chem. Lett., 1988, 1585–1588.

157 Hamilton, J.A. and Chen, L. (1988). J. Am. Chem. Soc., 110, 5833–5841.

158 Hamilton, J.A. and Chen, L. (1988). J. Am. Chem. Soc., 110, 4379–4391.
159 Vicens, J., Fujiwara, T., and Tomita, K. (1988). J. Incl. Phenom., 6, 577–581.
160 Giastas, P., Yannakopoulou, K., and Mavridis, I.M. (2003). Acta Crystallogr. Sect. B, 59, 287–299.
161 Stezowski, J.J., Jogun, K.H., Eckle, E., and Bartels, K. (1978). Nature, 274, 617–619.
162 Hursthouse, M.B., Smith, C.Z., Thornton-Pett, M., and Utley, J.H.P. (1982). J. Chem. Soc., Chem. Commun., 881–882.
163 Uekama, K., Hirayama, F., Imai, T., Otagiri, M., and Harata, K. (1983). Chem. Pharm. Bull., 31, 3363–3365.
164 Uekama, K., Imai, T., Hirayama, F., Otagiri, M., and Harata, K. (1984). Chem. Pharm. Bull., 32, 1662–1664.
165 Hamilton, J.A. and Sabesan, M.N. (1982). Acta Crystallogr. Sect. B, 38, 3063–3069.
166 Rontoyianni, A., Mavridis, I.M., Hadjoudis, E., and Duisenberg, J.M. (1994). Carbohydr. Res., 252, 19–32.
167 Shikun, M., Jinling, W., Aixiu, L., Ming, Y., Ming, S., Xiumei, S., and Fangming, M. (2001). Chinese Sci. Bull., 46, 390–392.
168 Caira, M.R., Griffith, V.J., Nassimbeni, L.R., and Oudtshoorn, B. (1996). Supramol. Chem., 7, 119–124.
169 Rontoyiammi, A. and Mavridis, I.M. (1996). Acta Crystallogr. Sect. C, 52, 2277–2281.
170 Makedomopoulou, S. and Mavridis, I.M. (2000). Acta Crystallogr. Sect. B, 56, 322–331.
171 Makedomopoulou, S. and Mavridis, I.M. (2001). Carbohydr. Res., 335, 213–220.
172 Bojinova, T., Gornitzka, H., Viguerie, N.L., and Rico-Lattes, I. (2003). Carbohydr. Res., 338, 781–785.
173 Yannakopoulou, K., Ripmeester, J.A., and Mavridis, I.M. (2002). J. Chem. Soc., Perkin Trans., 2, 1639–1644.
174 Brett, T.J., Alexander, J.M., and Stezowski, J.J. (2000). J.Chem. Soc., Perkin Trans., 2, 1105–1111.
175 Clark, J.L. and Stezowski, J.J. (2001). J. Am. Chem. Soc., 123, 9880–9888.
176 Clark, J.L., Booth, B.R., and Stezowski, J.J. (2001). J. Am. Chem. Soc., 123, 9889–9895.
177 Alexander, J.M., Clark, J.L., Brett, T.J., and Stezowski, J.J. (2002). Proc. Natl. Acad. Sci. USa, 99, 5115–5120.
178 Mentzafos, D., Mavridis, I.M., Bas, G.L., and Tsoucaris, G. (1991). Acta Crystallogr. Sect. B, 47, 746–757.
179 Steiner, T. and Saenger, W. (1998). Acta Crystallogr. Sect. B, 54, 450–455.
180 Ding, J., Steiner, T., and Saenger, W. (1991). Acta Crystallogr. Sect. B, 47, 731–738.
181 Kamitori, S., Hirotsu, K., and Higuchi, T. (1986). J. Chem. Soc., Chem. Commun., 690–691.
182 Kamitori, S., Hirotsu, K., and Higuchi, T. (1987). J. Am. Chem. Soc., 109, 2409–2414.
183 Akasaka, H., Endo, T., Nagase, H., Ueda, H., and Kobayashi, S. (2000). Chem. Pharm. Bull., 48, 1986–1989.
184 Nimz, O., Gebler, K., Uson, I., Laettig, S., Welfle, H., Sheldrick, G.M., and Saenger, W. (2003). Carbohydr. Res., 338, 977–986.
185 Harata, K. (1990). Bull. Chem. Soc. Jpn., 63, 2481–2486.
186 Harata, K. (1989). Carbohydr. Res., 192, 33–42 (1989).
187 Harata, K. (1998). J. Chem. Soc. Jpn., 1998, 285–297.
188 Aree, T., Hoier, H., Schulz, B., Reck, G., and Saenger, W. (2000). Carbohydr. Res., 323, 245–253.
189 Harata, K., Song, L.X., and Morii, H. (2000). Supramol. Chem., 11, 217–224.
190 Czugler, M., Eckle, E., and Stezowski, J.J. (1981). J. Chem. Soc. Chem. Commun., 1291–1292.
191 Glocker, P., Schollmeyer, D., and Pitter, H. (2002). Designed Monomers Polymers, 5, 163–172.
192 Harata, K. (1988). Bull. Chem. Soc. Jpn., 61, 1939–1944.
193 Tsorteki, F. and Mentzafos, D. (2002). Carbohydr. Res., 337, 1229–1233.

194 Selkti, M., Navaza, A., Villain, F., Charpin, P., and Rango, C. (1997). J. Incl. Phenom. Mol. Recogn. Chem., 27, 1–12.

195 Harata, K., Uekama, K., Otagiri, M., and Hirayama, F. (1982). Bull. Chem. Soc. Jpn., 55, 407–410.

196 Harata, K., Uekama, K., Otagiri, M., Hirayama, F., and Sugiyama, Y. (1982). Bull. Chem. Soc. Jpn., 55, 3386–3389.

197 Harata, K., Uekama, K., Otagiri, M., and Hirayama, F. (1982). Bull. Chem. Soc. Jpn., 55, 3904–3910.

198 Harata, K., Uekama, K., Otagiri, M., and Hirayama, F. (1984). J. Incl. Phenom., 1, 279–293.

199 Harata, K. (1990). J. Chem. Soc., Perkin Trans. 2, 1990, 799–804.

200 Harata, K., Uekama, K., Otagiri, M., and Hirayama, F. (1987). Bull. Chem. Soc. Jpn., 60, 497–502.

201 Harata, K. (1990). Minutes of the Fifth International Symposium on Cyclodextrins, Duchene, D. Ed., Editions de Sante (Paris), pp. 676–679.

202 Tsorteki, F., Bethanis, K., and Mentzafos, D. (2004). Carbohydr. Res., 339, 233–240.

203 Harata, K., Uekama, K., Imai, T., Hirayama, F., and Otagiri, M. (1988). J. Inclusion Phenom., 6, 443–460.

204 Harata, K., Hirayama, F., Arima, H., Uekama, K., and Miyaji, T. (1992). J. Chem. Soc. Perkin Trans. 2, 1992, 1159–1166.

205 Caira, M.R., Griffith, V.J., and Nassimbeni, L.R. (1995). J. Incl. Phenom. Mol. Recogn. Chem., 20, 277–290.

206 Brown, G.R., Caira, M.R., and Nassimbeni, L.R. (1996). J. Incl. Phenom. Mol. Recogn. Chem., 26, 281–294.

207 Mentzafos, D., Mavridis, I.M., and Schenk, H. (1994). Carbohydr. Res., 253, 39–50.

208 Harata, K. (1988). J. Chem. Soc., Chem. Commun., 1988, 928–929.

209 Cardinael, P., Peulon, V., Perez, G., Coquerel, G., and Toupet, L. (2001). J. Incl. Phenom. Mol. Recogn. Chem., 39, 159–167.

210 Grandeury, A., Petit, S., Gouhier, G., Agasse, V., and Coquerel, G. (2003). Tetrahedron: Asymmetry, 14, 2143–2152.

211 Chankvetadze, B., Burjanadze, N., Pintore, G., Bergenthal, D., Bergander, K., Muhlenbrock, C., Breitkreuz, J., and Blaschke, G. (2000). J. Chromatogr. A, 875, 471–484.

212 Makedonopoulou, S., Yannakopoulou, K., Mentzafos, D., Lamzin, V., Popov, A., and Mavridis, I.M. (2001). Acta Crystallogr. Sect. B, 57, 399–409.

213 Nicolis, I., Coleman, A.W., Selkti, M., Villain, F., Charpin, P., and Rango, C. (2001). J. Phys. Org. Chem., 14, 35–37.

214 Charpin, P., Nicolis, I., Villain, F., Rango, C., and Coleman, A.W. (1991). Acta Crystallogr. Sect. C, 47, 1829–1833.

215 Nicolis, I., Coleman, A.W., Charpin, P., and Rango, C. (1996). Acta Crystallogr. Sect. B, 52, 122–130.

216 Nicolis, I., Coleman, A.W., Charpin, P., and Rango, C. (1995). Angew. Chem. Int. Ed. Engl., 34, 2381–2383.

217 Klufers, P. and Schuhmacher, J. (1994). Angew. Chem. Int. Ed. Engl., 33, 1863–1865.

218 Alston, D.R., Slawin, A.M.Z., Stoddart, J.F., and Williams, D.J. (1985). Angew. Chem. Int. Ed. Engl., 24, 786–787.

219 Chen, Y., Luo, L.B., Chen, H.I., Hu, C.H., Chen, J., and Zheng, P.J. (2000). Bull. Chem. Soc. Jpn., 73, 1375–1378.

220 Stoddart, J.F. and Zarzycki, R. (1988). Recl. Trav. Chim. Pays-Bas, 107, 515–528.

221 Luo, L.B., Chen, Y., and Chen, H.L. (1998). Inorg. Chem., 37, 6147–6152.

222 Chen, Y., Xiang, P., Li, G., Chen, H.L., Chinnakali, K., and Fun, H.K. (2002) Supramol. Chem., 14, 339–346.

223 Chen, Y., Chen, H.L., Yang, Q.C., Song, X.Y., Duan, C.Y., and Mak, T.C.W. (1999). J. Chem. Soc., Dalton Trans., 629–633.

224 Chen, Y., Chen, H.L., Lian, H.Z.,

Mei, Y.H., Yang, Q.C., and Mak, T.C.W. (1999). Inorganic Chem. Commun., 2, 70–72.

225 Navaza, A., Iroulapt, M.G., and Navaza, J. (2000). J. Coord. Chem., 51, 153–168.

226 Odagaki, Y., Hirotsu, K., Higuchi, T., Harada, A., and Takahashi, S. (1990). J. Chem. Soc., Perkin Trans., 1, 1230–1231.

227 Klingert, B. and Rihs, G. (1991). J. Incl. Phenom. Mol. Recogn. Chem., 10, 255–265.

228 Klingert, B. and Rihs, G. (1991). J. Chem. Soc., Dalton Trans., 1, 2749–2760.

8
Microcalorimetry

Mikhail V. Rekharsky and Yoshihisa Inoue

8.1
Introduction

Modern microcalorimetry is a powerful and indispensable experimental technique that allows us to simultaneously determine the enthalpy change and equilibrium constant from a single experimental run [1]. Usually, in supramolecular chemistry, biochemistry, biotechnology, pharmacology, and medicinal chemistry, reactants are available in limited quantities and are very expensive, hence it is reasonable to increase the sensitivity of the instrument (and consequently reduce the amount of reactant required for a single calorimetric run), rather than employ less sensitive ones. Indeed, the sensitivity of calorimeters has been dramatically improved, particularly in recent years, to the level of a few microcalories or even less.

It should be noted that calorimetry is one of the oldest physicochemical experimental methods with a history of more than a century of scientific application for which numerous experimental devices and techniques have been designed, tested, and applied. Since the first extensive review of the use of microcalorimetry in the fields of biochemistry, biotechnology, and biology by Calvet and Prat in 1956 [2], a wide variety of different microcalorimeters have been developed and employed in various branches of the life sciences.

A detailed description of the most interesting and fruitful microcalorimetry designs over the next four decades (from 1950 to 1990) can be found in the book by Rekharsky and Egorov [3]. Until the 1970s, microcalorimeters were built and used predominantly by individual scientists or engineers as single instruments unique to the purpose of their particular laboratories. In the 1980s and particularly the 1990s, conventional microcalorimeters became commercially available, which certainly greatly facilitated thermodynamic approaches to the investigation of various areas of chemistry and biology.

Despite the great diversity in the design of microcalorimeters and the experimental procedures described in the literature [1–10], only two microcalorimetric methods have found widespread application in cyclodextrin (CyD) studies and drug-design research. These two methods are differential scanning calorimetry (DSC) and isothermal titration microcalorimetry (ITC). DSC and ITC can be con-

Cyclodextrins and Their Complexes. Edited by Helena Dodziuk

ISBN: 3-527-31280-3

sidered as complementary tools for investigating molecular recognition between a host and guest in both solution and the solid state [11]. Very briefly, we will mention the differences between these two methods. In DSC, the sample under investigation (liquid or solid) is gradually heated/cooled at a certain speed, and the molecular events caused by the temperature change (phase transitions, melting, decomposition, etc.) are recorded as heat release or absorption. In ITC, one component of the relevant species involved in the reaction (association, aggregation, or molecular recognition) is introduced in a stepwise fashion into a solution of the other component(s), with all the molecular events occurring upon this addition registered in a thermogram. By simultaneously using both methods, we can obtain complete thermodynamic information about the system under investigation at various molar ratios as well as at different temperatures [12, 13]. Reviews have been published [14–16], that give a comprehensive overview of the application of calorimetric methods to the investigation of CyDs as well as for drug-discovery research up to the beginning of 2000. Hence, in this chapter we will briefly discuss only the most pertinent experimental studies published after 2000. Although we can obtain complete thermodynamic features of the system under investigation by applying both DSC and ITC, as noted above, usually only one method (DSC or ITC) is employed in each specific study. Therefore, we will separately describe recent trends in the application of DSC and ITC.

8.2
Application of Differential Scanning Calorimetry in Cyclodextrin and Drug Studies

One recent trend is the use of DSC in combination with other experimental techniques, which include X-ray crystallography and various spectroscopic methods, such as IR, UV-Vis, circular dichroism, and ^{1}H and ^{13}C NMR spectroscopy (presented in Chapters 7, 9, and 10). Simultaneous analysis of complexation behavior using different methods obviously has a lot of advantages. Indeed, with the help of DSC one can obtain direct information about the thermal stability/phase transition(s) of CyD complexes (including those formed with drug molecules). Furthermore, DSC in combination with crystallographic and spectroscopic methods provides us with definitive experimental evidence for the formation of CyD–drug complexes and their stoichiometry and structure in the solution and solid states.

It is well known that in many cases native and modified CyDs dramatically improve the solubility of drugs that are less soluble in water via the formation of inclusion complexes (see Chapters 14 and 15), and their complexation behavior has been investigated by various microcalorimetric techniques [28–33]. There have been several recent experimental DSC studies aimed at applying the advantages of the highly soluble CyD–drug complexes to pharmacological use. Jain and Adeyeye [34] emphasized the importance of strong complex formation to increase the solubility of poorly soluble drugs. They reported that dissolution of the sulfobutylether(SBE)-β-CyD–danazol complex is significantly greater than that of the corresponding physical mixture. In their study, DSC was used to check for the disappearance of the melting peak of danazol, indicating complex formation, which, together with the X-ray diffraction method (that confirms the disappearance of the characteristic peaks for crystalline danazol), indicates the formation of an amorphous complex. Using DSC in combination with other methods, Naidu et al. [35]. demonstrated that the dissolution properties of meloxicam–CyD binary systems are superior to that of pure meloxicam. Similar conclusions were arrived at by Mura et al. [36, 37] when they found that the greater amorphizing properties (revealed by DSC and X-ray analysis) of CyD–ibuproxam [36] and CyD–naproxen [37] systems are mainly responsible for the enhanced dissolution efficiencies (by 5–15 times) in comparison to the pure drug. Perdomo-Lopez et al. [38] reported a direct correlation between CyD cavity size and the antimycotic activity of sertaconazole complexes. Maximum solubility, complex stability, and pharmacological activity were observed for sertaconazole–γ-CyD complexes and the lowest for α-CyD complexes. Furthermore, it was found that native γ-CyD exhibits better performance than hydroxypropyl-γ-CyD [38]. Similarly, Jambhekar et al. [39] demonstrated that the bioavailability of indomethacin is improved by complexation with β-CyD but not with hydroxyethyl- or hydroxypropyl-β-CyD. In some cases, it is necessary to use CyD dimers for accommodating a drug molecule of specific size and shape. Liu et al. [40] synthesized bis-β-CyDs linked with a tetraethylenepentaamino spacer for solubilizing the very poorly water-soluble drug paclitaxel. These novel bridged CyDs solubilize paclitaxel to levels as high as 2 mg mL^{-1}, with the CyD-drug complexes obtained characterized by various techniques, including DSC, NMR and FT-IR spectroscopy, TG-DTA, and X-ray diffraction. Importantly, these novel CyD–drug complexes [40] elicited an inhibitory effect on leukemia cells at drug concentrations of 10 pg mL^{-1}, much lower than the solubility of the complex.

It should be emphasized that drug complexation with CyDs can improve not only the drug solubility and pharmacological activity as exemplified above, but also the photostability, thus prolonging drug shelf-life. Bayomi et al. [41] reported that nifedipine (which is a highly photosensitive drug that requires strict protection from light during manufacturing, storage, and handling of its dosage forms) displays a dramatic enhancement of dissolution upon CyD inclusion, with a magnitude that depends on the type of CyD used. They found that inclusion complexes of nifedipine provide substantial protection against light exposure (to an extent that depends on the light source and wavelength), and eventually concluded that design of solid dosage forms of nifedipine, such as fast-dissolving nifedipine tablets, is possible taking advantages of the lower level of required light protection. Similar results were reported by Lutka [42]; that is the photostability of promethazine in solution is increased upon addition of β-CyD and its derivatives owing to the formation of inclusion complexes.

An interesting recent trend in DSC application can be found in the investigation of complexation systems that involve additional components other than drugs and CyDs. The major drive for doing such research is a desire to conveniently assess and/or understand the synergistic effect observed upon the addition of other components to a drug–CyD system. Mura et al. [43] studied ternary systems composed of naproxen, hydroxypropyl-β-CyD, and amino acids by DSC and other methods, finding that arginine is the most effective amino acid for improving drug solubility, and is the only one to exhibit a synergistic effect when used in combination with hydroxypropyl-β-CyD. Fatouros et al. [44] investigated the encapsulation behavior of prednisolone– and prednisolone–CyD complexes into liposomes. They found that release of prednisolone from liposomes is greatly accelerated when the drug is entrapped in the form of a complex with CyD, compare to the plain drug. However, it should be mentioned that inclusion of CyD and/or CyD complexes into liposomes is a rather complicated process which may cause damage to the liposome's lipid membrane upon direct interaction with CyD [45]. Another synthetic chemical approach was employed by Boudad et al. [46], who investigated the inclusion behavior of the hydroxypropyl-β-CyD–saquinavir complex into nanoparticles prepared by polymerization of *iso*-butyl and *iso*-hexyl cyanoacrylate monomers in aqueous solution, and found that this supramolecular system is promising for oral drug administration.

Certainly, DSC is a perfectly applicable technique for the detailed characterization of pharmacological formulations obtained by such traditional methods as co-grinding, as exemplified by Cirri et al. [47], who obtained and stabilized glyburide in the activated form as a complex with native and modified CyDs via the co-grinding technique.

8.3 Application of Isothermal Titration Calorimetry in Cyclodextrin and Drug Studies

In recent years, ITC has not been used as often in studies of CyD–drug interactions as in the previous two decades (see, for instance, the works by Guillory [48–50] and other groups [28, 29, 33] in the early 1990s). Generally speaking, the focus of ITC studies has overwhelmingly shifted from the microcalorimetry of CyD–drug interactions to the microcalorimetry of the direct interaction of drugs with proteins and nucleic acids [51–73]. Nevertheless, two recent experimental papers devoted to the thermodynamics of CyD–drug interactions should be briefly discussed. Using ITC in combination with UV, fluorescence, and circular dichroism spectroscopy, Koushik et al. [74] determined that the aromatic acid residues of the peptide drug deslorelin are included in the cavity of hydroxypropyl-β-CyD upon complexation. This inclusion complex causes steric hindrance for enzymatic hydrolysis by such enzymes as α-chymotrypsin and thus accounts for the greater drug stability in the presence of common hydrolytic enzymes. Kriz et al. [75] obtained consistent association constants from NMR (presented in Chapter 9) and ITC measurements of inclusion complexation of β-CyD with (+)-catechin, a polyphenolic compound of natural origin that exhibits antioxidant properties of interest for therapeutic and cosmetic uses. From NMR and computer simulation studies, they proposed a global complex architecture.

CyDs are intrinsically chiral macrocyclic hosts and this explains the recent trend of ITC application to elucidate the thermodynamics of chiral discrimination of enantiomeric and diastereomeric guest pairs by native and modified CyDs. It is rather surprising that until the beginning of this century only a relatively limited amount of effort had been devoted to such thermodynamic studies for determining not only the complex's stability constant but also the reaction enthalpy and entropy for inclusion complexation of chiral guests with CyDs. Before 2000 few experimental studies were published on chiral recognition by native CyDs using ITC, and there were no papers on the microcalorimetry of chiral recognition by modified CyDs. The early studies that appeared in the literature were done with the complexation of α-CyD with norleucine and norvaline [76], 1-ferrocenylethanol [77], phenylalanine, α-methylbenzylamine, mandelic acid, phenylfluoroethanol, and amphetamine [78]. However, the last two studies were performed under conditions that allowed the co-existence of different species of guest and/or CyD in the solution. In a similar vein the complexation thermodynamics of several carbohydrates with α- and β-CyDs were also investigated [79], and we also published our own work on the complexation thermodynamics of α- and β-CyDs with 2-alkanols [80], and ephedrines and pseudoephedrines [81].

The first comprehensive microcalorimetric study on chiral recognition by native β-CyD was published in 2000 [82]. In this study, microcalorimetric measurements were carried out in order to obtain accurate thermodynamic quantities for the inclusion complexation of 43 enantiomeric pairs of selected guests with β-CyD in aqueous buffer solutions. Based on the thermodynamic parameters obtained for several families of structurally related chiral guests, the relationship between the guest's structure and the enantioselectivity was elucidated, and the mechanisms and thermodynamic origin of the chiral recognition displayed by CyDs was discussed. The magnitude of chiral discrimination was not very high ($\leq$ 15–20% in affinity for enantiomeric guest pairs), which is most likely attributable to the chiral centers being very symmetrically distributed inside the β-CyD cavity. In this study, (1) a direct correlation was established between the mode of penetration and chiral recognition by β-CyD for aromatic amino acid derivatives (and for some other classes of organic chiral compounds); (2) chiral guests with a less symmetrical, nonpolar penetrating group and/or a greater distance between the chiral centre and the most hydrophilic, often charged, group are more likely to give better chiral recognition or enantio-discrimination; and (3) almost any alterations made to the guest molecule that result in stronger binding with β-CyD leads to smaller chiral recognition, since in almost all cases the additional weak interactions involved in the complexation process result in noncomplementarity between the chiral guest and the CyD cavity.

Further development of the above-mentioned ideas [82] has been reported [83, 84]. Mono- and diaminated β-CyDs were employed as chiral discriminators for a variety of chiral guest molecules [83, 84]. Thus, chiral centers less-symmetrically distributed inside the cavities of mono- and diaminated β-CyDs, compared to native β-CyD, do indeed lead to better chiral discrimination by the amino-modified hosts. The major findings [83, 84] may be summarized as follows: (1) the direct correlation between the mode of penetration and chiral recognition established above for β-CyD [82] holds for aminated β-CyD, and may now be considered a general rule for CyD complexation reactions. We may further conclude that the original chiral selectivity of each native CyD is preserved or "memorized" even after chemical modification; (2) negatively charged guests exhibit larger affinities toward monoamino-β-CyD than toward β-CyD. Flexible (e.g. hexahydromandelic acid) or less bulky (e.g. mandelic acid) guests often exhibit 3–5 times higher affinities toward amino-CyD, whilst bulky and/or rigid guests (e.g. camphanic acid) show only slightly higher (but <2 times) or even the same affinities; and (3) the enhanced chiral discrimination exhibited by amino-β-CyD (versus β-CyD) predominantly originates from the substantially different ability of each negatively charged enantiomeric guest to optimize the electrostatic attractive force by adjusting its position/location inside the chiral cavity, as typically observed upon complexation of α-methoxyphenylacetic acid. In accordance with this interpretation, neutral and positively charged guests do not show any significant changes in chiral discrimination upon complexation with amino-β-CyD.

Once the most essential rules that govern chiral discrimination by native and modified CyDs were established [82–84], the efforts of researchers in this field were directed predominantly in two directions, specifically: (1) the optimal guest structure for native CyDs, and (2) the best CyD modification to achieve the highest chiral discrimination towards a particular chiral guest. One successful example of the first approach is the ITC investigation of the chiral recognition of dipeptide derivatives containing two aromatic rings [85, 86]. It was shown that native γ-CyD gives unprecedented discrimination of several distereomeric pairs of dipeptide derivatives, affording equilibrium constants that differ by a factor of 6–7 (e.g. D,L-Cbz-Ala-Phe versus L,L-Cbz-Ala-Phe or D,L-Cbz-Pro-Phe versus L,L-Cbz-Pro-Phe (Cbz = benzyloxycarbonyl), the highest chiral discrimination reported for native CyDs [85, 86]. The second approach was also successfully applied by Kano [87, 88] and Liu [89, 90]. For instance, camphor displays a high enantioselectivity upon inclusion by 6-*O*-(4-chlorophenyl)-β-CyD [89], while phosphoryl-tethered β-CyDs effectively discriminate enantiomers of amino acid derivatives with hydrophobic substituents [90]. It should be noted that the chiral recognition of the enantiomers of camphor can be achieved not only by β-CyD dimers (phosphoryl-tethered β-CyDs [90]) but also with a large excess of native α-CyD through sandwich complexation [91]. (Caution should be exercised with the ITC data obtained for CyD complexation with complicated stoichiometries, such as 1:2 complexation, which will be discussed below in more detail). Finally, to highlight once again the importance and interest of calorimetric investigation for chiral discrimination by CyDs, it should be noted that a new calorimetric method has been developed for determining the thermodynamic parameters of enantioselective gas–surface interactions, such as absorption of two enantiomers of methyl-2-chloropropionate into a thin film of chiral receptor such as octakis(3-*O*-butanoyl-2,6-di-*O*-pentyl)-γ-CyD [92].

ITC has become a more and more attractive and indispensable tool for investigating complex supramolecular systems including CyDs. Mulder et al. [93] synthesized β-CyD dimers tethered by a rigid photoswitchable moiety, i.e. bis-(phenylthienyl)ethene. By irradiating with light of appropriate wavelengths, the dimers can be interconverted reversibly between a relatively flexible (open) form and a rigid (closed) form, thus showing promise for using such a supramolecular assembly as a switchable receptor system. ITC experiments revealed a large difference, by a factor of 8, in the binding affinity between the open and closed forms of the dimer towards a *meso*-tetrakis(4-sulfonatophenyl)porphyrin guest. The authors emphasized that this difference in binding affinity reflects the difference in enthalpy of binding between the two isomeric CyD dimers, indicating that the β-CyD cavities of the closed dimer are too far separated in space by the rigid closed bis(phenylthienyl)ethene tether to cooperatively bind the guest. The difference in binding affinity was large enough to trigger the release of the guest from the dimer upon irradiation. Liu et al. [94] reported supramolecular self-assembly of β-CyDs with an aromatic tether in a helical columnar rather than a linear channel fashion. ITC and NMR studies on this self-assembly behavior [94] in aqueous solution revealed that the dimerization step is the key to the formation of a linear polymeric supramolecular architecture, which is driven by favorable entropic contributions. Kano et al. [95, 96] characterized formation of 2:1 inclusion complexes of CyDs and charged porphyrins by ITC in aqueous and aqueous–organic media and used these thermodynamic data to develop a new approach to heteroporphyrin arrays in aqueous media.

One more very important trend should be addressed here briefly. Traditionally, by using conventional ITC methods, binding constants from 10^3 M^{-1} to 10^8 M^{-1} can be measured most accurately.[1] If binding constants significantly exceed 10^8 M^{-1}, microcalorimeter sensitivity becomes compromised as concentrations are lowered to the point where quantitative measurements of the heat released/absorbed approaches the limit of useful detection. However, binding constants substantially exceeding the 10^8 M^{-1} limit can be determined quantitative if such a strongly binding guest competes for the same binding site with a less-strongly binding second guest. Indeed, when the strongly binding guest is added stepwise into a solution of host and the weaker binder, the latter is gradually replaced by the stronger one. A convenient routine for performing competitive-binding ITC experiments has been developed by Sigurskjold [97]. Development of the competitive-binding model has greatly contributed to the accurate determination of thermodynamic parameters for complexes of very high affinity. However, the determination of precise thermodynamic parameters for complexation reactions with low affinity ($K < 10^3$ M^{-1}) is still considered as an even more challenging task.

Turnbull et al. [98] recently undertook a careful microcalorimetric study to evaluate the lower limit of association constant ($K < 10^3$ M^{-1}) that can be determined by the ITC method. From numerous experiments and computer simulations, they convincingly demonstrated that, in principle, there is no reason to doubt the association constant and reaction enthalpy obtained for low affinity complexation, provided that the experiment is well designed, by taking into account the effects of final receptor occupancy, the signal-to-noise levels and that the concentrations of ligand and receptor are accurately known; representative examples will be given for the ITC determination of $K < 10^3$ M^{-1} in the next section. By using their own experimental approach [98], the same authors successfully evaluated the complexation thermodynamic parameters for low-affinity interactions between cholera toxin and ganglioside GM1 [99].

8.4 Determination of Thermodynamic Parameters for Stoichiometric 1:1 Complexation by Cyclodextrin

In the case of 1:1 host–guest complexation, or a 1:1 binding model with a single binding site, we deal with the simultaneous determination of two parameters, i.e. the equilibrium constant (K) and the standard molar enthalpy (ΔH°), for the following complexation equilibrium of cyclodextrin (CyD) and guest (G) in solution:

$$\mathrm{CyD_{sln}} + \mathrm{G_{sln}} = \mathrm{CyD{\cdot}G_{sln}} \tag{1}$$

$$K = \gamma_{\mathrm{CyD\cdot G}}[\mathrm{CyD{\cdot}G}]/(\gamma_{\mathrm{CyD}}[\mathrm{CyD}] \times \gamma_{\mathrm{G}}[\mathrm{G}]) \tag{2}$$

where γ is the activity coefficient of relevant species. Nonideality corrections are often assumed unnecessary, particularly for the neutral species, since activity coefficients of neutral organic molecules are close to unity at low-to-moderate concentrations in aqueous solutions [81, 100]. This approximation holds reasonably well even when a charged guest, such as an organic ammonium salt, is involved, since reaction (1) is charge symmetric and the activity coefficients in the numerator and denominator of Eq. (2) should largely cancel each other at low-to-moderate ionic strengths [81, 100].

MicroCal's titration microcalorimeter is the most common instrument to have been used in many scientific studies on the complexation thermodynamics of CyDs. Several important issues which should be taken into account on using a titration microcalorimeter will be discussed below for investigation of CyD complexation and supramolecular association in general.

The apparent shape of the titration curve obtained in microcalorimetric titration experiments varies significantly, depending on the host and guest concentrations and the magnitude of the equilibrium constant. If the equilibrium constant is small and the initial reactant concentrations are low, only a small portion of the added guest can interact with the host in the cell to give a typical titration curve illustrated in Fig. 8.1.

When the initial concentrations of reactants (host and guest) are higher or the affinity is greater, the shape of titration curve varies substantially from a slowly saturating profile (Fig. 8.2 (a)) to a sigmoidal one (Fig. 8.2 (b)). From careful consideration of Eq. (2), one can readily understand the reason for such changes. Thus, if the initial reactant concentrations are sufficiently high or the complexation is strong, the added guest immediately interacts with the host in the cell from the beginning of titration until the stoichiometric point, efficiently consuming the host and producing an almost constant heat effect upon each injection (corresponding to the initial plateau in Fig. 8.2 (b)). After the stoichiometric point, the amount of free host remaining in the cell is rapidly reduced and hence the magnitude of the heat effect similarly decreases even upon addition of an excess amount of guest, giving a plateau at high guest/host ratios as in Fig. 8.2 (b).

In the case of 1:1 complexation, a typical microcalorimetric titration run consists of 15–25 successive injections of constant volume (5–10 μL per injection) of guest

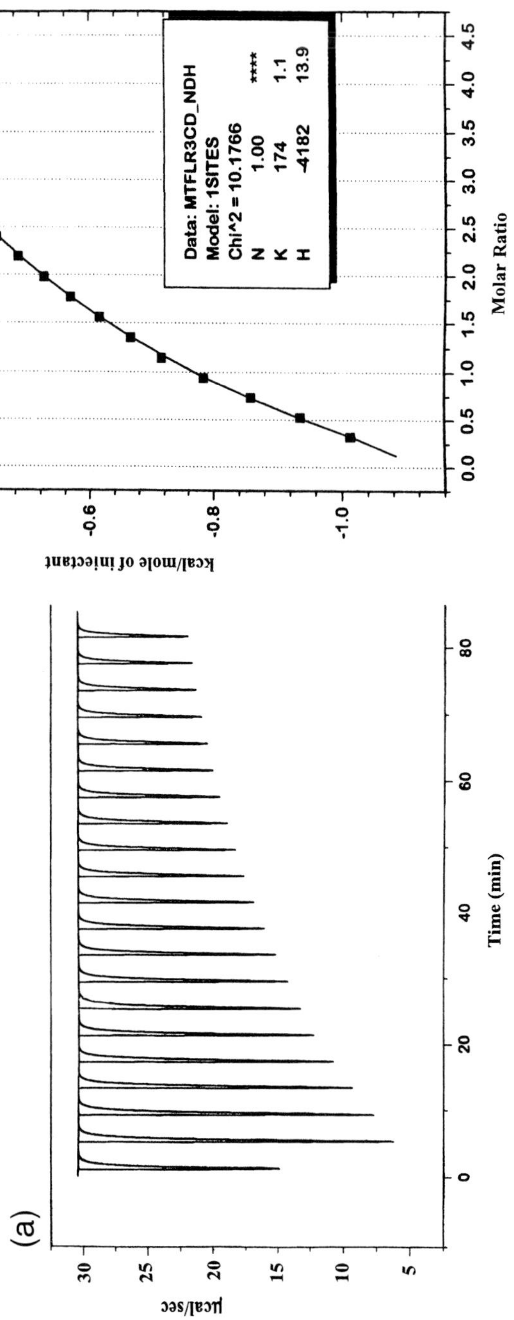

Fig. 8.1. Typical results obtained in a microcalorimetric titration experiment: (a) heat effect observed upon each injection of titrant solution and (b) non-linear curve fitting of the experimental results using a simple 1:1 model.

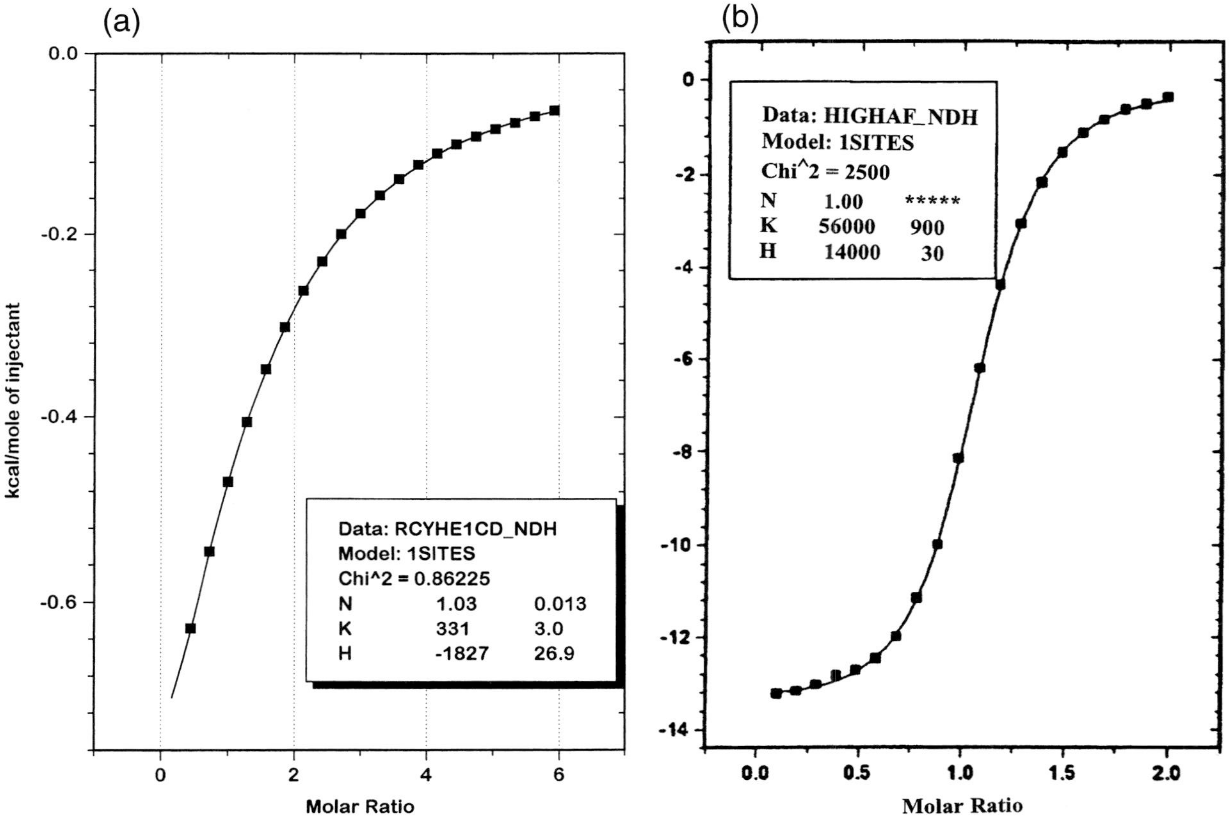

Fig. 8.2. Profiles of the microcalorimetric titration curve for (a) low affinity ($K = 331\ M^{-1}$) and (b) high affinity ($K = 56\,000\ M^{-1}$) complexation.

buffer solution into the reaction cell charged with a CyD solution (ca. 1.5 mL) in the same buffer. The heat effect obtained in the titration experiment must be corrected for the heat of dilution of the guest solution in the absence of a CyD host, which is determined in a separate experiment using the same number of injections and guest concentration as in the actual titration runs. The dilution enthalpies determined in the control experiments are algebraically subtracted from those obtained in the titration experiments. In the case of the MicroCal system, recording of all experimental data, calculation of heat effects from each injection, and data processing are performed by the Origin program (MicroCal), which is also used for calculation of the equilibrium constant and standard molar enthalpy of reaction from the titration curve. In addition, this software gives a standard deviation based on the scatter of the data points in a single titration curve. The issue of uncertainty is crucial in microcalorimetry, particularly when one compares the thermodynamic parameters obtained, e.g. for enantiomeric isomers, and will be discussed in further detail.

Firstly, there are two possible sources of random errors in microcalorimetric experiments using an ITC instrument. One of the errors is associated with a random deviation of experimental points from the theoretical curve on each individual microcalorimetric titration run. This originates from the baseline noise, the volume fluctuation of each injection, the measurement of the heat from each injection, and so on. Noninstrumental errors, arising from possible small deviations in sample concentration, pH, and ionic strength are also involved. The first source of error is known to be dominant when K is in the order of several hundreds [81]. When calorimetric titration is carefully repeated several times for the same reaction, with a new solution of different concentration each time, the standard deviation (σ) of the mean value is comparable to, or even smaller than, the σ value given in each run by the Origin fitting program [100]. When a guest shows a high affinity ($K > 1000$) and large heat effect, the experimental data points deviate only slightly from the fitted curve, affording a fairly small σ in each run, while the σ value arising from repeated measurements is not reduced and therefore often exceeds the σ value given by Origin in each independent run. Even for a moderately stable complex with a K value of several hundreds, the σ value arising from repeated experiments is occasionally two to three times larger than that given by Origin for each run. To be fair with your own data, it is necessary to report the larger of the two different types of uncertainty discussed above.

The major source of systematic errors lies in an inadequate application of the 1:1 model to more complicated systems, as we can illustrate below using our own experimental data. In previous papers, we reported two separate sets of data for the complexation of 3-phenylpropionic and 3-phenylbutyric acid with β-CyD. Slightly different values were reported for 3-phenylpropionic acid, i.e. (a) $K = 141 \pm 3$ M^{-1}, $\Delta H^{\circ} = -7.60 \pm 0.08$ kJ mol^{-181} and (b) $K = 149 \pm 4$ M^{-1}, $\Delta H^{\circ} = -7.32 \pm 0.08$ kJ mol^{-1} [101]; and also for 3-phenylbutyric acid, i.e. (a) $K = 379 \pm 10$ M^{-1}, $\Delta H^{\circ} = -9.41 \pm 0.10$ kJ mol^{-1100} and (b) $K = 387 \pm 6$ M^{-1}, $\Delta H^{\circ} = -9.16 \pm 0.04$ kJ mol^{-1} [101] (here and below in this chapter uncertainities of thermodynamic parameters are presented as a two standard deviation of the mean). More recently [82], we obtained a further set of microcalorimetric data with these two guests under the same physicochemical conditions as before [100]. Using a new ITC instrument also from MicroCal, a better fit of the experimental points was obtained, which allowed us to see small, but systematic, deviations of the data points from the theoretical curve particularly during the last part of titration with both guests. If one ignores this systematic deviation and executes the fitting by Origin using all of the experimental points, the calculations give the following values for 3-phenylpropionic acid: $K = 151 \pm 5\ \mathrm{M}^{-1}$, $\Delta H^\circ = -7.11 \pm 0.15\ \mathrm{kJ\ mol}^{-1}$. It should be noted that, although this result is in good agreement with the previous ones [101], the quality of the fit is considerably improved (by more than four times in χ^2) by introducing just one extra parameter, i.e. stoichiometry (n), to the fitting equation for the 1:1 model. We assigned the origin of this systematic deviation from the best-fit curve to the involvement of a complicated 1:n host–guest complex formation upon addition of excess amounts of guest to the CyD solution. In fact, when only the first half of the data points in the same experiment are used in the calculation in order to reduce the complicated final guest/host ratio, the Origin program gives a somewhat different result with a better fit: $K = 160 \pm 3\ \mathrm{M}^{-1}$, $\Delta H^\circ = -6.90 \pm 0.08\ \mathrm{kJ\ mol}^{-1}$. For further confirmation, we repeated the same experiments using host and guest solutions of half concentrations. In these runs, the quality of the fit (χ^2) to the theoretical curve was not appreciably improved by the use of more sophisticated models than 1:1 complex. We strongly recommend the execution of a "dilution test" as a standard procedure whenever sophisticated complex(es) are suspected of involvement, because their contribution can be reduced substantially by lowering the concentrations of guest and/or host. It is always necessary to make all possible attempts to avoid the involvement of 1:2 or more sophisticated complexes in solution, and all experiments should be performed at concentrations as low as possible in order to assure the sole formation of the 1:1 species. Certainly all the above discussion is valid only for cases when 1:1 complexation is the major target of the investigation; however, microcalorimetric study on higher-order CyD complexes, e.g. 1:2, 2:2, and so on does not necessarily require performance of the "dilution test".

It should be noted also that there is an illusive correlation between K and ΔH° observed in the erroneous and correct data presented above; thus a larger equilibrium constant is accompanied by a smaller heat effect and vice versa. Such a correlation between K and ΔH°, or more generally between two linearly correlated parameters, has been discussed in the literature from various points of view [1, 102–104]. Here we present a simple illustrative explanation for the source of $K - \Delta H^\circ$ correlation. For the simplest 1:1 model, a three-dimensional representation of the square sum of the deviation as a function of K and ΔH°, i.e. $\Sigma(\Delta Q_{\mathrm{exp}} - \Delta Q_{\mathrm{cal}})^2$, looks like an unsymmetrical well, where $(\Delta Q_{\mathrm{exp}} - \Delta Q_{\mathrm{cal}})$ is a difference between the observed experimental heat effect upon each injection (ΔQ_{exp}) and the theoretical curve (ΔQ_{cal}) [105, 106]. The side walls of this well become very steep if one tries to change one variable (for instance, K), keeping the others constant, and become even steeper if one tries to simultaneously alter two variables in the same direction. One can change the variables with an accompanying minimal increase of $\Sigma(\Delta Q_{\mathrm{exp}} - \Delta Q_{\mathrm{cal}})^2$ only by simultaneously decreasing one variable and increasing another. Hence, if some perturbation takes place (in the present case, the formation of a 1:2 species in addition to the 1:1 species) that affects the experimental data (K and/or ΔH°), it is most beneficial for K and ΔH° values to be adjusted in such a way as to minimize the change in $\Sigma(\Delta Q_{exp} - \Delta Q_{cal})^2$.

It is further advised that, even if the 1:1 stoichiometry is thought to be the most reasonable choice, the influence of removing some data points (up to 5 out of 20 points) from the initial and final parts of the titration curve on the overall quality of the fit should be routinely checked. In the initial stages of the titration experiment, the concentration of CyD in the cell far exceeds that of guest (G), and the occasionally observed systematic deviation of experimental points may be ascribed to the formation of 2:1 or higher-order CyD_n:G ($n > 1$) complex species. In the final stages of titration, the concentration of G is much higher than that of free CyD in the cell, which may lead to 1:2 or higher-order CyD:G_n ($n > 1$) complexes. When such systematic deviations are observed, the titration should be repeated using a 2–3 times less-concentrated guest and/or CyD solutions in order to reduce the contribution of these sophisticated host–guest complexes. In addition to calculation based on the 1:1 stoichiometry, other types of calculation, assuming 1:n and n:1 binding models, should also be performed whenever such higher-order complexes are suspected. If such calculations do not lead to any appreciable improvement of the overall fit (χ^2), then these more complicated models can be considered as irrelevant in the particular case under consideration and consequently the assumption of a 1:1 model with a single binding site becomes the only reasonable choice.

If one is suspicious of co-existing 1:1 and higher-order complexes, it is preferable in general to make all possible efforts to simplify the dominant complex species in the system rather than to apply more sophisticated models. We will present an example of such an approach employed in the analyses of titration calorimetry data obtained for the complexation of Cbz-protected aromatic amino acids with γ-CyD in aqueous buffer solutions. Thus, the data obtained by microcalorimetric titrations of 2-mM γ-CyD with 150–200-mM Cbz-Phe, Cbz-Tyr, Cbz-Trp, and Cbz-His showed significant deviations from the theoretical curve based on a simple 1:1 model [107]. The most logical explanation for these deviations is the co-existence of a 1:1 complex (in which two aromatic rings of one guest molecule are included in the same cavity) and various 2:1 complexes (in which two aromatic rings from two separate guest molecules are included in the same cavity) in the solution under the experimental conditions. Furthermore, we cannot logically exclude the involvement of the even more complex 2:2 species (in which two aromatic rings from two separate guest molecules are included in the same cavity and the two remaining aromatic rings of the same guest molecules are included in another cavity). To describe such a complicated reaction system, one needs at least six parameters, i.e. the reaction enthalpies and equilibrium constants for 1:1, 1:2, and 2:2 complexatons. Obviously, it is more reasonable to simplify the reaction mixture rather than to attempt the six-parametric fitting procedure involving a very complicated routine. Indeed, the deviations from the simple 1:1 model became smaller and finally vanished completely, when the host and guest solutions used in the titration experiment were diluted by 2–4 times from 2 mM and 150–200 mM to 0.5–1 mM and

50–70 mM, respectively [107]. We may conclude therefore that dilution is the most convenient and reliable standard procedure to avoid or eliminate the complications arising from the involvement of higher-order complexes [81, 82, 107].

8.5 Determination of Thermodynamic Parameters for 1:2 Guest–Host Complexation of Cyclodextrin

Complexation of Cbz-Gly with γ-CyD can serve as a typical example of the co-existence of 1:1 and 1:2 guest-host species in solution. In our study [107], we investigated the complexation behavior of Cbz-Gly with γ-CyD using a high guest concentration of 750 mM and high guest:host ratios of up to 40. Figure 8.3 illustrates a representative result obtained in a microcalorimetric titration performed under such conditions. Obviously, the experimental data points shown in Fig. 8.3 do not fit the simple 1:1 model, as a gradually increasing heat production is observed in the initial part of the titration curve. The simplest choice to give a satisfactory fit is the stepwise 1:2 host–guest complexation model.

A set of thermodynamic quantities obtained by the four parametric fit using the sequential binding model is shown in Fig. 8.3 (b). It should be mentioned that K_1 and K_2 in Fig. 8.3 (b) are intrinsic constants of the sequential binding model, which allow easy comparison with non-interacting expectations [105]. In the case of stepwise 1:2 host–guest complexation (Fig. 8.3), the value of the uncertainty given by the Origin program does not seem to be a good criterion for judging the accuracy of the thermodynamic parameters obtained, as several considerably different sets of parameters (i.e., $K_1{}^{\text{int}}$, ΔH_1; $K_2{}^{\text{int}}$, ΔH_2) can often give similarly good fits to the experimental data profile. Probably, the five-dimensional surface of χ^2 as a function of four parameters ($K_1{}^{\text{int}}$, ΔH_1; $K_2{}^{\text{int}}$, ΔH_2) is very "flat" (if this word is appropriate in a five-dimensional numerical universe) and therefore the scattering of the experimental data points does not allow the Origin program to find an absolute minimum [105, 106]. We repeated computer simulations many times with varying initial sets of the four parameters, using the consistency of χ^2 as a criterion for the reliability of the results of calculations, and finally we selected the following thermodynamic parameters for the complexation between Cbz-Gly and γ-CyD: $K_1 = 19 \pm 3$; $\Delta H_1 = -0.7 \pm 0.3$ kJ mol^{-1}; $K_2 = 8.5 \pm 1.5$ and $\Delta H_2 = -31 \pm 3$ kJ mol^{-1}. Practically, no other set of the parameters could better describe the experimental data without increasing the χ^2 value by two to three times.

It is interesting to compare the thermodynamic parameters obtained for the 1:1 complexation of Cbz-Gly with γ-CyD with those for the 1:1 complexation of Cbz-Gly with β-CyD. This comparison may also be taken as an independent assessment of the reliability of the above-mentioned four-parametric fit. The equilibrium constant for the 1:1 complex formation of Cbz-Gly with γ-CyD is about one-tenth as large as that for Cbz-Gly complexation with β-CyD [82], and the heat effect obtained for the γ-CyD complex is about one third as large as that for the β-CyD complex. It is reasonable to expect less-pronounced van der Waals interactions and therefore a smaller exothermic heat effect, when the cavity is too large to comfort-

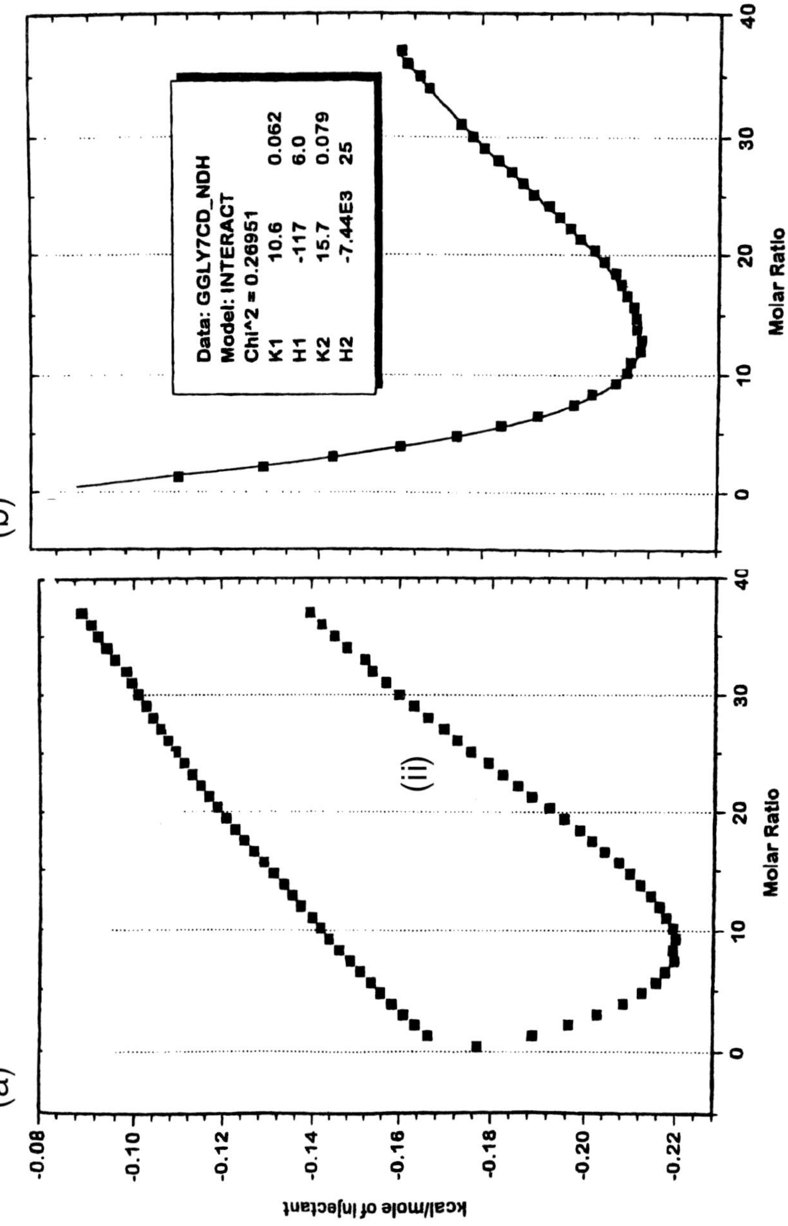

Fig. 8.3. Typical results of microcalorimetric titration experiment for 1:2 host–guest complexation: (a) plots of heat effects observed upon (i) dilution of a Cbz-Gly solution added to the buffer solution and (ii) titration of γ-CyD with Cbz-Gly both in the buffer solutions; (b) computer fitting of the experimental results to the theoretical curve for the 1:2 model.

ably accommodate the guest molecule, as demonstrated previously for the complexation of aliphatic alcohols with α- and β-CyDs [80], where the reaction enthalpy is always more exothermic for α-CyD than for β-CyD. Insertion of the second Cbz-Gly molecule into the same, half-occupied, γ-CyD cavity is expected to be accompanied by a large exothermic heat effect due to the more effective van der Waals contacts, and also by a large unfavorable entropy change due to the tight association of two Cbz-Gly molecules within the γ-CyD cavity. Indeed, the experimental thermodynamic data obtained agree well with those expected from such a prediction.

In principle, ternary complex formation is less favorable from the entropic point of view, and is consequently accompanied by a more negative entropy change than that observed for the binary counterpart [107]. In this context, it is reasonable to assume that the entropic loss associated with the inclusion of two separate aromatic guest molecules (such as Cbz-Gly) into the same cavity, giving a 1:2 complex [γ-CyD•(Cbz-Gly)$_2$], becomes larger than that caused by the simultaneous inclusion of two aromatic moieties of the same guest molecule (such as Cbz-Phe), forming a 1:1 complex with γ-CyD. This prediction is in good agreement with the experimental data. Thus, the inclusion of the second Cbz-Gly molecule into the γ-CyD cavity is accompanied by a very large loss of entropy ($T\Delta S^\circ = -26$ kJ mol^{-1}), which is in sharp contrast to the positive entropy change ($T\Delta S^\circ = 6.6$ kJ mol^{-1}) upon inclusion of the first Cbz-Gly molecule. This negative entropy change cancels the large exothermic enthalpy change ($\Delta H^\circ = -31$ kJ mol^{-1}) to a great extent, giving a smaller equilibrium constant for the second inclusion step: i.e. $K_1 = 19$ and $K_2 = 8.5$ M^{-1}.

In fact, the thermodynamics of 1:2 complexation is very sensitive to the structure of the guest, and even a small alteration in guest molecule such as the addition or removal of one methylene group has a large impact, as demonstrated in our study [107] of the complexation of ω-phenylalkanoic acids with γ-CyD. The heat evolution patterns upon injection of a guest solution in the titration microcalorimetric experiments are shown in Fig. 8.4 for the complexation reactions of γ-CyD with 4-phenylbutyric acid (Ph-C$_4$), 5-phenylvaleric acid (Ph-C$_5$), and 6-phenylhexanoic acid (Ph-C$_6$) under comparable conditions (2.7–2.9 mM γ-CyD and 190–230 mM guest in the initial solutions). Although the shape of each titration curve would appear to be qualitatively similar to that observed for Cbz-Gly, a close examination readily reveals a clear difference, namely that the curves for Ph-C$_4$, Ph-C$_5$, and Ph-C$_6$ cross the horizontal line at a different molar ratio in each case. The inversion of the sign of the heat evolution curve during the titration means that, in contrast to the Cbz-Gly case, the heat effect at the initial stage of the titration (or 1:1 complex formation) has a sign opposite to that in the second stage (or 2:1 complex formation); in other words, the first stage is endothermic, whilst the second is exothermic.

As was the case with Cbz-Gly, the stepwise 1:2 complexation model was used as the simplest, most reasonable, choice for the treatment of the microcalorimetric titration data obtained for the complexation of Ph-C$_4$, Ph-C$_5$, and Ph-C$_6$ with γ-CyD to afford a satisfactory match between the experimental data and the theoretical curve.

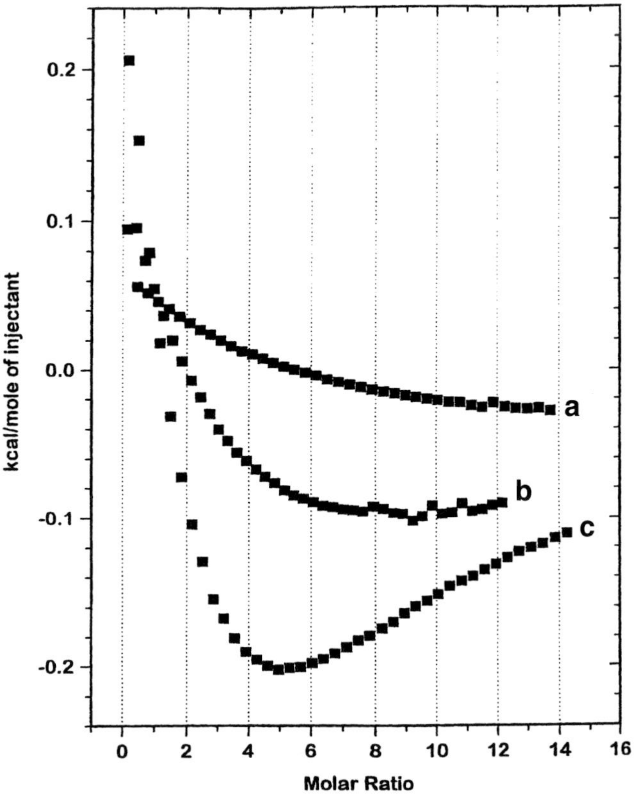

Fig. 8.4. Profiles of the heat effect observed for complexation reactions of γ-CyD with (a) 4-phenylbutyric acid, (b) 5-phenylvaleric acid, and (c) 6-phenylhexanoic acid under comparable experimental conditions (2.7–2.9 mM γ-CyD and 190–230 mM guest in the initial solutions).

8.6
A Thermodynamic View for Globally Understanding Molecular Recognition

Determination of the thermodynamic parameters is indispensable to a quantitative discussion and understanding of molecular recognition phenomena, involving various hosts, guests, and solvents, from the global viewpoint of supramolecular chemistry. For a wide variety of chemical and biological molecular recognition systems, the thermodynamic quantities have already been compiled, analyzed, and discussed in order to extract insights into the mechanistic and structural features of molecular recognition. However, most of the foregoing discussion and conclusions, although certainly self-consistent within particular and closely related systems, unfortunately seem *ad hoc* in many cases and are not always applicable to other molecular recognition systems composed of different categories of host–

guest combinations. Probably, the most general thermodynamic approach to a global understanding of molecular recognition is the comprehensive analysis of the compensatory enthalpy–entropy relationship observed in many supramolecular systems.

The enthalpy–entropy compensation effect has long been a hot topic in chemical literature, because in principle no explicit relationship between the enthalpy change and the entropy change can be derived from fundamental thermodynamics. Nevertheless, the compensatory enthalpy–entropy relationship has often been observed in both activation and thermodynamic quantities determined for a very wide variety of reactions and equilibria.

The linear ΔH° vs. ΔS° relationship observed experimentally leads to Eq. (3), where the proportional coefficient β has the dimension of temperature. From a combination of Eq. (3) and the differential form of the Gibbs–Helmholtz equation (Eq. (4)), we obtain Eq. (5).

$$\Delta H^\circ = \beta\Delta\Delta S^\circ \tag{3}$$

$$\Delta G^\circ = \Delta\Delta H^\circ - T\Delta\Delta S^\circ \tag{4}$$

$$\Delta G^\circ = (1 - T/\beta)\Delta\Delta H^\circ \tag{5}$$

Equation (5) clearly indicates that, at the critical point, or so-called "isokinetic" or "isoequilibrium" temperature (β), the rate or equilibrium constant is totally independent of the change in enthalpy caused by any alterations in substituent, solvent, and so on. It is interesting that such phenomena have been abundantly observed for various reactions and equilibria.

Certainly, one should be careful in selecting high-quality experimental data for an enthalpy–entropy compensation plot, particularly when limited sets of data are available, since an apparent compensation plot may arise from the experimental errors in measurement, as emphasized by Houk et al. [108]. However, the review of reliable experimental data by Williams et al. [109], as well as the theoretical consideration by Gilson et al. [110, 111] discussed below in detail, leave no doubts about the validity of the enthalpy–entropy compensation relationship itself.

8.7
Enthalpy–Entropy Compensation as a Practical Tool for a Global Understanding of Molecular Recognition

From the compensatory enthalpy–entropy relationship, the $T\Delta S^\circ$ values are linearly correlated with the ΔH° values to give Eq. (6). When integrated, this gives Eq. (7) and subsequent combination with Eq. (4) affords Eq. (8).

$$T\Delta\Delta S^\circ = \alpha\Delta\Delta H^\circ \tag{6}$$

$$T\Delta S^\circ = \alpha\Delta H^\circ + T\Delta S^\circ{}_0 \tag{7}$$

$$\Delta\Delta G^\circ = (1 - \alpha)\Delta\Delta H^\circ \tag{8}$$

Thus, the slope (α) of the $T\Delta S^\circ$-versus-ΔH° plot (Eq. (7)) indicates to what extent the enthalpic gain $\Delta\Delta H^\circ$ is canceled by the accompanying entropic loss $\Delta\Delta S^\circ$, both of which are induced by any alterations in host, guest, and solvent. In other words, only a fraction $(1 - \alpha)$ of the enthalpic gain can contribute to the enhancement of complex stability. On the other hand, the intercept $(T\Delta S^\circ_0)$ represents the inherent complex stability (ΔG°) obtained at $\Delta H^\circ = 0$, which means that the complex is stabilized even in the absence of enthalpic gain when the $T\Delta S^\circ_0$ term is positive. From comparative analyses of the thermodynamic data for cation binding by three types of ionophores with different topologies and rigidities (i.e. glyme, crown ether, and cryptand) [112–117], the slope (α) and the intercept $(T\Delta S^\circ_0)$ of the regression line were respectively related to the degree of conformational change and to the extent of desolvation upon complexation. Using the α and $T\Delta S^\circ_0$ values as quantitative measures for changes in conformation and desolvation of both host and guest, diverse chemical and biological supramolecular systems can be analyzed consistently, despite quite different weak interactions involved in each supramolecular system [112–117].

A simple illustration of the fundamental principles discussed above (α and $T\Delta S^\circ_0$ as quantitative measures for changes in conformation and desolvation, respectively, upon complex formation) is shown in Fig. 8.5. Obviously, the degree of

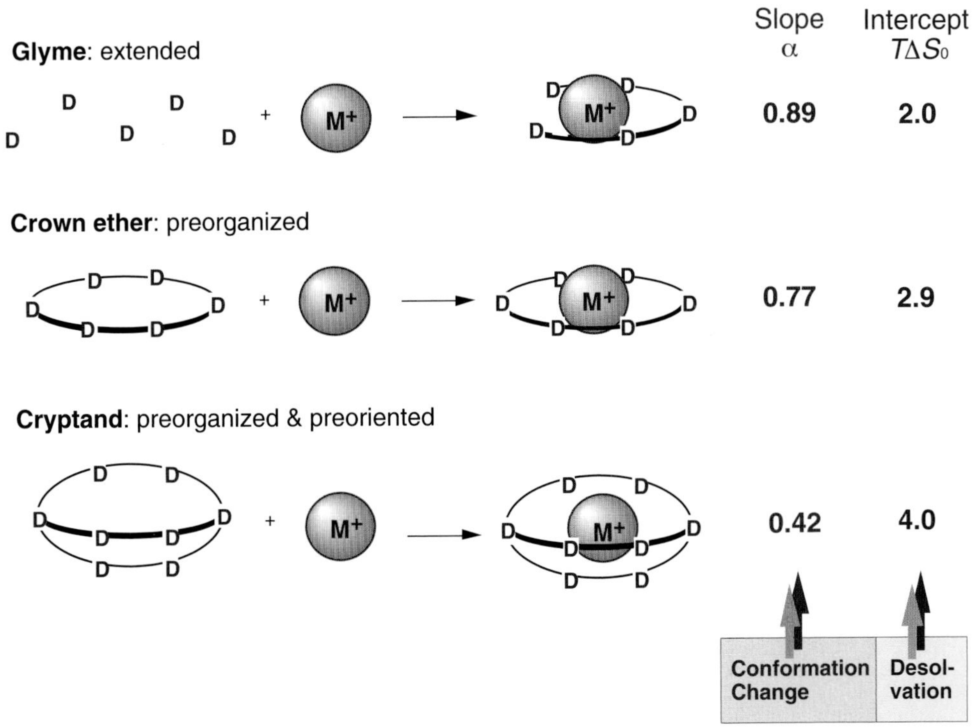

Fig. 8.5. Slope α as a measure of conformational changes and intercept $T\Delta S^\circ_0$ as a measure of desolvation upon complexation, initially proposed for cation binding by acyclic glyme, cyclic crown ether, and bicyclic cryptand, where D denotes a donor atom such as oxygen and nitrogen.

conformational change upon cation complexation decreases in the order: acyclic glyme > cyclic (preorganized) crown ether > bicyclic (preorganized and preoriented) cryptand. In line with this, the α value also decreases.

It is obvious that conformational changes and desolvation are much larger for the association reactions involving enzymes, antibodies, and DNA/RNAs than for the above simple cation binders. In accordance with that, the slope α and intercept $T\Delta S^\circ_0$ are much larger for biological supramolecular systems than for simple synthetic systems [112–117]. It should be emphasized that even closely related systems, such as CyDs of different sizes, can be consistently analyzed in terms of the slope α and intercept $T\Delta S^\circ_0$ as measures of the conformational changes and desolvation, respectively. Indeed, the slope α and the intercept $T\Delta S^\circ_0$ increase from α-CyD to β-CyD, and then to γ-CyD in accordance with the increasing conformational changes and extent of desolvation upon guest inclusion (Fig. 8.6) [16].

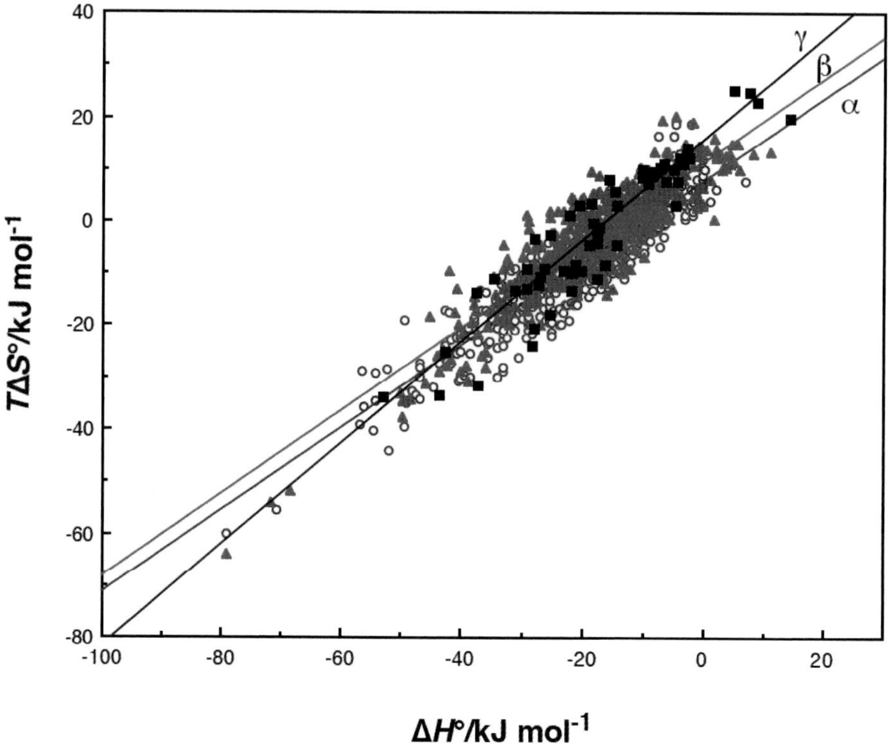

Fig. 8.6. Global enthalpy–entropy compensation plots for complexation of a wide variety of guests with α- (○), β- (▲), and γ-CyD (■).

8.8
Grunwald Theory on Enthalpy–Entropy Compensation

The general concept and methodology developed by Grunwald et al. [118, 119] provides us with a reliable tool for analyzing complexation thermodynamic parameters and particularly for predicting the existence or nonexistence of meaningful enthalpy–entropy compensation in a particular set of limited thermodynamic data. The idea is based on the separation of overall complexation thermodynamic parameters into two terms: *nominal* and *environmental*. The nominal part (ΔG_{nom}, ΔH_{nom}, and ΔS_{nom}) is associated with the complexation of a solvated host with a solvated guest to form a solvated host–guest complex, whilst the environmental part (ΔG_{env}, ΔH_{env}, and ΔS_{env}) is associated with water molecules involved in solvation/desolvation processes upon complexation. It was shown that ΔG_{env} is equal to zero in dilute solution and thus only ΔH_{env} and ΔS_{env} terms are subject to distinct enthalpy–entropy compensation [118, 119].

Recently, experimental confirmation of the Grunwald theory was achieved by comparison of the quality of differential enthalpy–entropy compensation plots for the exchange equilibrium between (*R*)- and (*S*)-enantiomers of chiral guests in β-CyD cavities (Eq. (9)) [82], and the exchange equilibrium between native β-CyD and 6-ammonio-6-deoxy-β-CyD (am-β-CyD) for chiral and achiral guests (Eq. (10)) [119]:

$$[\beta\text{-CyD}\cdot R] + [S] = [\beta\text{-CyD}\cdot S] + [R] \tag{9}$$

$$[\beta\text{-CyD}\cdot \mathrm{G}] + [\text{am-}\beta\text{-CyD}] = [\text{am-}\beta\text{-CyD}\cdot \mathrm{G}] + [\beta\text{-CyD}] \tag{10}$$

Differential thermodynamic parameters calculated for the hypothetical exchange equilibria (Eqs. (9) and (10)) are used to build the compensation plots shown in Fig. 8.7 (a) and (b), respectively. Interestingly, in spite of the same accuracy level and encompassing range of the original data, the two compensation plots show strikingly different scattering levels. These compensatory enthalpy–entropy relationships are a direct experimental confirmation of Grunwald's prediction that ΔG_{env} is equal to zero in dilute solution and thus only ΔH_{env} and ΔS_{env} are subject to enthalpy–entropy compensation. It is indeed obvious that a larger contribution from the nominal part (ΔG_{nom}, ΔH_{nom}, and ΔS_{nom}), associated with the particular complex structure, is expected for the host exchange from β-CyD to am-β-CyD than for the enantiomeric guest exchange in the same β-CyD cavity (Fig. 8.7 (a) and (b)).

8.9
Isotope Effect: Implications for the Comparison of Complexation Thermodynamics in H_2O versus D_2O

It was generally accepted for quite a long time that thermodynamic parameters (most often association constants) determined in H_2O and D_2O were not significantly different. However, no direct comparison of complex stability constants in D_2O and H_2O had been published until the late 1980s and the results reported

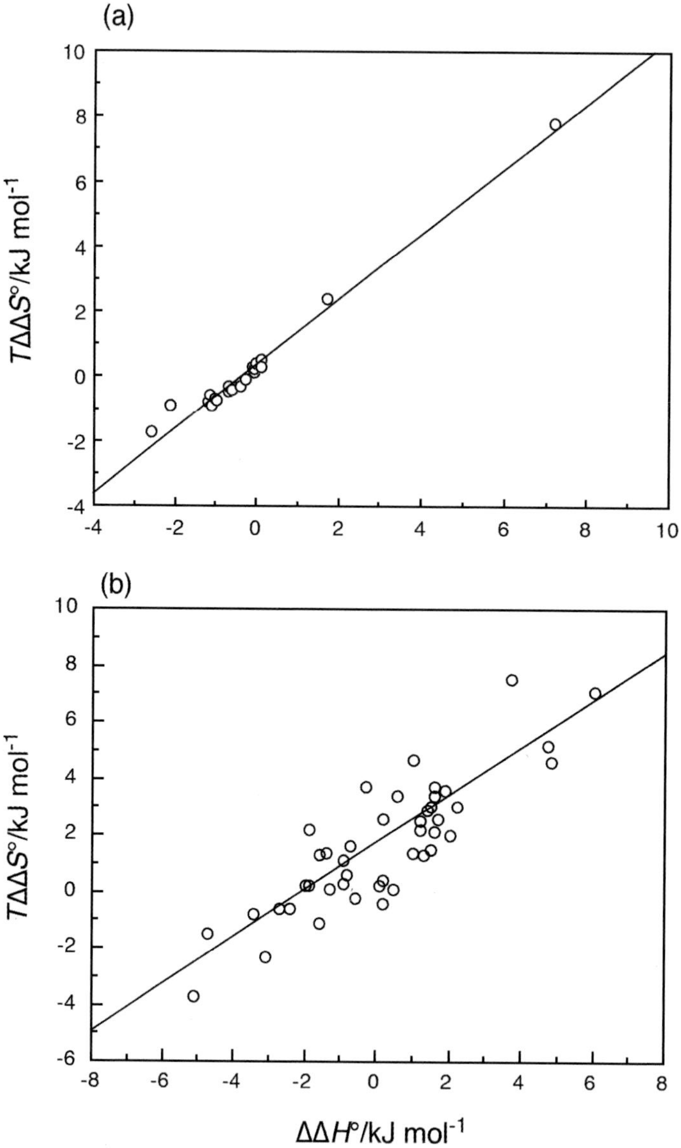

Fig. 8.7. Differential enthalpy–entropy compensation plots for (a) the guest-exchange equilibrium between (*R*)- and (*S*)-enantiomers of chiral guests in β-CyD's cavity (Eq. (9)), and (b) the host-exchange equilibrium between native β-CyD and 6-ammonio-6-dexoy-β-CyD (am-β-CyD) for chiral and achiral guests (Eq. (10)).

since then appear inconsistent [120–125]. There are some reported examples suitable for comparing the relevant data from NMR study in D_2O [120] and from calorimetric study in H_2O [124]. In the two studies, the $\Delta G°$ values for complexation of octanedioate in H_2O [121] and D_2O [124] clearly disagree with each other, whilst those for nonanedioate and decanedioate with α-CyD can be compared. However, the difference between the $\Delta H°$ values obtained in H_2O [120] and D_2O [124] is inconsistent in the alkanedioate series, varying from 2 to 8 kJ mol^{-1}. It should be emphasized that the literature data for the solvent isotope effect on $\Delta H°$ and $\Delta S°$ are very limited in general and our previous study was restricted to the comparison of complex stability ($\Delta G°$) in H_2O and D_2O [125]. Furthermore, a study by Schmidtchen is devoted to a careful examination of the complexation thermodynamics of only one chiral pair of camphor toward α-CyD in D_2O and H_2O [91].

In our study [126], a wide variety of guests (more than 30 charged and neutral guests, including seven totally or partially deuterated ones) were used to elucidate the global trend of thermodynamic behavior in D_2O versus H_2O. For general validity of the conclusions derived, we employed not only native α- and β-CyDs but also positively charged 6-amino-6-deoxy-β-CyD (am-β-CyD) as hosts in this microcalorimetric study. The main conclusions from this comprehensive ITC study are as follows: (1) Guest affinity toward CyDs is consistently enhanced by the use of D_2O as solvent. The quantitative affinity enhancement in D_2O versus H_2O directly correlates with the size and strength of the hydration shell around the charged/hydrophilic group of guests. For that reason, negatively/positively charged guests,

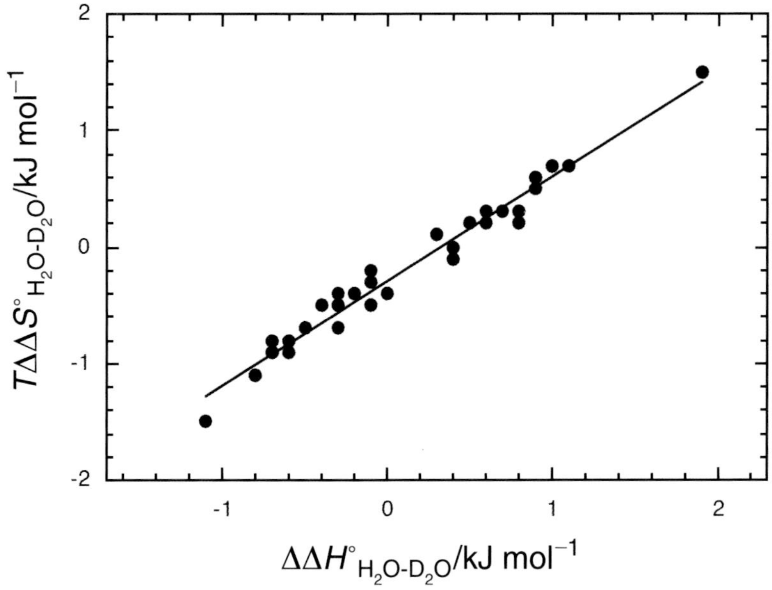

Fig. 8.8. Differential enthalpy–entropy compensation plots for the process of transfer of a CyD complex from H_2O to D_2O.

possessing relatively large and strong hydration shells, afford smaller K_{H_2O}/K_{D_2O} ratios than those for neutral guests with smaller and weaker hydration shells. (2) The enhanced affinity observed in D_2O for guests with rigid and bulky hydrophobic moieties is enthalpic in origin and attributable to the more favorable intracavity interactions in D_2O. In contrast, the increased affinity for flexible, less-bulky hydrophobic guests is entropic in origin and ascribable to the conformational adjustability of the flexible guest group upon inclusion in the CyD cavity. (3) Partial or total deuteration of guest leads to a reduced affinity toward CyDs in both H_2O and D_2O, which is probably ascribed to the lower ability of C–D bond to produce induced dipoles and thus results in reduced intracavity van der Waals interactions.

This study also revealed the validity of the enthalpy–entropy compensation effect even for such an exotic reaction as the transfer of a CyD complex from H_2O to D_2O (see Fig. 8.8):

$$[\beta\text{-CyD}\cdot\text{G}]_{H_2O} + n\ D_2O = [\beta\text{-CyD}\cdot\text{G}]_{D_2O} + n\ H_2O \tag{11}$$

and once again confirmed the very general nature of this compensation phenomenon.

8.10 A Theoretical Approach to Enthalpy–Entropy Compensation in Molecular Recognition

Gilson et al. [110, 111] have reported an interesting theoretical approach to enthalpy–entropy compensation, as commonly observed experimentally in a wide variety of molecular recognition phenomena [108–113]. By using the Mining Minima (M2) program, they calculated not only the affinities but also the changes in configurational energy and configurational entropy for associations of α-, β-, and γ-cyclodextrin with benzene, resorcinol, flurbiprofen, naproxen, and nabumetone [123], as well as for bindings of electron-accepting guests by molecular clip/tweezer and of barbiturate derivatives by a macrocyclic barbiturate receptor [124]. Quite interestingly, a strong correlation is found between the computed entropy changes and the computed potential-plus-solvation energy changes, resulting in near-linear compensation analogous to the enthalpy–entropy compensation [123, 124]. These results provide theoretical support for the extra-thermodynamic relationship and validate the global analyses of the enthalpy–entropy compensation plots [108–113].

8.11 Conclusions

Modern titration microcalorimetry is a powerful experimental method for the reliable determination of the thermodynamic parameters of complexation reactions in the fields of chemistry, biochemistry, pharmaceutics, and biology while differential scanning calorimetry is mainly used in drug studies. The main advantage of titration microcalorimetry is the simultaneous determination of a whole set of thermodynamic parameters (ΔG°, ΔH°, and ΔS°) from one single experiment. Thermodynamic data obtained with the help of this method can be used not only for describing the complete thermodynamic features but also for revealing the mechanisms and driving forces of a very wide variety of supramolecular interactions. Furthermore, the extra-thermodynamic enthalpy–entropy compensation relationship can contribute to a global analysis and understanding of the mechanisms and factors controlling a wide variety of molecular recognition phenomena in chemistry and biology.

References

1 Wiseman T, Williston S, Brandts J, Lin L, Rapid measurement of binding constants and heats of binding using a new titration calorimeter, Anal. Biochem. **1989**; 179:131–137.
2 Calvet E, Prat H, *Microcalorimetrie. Applications physicochimiques et biologiques*, Masson, Paris, **1956**.
3 Rekharsky MV, Egorov AM, *Thermodynamics of Biotechnological Processes*, Lomonosov Moscow State University Press, Moscow, **1992**.
4 Spink CH, Analytical calorimetry in biochemical and clinical applications, CRC Critical Rev. Anal. Chem. **1980**; 9:1–55.
5 Danielson B, Mathiasson B, Mosbash K, Enzyme thermistor devices and their analytical applications, Appl. Biochem. Bioeng. **1981**; 3:97–143.
6 Wadsö I, Design and testing of a micro reaction calorimeter, Acta Chem. Scand. **1968**; 22:927–937.
7 Wadsö I, A flow micro reaction calorimeter, Acta Chem. Scand. **1968**; 22:1842–1852.
8 Monk P, Wadsö I, Flow microcalorimetry as an analytical tool in biochemistry and related areas, Acta Chem. Scand. **1969**; 23:29–36.
9 Suurkuusk J, Wadsö I, Multichannel microcalorimetric system, Chim. Scrip. **1982**; 20:155–168.
10 Buzzell A, Sturtevant JM, A new calorimetric method, J. Am. Chem. Soc. **1951**; 73:2454–2458.
11 Jelesarov I, Bosshard HR, Isothermal titration calorimetry and differential scanning calorimetry as complementary tools to investigate the energetics of biomolecular recognition, J. Mol. Recogn. **1999**; 12:3–18.
12 Ladbury JE, Chowdhry BZ (editors), *Biocalorimetry: Applications of calorimetry in the biological sciences*, John Wiley & Sons, New York, **1998**.
13 Ackers GK, Johnson ML (editors), *Methods in Enzymology*, volume 295: Energetics of biological molecules. Part 2, Academic Press, New York-London, **1998**.
14 Raffa RB (editor), *Drug-receptor Thermodynamics: Introduction and Applications*, John Willey & Sons, New York, **2001**.
15 Ward WH, Holdgate GA, Isothermal titration calorimetry in drug discovery, Prog. Med. Chem. **2001**; 38:309–376.
16 Rekharsky MV, Inoue Y, Complexation thermodynamics of cyclodextrins, Chem. Rev. **1998**; 98:1875–1917.
17 Ficarra R, Tommasini S, Raneri D, Calabro ML, Di Bella MR, Rustichelli C, Gamberini MC, Ficarra P, Study of flavonoids/β-cyclodextrins inclusion complexes by NMR, FT-IR, DSC, X-ray investigation, J. Pharm. Biomed. Anal. **2002**; 29:1005–1014.

18 Li J, Guo Y, Zografi G, The solid state stability of amorphous quinapril in the presence of β-cyclodextrin, J. Pharm. Sci. **2002**; 91:229–243.
19 Granero G, Longhi M, Thermal analysis and spectroscopic characterization of interactions between a naphthoquinone derivative with HP-β-CyD or PVP, Pharm. Dev. Technol. **2002**; 7:381–390.
20 Hao X, Liang C, Jian-Bin C, Preparation and spectral investigation on inclusion complex of β-cyclodextrin with rutin, Analyst **2002**; 127:834–837.
21 Chao J, Li J, Meng D, Huang S, Preparation and study on the solid inclusion complex of sparfloxacin with HP-β-cyclodextrin, Spectrochim. Acta A, Mol. Biomol. Spectrosc. **2003**; 59:705–711.
22 Mura P, Faucci MT, Parrini PL, Effects of grinding with microcrystalline cellulose and cyclodextrin on ketoprofen physicochemical properties, Drug Dev. Ind. Pharm **2001**; 27:119–128.
23 Pralhad T, Rajendrakumar K, Study of freeze-dried quercetin-cyclodextrin binary systems by DSC, FT-IR, X-ray diffraction and SEM analysis, J. Pharm. Biomed. Anal. **2004**; 34:333–339.
24 Huh KM, Cho YW, Chung H, Kwon IC, Jeong SY, Ooya T, Lee WK, Sasaki S, You N, Supramolecular hydrogel formation based on inclusion complexation between poly(ethylene glycol)-modified chitosan and α-cyclodextrin, Macromol. Biosci. **2004**; 4:92–99.
25 Cerchiara T, Luppi B, Bigucci F, Zecchi V, Effect of chitosan on progesterone release from hydroxypropyl-β-cyclodextrin complexes, Int. J. Pharm. **2003**; 258:209–215.
26 Veiga MD, Merino M, Interaction of oxyphenbutazone with different cyclodextrins in aqueous medium and in the solid state, J. Pharm. Biomed. Anal. **2002**; 28:973–982.
27 Ribeiro LS, Ferreira DC, Veiga FJ, Physicochemical investigation of the effects of water-soluble polymers on vinpocetine complexation with β-cyclodextrin and its sulfobutylether derivative in solution and solid state, Eur. J. Pharm. Sci. **2003**; 20:253–266.
28 Orienti I, Fini A, Bertasi V, Zecchi V, Inclusion complexes between non steroidal antiinflammatory drugs and β-cyclodextrin, Eur. J. Pharm. Biopharm. **1991**; 37:110–112.
29 Menard FA, Dedhiya MG, Rhodes CT, Physico-chemical aspects of the complexation of some drugs with cyclodextrins, Drug Develop. Indust. Pharm. **1990**; 16:91–113.
30 Miyaji T, Kurono Y, Uekama K, Ikeda K, Simultaneous determination of complexation equilibrium constants for conjugated guest species by extended potentiometric titration Method: On barbiturate-β–cyclodextrin system. Chem. Pharm. Bull. **1976**; 24:1155–1159.
31 Min SH, Interaction of pharmaceuticals with β-cyclodextrin. I. Interaction with sulfonamides, J. Pharm. Soc. Korea **1971**; 15:8–15.
32 Hardee GE, Otagiri M, Perrin JH, Microcalorimetric investigations of pharmaceutical complexes. I. Drugs and β-cyclodextrin, Acta Pharm. Suec. **1978**; 15:188–199.
33 Danil de Namor AF, Tanaka DAP, Regueira LN, Gómez-Orellana I, Effect of β-cyclodextrin on the transfer of N^1-substituted sulfonamides from water to chloroform, J. Chem. Soc., Faraday Trans. **1992**; 88:1665–1668.
34 Jain AC, Adeyeye MC, Hydrophobicity, phase solubility and dissolution of various substituted sulfobutylether-β-cyclodextrin (SBE) and danazol-SBE inclusion complexes, Int. J. Pharm. **2001**; 212:177–186.
35 Naidu NB, Chowdary KPR, Murthy KVR, Satyanarayana V, Hayman AR, Becket G, Physicochemical characterization and dissolution properties of meloxicam–cyclodextrin binary systems, J. Pharm. Biomed. Anal. **2004**; 35:75–86.
36 Mura P, Zerrouk N, Faucci M, Maestrelli F, Chemtod C, Comparative study of ibuproxam complexation with amorphous β-cyclodextrin deriva-

tives in solution and in the solid state, Eur. J. Pharm. Biopharm. **2002**; 54:181–191.

37 Mura P, Faucci M, Maestrelli F, Furlanetto S, Pinzauti S, Characterization of physicochemical properties of naproxen systems with amorphous β-cyclodextrin-epichlorohydrin polymers, J. Pharm. Biomed. Anal. **2002**; 29:1015–1024.

38 Perdomo-Lopez I, Rodrigues-Perez AI, Yzquierdo-Peiro JM, White A, Estrada EG, Villa TG, Torres-Labandeira JJ, Effect of cyclodextrins on the solubility and antimycotic activity of sertaconazole: experimental and computational studies, J. Pharm. Sci. **2002**; 91:2408–2415.

39 Jambhekar S, Casella R, Maher T, The physicochemical characteristics and bioavailability of indomethacin from β-cyclodextrin, hydroxyethyl-β-cyclodextrin and hydroxypropyl-β-cyclodextrin complexes, Int. J. Pharm. **2004**; 270:149–166.

40 Liu Y, Chen GS, Li L, Zhang HY, Cao DX, Yuan YJ, Inclusion complexation and solubilization of paclitaxel by bridged bis(β-cyclodextrin)s containing a tetraethylenepentaamino spacer, J. Med. Chem. **2003**; 46:4634–4637.

41 Bayomi MA, Abanumay KA, Al Angary AA, Effect of inclusion complexation with cyclodextrins on photostability of nifedipine in solid state, Int. J. Pharm. **2002**; 243:107–117.

42 Lutka A, Investigation of interaction of promethazine with cyclodextrins, Acta Pol. Pharm. **2002**; 59:45–51.

43 Mura P, Maestrelli F, Cirri M, Ternary systems of naproxen with hydroxypropyl-β-cyclodextrin and aminoacids, Int. J. Pharm. **2003**; 260:293–302.

44 Fatouros DG, Hatzidimitriuou K, Antimisiaris SG, Liposomes encapsulating prednisolone and prednisolone-cyclodextrin complexes: comparison of membrane integrity and drug release, Eur. J. Pharm. Sci. **2001**; 13:287–296.

45 Anderson TG, Tan A, Ganz P, Seelig J, Calorimetric measurement of phospholipid interaction with methyl-β-cyclodextrin, Biochemistry **2004**; 43:2251–2261.

46 Boudad H, Legrand P, Lebas G, Cheron M, Duchene D, Ponchel G, Combined hydroxypropyl-β-cyclodextrin and poly(alkylcyanoacrylate)-nanoparticles intended for oral administration of saquinavir, Int. J. Pharm. **2001**; 218:113–124.

47 Cirri M, Maestrelli F, Furlanetto S, Mura P, Solid state characterization of glyburide-cyclodextrin co-ground products, J. Therm. Anal. Calorim. **2004**; 77:413–422.

48 Tong WQ, Lach JL, Chin TF, Guillory JK, Structural effects on the binding of amine drugs with the diphenylmethyl functionality to cyclodextrins. I. A microcalorimetric study, Pharm. Res. **1991**; 8:951–957.

49 Tong WQ, Lach JL, Chin TF, Guillory JK, Structural effects on the binding of amine drugs with the diphenylmethyl functionality to cyclodextrins. II. A molecular modeling study, Pharm. Res. **1991**; 8:1307–1312.

50 Tong WQ, Lach JL, Chin TF, Guillory JK, Microcalorimetric investigation of the complexation between 2-hydroxypropyl-β-cyclodextrin amine drugs with the diphenylmethyl functionality, J. Pharm. Biomed. Anal. **1991**; 9:1139–1146.

51 Arabzadeh A, Bathaie SZ, Farsam H, Amanlou M, Saboury AA, Shockravi A, Moosavi-Movahedi AA, Studies on mechanism of 8-methoxypsoralen-DNA interaction in the dark, Int. J. Pharm. **2002**; 237:47–55.

52 Arya DP, Micovic L, Charles I, Coffee RL, Willis B, Xue L, Neomycin binding to Watson-Hoogsteen (W-H) DNA triplex groove: a model, J. Am. Chem. Soc. **2003**; 125:3733–3744.

53 Barcelo F, Ortiz-Lombardia M, Portugal J, Heterogeneous DNA binding modes of berenil, Biochim. Biophys. Acta **2001**; 1519:175–184.

54 Barcelo F, Capo D, Portugal J, Thermodynamic characterization of the multivalent binding of chartreusin to DNA, Nucleic Acids Res. **2002**; 30:4567–4573.

55 Barbieri CM, Li TK, Guo S, Wang G, Shallop AJ, Pan W, Yang G, Gaffney BL, Jones RA, Pilch DS, Aminoglycoside complexation with a DNAoRNA hybrid duplex: the thermodynamics of recognition and inhibition of RNA processing enzymes, J. Am Chem. Soc. **2003**; 125:6469–6477.

56 Barcelo F, Ortiz-Lombardia M, Portugal J, Heterogeneous DNA binding modes of berenil, Biochim. Biophys. Acta **2001**; 1519:175–184.

57 Barcelo F, Capo D, Portugal J, Thermodynamic characterization of the multivalent binding of chartreusin to DNA, Nucleic Acids Res. **2002**; 30:4567–4573.

58 Carrasco C, Vezin H, Wilson WD, Ren J, Chaires JB, Bailly C, DNA binding properties of the indolocarbazole antitumor drug NB-506, Anticancer Drug. Des. **2001**; 16:99–10.

59 Chaires JB, Analysis and interpretation of ligand-DNA binding isotherms, Methods Enzymol. **2001**; 340:3–22.

60 Fleck SL, Birdsall B, Babon J, Dluzewski AR, Martin SR, Morgan WD, Angov E, Kettleborough CA, Feeney J, Blackman MJ, Holder AA, Suramin and suramin analogues inhibit merozoite surface protein-1 secondary processing and erythrocyte invasion by the malaria parasite *Plasmodium falciparum*, J. Biol. Chem. **2003**; 278:47670–47677.

61 Huang HM, Chen WY, Ruaan RC, Microcalorimetric studies of the mechanism of interaction between designed peptides and hydrophobic adsorbents, J. Colloid. Interface Sci. **2003**; 263:23–28.

62 Kaul M, Pilch DS, Thermodynamics of aminoglycoside-rRNA recognition: the binding of neomycin-class aminoglycosides to the A site of 16S rRNA, Biochemistry **2002**; 41:7695–7706.

63 Kaul M, Barbieri CM, Kerrigan JE, Pilch DS, Coupling of drug protonation to the specific binding of aminoglycosides to the A site of 16 S rRNA: elucidation of the number of drug amino groups involved and their identities, J. Mol. Biol. **2003**; 326:1373–1387.

64 Magnet S, Smith TA, Zheng R, Nordmann P, Blanchard JS, Aminoglycoside resistance resulting from tight drug binding to an altered aminoglycoside acetyltransferase, Antimicrob. Agents Chemother. **2003**; 47:1577–1583.

65 Muzammil S, Ross P, Freire E, A major role for a set of non-active site mutations in the development of HIV-1 protease drug resistance, Biochemistry **2003**; 42:631–638.

66 Nezami A, Luque I, Kimura T, Kiso Y, Freire E, Identification and characterization of allophenylnorstatine-based inhibitors of plasmepsin II, an antimalarial target, Biochemistry **2002**; 41:2273–2280.

67 Nezami A, Kimura T, Hidaka K, Kiso A, Liu J, Kiso Y, Goldberg DE, Freire E, Highaffinity inhibition of a family of Plasmodium falciparum proteases by a designed adaptive inhibitor, Biochemistry **2003**; 42:8459–8464.

68 Ohtaka H, Velazquez-Campoy A, Xie D, Freire E, Overcoming drug resistance in HIV-1 chemotherapy: the binding thermodynamics of Amprenavir and TMC-126 to wild-type and drug-resistant mutants of the HIV-1 protease, Protein. Sci. **2002**; 11:1908–1916.

69 Ohtaka H, Schon A, Freire E, Multidrug resistance to HIV-1 protease inhibition requires cooperative coupling between distal mutations, Biochemistry **2003**; 42:13659–13666.

70 Velazquez-Campoy A, Kiso Y, Freire E, The binding energetics of first- and second-generation HIV-1 protease inhibitors: implications for drug design, Arch. Biochem. Biophys. **2001**; 390:169–175.

71 Velazquez-Campoy A, Vega S, Freire E, Amplification of the effects of drug resistance mutations by background

polymorphisms in HIV-1 protease from African subtypes, Biochemistry **2002**; 41:8613–8619.
72 Veselkov AN, Maleev VY, Glibin EN, Karawajew L, Davies DB, Structure-activity relation for synthetic phenoxazone drugs: evidence for a direct correlation between DNA and pro-apoptotic activity. Eur. J. Biochem. **2003**; 270:4200–4207.
73 Weber PC, Salemme FR, Applications of calorimetric methods to drug discovery and the study of protein interactions, Curr. Opin. Struct. Biol. **2003**; 13:115–121.
74 Koushik KN, Bandi N, Kompella UB, Interaction of [D-Trp6, Des-Gly10]-LHRH ethylamide and hydroxypropyl-β-cyclodextrin: thermodynamics of interaction and protection from degradation by alpha-chymotrypsin, Pharm. Dev. Technol. **2001**; 6:595–606.
75 Kriz Z, Koca J, Imberty A, Charlot A, Auzely-Velty R, Investigation of the complexation of (+)-catechin by β-cyclodextrin by a combination of NMR, microcalorimetry and molecular modeling techniques, Org. Biomol. Chem. **2003**; 1:2590–2595.
76 Barone G, Castronuovo G, di Ruocco V, Elia V, Giancola C, Inclusion compounds in water: Thermodynamics of the interaction of cyclomaltohexaose with amino acids at 25 °C, Carbohydr. Res. **1989**; 192:331–341.
77 Harada A, Saeki K, Takahashi S, Optical resolution of ferrocene derivatives by liquid chromatography using aqueous cyclomaltohexaose as the mobile phase and polyamide as the stationary phase, Carbohydr. Res. **1989**; 192:1–7.
78 Cooper A, MacNicol DD, Chiral host–guest complexes: interaction of α-cyclodextrin with optically active benzene derivatives, J. Chem. Soc., Perkin Trans. 2 **1978**:760–763.
79 Danil de Namor AF, Blackett PM, Cabaleiro MC, Al Rawl JMA, Cyclodextrin–monosaccharide interactions in water, J. Chem. Soc., Faraday Trans. **1994**; 90:845–847.
80 Rekharsky MV, Schwarz FP, Tewari YB, Goldberg RN, A thermodynamic study of the reactions of cyclodextrins with primary and secondary aliphatic alcohols, with *d*- and *l*-phenylalanine, and with *l*-phenylalanine amide, J. Phys. Chem. **1994**; 98:10282–10288.
81 Rekharsky MV, Goldberg RN, Schwarz FP, Tewari YB, Ross PD, Yamashoji Y, Inoue Y, A thermodynamic and nuclear magnetic resonance study of the interactions of α- and β-cyclodextrin with model substances: phenethylamine, ephedrines, and related substances, J. Am. Chem. Soc. **1995**; 117:8830–8840.
82 Rekharsky MV, Inoue Y, Chiral recognition thermodynamics of β-cyclodextrin: The thermodynamic origin of enantioselectivity and the enthalpy-entropy compensation effect, J. Am. Chem. Soc. **2000**; 122:4418–4435.
83 Rekharsky MV, Inoue Y, Complexation and chiral recognition thermodynamics of 6-amino-6-deoxy-β-cyclodextrin with anionic, cationic, and neutral chiral guests: Counterbalance between van der Waals and Coulombic interactions, J. Am. Chem. Soc. **2002**; 124:813–826.
84 Rekharsky MV, Yamamura H, Kawai M, Inoue Y, Critical difference in chiral recognition of *N*-benzyloxycarbonyl-D/L-aspartic and -glutamic acids by mono- and bis(trimethylammonio)-β-cyclodextrins: A calorimetric and NMR study, J. Am. Chem. Soc. **2001**; 123:5360–5361.
85 Rekharsky MV, Yamamura H, Kawai M, Inoue Y, Complexation and chiral recognition thermodynamics of γ-cyclodextrin with *N*-acetyl- and *N*-carbobenzyloxy-dipeptides possessing two aromatic rings, J. Org. Chem. **2003**; 68: 5228–5235.
86 Yamamura H, Rekharsky MV, Ishihara Y, Kawai M, Inoue Y, Factors controlling the complex architecture of native and modified cyclodextrins with dipeptide (Z-Glu-Tyr) studied by microcalorimetry and NMR spectroscopy. Critical effects of peripheral bis-trimethylamination and

cavity size J. Am. Chem. Soc. **2004**; 126:14224.

87 Kano K, Nishiyabu R, Chiral recognition by cyclodextrins: a general mechanism, in *Advances in Supramolecular Chemistry*, vol. 9 (Gokel GW, ed.), Cerberus Press, Miami **2003**, pp39–69.

88 Kano K, Hasegawa H, Chiral recognition of helical metal complexes by modified cyclodextrins, J. Am. Chem. Soc. **2001**; 123:10616–10627.

89 Liu Y, Yang EC, Yang YW, Zhang HY, Fan Z, Ding F, Cao R, Thermodynamics of the molecular and chiral recognition of cycloalkanols and camphor by modified β-cyclodextrins possessing simple aromatic tethers, J. Org. Chem. **2004**; 69:173–180.

90 Liu Y, Li XY, Zhang HY, Wada T, Inoue Y, Synthesis of phosphoryl-tethered-β-cyclodextrins and their molecular and chiral recognition thermodynamics, J. Org. Chem. **2003**; 68:3646–3657.

91 Schmidtchen FP, The anatomy of the energetics of molecular recognition by calorimetry: chiral discrimination of camphor by α-cyclodextrin, Chem. Eur. J **2002**; 8:3522–3528.

92 Lerchner J, Kirchner R, Seidel J, Waehlisch D, Wolf G, Determination of molar heats of absorption of enantiomers into thin chiral coatings by combined IC calorimetric and microgravimetric (QMB) measurements. I. IC calorimetric measurements of heats of absorption, Thermochim. Acta **2004**; 415:27–34.

93 Mulder A, Jukovic A, Huskens J, Reinhoudt DN, Bis(phenylthienyl)-ethane-tethered β-cyclodextrin dimers as photoswitchable hosts, Org. Biomol. Chem. **2004**; 2:1748–1755.

94 Liu Y, Fan Z, Zhang HY, Yang YW, Ding F, Liu SX, Wu X, Wada T, Inoue Y, Supramolecular self-assemblies of β-cyclodextrins with aromatic tethers: factors governing the helical columnar versus linear channel superstructures, J. Org. Chem. **2003**; 68:8345–8352.

95 Kano K, Nishiyabu R, Asada T, Kuroda Y, Static and dynamic behavior of 2:1 inclusion complexes on cyclodextrins and charged porphyrins in aqueous organic media, J. Am. Chem. Soc. **2002**; 124:9937–9944.

96 Kano K, Nishiyabu R, Yamazaki T, Yamazaki I, Convenient scaffold for forming heteroporphyrin arrays in aqueous media, J. Am. Chem. Soc. **2003**; 125:10625–10634.

97 Sigurskjold BW, Exact analysis of competition ligand binding by displacement isothermal titration calorimetry, Anal. Biochem. **2000**; 277:260–266.

98 Turnbull WB, Daranas AH, On the value of c: can low affinity systems be studied by isothermal titration calorimetry? J. Am. Chem. Soc. **2003**; 125:14859–14866.

99 Turnbull WB, Precious BL, Homans SW, Dissecting the cholera toxin-ganglioside GM1 interaction by isothermal titration calorimetry, J. Am. Chem. Soc. **2004**; 126:1047–1054.

100 Rekharsky MV, Mayhew MP, Goldberg RN, Ross PD, Yamashoji Y, Inoue Y, A Thermodynamic and nuclear magnetic resonance study of the reaction of α- and β-cyclodextrin with acids, aliphatic amines, and cyclic alcohols, J. Phys. Chem. B **1997**; 101:87–100.

101 Ross PD, Rekharsky MV, Thermodynamics of hydrogen bond and hydrophobic interactions in cyclodextrin complexes, Biophys. J. **1996**; 71:2144–2154.

102 Connors KA, The stability of cyclodextrin complexes in solution, Chem. Rev. **1997**; 97:1325–1357.

103 Hallén D, Data-treatment: Considerations when applying binding reaction data to a model, Pure Appl. Chem. **1993**; 65:1527–1532.

104 Eatough DJ, Lewis EA, Hansen LD, in *Analytical Solution Calorimetry* (Grime JK, ed.), J. Wiley & Sons, New York, **1985**, pp 137–161.

105 Yang CP, ITC Data Analysis in Origin v.2.9. MicroCal Inc., Northampton, MA, **1993**.

106 Rekharsky MV, *Application of*

Microcalorimetry in Biochemistry, Thesis Doctor of Science, Institute of Biological and Medical Chemistry, Russian Academy of Medical Sciences, Moscow, **1997**.

107 Rekharsky M, Inoue Y, 1:1 and 1:2 complexation thermodynamics of γ-cyclodextrin with *N*-carbobenzyloxy-aromatic amino acids and ω-phenylalkanoic acids, J. Am. Chem. Soc. **2000**; 122:10949–10955.

108 Houk KN, Leach AG, Kim SP, Zhang X, Binding affinities of host-guest, protein-ligand, and protein-transition-state complexes, Angew. Chem. Int. Ed. **2003**; 42:4872–4897.

109 Williams DH, Stephens E, O'Brien DP, Zhou M, Understanding noncovalent interactions: ligand binding energy and catalytic efficiency from ligand-induced reductions in motion within receptors and enzymes, Angew. Chem. Int. Ed. **2004**; 43:6596–6616.

110 Chen W, Chang CE, Gilson MK, Calculation of cyclodextrin binding affinities, energy, entropy, and implications for drug design, Biophys. J. **2004**; 87:3035–3049.

111 Chang CE, Gilson MK, Free energy, entropy, and induced fit in host-guest recognition: calculations with the second-generation mining minima algorithm, J. Am. Chem. Soc. **2004**; 126:13156–13164.

112 Inoue Y, Hakushi T, Liu Y, Tong LH, Shen BJ, Jin DS, Thermodynamics of molecular recognition by cyclodextrins. 1. Calorimetric titration of inclusion complexation of naphthalenesulfonates with α-, β-, and γ-cyclodextrins: enthalpy-entropy compensation, J. Am. Chem. Soc. **1993**; 115:475–481.

113 Inoue Y, Liu Y, Tong LH, Shen BJ, Jin DS, Calorimetric titration of inclusion complexation with modified β-cyclodextrins. Enthalpy-entropy compensation in host-guest complexation: from ionophore to cyclodextrin and cyclophane, J. Am. Chem. Soc. **1993**; 115:10637–10644.

114 Inoue Y, Wada T, in *Advances in Supramolecular Chemistry* (Gokel GW, ed.), Molecular recognition in chemistry and biology as viewed from enthalpy-entropy compensation effect, JAI Press, Greenwich, CT, **1997**, pp 55–96.

115 Inoue Y, Hakushi T, Enthalpy-entropy compensation in complexation of cations with crown ethers and related ligands, J. Chem. Soc., Perkin Trans. 2 **1985**; 935–946.

116 Inoue Y, Hakushi T, Liu Y, in *Cation Binding by Macrocycles* (Inoue Y, Gokel GW, ed.), Marcel Dekker, New York, NY, **1990**, Chapter 1.

117 Inoue Y, Wada T, in *Molecular Recognition Chemistry* (Tsukube H, ed.), Sankyo Shuppan, Tokyo, **1996**; Chapter 2 (in Japanese).

118 (a) Grunwald E, Steel C, Solvent reorganization and thermodynamic enthalpy-entropy compensation, J. Am. Chem. Soc. **1995**; 117:5687–5692. (b) Grunwald, E, *Thermodynamics of Molecular Species*, Wiley-Interscience, New York, **1996**.

119 Rekharsky MV, Inoue Y, Complexation and chiral recognition thermodynamics of 6-amino-6-deoxy-β-cyclodextrin with anionic, cationic, and neutral chiral guests: counterbalance between van der Waals and Coulombic interactions, J. Am. Chem. Soc. **2002**; 124:813–826.

120 Matsui Y, Fujie M, Hanaoka K, Host-guest complexation of mono[6-(1-pyridinio)-6-deoxy]-α-cyclodextrin with several inorganic anions, Bull. Chem. Soc. Jpn. **1989**; 62:1451–1457.

121 Watanabe M, Nakamura H, Matsuo T, Formation of through-ring α-cyclodextrin complexes with α,ω-alkanedicarboxylate anion. Effects of the aliphatic chain length and electrostatic factors on the complexation behavior, Bull. Chem. Soc. Jpn. **1992**; 65:164–169.

122 Mu P, Okada T, Iwami N, Matsui Y, ^{1}HNMR response of mono[6-(1-pyridinio)-6-deoxy]-α-cyclodextrin to inorganic anions, Bull. Chem. Soc. Jpn. **1993**; 66:1924–1928.

123 Wang AS, Matsui Y, Solvent isotope effect on the complexation of cyclodextrins in aqueous solutions,

Bull. Chem. Soc. Jpn. **1994**; 67:2917–2920.

124 Castronuovo G, Elia V, Velleca F, Viscardi G, Thermodynamics of the interaction of α-cyclodextrin with α,ω-dicarboxylic acids in aqueous solutions. A calorimetric study at 25 °C, Thermochim. Acta **1997**; 292:31–37.

125 Inoue Y, Yamamoto K, Wada T, Everitt S, Gao XM, Hoe ZJ, Tong LH, Jiang SK, Wu HM, Inclusion complexation of (cyclo)alkanes and (cyclo)alkanols with 6-*O*-Modified cyclodextrins, J. Chem. Soc., Perkin Trans. 2 **1998**:1807.

126 Rekharsky MV, Inoue Y, Solvent and guest isotope effects on complexation thermodynamics of α-, β-, and 6-amino-6-deoxy-β-cyclodextrins, J. Am. Chem. Soc. **2002**; 124:12361–12371.

9
NMR of Cyclodextrins and Their Complexes

Andrzej Ejchart and Wiktor Koźmiński

Nuclear magnetic resonance (NMR) spectroscopy is one of the most powerful and versatile methods for the elucidation of molecular structure and dynamics. It is also very well suited to study molecular complexes and their properties [1]. Therefore, it has been widely used for studying inclusion complexes formed by cyclodextrins (CyD) [2–4]. Some examples of the applications of NMR in conjunction with other techniques are presented in other chapters, in particular in Chapter 6. The success of NMR spectroscopy in this field is due to its ability to study complex chemical systems and to determine stoichiometry, association constants, and conformations of molecular complexes, as well as to provide information on their symmetry and dynamics. Furthermore, compared to other techniques, NMR spectroscopy provides a superior method to study complexation phenomena, because guest and host molecules are simultaneously observed at the atomic level.

9.1
NMR Methodology

9.1.1
Standard NMR Spectra

Isotopes possessing nuclei with spins, and thus magnetic moments, are magnetically active. Almost all elements possess at least one stable, magnetically active isotope. When the system of nuclear spins is placed in an external magnetic field, macroscopic magnetization is generated. Its perturbation allows us to measure the NMR spectrum of the perturbed nuclei. The resonance frequency of a nucleus is given by Eq. (1):

$$f = \gamma B_0(1 - \sigma)/2\pi \tag{1}$$

where B_0 is the strength of the spectrometer magnetic field and the magnetogyric ratio, γ, is a constant characteristic for a given isotope. NMR resonance frequen-

Cyclodextrins and Their Complexes. Edited by Helena Dodziuk

ISBN: 3-527-31280-3

cies of all nuclei fall in the radio-frequency range: e.g., $f(^1\text{H}) \approx 500$ MHz or $f(^{13}\text{C}) \approx 125.7$ MHz at $B_0 = 11.7$ T. The shielding constant, σ, strongly depends on the distribution of the electronic charge in the vicinity of the nucleus, and therefore on the local molecular structure, and typically it is of the order of 10^{-4}–10^{-5}. Since the use of shielding constants is not convenient, a related parameter–chemical shift–has been introduced. It is given by the difference in resonance frequencies between the nucleus of interest and a reference nucleus and expressed as a dimensionless parameter δ:

$$\delta = 10^6(f - f_{ref})/f_{ref} \approx 10^6(\sigma_{ref} - \sigma) \qquad (2)$$

The factor 10^6 is introduced for convenience and, therefore, the chemical shift is expressed in parts per million (ppm).

^{1}H and ^{13}C NMR spectra of α-CyD in D_2O are shown in Figs. 9.1 and 9.2. Signal shapes in the spectra of these two isotopes differ significantly due to the following reason. Besides the chemical shifts generated by interactions of nuclear spins with the external magnetic field, another interactions strongly influence the appearance of NMR spectra. These are interactions transmitted through chemical bonds, the spin–spin (scalar) interactions, which are generally observed between nuclei separated by four or fewer chemical bonds. They depend on the type of elements and chemical bonds in the pathway. The spin–spin interaction manifests itself in the spectrum as a signal splitting. Its strength can be read out directly from the spectrum if the frequency difference between the coupled nuclei is large enough. Then

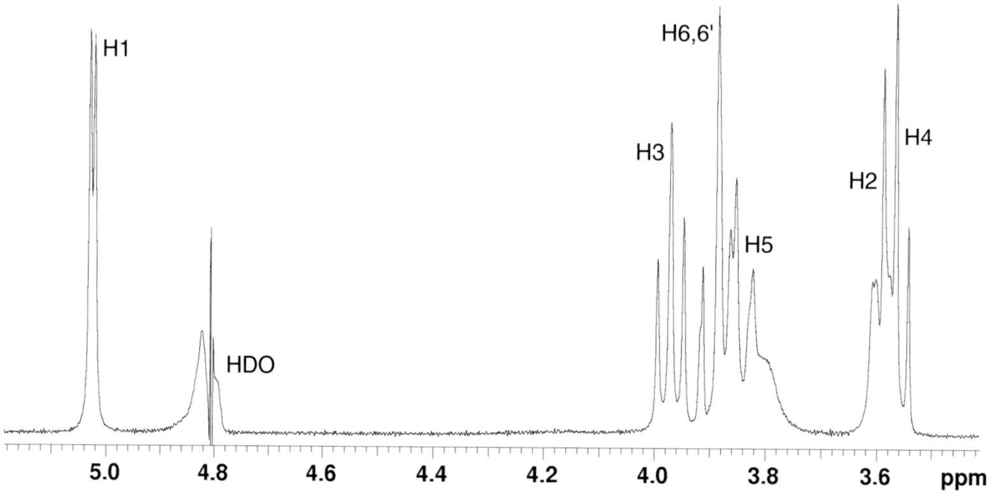

Fig. 9.1. ^{1}H NMR spectrum of α-CyD in D_2O. Assignments of all protons are shown in the spectrum. The residual solvent HDO signal is suppressed by presaturation.

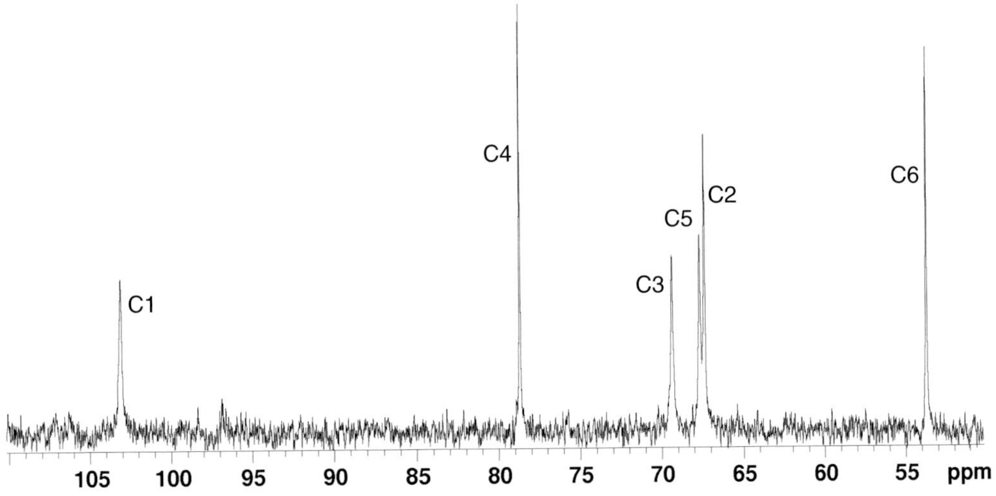

Fig. 9.2. ^{13}C NMR spectrum of α-CyD in D_2O. Owing to ^1H decoupling and low (1.1%) natural ^{13}C abundance all signals are singlets. Such a spectrum contains no spin–spin coupling information.

signal components are separated by the spin–spin coupling constant, J, which contains structural information [5]. Vicinal coupling constants between nuclei separated by three bond, ^3J, are particularly important from this standpoint, because they depend on the dihedral angle defined by the intervening bonds according to so called Karplus relation [6, 7]. Beside the structural information they provide, spin–spin coupling constants are utilized for the polarization transfer – the technique used for the sensitivity enhancement in the NMR spectroscopy [8]. Large, heteronuclear one-bond couplings are especially useful for this purpose (e.g., $^1J(^1H,^{13}C) > 120$ Hz).

In the conventional, one-dimensional (1D) experiment, generation of the initial nonequilibrium state of the spin system (preparation) is followed by the optional magnetization transfer between different spins (mixing) and the detection of the response of a spin system (acquisition). So the obtained signal of free induction decay (FID) has to be Fourier transformed in order to change the time domain of the FID signal to the frequency domain of the spectrum. There is a difference in the character of signals depending on the natural abundance of the studied isotope. At high natural abundance most of the signals are split owing to the homonuclear spin–spin coupling constants as is the case for ^1H nuclei (see Fig. 9.1). The opposite is true for low natural abundance nuclei. Their spectra usually contain singlets owing to the lack of homonuclear spin–spin couplings and the removal of heteronuclear ones by the use of decoupling techniques (see Fig. 9.2) [8].

9.1.2
Nuclear Magnetic Relaxation

A perturbed system of nuclear spins restores its thermodynamic equilibrium by the process known as relaxation. Longitudinal relaxation, characterized by the rate R_1, restores the static magnetization parallel to B_0, whereas transverse relaxation causes the decay of transverse magnetization, which is equal to zero at equilibrium. The transverse relaxation rate, R_2, is the main factor determining line width in NMR spectroscopy. Both R_2 and line width increase with the molecular mass and viscosity of the solution. Relaxation rates of many magnetically active nuclei are important structural parameters. Moreover, they can provide valuable information on the dynamics of a system under study. Among a number of relaxation mechanisms which take place in the relaxation of nuclei with spin $I = \frac{1}{2}$, such as ^1H or ^{13}C nuclei, the dipolar (DD) mechanism is of special importance [8]. Relaxation rates due to the dipolar mechanism depend on an effective correlation time characterizing overall molecular tumbling and intramolecular motions as shown in Fig. 9.3.

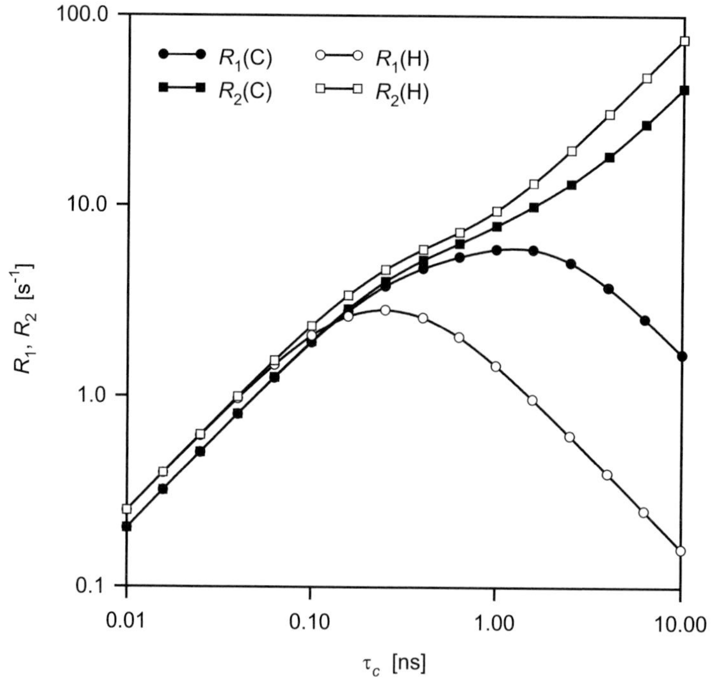

Fig. 9.3. Longitudinal (R_1) and transverse (R_2) relaxation rates for two dipolar interacting nuclei, C–H ($r_{CH} = 0.11$ nm) and H–H ($r_{HH} = 0.18$ nm) as a function of molecular correlation time τ_c in a magnetic field $B_0 = 9.4$ T.

One can distinguish three regions for such a relationship; (i) an extreme narrowing limit ($2\pi f\tau_c \ll 1$) on the left of Fig. 9.3, for which the equality $R_1 = R_2$ is valid, (ii) a spin diffusion limit ($2\pi f\tau_c \gg 1$) on the right of Fig. 9.3, for which $R_1 \ll R_2$ and NMR line widths are very broad, often too broad to be observed, (iii) a dispersion region lying between the extreme narrowing and spin diffusion limits. Owing to their molecular masses, inclusion complexes of medium-sized organic molecules with CyDs are characterized by correlation times typical for the dispersion region at the magnetic fields available at present.

Nuclear relaxation rates have been used in studies of CyD complexes for the determination of correlation times, thus giving insight into the dynamics of the host and/or guest in the complex. ^1H selective and nonselective longitudinal relaxation rates enable us to characterize the dynamics of the acridine–β-CyD complex [9]. Proton longitudinal, $R_1(^1H)$, and transverse, $R_2(^1H)$, relaxation rates were used to determine the motion of ($\pm$)camphor guest molecules in both diastereomeric complexes with α-CyD [10]. The $R_1(^{13}C)$ rates were used as early as 1976 to obtain correlation times for guest and host molecules in complexes formed by α-CyD with three aromatic compounds [11]. The $R_1(^{13}C)$ rates were also used for the determination of correlation times of constituents and complexes of several CyDs with azo dyes [12].

Nuclei with spin $I > \frac{1}{2}$ possess electric quadrupolar moments besides the magnetic ones. The interaction of a nuclear quadrupolar moment with molecular electric field gradients is usually much stronger than all other interactions of nuclei with their surroundings and quadrupolar mechanisms dominate relaxation processes. For this very reason transverse relaxation rates of quadrupolar nuclei are much faster than the corresponding rates of $\frac{1}{2}$-spin nuclei, which lack a quadrupolar moment. Fast relaxation causes significant broadening of NMR signals and often makes their observation impossible. If relaxation rate measurements for quadrupolar nuclei, however, are experimentally feasible, the results can allow the determination of corresponding correlation times [11]. Nevertheless, for the aforementioned reason, the main interest of NMR spectroscopists is focused on the $\frac{1}{2}$-spin nuclei that display narrow signals. Luckily, for all chemically important elements except oxygen, isotopes with $\frac{1}{2}$-spin nuclei exist. For instance, ^1H and ^{31}P are isotopes with large natural abundance (99.98% and 100%, respectively) whereas the opposite is true for ^{13}C and ^{15}N isotopes (1.1% and 0.37%, respectively). Therefore, the enrichment of the latter isotopes can sometimes be necessary for successful NMR study.

9.1.3
Dipolar Interactions and NOE

Nuclear spins can mutually interact in two ways, through chemical bonds (spin–spin coupling) and through space as magnetic dipoles (dipolar coupling). The static dipolar coupling constant, D_{ij}, depends on both the distance between the interacting nuclei and their magnetogyric ratios; $D_{ij} \sim \gamma_i\gamma_j r_{ij}^{-3}$. Typical D_{ij} values are several orders of magnitude larger than spin–spin coupling constants (e.g.

$|D_{C,H}| > 24$ kHz). However, dipolar interactions, which depend on the orientation of internuclear vectors relative to B_0, do not split NMR signals in isotropic solutions, because they are averaged to zero by the fast diffusional tumbling of solute molecules. Even so, dipolar interactions manifest themselves in isotropic solutions indirectly and generate the nuclear Overhauser effect (NOE) which is the change in the intensity of the NMR signal of a nuclear spin when the thermodynamic equilibrium of another nuclear spin interacting with a given one is perturbed. The NOE arises from the process called cross relaxation. It can be detected during appropriately designed and performed experiments, which result in the transfer of magnetization between dipolarly interacting nuclei. The most important dependence of the cross relaxation rate σ_{ij} between two nuclei i and j from a structural standpoint is on the inverse sixth power of the internuclear distance, $\sigma_{ij} \sim D_{ij}{}^2 \sim r_{ij}{}^{-6}$. Therefore, NOEs, especially those among the protons of structurally peripheral ^{1}H nuclei, bear structural information about interatomic distances in molecules [13], and thus provide information about the molecular conformation. Equally importantly intermolecular NOEs provide information about complexation as well as the distances and relative orientation of the host and guest molecules in a complex. In particular, NOEs occurring between host protons of the inner surface, H3 and H5 in the case of CyDs, and guest protons are indicators of inclusion complex formation [3]. One should be aware, however, that NOEs in dynamic complexes, like other spectral parameters, are averaged over all species and/or conformations in the solution, and owing to the r^{-6} dependence, even a small fraction of a species with a short interproton distance gives rise to a substantial NOE.

NOE magnitudes depend, among other things, on the effective correlation times τ_{eff} characterizing overall molecular tumbling and intramolecular motions. CyD complexes with molecular masses around 1–3 kDa display τ_{eff} of a fraction of nanosecond and are placed in the close to zero dispersion region near the transition between positive and negative NOEs often precluding detection of dipolar interactions. However, NOE in the rotating frame (ROE) is always positive (Fig. 9.4) becoming a method of choice in studies of medium-sized molecules and complexes [3, 4, 13]. Some examples of the application of NOE/ROE techniques with the aim of rationalizing differences in elution order and enantioselectivity are presented in Chapter 6.

9.1.4
Molecular Dynamics

NMR spectroscopy is sensitive to dynamic processes covering a broad range of frequencies. Therefore, the definition of the NMR time scale is crucial for understanding the influence of molecular motions on the NMR parameters. If a given motion is fast in the NMR time scale it will average an NMR parameter, resulting in a population-weighted average of the individual species. It takes place when lifetimes, τ, of those species fulfil the condition: $\tau_i, \tau_j \ll 1/\Delta P_{ij}$, where ΔP_{ij} represents the parameter difference between species i and j. Otherwise, exchange-broadened signals appear for $\tau_i, \tau_j \approx 1/\Delta P_{ij}$, or separate spectra are observed for each species if

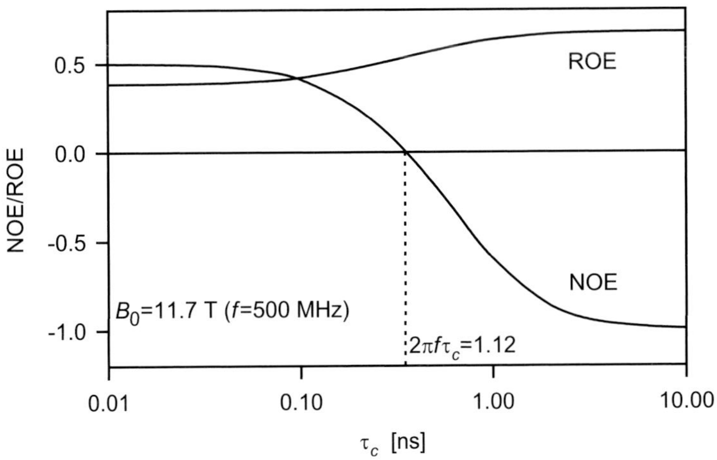

Fig. 9.4. Homonuclear (e.g. $^1H \cdots {}^1H$) Overhauser effect in laboratory (NOE) and rotating (ROE) frame as a function of molecular correlation time τ_c. If the product $2\pi f\tau_c \approx 1.12$ the NOE vanishes and dipolar interactions cannot be detected in this way.

$\tau_i, \tau_j \gg 1/\Delta P_{ij}$ (see Fig. 9.5). Chemical shifts are typically averaged by processes in which the lifetimes are 10^{-3} s or shorter [14]. Since the major part of conformational motions is much faster, usually conformationally averaged NMR spectra are observed. It means that most often NMR derived structural information is conformationally averaged. It should be pointed out that other spectroscopic techniques, like IR or UV ones, are characterized by different time scales.

Rates of association and dissociation of the complexes formed by CyDs are predominantly fast in the NMR time scale. Nevertheless, if the free enthalpy of activation for the complexation, $\Delta G^{\#}$, is large, separate spectra of free and bound host and/or guest molecules can be observed [15, 16]. Another behavior was observed for the complexes of acenaphthene with α-CyD [17] and dimethylnaphthalene with β-CyD [18]. The authors claim that they observed different sets of signals belonging to a 1:2 complex and a 1:1 complex remaining in fast exchange with free host and guest molecules at low temperature. The role of NMR spectra in establishing the nonrigidity of CyDs and the dynamic character of their complexes is briefly presented in Chapter 1.

9.1.5 2D Spectra

In two-dimensional (2D) spectroscopy, after the preparation period the perturbed spin system is left to evolve at its characteristic frequencies (evolution). Owing to the mixing, these frequencies can modulate the oscillations detected during acqui-

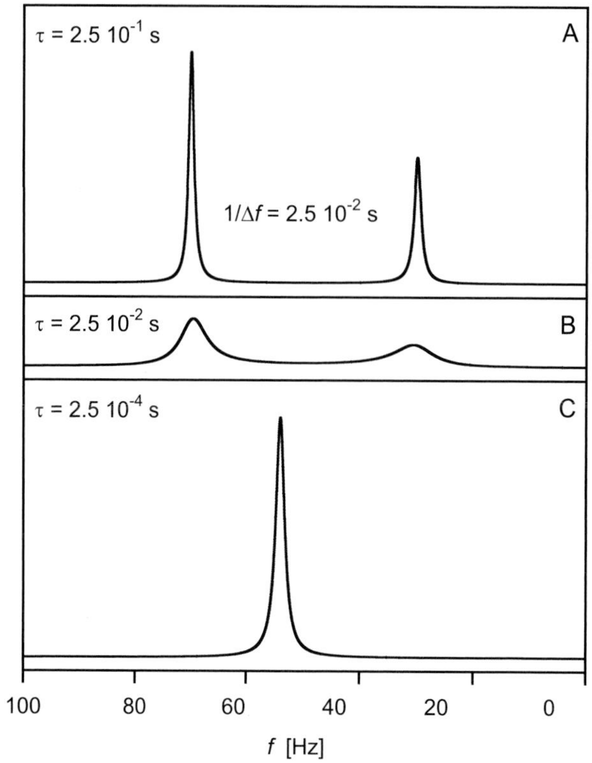

Fig. 9.5. Lineshapes representing exchange processes between two unequally populated sites, $x_1 = 0.6$ at $f_1 = 70$ Hz and $x_2 = 0.4$ at $f_2 = 30$ Hz. Depending on the lifetimes, governed by the free enthalpy of activation of the exchange process, the exchange is slow, $\tau \gg 1/\Delta f$ (A), intermediate, $\tau = 1/\Delta f$ (B) or fast, $\tau \ll 1/\Delta f$ (C) in the NMR timescale. The frequency of the averaged signal in spectrum C is a population-weighted average of signals in spectrum A.

sition. If the evolution period is systematically incremented, a set of 1D data, each differently modulated, will be obtained forming a second time domain. Two subsequent Fourier transformations applied to both time domains move the data to the two-dimensional frequency domain. If evolved and detected spins belong to the same isotope, the resulting homonuclear 2D spectrum will display two types of signal: diagonal peaks essentially corresponding to those appearing in a 1D spectrum and off-diagonal cross peaks whose coordinates $f1 \neq f2$ represent chemical shifts of interacting nuclear spins. The type of interaction, either scalar or dipolar, is chosen by the appropriate sequence of mixing periods. The corresponding techniques are known as COSY (correlated spectroscopy) (Fig. 9.6) or NOESY/ROESY (nuclear Overhauser effect spectroscopy in laboratory/rotating frame) spectra. If evolved and detected spins belong to different isotopes, the corresponding heteronuclear 2D spectrum will contain only cross peaks (Fig. 9.7) [5, 8].

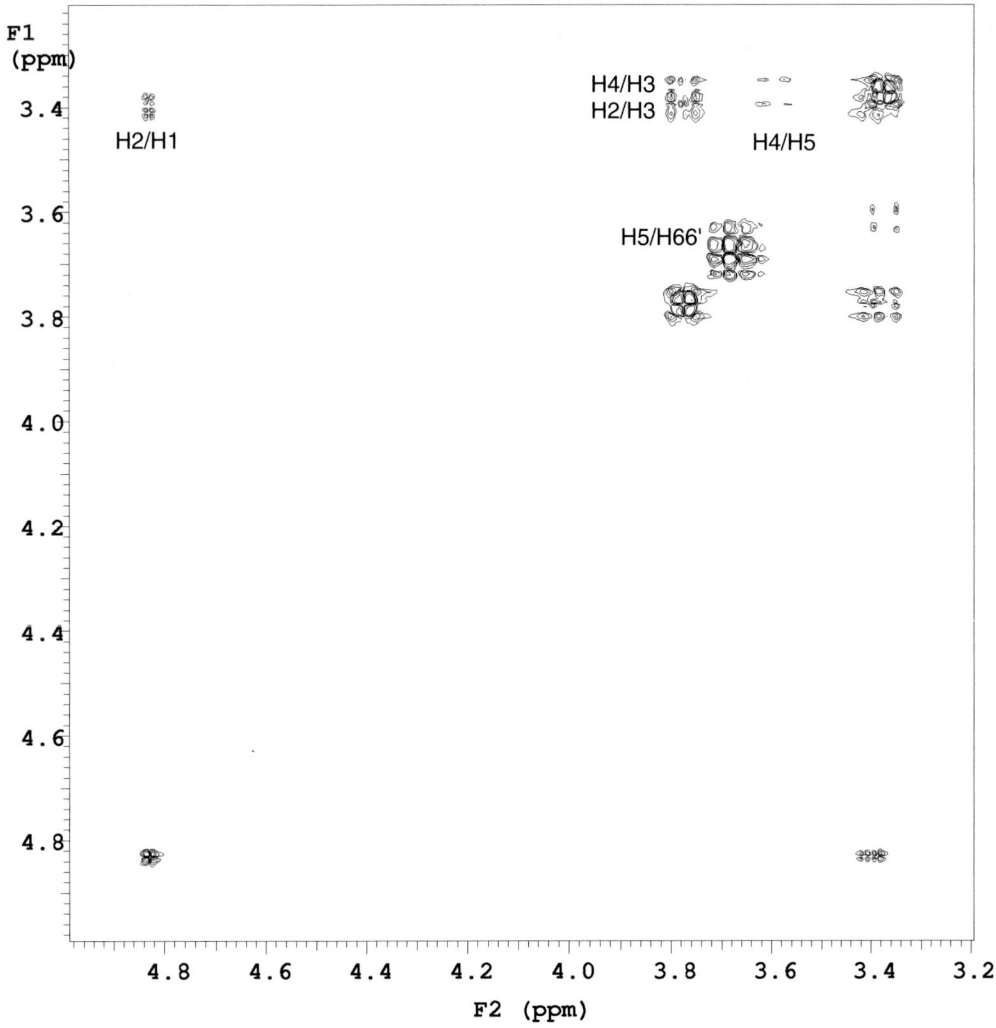

Fig. 9.6. Two-dimensional (2D) COSY spectrum of α-CyD in D_2O. Off-diagonal cross peaks allow us to follow the spin–spin coupling network and assign subsequent signals starting at the anomeric H-1 proton characteristic of its large chemical shift.

As previously mentioned, the highest theoretically possible sensitivity could be achieved when the spin states of large γ nuclei are initially perturbed and their magnetization is detected. Therefore, among many possible heteronuclear correlation schemes, those exploiting excitation and detection of sensitive nuclei (e.g. ^1H) are of great importance. Such methods are characterized by two-fold coherence

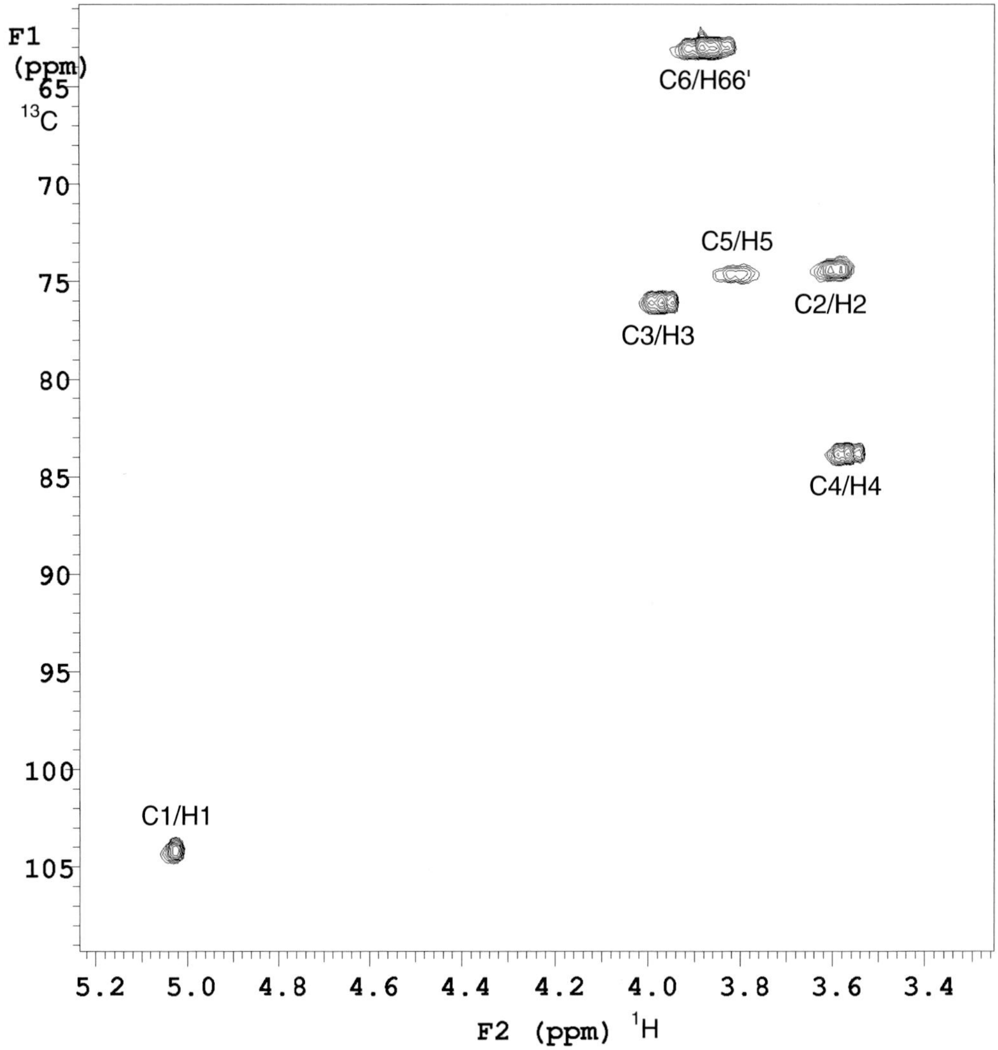

Fig. 9.7. Two-dimensional HSQC spectrum correlating ^{1}H (F2) and ^{13}C (F1) chemical shifts in α-CyD. Correlations corresponding to signals superposed in 1D and 2D ^{1}H spectra, H-2, H-4 and H-5, H-6,6′ (Figs. 9.4 and 9.6) are dispersed along the ^{13}C axis.

transfer between two kinds of nuclei and sampling the insensitive nuclei during the evolution period.

There are two basic types of such techniques, heteronuclear single quantum correlation (HSQC) (see Fig. 9.7) and heteronuclear multiple quantum correlation

(HMQC). The first is mostly applied for one-bond correlations using usually large heteronuclear ^{1}J couplings for the coherence transfer. It could be also used for multiple-bond correlations, although, when the magnitudes of homonuclear couplings (involving spins of larger γ) and heteronuclear couplings are comparable, the resulting signal is strongly attenuated. On the other hand, the sensitivity of HMQC experiment does not depend on possible homonuclear interactions between sensitive nuclei. Therefore, this technique is ideally suited for long-range correlations using small heteronuclear couplings [8, 19].

In all heteronuclear experiments with ^{1}H detection performed at the natural isotopic abundance of heteronuclei, weak signals of isotopomers of interest (for example $^{1}H/^{13}C$ and $^{1}H/^{15}N$), are masked by intense signals from abundant isotopomers (i.e. involving ^{12}C or ^{14}N), which have to be efficiently suppressed [8].

9.1.6 Translational Self-diffusion

NMR spectroscopy is well suited for the measurements of self-diffusion coefficients (D) of pure liquid compounds as well as of complex mixtures [20, 21]. This type of experiment was made possible by the introduction of pulsed field gradients (PFGs) which were used for labeling the position of observed species in the sample. All molecules in solution are subject to random translational motions. According to the Einstein equation, the root mean square displacement is equal to $(6Dt_{\mathrm{d}})^{1/2}$, where t_{d} is the diffusion time of three-dimensional motion. A PFG spin-echo experiment allows us to dephase magnetization by introducing a short, intense gradient of B_0 and to rephase it again after time Δ. Only those molecules that, after time Δ, remain at the same spatial coordinates, contribute to the acquired signal. The observed relative attenuation I/I_0, where I and I_0, are the experimental signal intensities with and without PFG, is dependent on Δ, D, gradient strength G, and gradient length δ.

$$I/I_0 = \exp[-\gamma^2 G^2 \delta^2 (\Delta - \delta/3) D] \tag{3}$$

For analysis of complex mixtures the quasi 2D diffusion-ordered spectroscopy (DOSY) method could be applied [22, 23].

The self-diffusion coefficients reflect the molecular mobility in solution and are sensitive to temperature, solvent viscosity, and molecular mass. Similarly to other spectral parameters, the apparent self-diffusion coefficient is the weighted average for all species remaining in the equilibrium. Thus, when a small guest molecule interacts with a bigger host molecule its apparent diffusion coefficient decreases, allowing us to detect the formation of an inclusion complex. Moreover, the dependence of the self-diffusion coefficient of guest on the host molar fraction allows us to determine the association constant similarly to the chemical shift titration.

There are numerous examples of PFG diffusion measurements in studies of CyD complexes. For example, diffusion measurement was used for tracing the interaction between β-CyD and Alzheimer amyloid β-peptide [24]. It revealed that β-CyD binding occurred at two sites via interaction of β-CyD with the aromatic side chains. Diffusion measurements were also applied for evaluation of the association constants of β-CyD with cyclohexylacetic and cholic acids [25] and those of the complexes of α-CyD with L-phenylalanine, L-leucine, and L-valine [26]. In the latter example, the methodology and possible sources of errors in the determination of association constants were discussed.

Diffusion measurements make possible the determination of association constants for complexes with small chemical shift changes induced by the association. A representative example of such application is the evaluation of association constants for the complexes of γ-CyD with 12-crown-4 and its tetraaza and tetrathia analogues [27]. A titration experiment with self-diffusion coefficient detection was also used for determination of the association constants of the 1-adamantanecarboxylic acid–β-CyD complex at two temperatures [28].

9.2 Application of NMR to Cyclodextrins and Their Complexes

9.2.1 NMR Parameters and the Complexing Ability of Cyclodextrins and Their Derivatives

The unequivocal assignment of as many signals in NMR spectra as possible is a prerequisite of a successful structural and/or conformational analysis. Therefore, a knowledge of ^{1}H and ^{13}C chemical shifts of cyclodextrin molecules is crucial in any study of their complexes. Besides the data for the smallest, most often used α-, β-, and γ-CyDs [3], resonance assignments were obtained for the larger CyDs composed of more than eight α-glucose units discussed in Chapter 13 [29–32].

Chemical modifications of CyDs have advantages and drawbacks. Usually, CyDs are modified to be better soluble and more resistant to enzymatic degradation than the native CyDs. Modified CyDs are also obtained to bind guest molecules more tightly and/or to display improved chiral recognition abilities [33]. On the other hand, NMR spectra of modified CyDs are often very complicated owing to low average molecular symmetry and/or the large number of functional groups, making quantitative analysis of host spectra difficult or impossible. Branched CyDs, a group of CyDs to which one or more glucopyranosyl residues are covalently linked, are good examples of such molecules [34, 35]. Moreover, a flexible substituent can penetrate the cavity of the CyD molecule (self-inclusion) competing with guest molecules in the inclusion process [36–39]. On the contrary, native CyD molecules display relatively simple NMR spectra, because all glucopyranosyl residues are chemically equivalent in the NMR time scale owing to fast conformational averaging. The same is true for a number of groups of symmetrically substituted CyDs, such as *O*-methyl- [40], *O*-acetyl- [41], *O*-nitro- [42], and sulfated [43] cyclodextrins, which have been thoroughly studied and successfully used in many applications. One should be aware, however, that even symmetrically located substituents introduce strain into the CyD molecule and/or reorganize a network of intramolecular hydrogen bonds, which may lower molecular symmetry [44] and significantly hinder interpretation of NMR spectra. A good example of such behavior was reported for hexakis(2,3-di-*O*-benzoyl)-α-CyD [45]. At low temperature its conformationally averaged symmetry was lowered from C_6 to C_3. The effect of chemical modification on the complexation ability can depend on the size of CyD molecule. It has been reported that tri-*O*-methylation significantly enhanced the complexing ability of α-CyD, moderately reduced it for β-CyD and largely diminished it for γ-CyD in complexes with 1,7-dioxaspiro[5.5]undecane, nonanal, and ethyl dodecanoate [46].

9.2.2 Stoichiometry

The first step in the study of a molecular complex is to determine its stoichiometry. This is frequently obtained by applying the method of continuous variations (Job's plot) [1, 4]. Values of the experimentally observed spectral parameter (usually chemical shifts that are sensitive to complex formation) are determined for varying host (H) to guest (G) mole ratios where the sum of their concentrations is kept constant. The data are plotted as the product of guest mole fraction and its complexation induced shift (CIS) vs. host mole fraction. The ratio of mole fractions x_G:x_H at the position of the maximum indicates the complex stoichiometry as shown in Fig. 9.8 (A). Interpretation of the continuous variations data becomes ambiguous in the case of two or more competing complexations, e.g. H:G = 1:1 and H:G = 2:1. In such a case the maximum can be found somewhere between positions corresponding to a single stoichiometry (Fig. 9.8 (B)).

For CyD complexes a number of stoichiometric ratios has been observed [2]. The most commonly reported ratios are H:G = 1:1 and H:G = 2:1. However, other stoichiometries as well as ternary CyD-containing complexes [47] are known. An example of 2:1 stoichiometry is the camphor–α-CyD complex in which the guest molecule is embedded inside a capsule formed by two host molecules [48]. Fenbufen (γ-oxo-[1,1′-biphenyl]-4-butanoic acid) is an interesting example of a compound which shows stoichiometry dependence on the CyD cavity size. It does not form an inclusion complex with α-CyD, but displays H:G = 1:1 stoichiometry with β-CyD and H:G = 1:2 stoichiometry with γ-CyD [49, 50]. Metoprolol is another such compound which forms 1:1 complexes with α-CyD and β-CyD but with γ-CyD it forms an H:G = 1:2 complex [51]. A similar phenomenon detected using HPLC for a complex with a first-generation dendrimer is presented in Chapter 5 [52]. On the other hand, 1-adamantanecarboxylic acid and β-CyD form a complex with temperature-dependent stoichiometry, H:G = 1:1 at 25 °C and H:G = 1:2 at 0 °C [28]. For the complexation of dodecyltrimethylammonium bromide with α-CyD two competing associations with stoichiometries of H:G = 1:1 and H:G = 2:1 have been reported [53]. Use of the method of continuous variations in such situations becomes questionable and information about the complex stoichiometry is revealed directly from the titration measurement described in Section 9.2.3.

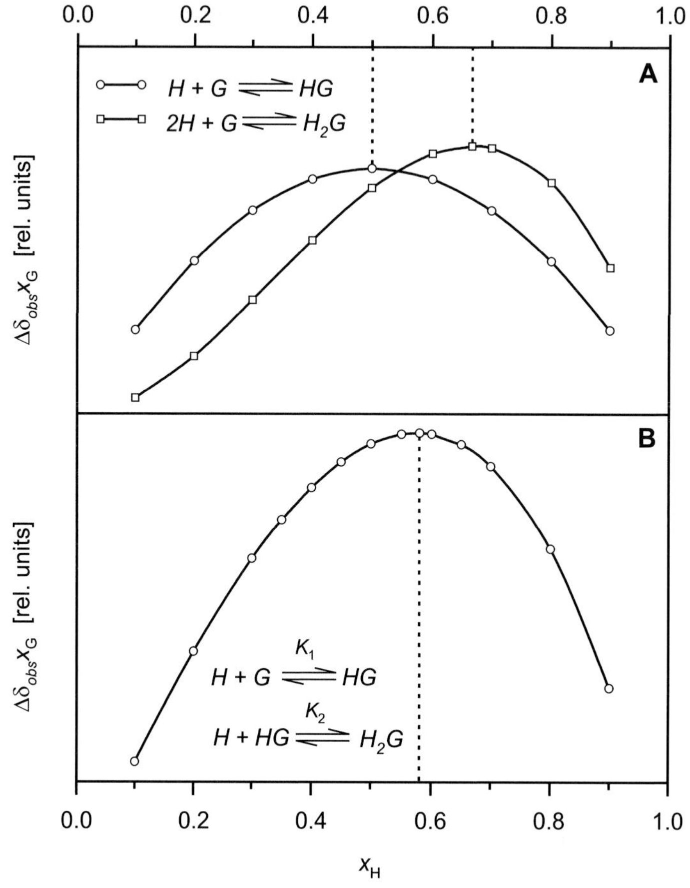

Fig. 9.8. (A) The continuous variation method enables us to determine complexation stoichiometry on the basis of the maximum plot position. (B) Two simultaneous, competing complexations give an ambiguous result ($K_1 = 100\ M^{-1}$, $CIS_1 = 50$ Hz, $K_2 = 100\ M^{-1}$, $CIS_2 = 80$ Hz).

9.2.3
Determination of Association Constants

The association constant (the synonymous expressions binding constant, stability constant, or complex constant can be found in the literature) K for complex formation of guest molecules with host CyD can be determined from NMR titration experiments. Assuming a given stoichiometry of the complex:

$$mG + nH \rightleftharpoons G_mH_n \tag{4}$$

the equation expressing the association constant in terms of initial concentrations of the host, $[H_0]$, guest $[G_0]$, and unknown complex concentration $[C]$ can be applied:

$$K = [C]/[G_0 - mC]^m [H_0 - nC]^n \tag{5}$$

Since the rates of complex formation and decomposition are usually faster than the NMR time scale, the chemical shift averaged signals δ_{obs} are observed:

$$\delta_{obs} = x_c\delta_c + x_f\delta_f \tag{6}$$

where δ and x are the chemical shifts and the mole fractions, respectively, and c and f indices denote complex and free component, respectively. The complexation-induced shift (CIS), $\Delta\delta$, is defined as:

$$\Delta\delta = \delta_f - \delta_{obs} \tag{7}$$

and can be observed either for guest, or host, or both components of a complex. K and δ_c parameters are calculated by fitting the experimental dependence of $\Delta\delta$ vs. concentration obtained for a given concentration of the host or guest at various concentrations of the other component [1]. Any approximation, such as the Benesi–Hildebrand equation [54] or other linearization approaches, should be better avoided. Examples of chemical shift titrations are shown in Fig. 9.9.

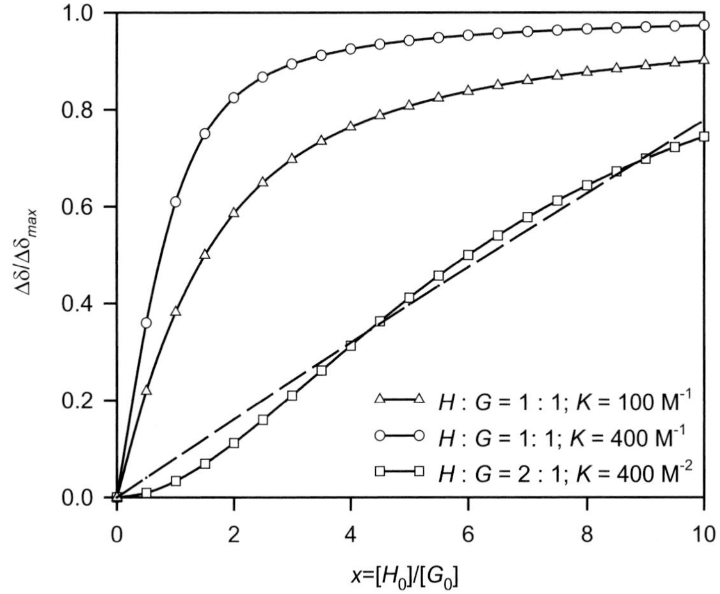

Fig. 9.9. Chemical shift titration curves depend not only on the association constant K, but also on the stoichiometry of the complexation. Presented data are calculated for $[G_0] = 10^{-2}$ mol L^{-1}. An attempt to fit titration curve representing 1:1 complexation model (dotted line) to the 2:1 data failed, resulting in a poor quality fit and unreasonable values of fitted parameters: $K = 0.44$ M^{-1}, $\Delta\delta/\Delta\delta_{\mathrm{max}} = 18.4$.

Owing to their accuracy and simplicity of determination, ^{1}H chemical shift titrations make up the vast majority of NMR titration experiments. A number of other $\frac{1}{2}$-spin nuclei, ^{13}C, ^{31}P, ^{19}F, or ^{15}N, are also used for this purpose [3, 4]. One should mention, however, that the lack of CIS observation does not necessarily correspond to a lack of complexation. In such cases any NMR observables, such as relaxation rates [55] or diffusion coefficients [26, 28], can be utilized for the determination of association constants in a titration experiment.

9.2.4
Host–Guest Orientation and Geometry of CyD Inclusion Complexes

A schematic representation of an average CyD molecule structure, which resembles a truncated cone, is shown in Fig. 1.2. Differences among native cyclodextrins arise from the size of the hydrophobic cavity, which depends on the number of glucopyranosyl residues constituting a whole molecule. Orientations of CyD protons in the molecule are crucial for understanding any spectral changes induced by the formation of inclusion complexes. Chemical shift changes of H-3 and H-5 protons pointing inside the cavity are indicators of complexation. Observation of their complexation shifts or NOEs to guest protons are the best hallmarks of the complex formation. Complexes of three noncarbohydrate-binding proteins (chymotrypsin inhibitor 2, S6, and insuline mutant) with β-CyD are convincing examples of this behavior [56]. NOEs between cyclodextrin H-3, H-5 protons and solvent-accessible aromatic side-chain protons on the surface of proteins have been observed resulting in the suppression of protein aggregation. Similarly, in the complex of acridine **1** with β-CyD H-3, H-5 host protons give correlation in the ROESY spectrum with all acridine protons, proving inclusion complex formation [9]. Restrained molecular dynamics calculations with interproton distance constraints obtained from the quantitative ROESY data were used to position the acridine in the β-CyD cavity.

In the complex of carbazol-viologen-linked compound **2** with α-CyD, NOEs have been observed between H-5 and one H-6 host protons and the spacer methylene protons [15]. Orientation of the guest molecule could be revealed from the NOE correlations between another host methylene proton H-6′ and carbazole **a** and **d** ones, strongly suggesting that the carbazole moiety protrudes from the narrow rim of the host cavity. In the NOESY spectrum of the doxorubicin (**3**)–γ-CyD complex, insertion of the D ring into the host cavity is confirmed by the H-1(*G*)/H-

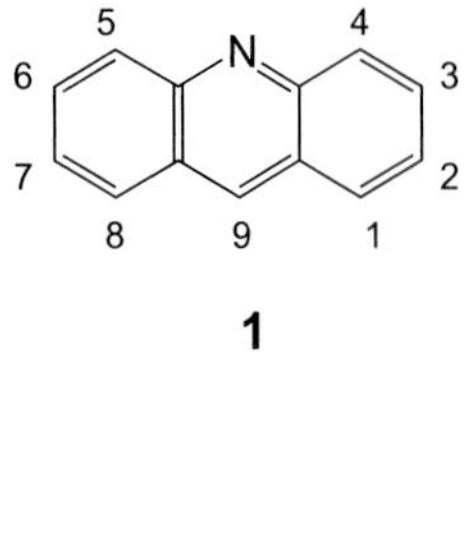

1

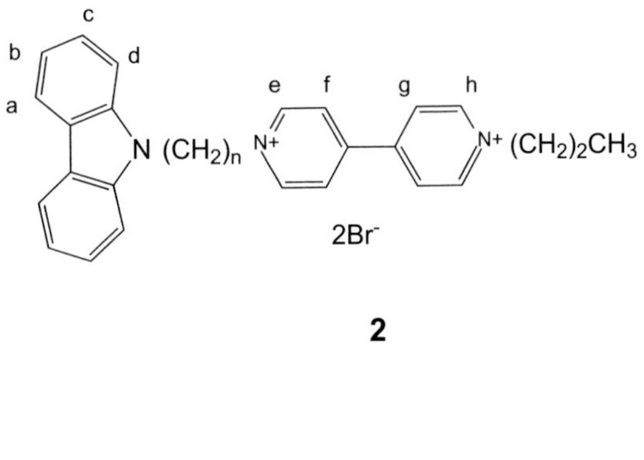

2

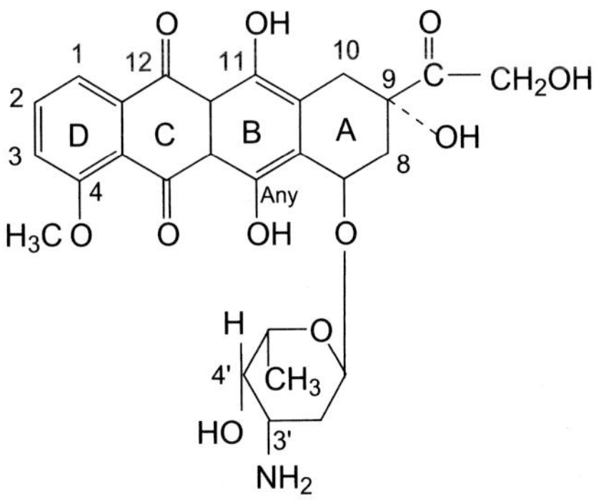

3

5(*H*) and H-3(*G*)/H-3(*H*) cross peaks. The sugar residue of doxorubicin is hanging outside the γ-CyD giving correlation H-4′(*G*)/H-2(*H*) [57].

Quantitative analysis of relaxation rates can also be used for the elucidation of inclusion complex geometry. Longitudinal and transverse ^{1}H relaxation rates were exploited to determine the orientation of the camphor **4** molecule inside the CyD capsule in the ($\pm$)camphor–α-CyD 1:2 complexes applying the model of anisotropic tumbling of the guest molecules [10]. It is noteworthy that in this particular case the complex geometry could not be obtained from analysis of NOE correlations.

Most of the inclusion complexes of organic molecules with CyDs, for which geometry has been investigated by NOESY/ROESY experiments and/or molecular modeling, showed unique relative orientations of host and guest molecules in the NMR time scale. Specific compounds, however, can form inclusion complexes with identical stoichiometry, but different geometry – a phenomenon termed multimodal complexation. For instance, in the 1:1 complex of salbutamol **5** with β-CyD, either the aromatic ring of the guest molecule or its alkyl chain is located in the host cavity [58]. Lipoic acid **6** in complex with β-CyD also shows bimodal complexation. One of its terminal groups, the carboxyl group, protrudes from either the wider side of the host cavity or its narrower side. The alkyl linker is located inside the host cavity and the dithiolane ring on the opposite side to the carboxyl group [59]. For the multimodal complexation of ampicillin **7** with β-CyD, association con-

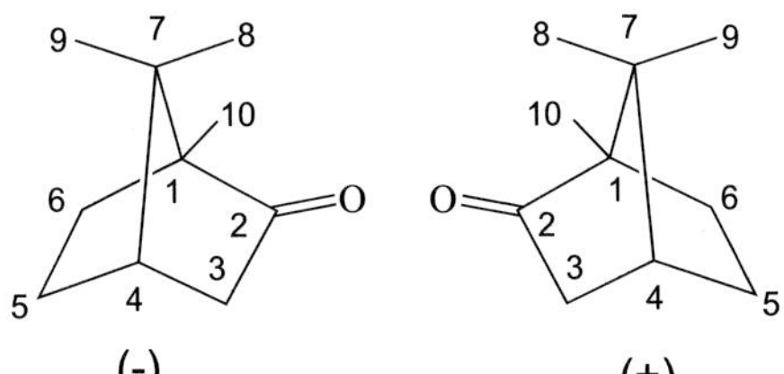

4

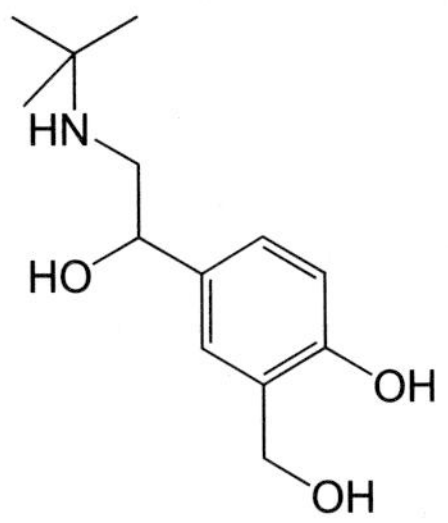

5

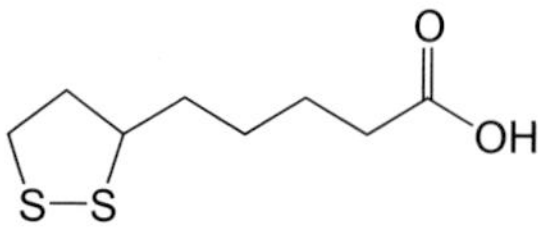

6

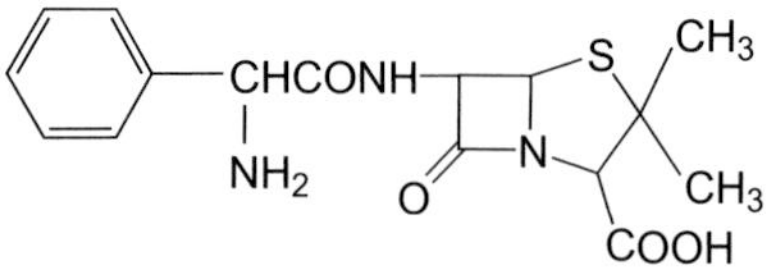

7

stants were determined for two types of complex geometries with either penam ring (higher *K*) or phenyl ring (lower *K*) inserted into the *β*-CyD cavity [60].

9.2.5 Chiral Recognition

It has been long appreciated that a chiral environment may differentiate any physical property of enantiomeric molecules. NMR spectroscopy is a sensitive probe for the occurrence of interactions between chiral molecules [4]. NMR spectra of enantiomers in an achiral medium are identical because enantiotopic groups display the same values of NMR parameters. Enantiodifferentiation of the spectral parameters (chemical shifts, spin–spin coupling constants, relaxation rates) requires the use of a chiral medium, such as CyDs, that converts the mixture of enantiomers into a mixture of diastereomeric complexes. Other types of chiral systems used in NMR spectroscopy include chiral lanthanide chemical shift reagents [61, 62] and chiral liquid crystals [63, 64]. These approaches can be combined. For example, CyD as a chiral solvating medium was used for chiral recognition in the analysis of ^{2}H residual quadrupolar splittings in an achiral lyotropic liquid crystal [65].

The application of NMR spectroscopic methods to chiral recognition by cyclodextrins has been reviewed [4] and some of the key results are now discussed.

The chirality of guest molecules is not always sufficient to differentiate NMR parameters of their diastereomeric CyD complexes. If such complexes are weak, no differences in the NMR parameters of enantiomeric guest molecules in complexes with CyDs can be observed [66]. Modern NMR techniques, especially correlation spectroscopy, greatly improve signal dispersion and enable observation of CIS in two separate dimensions. Good examples of such an approach are HSQC spectra of *α*-CyD complexes with (±)camphor [48] (Fig. 9.10) and (±)pinene [64]. For these systems measurements of the HSQC spectra not only demonstrated chiral recognition but also made possible determination of ^{13}C chemical shifts in diluted solutions at the natural ^{13}C abundance.

Most often, qualitative observation of chiral recognition is based on the changes of CIS [4]. Typical examples of ^{1}H and ^{13}C CIS-based chiral recognition are spectra of (±)*α*-pinene complexes with *α*-CyD in D_2O [64] and $(CD_3)_2SO$ [67]. The differentiation of signals arising from both diastereomeric complexes in ^{1}H spectra was observed in D_2O solution while in $(CD_3)_2SO$ the CIS differences were too small to be observed. On the other hand, owing to the larger chemical shift range, ^{13}C

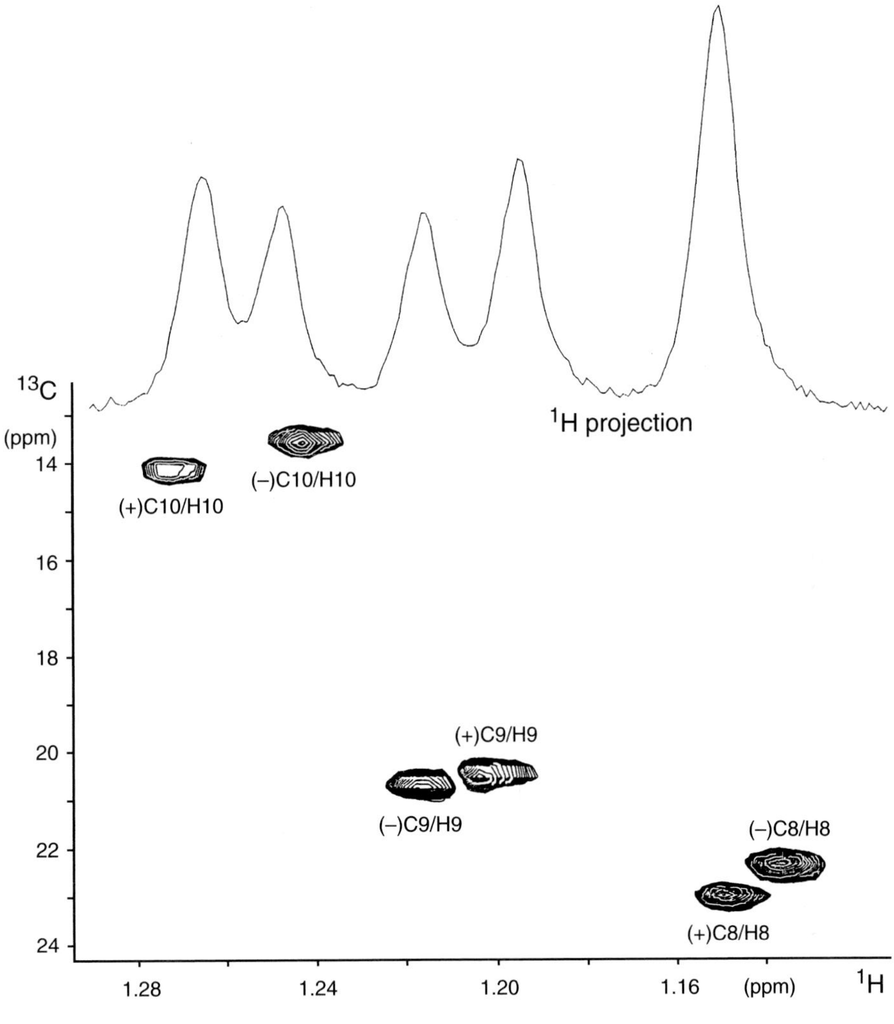

Fig. 9.10. Part of the $^{1}H/^{13}C$ HSQC spectrum showing enantiodifferentiation of methyl correlations in two diastereomeric (±)camphor–α-CyD complexes. Much better signal dispersion is evident in the 2D spectrum.

CIS differentiation was detected in both solvents with the differentiation effect scaled down by an order of magnitude for the $(CD_3)_2SO$ solution.

An interesting example of the assessment of drug enantiomeric composition based on ^{1}H CIS differences was observed in the complexes with heptakis(2,3-di-*O*-acetylated)-β-CyD [68]. The CIS differences could be also conveniently observed for other $\frac{1}{2}$-spin nuclei with a large chemical shift range. An example of enantiodifferentiation in ^{31}P spectra was observed for complexes of α-aminophosphonic and α-aminophosphinic acids with α- and β-CyD [69]. An additional advantage of observing ^{31}P or other less common nuclei is the much simpler NMR spectrum.

The association constants for CyD complexes with chiral guests are generally different and mostly quantitatively determined by the chemical shift titration experiments. However, as mentioned above, other NMR parameters, such as relaxation rates [10, 70] or self-diffusion coefficients, may also be used. Both those parameters were successfully applied for the enantiodifferentiation and determination of association constants in complexes of the trifluoroacetate salts of the enantiomers of amphetamine, ephedrine, and propranolol with 2,6-di-*O*-dodecyl-α-CyD and its β analogue [71]. The DOSY technique was employed for the determination of diffusion coefficients of enantiomers of cyclohexanone derivatives complexed with α-, β- and γ-CyDs as well as with their per-*O*-methylated analogues [72].

Systematic studies of relaxation rates, preferably at more than one magnetic field strength, provide, in addition to a simple differentiation of enantiomers, information about complex formation and its dynamics. For example, ^{1}H longitudinal relaxation rates were determined in the study of the tryptophan–α-CyD complex in D_2O [70]. A difference in the anisotropic tumbling of ($\pm$)camphor enantiomers inside the capsule formed by two α-CyD molecules in D_2O was elucidated from the analysis of longitudinal and transverse relaxation rates at two magnetic field strengths [10].

Chiral recognition by forming the inclusion complexes with CyDs could be also observed using NOE. As already mentioned, NOE/ROE experiments rely on the dependence of dipolar interactions on the distances between interacting nuclei. Thus, the mutual interactions of host and guest protons reflect information about the difference in relative orientation of host and enantiomeric guests in both diastereomeric complexes as can be seen in Fig. 9.11. Typical examples of the use of ROESY spectra to investigate the enantiodifferentiation of CyD-complexed molecules include observation of different complexation modes for (*R*)- and (*S*)-enantiomers [73] and structural studies of diastereomeric complexes [54, 74]. The ROESY-derived data are often complemented by molecular modeling calculations [75, 76] (see, however, the discussion on the reliability of CyD modeling in Chapter 11).

9.2.6
NMR Experiments in the Solid State

NMR spectra in the solid state are dominated by anisotropic interactions: dipolar couplings and chemical shift anisotropy. Since the 1980s significant progress has been made in methods of solid-state NMR spectroscopy. The modern techniques of cross-polarization (CP) and magic angle spinning (MAS) allow the acquisition of high-resolution chemical shift spectra and the evaluation of anisotropic parameters [77].

The relationship between glycosidic linkage conformation and ^{13}C chemical shifts in the solid state for a number of $(1 \rightarrow 4)$-linked polysaccharide polymers and oligomers including CyDs was studied by the CP MAS method [78–80]. The CyD hydration process was investigated using ^{13}C CP MAS spectra of anhydrous

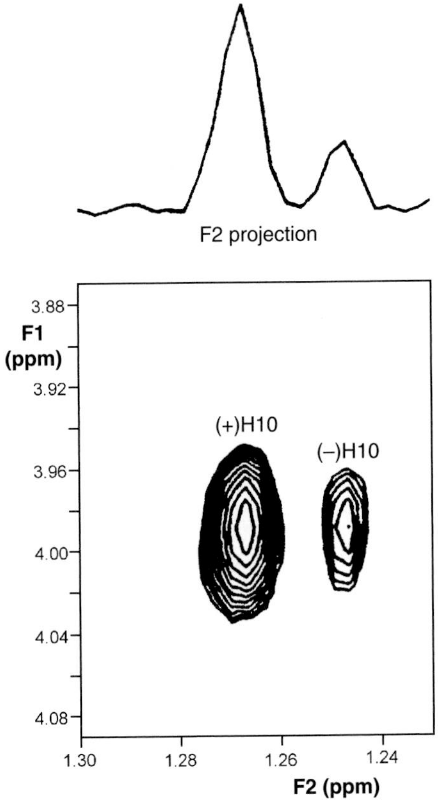

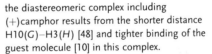

Fig. 9.11. Enantiodifferentiation of H10(*G*)/H3(*H*) NOE correlations in complexes of a racemic mixture of camphor enantiomers with α-CyD. Much stronger interaction in the diastereomeric complex including (+)camphor results from the shorter distance H10(*G*)–H3(*H*) [48] and tighter binding of the guest molecule [10] in this complex.

and hydrated α-, β-, and γ-CyD, and revealed that conformation of the CyD cone seemed to be determined by the hydration water [81]. The structure and molecular dynamics in the solid state of benzaldehyde and α-, β-, and permethylated α-cyclodextrin was studied by the CP MAS technique and revealed that the guest molecule undergoes motion in the cavity with a rate dependent on the CyD size [82]. The mobility of benzaldehyde in α-, β-, and γ-CyD cavities was also studied by means of CP dynamics, $R_1(^{13}C)$ and $R_1(^{1}H)$ in rotating frame relaxation rates [83]. Other interesting examples of solid-state investigation of CyDs and their complexes include study of tautomeric proton-transfer kinetics in tropolone complexes with α- and β-CyD [84], the γ-CyD inclusion complex of C_{60} fullerene obtained by its solubilization in water [85] and the β-CyD inclusion compounds containing the lanthanide tris(β-diketonates) complex hydrates [86].

References

1 L. Fielding, *Tetrahedron* **2000**, *56*, 6151–6170.
2 K. A. Connors, *Chem. Rev.* **1997**, *97*, 1325–1357.
3 H.-J. Schneider, F. Hacket, V. Rüdiger, H. Ikeda, *Chem. Rev.* **1998**, *98*, 1755–1785.
4 H. Dodziuk, W. Koźmiński, A. Ejchart, *Chirality* **2004**, *16*, 90–105.
5 J. K. M. Sanders, B. K. Hunter, *Modern NMR Spectroscopy. A Guide for Chemists*, 2nd ed., Oxford University Press, Oxford, **1993**.
6 M. Karplus, *J. Am. Chem. Soc.* **1963**, *85*, 2870–2871.
7 K. Imai, E. Osawa, *Magn. Reson. Chem.* **1990**, *28*, 668–674.
8 F. J. M. van de Ven, *Multidimensional NMR in Liquids*, VCH Publishers Inc., New York, **1995**.
9 I. Correia, N. Bezzenine, N. Ronzani, N. Platzer, J.-C. Beloeil, B.-T. Doan, *J. Phys. Org. Chem.* **2002**, *15*, 647–659.
10 W. Anczewski, H. Dodziuk, A. Ejchart, *Chirality* **2003**, *15*, 654–659.
11 J. P. Behr, J. M. Lehn, *J. Am. Chem. Soc.* **1976**, *98*, 1743–1747.
12 M. Suzuki, H. Takai, J. Szejtli, E. Fenyvesi, *Carbohydr. Res.* **1990**, *201*, 1–14.
13 D. Neuhaus, M. P. Williamson, *The Nuclear Overhauser Effect in Structural and Conformational Analysis*, VCH Publishers Inc., New York, **1989**.
14 J. Sandström, *Dynamic NMR Spectroscopy*, Academic Press, London, **1982**.
15 H. Yonemura, M. Kasahara, H. Saito, H. Nakamura, T. Matsuo, *J. Phys. Chem.* **1992**, *96*, 5765–5770.
16 U. Berg, M. Gustavsson, N. Åström, *J. Am. Chem. Soc.* **1995**, *117*, 2114–2115.
17 H. Dodziuk, J. Sitkowski, L. Stefaniak, D. Sybilska, J. Jurczak, *Supramol. Chem.* **1996**, *7*, 33–35.
18 H. Dodziuk, J. Sitkowski, L. Stefaniak, D. Sybilska, J. Jurczak, K. Chmurski, *Pol. J. Chem.* **1997**, *71*, 757–766.
19 A. Bax, M. F. Summers, *J. Am. Chem. Soc.* **1986**, *108*, 2093–2094.
20 L. H. Lucas, C. K. Larive, *Concepts Magn. Reson.* **2004**, *20A*, 24–41.
21 Y. Cohen, L. Avram, L. Frish, *Angew. Chem. Int. Ed.* **2005**, *44*, 520–554.
22 (a) K. F. Morris, C. S. Johnson Jr., *J. Am. Chem. Soc.* **1992**, *114*, 3139–3141; (b) *ibid.* **1993**, *115*, 4291–4299.
23 A. R. Waldeck, P. W. Kuchel, A. J. Lennon, B. E. Chapman, *Prog. NMR Spectrosc.* **1997**, *30*, 39–68.
24 J. Danielsson, J. Jarvet, P. Damberg, A. Gräslund, *Biochemistry* **2004**, *43*, 6261–6269.
25 K. S. Cameron, L. Fielding, *J. Org. Chem.* **2001**, *66*, 9891–6895.
26 R. Wimmer, F. L. Aachmann, K. L. Larsen, S. B. Petersen, *Carbohydr. Res.* **2002**, *337*, 841–849.
27 A. Gafni, Y. Cohen, *J. Org. Chem.* **1997**, *62*, 120–125.
28 Z. Tošner, *PhD Thesis*, Stockholm University, Sweden, **2005**.
29 I. Miyazawa, H. Ueda, H. Nagase, T. Endo, S. Kobayashi, T. Nagai, *Eur. J. Pharm. Sci.* **1995**, *3*, 153–162.
30 H. Ueda, T. Endo, H. Nagase, S. Kobayashi, T. Nagai, *J. Incl. Phenom.* **1996**, *25*, 17–20.
31 T. Endo, H. H. Nagase, Ueda, S. Kobayashi, T. Nagai, *Chem. Pharm. Bull.* **1997**, *45*, 532–536.
32 T. Endo, H. Nagase, H. Ueda, A. Shigihara, S. Kobayashi, T. Nagai, *Chem. Pharm. Bull.* **1997**, *45*, 1856–1859.
33 K. Harata, K. Uekama, M. Otagiri, F. Hirayama, *Bull. Chem. Soc. Jpn.* **1987**, *60*, 497–502.
34 Y. Yamamoto, Y. Inoue, *J. Carbohydr. Chem.* 1989, *8*, 29–46.
35 Y. Ishizuka, T. Nemoto, K. Kanazawa, H. Nakanishi, *Carbohydr. Res.* **2004**, *339*, 777–785.
36 S. R. McAlpine, M. A. Garcia-Garibay, *J. Am. Chem. Soc.* **1996**, *118*, 2750–2751.
37 S. R. McAlpine, M. A. Garcia-Garibay, *J. Org. Chem.* **1996**, *61*, 8307–8309.
38 H. Ikeda, M. Nakamura, N. Ise, N.

Oguma, A. Nakamura, T. Ikeda, F. Toda, A. Ueno, *J. Am. Chem. Soc.* **1996**, *118*, 10980–10988.
39 N. Birlirakis, B. Henry, P. Berthault, F. Venema, R. J. M. Nolte, *Tetrahedron* **1998**, *54*, 3513–3522.
40 Y. Yamamoto, M. Onda, Y. Takanashi, Y. Inoue, R. Chûjô, *Carbohydr. Res.* **1987**, *170*, 229–234.
41 G. U. Barretta, G. Sicoli, F. Balzano, P. Salvadori, *Carbohydr. Res.* **2003**, *338*, 1103–1107.
42 S. Bulusu, T. Axenrod, B. Liang, Y. He, L. Yuan, *Magn. Reson. Chem.* **1991**, *29*, 1018–1023.
43 T. J. Wenzel, E. P. Amonoo, S. S. Shariff, S. E. Aniagyei, *Tetrahedron: Asymmetry* **2004**, *14*, 3099–3104.
44 G. Uccello-Barretta, F. Balzano, A. Cuzzola, R. Menicagli, P. Salvadori, *Eur. J. Org. Chem.* **2000**, 449–453.
45 P. Ellwood, C. M. Spencer, N. Spencer, J. F. Stoddart, R. Zarycki, *J. Incl. Phenom.* **1992**, *12*, 121–150.
46 A. Botsi, K. Yannakopoulou, B. Perly, E. Hadjoudis, *J. Org. Chem.* **1995**, *60*, 4017–4023.
47 Y. Liu, Z.-Y. Duan, Y. Chen, J.-R. Han, H. Cui, *Org. Biomol. Chem.* **2004**, *2*, 2359–2364.
48 H. Dodziuk, A. Ejchart, O. Lukin, M. O. Vysotsky, *J. Org. Chem.* **1999**, *64*, 1503–1507.
49 I. Bratu, J. M. Gavira-Vallejo, A. Hernanz, M. Bogdan, Gh. Bora, *Biopolymers* **2004**, *73*, 451–456.
50 I. Bratu, J. M. Gavira-Vallejo, A. Hernanz, *Biopolymers* **2005**, *77*, 361–367.
51 Y. Ikeda, F. Hirayama, H. Arima, K. Uekama, Y. Yoshitake, K. Harano, *J. Pharm. Sci.* **2004**, *93*, 1659–1671.
52 H. Dodziuk, O. M. Demchuk, A. Bielejewska, W. Koźmiński, G. Dolgonos, *Supramol. Chem.* **2004**, *16*, 287–292.
53 N. Funasaki, S. Ishikawa, S. Neya, *J. Phys. Chem. B* **2004**, *108*, 9593–9598.
54 H. A. Benesi, J. H. Hildebrand, *J. Am. Chem. Soc.* **1949**, *71*, 2703–2707.
55 J. Ahmed, T. Yamamoto, Y. Matsui, *J. Incl. Phenom.* **2000**, *38*, 267–276.
56 F. L. Aachmann, D. E. Otzen, K. L. Larsen, R. Wimmer, *Protein Eng.* **2003**, *16*, 905–912.
57 O. Bekers, J. J. Kettenes-van den Bosh, S. P. van Helden, D. Seijkens, J. H. Beijnen, A. Bult, W. J. M. Underberg, *J. Incl. Phenom.* **1991**, *11*, 185–193.
58 E. Estrada, I. Perdomo-López, J. J. Torres-Labandeira, *J. Org. Chem.* **2000**, *65*, 8510–8517.
59 T. Carofiglio, R. Fornasier, L. Jicsinszky, G. Saielli, U. Tonellato, R. Vetta, *Eur. J. Org. Chem.* **2002**, 1191–1196.
60 H. Aki, T. Niiya, Y. Iwase, Y. Kawasaki, K. Kumai, T. Kimura, *Thermochem. Acta* **2004**, *416*, 87–92.
61 G. R. Sullivan, *Topics Stereochem.* **1978**, *10*, 287–329.
62 J. A. Peters, J. Huskens, D. J. Raber, *Prog. NMR Spectrosc.* **1996**, *28*, 283–350.
63 J. Courtieu, P. Lesot, A. Meddour, D. Merlet, C. Aroulanda, *Encyclopedia of Nuclear Magnetic Resonance*, D. M. Grant and R. K. Harris, Eds., J. Wiley & Sons, Chichester, **2002**, *Vol. 9*, 497–505.
64 H. Dodziuk, W. Koźmiński, O. Lukin, D. Sybilska, *J. Mol. Struct.* **2000**, *523*, 205–212.
65 J.-M. Péchiné, A. Meddour, J. Courtieu, *Chem. Commun.* **2002**, 1734–1735.
66 E. Butkus, J. C. Martins, U. Berg, *J. Incl. Phenom.* **1996**, *26*, 209–218.
67 H. Dodziuk, J. Sitkowski, L. Stefaniak, D. Sybilska, *Pol. J. Chem.* **1996**, *70*, 1361–1364.
68 M. Thunhorst, U. Holzgrabe, *Magn. Reson. Chem.* **1998**, *36*, 211–216.
69 Ł. Berlicki, E. Rudzińska, P. Kafarski, *Tetrahedron: Asymmetry* **2003**, *14*, 1535–1539.
70 K. B. Lipkowitz, S. Raghothama, J. Yang, *J. Am. Chem. Soc.* **1992**, *114*, 1554–1562.
71 A. Gafni, Y. Cohen, R. Kataky, S. Palmer, D. Parker, *J. Chem. Soc., Perkin Trans. 2* **1998**, 19–23.

72 A. Laverde, Jr., G. J. A. da Conceiao, S. C. N. Queiroz, F. Y. Fujiwara, A. J. Marsaioli, *Magn. Res. Chem.* **2002**, *40*, 433–442.
73 S. Negi, K. Terai, K. Nakamura, *J. Chem. Res. (S)* **1998**, 750–751.
74 P. K. Owens, A. F. Fell, M. W. Coleman, J. C. Berridge, *J. Incl. Phenom. Macro.* **2000**, *38*, 133–151.
75 M. E. Amato, G. M. Lombardo, G. C. Pappalardo, G. Scarlata, *J. Mol. Struct.* **1995**, *350*, 71–82.
76 E. Redenti, P. Ventura, G. Fronza, A. Selva, S. Rivara, P. V. Plazzi et al., *J. Pharm. Sci.* **1999**, *88*, 599–607.
77 F. Engelke, *Encyclopedia of Nuclear Magnetic Resonance*, D. M. Grant and R. K. Harris, Eds., J. Wiley & Sons, Chichester, **1996**, *Vol. 3*, 1529–1535.
78 R. P. Veregin, C. A. Fyfe, R. H. Marchessault, M. G. Taylor, *Carbohydr. Res.* **1987**, *160*, 41–56.
79 M. J. Gidley, S. M. Bociek, *J. Am. Chem. Soc.* **1988**, *110*, 3820–3829.
80 M. C. Jarvis, *Carbohydr. Res.* **1994**, *259*, 311–318.
81 I. Furó, I. Pócsik, K. Tompa, R. Teeäär, E. Lippmaa, *Carbohydr. Res.* **1987**, *166*, 27–33.
82 F.-H. Kuan, Y. Inoue, R. Chûjô, *J. Incl. Phenom.* **1986**, *4*, 281–290.
83 F. O. Garces, V. Pushkara Rao, M. A. Garcia-Garibay, N. J. Turro, *Supramolecular. Chem.* **1992**, *1*, 66–72.
84 K. Takegoshi, K. Hikichi, *J. Am. Chem. Soc.* **1993**, *115*, 9747–9749.
85 Á. Buvári-Barcza, J. Rohonczy, N. Rozlosnik, T. Gilányi, B. Szabó, G. Lovas, T. Braun, J. Samu, L. Barcza, *J. Chem. Soc., Perkin Trans. 2* **2001**, *338*, 1191–1196.
86 S. S. Braga, R. A. Sá Ferreira, I. S. Gonçalves, M. Pillinger, J. Rocha, J. J. C. Teixeir-Dias, L. D. Carlos, *J. Phys. Chem. B* **2002**, *106*, 11430–11437.

10
Other Physicochemical Methods

10.1
Introductory Remarks

Helena Dodziuk

In addition to chromatography, X-ray, NMR spectroscopy, and calorimetry presented in Chapters 5–9, respectively, several other techniques are applied in CyD studies. Mass spectroscopy, electrochemistry, UV-Vis absorption and emission spectra, circular dichroism, and scanning probe microscopy methods will be discussed in some detail in this chapter while the use of other techniques deserves a few words here. Solubility studies were significant in the early stages of development of supramolecular chemistry [1]. Today they are of importance only in specific applications, e.g. in CyD usage as drug carriers presented in Chapters 14 and 15. In this respect it is worth mentioning the quite unusual higher solubility below 50 °C of per-2,6-*O*-dimethyl-*β*-CyD **2** (in which 14 OH groups are substituted by OMe ones) than that of the parent **1**. The decrease in solubility of the former molecule as the temperature rises is also not typical [2]. Diffusion, viscosity, and dielectric CyD properties as well as ultrasonic absorption studies have been briefly reviewed by Szejtli [3], while other analytical methods including IR, Raman, and reflectance spectroscopies, the positron annihilation technique, electron microscopy, membrane permeation studies, and measurements of surface tension as well as electron spin resonance and some other methods were briefly discussed by Szente [4]. The most interesting novel applications of these techniques seem not to aim at studying simple inclusion CyD complexes using a single method but are rather devoted to the inspection of higher-order aggregates using several techniques simultaneously [5–9] with a special emphasis on prospective applications (presented in Chapter 16) [10–12]. Very few novel techniques, such as surface-plasmon resonance [11], or Fourier-transform infrared-reflection absorption spectroscopy (FTIR-RAS) [9] seem to appear in CyD studies.

Two important points discussed in detail in Section 1.3 should be remembered here: the fact that CyDs and their complexes are highly dynamic systems and that

Cyclodextrins and Their Complexes. Edited by Helena Dodziuk

ISBN: 3-527-31280-3

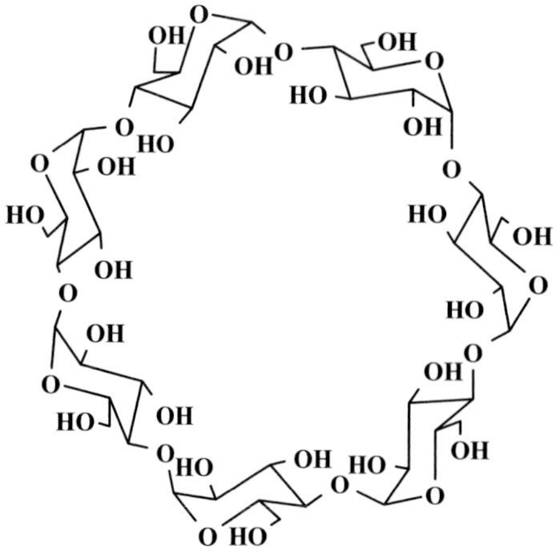

1

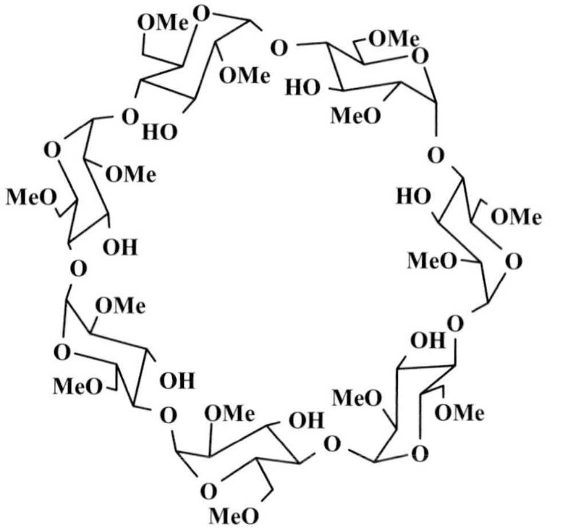

2

the latter are composed of weakly interacting fragments. Thus, the dependence of experimental results for these systems on the technique applied and, consequently, the necessity of studying them by applying several experimental methods simultaneously as emphasized there is of considerable importance.

References

1 Pedersen, C. J. *J. Am. Chem. Soc.* **1967**, *89*, 7017.

2 Nagai, I.; Ueda, H. In *Comprehensive Supramolecular Chemistry*; Szejtli, J., Ed.; Pergamon: Oxford, 1996; Vol. 3, p 441.

3 Szejtli, J. In *Comprehensive Supramolecular Chemistry*; Szejtli, J., Ed.; Pergamon: Oxford, 1996; Vol. 3, p 5.

4 Szente, L. In *Comprehensive Supramolecular Chemistry*; Szejtli, J., Ed.; Pergamon: Oxford, 1996; Vol. 3, p 253.

5 Mohanambe, L.; Vasudevan, S. *Inorg. Chem.* **2005**, *44*, 2128–2130.

6 Ponchel, A.; Abramson, S.; Quartararo, J.; Bormann, D.; Barbaux, Y.; Monflier, E. *Microporous and Nanoporous Mat.* **2004**, *75*, 261–272.

7 Liu, Y. L.; Male, K. B.; Bouvrette, P.; Luong, J. H. T. *Chem. Mater.* **2003**, *15*, 4172–4180.

8 Hirose, H.; Sano, H.; Mizutani, G.; Eguchi, M.; Ooya, T.; Yui, N. *Langmuir* **2004**, *20*, 2852–2854.

9 Choi, S. W.; Jang, J. H.; Kang, Y. G.; Lee, C. J.; Kim, J. H. *Colloids Surfaces A – Physicochem. Eng. Aspects* **2005**, *257–258*, 31–36.

10 Hou, Y.; Kondoh, H.; Shimojo, M.; Sako, E. O.; Ozaki, N.; Kogure, T.; Ohta, T. *J. Phys. Chem. B* **2005**, *109*, 4845–4852.

11 Morokoshi, S.; Ohhori, K.; Mizukami, K.; Kitano, H. *Langmuir* **2004**, *20*, 8897–8902.

12 Nakanishi, T.; Ohwaki, H.; Tanaka, H.; Murakami, H.; Sagara, T.; Nakashima, N. *J. Phys. Chem. B* **2004**, *108*, 77654–77762.

10.2 Mass Spectrometry of CyDs and Their Complexes

Witold Danikiewicz

10.2.1 Introduction – Mass Spectrometry Techniques Used for Studying Large Molecules and Their Complexes

Mass spectrometry (MS) has been one of the most important analytical methods in chemistry for many decades but its use was originally limited to relatively small and volatile organic molecules. This situation changed dramatically in 1981 when Barber and coworkers [1, 2] introduced a method named Fast Atom Bombardment (FAB). This was the first practical technique which made possible the generation and transfer to the gas phase of ions of polar, nonvolatile, and thermally unstable compounds, including the most important biomolecules such as peptides, nucleic acids, and sugars. Such methods are called "mild" or "soft" ionization techniques. Since that time MS has been constantly gaining importance as the major analytical tool in biochemistry as well as in polymer and supramolecular chemistry.

The "classical" FAB ionization technique consists in the bombardment of the sample dissolved in a nonvolatile, viscous, and polar liquid (the matrix) with neutral xenon atoms. However, a much more popular version is Liquid-matrix Secondary-ion Mass Spectrometry (LSIMS) in which the sample solution is bombarded with heavy ions, usually Cs^+ or Xe^+. Basic information about FAB and LSIMS can be found in mass spectrometry monographs [3, 4]. Comprehensive reviews on the application of FAB for analyzing biomolecules [5–7] and, specifically, saccharides and glycoconjugates [8] are also available.

To overcome the drawbacks of FAB, discussed later, two new methods, electrospray ionization (ESI, also known as Ion Spray, IS) and Matrix Assisted Laser Desorption/Ionization (MALDI) were developed, making MS one of the most important tools in chemistry, including biochemistry, supramolecular chemistry, and material chemistry. Acknowledging the development of these methods, Professor John B. Fenn from Yale University and Dr Koichi Tanaka from the Schimadzu Corporation were awarded the Nobel Prize in Chemistry in 2002, Professor Fenn for the development of electrospray and Dr Tanaka for the application of the MALDI technique to the analysis of proteins.

In the ESI technique [9, 10] a solution of the sample in a polar, volatile solvent (H_2O, MeOH, CH_3CN, etc.) is sprayed at atmospheric pressure into the ion source. During evaporation of the droplets, ions are liberated and transferred into mass spectrometer. The main advantages of ESI are: simplicity of sample preparation and measurement of the spectrum, relatively low cost, and ability to transfer even weakly bonded complexes from the liquid to the gas phase. Importantly, the ESI ion source can be coupled directly to the outlet of an HPLC column making possible on-line HPLC-MS analyses of complex mixtures (HPLC studies of CyDs are presented in Chapter 5). The only requirement for the successful acquisition of the ESI mass spectrum is the solubility of the analyzed compound in a polar solvent, or, for less-polar analytes, in solvent mixtures containing, most commonly, CH_2Cl_2 or $CHCl_3$. Owing to the extremely high sensitivity of MS, the solubility of the studied compounds can be very low, because even concentrations of about 1×10^{-5} mol dm^{-3} give reasonable signal-to-noise ratios. Taking into account that a single experiment with a standard ESI ion source requires about 100 μL of sample solution, only about 1 femtomole of the sample is required. Using ion sources with reduced flow (microspray and nanospray), the amount of the sample can be reduced even further.

The nanospray technique has one more significant advantage. The "classical" ESI ionization of free oligosaccharides, including CyDs, is known to be significantly less sensitive than the ionization of peptides and many other classes of compounds. The reason is the very strong solvation of oligosaccharides having numerous OH groups. This drop in sensitivity can be overcome by using the nanospray technique, in which extremely small droplets are formed. This results in better efficiency of ion transfer from the liquid to the gas phase, and, consequently, in a significant sensitivity gain.

Comprehensive information about ESI can be found in two monographs [11, 12]. A review of the mass spectrometry of oligosaccharides, focused especially on ESI and MALDI, was published by Zaia [13].

The MALDI technique was developed mainly by Karas and Hillenkamp [14–16], although it was Koichi Tanaka who was awarded the Nobel Prize in 2002 [17]. MALDI differs from FAB and ESI because the ionization process in this method takes place in a solid, rather than a liquid, sample upon irradiation with a very short laser pulse. The sample is prepared by evaporation of a solution of the analyte with a matrix compound on the stainless steel target plate. The role of the matrix is to absorb the laser radiation and transfer energy to the analyte. It is possible also to use a liquid matrix in MALDI, e.g. glycerol for IR- and glycerol with *p*-nitroaniline for UV-based MALDI [18]. It has to be noted that owing to the pulse character of MALDI ionization, MALDI ion sources are usually coupled with time-of-flight (TOF) mass analyzers, so the MALDI/TOF acronym is used most frequently to describe this MS method. Basic information about MALDI can be found in mass spectrometry monographs [3, 4]. More detailed discussion about the ion formation processes in MALDI was published by Zenobi and Knochenmuss [19]. An application of MALDI for the analysis of oligosaccharides was reviewed by Zaia [13].

FAB, ESI, and MALDI are extensively used for studying CyDs and their complexes. Of course, the most common application of MS in CyD research is the analytical one, providing the exact value of the molecular mass of the studied compound thus confirming its elemental composition. Nowadays there are no technical problems with acquiring high-quality mass spectra of both native CyDs and their various derivatives. However, much more interesting information about the studied compounds can be obtained from MS. For instance, Collision Induced Dissociation (CID) of the molecular ions gives insight into the structure of the studied molecules.

The most important application of MS in CyD research is in studying inclusion complexes. In many cases FAB, ESI, and MALDI allow transfer of the unchanged complex (in the ionized form) from the liquid or solid phase into the gas phase, where it can be studied by several MS methods. This field of research is often called "gas-phase supramolecular chemistry" and is one of the most challenging applications of modern MS. For instance, it is possible not only to establish the stoichiometry of an inclusion complex but also to estimate the energy of the host–guest interaction [20, 21] and even to study the gas-phase reactions, such as ligand exchange processes [22–26].

In this chapter a representative selection of MS applications in CyD chemistry will be presented, concentrating on the studies of their inclusion complexes.

10.2.2 Mass Spectrometry of Free CyDs

10.2.2.1 Fast Atom Bombardment (FAB) and Liquid-matrix Secondary-ion Mass Spectrometry (LSIMS)

The most important factor determining the quality of FAB*) spectra is the correct choice of liquid matrix (Scheme 10.2.1).

$HO\text{-}CH_2\text{-}CH(OH)\text{-}CH_2\text{-}OH$

glycerol (GLY)

$HO\text{-}CH_2\text{-}CH(OH)\text{-}CH_2\text{-}SH$

thioglycerol (TGL)

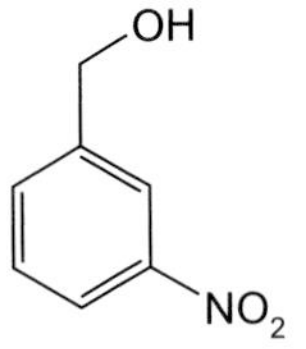

m-nitrobenzyl alcohol (NBA)

$HO\text{-}(\text{-}CH_2\text{-}CH_2\text{-}O\text{-})_n\text{-}H$

polyethylene glycol (PEG)
n = 3 - 10

Scheme 10.2.1. Common matrices for FAB analysis of CyDs.

For native CyDs the most frequently used matrix is thioglycerol (TGL) [27–30]. In this matrix both protonated molecular ions and ions formed by complexation of other cations, such as Na^+, K^+, NH_4^+, etc. are formed. The formation of $[M + H]^+$ ions can be enhanced by acidification of the matrix with trifluoroacetic acid (TFA). On the other hand, addition of the appropriate salt enhances the for-

*) In the following text the acronym FAB will also be used for LSIMS measurements, which is a common practice in the literature.

mation of $[M + cation]^+$ ions, e.g. $[M + Na]^+$ after addition of NaOAc. The second common matrix, used mostly for the less-polar substituted CyDs, is *m*-nitrobenzyl alcohol (NBA) [31, 32]. In this matrix only $[M + Na]^+$ ions are observed and the sensitivity of their detection can be strongly enhanced by addition of NaOAc [33]. In some individual cases other matrices or their mixtures are used (Scheme 10.2.1) [34, 35]. CyDs in FAB are usually recorded as the positive ions, although compounds containing free OH groups can also form $[M - H]^-$ ions [34].

The main advantage of the FAB technique compared to ESI and MALDI is, in fact, not the ionization method itself but the type of mass analyzer it works with. FAB ion sources are usually attached to magnetic sector, double-focusing mass spectrometers, enabling accurate mass measurements and, which is even more important, allowing the study of high-energy collision-induced dissociation (CID) reactions which provide information about the structure of the studied ion (see next section).

10.2.2.2
Electrospray Ionization (ESI)

There is no doubt that the most important ionization method used for studying CyDs and other complex and heavy biomolecules is electrospray.

Typical positive and negative ion ESI spectra of commercial *β*-CyD are presented in Fig. 10.2.1. In the positive mode, the $[M + Na]^+$ ion ($m/z = 1157.4$) is the main one; no $[M + H]^+$ ions are observed in this experiment. However, there are reports in the literature [22] that $[M + H]^+$ ions can be generated using a MeOH:H_2O:AcOH (48.5:48.5:3) solvent system. In the negative ion mode the most intense peak corresponds to the $[M - H]^-$ ion ($m/z = 1133.3$). There is also an $m/z = 1169.2$ peak visible, which corresponds to the $[M + Cl]^-$ ion. Other peaks correspond to clusters with different anions. For example, the isotope pattern around the $m/z = 1182.8$ peak corresponds to the doubly charged $[2\,CyD + SO_4^{2-}]^{2-}$ ion. This result shows clearly that, owing to the very strong complexing ability of CyDs, many strange peaks can appear in the spectrum reflecting the presence and content of the impurities in the sample. Some metal cations, especially Li^+, can not only form complexes with CyDs but can also exchange protons in OH groups yielding very complex spectra [36].

ESI mass spectrometry is used now as a routine tool for identification of modified CyDs. For example, Péan et al. [37] determined the molecular mass of some peptidyl-CyDs such as **1** using ESI in both positive and negative ion modes.

Only the negative ion mode was found suitable for the analysis of strongly acidic sulfated CyDs (e.g. **2**) [38] and sulfobutyl ether derivatives of *β*-CyD (e.g. **3**) [39]. In the latter case, a mixture of differently substituted *β*-CyD derivatives was separated on HPLC directly coupled to an ESI-equipped mass spectrometer. The HPLC/MS technique can also be used for quantitative determination of CyDs in complex matrices. Hammes et al. [40] developed a method for quantitative determination of *α*-CyD in human plasma using *β*-CyD as an internal standard. Linear calibration curve was obtained over the concentration range 5–1000 ng mL^{-1}.

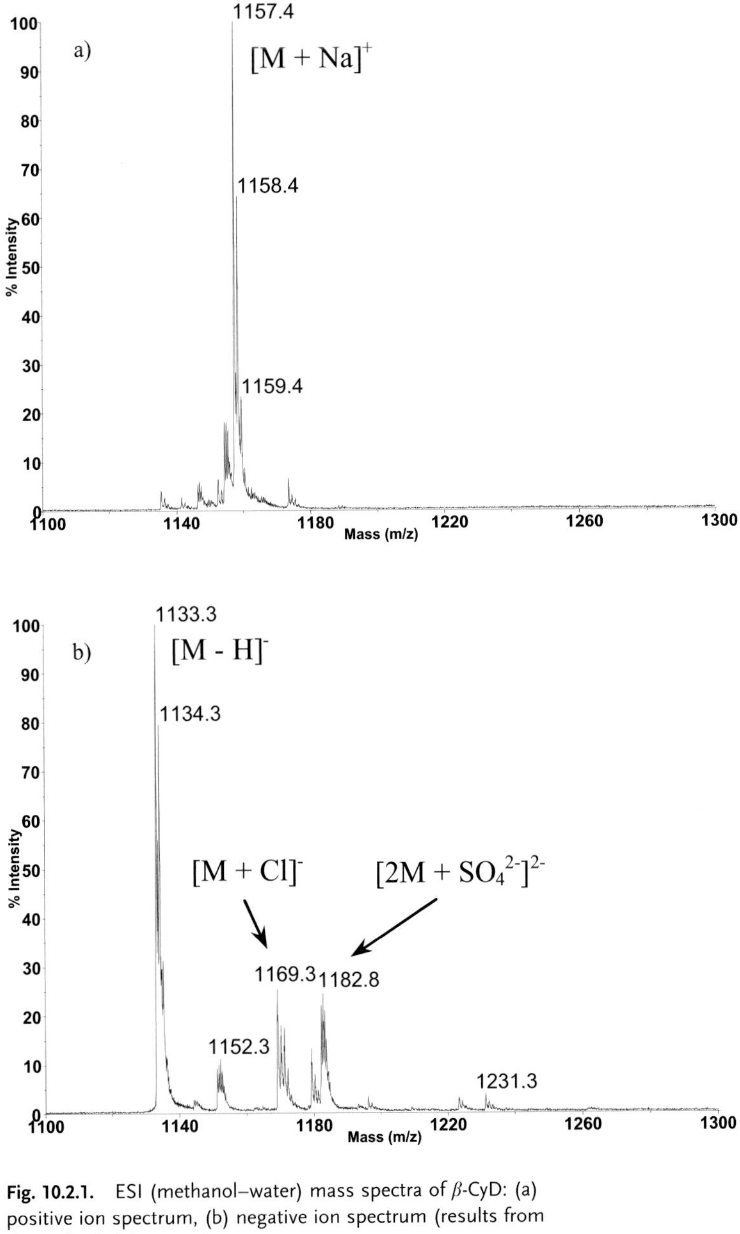

Fig. 10.2.1. ESI (methanol–water) mass spectra of β-CyD: (a) positive ion spectrum, (b) negative ion spectrum (results from the author's laboratory).

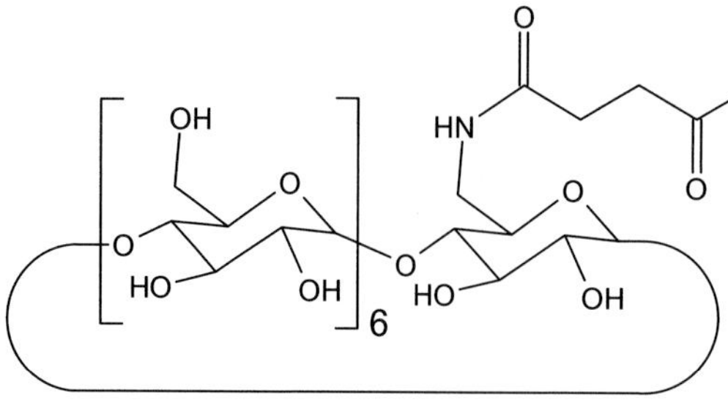

1

R = peptidyl chain, e.g. Gln-Gln-Phe-Phe-Gly-Leu-Met-NH_2

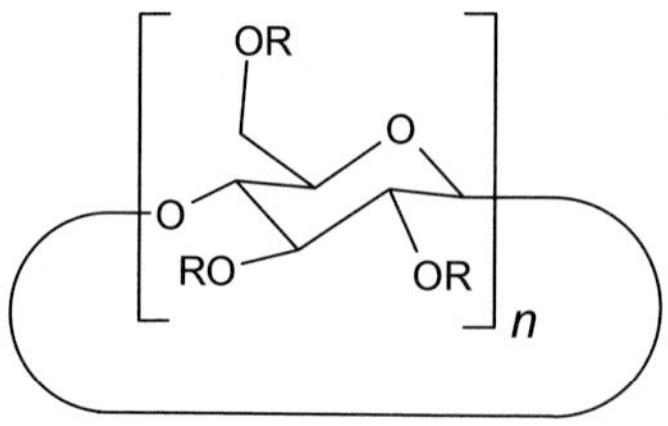

2

R = -SO_2ONa or H, *n* = 6, 7 or 8

3

R = -$CH_2(CH_2)_3OSO_3Na$ or H, *n* = 7

Electrospray ionization yields very stable ions, so fragmenting them requires additional energy which is delivered usually by collisions with molecules of a neutral gas. Native CyDs form such stable ions that, even under CID conditions, it is difficult to fragment them. Fragmentation of substituted CyDs is substantially easier. Salvador et al. [41] described the use of several chromatographic techniques together with the ESI triple-quadrupole mass spectrometer for the analysis of five commercial dimethylated *β*-CyDs, i.e. compounds with statistically two methyl groups on each sugar ring. Using the CID technique they were able to distinguish among symmetrically and unsymmetrically substituted $(Me)_{14}$-*β*-CyDs, i.e. *β*-CyD with two methyl groups on each ring and compounds with three Me groups on one ring, one Me group on the other ring and two Me groups on the rest of the rings. Sforza et al. [42] used a cone-voltage fragmentation technique together with statistical analysis for distinguishing between isomeric disubstituted *β*-CyDs (Scheme 10.2.2). These authors proved that their method gives reliable results independently of the type of substituents.

10.2.2.3
Matrix-Assisted Laser Desorption Ionization (MALDI)

MALDI is less popular than ESI, but the number of reports in the literature describing its use for CyDs is quite significant. Using 2,5-dihydroxybenzoic acid (DHB or DHBA), which is the most common matrix for sugars, abundant peaks corresponding to $[M + Na]^+$ ions, often accompanied by peaks of $[M + K]^+$ ions are obtained in the positive ion mode [43–45]. Owing to the high complexing capabilities of CyDs, cations with other metal ions can also be observed [46]. MALDI spectra of CyDs and their complexes were also acquired in sinapinic acid (*trans*-3,5-dimethoxy-4-hydroxycinnamic acid, SA) [47] and *p*-nitroaniline (PA) [48]. Raju et al. [49] proposed 5-ethyl-2-mercaptothiazole (EMT) as the matrix of choice for CyDs, showing high-quality results both in the positive ($[M + Na]^+$ ions) and negative ion mode ($[M - H]^-$ ions) (Fig. 10.2.2).

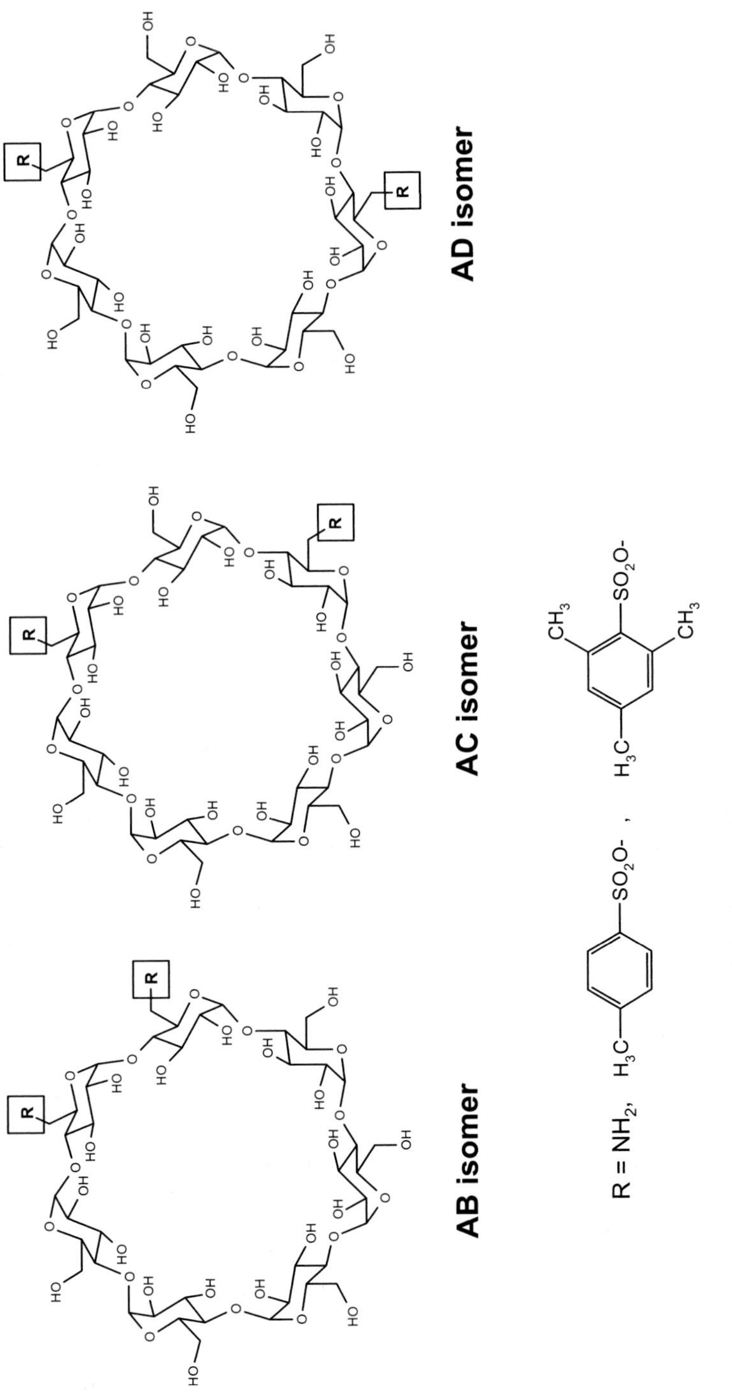

Scheme 10.2.2. AB-, AC- and AD-disubstituted derivatives of β-CyD studied using an ESI-MS/MS technique [42].

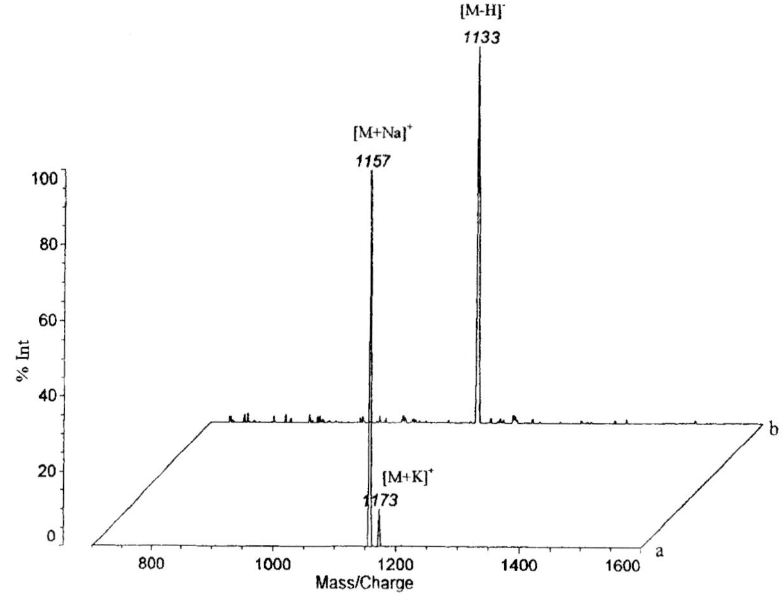

Fig. 10.2.2. MALDI-TOF mass spectra of β-CyD obtained using EMT as matrix in (a) positive and (b) negative ion mode (reproduced with permission from Ref. [49]).

It is worth mentioning that no $[M + H]^+$ ions can be observed except from CyD derivatives containing basic residues [50]. The original approach was proposed by Williams and Fenselau [18] who demonstrated that using a mixture of glycerol with *p*-nitroaniline as a liquid matrix for MALDI can give highly reproducible results for many classes of biomolecules, including CyDs. More information about designing MALDI experiments can be found in a review by Zenobi and Knochenmuss [19].

Mele and Malpezzi [51] observed, that in the case of methylated CyDs abundant peaks are recorded in MALDI spectra corresponding to CyD adducts with DHB and metal cations (Na^+, K^+). No such peaks are present for native CyDs. This observation indicates that the MALDI technique should be used with caution in the analysis of CyDs because of possible clustering of the matrix with the analyte.

The use of MALDI/TOF mass spectrometry with post-source decay (PSD) fragmentation for distinguishing among isomeric glycosylated CyDs was described in a series of papers by Yamagaki et al. [52–54]. For example, Yamagaki and Nakanishi described the PSD fragmentation of isomeric 6-*O*-glucosyl-β-CyD and 6,6′-di-*O*-glucosyl-α-CyD, showing that it is possible to distinguish between these two compounds by changing the acceleration voltage [44].

Interestingly, MALDI/TOF was also used for studying self-assembled mono-

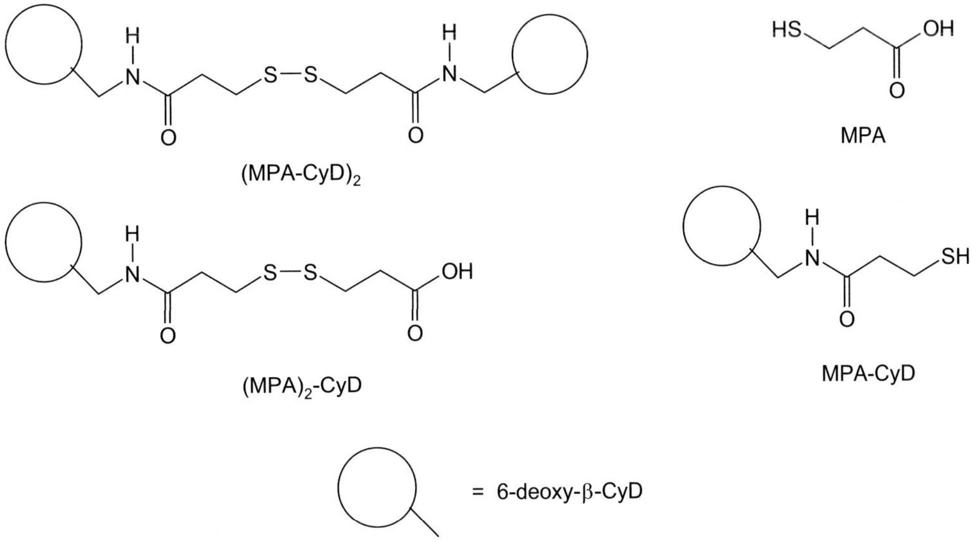

Scheme 10.2.3.

layers of monofunctionalized CyDs on gold surfaces. Henke et al. [55] used this technique to prove that 6-deoxyamino-β-CyD modified with mercaptopropionic acid (MPA) (Scheme 10.2.3) is chemisorbed on the gold surface by Au–S bonds.

They found that after evaporation of the $(MPA\text{-}CyD)_2$ solution on a standard stainless steel MALDI target, followed by evaporation of the DHB matrix solution, an intense peak corresponding to $[(MPA\text{-}CyD)_2 + K]^+$ is observed in a MALDI/TOF mass spectrum. This peak was missing when the self-assembled monolayer of MPA-CyD prepared from $(MPA\text{-}CyD)_2$ on the gold surface was used as a target for MALDI (after covering with DHB). After the addition of MPA, a peak corresponding to $[MPA\text{-}CyD + Na]^+$ appeared, indicating that MPA-CyD molecules were liberated from the surface and replaced with MPA. Similar results were obtained for $(MPA)_2$-CyD.

10.2.3 Mass Spectrometry of Inclusion Complexes of CyDs

Much more challenging than recording mass spectra of pure CyDs are applications of MS for characterization of stoichiometry, physicochemical properties, and even gas-phase reactions of the inclusion complexes of CyDs with a broad variety of ligands, both organic and inorganic. Since the mid-1990s several extensive review articles [20, 21, 56–58] and the chapter in Vol. 8 of the *Comprehensive Supramolecular Chemistry* [59] have been published on these subjects. Almost all of them contain data concerning complexes of CyDs. The review by Lebrilla [26] is devoted specifically to the gas-phase chemistry of CyD complexes.

All three leading soft ionization techniques, FAB, ESI, and MALDI, are used for generating gas-phase ions of CyD inclusion complexes. However, there is still an open question about how accurately mass spectra reflect the solution-phase chemistry. There is no general answer to this question and each system has to be treated separately. A few helpful hints, however, have to be borne in mind. Polar interactions, such as hydrogen bonds and electrostatic attraction, are usually stronger in the gas phase than in solution, especially in polar solvents. Therefore, it is possible to observe some cluster ions in the gas phase which cannot be detected in solution. On the other hand, nonpolar interactions are usually weakened in the gas phase. Thus, some complexes that do exist in solution, as confirmed by spectral methods such as NMR spectroscopy (presented in Chapter 9), cannot be transferred to the gas phase without decomposition. Applying these rules to complexes of CyDs, the conclusion can be drawn that only complexes with relatively polar compounds can be observed in the gas phase [58, 60].

Another important question concerns the structure of ions with the composition corresponding to the expected inclusion complexes. Loosely and nonspecifically bonded clusters are common in MS, especially for such polar compounds as CyDs. Therefore, there is always a question whether the "real" inclusion complex or just a nonspecific cluster ion is observed (Fig. 10.2.3).

In spite of some attempts to answer this question, the results are not clear. It seems that, in principle, it is possible to transfer intact inclusion complex ions to the gas phase, but, as stated above, any individual case requires careful examination. One possible method of solving this problem consists in a comparison between CyDs and linear oligosaccharides as host molecules for the same guest. If the complex ion is observed with similar abundance for both CyD and the linear ligand, which cannot form an inclusion complex typical of a CyD, there is a good chance that nonspecific complexation dominates for the CyD also. On the other

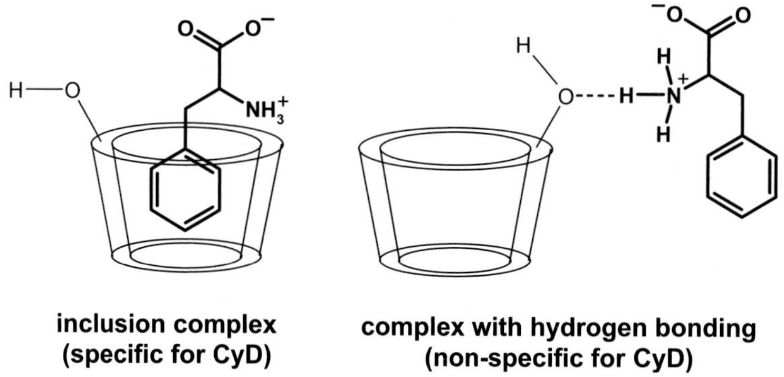

Fig. 10.2.3. Structures of the inclusion complex (interaction specific for the CyD structure) and the complex with hydrogen bonding (interaction nonspecific for the CyD structure).

hand, observation of an intense peak corresponding to the complex ion only in the case of CyD proves that we are observing a specific inclusion complex [61]. The second method of distinguishing between an inclusion and a nonspecific structure of a CyD complex consists in measuring the relative intensity of the peak corresponding to the complex ion as a function of the collision energy under CID conditions. The resulting plots (named breakdown curves) make it possible to compare the noncovalent binding energies and, therefore, to distinguish between inclusion complexes and less-specific structures [61].

10.2.3.1 Fast Atom Bombardment (FAB) and Liquid-matrix Secondary Ion Mass Spectrometry (LSIMS)

Although FAB is used less frequently today, it is still a very valuable method. It is widely assumed that FAB mass spectra reveal solution-phase chemistry quite accurately, while this statement is still controversial in ESI/MS (*vide infra*).

In only in a few cases has the application of FAB and ESI techniques for analyzing CyDs inclusion complexes been compared directly. Bongiorno et al. [30] studied the complex of melatonin with β-CyD by both methods and obtained comparable results, as shown by the similar CID spectra of the protonated 1:1 complex ions generated by FAB and ESI. Thioglycerol (TGL) was the matrix of choice in this study as well as in several other studies of complexes of simple organic molecules with unmodified CyDs [27–29, 47]. Progesterone complexes with two less polar, amphiphilic β-CyDs acylated with long-chain fatty acids (Scheme 10.2.4) were studied using *m*-nitrobenzyl alcohol (NBA) as the matrix [32]. For both compounds studied, peaks corresponding to 1:1 complexes were observed.

FAB was also the first MS method for studying the chiral recognition exhibited by native CyDs and their derivatives, but the observed chiral selectivities were very low [31, 62, 63]. The best result was obtained for the complex of permethylated α-CyD with the isopropyl ester of tryptophan. An intensity ratio of 1.29 was found for the peaks corresponding to the complexes of *R* and *S* enantiomers [63]. Distinction between the complexes formed by *R* and *S* enantiomers was possible because the *S* enantiomer of tryptophane was used as a d_7-isopropyl ester (isotope labeling method). Much higher chiral recognition values, up to 10:1, were reported by Davey et al. [35] who studied the complexation of the methyl esters of tryptophan and phenylalanine by permethylated α- and β-CyDs. Unfortunately, their method, based on a comparison of the results of two separate measurements, seems to be less reliable than the isotope labeling method. There are no other reports confirming these results, so they, as well as the entire method, have to be treated with caution.

10.2.3.2 Electrospray Ionization (ESI)

ESI, as the "mildest" ionization method, is able to transfer even relatively weakly bonded complex ions from the liquid to the gas phase, which makes it the most

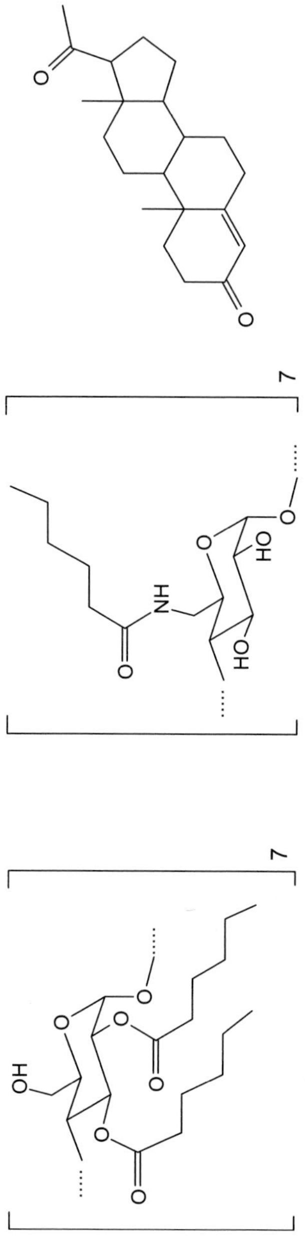

progesterone

6-*N*-Capro-β-CyD

β-CyD-$(C6)_2$

Scheme 10.2.4.

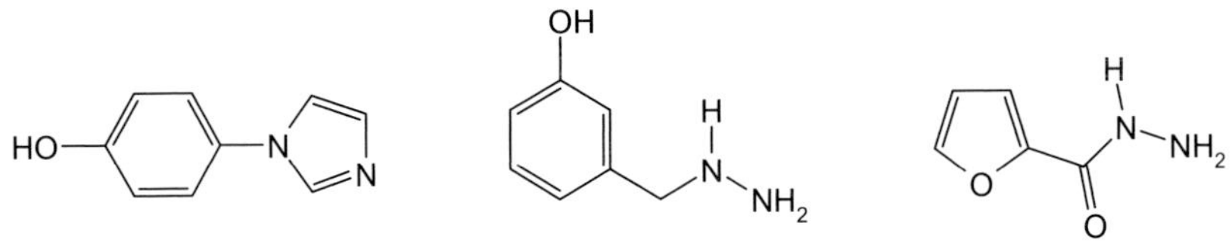

	4-IMP	3-HBH	2-FUH
K_{st}(MS):	1438(±75)	1084(±58)	906(±43)
K_{st}(UV):	1281(±61)	1112(±55)	820(±38)

Scheme 10.2.5. Stability constants of the complexes of β-CyD with three simple organic molecules measured by MS and UV spectrometry [64].

important MS tool for studying CyD inclusion complexes. However, the structures of these complexes in solution and in the gas phase can differ. The question of how accurately the relative intensities of peaks in an ESI mass spectrum reflect the concentration ratios of the corresponding ions in solution is still open, but there are already many well-established methods which give semiquantitative, or even fully quantitative results allowing us to determine stability constants for some complex ions in solution. For example, Dotsikas and Loukas [64] determined the stability constants of the complexes of β-CyD with three simple organic molecules and obtained results very similar to those resulting from UV spectroscopic measurements (Scheme 10.2.5). Taking into account differences in experimental conditions, the similarity is striking. It has to be noted, however, that the authors present their results with an accuracy which is not justified by the uncertainty level of the measurements.

Representative examples of ESI measurements of CyD inclusion complexes selected from the literature (starting from 1996) are collected in Table 10.2.1.

It must be repeated again that the presence of a peak in the MS spectrum with an m/z value corresponding to the expected cationized or anionized complex ion does not necessarily mean that we are observing an inclusion complex that is also present in solution.

There are several papers dealing with the application of the ESI/MS technique for studying chiral recognition by CyDs. The most accurate, but relatively complex, method was introduced by Lebrilla and coworkers [65]. Using a Fourier-transform Ion Cyclotron Resonance (FT/ICR) mass spectrometer it is possible to measure the reaction rates between the CyD–amino acid inclusion complex and an aliphatic amine (usually n-propylamine) in the gas phase. For the majority of amino acids, the reaction rate of a D-amino acid complex is lower (by a factor of up to 3) than the reaction rate of the L-enantiomer complex, indicating the higher stability of the D-amino acids complexes. Lebrilla's group [23–25, 65–67] applied and critically evaluated this method for many host–guest systems (see, however, critical remarks on these and other authors' use of the three-point model of chiral recognition in Section 1.4). FT/ICR mass spectrometers are very expensive and because of this

Tab. 10.2.1. Selected examples of CyDs inclusion complexes studied by ESI.

CyD	Guest	Stoichiometry (CyD:guest)	Solvent	Notes	Reference
α-CyD, β-CyD	lycopene (ψ,ψ-carotene)	1:1	H_2O/ACN 1:1	(−) mode	72
DM-β-CyD TM-β-CyD	1-anilinonaphthalene-8-sulfonate 2-*p*-toluidinylnaphthalene-6-sulfonate	1:1	H_2O/ACN 1:1	+NH_4OAc in (+) mode	73
α-CyD, β-CyD, γ-CyD	1-anilinonaphthalene-8-sulfonate 2-*p*-toluidinylnaphthalene-6-sulfonate	1:1	H_2O/ACN 1:1	+NH_4OAc in (+) mode +NH_3 in (−) mode	74
β-CyD, γ-CyD	cinchonine	1:1, 2:1	H_2O	(+) mode	75
β-CyD	egg-white lysozyme	1:1, 2:1, 3:1	H_2O	(+) mode	76
α-CyD, β-CyD, DM-β-CyD	10 different aromatic compounds (aldehydes and phenol esters)	1:1, 2:1 (depending on the guest)	H_2O/ACN 1:4	+NaOAc, (+) mode	77
DM-β-CyD	Euplotin C	1:1	H_2O	(+) mode	78
β-CyD, γ-CyD	several surfactants (aliphatic compounds with long chains)	1:1, 2:1	H_2O/ACN 1:1	(−) mode	79
α-CyD, β-CyD, γ-CyD	6 primary amines	1:1	H_2O/MeOH/AcOH (48.5:48.5:3)	(+) mode; ligand exchange studies	22
DM-β-CyD TM-β-CyD M-β-CyD	proparacaine, lidocaine, prilocaine, mepricaline, dibucaine, dyclonine	1:1	100 mM NH_4OCHO_{aq}	(+) mode	80

α-CyD	7 linear α,ω-dicarboxylic acids	1:1, 2:1	NH_4OH_{aq} (pH = 9)	(−) mode	61
β-CyD	Gly, Phe, Trp	1:1	NH_4OAc_{aq}	(+) and (−) mode	81
M-β-CyD	5 steroid hormones	1:1, 2:1	H_2O/MeOH 4:1	(+) mode	82
β-CyD, γ-CyD, HP-β-CyD, HP-γ-CyD	6 barbiturates	1:1	NH_4OAc buffer (pH = 9)	(−) mode	83
α-CyD, β-CyD, γ-CyD	oleanolic acid	1:1	MeOH	(−) mode	84
α-CyD, β-CyD, γ-CyD	rutin	1:1	MeOH	(+) and (−) mode	85
γ-CyD	Nile Red	no complex observed	EtOH, H_2O	(+) mode	86
β-CyD	sodium salt of tri(*m*-sulfonylphenyl)phosphine	1:1, 2:1, 3:1	H_2O/ACN 1:1 + NH_4OAc	(+) mode	87
α-CyD, β-CyD, γ-CyD	titanocene dihalides	1:1	H_2O	(+) mode	88
α-CyD, β-CyD	ferrocene and its derivatives	1:1, 2:1	NH_4OAc_{aq}	(+) mode	89
α-CyD, β-CyD, γ-CyD	meloxicam	1:1	H_2O/MeOH/AcOH (49.5:49.5:1)	(+) mode	90
β-CyD	piroxicam, terfenadine	1:1	H_2O/ACN 1:1	(+) mode	91
β-CyD	melatonin	1:1	H_2O/ACN 1:1	(+) mode	30
α-CyD, γ-CyD, HP-α-CyD, HP-β-CyD, HP-γ-CyD	lorazepam	1:1, 2:1, 3:1, 3:2	H_2O	(+) mode	92

Tab. 10.2.1 *(continued)*

CyD	Guest	Stoichiometry (CyD:guest)	Solvent	Notes	Reference
β-CyD	glybenclamide and furosemide with diethanolamine in ternary complex	1:1:1	H_2O/ACN 1:1	$+NH_4OAc$ in (+) mode	93
β-CyD	ketoconazole with tartaric acid in ternary complex	1:1:1	H_2O/ACN 1:1	(+) and (−) mode	94
β-CyD	toluene and other aromatic compounds with Fe^{2+} or Mg^{2+} in ternary complexes	1:1:1	H_2O	(+) mode	95

Acronyms used in the table: DM-β-CyD – heptakis-(2,6-di-*O*-methyl)-β-CyD; TM-β-CyD – heptakis-(2,3,6-tri-*O*-methyl)-β-CyD; M-β-CyD – mixture of methylated β-CyDs with different numbers of Me groups; HP-x-CyD ($x = \alpha, \beta$, or γ) – hydroxypropyl-x-CyDs with different numbers of hydroxypropyl groups; ACN – acetonitrile.

not in widespread use, so the same authors tried to use the much more easily available ion trap mass spectrometer instead of FT/ICR. However, the results were not satisfactory [68].

Other researchers also studied this subject and obtained different results. Franchi et al. reported that, using ESI, they were not able to observe any discrimination between the enantiomeric pairs of chiral nitroxides by modified CyD, while they observed such discrimination in solution using other methods [69]. On the other hand, Lu et al. [70] presented positive results concerning chiral recognition of pseudoephedrine with α-, β- and γ-CyDs while Cheng and Hercules [71] succeeded in discriminating enantiomers of tryptophan and tyrosine with α- and β-CyDs. There is no doubt that ESI/MS can be a valuable method for studying chiral recognition but this subject requires more investigation.

10.2.3.3
Matrix-assisted Laser Desorption Ionization (MALDI)

The "classical" MALDI ionization technique differs from the two other soft ionization methods – FAB and ESI – because the ions are transferred to the gas phase from a solid, rather than a liquid phase. This should make possible, in principle, the study of inclusion complexes of CyDs in the solid state. Some positive results, in which peaks corresponding to host–guest complex ions have been observed, have been reported [47]. Lehmann et al. [48] discussed critically the problem of how MALDI mass spectra reflect solution-phase formation of CyD inclusion complexes. They came to the conclusion that there is a high probability that preparation of the MALDI target destroys relatively weakly bonded inclusion complexes, so this technique is not suitable for studying such species. In contrast to these results, So et al. [43] found that MALDI/MS can be useful for differentiating the enantiomers of amino acids. It is also possible that the use of a liquid matrix for MALDI, as proposed by Williams and Fenselau [18] could make this technique more appropriate for studying inclusion complexes of CyDs, but this path has not been explored yet. An example of the application of MALDI/TOF to the determination of a rotaxane structure involving CyD is presented in Section 3.1.9.

References

1 Barber, M.; Bordoli, R. S.; Sedgwick, R. D.; Tyler, A. N. *J. Chem. Soc., Chem. Commun.* 1981, 325–326.

2 Barber, M.; Bordoli, R. S.; Elliott, G. J.; Sedgwick, R. D.; Tyler, A. N. *Anal. Chem.* 1982, *54*, 645–657.

3 Chapman, J. R. *Practical Organic Mass Spectrometry: A Guide for Chemical and Biochemical Analysis*, 2nd Edition; J. Wiley & Sons: New York, 1994.

4 De Hoffmann, E.; Stroobant, V. *Mass Spectrometry: Principles and Applications, Second Edition*; Wiley Interscience: New York, 2001.

5 Fenselau, C.; Cotter, R. J. *Chem. Rev.* 1987, *87*, 501–512.

6 Tomer, K. B. *Mass Spectrom. Rev.* 1989, *8*, 445–482.

7 Tomer, K. B. *Mass Spectrom. Rev.* 1989, *8*, 483–511.
8 Egge, H.; Peter-Katalinić, J. *Mass Spectrom. Rev.* 1987, *6*, 331–393.
9 Yamashita, M.; Fenn, J. B. *J. Phys. Chem.* 1984, *88*, 4451–4459.
10 Yamashita, M.; Fenn, J. B. *J. Phys. Chem.* 1984, *88*, 4671–4675.
11 *Electrospray Ionization Mass Spectrometry: Fundamentals, Instrumentation, and Applications*; Cole, R. B., Ed.; Wiley Interscience: New York, 1997.
12 *Applied Electrospray Mass Spectrometry*; Pramanik, B. N.; Ganguly, A. K.; Gross, M. L., Eds.; Marcel Dekker: New York, 2002; Vol. 32.
13 Zaia, J. *Mass Spectrom. Rev.* 2004, *23*, 161–227.
14 Karas, M.; Bachmann, D.; Hillenkamp, F. *Anal. Chem.* 1985, *57*, 2935–2939.
15 Karas, M.; Bachmann, D.; Bahr, U.; Hillenkamp, F. *Int. J. Mass Spectrom Ion Processes* 1987, *78*, 53.
16 Karas, M.; Hillenkamp, F. *Anal. Chem.* 1988, *60*, 2299–2301.
17 Tanaka, K.; Waiki, H.; Ido, Y.; Akita, S.; Yoshida, Y.; Yoshida, T. *Rapid Commun. Mass Spectrom.* 1988, *2*, 151.
18 Williams, T. L.; Fenselau, C. *Eur. Mass Spectrom.* 1998, *4*, 379–383.
19 Zenobi, R.; Knochenmuss, R. *Mass Spectrom. Rev.* 1998, *17*, 337–366.
20 Daniel, J. M.; Friess, S. D.; Rajagopalan, S.; Wendt, S.; Zenobi, R. *Int. J. Mass Spectrom.* 2002, *216*, 1–27.
21 Brodbelt, J. S. *Int. J. Mass Spectrom.* 2000, *200*, 57–69.
22 Kellersberger, K. A.; Dejsupa, C.; Liang, Y. J.; Pope, R. M.; Dearden, D. V. *Int. J. Mass Spectrom.* 1999, *193*, 181–195.
23 Grigorean, G.; Ramirez, J.; Ahn, S. H.; Lebrilla, C. B. *Anal. Chem.* 2000, *72*, 4275–4281.
24 Grigorean, G.; Lebrilla, C. B. *Anal. Chem.* 2001, *73*, 1684–1691.
25 Grigorean, G.; Cong, X.; Lebrilla, C. B. *Int. J. Mass Spectrom.* 2004, *234*, 71–77.
26 Lebrilla, C. B. *Acc. Chem. Res.* 2001, *34*, 653–661.
27 Mele, A.; Selva, A. *J. Mass Spectrom.* 1995, *30*, 645–647.
28 Mele, A.; Panzeri, W.; Selva, A. *J. Mass Spectrom.* 1997, *32*, 807–812.
29 Madhusudanan, K. P.; Katti, S. B.; Dwivedi, A. K. *J. Mass Spectrom.* 1998, *33*, 1017–1022.
30 Bongiorno, D.; Ceraulo, L.; Mele, A.; Panzeri, W.; Selva, A.; Liveri, V. T. *J. Mass Spectrom.* 2001, *36*, 1189–1194.
31 Shizuma, M.; Adachi, H.; Amemura, A.; Takai, Y.; Takeda, T.; Sawada, M. *Tetrahedron* 2001, *57*, 4567–4578.
32 Memisoglu, E.; Bochot, A.; Sen, M.; Duchene, D.; Hincal, A. A. *Int. J. Pharm.* 2003, *251*, 143–153.
33 Ashton, P. R.; Balzani, V.; Clemente-Leon, M.; Colonna, B.; Credi, A.; Jayaraman, N.; Raymo, F. M.; Stoddart, J. F.; Venturi, M. *Chem. Eur. J.* 2002, *8*, 673–684.
34 Bansal, P. S.; Francis, C. L.; Hart, N. K.; Henderson, S. A.; Oakenfull, D.; Robertson, A. D.; Simpson, G. W. *Australian J. Chem.* 1998, *51*, 915–923.
35 Davey, S. N.; Leigh, D. A.; Smart, J. P.; Tetler, L. W.; Truscello, A. M. *Carbohydrate Res.* 1996, *290*, 117–123.
36 Madhusudanan, K. P. *J. Mass Spectrom.* 2003, *38*, 409–416.
37 Pean, C.; Creminon, C.; Wijkhuisen, A.; Grassi, J.; Guenot, P.; Jehan, P.; Dalbiez, J. P.; Perly, B.; Djedaini-Pilard, F. *J. Chem. Soc., Perkin Trans. 2* 2000, 853–863.
38 Chen, F. T. A.; Shen, G.; Evangelista, R. A. *J. Chromatog. A* 2001, *924*, 523–532.
39 Grard, S.; Elfakir, C.; Dreux, A. *J. Chromatog. A* 2001, *925*, 79–87.
40 Hammes, W.; Bourscheidt, C.; Buchsler, U.; Stodt, G.; Bokens, H. *J. Mass Spectrom.* 2000, *35*, 378–384.
41 Salvador, A.; Herbreteau, B.; Dreux, M. *J. Chromatog. A* 1999, *855*, 645–656.
42 Sforza, S.; Galaverna, G.; Corradini, R.; Dossena, A.; Marchelli, R. *J. Am. Soc. Mass Spectrom.* 2003, *14*, 124–135.
43 So, M. P.; Wan, T. S. M.; Chan, T. W. D. *Rapid Commun. Mass Spectrom.* 2000, *14*, 692–695.

44 Yamagaki, T.; Nakanishi, H. *Rapid Commun. Mass Spectrom.* 1999, *13*, 2199–2203.
45 Janshoff, A.; Steinem, C.; Michalke, A.; Henke, C.; Galla, H.-J. *Sensors Actuators B Chem.* 2000, *70*, 243–253.
46 Wong, C. K. L.; Chan, T. W. D. *Rapid Commun. Mass Spectrom.* 1997, *11*, 513–519.
47 Gallagher, R. T.; Ball, C. P.; Gatehouse, D. R.; Gates, P. J.; Lobell, M.; Derrick, P. J. *Int. J. Mass Spectrom.* 1997, *165*, 523–531.
48 Lehmann, E.; Salih, B.; Gomez-Lopez, M.; Diederich, F.; Zenobi, R. *Analyst* 2000, *125*, 849–854.
49 Raju, N. P.; Mirza, S. P.; Vairamani, M.; Ramulu, A. R.; Pardhasaradhi, M. *Rapid Commun. Mass Spectrom.* 2001, *15*, 1879–1884.
50 Carofiglio, T.; Fornasier, R.; Lucchini, V.; Simonato, L.; Tonellato, U. *J. Org. Chem.* 2000, *65*, 9013–9021.
51 Mele, A.; Malpezzi, L. *J. Am. Soc. Mass Spectrom.* 2000, *11*, 228–236.
52 Yamagaki, T.; Nakanishi, H. *Rapid Commun. Mass Spectrom.* 1998, *12*, 1069–1074.
53 Yamagaki, T.; Ishizuka, Y.; Kawabata, S.; Nakanishi, H. *Rapid Commun. Mass Spectrom.* 1996, *10*, 1887–1890.
54 Yamagaki, T. *Bunseki Kagaku* 1999, *48*, 949–950.
55 Henke, C.; Steinem, C.; Janshoff, A.; Steffan, G.; Luftmann, H.; Sieber, M.; Galla, H.-J. *Anal. Chem.* 1996, *68*, 3158–3165.
56 Vincenti, M. *J. Mass Spectrom.* 1995, *30*, 925–939.
57 Schalley, C. A. *Int. J. Mass Spectrom.* 2000, *194*, 11–39.
58 Schalley, C. A. *Mass Spectrom. Rev.* 2001, *20*, 253–309.
59 *Physical Methods in Supramolecular Chemistry*; Davies, J. E. D.; Ripmeester, J. A., Eds.; Elsevier Science: New York, 1996; Vol. 8.
60 Cunniff, J. B.; Vouros, P. *J. Am. Soc. Mass Spectrom.* 1995, *6*, 437–447.
61 Gabelica, V.; Galic, N.; De Pauw, E. *J. Am. Soc. Mass Spectrom.* 2002, *13*, 946–953.
62 Sawada, M. *Mass Spectrom. Rev.* 1997, *16*, 73–90.
63 Sawada, M.; Shizuma, M.; Takai, Y.; Adachi, H.; Takedab, T.; Uchiyamad, T. *Chem. Commun.* 1998, 1453–1454.
64 Dotsikas, Y.; Loukas, Y. L. *J. Am. Soc. Mass Spectrom.* 2003, *14*, 1123–1129.
65 Ramirez, J.; He, F.; Lebrilla, C. B. *J. Am. Chem. Soc.* 1998, *120*, 7387–7388.
66 Ramirez, J.; Ahn, S. H.; Grigorean, G.; Lebrilla, C. B. *J. Am. Chem. Soc.* 2000, *122*, 6884–6890.
67 Ahn, S.; Ramirez, J.; Grigorean, G.; Lebrilla, C. B. *J. Am. Soc. Mass Spectrom.* 2001, *12*, 278–287.
68 Grigorean, G.; Gronert, S.; Lebrilla, C. B. *Int. J. Mass Spectrom.* 2002, *219*, 79–87.
69 Franchi, P.; Lucarini, M.; Mezzina, E.; Pedulli, G. F. *J. Am. Chem. Soc.* 2004, *126*, 4343–4354.
70 Lu, H. J.; Yu, C. T.; Guo, Y. L. *Acta Chim. Sinica* 2002, *60*, 882–885.
71 Cheng, Y.; Hercules, D. M. *J. Mass Spectrom.* 2001, *36*, 834–836.
72 Mele, A.; Mendichi, R.; Selva, A.; Molnar, P.; Toth, G. *Carbohydrate Res.* 2002, *337*, 1129–1136.
73 Cescutti, P.; Garozzo, D.; Rizzo, R. *Carbohydrate Res.* 1997, *302*, 1–6.
74 Cescutti, P.; Garozzo, D.; Rizzo, R. *Carbohydrate Res.* 1996, *290*, 105–115.
75 Wen, X. H.; Liu, Z. Y.; Zhu, T. Q.; Zhu, M. Q.; Jiang, K. Z.; Huang, Q. Q. *Bioorg. Chem.* 2004, *32*, 223–233.
76 Yu, C. T.; Guo, Y. L.; Zhang, Z. J.; Xiang, B. R.; An, D. K. *Acta Chim. Sinica* 2001, *59*, 615–618.
77 Lamcharfi, E.; Chuilon, S.; Kerbal, A.; Kunesch, G.; Libot, F. *J. Mass Spectrom.* 1996, *31*, 982–986.
78 Guella, G.; Callone, E.; Mancini, I.; Uccello-Barretta, G.; Balzano, F.; Dini, F. *Eur. J. Org. Chem.* 2004, 1308–1317.
79 Guernelli, S.; Lagana, M. F.; Mezzina, E.; Ferroni, F.; Siani, G.; Spinelli, D. *Eur. J. Org. Chem.* 2003, 4765–4776.
80 Al-nouti, Y.; Bartlett, M. G. *J. Am. Soc. Mass Spectrom.* 2002, *13*, 928–935.

81 Sun, W. X.; Cui, M.; Liu, S. Y.; Song, F. R.; Elkin, Y. N. *Rapid Commun. Mass Spectrom.* 1998, *12*, 2016–2022.
82 Bakhtiar, R.; Hop, C. *Rapid Commun. Mass Spectrom.* 1997, *11*, 1478–1491.
83 Srinivasan, K.; Bartlett, M. G. *Rapid Commun. Mass Spectrom.* 2000, *14*, 624–632.
84 Guo, M. Q.; Zhang, S. Q.; Song, F. R.; Wang, D. W.; Liu, Z. Q.; Liu, S. Y. *J. Mass Spectrom.* 2003, *38*, 723–731.
85 Guo, M. Q.; Song, F. R.; Liu, Z. Q.; Liu, S. Y. *J. Mass Spectrom.* 2004, *39*, 594–599.
86 Wagner, B. D.; Stojanovic, N.; Leclair, G.; Jankowski, C. K. *J. Inclusion Phenom. Macrocyclic Chem.* 2003, *45*, 275–283.
87 Caron, L.; Tilloy, S.; Monflier, E.; Wieruszeski, J. M.; Lippens, G.; Landy, D.; Fourmentin, S.; Surpateanu, G. *J. Inclusion Phenom. Macrocyclic Chem.* 2000, *38*, 361–379.
88 Turel, I.; Demsar, A.; Kosmrlj, J. *J. Inclusion Phenom. Macrocyclic Chem.* 1999, *35*, 595–604.
89 Bakhtiar, R.; Kaifer, A. E. *Rapid Commun. Mass Spectrom.* 1998, *12*, 111–114.
90 Naidu, N. B.; Chowdary, K. P. R.; Murthy, K. V. R.; Satyanarayana, V.; Hayman, A. R.; Becket, G. *J. Pharm. Biomed Anal.* 2004, *35*, 75–86.
91 Selva, A.; Redenti, E.; Casetta, B. *Org. Mass Spectrom.* 1993, *28*, 983–986.
92 Kobetic, R.; Jursic, B. S.; Bonnette, S.; Tsai, J. S. C.; Salvatore, S. J. *Tetrahedron Lett.* 2001, *42*, 6077–6082.
93 Selva, A.; Redenti, E.; Ventura, P.; Zanol, M.; Casetta, B. *J. Mass Spectrom.* 1996, *31*, 1364–1370.
94 Selva, A.; Redenti, E.; Ventura, P.; Zanol, M.; Casetta, B. *J. Mass Spectrom.* 1998, *33*, 729–734.
95 Cai, Y.; Tarr, M. A.; Xu, G. X.; Yalcin, T.; Cole, R. B. *J. Am. Soc. Mass Spectrom.* 2003, *14*, 449–459.

10.3
Studies of Cyclodextrin Inclusion Complexes by Electronic (UV-Vis Absorption and Emission) Spectroscopy

Gottfried Grabner

10.3.1
Introduction

Electronic (absorption and emission) spectroscopies are among the most widely applied experimental techniques in supramolecular chemistry [1]. This section provides a condensed overview of the principles and uses of UV-Vis absorption and emission (fluorescence and phosphorescence) spectroscopies in the study of cyclodextrin (CyD) inclusion complexes. The emphasis will be on a presentation of the main effects of complex formation on measured spectra, quantum yields, and kinetics. This latter point will be treated in a separate section as it exemplifies the power of spectroscopic techniques in supramolecular studies. Only nonderivatized CyDs will be discussed. This is not a comprehensive review; cited references, taken from the literature of the literature of the past ten years, are mainly intended to provide illustrative examples.

As in all chemistry, the ultimate goal of supramolecular chemistry is to relate structure to function. Electronic spectroscopy can be valuable on both sides of the game when it comes to the study of CyD inclusion complexes. On one hand, it is of virtually general use in monitoring the formation of complexes by their influence on measured spectra, and is on a par with other techniques in the quantitative study of association equilibria [2–4]; this is widely used in pharmaceutical chemistry in the development of drug-delivery technology [5]. On the other hand, CyDs are valuable as models of enzyme function (briefly presented in Section 1.1 and in Chapter 4) [6–8] and as functional components of various optical sensor architectures [9–13]. Frequently, electronic spectroscopies play a major role in this research.

The value of absorption and emission spectroscopies in these studies relies on three distinctive properties: sensitivity, time resolution, and spatial resolution. Fluorescence spectroscopy is one of the most sensitive techniques available; this constitutes the basis of its application for the sensing of fluorescent analytes, but in a more general perspective provides the almost unique possibility of studying stable inclusion complexes in solution, i.e. in a situation which requires the use of very low guest concentrations [14]. The ability of optical spectroscopies to provide time resolutions down to the femtosecond range opens the way to investigations of the kinetics and dynamics of CyD complexes [15, 16]. Finally, fluorescence microscopy is one of the fundamental tools in the fast-developing area of surface patterning towards supramolecular lithography, which includes the preparation and study of CyD-derivatized surfaces [17].

The other side of the coin is the inability of electronic spectroscopies to provide direct structural information (circular dichroism spectroscopy, which to some extent is able to do so, is treated in Section 10.4). For this reason, a comprehensive investigation of any given CyD guest–host system will generally involve additional, complementary approaches. On the experimental side, these may include techniques providing direct structural information, such as vibrational and nuclear magnetic resonance spectroscopies [18], or thermodynamic quantities, such as calorimetry [19]. A further possibility is the combination of electronic spectroscopy with theoretical modeling of complex structures to guide the interpretation of experimental results [20]. In the following, we will draw some illustrative examples from work applying this last strategy, namely an investigation of naphthalene/CyD complexes [21].

10.3.2
Steady-state Spectroscopies

10.3.2.1
UV-Vis Absorption Spectroscopy

UV-Vis spectroscopy is a technique widely applicable to CyD complex formation except when the guest molecules are weakly absorbing. In principle, the method is well suited for the quantitative study of association equilibria since the measured absorbances are proportional to the respective concentrations by virtue of the Lambert–Beer law. Unfortunately, the effects of CyD inclusion on the absorption spectra of guest molecules are often weak, frequently characterized by peak shifts of the order of a few nm, and changes of the extinction coefficients seldom exceeding ten percent. These effects may reflect the altered polarity of the cavity microenvironment and/or specific guest–host interactions.

Distinct polarity effects on absorption were encountered for guest molecules exhibiting polarity-dependent vibronic fine structure in their spectra, as for instance naphthalene [21], which showed a marked enhancement of fine structure in the presence of 0.05-M α-CyD but not in that of 0.01-M β-CyD (Fig. 10.3.1). In such favorable cases, the stoichiometry of inclusion could be probed using a suitably modified Benesi–Hildebrand treatment based on the absorption of a vibronic subband whose intensity is specifically enhanced by CyD inclusion. This is the case for the 290-nm transition of the ^1L_a band of naphthalene which is threefold more intense in the presence of a high α-CD concentration than in neat aqueous solution

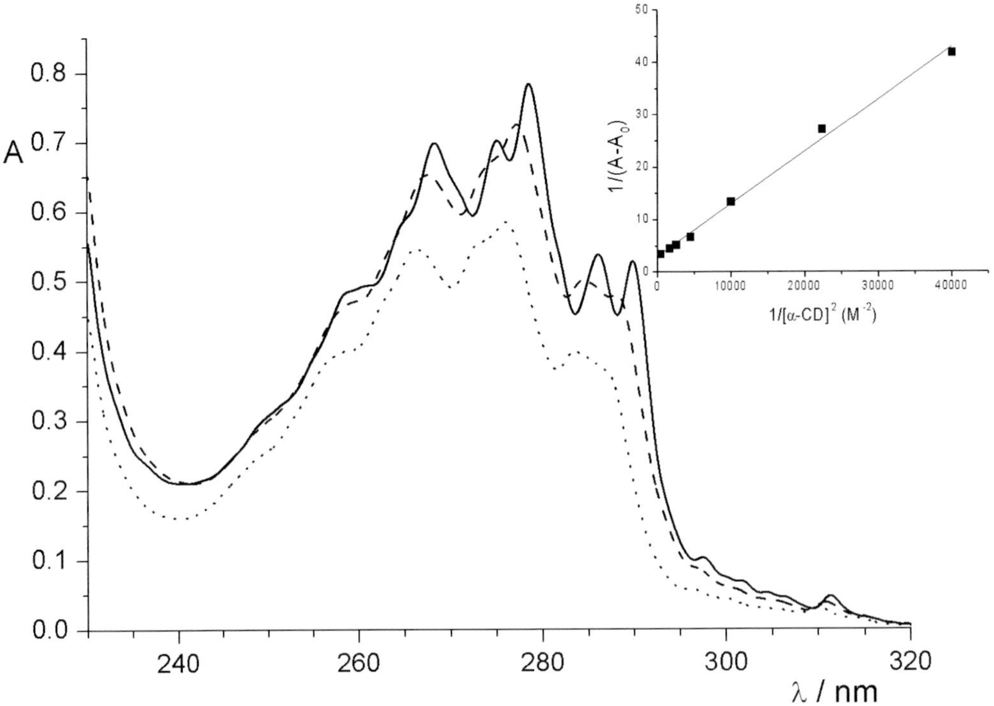

Fig. 10.3.1. Absorption spectra of naphthalene (1.1×10^{-4} M) in water (····) and in the presence of 0.01-M β-CyD (––––) and 0.05-M α-Cyd (———). Insert: Benesi–Hildebrand treatment for 290-nm absorption data according to a model of 1:2 complex formation (adapted from Ref. [21]).

(Fig. 10.3.1). Various models of complex stoichiometries could then be tested against the α-CD concentration-dependent experimental data. The insert in Fig. 10.3.1 shows consistency with a model assuming formation of a 1:2 guest:host complex with formally simultaneous addition of both host molecules (i.e. two complexation steps with $K_2 \gg K_1$). The vibronic structure observed in the naphthalene/α-CyD absorption could thus be linked to the shielding from the aqueous solvent provided by the host cage. In contrast, the absorption spectrum of naphthalene in the β-CyD complex was no more structured than that in water, in agreement with the finding that a 1:1 stoichiometry that does not prevent guest–solvent interaction prevailed in this case. The prevalence of 1:2 α-CyD and 1:1 β-CyD stoichiometries was in agreement with the predictions of theoretical modeling [21].

Specific guest/host interactions, reflecting either the weak electrostatic and van der Waals forces that are responsible for complexation [22] or steric hindrance of guest motion, may influence the transition dipole moments in absorption as well as in emission. In many cases, the absorbance of the guest is found to increase upon complexation (compare Fig. 10.3.1), but the opposite trend may also be encountered, as, for example, in the system 4-hydroxybiphenyl/β-Cyd. In this latter case, steric hindrance of the excited-state planarization of the guest has been put forward as a possible explanation [23].

In spite of the small effects encountered in absorption, UV-Vis spectrometry is nevertheless frequently employed as an easily performed first test of the occurrence of complexation, in particular in nonfluorescing systems; to cite an example at random, UV-Vis absorption gave the first hint of the possibility of an interaction between γ-CyD and carbon nanotubes, which was subsequently confirmed by Raman spectroscopy and differential scanning calorimetry (described briefly in Chapter 8) [24]. In those frequent cases when single-wavelength absorbance changes are too small to allow a reliable quantitative analysis of association equilibria (binding stoichiometries and association constants), state-of-the art global fitting may still achieve discrimination between different binding models [25]. In fact, the power of modern chemometric techniques allows valuable analytical applications of small effects of CyD inclusion on UV-Vis spectra; for instance, a method based on the UV-Vis absorption of β-CyD inclusion complexes of chiral guests has been proposed for the determination of enantiomeric purity of the guest molecules [26]. An example of a system transducing an optical system to an electrochemical one is described in Section 12.4.

10.3.2.2
Fluorescence Emission Spectroscopy

Organic fluorescing guests frequently show a marked increase of fluorescence quantum yield and lifetime with increasing host concentration, reflecting the formation of inclusion complexes [2]. This effect is usually much larger than that observed in absorption, and has therefore been used more frequently for determining complexation stoichiometries and association constants. The sensitivity of the emission measurement is such that a very broad range of guest and host concentrations is usually accessible; this is a virtually unique advantage over other techniques which may be unable to monitor the first stage of a sequence of consecutive complexation steps [27]. It should be kept in mind, however, that the experimental output of an emission measurement, i.e. quantum yield or lifetime of emission, is related to, but not strictly proportional to, the concentration of the emitter. This may lead to problems in data evaluation [28].

Comprehensive photophysical studies show in many cases that the rate constant of internal conversion (k_{ic}) is reduced on complexation, whereas the rate constants of radiative deactivation and intersystem crossing are less affected; formation of higher-order complexes, such as 1:2 or 2:2 guest:host, may have a particularly large impact. To give an example, k_{ic} for 1,4-dimethoxybenzene decreased from 2.5×10^8 s^{-1} in neat aqueous solution to 1.4×10^8 s^{-1} in the β-CyD 1:1 complex, but in the α-CyD 1:2 complex the sum of the fluorescence and intersystem crossing yields was unity within experimental error, implying $k_{ic} < 2 \times 10^7$ s^{-1} [29]. Steric influences on the vibrational interactions, promoting deactivation by internal conversion, have been put forward as the main reason for this effect [29, 30]. The reduction of internal conversion is therefore not a universal effect, but depends on the specific inclusion configuration as well as on the possible co-inclusion of solvent molecules; for instance, nonradiative deactivation was hardly affected by β-CyD complexation for the guests phenol and *p*-cresol, but strongly so for dimethylated and trimethylated phenols [30].

If the guest molecules are prone to form excimers, occurrence of the corresponding emission band in the presence of CyDs can be a source of information about the complex stoichiometry. The naphthalene/β-CyD system gives a good example [21, 31]. Figure 10.3.2 shows the fluorescence spectra of naphthalene in water and with added 0.01-M β-CyD; the band at 410 nm has been assigned to the naphthalene excimer in a 2:2 guest:host complex.

The effect of CyD inclusion on guest fluorescence may be still more pronounced for those guest molecules that undergo excited-state intramolecular charge-transfer (ICT) or proton-transfer (ESPT) processes. The dual fluorescence emission observed in these cases is usually strongly affected by CyD complex formation. An extensively studied fluorescence is that of the ICT molecule 4-*N*,*N*-dimethylaminobenzonitrile (DMABN) [32]. The dual fluorescence of this molecule, characteristic for polar solvents, involving a locally excited (LE) and a charge transfer (CT) band, was found to be strongly affected by CyD inclusion [33]. The CT emission was promoted in the α-CyD complexes, with band maxima at 510 nm for the 1:1 and 440 nm for the 1:2 complexes, respectively, demonstrating the sensitivity to the microenvironmental polarity. Conversely, the 2:2 complex with β-CyD showed no CT emission, but an enhanced LE band (350 nm) and a shoulder at 420 nm as a new feature, assigned to two different guest:host conformations as suggested by model calculations [25, 33].

The sensitivity of the emission of ICT and ESPT systems has been exploited for studies of CyD complex properties using these systems as static or dynamic probes. The DMABN/β-CyD system has been proposed as a temperature sensor [34]. ESPT has the potential to probe details of the inclusion process. It has been

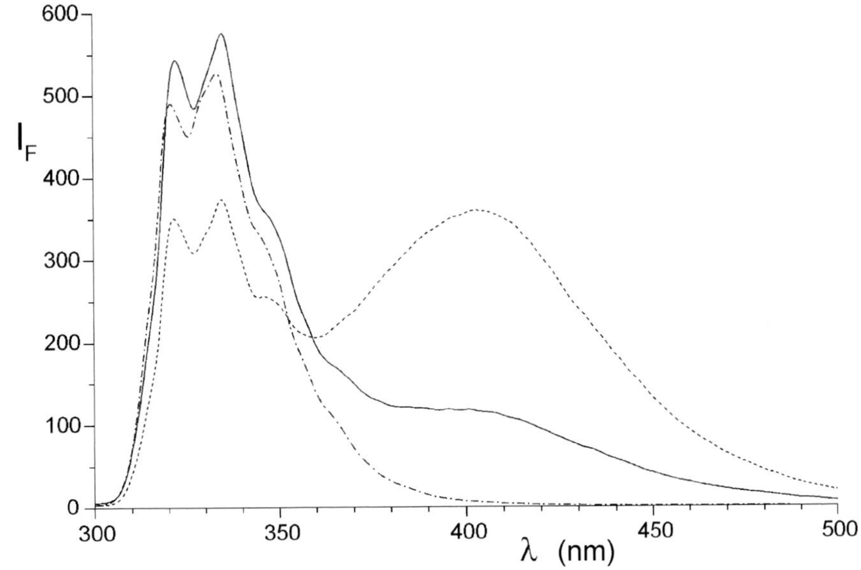

Fig. 10.3.2. Fluorescence spectra of naphthalene (2×10^{-4} M) in aqueous solution at 8 °C (–·–·–·), and in the presence of 0.01 M β-CyD at 25 °C (———) and at 8 °C (- - - -) (from Ref. [21]).

found to be inhibited by the co-inclusion of organic solvent molecules in complexes involving α- or β-CyD and 4-methyl-2,6-dicarbomethoxyphenol, and thereby provides a means to monitor the formation of ternary complexes [35]. Another ESPT probe molecule, 2-(2′-hydroxyphenyl)-4-methyloxazole, has been proposed as a cavity size ruler because of the occurrence of an intramolecular twisting motion in the phototautomer, which was shown to be sensitive to the steric properties of the inclusion complex [36]. Some examples of fluorescence enhancement for rotaxanes involving CyDs are presented in Sect. 12.4.

A further interesting application of fluorescence spectroscopy is its potential enantioselectivity. Chiral discrimination has been demonstrated for CyD inclusion of camphorquinone [37]. The measurement of fluorescence anisotropy has been proposed as a method to determine the enantiomeric composition of samples, and validated using 1,1′-binaphthyl-2,2′-diyl hydrogen phosphate [38].

10.3.2.3
Room-temperature Phosphorescence (RTP)

Phosphorescence is usually not observable in fluid solution even from compounds with high triplet yields, being overruled by nonradiative triplet deactivation and/or quenching by co-solutes such as O_2. Cyclodextrin inclusion is a way to overcome these limitations [39, 40], provided a heavy atom-containing molecule is present which functions to increase the rate constants of both intersystem crossing and radiative triplet deactivation. This can be the guest molecule itself, such as 6-bromo-2-naphthol which exhibited RTP in the 1:2 α-CyD complex even in non-deoxygenated solution owing to protection from O_2 quenching, although nitrogen purging increased the emission intensity 13-fold [41]. From the point of view of analytical application, detection of analytes by means of RTP could be accomplished by co-inclusion of heavy atom-containing compounds such as 3-bromo-1-propanol [42], or by using heavy atom-derivatized CyDs such as 6-deoxy-6-iodo-β-CyD [43]. The use of RTP in actual analytical applications is still at the proof-of-principle stage [40].

10.3.3
Time-resolved Spectroscopies

10.3.3.1
Excited-state Absorption

The power of optical spectroscopic techniques becomes obvious when it comes to studies of the dynamic behavior of CyD complexes. The characteristic time window of a time-resolved measurement is defined by the intrinsic lifetime of the excited state under consideration, which usually belongs to the included guest molecule. Depending on whether the guest exhibits ICT or ESPT phenomena or unreactive excited states, the relevant time windows range from the picosecond/subpicosecond range in the former to the nanosecond (singlet states) and microsecond/millisecond (triplet states and photoproduced radicals) ranges in the latter case.

Besides fluorescence spectroscopy, time-resolved spectroscopy can rely on the measurement of excited (singlet or triplet) state absorption. Similarly to ground-state absorption, the spectral and absorbance properties may be altered by CyD complexation and yield information about the behavior of the complex in the excited state; in addition, the time dependence (formation and decay) of the excited state absorption yields information about the kinetics and dynamics of the system. This is illustrated by the behavior of the lowest triplet state of naphthalene as measured by nanosecond spectroscopy using a Q-switched Nd:YAG laser at 266 nm for excitation [21]. The triplet–triplet absorption spectra were measured in neat solvents (water and ethanol) and in the presence of α- and β-CyD (Fig. 10.3.3). The spectra in ethanol and H_2O had the same absorption maximum, but the transition was considerably weaker and broadened in H_2O. Both CyDs induced a red shift, and α-CyD additionally narrowed the main band considerably. Fig. 10.3.4 shows the effect of α-CD concentration on the time evolution of the triplet-triplet absorption at 416 nm in the microsecond range. Triplet decay was caused by O_2 quenching; a detailed kinetic analysis of the time dependence yielded two main components which could be assigned to the free guest and the 1:2 complex, in full

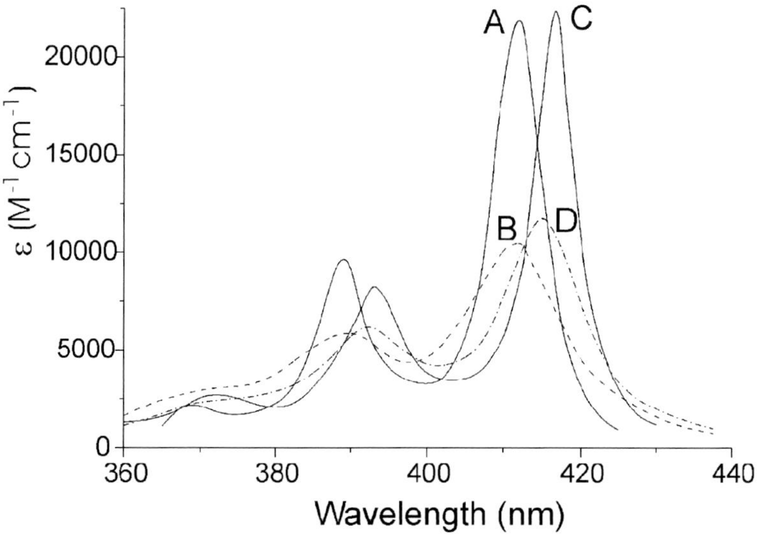

Fig. 10.3.3. Triplet–triplet absorption spectra of naphthalene (10^{-4} M) at 25 °C in ethanol (A), in H_2O (B), in H_2O + 0.05-M α-CD (C), and in H_2O + 0.01-M β-CD (D) (from Ref. [21]).

agreement with the results of ground-state absorption (see above). Again, the inhibition of triplet quenching in the 1:2 complex was explained by shielding of the guest, this time from access by the triplet quencher O_2 [21].

The frequently quite large effect of CyD inclusion on triplet lifetimes can be used as a further spectroscopic approach to ascertain and study the formation of inclusion complexes; this can be advantageous when the variations of other parameters caused by inclusion are small [44]. In addition, transient spectroscopy is valuable in studies of the photochemistry of CyD/drugs systems, which are of great importance in ascertaining the photostability and photodegradation pathways of photosensitive drugs when CyDs are used for drug delivery [45]. The measurement of excited states and photoproducts can also be performed on solid surfaces by means of time-resolved diffuse reflectance spectroscopy [46–48].

10.3.3.2
Complexation Kinetics

The study of the dynamics of host–guest interaction can be subdivided into three aspects: the kinetics of complexation, the flexibility of the host, and the dynamics of the guest. A method based on high-pressure stopped-flow kinetics/UV-Vis absorption spectroscopy has been described for the investigation of the kinetics and thermodynamics of the formation of inclusion complexes between phenylazo dyes and α-CyD [49]. Time-resolved measurements of excited-state evolution could be

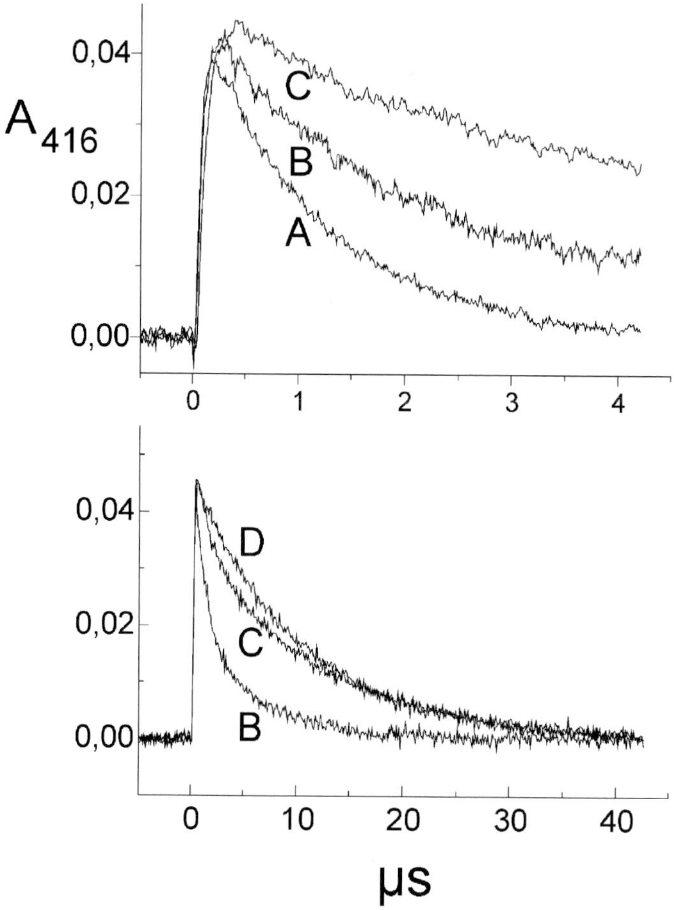

Fig. 10.3.4. Decay of the naphthalene triplet–triplet absorption in solutions containing O_2 (2.5×10^{-4} M) at 25 °C in the presence of different concentrations of α-CD (A) 0, (B) 0.015 M, (C) 0.025 M, (D) 0.05 M (from Ref. [21]).

exploited for studies of the kinetics of both formation and dissociation of CyD inclusion complexes in solution. A quite general approach is based on the quenching of the triplet state of guest molecules by quenchers which reside exclusively in the solvent phase, and therefore have widely different rate constants for interaction with either complexed or free guest molecules. This method was applied to the study of the complexation kinetics of CyDs with partner molecules such as naphthylethanols [50] or flavones and chromones [51] using Mn^{2+}, Cu^{2+}, or nitrite as ionic quenchers in aqueous solution. This method was complemented by specific ones, applicable for guest compounds with solvent polarity-dependent triplet–

triplet absorption [52, 53], or for systems exhibiting excimer formation, such as the pyrene/γ-CyD system, which could be investigated by means of stopped-flow/fluorescence kinetics [54]. These studies yielded some general results: for β-CyD1:1 complexes, the entry rate constants were found to be of the order of 10^9 M^{-1} s^{-1} with variations of about a factor of ten among the studied molecules. The exit rate constants showed more pronounced differences; their values range from 1.8×10^5 s^{-1} for 2-napthylethanol to 2.1×10^7 s^{-1} for chromone. Molecular size and excited-state basicity of the guest have been suggested as determinants of this behavior [51]. The exit of the guest from higher-order complexes was much slower. For the 2:2 complex of pyrene with γ-CyD, rate constants of 73 s^{-1} for dissociation into 1:1 complexes and less than 3 s^{-1} for exit of one guest to yield a 1:2 complex could be determined [54]; using RTP, the 6-bromo-2-naphthol triplet was found to exit from the 1:2 complex with α-CyD with $k = 8.4 \times 10^3$ s^{-1} [55]. The kinetics of dissociation have also been studied using the photochemical production of radicals within CyD cavities; the exit of the radical was generally found to be in the nanosecond or subnanosecond range [56–58].

10.3.3.3
Host Structural Flexibility

One of the factors that influence the kinetics of association/dissociation is the conformational flexibility of the CyD host discussed in detail in Section 1.3. This is of particular importance for higher-order complexes stabilized by hydrogen bonds that must be broken to allow exit of the included molecule [59]. Molecular dynamics simulations indicate that guest molecules such as *p*-chlorophenol and *p*-hydroxybenzoic acid exhibit transient exit-reentry dynamics in α-CyD complexes on the 100-ps timescale [60]. Experimental tests of such predictions are difficult, and few studies devoted to them have been performed. The available experimental evidence, mainly from studies using NMR spectroscopy, and related theoretical modeling work, has been reviewed [61]. Again, the contribution of optical spectroscopies is the potential access to real-time dynamics of configurational changes in fluid solutions; however, direct measurements are obviously difficult owing to the absence of any easily accessible signature from the CyD molecules themselves. For this reason, experimental results on the flexibility of CyD complexes obtained by means of optical spectroscopic work are mainly based on indirect evidence. The technique of Raman optical activity, which is able to directly sense the skeletal mobility of glycosidic bonds, constitutes an exception; using this approach, there was evidence of an inverse correlation of conformational flexibility and tightness of binding in complexes of benzoate and benzoic acid with unmodified and modified β-CyDs in aqueous solution [62]. Raman spectroscopy was also employed to study H/D and D/H exchange rates of hydroxy groups in crystalline α-CyD; it was shown that the exchange rates are much faster in the presence of included guest molecules than in their absence, implying a guest-induced decrease of structural rigidity [63]. The kinetics of guest triplet decay have also been exploited to gain information about host flexibility. It has been noted that inclusion of α-terthiophene in a 1:1 β-CyD complex reduced the intensity of the delayed fluorescence originating from triplet–triplet annihilation, but did not eliminate it; this was tentatively explained by incomplete guest inclusion [64]. However, real-time kinetic measurements of triplet decay in 1:2 α-CyD complexes of naphthalene demonstrated that triplet–triplet annihilation is observed even in this case, although the calculated structure gave no hint of the possibility of escape of the guest from the host cavity [21]. A temperature-dependent study of triplet decay kinetics yielded activation energies that were equal within error limits for both triplet quenching by O_2 and triplet–triplet annihilation; the measured values were only slightly higher for the 1:1 β-Cyd complex (26 kJ mol^{-1}) than for a pure aqueous solution (16 kJ mol^{-1}), but distinctly higher for the 1:2 α-CyD complex (66.5 kJ mol^{-1}). It has been speculated that this additional activated process might be related to a transient opening of the cavity which is necessary to provide the close contact between two naphthalene molecules to allow triplet–triplet quenching or annihilation. This conclusion concerning conformational flexibility was supported by the observation that the fluorescence spectrum of the 1:2 α-CyD complex exhibited a gradual loss of vibrational fine structure with increasing temperature, suggesting an enhanced accessibility to the aqueous solvent [21].

10.3.3.4
Ultrafast Studies of Guest Dynamics

Investigations of the time dependence of spectroscopic properties of CyD inclusion complexes in the pico- and femtosecond time domains mainly probe the dynamics of the included guest, as the host can be assumed to be quasi-rigid during these short time periods. Rotation of guest molecules within the host cavity, as opposed to overall rotation of the complex, can conveniently be studied using picosecond time-resolved fluorescence anisotropy. This has mainly been demonstrated on the example of suitably strongly fluorescing included dyes [65, 66]. On the low-picosecond/femtosecond timescales, the studied phenomena include the influence of guest inclusion of photodissociation/recombination/vibrational cooling in I_2 [67], the dynamics of excited-state double proton transfer in 7-hydroxyquinoline [68], and of bond rotation and twisting connected to ESPT and ICT [69–71]. This work has been reviewed [16].

References

1 Johnston, L. J.; Wagner, B. D. *Comprehensive Supramolecular Chemistry*, Davies, J. E. D.; Ripmeester, J. A., Eds.; Elsevier: Oxford, Vol. 8, p. 537, **1996**.

2 Bortolus, P.; Monti, S. *Adv. Photochem.* **1996**, *21*, 1.

3 Connors, K. A. *Chem. Rev.* **1997**, *97*, 1567.

4 Hamai, S. *Handbook of Photochemistry*

and Photobiology, Nalwa, H. S., Ed.; Am. Scientific Publ., Stevenson Ranch, Vol. 3, p. 59, **2003**.

5 Davis, M. E.; Brewster, M. E. *Nat. Rev. Drug Discovery* **2004**, *3*, 1023.

6 Kirby, A. J. *Angew. Chem. Int. Ed. Engl.* **1996**, *36*, 707.

7 Komiyama, M.; Shigekawa, H. *Comprehensive Supramolecular Chemistry*, Davies, J. E. D.; Ripmeester, J. A., Eds.; Elsevier: Oxford, Vol. 3, p. 401, 1996.

8 Breslow, R.; Dong, D. D. *Chem. Rev.* **1998**, *98*, 1997.

9 Ueno, A. *Supramol. Sci.* **1996**, *3*, 31.

10 Rudzinski, C. M.; Hartmann, W. K.; Nocera, D. G. *Coord. Chem. Rev.* **1998**, *171*, 115.

11 Liu, Y.; Han, B. H.; Zhang, H. Y. *Curr. Org. Chem.* **2004**, *8*, 35.

12 Féry-Forgues, S.; Dondon, R.; Bertorelle, F. *Handbook of Photochemistry and Photobiology*, Nalwa, H. S., Ed.; Am. Scientific Publ., Stevenson Ranch, Vol. 3, p. 121, **2003**.

13 Hayashita, T.; Yamauchi, A.; Tong, A. J.; Lee, J.; Smith, B.; Teramae, N. J. *Inclusion Phenom. Macrocycl. Chem.* **2004**, *50*, 87.

14 Kano, K.; Nishiyabu, R.; Asada, T.; Kuroda, Y. *J. Am. Chem. Soc.* **2002**, *124*, 9937.

15 Bhattacharyya, K. *Acc. Chem. Res.* **2003**, *36*, 95.

16 Douhal, A. *Chem. Rev.* **2004**, *104*, 1955.

17 Mulder, A.; Onclin, S.; Peter, M.; Hoogenboom, J. P.; Beijleveld, H.; ter Maat, J.; García-Parajó, M. F.; Ravoo, B. J.; Huskens, J.; van Hulst, N. F.; Reinhoudt, D. N. *Small* **2005**, *1*, 242.

18 Schneider, H. J.; Hacket, F.; Rüdiger, V.; Ikeda, H. *Chem. Rev.* **1998**, *98*, 1755.

19 Rekharsky, M. V.; Inoue, Y. *Chem. Rev.* **1998**, *98*, 1875.

20 Lipkowitz, K. B. *Chem. Rev.* **1998**, *98*, 1829.

21 Grabner, G.; Rechthaler, K.; Mayer, B.; Köhler, G.; Rotkiewicz, K. *J. Phys. Chem. A* **2000**, *104*, 1365.

22 Liu, L.; Guo, Q. X. *J. Inclusion Phenom. Macrocycl. Chem.* **2002**, *42*, 1.

23 Bortolus, P.; Marconi, G.; Monti, S.; Mayer, B.; Köhler, G.; Grabner, G. *Chem. Eur. J.* **2000**, *6*, 1578.

24 Chambers, G.; Carroll, C.; Farrell, G. F.; Dalton, A. B.; McNamara, M.; In Het Panhuis, M.; Byrne, H. J. *Nano Lett.* **2003**, *3*, 843.

25 Monti, S.; Marconi, G.; Manoli, F.; Bortolus, P.; Mayer, B.; Grabner, G.; Köhler, G.; Boszczyk, W.; Rotkiewicz, K. *Phys. Chem. Chem. Phys.* **2003**, *5*, 1019.

26 Busch, K. W.; Swamidoss, I. M.; Fakayode, S. O.; Busch, M. A. *J. Am. Chem. Soc.* **2003**, *125*, 1690.

27 Dodziuk, H.; Nowinski, K. S.; Kozminski, W.; Dolgonos, G. *Org. Biomol. Chem.* **2003**, *1*, 581.

28 Pistolis, G.; Malliaris, A. *Chem. Phys. Lett.* **1999**, *310*, 501.

29 Grabner, G.; Monti, S.; Marconi, G.; Mayer, B.; Klein, C.; Köhler, G. *J. Phys. Chem.* **1996**, *100*, 20068.

30 Monti, S.; Köhler, G.; Grabner, G. *J. Phys. Chem.* **1993**, *97*, 13011.

31 Sau, S.; Solanki, B.; Orprecio, R.; van Stam, J.; Evans, C. H. *J. Inclusion Phenom. Macrocycl. Chem* **2004**, *48*, 173.

32 Grabowski, Z. R.; Rotkiewicz, K.; Rettig, W. *Chem. Rev.* **2003**, *103*, 3899.

33 Monti, S.; Bortolus, P.; Manoli, F.; Marconi, G.; Grabner, G.; Köhler, G.; Mayer, B.; Boszczyk, W.; Rotkiewicz, K. *Photochem. Photobiol. Sci.* **2003**, *2*, 203.

34 Figueroa, I. D.; El Baraka, M.; Quinones, E.; Rosario, O. *Anal. Chem.* **1998**, *70*, 3974.

35 Mitra, S.; Das, R.; Mukherjee, S. *J. Phys. Chem. B* **1998**, *102*, 3730.

36 García-Ochoa, I.; López, M. A. D.; Viñas, M. H.; Santos, L.; Ataz, E. M.; Amat-Guerri, F.; Douhal, A. *Chem. Eur. J.* **1999**, *5*, 897.

37 Bortolus, P.; Marconi, G.; Monti, S.; Mayer, B. *J. Phys. Chem. A* **2002**, *106*, 1686.

38 Xu, Y. F.; McCarroll, M. E. *J. Phys. Chem. A.* **2004**, *108*, 6929.

39 Muñoz de la Peña, A.; Mahedero, M. C.; Sánchez, A. B. *Analusis* **2000**, *28*, 670.

40 Kuijt, J.; Ariese, F.; Brikman, U. A. T.; Gooijer, C. *Anal. Chim. Acta* **2003**, *488*, 135.

41 Muñoz de la Peña, A.; Rodriguez, M. P.; Escandar, G. M. *Talanta* **2000**, *51*, 949.

42 González, M. A.; Hernández-López, M. *Analyst* **1998**, *123*, 2217.

43 Hamai, S. *Bull. Chem. Soc. Jpn.* **1998**, *71*, 1549.

44 Lang, K.; Kubat, P.; Lhotak, P.; Mosinger, J.; Wagnerova, D. M. *Photochem. Photobiol.* **2001**, *74*, 558.

45 Monti, S.; Sortino, S. *Chem. Soc. Rev.* **2002**, *31*, 287.

46 Barra, M.; Agha, K. A. *J. Photochem. Photobiol. A: Chem.* **1997**, *109*, 293.

47 Mir, M.; Wilkinson, F.; Worrall, D. R.; Bourdelande, J. L.; Marquet, J. *J. Photochem. Photobiol. A: Chem.* **1997**, *111*, 241.

48 Vieira Ferreira, L. F.; Machado, I. F.; da Silva, J. P.; Oliveira, A. S. *Photochem. Photobiol. Sci.* **2004**, *3*, 174.

49 Abou-Hamdan, A.; Bugnon, P.; Saudan, C.; Lye, P. G.; Merbach, A. E. *J. Am. Chem. Soc.* **2000**, *122*, 592.

50 Barros, T. C.; Stefaniak, K.; Holzwarth, J. F.; Bohne, C. *J. Phys. Chem. A* **1998**, *102*, 5639.

51 Christoff, M.; Okano, L. T.; Bohne, C. *J. Photochem. Photobiol. A: Chem.* **2000**, *134*, 169.

52 Liao, Y.; Bohne, C. *J. Phys. Chem.* **1996**, *100*, 734.

53 Okano, L. T.; Barros, T. C.; Chou, D. T. H.; Bennet, A. J.; Bohne, C. *J. Phys. Chem. B* **2001**, *105*, 2122.

54 Dyck, A. S. M.; Kisiel, U.; Bohne, C. *J. Phys. Chem. B* **2003**, *107*, 11652.

55 Brewster, R. E.; Teresa, B. F.; Schuh, M. D. *J. Phys. Chem. A* **2003**, *107*, 10521.

56 Murphy, R. S.; Bohne, C. *Photochem. Photobiol.* **2000**, *71*, 35.

57 Takamori, D.; Aoki, T.; Yashiro, H.; Murai, H. *J. Phys. Chem. A* **2001**, *105*, 6001.

58 Hapiot, P.; Lagrost, C.; Aeiyach, S.; Jouini, M.; Lacroix, J. C. *J. Phys. Chem. B* **2002**, *106*, 3622.

59 Saenger, W.; Jacob, J.; Gessler, K.; Steiner, T.; Hoffmann, D.; Sanbe, H.; Koizumi, K.; Smith, S. M.; Takaha, T. *Chem. Rev.* **1998**, *98*, 1787.

60 Van Helden, P.; van Eijck, B. P.; Janssen, L. H. M. *J. Biomol. Struct. Dyn.* **1992**, *9*, 1269.

61 Dodziuk, H. *J. Mol. Struct.* **2002**, *614*, 33.

62 Bell, A. F.; Hecht, L.; Barron, L. D. *Chem. Eur. J.* **1997**, *3*, 1292.

63 Amado, A. M.; Ribeiro-Claro, P. J. *J. Chem. Soc., Faraday Trans.* **1997**, *93*, 2387.

64 Evans, C. H.; De Feyter, S.; Viaene, L.; van Stam, J.; De Schryver, F. C. *J. Phys. Chem.* **1996**, *100*, 2129.

65 Balabai, N.; Linton, B.; Napper, A.; Priyadarshy, S.; Sukharevsky, A. P.; Waldeck, D. H. *J. Phys. Chem. A.* **1998**, *102*, 9617.

66 Singh, M. K.; Pal, H.; Koti, A. S. R.; Sapre, A. V. *J. Phys. Chem. A.* **2004**, *108*, 1465.

67 Chachisvilis, M.; García-Ochoa, I.; Douhal, A.; Zewail, A. H. *Chem. Phys. Lett.* **1998**, *293*, 153.

68 García-Ochoa, I.; Díez-Lopez, M. A.; Viñas, M. H.; Santos, L.; Ataz, E. M.; Sánchez, F.; Douhal, A. *Chem. Phys. Lett.* **1998**, *296*, 335.

69 Organero, J. A.; Tormo, L.; Douhal, A. *Chem. Phys. Lett.* **2002**, *363*, 409.

70 Fayed, T. A.; Organero, J. A.; García-Ochoa, I.; Tormo, L.; Douhal, A. *Chem. Phys. Lett.* **2002**, *364*, 108.

71 Organero, J. A.; Douhal, A. *Chem. Phys. Lett.* **2003**, *373*, 426.

10.4
Circular Dichroism of Cyclodextrin Complexes

Daniel Krois and Udo H. Brinker

10.4.1
General Introduction to Circular Dichroism

Circular dichroism (CD) [1] is a spectroscopic technique that detects the differential absorption of circularly polarized light passing through a chiral substance. It is defined as the difference of the molar extinction coefficients for the left and right circularly polarized light $\Delta\varepsilon = \varepsilon_l - \varepsilon_r$. Consequently, nothing results other than an absorption spectrum with positive and/or negative signs. If it observed in the region of infrared radiation (excitation of molecular vibration) then it is called vibrational circular dichroism (VCD) [2, 3]. The term circular dichroism (CD) itself is used for the phenomenon in the region of UV-Vis radiation (i.e. electronic absorption) [4]. In the following we will only discuss the latter type, which is of widespread use.

CD spectra are recorded as $\Delta A = A_l - A_r$ [absorption $A = \log(I_0/I)$]. In analogy to the Lambert–Beer law the circular dichroism then is: $\Delta\varepsilon = \Delta A/(c.d)$ [where c stands for the molar concentration (M) and d for the path length (nm)]. But very often it is also given as the ellipticity ψ. When linearly polarized light – a racemate of left and right-handed circularly polarized light – passes through solutions of chiral non-racemic compounds, it becomes elliptically polarized within their absorption bands. Accordingly, it is mathematically characterized by the ellipticity, the ratio of the half-axes b/a. There is a simple relationship between the molar ellipticity $[\psi]$ and $\Delta\varepsilon$: $[\psi] = 3300\Delta\varepsilon$. As most chromophores absorbing in the easily accessible UV-Vis region (200–1000 nm) are achiral themselves,*) circular dichroism is only observed due to perturbations by the chiral σ-bond skeleton of the molecule.

10.4.2
Types of CD Observed for Cyclodextrin Complexes

α-, β-, and γ-Cyclodextrins (CyDs) themselves do not absorb in the above-mentioned wavelength range and, therefore, show no CD, in spite of being chiral compounds. But, of course, CyDs can induce intermolecularly (or more precise innermolecularly) CD if they form complexes with achiral compounds bearing observable chromophoric groups [5–7]. This phenomenon is known as induced circular dichroism (ICD) (Scheme 10.4.1, type A). The appearance of ICD is a measure of the forma-

*) In this case the magnetic $[M]$ and electric dipole transition moments $[\mu]$ are orthogonal and thus the rotatory strength R, which is proportional to the area of the CD band of the specific transition, vanishes according to: $R = \vec{\mathbf{M}} \circ \vec{\mu} = \bar{M}.\bar{m}.\cos\theta$.

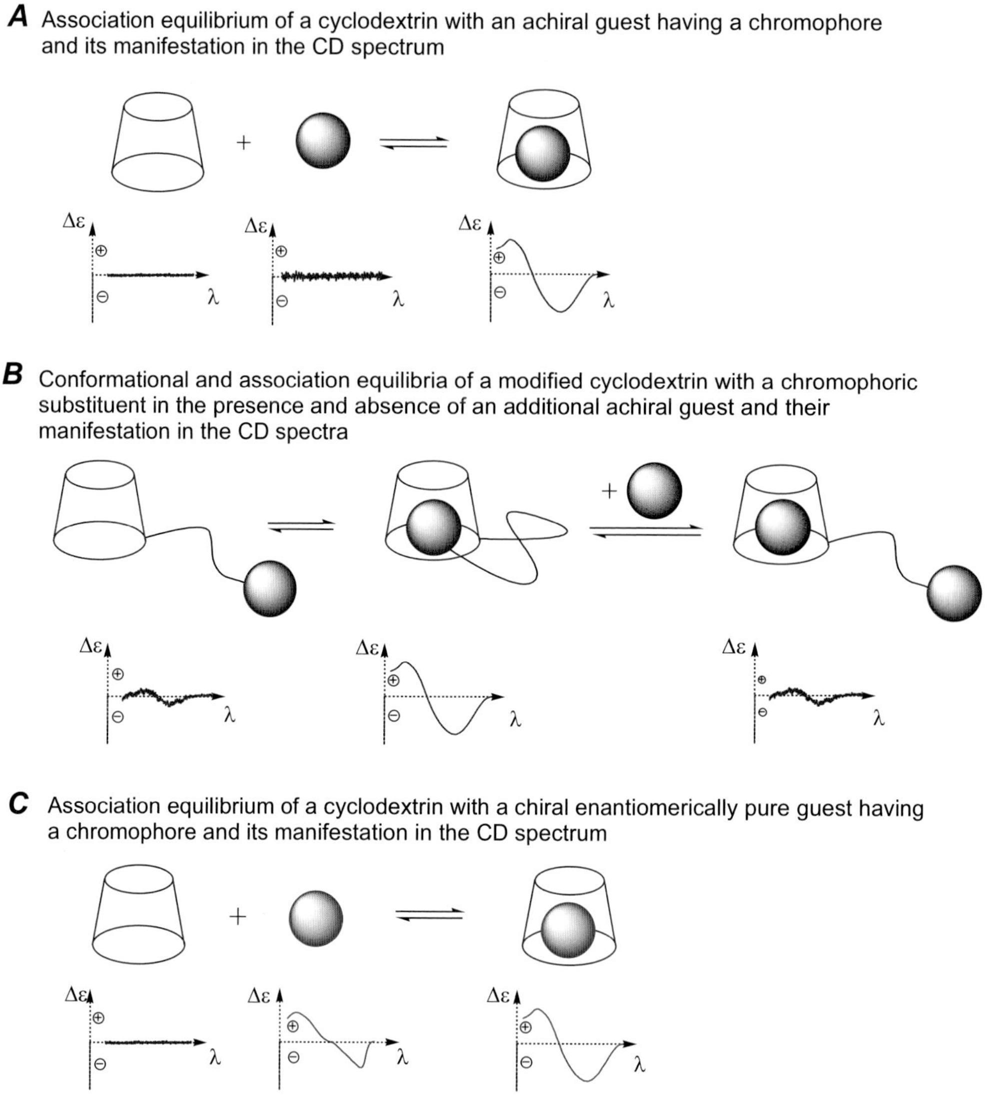

A Association equilibrium of a cyclodextrin with an achiral guest having a chromophore and its manifestation in the CD spectrum
B Conformational and association equilibria of a modified cyclodextrin with a chromophoric substituent in the presence and absence of an additional achiral guest and their manifestation in the CD spectra
C Association equilibrium of a cyclodextrin with a chiral enantiomerically pure guest having a chromophore and its manifestation in the CD spectrum
+
Δε
⊕
⊖
λ

Scheme 10.4.1. Blue: chiral compound (cyclodextrin) with no chromophore in the wavelength range of 200–1000 nm; red: achiral compound or "achiral" substituent containing a chromophore; magenta: chiral enantiomerically pure guest with a chromophore; black: achiral nonchromophoric compound.

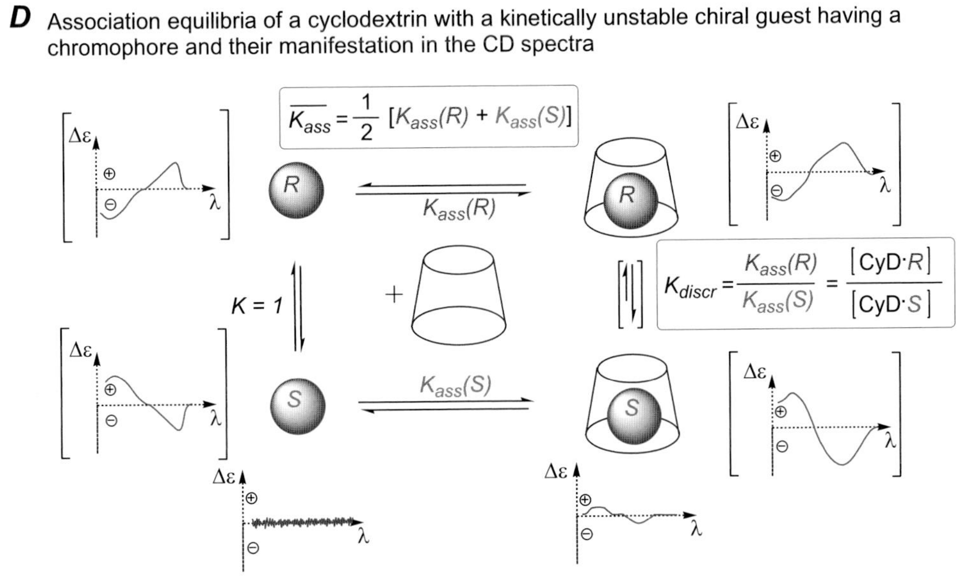

Scheme 10.4.1. *(continued)*

tion of the complex and allows for determination of the equilibrium constant for association (also called the association constant or stability constant) or dissociation.

A special type of CD is shown by modified CyDs (Scheme 10.4.1, type B), that comprise chromophoric group(s) covalently bound to the CyD. It can be regarded as true CD or as ICD, because very often it only can be observed if the chromophoric group is intramolecularly complexed by the CyD, i.e., it is self-included. Accordingly, this CD is strongly conformation dependent. But, of course, a concentration dependence can also occur when intermolecular and intramolecular complexation are competing. The intermolecular complexation behavior of such modified CD-active CyDs with nonchromophoric guests can then be observed by the decrease of CD with growing concentration of guest.

With chiral guests, diastereoisomeric complexes are formed. Therefore, association constants can only be determined for enantiomerically pure guests [8, 9]. If they also carry a chromophore (Scheme 10.4.1, type C), the guest alone exhibits a CD spectrum, which is altered to some degree by complexation with a CyD. The change in CD can be used to follow complexation in a similar way to the UV-Vis spectrum (see Section 10.3) or any other spectroscopic technique (for instance, NMR spectroscopy, discussed in Chapter 9). This is in contrast to type A, where the absence of complexation corresponds to a zero line in the CD spectrum.

In case of a kinetically labile racemate as guest, a chiral discrimination can be observed during the course of complexation (Scheme 10.4.1, type D), reflecting the different association constants for the two enantiomeric species that inter-convert quite rapidly. The CD and ICD observed may exhibit a complex shape, because they represent a difference spectrum of the CD of two diastereoisomeric complexes, which are present in different amounts according to K_{discr} [8].

If both modified CyD and chiral guest exhibit a CD, this method is less suited for detecting complex formation (not shown in Scheme 10.4.1).

10.4.3
Application of CD Spectroscopy for the Characterization of Cyclodextrin Complexes

CD spectroscopy can provide two kinds of information about CyD complexes.

- As the ICD or CD changes in the course of complexation of the guest, the concentration dependence of the CD allows us to estimate association constants. This can be achieved for types A–D, but most straightforwardly for type A.
- The sign, shape, and intensity of the ICD of complexes of type A are, in principle, structure dependent and can thus be used to identify possible arrangements of guest and host. On the other hand, the CD of CyDs of type B can be used to characterize the structure of these modified host molecules.

Binding or association constants in principle can be obtained by any spectroscopic method, which permits the analysis of solutions, i.e., ^1H NMR (Chapter 9), UV-Vis and fluorescence spectra (Section 10.3), and CD, and also by microcalorimetric methods (Chapter 8), potentiometric titrations (Section 10.5), etc. [5, 10]. Information about the supramolecular structure in solution, on the other hand, only can be gained by NMR (NOESY or ROESY) or CD spectroscopy – preferably in combination with computational methods [11–17] (see, however, Chapter 11, which details the limitations of CyD modeling).

In the following we present a short overview of the achievements obtained by CD as documented in publications in the period from 1995 to 2005, with emphasis on both types of information (*vide supra*). Special focus will be put on the comparison of results achieved by other methods with the aim of providing some general guidelines, where CD indeed provides not only complementary but also more precise and reliable information than other types of measurements.

10.4.3.1
Determination of Complex Stability and Stoichiometry by CD

The concentration dependence of ICD has widely been used to investigate intermolecular interactions in solution [5, 7, 18]. Several techniques for the calculation of association constants (K_{ass}) have been applied, for instance Benesi–Hildebrand [19], Scatchard plots, [20, 21] and, with the growing capacities of computer calculations, nonlinear curve-fitting procedures [22, 23]. As already mentioned, CyD complexes of type A seem especially well suited for determination of complex stabilities by ICD. Of course, the reliability of the calculated K_{ass} largely depends on the signal/noise ratio of the data obtained.*) This in turn is determined by the extinction coefficients ε of the UV-Vis absorption band, the range of K_{ass} for the complex considered, and the anisotropy factor $g = (\Delta\varepsilon/\varepsilon)$ of the CD and can only in part be compensated by using different path lengths of cuvettes (practically accessible range from 0.01 to 10 cm) and longer measurement times (more accumulated scans). A further restriction is imposed by the condition that the UV-Vis absorption must not exceed $A < 1.2$ for a reliable CD measurement. Generally, higher complex stabilities can better be determined by spectrophotometric methods (UV-Vis, ICD, fluorescence) whereas NMR spectroscopy is more suitable for less stable complexes. There are only a few examples in the literature where K_{ass} determined by ICD is compared with values obtained by other methods [23–26]. In these cases the agreement is quite good. Landy et al. in their very accurate work [27] come to the conclusion that in the case of substituted phenols, NMR and UV titrations give much more accurate results than ICD, due to the quite unfavorable g factor for these aromatic systems. In general, ICD and CD have more favorable g factors, if magnetically allowed but weak transitions are involved. This holds true for the $n \rightarrow \pi^*$ transition of carbonyl compounds, and also for the $n \rightarrow \pi^*$ transition of the azo chromophore in diazirines and other cyclic azo-compounds. Indeed, in the latter cases exceptionally good results were obtained by ICD [17, 21, 23, 28, 29].

The shape of normalized ICD spectra (type A) should not change with varying CyD concentration and thus show isodichroic points, if only 1:1 complexes (or 2:2) complexes are formed in the concentration range investigated. Thus ICD spectra give additional information about the number of species involved [17, 21, 23, 28–33]. On the other hand, the occurrence of different types of 1:1 complexes cannot be excluded, in contrast to the statement in Ref. [34]. As the proportion of different complexes must always remain constant (determined by K_{XY} in Scheme 10.4.2) a weighted ICD spectrum is observed that increases according to the apparent K_{ass} and thus isodichroic points must occur.

If 1:1 and 2:1 complexes are formed, a fact frequently noticed for α-CyD complexes, a change in phenotype (i.e. the shape of the spectrum) with increasing concentration of CyD should be expected. But in certain instances no such changes occur. This is the case for the α-CyD complexes of 2-aziadamantane, as the ICD for the 1:1 complex is too weak to be observed at all, when compared with the very strong ICD for the 2:1 complex [21]. In the complexation of 2,6-diaziadamantane neither phenotype changes, as both diazirine chromophores experience similar effects and, therefore, an additive model can explain the experimental data [28].

*) CD measurements are especially sensitive to any type of aggregation and thus to slight turbidities of solutions. Of course, determination of association equilibria makes no sense unless all components are completely dissolved. However, in the case of CD not only would the estimated concentrations be wrong, but turbidities would provoke scattering, which could lead to dominating artifacts in the recorded spectrum. Thus, it is extremely important to measure only entirely clear solutions; eventually filtration through membrane filters can help to assess the absence of artifacts due to scattering.

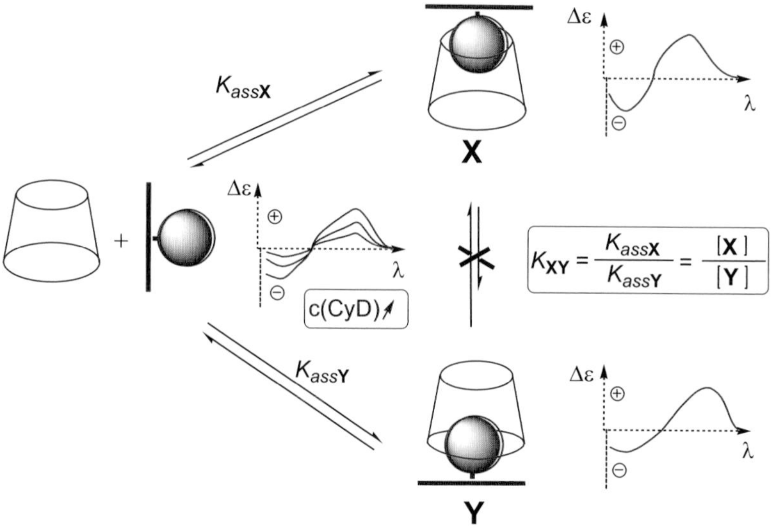

Scheme 10.4.2. Blue: cyclodextrin (chiral) bearing no chromophore in the wavelength range 200–1000 nm; red: achiral compound containing a chromophore.

For CyDs of type B, that is with an attached chromophore, pronounced CD spectra occur due to self-inclusion (see Scheme 10.4.1). These spectra can help to elucidate the conformation of these modified CyDs [33, 35–40]. Complexation of a non-chromophoric guest generally results in a decrease of ICD, although sometimes an enhancement is observed instead. The latter effect could be interpreted in terms of a lid-type complexation, where the guest provokes even tighter contact of the appended chromophore with the CyD [35]. But besides self-inclusion, a stronger CD for type B CyDs could also arise from dimerization, whereby the appended chromophore is included as guest by a second CyD molecule. These two possibilities can only be distinguished by means of a concentration dependence – a solvent dependence, on the other hand, allows no decision [38]. If a guest carrying a chromophore, that is not overlapping with the chromophore of a type B CyD, forms a complex with this modified CyD, the effects of type A and B can be observed, independently [36].

For complexes of type C the measurement of racemates makes no sense, but only of pure enantiomers [8, 9]. In any case, the CD observed is very complex, as stated in Refs. [41, 42]. If the racemization reaction is quite rapid, as for bilirubins [8, 43] or binaphthyl [44], type D applies, which can give rise to interesting effects. In Ref. [44] the very fast association reaction accompanied by the appearance of an ICD (of type A) is followed by the slower diastereoisomerization reaction, yielding a partially resolved racemate complexed by the CyD and thus a strongly modified overall CD spectrum.

Even if ICD is not applied for calculations of K_{ass}, it is often used to assess the occurrence of true inclusion complexes, because other kinds of guest association have never shown measurable ICD effects – as far as stated in the literature. Furthermore, it is used to investigate the stoichiometry (Job plot) [23, 30, 31, 34] and qualitatively the kind of complex [21, 28–34, 45]. If the guest molecule contains more than one chromophore, interpretation of ICD spectra may become quite difficult. In the 1:1 complex of β-CyD with symmetrical 2,6-diaziadamantane the identical chromophores of the guest experience totally different environments. Thus the ICD spectrum is composed of two overlapping ICD spectra of opposite sign. The resulting unusual overall phenotype can only be interpreted properly with additional information about the corresponding monochromophoric guest [28]. In our research the information obtained by ICD spectra was indeed of high value in giving insight into the supramolecular structure of diazirines, which can be further correlated with their reactivity [21, 23, 28, 46].

10.4.3.2 Determination of Complex Geometry by CD

Besides NOESY and ROESY NMR experiments, the ICD method is the only one that can provide information concerning the three-dimensional structure of the complexes in solution. Several rules have been proposed, that have been theoretically deduced especially for aromatic chromophores under certain approximations [11–13]. The calculations are based on the Kirkwood–Tinoco approximation, which accounts for the dipole–dipole interactions of the electronic transition observed with all dipoles in the chiral molecule, often approximated by the polarizabilities of the individual bonds. But, of course, slight distortions of the geometry of the guest's chromophore or the CyD's structure may lead to quite strong deviations. As already known for rules of optical activity in general, these may apply with rather high incidence, but may fail in some important cases. Especially in the case of CyD complexes, the general orientation (axial or equatorial) of the guest is very often obvious without the application of any rule. Therefore, the application of theoretical models is only justified if it does indeed yield important, highly reliable and detailed structural information. Marconi and Mayer point out quite clearly in their review [15], that more precise calculation methods can provide more accurate results, which can be better matched with experimental spectra (see, however, the discussion in Chapter 11). They propose a calculation scheme in three stages: (1) calculation of geometries of the guest by molecular mechanics methods, (2) determination of possible complex conformations by dynamic Monte Carlo simulations (DMC, MCS), whereby a solvent continuum model accounts for the important influence of the solvent, and (3) the calculation of ICD for the major conformer by semi-empirical methods using the Kirkwood–Tinoco approximation. This method has been successfully applied in several cases [14, 41, 47–50]. Another calculation method was proposed by Adeagbo et al. [16]. Taking into account the improvement of *ab initio* methods, Marconi et al. [51] applied a B3LYP/6-31G* method for the calculation of ground state geometries and time-dependent density functional theory (TDDFT) for the calculation of the excited states of the electronic transition and of the rotational strength. The calculated CD spectra were then compared with carefully measured experimental spectra.

It is obvious that information on a particular complex can only be gained if the ICD spectrum for this complex has been determined or extrapolated using the appropriate K_{ass} [23, 25, 28, 33, 34, 49]. Thus, an ICD spectrum measured at only one concentration is of no significance [52], especially as not only the sign but also the strength of the ICD is determined by the geometry. A faint ICD spectrum very often originates from contributions from more than one conformer and in such a case an interpretation might be quite misleading.

The CD or ICD observed for type B cyclodextrins is often used to gather information about the relative orientation of the appended chromophore and CyD moiety in the uncomplexed and complexed state. Application of the rules mentioned above [11–13] can give only very limited information, in our opinion. In these cases the calculation of possible conformations should be quite straightforward and would significantly reduce possible arrangements that must be considered. This would facilitate the drawing of correct conclusions from the experimental CD.

10.4.4 Conclusions

In Section 10.4.1, the ICD (CD) due to the formation of complexes and the possibility of determining complex stabilities was discussed. As already mentioned, other methods can often give more accurate results depending in the first place on the chromophores that are observed. In fact, most K_{ass} values have been obtained by calorimetry, NMR, UV-Vis, fluorescence, or other methods. In part the complex stabilities determined by different research groups with different methods do not compare well [5, 27]. Such a discrepancy is, at least partly due to the differences in experimental conditions. Thus, the use of two or even three diverse methods applied to the same complex would be preferable. Accordingly, ICD as an additional method can be recommended for complexes of type A, even if applied only semi-quantitatively. For detailed information on the structure in solution (see Section 10.4.2), ICD can only compete with NMR methods (applying 2D ROESY or NOESY discussed in Chapter 9), if combined with theoretical calculations for the case in question. Owing to advances in computational calculations in recent years, these are quite easily performed nowadays, at least at a not too high level of theory. Such procedures should definitely give results more accurate than simple geometric rules. (See, however, the critical presentation of CyDs modeling in Chapter 11).

References

1 F. Ciardelli, P. Salvadori, *Fundamental Aspects and Recent Developments in Optical Rotatory Dispersion and Circular Dichroism.* Heyden, London, **1973**.

2 P. L. Polavarapu, *Vibrational Spectra:*

Principles and Applications with Emphasis on Optical Activity, Elsevier, New York, **1998**.

3 (a) P. K. Bose, P. L. Polavarapu, *Carbohydrate Res.*, **2000**, *323*, 63–72; (b) H. Pham-Tuan, C. Larsson, F. Hoffmann, A. Bergman, M. Fröba, H. Hühnerfuss, *Chirality*, **2005**, *17*, 266–280; (c) P. Lesot, M. Sarfati, J. Courtieu, *Chem. Eur. J.*, **2003**, *9*, 1724–1745; (d) V. Setnicka, M. Urbanova, V. Kral, K. Volka, *Spectrochim. Acta, Part A*, **2002**, *58*, 2983–2989.

4 (a) N. Berova, K. Nakanishi, R. W. Woody, *Circular Dichroism. Principles and Applications*, 2nd edition, Wiley-VCH, New York, **2000**; (b) D. A. Lightner, J. E. Gurst, *Organic Conformational Analysis and Stereochemistry from Circular Dichroism Spectroscopy*, Wiley-VCH, New York, **2000**.

5 (a) K. A. Connors, *Chem. Rev.*, **1997**, *97*, 1325–1357; (b) R. R. Burnette, K. A. Connors, *J. Pharm. Sci.*, **2000**, *89*, 1389–1394; (c) K. A. Connors, *J. Pharm. Sci.*, **1995**, *84*, 843–848.

6 M. V. Rekharsky, Y. Inoue, *Chem. Rev.*, **1998**, *98*, 1875–1917.

7 Y. A. Zhdanov, Y. E. Alekseev, E. V. Kompantseva, E. N. Vergeychik, *Russ. Chem. Rev. (Engl. Transl.)*, **1992**, *61*, 563–575.

8 D. Krois, H. Lehner, *J. Chem. Soc., Perkin Trans. 2*, **1995**, 489–494.

9 J. Gawroński, J. Grajewski, *Org. Lett.*, **2003**, *5*, 3301–3303.

10 K. N. Houk, A. G. Leach, S. P. Kim, X. Zhang, *Angew. Chem. Int. Ed.*, **2003**, *42*, 4872–4897.

11 K. Harata, H. Uedaira, *Bull. Chem. Soc. Jpn.*, **1975**, *48*, 375–378.

12 M. Kajtar, C. Horvath-Toro, E. Kuthi, J. Szejtli, *Acta Chim. Acad. Sci. Hung.*, **1982**, *110*, 327–355.

13 M. Kodaka, *J. Phys. Chem.*, **1998**, *102*, 8101–8103.

14 G. Marconi, S. Monti, B. Mayer, G. Köhler, *J. Phys. Chem.*, **1995**, *99*, 3943–3950.

15 G. Marconi, B. Mayer, *Pure Appl. Chem.*, **1997**, *69*, 779–783.

16 W. A. Adeagbo, V. Buß, P. Entel, *J. Incl. Phenom.*, **2002**, *44*, 203–205.

17 H. Bakirci, X. Zhang, W. M. Nau, *J. Org. Chem.*, **2005**, *70*, 39–46.

18 D. Krois, *Tetrahedron*, **1993**, *49*, 8855–8864.

19 H. A. Benesi, J. H. Hildebrand, *J. Am. Chem. Soc.*, **1949**, *71*, 2703–2707.

20 (a) C. R. Cantor, P. R. Schimmel, *Behavior of Biological Macromolecules. Part III*, W. H. Freeman, San Francisco, **1980**; (b) G. Scatchard, *Ann. N. Y. Acad. Sci.*, **1949**, *51*, 660–672.

21 D. Krois, U. H. Brinker, *J. Am. Chem. Soc.*, **1998**, *120*, 11627–11632.

22 (a) J. O. Machuca-Herrera, *J. Chem. Ed.*, **1997**, *74*, 448–449; (b) G. González-Gaitano, G. Tardajos, *J. Chem. Ed.*, **2004**, *81*, 270–274; (c) T. J. Zielinski, R. D. Allendoerfer, *J. Chem. Ed.*, **1997**, *74*, 1001–1007.

23 J.-L. Mieusset, D. Krois, M. Pacar, L. Brecker, G. Giester, U. H. Brinker, *Org. Lett.*, **2004**, *6*, 1967–1970.

24 J. M. Schuette, T. T. Ndou, I. M. Warner, *J. Phys. Chem.*, **1992**, *96*, 5309–5314.

25 J. Dey, E. L. Roberts, I. M. Warner, *J. Phys. Chem. A*, **1998**, *102*, 301–305.

26 L. Yang, N. Takisawa, T. Kaikawa, K. Shirahama, *Langmuir*, **1996**, *12*, 1154–1158.

27 D. Landy, S. Fourmentin, M. Salome, G. Surpateanu, *J. Incl. Phenom.*, **2000**, *38*, 187–198.

28 M. M. Bobek, D. Krois, U. H. Brinker, *Org. Lett.*, **2000**, *2*, 1999–2002.

29 X. Zhang, W. M. Nau, *Angew. Chem. Int. Ed.*, **2000**, *39*, 544–547.

30 A. Nakamura, Y. Inoue, *J. Am. Chem. Soc.*, **2003**, *125*, 966–972.

31 S. Hamai, *J. Phys. Chem.*, **1995**, *99*, 12109–12114.

32 Y. Sueishi, T. Miyakawa, *J. Phys. Org. Chem.*, **1999**, *12*, 541–546.

33 J. W. Park, H. E. Song, S. Y. Lee, *J. Phys. Chem. B*, **2002**, *106*, 7186–7192.

34 Y. Sueishi, H. Hishikawa, *Int. J. Chem. Kinet.*, **2002**, *34*, 481–487.

35 X.-M. Gao, Y.-L. Zhang, L.-H. Tong, Y.-H. Ye, X.-Y. Ma, W.-S. Liu, Y. Inoue, *J. Incl. Phenom.*, **2001**, *39*, 77–80.

36 Y. Liu, Y. Chen, L. Li, H.-Y. Zhang, S.-X. Liu, X.-D. Guan, *J. Org. Chem.*, **2001**, *66*, 8518–8527.
37 Y. Liu, Y.-W. Yang, L. Li, Y. Chen, *Org. Biomol. Chem.*, **2004**, *2*, 1542–1548.
38 Y. Liu, Y.-L. Zhao, H.-Y. Zhang, Z. Fan, G.-D. Wen, F. Ding, *J. Phys. Chem. B*, **2004**, *108*, 8836–8843.
39 M. Narita, J. Itoh, T. Kikuchi, F. Hamada, *J. Incl. Phenom.*, **2002**, *42*, 107–114.
40 A. Ueno, F. Moriwaki, T. Osa, F. Hamada, K. Murai, *J. Am. Chem. Soc.*, **1988**, *110*, 4323–4328.
41 R. S. Murphy, T. C. Barros, B. Mayer, G. Marconi, C. Bohne, *Langmuir*, **2000**, *16*, 8780–8788.
42 M. Blanco, J. Coello, H. Iturriaga, S. Maspoch, C. Pérez-Maseda, *Anal. Chim. Acta*, **2000**, *407*, 233–245.
43 K. Kano, K. Imaeda, K. Ota, R. Doi, *Bull. Chem. Soc. Jpn.*, **2003**, *76*, 1035–1041.
44 S. Y. Yang, M. M. Green, G. Schultz, S. K. Jha, A. H. E. Müller, *J. Am. Chem. Soc.*, **1997**, *119*, 12404–12405.
45 T. Masuda, J. Hayashi, S. Tamagaki, *J. Chem. Soc., Perkin Trans. 2*, **2000**, 161–167.
46 D. Krois, L. Brecker, A. Werner, U. H. Brinker, *Adv. Synth. Catal.*, **2004**, *346*, 1367–1374; M. G. Rosenberg, U. H. Brinker, *J. Org. Chem.*, **2003**, *68*, 4819–4832; W. Knoll, M. M. Bobek, G. Giester, U. H. Brinker, *Tetrahedron Lett.*, **2001**, *42*, 9161–9165; D. Krois, M. M. Bobek, A. Werner, H. Kählig, U. H. Brinker, *Org. Lett.*, **2000**, *2*, 315–318.
47 B. Mayer, G. Marconi, C. Klein, G. Köhler, P. Wolschann, *J. Incl. Phenom.*, **1997**, *29*, 79–93.
48 R. S. Murphy, T. C. Barros, J. Barnes, B. Mayer, G. Marconi, C. Bohne, *J. Phys. Chem. A*, **1999**, *103*, 137–146.
49 B. Mayer, X. Zhang, W. M. Nau, G. Marconi, *J. Am. Chem. Soc.*, **2001**, *123*, 5240–5248.
50 S. Monti, S. Encinas, A. Lahoz, G. Marconi, S. Sortino, J. Perez-Prieto, M. A. Miranda, *Helv. Chim. Acta*, **2001**, *84*, 2452–2466.
51 (a) G. Marconi, S. Monti, F. Manoli, A. Degli Esposti, A. Guerrini, *Helv. Chim. Acta*, **2004**, *87*, 2368–2377; (b) G. Marconi, S. Monti, F. Manoli, A. Degli Esposti, B. Mayer, *Chem. Phys. Lett.*, **2004**, *383*, 566–571.
52 M. Kawamura, M. Higashi, *Helv. Chim. Acta*, **2003**, *86*, 2342–2348.

10.5 Electrochemistry of Cyclodextrins

Renata Bilewicz and Kazimierz Chmurski

10.5.1 Introduction

Cyclodextrins (CyDs) with their largely hydrophobic cavities of variable size and numerous ways of chemical modification are the subject of intensive electrochemical research including both their behavior in homogeneous solutions and in thin films attached to the electrode surfaces [1–8]. Electroanalytical methods measuring the current response to the potential applied, linear scan, staircase, and pulse voltammetries, and potential-step techniques such as chronoamperometry and chronocoulometry, allow us to trigger or to monitor changes in the redox states of the electroactive sites in CyD-based systems. The potential waveforms for these techniques are collected in Fig. 10.5.1. Linear scan (LSV) or staircase (SCV) cyclic voltammetries are the most frequently employed electrochemical techniques for studies of electroactive CyD systems. Thanks to the well-developed theory and easily accessible experimental data they provide useful information on the nature of the reduced and oxidized forms of the studied compounds, and on the mechanistic aspects of the electrode processes. They also allow us to monitor even very subtle changes of the molecular environment of the redox centers by following their redox potentials.

If the concentrations of the redox compounds in the solution or at the electrode surface are low, and better sensitivities are needed than those optimal for LSV and SCV, differential pulse (DPV), normal pulse (NPV) and Osteryoung square-wave voltammetries (OSWV) are more suitable [9a, 9b]. They allow better elimination of the capacitive/background currents and, therefore, the measurement of smaller faradic signals becomes easier. This is achieved either by sampling the current at the end of each pulse (OSWV, NPV) (Fig. 10.5.1E and F) or twice at the end and before pulse application (DPV) (Fig. 10.5.1D).

Normal and reverse pulse voltammetries are often employed for the studies of stepwise or chemically complicated processes where the direct product of electrode reaction is unstable and undergoes transformations into other species – either electroinactive or active in a different potential range. Reverse pulse voltammetry allows a direct examination of the product generated between pulse applications; thus it is simply a short electrolysis (seconds) followed by recording the product of electrooxidation or reduction on the time-scale of milliseconds.

The relationship between electrode potential and current is determined by the electrochemical reaction taking place at the working electrode. Measurements are done usually in a three-electrode arrangement with a reference electrode to control the potential of the working electrode (typically no current is in practice allowed to flow through the reference electrode and its potential is constant) and a counter (auxiliary) electrode where a counterbalancing but not rate-determining electrode process takes place. In cyclic voltammetry for a reversible electrode reaction, the cathodic, E_{pc} and anodic, E_{pa} peak potentials depend on the formal potential, $E^{0'}$ [10]:

$$E_{pc} = E_c^{0'} - 1.1\frac{RT}{2nF}\ln\frac{D_{Ox}^{1/2}}{D_{red}^{1/2}} \tag{10.5.1}$$

$$E_{pa} = E_c^{0'} + 1.1\frac{RT}{2nF}\ln\frac{D_{Ox}^{1/2}}{D_{red}^{1/2}} \tag{10.5.2}$$

where R is the gas constant, J mol^{-1} K^{-1}, T is temperature, K, n is the number of electrons transferred in the electrode reaction, F is the Faraday constant 96 484.6 C mol^{-1}, and D_{ox}, D_{red} are diffusion coefficients for the oxidized and reduced forms, m^2 s^{-1}, respectively. Thus the mid-peak potential can be calculated:

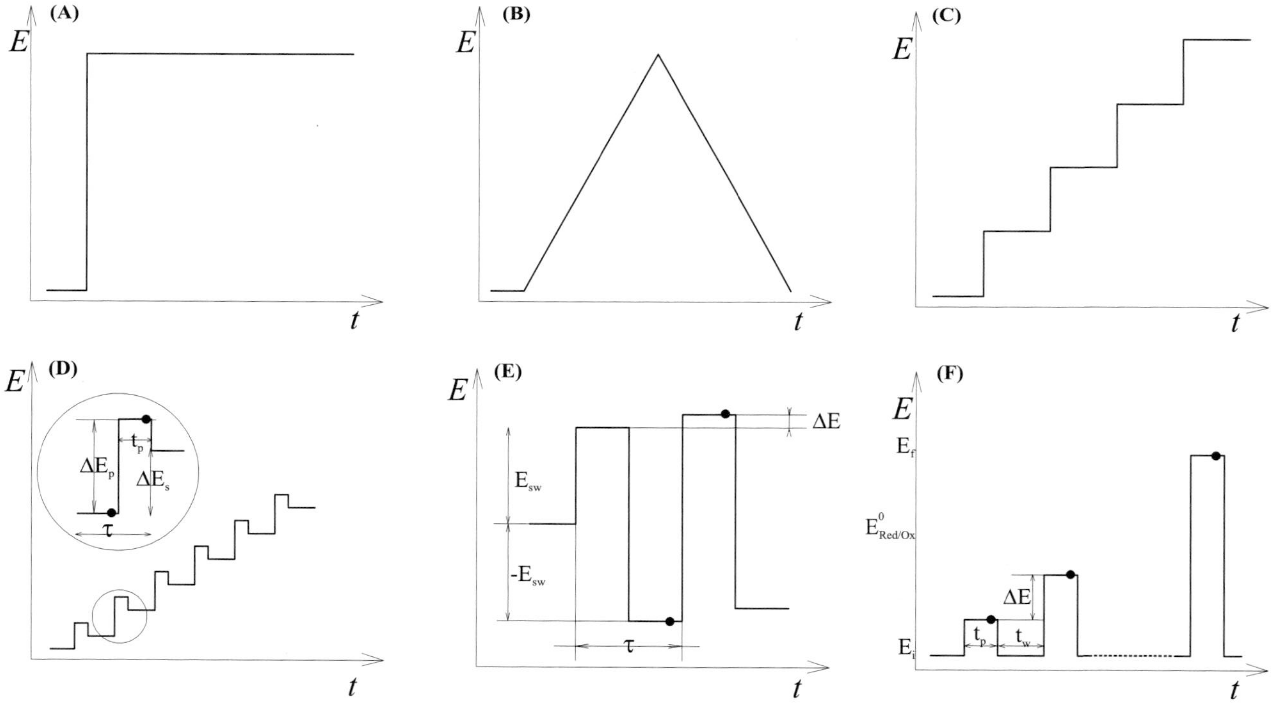

Fig. 10.5.1. (A), Chronoamperometry; (B), Linear scan cyclic voltammetry; (C), Staircase voltammetry; (D), Differential pulse voltammetry; (E), Osteryoung square-wave voltammetry; (F), Normal pulse voltammetry. Bold points indicate sampling of current during or before the pulse.

$$E_c^{0'} = \frac{E_{pa} + E_{pc}}{2} \tag{10.5.3}$$

This equation is often used to determine the formal potential of a given redox system with the help of cyclic voltammetry. However, the assumption that mid-peak potential is equal to formal potential holds only for a reversible electrode reaction. The diagnostic criteria and characteristics of cyclic voltammetric responses for solution systems undergoing reversible, quasi-reversible, or irreversible heterogeneous electron-transfer process are discussed, for example in Ref. [9c]. An electrochemically reversible process implies that the anodic to cathodic peak current ratio, I_{pa}/I_{pc} is equal to 1 and $E_{pc} - E_{pa}$ is $2.218RT/nF$, which at 298 K is equal to $57/n$ mV and is independent of the scan rate. For a diffusion-controlled reduction process, I_p should be proportional to the square root of the scan rate v, according to the Randles–Sevčik equation [10]:

$$I_p = -0.446nFA[C]_{bulk}\sqrt{\frac{nFvD}{RT}} \tag{10.5.4}$$

where, A is electrode area and $[C]_{bulk}$ is bulk concentration of the depolarizer.

In the case of redox reagents adsorbed on the electrode surface or confined to the electrode by means of a thin film, the expression for the peak current for a reversible process is [11]:

$$I_p = \frac{n^2F^2}{4RT}vA\Gamma_0 \tag{10.5.5}$$

where Γ_0 is the surface concentration of the species undergoing the electrode process.

The width of the peak at half-height is equal to:

$$3.53\frac{RT}{nF} = \frac{90.6}{n}\ [\text{mV}] \tag{10.5.6}$$

for the ideal case of a Langmuir-type adsorption without any electrostatic interactions between redox species in the layer. Larger values point to repulsive and smaller ones to attractive interactions in the layer. Changes of the shape together with an increasing difference of E_{pa} and E_{pac} with increasing scan rate indicate departure from electrochemical reversibility.

When microelectrodes are used instead of the normal-sized electrode in cyclic voltammetry experiments, voltammograms with the steady-state wave shape are obtained (Fig. 10.5.2). The half-wave potential, $E_{1/2}$ is equal to the mid-peak potential of the cyclic voltammetric curve recorded using a large electrode.

Voltammetric experiments covering a certain time domain are often employed to detect the rate constants of homogeneous chemical reactions accompanying the electron-transfer steps. For processes that are not fully chemically reversible

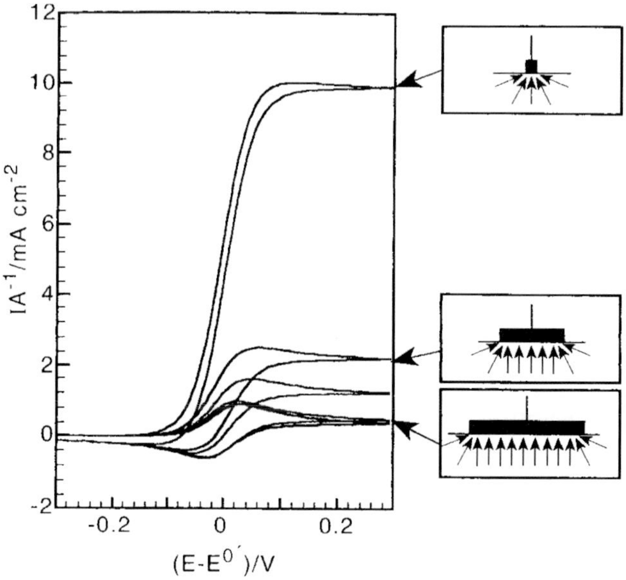

Fig. 10.5.2. The effect of electrode geometry on the shape of reversible cyclic voltammmograms (adapted from Ref. [9c]).

(decomposition of the oxidized or reduced form), deviation of the peak ratio, I_{pa}/I_{pc} from 1 and dependence of this ratio on the scan rate (time scale) is observed. Various combinations of homogeneous reversible and irreversible chemical processes coupled to electron transfer at the electrode surface have been discussed by Marken et al. [9c].

In the case of a reversible electroreduction process, in which a complex is formed between an electroactive guest G and a host ligand H according to a general equation:

$$\mathrm{G} + p\mathrm{H} \rightleftharpoons \mathrm{GH}_p \tag{10.5.7}$$

the stability constant K and stoichiometric coefficient p can be determined from Eq. (10.5.8):

$$E_{1/2}^{\mathrm{G(H)}} = E_{1/2}^{\mathrm{G}} - \frac{RT}{nF} \ln K - \frac{RT}{nF} \ln a_{\mathrm{H}}^{p} \tag{10.5.8}$$

where a_{H}^{p} is the activity of the host ligand.

From the plot of $E_{1/2}^{\mathrm{G(H)}}$ vs. logarithm of the concentration of the host H (assuming no strong deviation from its activity) the value of p can be obtained from the slope and K from the intercept. This equation assumes that only the oxidized form of the guest is complexed and the ligand concentration greatly exceeds that of the guest.

When the guest upon reduction to a lower oxidation state still forms a complex with the host, the $E_{1/2}$ dependence on ligand concentration gives only the difference in p values and the ratio of K values of the two complexes of the guest in the two oxidation states.

When an electroinactive guest is added to a solution of an electroactive host, a titration curve is obtained of $E_{1/2}$ vs. concentration of G [12]:

$$E_{1/2}[\mathrm{G(H)}] = E_{1/2}[\mathrm{G}] + \frac{RT}{2nF} \ln \frac{(1 + K_{m+}[\mathrm{G}])}{(1 + K_{n+}[\mathrm{G}])} \tag{10.5.9}$$

where K_{m+} and K_{n+} are the binding constants of the guest G with the host H in its $m+$ and $n+$ oxidation states. Recording the half-wave or mid-point potential value before and after the addition of large excess of guest to an electroactive host solution has often been used to evaluate the ratio of binding constants in the oxidized and reduced form of the host [13].

Of special interest is the evaluation of the binding affinity of a guest to a ligand immobilized on the electrode surface. Cyclic voltammograms recorded for increasing concentration of guest in the solution using a CyD-modified electrode as host allows us to calculate the binding constant to the surface-immobilized cyclodextrin. The association constant of the guest with surface-confined CyD, K_1 is usually determined by fitting the experimental data to the following equation:

$$\frac{[\mathrm{G}]}{I_{pa}} = \frac{1}{K_1 I_{max}} + \frac{[\mathrm{G}]}{I_{max}} \tag{10.5.10}$$

where I_{pa} is the anodic peak current for a given concentration of guest [G] in the linear part of the I_{pa} vs. [G] plot, I_{max} is the maximum peak current when the current attains constancy. K_1 is calculated from the x-intercept of the linear part of the plots for different scan rates [14–16]. The association constant for, for example, ferrocene (Fc) and the β-CyD mono(6-deoxy-6-lipoylamide)-β-cyclodextrin is $3.3 \times 10^4\ \mathrm{M}^{-1}$ [16], similar to the value $3.9 \times 10^4\ \mathrm{M}^{-1}$ obtained by Rojas et al. [5] for Fc and per(6-deoxy-6-thio)-β-CD anchored at gold electrode surfaces.

The formation of host–guest complexes with CyDs is the subject of extensive electrochemical studies allowing us to understand the thermodynamic, kinetic, and molecular-recognition properties of these complexes. Based on this knowledge, CyDs are employed in electrochemical sensing devices for the determination of selected analytes. Electrochemically controlled binding and removal of organic compounds, and CyD-based drug delivery systems (discussed in Chapters 14 and 15) are actively studied and the well-defined and X-ray-rigid structure of CyDs (see, however, Section 1.3 on the nonrigidity of CyDs) makes them especially interesting as potential biomimetic molecular and ion channels [17–20].

Challenges for supramolecular chemists and electrochemists include electrochemically switchable molecules with mechanically interlocked components, such as rotaxanes and catenanes (presented in Chapter 12). Since they show electrically triggered internal movement they are expected to lead in a bottom-up approach to the development of microelectromechanical systems and novel information storage devices (briefly discussed in Chapter 16) [21, 22].

10.5.2
Electrochemistry in Homogeneous Solutions

The formation of inclusion complexes and intertwined molecules, like the catenanes and rotaxanes discussed in Chapter 12, is the point of interest in most electrochemical investigations of CyDs in solution [23]. The internal hydrophobicity of the cavity with the hydrophilic bases of CyD allows us to increase the solubility of aromatic electroactive guests as shown for example in the use of azobenzene [24], ferrocene [25–28], or viologen [29], so that their electrochemistry can be carried out in an aqueous medium.

Binding constants in solution are usually determined from potentiometric titrations or plots of cyclic voltammetry peak potentials vs. CyD concentration after assessing the guest/host ratio in the complex [3, 4]. Potentiometric measurements are more frequently used, and ion-selective electrodes are employed, for the direct measurement of the guest activity in the solution. Measurements of pH allow the evaluation of concentrations of several reaction components.

Hydrophobic interactions are one of the major forces in selective binding of organic analytes. However, inclusion of small inorganic ions has been also demonstrated and explained by electrostatic forces including the charged electrode surface. The CyDs act as second-sphere ligands for the cyclopentadienyl moieties of ferrocenes, altering their solution electrochemistry behavior. Decreased peak currents in cyclic voltammetry and shifts in the oxidation and reduction potentials are observed on complexation, owing to the smaller diffusion coefficients of the complexes than that of the guest alone, and to the relatively high stability of these complexes [26, 27, 30].

Common inorganic anions are strongly hydrated and bind only very weakly to α- and β-CyDs ($K_{ass} = 1$–3 and 0.2–$5.5\ M^{-1}$, respectively) [21]. Only the less-hydrated ClO_4^- ions were reported to bind more strongly ($K_{ass} = 16.9$–66 and 9.0–$16.7\ M^{-1}$, respectively) [31]. The cavities of native α-CyDs were found too small to bind adamantyl groups in homogeneous aqueous solutions while smaller guests, e.g. 1-hexanol, are complexed much more strongly ($K_{ass} = 8 \times 10^2\ M^{-1}$) [32].

The stabilizing effect of permethylation in the binding of some guests to α-CyDs was explained by the enlargement of the hydrophobic cavity [33].

10.5.3
Electrochemistry of Cyclodextrin Thin Films

Numerous electrochemical papers concern complexation abilities, electrochemical properties, and analytical applications of thin films of CyDs attached to the electrode surfaces, since only when the systems have been studied can new routes be opened to solid-state devices and to their miniaturization [6, 16, 31, 34–46].

Monolayer or sequential monolayer modifications of electrodes are performed using either the Langmuir–Blodgett (L–B) technique, in which the monolayer is formed at an air–water interface and transferred onto solid electrode substrates, or the self-assembly method directly from the solution.

Langmuir monolayers at the air–water interface and L–B monolayers obtained on transfer of these films onto solid substrates require that the molecules are amphiphilic and organize into ordered layers, so several modifications of the CyDs have been proposed [17]. Biomimetic intramolecular channels gated by recognition of chemical species by the CyDs were first described by Umezawa et al. [18] who investigated monolayers of CyDs at the air–water interface. By contacting *in situ* an electrode with the monolayer, the authors could show binding of selected solution species, since it was easily detected using cyclic voltammetry. Blocking of the CyD channels with an appropriate guest led to a decrease in the reduction current of an electrochemical probe in the solution. This observation opened a new route for the electrochemical detection of selected nonelectroactive analytes.

Various amphiphilic α- and β-CyDs functionalized with alkyl chains at the primary hydroxy positions or modified on the secondary face form stable and well-organized monomolecular layers at the air–water interface [47, 48]. They can then be used for the preparation of L–B films on solid substrates. Electroactive CyDs persubstituted with electroactive groups, e.g. terathiafulvalene (TTF) moieties [49], were transferred as L–B films from the water surface with a transfer ratio of 1 to the solid substrate (Fig. 10.5.3). Charge-transfer interactions in the

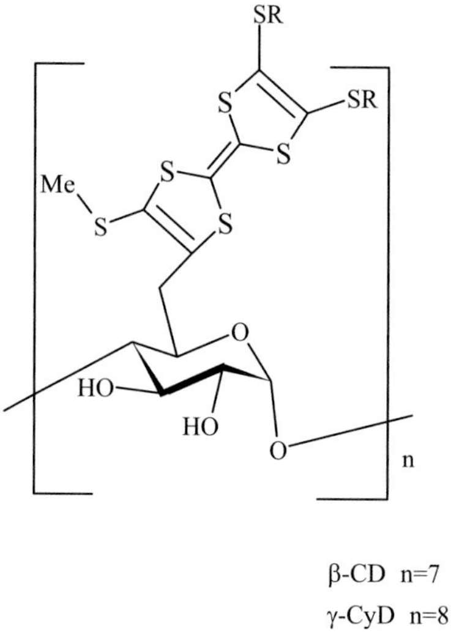

Fig. 10.5.3. CyD derivatives with covalently linked TTF (adapted from Ref. [49]).

L–B films were then studied and new sensing devices were proposed [50–56]. More information on CyD-based devices is presented in Chapter 16.

By controlling the electrode potential, a delicate balance of various interactions can be achieved, resulting in the self-organization of molecules on the surface [57–59].

The formation and successful *in situ* imaging of "nanotube" structures of β-CyD at controlled electrode potentials was described by Ohira et al. [59]. The self-organization of β-CyD into a "nanotube" structure similar to that of CyD-polyrotaxane was found to be induced by potential-controlled adsorption on Au(111) surfaces in sodium perchlorate solution. *In situ* STM revealed that the cavities of β-CyD faced sideways not upward in the tubes (Fig. 10.5.4). This ordered structure can form only under conditions where the appropriate potential is applied to the surface. β-CyD molecules were in a disordered state on bare Au(111) surfaces without potential control. AFM and STM techniques, as applied to CyDs and their complexes, are discussed in Section 10.6.

Self-assembled monolayers (SAMs) provide a simple way to functionalize coinage metal electrode surfaces by organic molecules containing anchor groups such as thiols, disulfides, or silanes. The monolayers are usually well packed, ordered, and stable and can be easily tailored for certain applications by changing the functional terminal group in contact with the solution. Alkanethiol monolayers can

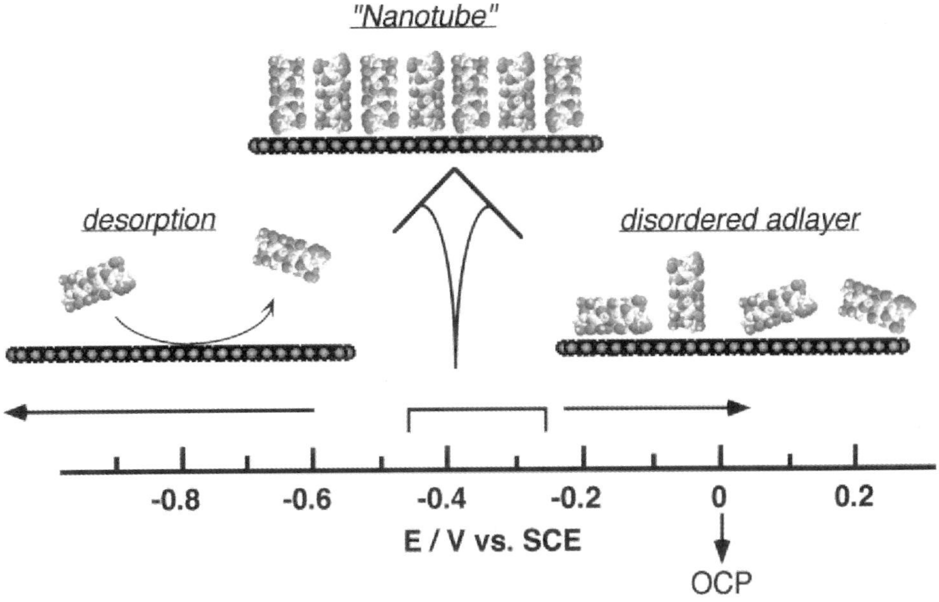

Fig. 10.5.4. Schematic representation of the adsorption behavior of β-CyD on Au(111) surfaces with changing electrode potential [59]. Reprinted with permission of Elsevier.

provide a membrane matrix useful for immobilizing structures such as cylodextrins [16, 45–47].

A host–guest equilibrium involving the redox guest 11-(ferrocenylcarbonyloxy)undecanethiol (FcSH) in a self-assembled monolayer on gold electrode surfaces and a CyD host in solution was investigated by Ju and Leech [41]. The effect of inclusion on the electrochemical characteristics of the ferrocene/ferricenium redox process can be examined using cyclic voltammetry and chronoamperometry. α- and γ-CyD has a negligible effect on both the thermodynamics and kinetics of the redox reaction. The electrochemical and ion-pairing behavior of FcSH-SAMs in solutions containing β-CyD suggests that an inclusion complex is formed and that the included form of the ferrocenyl group is electroinactive in the potential window investigated. Electrodes can be prepared by the electrostatic immobilization of an adlayer of positively charged per(6-amino-6-deoxy)-β-CyD on an SAM of thioctic acid. Studies of this assembly proved that the water-soluble guest ferrocenecarboxylate is effectively bound by the host with a binding constant of ca. 2×10^4 M^{-1} [42].

Single-component monolayers of CyDs are conveniently deposited on electrode surfaces using the self-assembly method, provided that the CyD derivative has one or more anchoring units, such as the thiol group, responsible for the attachment of the molecule to the metal electrode surface [16, 38, 45, 46, 60, 61].

The monolayers are not perfectly ordered; intermolecular voids are present and, therefore, no efficient inhibition of electron transfer across such layers was seen, as reported by Kaifer et al. [36]. There have been several approaches to the provision of anchoring groups of thiol, sulfide, and disulfide functionalities with which CyDs would be deposited on the gold surface as a better ordered monolayer. In the case of β-CyD, full substitution of the seven primary hydroxyl groups by thiol- or sulfide-containing groups results in submonolayers with relatively large defect sites [36]. The presence of the long dialkylsulfide moiety is likely to be essential for CyDs to form a well-covered monolayer, if persubstituted CyDs are used as membrane components [62, 63]. Contrary to this, β-CyD having only one thiol or disulfide group can perfectly cover a gold electrode surface [37, 39, 60–62]. This difference may be attributable to the mobility of the CyD rings in the resulting monolayers. CyD derivatives possessing seven thiol groups with no flexible spacer alkyl chain could not achieve the free movement with respect to the CyD cavities needed to convert a pre-formed submonolayer into a defect-free monolayer. Multiple anchoring of one molecule may strongly restrict the lateral diffusion process requisite for a defect-free monolayer [36, 63]. The presence of a single relatively long alkyl chain should make monosubstituted CyD derivatives more flexible. With the assistance of this flexibility, CyD cavities would be able to align at the gold surface with monolayer order.

Attempts to remove defects in the monolayers included backfilling the layer with alkanethiols or substitution of CyDs with long alkyl chains [16, 36, 37, 45–47, 61, 64–66].

Different problems were reported, e.g. failure to exclude alkanethiol from the cavities. An alternative successful sequential method has been proposed [16, 36,

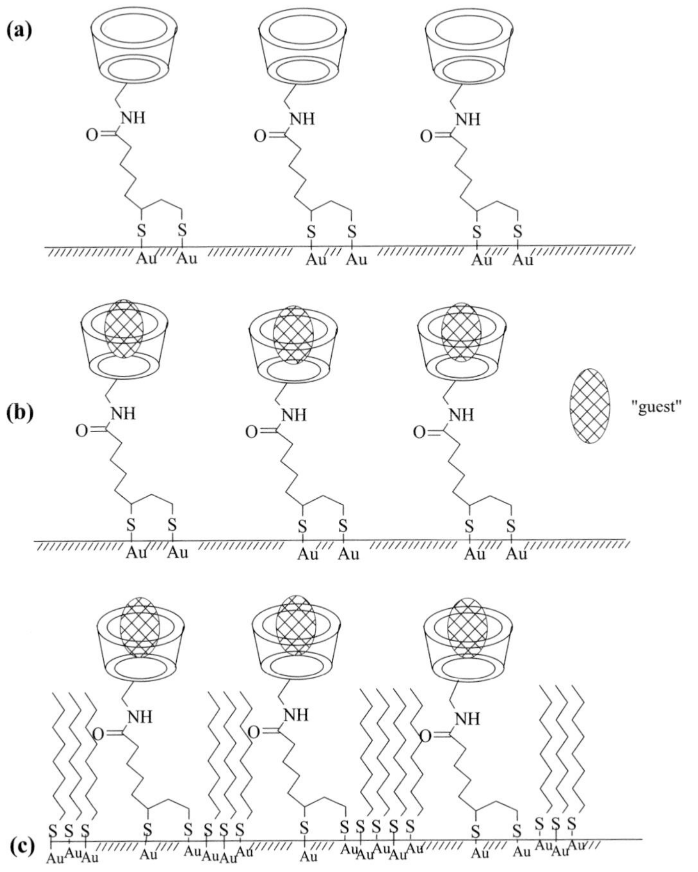

Fig. 10.5.5. Schematic representation of the sequential modification of a gold electrode.

45, 46]. It involves a three-step modification procedure leading to stable and ordered monolayers on a gold electrode (Fig. 10.5.5) [16, 45, 46]. The approach relies on the formation of alkanethiol and lipoylamide-β-CyD monolayers, with the thiol component responsible for blocking the electrode surface and lipoylamide-β-CyD for controlled opening of the access of the electroactive probe to the electrode. The response towards guest (ferrocene) was found to be highly improved when ferrocene was added to the solution following self-assembly of cyclodextrin but prior to the thiol self-assembly step (Fig. 10.5.5b). The sequential monolayer formation scheme led to well-organized and stable modified electrode surfaces with improved sensitivity towards solution species compared to other procedures of electrode modification with CyD derivatives [16, 36, 67, 68].

Two derivatives of lipoylamide-β-CyD of different hydrophobicity were used to prepare gold electrodes responsive towards selected drugs. The electrode, modified

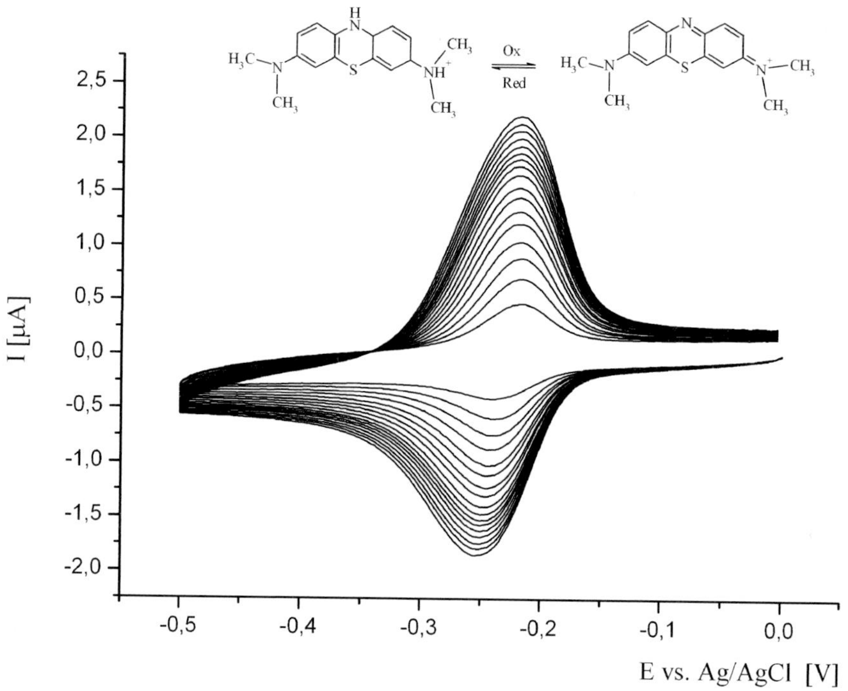

Fig. 10.5.6. Cyclic voltammetry curves for an Au electrode modified with the MB complex with β-CyD. The peak current increases linearly with increasing MB concentration in the solution [46].

(using the three-step sequential procedure) with the CyD derivative, binds ferrocene to form an electroactive complex with the ferrocene oxidation current decreasing in the presence of ibuprofen in the solution. The competition of ferrocene and ibuprofen for the CyD cavities in the monolayer provided a means for the determination of the binding constants of ibuprofen to CyD [45].

A similar three-step sequential self-assembly procedure was applied to prepare gold electrodes modified in a stable and controlled way by a monolayer of per(6-dexy-6-thio-2,3-di-OMe)-β-CyD, with methylene blue (MB) in its cavity as the active component of the monolayer (Fig. 10.5.6) [46].

Methylene blue acted as a mediator of electrons providing electrical contact between the electrode and the solution-resident enzyme laccase (Fig. 10.5.7) catalyzing reduction of oxygen to water [46]. The catalytic effect is revealed by the wave shape of the cyclic voltammetry curve in the oxygen-saturated solutions and its increased limiting current.

Disubstituted CyDs form a family of positional isomers. The organization of the monolayer of CyDs is affected by the type of isomer as discussed by Suzuki et al. [62]. Studies of four positional isomers of γ-CyD derivatives possessing two lipoyl

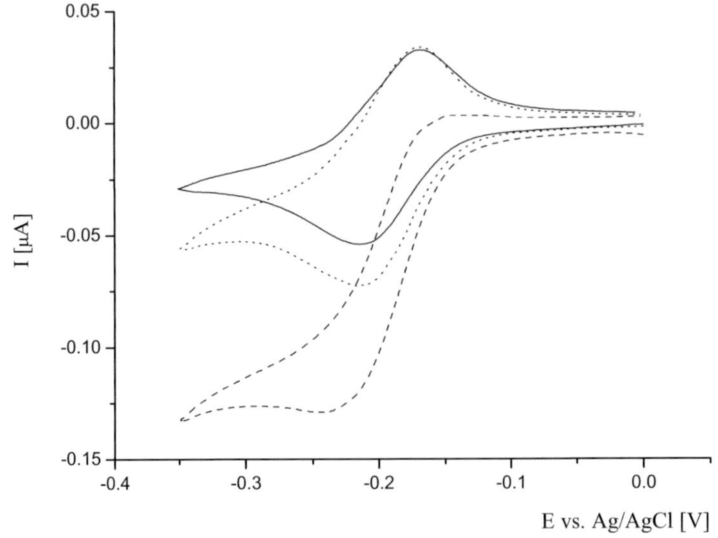

Fig. 10.5.7. Cyclic voltammograms for MB @ β-CyD-modified electrode in deoxygenated (———); oxygen-saturated solution (···), and in oxygen saturated solution containing 1 mg mL^{-1} laccase (---) [46].

moieties at the primary hydroxyl side showed that only isomer 4 (A–E) can form a high-quality γ-CyD monolayer with molecular recognition ability (Fig. 10.5.8). Interestingly, other isomers also retained a molecular recognition ability similar to that in the solution phase; however, large intermolecular voids were also present on the electrode.

The disorder usually found in single-component CyD layers appears to be connected with the surface roughness of the electrode substrates and with the limited packing ability of the molecules. More-ordered layers were reported for mercury substrates and this was explained by Majda et al. [31] as due to both the larger lateral mobility of the molecules on the liquid mercury and to the atomically uniform surface of this electrode. Higher packing densities relative to those for self-assembly on gold were noted in this case. Majda et al. studied inclusion of both inorganic ions and uncharged hydrophobic guests into these monolayers using capacitance measurements [31].

Interestingly, double-layer electrostatic forces induced inclusion of inorganic cations and anions into CyD monolayers leading to an increase in the interfacial capacitance. Anion binding constants were obtained from the dependence of capacitance current on anion concentration. The charge density in the plane of nitrate inclusion is a function of total surface concentration of CyD and the nitrate occupancy. The electrostatic contribution to the potential was determined and used for the calculation of corrected binding constant for nitrates (0.11 M^{-1} for β-

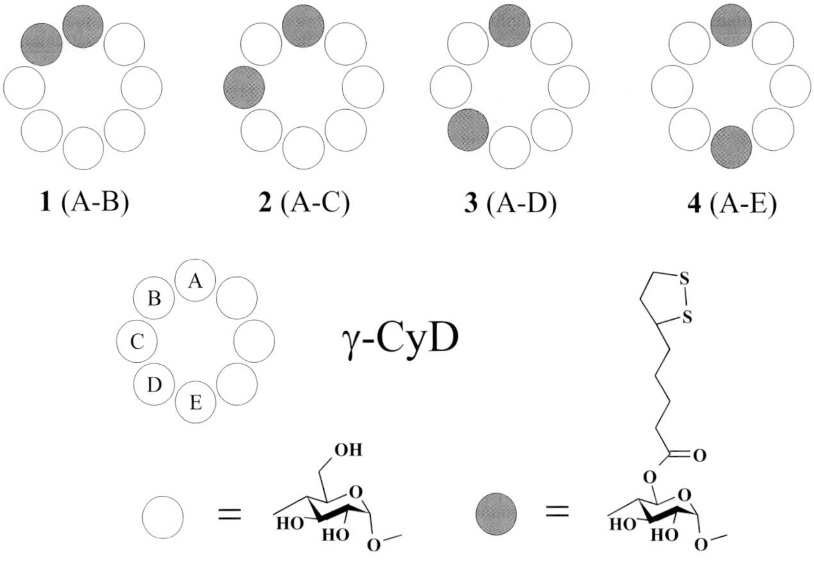

Fig. 10.5.8. Positional isomers of γ-CyD derivatives used for modification of gold electrodes [62].

CyD). Competitive binding of hydrophobic guests decreased, in turn, the interfacial capacitance [69]. Fitting Langmuir isotherms to the plots of interfacial capacitance vs. organic guest concentration yielded binding constants (e.g. 1.0×10^4 M^{-1} for adamantanol) [31, 39, 42].

Changes in capacitances of the CyD monolayers on gold surfaces in the presence of selected analytes may be used as a signal transduction mode alternative to the more commonly used voltammetric and amperometric techniques and impedance analysis for monitoring host–guest interactions in monolayers.

CyDs with both electroactive and anchoring units can be prepared by covalent bonding of, for example, ferrocene derivatives by means of a tether of appropriate length to secondary OH groups of β-CyD [70]. The models of these compounds adsorbed onto a gold electrode, show that in the reduced form the ferrocene derivative fits well to the cavity size while in the oxidized form it is pushed outside (Fig. 10.5.9).

The observed electron-transfer mediating role of the ferrocenes tethered to CyDs is of importance for biosensor applications (Fig. 10.5.10). The presence of several ferrocene units on one CyD molecule may enhance the functioning of these molecules as electron mediators. Electroactive monolayers of CyDs to be used as surface binding sites are usually prepared by attaching a ferrocene unit far from the cavity and close to the electrode surface [43, 71]. Compounds **1** and **2** (Fig. 10.5.10) are examples of such systems [71].

Willner and coworkers described an exciting light-driven molecular shuttle organized on the electrode surface [72]. The assembly consists of a ferrocene-

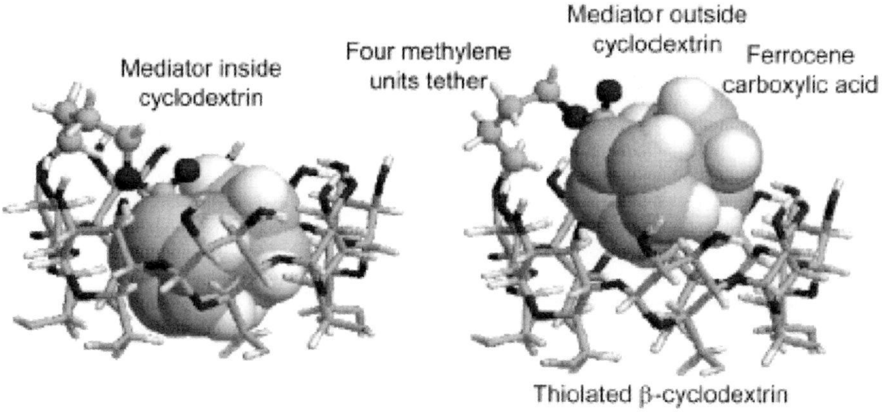

Fig. 10.5.9. Oxidation-state-dependent inclusion of the ferrocene moiety into CyD. Reprinted with permission of Elsevier [70].

functionalized β-CyD (Fc-β-CyD) threaded on a monolayer-immobilized long alkyl component containing a photoisomerizable azobenzene unit, and terminated with a bulky anthracene group (Fig. 10.5.11). The assembly functions as a molecular optoelectronic system since it records optical information and transduces it by an electronic signal.

The CyD resides preferentially on the *trans*-azobenzene component and photoisomerization to the *cis*-azobenzene state causes translocation of the Fc-β-CyD to the alkyl chain component of the assembly. This light-driven translocation is reversible and proceeds by isomerization of azobenzene between the *trans* and *cis* states. The chronoamperometric response of the redox-active ferrocene group associated with a CyD unit reflects a change of position on the molecular array.

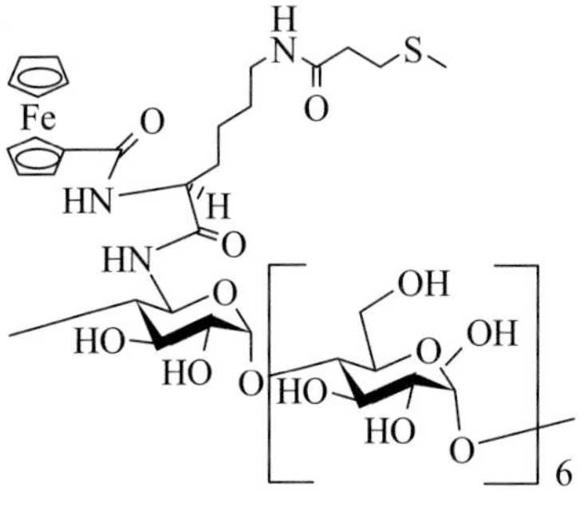

1

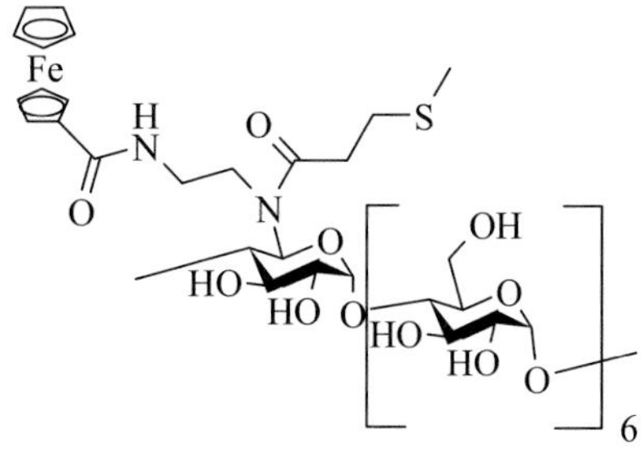

2

Fig. 10.5.10. Examples of β-CyD derivatives with a ferrocene moiety linked directly to the CyD ring [71].

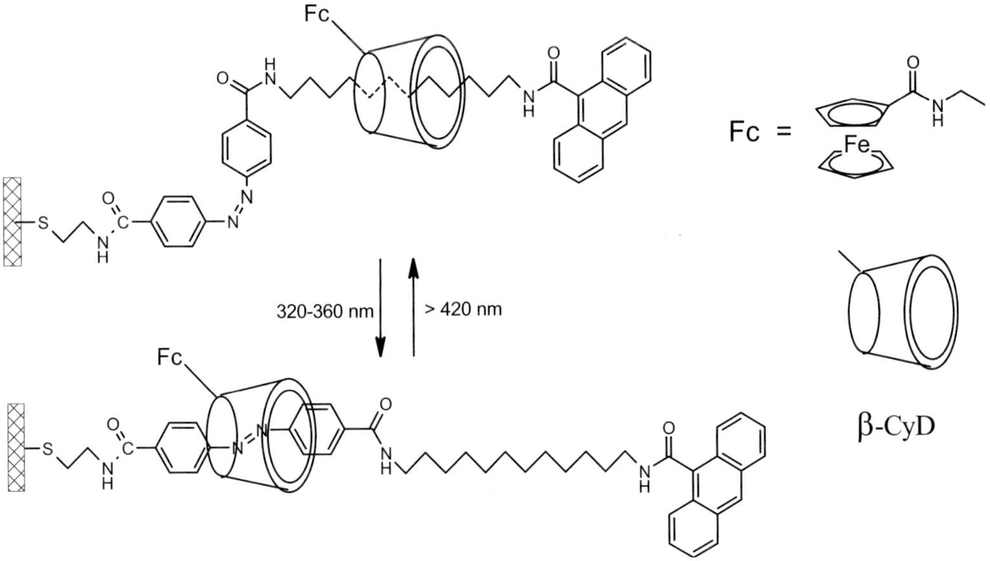

Fig. 10.5.11. Willner's molecular train (adapted from Ref. [72]).

Various matrices are used to immobilize CyDs on electrodes. Polypyrrole-sulfated β-CyD films can be prepared electrochemically using a mixture of pyrrole monomer and sulfated CyD. The presence of the CyD in the film is proved voltammetrically and by dispersive X-ray analysis. CyD preferentially dopes the polymer even in the presence of large concentrations of perchlorate [73].

Thin polymer films of CyDs can be prepared on an electrode surface by polycondensation with dialdehyde [26, 74–76] or by electropolymerization following derivatization [26]. In some cases the stability of complexes embedded in the polymer matrix was found to be enhanced [2, 77–80].

The accumulating properties of a thin film of condensation polymer of β-CyD are used to determine selected drugs. The procedure involves a preconcentration step followed by differential pulse voltammetry quantification. The detection limits for tricyclic and antidepressive drugs was reported to be 4×10^{-8} M [81, 82].

The "molecular filter approach" presented by Crooks et al. allowed them to enhance the selectivity of sensors based on β-CyD – functionalized hyperbranched polyacrylic acid (PAA) films capped with a chemically grafted ultrathin polyamine layer [83]. The latter served as a pH-sensitive filter, passing only suitably charged analytes. At low pH, the high charge of this protonated layer did not allow positively charged voltammetric probes to permeate through the CyD receptor. Thin films prepared by polycondensation on the electrode surface using prepolymers (β-CyDP) or carboxymethylated prepolymers (β-CyDPA) in glutaric dialdehyde solution appear to be useful as amperometric glucose biosensors [74, 84].

Glucose oxidase (GOx) was covalently immobilized in the membrane and the

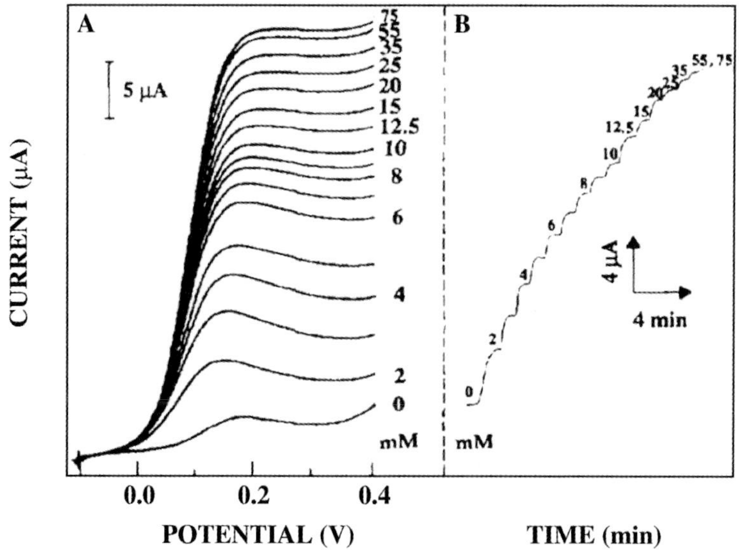

Fig. 10.5.12. Voltammetric and amperometric responses of the β-CyDPA-GOx–TTF membrane-coated electrode to increasing glucose concentration in solution. Reprinted with permission of Elsevier [74].

tetrathiafulvalenium/tetrathiafulvalene (TTF^+/TTF) mediating couple was retained in the β-CyDP membrane by complex formation or in the case of β-CyDPA by both supramolecular complex formation and ion exchange by the pendant carboxymethyl groups (Fig. 10.5.12). The biosensor displays high selectivity towards glucose in the presence of commonly interfering substances such as uric acid, acetaminophen, or ascorbic acid and was employed for the fabrication of glucose biosensors based on a disposable screen-printed Ag–carbon strip two-electrode transducer [74, 75].

Enzymes and ferrocene in β-CyD can also be immobilized in polymer via cross-linking with glutaraldehyde. This configuration decreases the distance between redox centers of the enzyme and ferrocene – the electron-transfer mediator – and was shown to improve communication between the electrode and the redox centers of GOx [76, 81, 84].

Polypyridine complexes of ruthenium are also used as charge mediators in second-generation glucose biosensors. Anionic carboxymethylated β-CyD polymer films are cast on the electrode and GOx is next immobilized in the films by covalent bonding. The Ru complex is incorporated by ion exchange at the fixed carboxymethyl cation-exchange sites. Enzymatic glucose oxidation is mediated by the Ru complex at its redox potential. The biosensor response to glucose is reversible with a detection limit of 1 mM, but the concentration range is smaller than one order of magnitude [85]. Sensors based on cobalt tetraphenylporphyrins hosted by

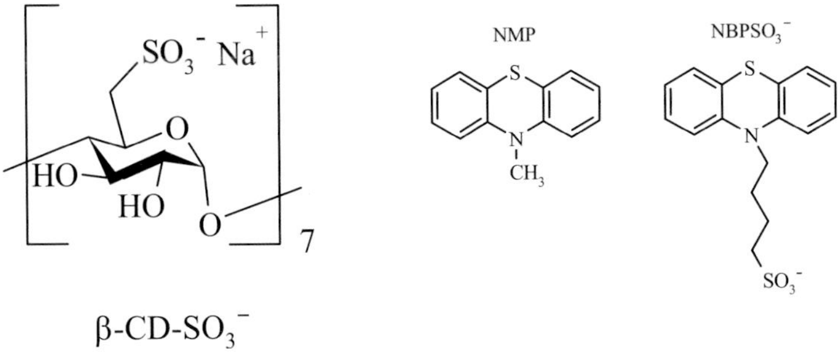

Fig. 10.5.13. Persulfonated β-CyD derivative and the drug guests used by Bidan et al. (adapted from Ref. [87]).

CyD polymer films were proposed by D'Souza et al. [86] as useful for electrocatalytic 2*e* reduction of oxygen to hydrogen peroxide, since the catalytic rate constants were higher than the corresponding values for adsorbed or electropolymerized catalysts at the electrode. Other examples of CyD catalysis are presented in Chapter 4.

The electrocontrolled delivery of drugs may be based on a doping mechanism where doping relies on interaction with modified β-CyDs. The binding of methylphenothiazine, NMP, drugs to persulfonated β-CyD-SO_3^- (Fig. 10.5.13), which has been tailor-made to dope polypyrrole film, was reported by Bidan et al. [87]. The release of NMP^+ from guest-preloaded polymer is triggered electrochemically by applying a suitable electrode potential, which provides a convenient way to regain the initial properties of the electrode surface.

10.5.4 Summary and Prospects

A significant number of electrochemical investigations have been directed towards understanding the relation between structural features (geometry, substituents, charge) and the complexing or recognition abilities of CyDs both in solution and immobilized on electrodes. Most of the papers provide information on the selectivity and stability of the complexes. The electrochemical properties of the complexes have been assessed, especially in relation to their possible electrocatalytic properties.

Another widely described role for electrochemical methods in the field of CyDs is the external on/off switching of binding events or opening/closing biomimetic CyD channels. Electrochemical measurements are often employed to explain the chemical reactions taking place inside the cavity and to rationalize the catalytic properties of the CyD microenvironment.

We have mentioned selected examples from the vast literature on electroanalysis based on CyDs which promises new electrochemical sensors and drug-delivery systems. In connection with the latter subject, an important goal is to gain a better understanding of the interactions of CyDs with biological membranes and with their components, and their possible influence on the structure and integrity of the biomembrane. Surface electrochemistry combined with scanning probe microscopy (presented in Section 10.6) should bring such information together with in-depth studies of biomimetic model membrane systems. Miniaturized electrochemical sensing/biosensing devices based on CyDs will certainly remain a very active field of research and of applications in environmental, medicinal and forensic chemistry.

The construction of externally switched solid-state devices awaits development of molecular junctions sandwiched between two electrodes [89]. They may, for example, consist of CyD-based meccano molecules such as rotaxanes (discussed in Chapter 12) for which switching in the solution phase or in monolayers has been demonstrated [21].

Functional electrochemically controlled supramolecular machines based on CyD units now being actively studied and already presenting motor-like behavior on the molecular level, following new routes in the bottom-up approach, should soon enter the world of useful micro- and mesoscale electromechanical devices. This subject is discussed in Chapter 16.

References

1 M. L. Bender, M. Komiyama, *Cyclodextrin Chemistry*, Springer, Berlin, 1978.

2 J. Szejtli, *Cyclodextrin Technology*, Kluwer Academic Press, Dordrecht, 1988.

3 K. A. Connors, *Chem. Rev.* 1997, *97*, 1325–1358.

4 M. V. Rekharski, Y. Inoue, *Chem. Rev.* 1998, *98*, 1875–1917.

5 A. E. Kaifer, M. Gómez-Kaifer, In *Supramolecular Electrochemistry: Self-Assembled Monolayers*; Wiley-VCH: Weinheim, 1999, 91–206.

6 A. Ferancova, J. Labuda, *Fresenius' J. Anal. Chem.* 2001, *370*, 1–10.

7 M. Singh, R. Sharma, U. C. Banerjee, *Biotechnology Advances*, 2002, *20*, 341–359.

8 S. Li, W. C. Purdy, *Chem. Rev.* 1992, *92*, 1457–1470.

9 E. Scholz (Ed.), Electroanalytical Methods, Guide to Experiments and Applications, Springer-Verlag, Berlin, Heidelberg, 2002, (a) Z. Stojek in [9] pp. 100–110, (b) M. Lovrić in [9] pp. 111–136, (c) E. Marken, A. Neudeck, A. M. Bond in [9] pp. 51–97.

10 Z. Galus, *Fundamentals of Electrochemical Analysis*, Ellis Harwood and Polish Scientific Publishers PWN, New York, 1994.

11 J. O'M. Bockris, *Surface Electrochemistry: a molecular level approach*, Plenum, New York, 1993.

12 J. W. Steed, J. L. Atwood, *Supramolecular Chemistry*, J. Wiley & Sons, Chichester, New York, 2000.

13 A. E. Kaifer in *Comprehensive Supramolecular Chemistry*, J. L. Atwood, J. E. D. Davies, D. D. MacNicol, F. Vőgtle (Eds) Pergamon, Oxford, 1996, pp. 499–535.

14 Y. Maeda, T. Fukuda, H. Yamamoto, H. Kitano, *Langmuir* 1997, *13*, 4187–4189.

15 T. Fukuda, Y. Maeda, H. Kitano, *Langmuir* 1999, *15*, 1887–1890.

16 K. Chmurski, A. Temeriusz, R. Bilewicz, *Anal. Chem.* 2003, *75*, 5687–5691.

17 S. Nagase, M. Kataoka, R. Naganawa, R. Komatsu, K. Odashima, Y. Umezawa, *Anal. Chem.* 1990, *62*, 1252–1259.

18 K. Odashima, M. Kotato, M.

Sugawara, Y. Umezawa, *Anal. Chem.* 1993, *65*, 927–936.
19 K. Toda, S. Amemiya, T. Ohki, S. Nagahora, S. Tanaka, P. Buhlmann, Y. Umezawa, *Isr. J. Chem.* 1997, *37*, 267–275.
20 T. M. Flyes, W. F. van Straaten-Nijenhuis, In *Comprehensive Supramolecular Chemistry*, D. N. Reinhoudt (Ed.) Pergamon, Oxford, 1996.
21 A. H. Flood, R. J. A. Ramirez, W.-Q. Deng, R. P. Muller, W. A. Goddard III, J. F. Stoddart, *Aust. J. Chem.* 2004, *57*, 301–322.
22 A. H. Flood, J. F. Stoddart, D. W. Steuerman, J. R. Heath, *Science* 2004, *306*, 2055–2056.
23 L. A. Godinez, J. Lin, M. Munoz, A. W. Coleman, A. E. Kaifer, *J. Chem. Soc. Faraday Trans.* 1996, *92*, 645–650.
24 M. Tanaka, Y. Ishizuka, M. Matsumoto, T. Nakamura, A. Yabe, H. Nakanishi, Y. Kawabata, H. Takahashi, S. Tamura, W. Tagaki, H. Nakahara, K. Fukuda, *Chem. Lett.* 1987, 1307–1310.
25 M. Opałło, N. Kobayashi, T. Osa, H. Yamada, S. Onodera, *Bull. Chem. Soc. Soc. Jpn.* 1989, *62*, 2995–2997.
26 W. Kutner, K. Doblhofer, *J. Electroanal. Chem.* 1992, *326*, 139–160.
27 S. McCormack, N. Russel, J. Cassidy, *Electrochim. Acta* 1992, *37*, 1939–1944.
28 I. Suzuki, Q. Chen, Y. Kashiwagi, T. Osa, A. Ueno, *Chem. Lett.* 1993, 1719–1722.
29 U. Sivagnanam, M. Palaniandavar, *J. Electroanal. Chem.* 1992, *341*, 197–207.
30 T. Matsue, D. H. Evans, T. Osa, N. Kobayashi, *J. Am. Chem. Soc.* 1985, *107*, 3411–3417.
31 R. V. Chamberlain II, K. Słowińska, M. Majda, *Langmuir* 2000, *16*, 1388–1396.
32 Y. Matsui, K. Mochida, *Bull. Chem. Soc. Jpn.* 1979, *52*, 2808–2814.
33 A. Botsi, K. Yannakopoulou, B. Perly, E. Hadjoudis, *J. Org. Chem.* 1995, *60*, 4017–4023.
34 D. Bouchta, N. Izaoumen, H. Zejli, M. E. Kaoutit, K. R. Temsamani, *Biosensors Bioelectronics*, 2005, *20*, 2228–2235.
35 D. A. Reece, S. F. Ralph, G. G. Wallace, *J. Membrane Sci.* 2005, *249*, 9–20.
36 M. T. Rojas, R. Königer, J. F. Stoddart, A. E. Kaifer, *J. Am Chem. Soc.* 1995, *117*, 336–343.
37 G. Nelles, M. Weisser, R. Back, P. Wohlfart, G. Wenz, S. Mittler-Neher, *J. Am Chem. Soc.* 1996, *118*, 5039–5046.
38 M. Weisser, G. Nelles, G. Wenz, S. Mittler-Neher, *Sensors Actuators B*, 1997, *38*, 58–67.
39 C. Henke, C. Steinem, A. Janshoff, G. Steffan, H. Luftmann, M. Sieber, H.-J. Galla, *Anal. Chem.* 1996, *68*, 3158–3165.
40 M. Lahav, K. T. Ranjit, E. Katz, I. Willner, *Isr. J. Chem.* 1997, *37*, 185–195.
41 H. X. Ju, D. Leech, *Langmuir* 1998, *14*, 300–306.
42 Y. Wang, A. E. Kaifer, *J. Phys. Chem. B* 1998, *102*, 9922–9927.
43 K. J. Stine, D. M. Andrauskas, A. R. Khan, P. Forgo, V. T. D'Souza, *J. Electroanal. Chem.* 1999, *465*, 209–218.
44 M. Hromadova, R. J. de Levie, *J. Electroanal. Chem.* 1999, *465*, 51–62.
45 K. Chmurski, A. Temeriusz, R. Bilewicz, *J. Incl. Phen. Macrocyc. Chem.* 2004, *49*, 187–191.
46 K. Chmurski, A. Koralewska, A. Temeriusz, R. Bilewicz, *Electroanalysis* 2004, *16*, 1407–1412.
47 K. Chmurski, R. Bilewicz, J. Jurczak, *Langmuir* 1996, *12*, 6114–6118.
48 M.Ważyńska, A. Temeriusz, K. Chmurski, R. Bilewicz, J. Jurczak, *Tetrahedron Lett.* 2000, *41*, 9119–9123.
49 Y. LeBras, M. Salle, P. Leriche, C. Mingotaud, P. Richomme, J. Moller, *J. Mater. Chem.* 1997, *7*, 2393–2396.
50 Y. Kawabata, M. Matsumoto, M. Tanaka, H. Takahashi, Y. Irinatsu, S. Tamura, W. Tagaki, H. Nakahara, K. Fukuda, *Chem. Lett.* 1986, 1933–1934.
51 Y. Kawabata, M. Matsumoto, T. Nakamura, M. Tanaka, E. Manda, H.

Takahashi, S. Tamura, W. Tagaki, H. Nakahara, K. Fukuda, *Thin Solid Films* 1988, *159*, 353–358.
52 P. Zhang, H. Parrot-Lopez, P. Tchoreloff, A. Baszkin, C. C. Ling, C. de Rango, A. W. Coleman, *J. Phys. Org. Chem.* 1992, *5*, 518–528.
53 H. Parrot-Lopez, C. C. Ling, P. Zhang, A. Baszkin, G. Albrecht, C. De Rango, A. W. Coleman, *J. Am. Chem. Soc.* 1992, *114*, 5479–5480.
54 M. H. Greenhall, P. Lukes, R. Kataky, N. E. Agbor, J. P. S. Badyal, J. Yarwood, D. Parker, M. C. Petty, *Langmuir* 1995, *11*, 3997–4000.
55 D. P. Parazak, A. R. Khan, V. T. D'Souza, K. J. Stine, *Langmuir* 1996, *12*, 4046–4049.
56 M. Kunitake, K. Kotoo, O. Manabe, T. Muramatsu, N. Nakashima, *Chem. Lett.* 1993, 1033–1036.
57 A. Harada, *Acta Polym.* 1998, *49*, 3–17.
58 S. A. Nepogodiev, J. F. Stoddart, *Chem. Rev.* 1998, *98*, 1959–1976.
59 A. Ohira, T. Ishizaki, M. Sakata, I. Taniguchi, C. Hirayama, M. Kunitake, *Colloids Surfaces* 2000, *169*, 27–33.
60 P. He, J. Ye, Y. Fang, I. Suzuki, T. Osa, *Anal. Chim. Acta* 1997, *337*, 217–223.
61 P. He, J. Ye, Y. Fang, I. Suzuki, T. Osa, *Electroanalysis* 1997, *9*, 68–73.
62 I. Suzuki, K. Murakami, J. Anzai, *Mater. Sci. Eng. C* 2001, *17*, 143–148.
63 I. Suzuki, K. Murakami, J. Anzai, T. Osa, P. G. He, Y. Z. Fang, *Mater. Sci. Eng. C* 1998, *6*, 19–25.
64 S. Minato, T. Osa, A. Ueno, *J. Chem. Soc. Chem. Commun.* 1991, 107–108.
65 A. Ueno, S. Minato, T. Osa, *Anal. Chem.* 1992, *64*, 1154–1157.
66 S.-W. Choi, J.-H. Jang, Y.-G. Kang, C.-J. Lee, J.-H. Kim, *Colloids Surfaces* A 2005, *257*, 31–36.
67 J.-Y. Lee, S.-M. Park, *J. Phys. Chem. B* 1998, *102*, 9940–9945.
68 M. Weisser, G. Nelles, P. Wohlfart, G. Wenz, S. Mittler-Neher, *J. Phys. Chem.* 1996, *100*, 17893–17900.
69 K. Słowinski, R. V. Chamberlain, C. J. Miller, M. Majda, *J. Am. Chem. Soc.* 1997, *119*, 11910–11919.
70 G. Favero, L. Campanella, A. D'Annibale, T. Ferri, *Microchemical J.* 2004, *76*, 77–84.
71 I. Suzuki, K. Murakami, J. Anzai, *Mater. Sci. Eng. C* 2001, *17*, 149–154.
72 I. Willner, S. Marx, Y. Eichen, *Angew. Chem. Int. Ed. Engl.* 1992, *31*, 1243–1244.
73 K. R. Temsamani, O. Ceylan, B. J. Yates, S. Oztemiz, T. B. Gbatu, A. M. Stalcup, H. B. Mark, W. Kutner, *J. Solid State Electrochem.* 2002, *6*, 494–497.
74 Q. Chen, P. V. A. Pamidi, J. Wang, W. Kutner, *Anal. Chim. Acta* 1995, *306*, 201–208.
75 W. Kutner, W. Storck, K. Doblhofer, *J. Incl. Phen.* 1992, *13*, 257–265.
76 W. Kutner, *Electrochim. Acta* 1992, *37*, 1109–1117.
77 A. Harada, A. Furue, S.-I. Nozakura, *Polymer J.* 1981, *13*, 777–779.
78 T. Cserhati, G. Oros, E. Fenyvesi, J. Szejtli, *J. Incl. Phenom.* 1984, *1*, 395–402.
79 J. Szeman, E. Fenyvesi, J. Szejtli, H. Ueda, Y. Machida, T. Nagai, *J. Incl. Phenom.* 1987, *5*, 427–431.
80 D. Koradecki, W. Kutner, *J. Incl. Phenom.* 1991, *10*, 79–96.
81 A. Ferancova, E. Korgova, T. Buzinkaiova, W. Kutner, I. Stepanek, J. Labuda, *Anal. Chim. Acta* 2001, *447*, 47–54.
82 A. Ferancova, J. Labuda, W. Kutner, *Electroanalysis* 2001, *13* (17), 1417–1423.
83 D. L. Dermody, R. F. Peez, D. E. Berg-Breiter, R. M. Crooks, *Langmuir* 1999, *15*, 885–890.
84 W. Kutner, H. Wu, K. M. Kadish, *Electroanalysis* 1994, *6*, 934–944.
85 E. Kosela, H. Elzanowska, W. Kutner, *Anal. Bioanal. Chem.* 2002, *373*, 724–734.
86 F. D'Souza, Y. Y. Hsieh, H. Wickman, W. Kutner, *Electroanalysis* 1997, *9*, 1093–1101.
87 G. Bidan, C. Lopez, F. Mendes-Viegas, E. Veil, A. Gadelle, *Biosensors Bioelectronics* 1994, *9*, 219–229.
88 S. Sęk, R. Bilewicz, K. Słowinski, *Chem. Comm.* 2004, 404–405.

10.6 Visualization of Cyclodextrins in Supramolecular Structures by Scanning Probe Microscopy

Masashi Kunitake and Akihiro Ohira

10.6.1 Introduction

Cyclodextrins (CyDs) possess a hydrophobic cavity that can accommodate a wide variety of guest species [1]. Since the mid-1990s, supramolecular chemistry has attracted increasing interest in organic chemistry, and CyDs have also been used as components of inclusion complexes and other self-assembled species [2, 3–6], particularly the rotaxanes, catenanes, and polyrotaxanes discussed in detail in Chapter 12. Organic chemists interested in supramolecular chemistry wish to observe the real shape of supramolecular structures. Much attention has been focused on scanning probe microscopy (SPM) [7], a technique allowing visualization of arrangements, orientations, and even the inner structures of organic molecules existing in air [8], in ultrahigh vacuum (UHV) [9], and in solution [10–14]. Visualization by SPM may be the best method that can directly confirm the architectures of supramolecular arrangements.

The scanning probe microscope skims the surface of an object while the distance (or interaction such as force or tunneling current) between the apex of the probe and the surface is kept constant by a feedback mechanism. From traces of the tip probe, images in which surface morphologies are visualized are created on a computer screen. Tunneling currents and direct force between the apex of the probe and the surface are used as a monitoring feedback in scanning tunneling microscopy (STM) and atomic force microscopy (AFM), respectively. The use of an atomically sharpened probe tip allows the production of molecular or atomic resolutions of images for AFM and STM. In terms of resolution, STM is more advantageous than AFM. However, an electro-conductive substrate is always necessary for STM, because bias has to be applied. Therefore, suitable sample types for STM measurements are relatively limited, and STM observations of organic molecules are regularly conducted at the level of monolayer or molecular adsorbates on electro-conductive substrates such as single-crystal metals, semiconductors, and highly ordered pyrolytic graphite (HOPG). On the other hand, there are no such limitations for AFM, although resolution is limited. STM and AFM are widely used to visualize CyD molecules in supramolecular systems such as polyrotaxane, bis(molecular tube)s, inclusion compounds, self-assembled monolayers (SAMs) and one-dimensional arrangements. In this section, we introduce and focus on SPM imaging of CyD molecules and their inclusion complexes.

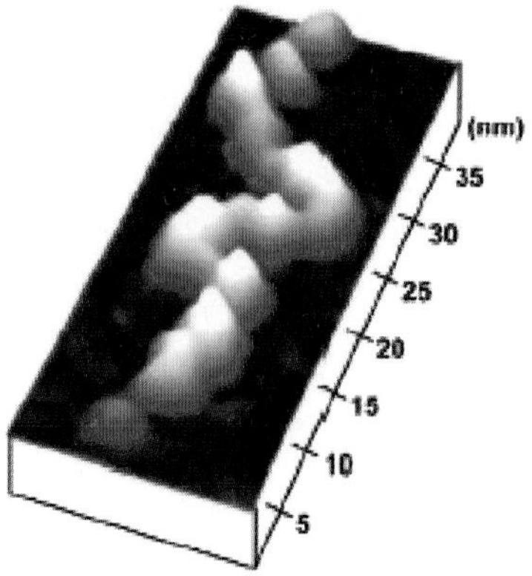

Fig. 10.6.1. STM image of a molecular necklace formed by polyethylene glycol and α-CyDs. Reprinted with permission from ref 15.
Copyright by American chemical society.

10.6.2
STM or AFM Observations of a CyD Nanotube in Air

A real spacing image of a polyethylene glycol (PEG)-α-CyD molecular necklace, which was synthesized for the first time by Harada and co-workers, has been observed using STM in air [15, 16], as shown in Fig. 10.6.1. The image clearly reveals a molecular necklace meandering like a snake on the HOPG substrate, and each individual CyD molecule can be recognized in the string.

STM allows us not only to visualize the real shape of polyrotaxane molecules (also discussed in Section 12.4) but also to manipulate an individual CyD molecule or a polymer chain itself. Shigekawa and co-workers demonstrated that a selected α-CyD molecule in polyrotaxane was reversibly shuttled using a STM tip resulting in a "molecular abacus" [16]. Figures 10.6.2a–c show STM imaging of simple (one molecule) shuttling: one CyD molecule is moved by the tip sweeping from right (Fig. 10.6.2a) to left (Fig. 10.6.2b); then, CyD moved by the tip is returned to its original position by the tip sweeping in the reverse direction (Fig. 10.6.2c). Shigekawa and co-workers also achieved actuation of a molecular necklace chain as shown in Figs. 10.6.2d–f. When molecules were pushed laterally by the tip sweeping in a perpendicular direction (Fig. 10.6.2d), the molecular necklace bent (Fig. 10.6.2e). Interestingly, a hook-shaped structure seemed to form by synchronized repositioning of several α-CyDs (Fig. 10.6.2f). This indicated that CyDs bound noncovalently along the chain in a cooperative motion, and could be observed using STM.

Other polyrotaxane molecules with CyD besides PEG-CyD were also observed using SPM. STM observation of polypseudorotaxane, in which β-CyD threaded

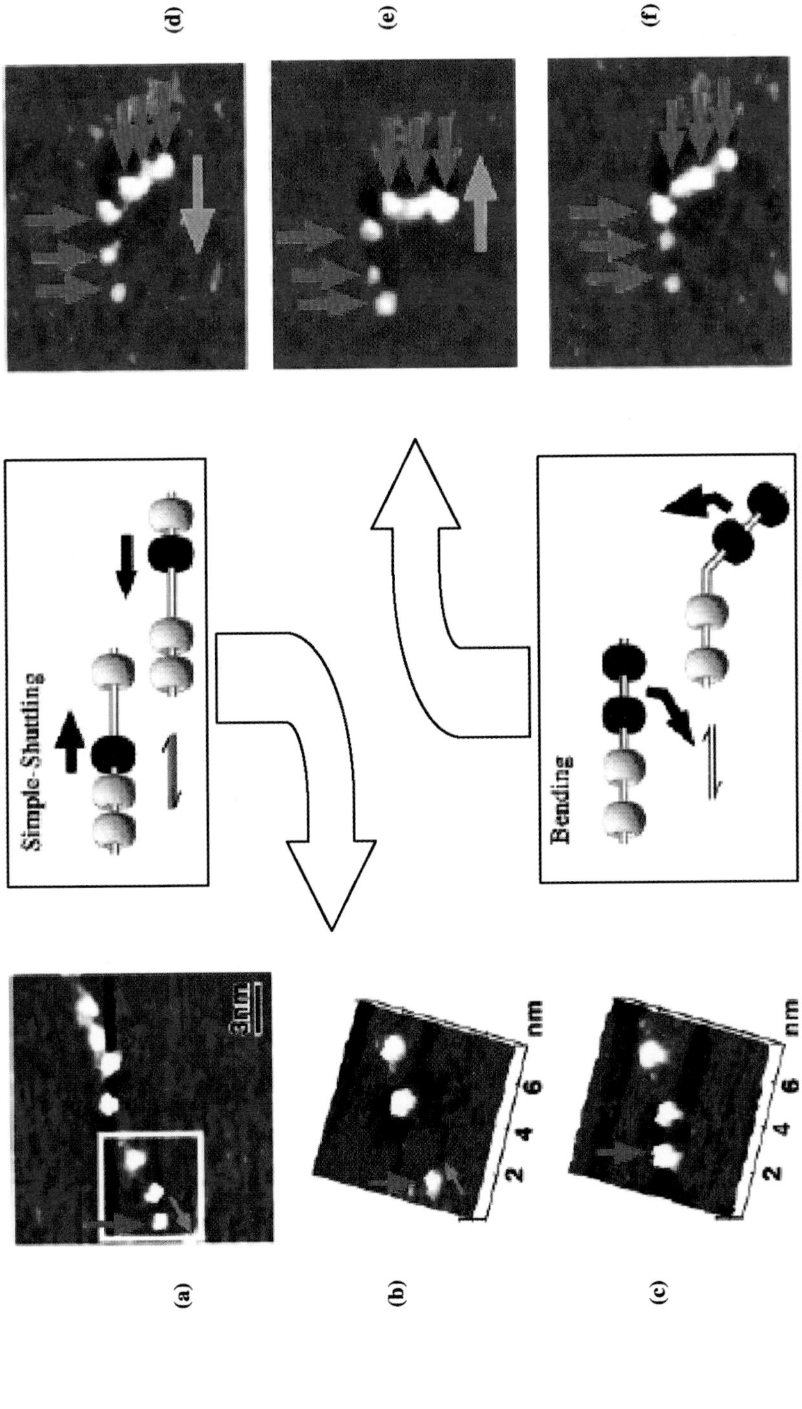

Fig. 10.6.2. STM images of simple-shuttling (a), (b), and (c) and bending (d), (e), and (f). Reprinted with permission from ref 16. Copyright by American chemical society.

with polyazomethine, was reported [17]. Polypseudorotaxane with polyazomethine was synthesized by polycondensation of two simple inclusion complexes of *o*-toluidine@β-CyD and *p*-phthaldehyde@β-CyD. High-resolution STM images revealed several nanometer-sized molecular wires approximately 1.5 nm in diameter on a graphite (HOPG) substrate. Nano-wires consisted of joint-like features, which might be attributed to a "tail-to-tail" coupling of CyDs with longitudinal axes parallel to the HOPG surface.

Liu and co-workers reported the synthesis of several bis(polypseudorotaxane)s which were formed from metallo-bridged β-CyDs. They synthesized CyD derivatives with a ligand moiety as a building block. Bis(polypseudorotaxane) was formed from metallo-bridged bis-(CyD)s and poly(propylene glycol) (PPG) [18–20]. Structural identification was conducted by AFM and STM. Compared with indistinct AFM imaging, STM imaging obviously succeeded in identifying the unique "double line" structure of bis(polypseudorotaxane)s on HOPG [19]. In STM images, four bright dots would represent a block of metallo-bridged bis(β-CyD)s, and such blocks were lined in a regular double alignment to form bis(polypseudorotaxane). The length and width of bis(polypseudorotaxane) were ca. 20 and 8 nm, respectively. The STM image and one of the possible schematic structures of bis(polypseudorotaxane) are shown in Fig. 10.6.3.

Hadziioannou and co-workers reported AFM observation and analysis of CyD-polyrotaxanes based on semiconducting conjugated polymers such as polythiophene and polyfluorene [22]. The polyrotaxane architecture with a single-chain conjugated polymer could be a very interesting alternative for the synthesis of soluble insulated single-chain conjugated polymers, but not by side-chain substitution. Hadziioannou et al. estimated the average length of polyrotaxane to be typically 15–20 nm by AFM observation of isolated short strings. Relatively long nanotube structures (>100 nm) are easy to observe even by AFM. However, limited AFM resolution did not allow discrimination of each CyD unit in the AFM images. The formation of inclusion complexes between polyaniline (PANI) with emeraldine base resulting in a molecular nanotube (length 25 nm) synthesized from α-CyD has been studied by AFM [23]. A rod-like inclusion complex about 300 nm long was clearly observed, confirming the formation of the inclusion complex. Since features of the rotaxane in the AFM image were uniform, it was suggested that a conducting wire of PANI was fully covered by molecular nanotubes to form an insulator. To judge from the length histogram estimated from more than 200 images, PANI-nanotube wires that are much longer than PANI-CyD inclusion complexes have also been found. SPM measurements demonstrated that longer wires could be formed not by CyD molecules, but by linking some PANI chains together as nanotubes.

Furthermore, Baglioni reported that the inclusion compound of β-CyD and semi-fluorinated-*n*-alkanes (short-chain block copolymers where a fluorinated chain is covalently bound to a hydrogenated tail) formed long supramolecular tubular structures [24]. Very long and relatively straight rigid tubular structures were clearly observed with many clusters, using AFM, to confirm the presence of a long-range

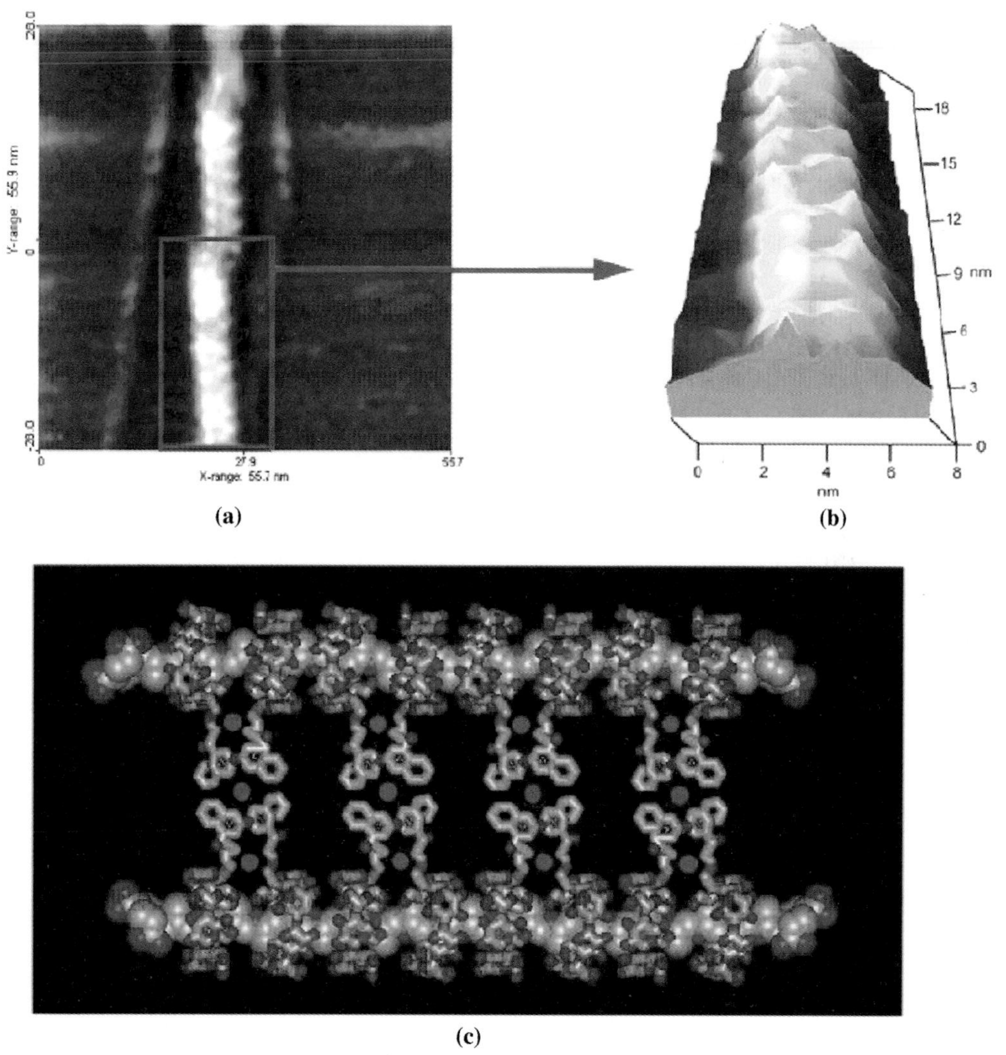

Fig. 10.6.3. (a) STM image of inclusion complexes on HOPG surfaces. (b) Sectional image in 3D mode. (c) Possible schematic structure of bis(polypseudorotaxane). Reprinted with permission from ref 19. Copyright by American chemical society.

aggregated structure (Fig. 10.6.4). The width and the length of the inclusion compounds were about 20 nm and 1.3 μm, respectively. It is worth noting that the width of nanotubes observed by AFM might not reflect the real van der Waals diameter of CyDs but the apex radius of the probe tip.

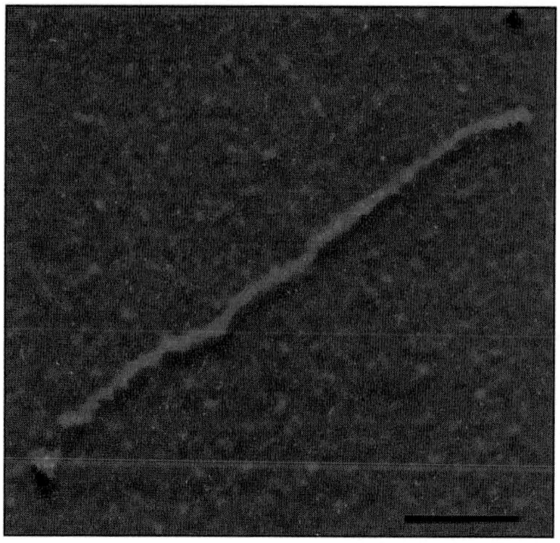

Fig. 10.6.4. AFM images (non-contact) for F_8H_{16} (semifluorinated alkane)@β-CyD inclusion complex scale bar = 200 nm.
Reprinted with permission from ref 24.
Copyright by American chemical society.

10.6.3
STM Observation of CyDs in Aqueous Solution

Observation of intact CyD molecules was achieved by STM not only in air but also in aqueous solution in the presence of CyDs using electrochemical STM (EC-STM) [13d]. (Other electrochemical CyD studies are presented in Section 10.5.) *In situ* EC-STM, which can be conducted in aqueous solution with electrochemical potential control, was developed for real-time, atomic scale, *in situ* characterization of electrode surfaces, and nowadays is also frequently applied for observation of molecular adsorption states, mainly for 2D supramolecular arrays. Thermodynamic adsorption behaviors of native α-, β- and γ-CyDs on Au(111) single-crystal surfaces were investigated by means of EC-STM. It was found that 2D arrays consisted of "nanotube" structures of α-, β-, and γ-CyDs which were similar to those of CyD-polyrotaxane. They were constructed by potential-controlled adsorption onto Au(111) surfaces in an electrolyte solution without a threaded polymer. Adsorption behaviors, including formation of nanotube structures of CyDs, significantly depended on the electrode potential. Figure 10.6.5 shows typical STM images of adsorption conditions of α-CyD, and schematically represents the electrode potential dependence for adsorption of CyDs. Essentially the same adsorption states, i.e. "desorption", "self-organization to nanotube" and "disordered adsorption" processes, were observed for α-, β-, and γ-CyDs. At the open-circuit potential which corresponded to no potential control, the surface was totally covered by randomly

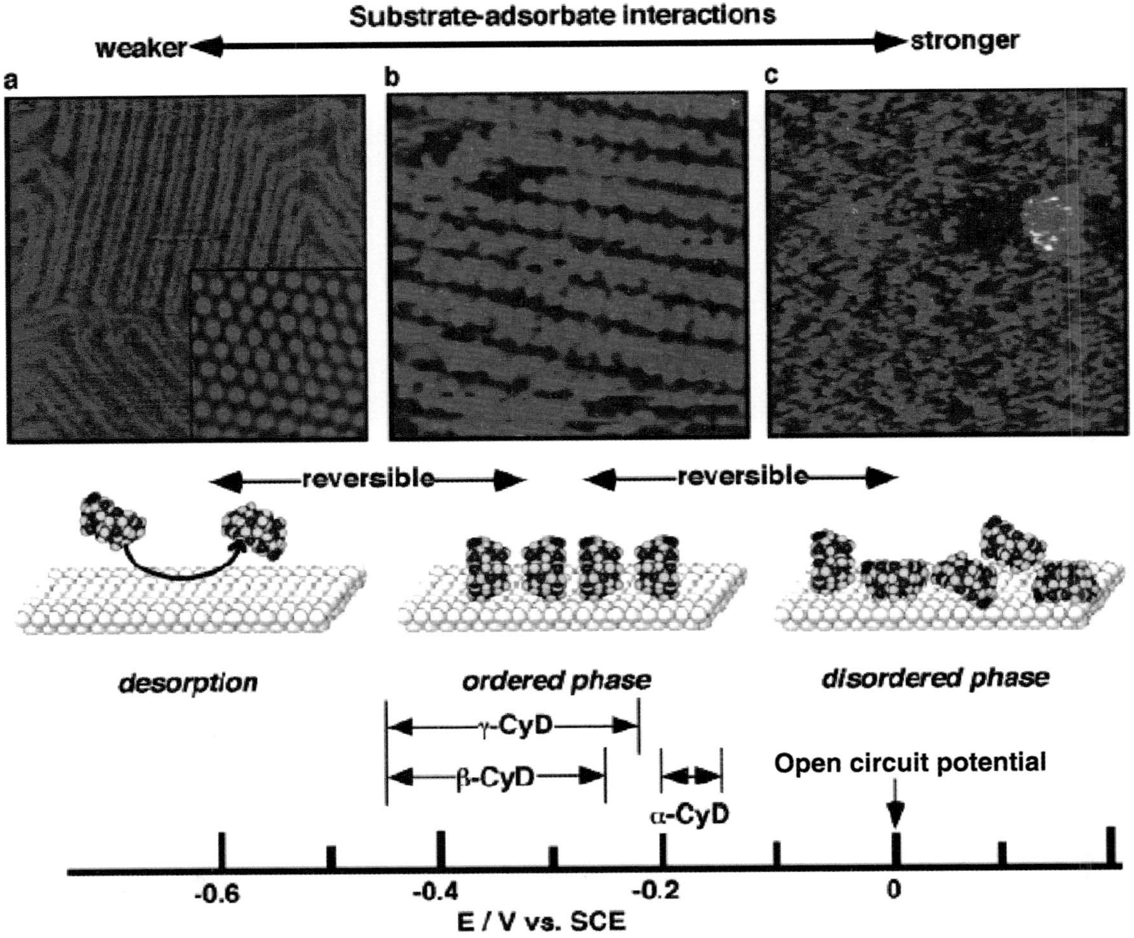

Fig. 10.6.5. Electrode potential diagrams for three phases: (a) "desorption" phase, (b) "ordered phase", and (c) "disordered phase", with corresponding schematic illustrations and typical STM images. (a) electrode potential $E_s = -0.60$ V. (b) $E_s = -0.45$ V. (c) $E_s = +0.10$ V. Each AISO potential range for nanotube formation of α-, β-, and γ-CyDs is marked by an arrow.
Reprinted with permission from ref 13 (d).
Copyright by American chemical society.

adsorbed CyDs (Fig. 10.6.5c). In the negative potential region lower than −0.5 V for β- and γ-CyDs, an intact atomic lattice of Au(111) was observed in spite of the presence of CyDs, suggesting that CyD molecules were desorbed at negative potentials. Adsorption and desorption of CyD molecules were essentially reversible for all CyDs. Generally, coverage of neutral molecules was greatest at a potential of zero charge (PZC), and adsorption tended to be prevented onto polarized surfaces [25]. At a specified mid-range situated between these potentials, ordered molecular arrays with nanotube structures formed on the Au(111) surface (Fig. 10.6.5b). Formation of these ordered arrays under conditions where substrate–adsorbate interactions were neither too strong nor too weak, was termed "adsorption-induced self-organization (AISO)".

Interestingly, the potential range for the formation of nanotubes in AISO differed for each CyD, as indicated by arrows in the potential diagram in Fig. 10.6.5. AISO potential ranges for α-CyD were positive relative to those for β- and γ-CyDs. This suggested that adsorption strengths of β- and γ-CyDs were stronger than that for α-CyDs, because β- and γ-CyDs required a more negative potential for AISO than α-CyDs in order to weaken adsorption and induce self-organization. This is likely attributable to the much bigger CyD ring sizes. These results indicate that control of the electrode potential facilitates management of the delicate figure balance of various interactions, resulting in the formation of 2D supramolecular structures on substrates.

The key to inducing self-organization onto water–solid substrate interfaces is to achieve mild adsorption under controlled conditions. If adsorbate–substrate interactions are too strong, molecules cannot move around on the substrate surface. On the other hand, when adsorbate–substrate interactions are too weak, molecules desorb from surfaces. Relatively mild adsorption conditions between these extreme states leads to induction of 2D self-organization of molecules via rapid surface diffusion and acceleration of the adsorption/desorption equilibrium. Electrochemical potential management would be convenient for AISO, because it allows for precise control of adsorption strength in units of mV [11, 13, 14].

Figures 10.6.6a–c show STM images of self-organized nanotube structures of α-, β-, and γ-CyDs under AISO conditions. CyD molecules, in the form of nanotube structures, covered the surfaces and were clearly individually visualized over a relatively wide region. Interestingly, for α-CyDs the surface morphology of the nanotube was different from that for β- and γ-CyDs. Typical α-CyD-nanotube structures with triangular arrangements indicate the epitaxial arrangement of nanotubes against underlying Au(111) lattices. On the other hand, nanotubes formed by β- and γ-CyDs were longer than those formed from α-CyDs. Furthermore, the direction of β- and γ-CyD nanotubes did not seem to be regulated by the underlying Au(111) lattices in contrast with the arrangements of the α-CyD-nanotubes. This suggests that intermolecular interactions for β- and γ-CyDs are stronger than those for α-CyDs. Similarly to the formation of polyrotaxanes discussed in Chapter 12, the dominant intermolecular interactions that form nanotubes are hydrogen bonds between hydroxyl groups on both bases of CyDs. Intermolecular interactions are therefore, roughly proportional to the number of glucose units in each CyD

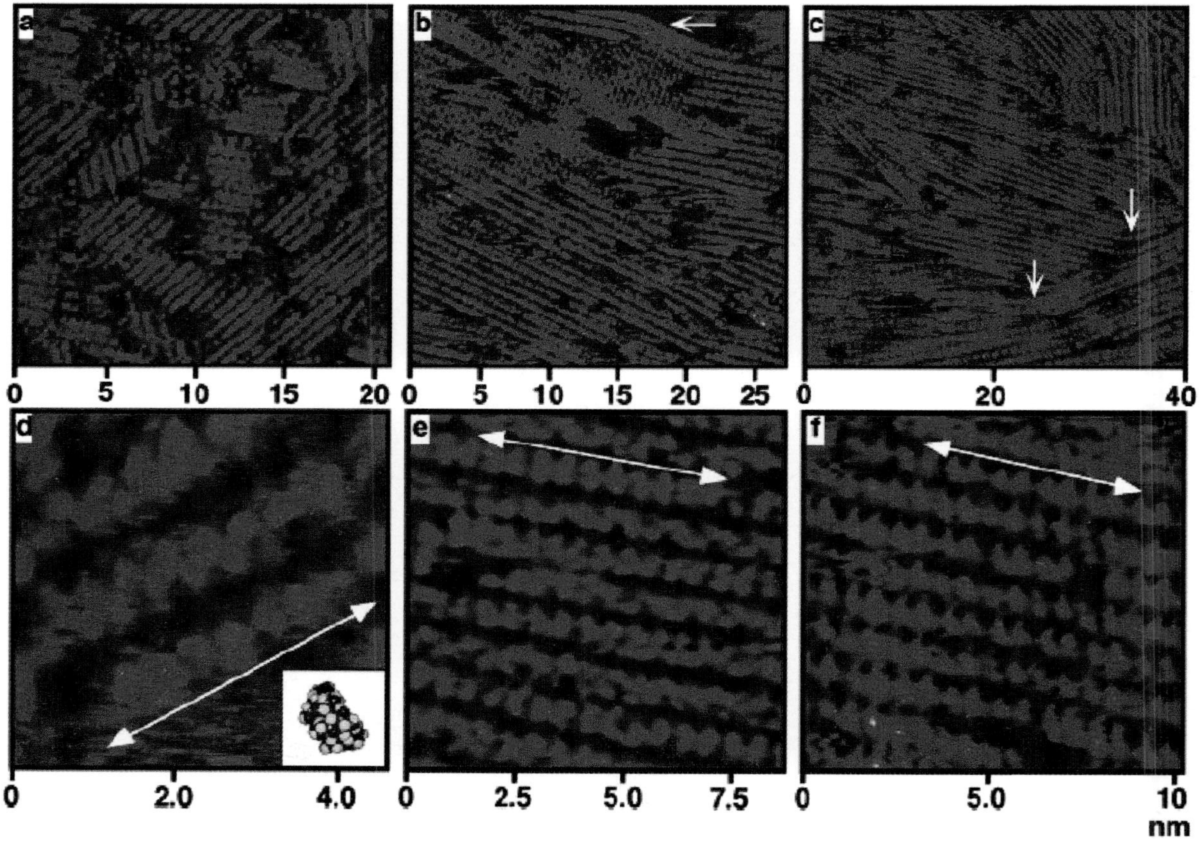

Fig. 10.6.6. High-resolution STM images of CyD-nanotube arrays on Au(111) surfaces for α-CyD (a) 22 nm × 22 nm, $E_s = -0.20$ V, (d) 4.6 nm × 4.6 nm, $E_s = -0.15$ V; for β-CyD (b) 28 nm × 28 nm, $E_s = -0.45$ V, (e) 8.7 nm × 8.7 nm, $E_s = -0.45$ V; and for γ-CyD (c) 40 nm × 40 nm, $E_s = -0.35$ V, (f) 10.2 nm × 10.2 nm, $E_s = -0.35$ V. Inset in (d) shows a space-filling model of α-CyD, which possesses the same orientation as that in the STM image. Arrows in (d–f) illustrate the running direction of the nanotubes.
Reprinted with permission from ref 13 (d). Copyright by American chemical society.

molecule. Interestingly, bendings of the nanotubes can be frequently seen in Figs. 10.6.5b and c (arrows). Such bendings were rarely observed for α-CyDs. This may also be due to the stronger intermolecular interactions and softness of β- and γ-CyDs relative to those of α-CyDs. High-resolution STM imaging clearly provides information on the molecular shape and arrangement of individual molecules within nanotubes, as shown in Figs. 10.6.6d, e, and f. Typically, each CyD molecule can be recognized as a rectangular bright spot, probably indicating that molecules are connected to each other in "head-to-head" and "tail-to-tail" arrangements to form the nanotube. These results indicate that the dominant driving forces for the formation of the tube structure are intermolecular H-bond interactions between (since these are pairwise interactions) hydroxyl groups on the same bases of CyD molecules. The distances observed between rectangular features along the tube were 0.6 0.1 nm were essentially constant for the series of CyDs. These values were consistent with the expected height of CyDs along the cavity but were too short for the diameter of CyDs with an upward orientation. In contrast, interval distances between regularly aligned parallel tubes were estimated as 0.9 ± 0.1 nm, 1.0 ± 0.1 nm and 1.1 ± 0.1 nm, for α-, β- and γ-CyDs, respectively. These sizes were consistent with values expected from chemical models. High-resolution images occasionally obtained allowed visualization of the inner structure of CyD as shown in Fig. 10.6.6d. Here, three α-CyD-nanotubes are shown running from the upper right direction to the lower left. Each molecule seems to consist of three spots, which might correspond to glucose moieties.

Not only static high-resolution STM images of CyD molecules but also dynamic phase transition processes between three phases, i.e. "desorption", "self-organization to nanotube", and "disordered" states of CyDs, have been observed by means of EC-STM. The possibilitiers of determining not only static structures but also dynamic processes are significant advantages of electrochemical STM imaging compared to ordinary procedures using casting of sample and observation exposed to air.

Note that phase transitions are significantly controlled by delicate potential management, and observation of dynamic processes for 2D molecular aggregation would allow us to visually understand the thermodynamic balance based on intermolecular interactions and perpendicular interactions between adsorbates and substrates.

10.6.4
Dynamic Force Spectroscopy

Another interesting attempt using chemical force microscopy based on an AFM system has also been reported [26, 27] consisting not in the visualization of molecules in a supramolecular system but allowing the detection of interactions operating among the host–guest complexes. Huskens and co-workers successfully demonstrated direct measurement of single host–guest interactions within the complex between β-CyD SAMs on substrates as a host and a guest molecule con-

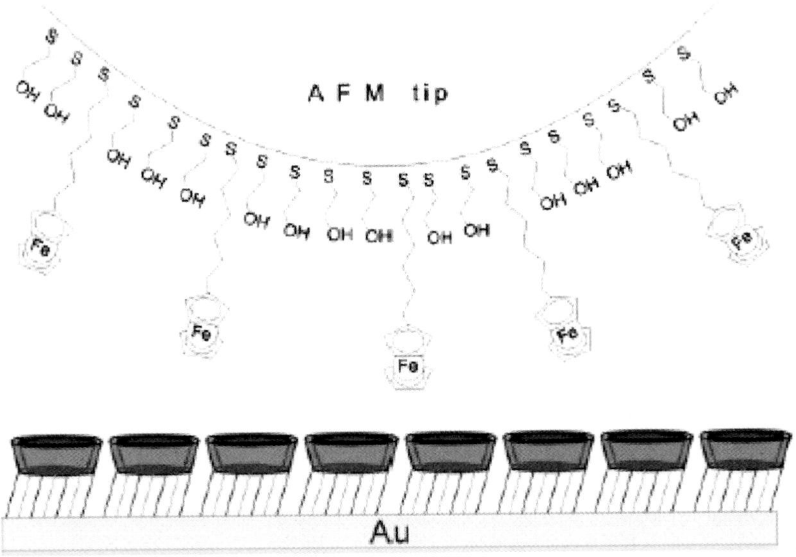

Fig. 10.6.7. Schematic representation of AFM-based single molecule force spectroscopy of a ferrocene guest immobilized on an AFM tip and the host CyD-SAM on Au(111) surfaces. Reprinted with permission from ref 26 (a). Copyright by American chemical society.

nected to the surface of a gold-coated AFM tip (Fig. 10.6.7) [26]. Since binding kinetics are independent of the AFM measurement time scale, the resulting force in the host–guest complex indicates interactions under thermodynamic control in this experimental system. This approach could potentially be applied to other host–guest complex systems such as calixarenes, crown ethers, and other organic molecules as well as in natural and/or artificial proteins in biological systems.

10.6.5 Conclusion

As this section has indicated, scanning probe microscopy has already shown remarkable capacities in visualizing molecules forming nanostructures, and it has now become an indispensable method in supramolecular chemistry. Similarly to the imaging of CyDs, high-resolution STM imaging of various molecules, which has attracted attention in the field of supramolecular chemistry, has been achieved. SPM including high-resolution imaging and force spectroscopy will be extremely valuable for future applications in many fields. However, researchers who want to use SPM to view their favorite molecule should pay attention to unexpected observations of artifacts resulting from the "tip-broadening effect" [28] or "multiple tip effect," and to overestimations from the image obtained. For instance, artifact images on HOPG surfaces [29] have sometimes been reported. Therefore, experimental reproducibility and various combinations of research methodologies are necessary and crucial to clarify the images.

References

1 Bender, M.; Komiyama, M. *Cyclodextrin Chemistry*; Springer, Berlin, 1976.

2 Lehn, J.-M. *Supramolecular Chemistry*; VCH, Weinheim, 1995.

3 (a) Sauvage, J.-P. *Acc. Chem. Res.* 1998, *31*, 611–619. (b) Balzani, V.; Gomez-Lopez, M.; Stoddart, J. F. *Acc. Chem. Res.* 1998, *31*, 405–411. (c) Nepogodiev, S. A.; Stoddart, J. F. *Chem. Rev.* 1998, *98*, 1959–1976. (d) Hoshino, T.; Miyauchi, M.; Kawaguchi, Y.; Yamaguchi, H.; Harada, A. *J. Am. Chem. Soc.* 2000, *122*, 9876–9877.

4 (a) Kunitake, M.; Kotoo, K.; Manabe, O.; Muramatsu, T.; Nakashima, N. *Chem. Lett.* 1993, 1033–1036. (b) Harada, A.; Li, J.; Kamachi, M. *Chem. Commun.* 1997, 1413–1414. (c) Murakami, H.; Kawabuchi, A.; Kotoo, K.; Kunitake, M.; Nakashima, N. *J. Am. Chem. Soc.* 1997, *119*, 7605–7606.

5 (a) Harada, A.; Kamachi, M. *Chem. Commun.* 1990, 1322–1323. (b) Harada, A.; Kamachi, M. *Macromolecules* 1990, *23*, 2821–2823. (c) Harada, A.; Li, J.; Kamachi, M. *Nature* 1992, *356*, 325–327. (d) Harada, A.; Li, J.; Kamachi, M. *J. Am. Chem. Soc.*, 1994, *116*, 3192–3196. (e) Harada, A.; Li, J.; Kamachi, M.; Kitagawa, Y.; Katsube, Y. *Carbohydrate Research*, 1998, *305*, 127–129. (f) Harada, A. *Acta Polymer.* 1998, *49*, 3–17. (g) Okada, M.; Kamachi, M.; Harada, A. *J. Phys. Chem. B* 1999, *103*, 2607–2613. (h) Kamitori, S.; Matsuzaka, O.; Kondo, S.; Muraoka, S.; Okuyama, K.; Noguchi, K.; Okada, M.; Harada, A. *Macromolecules* 2000, *33*, 1500–1502. (i) Okada, M.; Kawaguchi, Y.; Okumura, H.; Kamachi, M.; Harada, A. *J. Polym. Sci., A: Polym. Chem.* 2000, *38*, 4839–4849.

6 (a) Harada, A.; Li, J.; Kamachi, M. *Nature* 1993, *364*, 516–518. (b) Harada, A.; Li, J.; Kamachi, M. *Nature* 1994, *370*, 126–128. (c) Harada, A. *Acc. Chem. Res.* 2001, *34*, 456–464.

7 (a) Itaya, K. *Prog. Surf. Sci.* 1998, *58*, 121–247. (b) Ikai, A. *Surf. Sci. Rep.* 1997, *26*, 261–332. (c) Jandt, K. D. *Surf. Sci.* 2001, *491*, 303–332.

8 (a) Heckl, W. M.; Kallury, K. M. R.; Thompson, M.; Gerber, C.; Horber, H. J. K.; Binnig, G. *Langmuir* 1989, *5*, 1433–1435. (b) Snyder, S. R.; White, H. S.; Lopez, S.; Abruna, H. D.; *J. Am. Chem. Soc.* 1990, *112*, 1333–1337. (c) Widrig, C. A.; Alves, C. A.; Porter, M. D. *J. Am. Chem. Soc.* 1991, *113*, 2805–2810. (d) Kim, Y.-T.; Bard, A. J. *Langmuir* 1992, *8*, 1096–1102. (e) Luttrull, D. K.; Graham, J.; DeRose, J. A.; Gust, D.; Moore, T. A.; Lindsay, S. M. *Langmuir* 1992, *8*, 765–768. (f) Kim, Y.-T.; McCarley, R. L.; Bard, A. J. *J. Phys. Chem.* 1992, *96*, 7416–7421. (g) Viswanathan, R.; Zasadzinski, J. A.; Schwartz, D. K.; *Science*, 1993, *261*, 449–452. (h) Li, M. Q. *Applied Phys. A: Materials Sci. Processing* 1999, *A68*, 255–258.

9 (a) Poirier, G. E.; Pylant, E. D. *Science* 1996, *272*, 1145–1148. (b) Chiang, S. *Chem. Rev.* 1997, *97*, 1083–1096. (c) Jung, T. A.; Schlittler, R. R.; Gimzewski, J. K. *Nature* 1997, *386*, 696–699. (d) Soriaga, M. P.; Itaya, K.; Stickney, J. L. *Electrochem. Nanotechnol.* 1998, 267–276. (e) Gimzewski, J. K.; Joachim, C.; Schlittler, R. R.; Lauglais, V.; Tang, H.; Johannsen, I. *Science* 1998, *281*, 531–533. (f) Gimzewski, J. K.; Joachim, C. *Science*

1999, *283*, 1683–1688. (g) Tanaka, H.; Hamai, C.; Kanno, T.; Kawai, T. *Surf. Sci.* 1999, *432*, L611–L616. (h) Taniguchi, M.; Nakagawa, H.; Yamagishi, A.; Yamada, K. *Surf. Sci.* 2000, *454–456*, 1005–1009. (i) Bohringer, M.; Schneider, W.-D.; Berndt, R. *Angew. Chem., Int. Ed. Engl.* 2000, *39*, 792–795. (j) Hla, S.-W.; Meyer, G.; Rieder, K.-H. *Chem. Phys. Chem.* 2001, *2*, 361–366, (k) Yokoyama, T.; Yokoyama, S.; Kamikado, T.; Okuno, Y.; Mashiko, S. *Nature* 2001, *413*, 619–621. (l) Poirier, G. E. *Phys. Rev. Lett.* 2001, *86*, 83–86. (m) Weckesser, J.; Vita, A. De.; Barth, J. V.; Cai, C.; Kern, K. *Phys. Rev. Lett.* 2001, *87*, 096101-1–096101-4. (n) Ackinger, M.; Griessl, S.; Heckl, Wolfgang M.; Hietschold, M. *J. Phys. Chem. B* 2002, *106*, 4482–4485. (o) Barth, J. V.; Weckesser, J.; Trimarchi, G.; Vladimirova, M.; Vita, A. De.; Cai, C.; Brune, H.; Gunter, P.; Kern, K. *J. Am. Chem. Soc.* 2002, *124*, 7991–8000.

10 (a) Rabe, J. P.; Buchholz, S. *Science* 1991, *253*, 424–427. (b) Schwinn, T.; Gaub, H. E.; Rabe, J. P. *Supramol. Sci.* 1994, *1*, 85–90. (c) Cincotti, S.; Rabe, J. P. *Supramol. Sci.* 1994, *1*, 7–10. (d) Heinz, R.; Stabel, A.; DeSchryver, F. C.; Rabe, J. P. *J. Phys. Chem.* 1995, *99*, 505–507. (e) Heinz, R.; Stabel, A.; DeSchryver, F. C.; Rabe, J. P. *J. Phys. Chem.* 1995, *99*, 8690–8697. (f) Eichhorst-Gerner, K.; Stabel, A.; Moessner, G.; Declerq, D.; Valiyaveettil, S.; Enkelmann, V.; Müllen, K.; Rabe, J. P. *Angew. Chem., Int. Ed. Engl.* 1996, *35*, 1492–1495. (g) DeFeyter, S.; Gesquiere, A.; Abdel-Mottaleb, M. M.; Grim, P. C. M.; DeSchryver, F. C. *Acc. Chem. Res.* 2000, *33*, 520–531. (h) Samorí, P.; Jäckel, F.; Ünsal, Ö.; Godt, A.; Rabe, J. P. *Chem. Phys. Chem.* 2001, *7*, 461–464. (i) Giancarlo, L. C.; Fang, H.; Rubin, S. M.; Bront, A. A.; Flynn, G. W. *J. Phys. Chem. B* 1998, *102*, 10255–10263.

11 Ishikawa, Y.; Ohira, A.; Sakata, M.; Hirayama, C.; Kunitake, M. *Chem. Commun.* 2002, 2652–2653.

12 Hooks, D. E.; Fritz, T.; Ward, M. D. *Adv. Mater.* 2001, *13*, 227–241.

13 (a) Sakai, T.; Ohira, A.; Sakata, M.; Hirayama, C.; Kunitake, M. *Chem. Lett.* 2001, 782–783. (b) Pan, G.-B.; Wan, L.-J.; Zheng, Q.-Y.; Bai, C.-L.; Itaya, K. *Chem. Phys. Lett.* 2002, *359*, 83–88. (c) Ohira, A.; Sakata, M.; Hirayama, C.; Kunitake, M. *Org. Biomolec. Chem.* 2003, *1*, 251–253. (d) Ohira, A.; Sakata, M.; Taniguchi, I.; Hirayama, C.; Kunitake, M. *J. Am. Chem. Soc.* 2003, *125*, 5057–5065.

14 (a) Cunha, F.; Tao, N. J. *Phys. Rev. Lett.* 1995, *75*, 2376–2379. (b) Wandlowski, T.; Lampner, D.; Lindsay, S. M. *J. Electroanal. Chem.* 1996, *404*, 215–226. (c) Cunha, F.; Tao, N. J.; Wang, X. W.; Jin, Q.; Duong, B.; D'Agnese, J. *Langmuir* 1996, *12*, 6410–6418. (d) Wandlowski, Th.; Dretschkow, Th.; Dakkouri, S. A. *Langmuir* 1997, *13*, 2843–2856. (e) Cunha, F.; Jin, Q.; Tao, N. J.; Li, C. Z. *Surf. Sci.* 1997, *389*, 19–28. (f) Schweizer, M.; Hagenstrom, H.; Kolb, D. M. *Surf. Sci.* 2001, *490*, L627–636.

15 Miyake, K.; Yasuda, S.; Harada, A.; Sumaoka, J.; Komiyama, M.; Shigekawa, H. *J. Am. Chem. Soc.* 2003, *125*, 5080–5085.

16 Shigekawa, H.; Miyake, K.; Sumaoka, J.; Harada, A.; Komiyama, M. *J. Am. Chem. Soc.* 2000, *122*, 5411–5412.

17 Liu, Y.; Zhao, Y.-L.; Zhang, H.-Y.; Li, X.-Y.; Liang, P.; Zhang, X.-Z.; Xu, J.-J. *Macromolecules* 2004, *37*, 6362–6369.

18 Liu, Y.; Li, L.; Zhang, H.-Y.; Zhao, Y.-L.; Wu, X. *Macromolecules* 2002, *35*, 9934–9938.

19 Liu, Y.; Song, Y.; Wang, H.; Zhang, H.-Y.; Li, X.-Q. *Macromolecules* 2004, *37*, 6370–6375.

20 Liu, Y.; You, C.-C.; Zhang, H.-Y.; Kang, S.-Z.; Zhu, C.-F.; Wang, C. *Nano Lett.* 2001, *11*, 613–616.

21 Miyauchi, M.; Takashima, Y.; Yamaguchi, H.; Harada, A. *J. Am. Chem. Soc.* 2005, *127*, 2984–2989.

22 van den Boogaard, M.; Bonnet, G.; van 't Hof, P.; Wang, Y.; Brochon, C.; van Hutten, P.; Lapp, A.;

Hadziioannou, G. *Chem. Mater.* 2004, *16*, 4383–4385.

23 Shimomura, T.; Akai, T.; Abe, T.; Ito, K. *J. Chem. Phys.* 2002, *116*, 1753–1756.

24 Lo Nostro, P.; Santoni, I.; Bonini, M.; Baglioni, P. *Langmuir* 2003, *19*, 2313–2317.

25 Lipokowski, J.; Ross, P. N. (Eds.) *Adsorption of Molecules at Metal Electrodes*, VCH, New York, 1992.

26 (a) Auletta, T.; de Jong, M. R.; Mulder, A.; van Veggel, F. C. J. M.; Huskens, J.; Reinhoudt, D. N.; Zou, S.; Zapotoczny, S.; Schönherr, H.; Vancso, G. J.; Kuipers, L. *J. Am. Chem. Soc.* 2004, *126*, 1577–1584. (b) Zapotoczny, S.; Auletta, T.; de Jong, M. R.; Schönherr, H.; Huskens, J.; van Veggel, F. C. J. M.; Reinhoudt, D. N.; Vancso, G. J. *Langmuir* 2002, *18*, 6988–6994. (c) Schönherr, H.; Beulen, M. W. J.; Buügler, J.; Huskens, J.; van Veggel, F. C. J. M.; Reinhoudt, D. N.; Vancso, G. J. *J. Am. Chem. Soc.* 2000, *122*, 4963–4967.

27 (a) Frisbie, C. D.; Rozsnyai, L. F.; Noy, A.; Wrighton, M. S.; Lieber, C. M. *Science* 1994, *265*, 2071–2074. (b) Noy, A.; Vezenov, D. V.; Lieber, C. M. *Annu. Rev. Mater. Sci.* 1997, *27*, 381–421.

28 Samori, P.; Francke, V.; Mangel, T.; Mullen, K.; Rabe, J. *Opt. Mater.* 1998, *9*, 390–393.

29 Clemmer, C. R.; Beebe, T. P. Jr. *Science* 1991, *251*, 640–642.

11
Modeling of CyDs and Their Complexes

Helena Dodziuk

11.1
Introduction

Molecular modeling is trendy and in a large number of studies of isolated molecules of small or medium size it can provide useful information using well-tested methods and established parameter sets. Calculations on large flexible molecules and supramolecular systems, in particular those involving CyDs, are much more problematic [1]. The situation is influenced by at least five factors.

- There is a real need of models that allow one to better understand the observed phenomena, especially on chiral recognition, and thus be able to make predictions and use modeling to develop chiral phases for efficient chromatographic separations. One can feel this need by reading, for example, Chapter 6 of this book.
- As will be shown below, a competent evaluation of the pertinent experimental data for comparison with the calculated data presents a serious problem.
- The enormous complexity of CyDs and their complexes from the theoretical point of view is mostly overlooked. For instance, the nonrigidity of CyDs (discussed in Section 1.3) and its implications are totally neglected in most theoretical CyD studies. A good example of such neglect is the frequently unformulated assumption on the equal energy of the CyD host in complexes with different isomer guests, e.g. enantiomers, which ignores the significant effects of the calculated energy differences, under the influence of the induced fit mechanism of complexation, on the CyD geometry and energy. (See, for instance, Section 6.5 and the discussion of chiral recognition modeling there.)
- The (relative) ease of carrying out the calculations, treating relevant programs as black boxes, takes its toll. In consequence, the accuracy of the calculations and the dependence of the results obtained on the assumed parameter values have almost never been tested.
- Last but not least, beautiful computer models attract attention and convey the feeling that they can disclose some valuable information. As we shall show in the discussion of so-called lipophilic surfaces [2], this is not always the case.

Cyclodextrins and Their Complexes. Edited by Helena Dodziuk

ISBN: 3-527-31280-3

As we discuss below, most existing reviews [3, 4, 5] neglect the above and some other objections. Describing modeling of chiral recognition (usually involving very small energy differences) by CyDs, proteins and synthetic receptors Lipkowitz [4] states "This Account ... explains why differential free energies of binding can be computed so accurately. The review focuses on chiral recognition in chromatography, emphasizing binding and enantiodiscriminating forces responsible for chiral recognition ... in cyclodextrins, proteins, and synthetic receptors." He also claims [3] that "One can very accurately predict an unknown molecule's property or anticipated response by computing its molecular descriptors and substituting those values into the model" and states that "... only recently has the average chemist had the tools available to carry out these calculations (i.e. CyD modelling in 1998, HD) in a reasonable time period." We do not share this optimism, which triggered numerous subsequent applications of molecular modeling to CyDs and their complexes. The difference between our and Lipkowitz's approaches to the modeling of such large and mobile systems can be seen from his conclusions [3] concerning the discrepancies between the results of the calculations by his and Lichtenthaler's groups [6, 2] declaring that "The beauty of science is to discover the truth and neither group claims their method to be right." We believe that the areas of applicability and limitations should be explored and only the results that do not greatly depend on a reasonable choice of parameters and that are carefully checked against experimental data deserve publishing.

Unfortunately, in numerous cases, the results of a modeling of a single molecule or of a complex with a single guest, instead of a large series, are presented as "hard findings" without confronting them with experimental data and without looking for trends rather than "absolute" values. Moreover, the experimental data are often misinterpreted. For instance, as discussed in Section 1.4, comparison of X-ray geometry with the calculated one can be carried out in two ways. One can either choose a very laborious approach and try to compare all geometrical parameters, e.g. all $O4_n$–$O4_{n+1}$ distances, in the CyD under study or compare only the average values. As the data collected in Table 1.2 shows, much useful information gets lost if the latter method of data evaluation is used. Therefore, comparability of the average values of geometrical parameters cannot be considered as a sufficient condition of agreement between the computed and experimental results.

Another example showing the misinterpretation of experimental data is provided by the modeling of chiral recognition of *cis*-decalin invertomers by β-CyD presented briefly below. In this case, the calculations [7] did not yield any reliable information and the authors misinterpreted our experimental results [8] to justify them. However, these calculations were cited [3] as one of numerous examples of the successful application of modeling in CyD studies.

Before we discuss the calculations, it is appropriate to list the acronyms denoting methods used in this field.

AM1 – Austin Model 1 [9]
AMBER – molecular mechanics program and force field developed by Kollman's group [10]
BLYP – Becke, Lee, Yang, Parr [11, 12]

CFF91, CVFF – force fields used in molecular mechanics calculations
CNDO – Complete Neglect of Differential Overlap
DFT – Density Functional Theory
DS – Dynamic Simulations
MC – Monte Carlo Simulations
MD – Molecular Dynamics
MM – Molecular Mechanics
MNDO – Modified Neglect of Differential Overlap [13]
MM2, MMX – versions of the MM method
ONIOM – our Own N-layered Integrated molecular Orbital and molecular Mechanics method Morokuma [121, 122]
PDDG – Pairwise Distance Directed Gaussian [14]
PM3 – Parametric Method number 3 [15]
QC – Quantum Chemical Calculations
QSAR – Quantitative Structure–Activity (or Property) Relation
RHF – Restricted Hartree Fock
SCF – Self-Consistent Field Method
STO – Slater-type Orbital

One more general remark. Modeling of supramolecular complexes is a relatively new field. Until now, it has consisted mainly in the application of methods well established for relatively rigid isolated molecules to systems that are dynamic and held together by much weaker forces than those operating in binding organic molecules. Therefore, we believe that most of the methods used in CyD modeling are improperly applied and those acceptable can yield only qualitative results.

The beginning of CyD modeling was not very promising. In the earliest work in this domain that we are aware of, in which a simplified MM model was applied [16], CyDs smaller than α-CyD, with fewer than six glucopyranose units, were found to be too strained to exist. This opinion was repeated for more than 20 years until the synthesis of such a macrocycle by Nakagawa et al. [17]. Tabushi [18] developed an elaborate model that included not only host–guest interactions but also the energy portion associated with the redistribution that accompanied the collapse of the cavity in water arising from the removal of the guest from the solvent as a result of complex formation. This model included both enthalpy and entropy terms. Tabushi's group also tried to analyze the significance of guest polarity in the stabilization of CyD complexes on the basis of a MM model [19]. These works are of historical interest today. Considerable increases in computer power paved the way to massive applications of theoretical modeling to CyDs and their complexes starting in the early 1990s.

Our own interest in CyD modeling has evolved from the experimental study of molecular and chiral recognition of decalin isomers **1** by β-CyD **2** [20, 8, 21]. These molecules provide a unique set for such studies since at room temperature the *cis*-isomer undergoes rapid ring inversions between the **1b** and **1c** forms, which are enantiomers as well. Thus, decalin isomers allow us to study both molecular and chiral recognition. Chromatographic analysis by Sybilska's group [20] has shown that the *trans* isomer **1a** forms weak complexes with **2** while the complexes with the *cis* **1b** and **1c** ones are of considerable stability. Our measurements of NMR spectra of the latter complex at low temperatures, when the ring inversion is frozen, have shown that chiral recognition of **1b** and **1c** by β-CyD manifests itself by splitting or broadening of all signals in ^{13}C NMR spectra [8] and by unusually complicated patterns of signals in ^{1}H spectra [21].

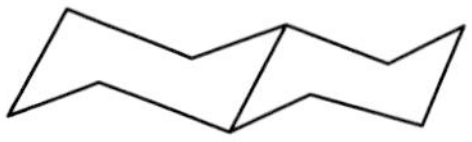

trans-**1a**

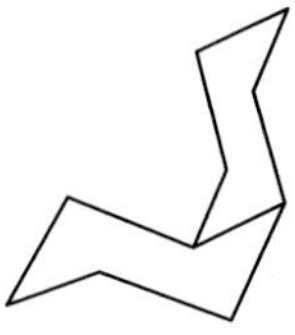

cis-**1b**

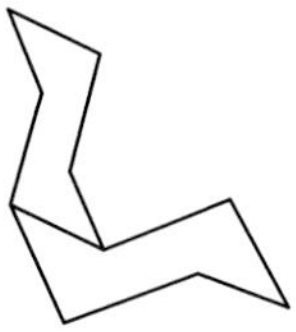

cis-**1c**

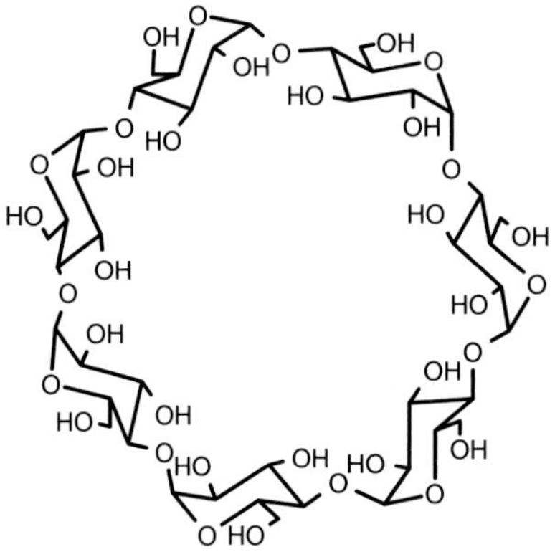

2

Having experience in MM calculations for small molecules [22, 23], we have carried out the modeling of the complexes of isomers of **1** with **2** and obtained, contrary to experimental findings, a higher value of stabilization energy (defined in Section 11.3) for the complex with the *trans* isomer than for complexes with the *cis* ones [24]. Therefore, we have not continued the computations. However, to our surprise, an MM study of the latter complexes has been published [7] claiming (1) that there is a small energy difference between the complexes with **1b** and **1c** and (2) that the results of this modeling agree with our ^{13}C spectra [8]. This was not the case, since the small energy difference for the pair of complexes with enantiomeric species was to be expected without any computation, although no assignment of the split signals to **1b** and **1c** could be carried out. Therefore, NMR data did not indicate which invertomer was more stable. The well-known difficulties in estimating the relative contents of the species on the basis of the signal intensities in ^{13}C NMR spectra did not allow us to draw any qualitative conclusion concerning the relative stability of the complexes under investigation. However, the MM study of *cis*-decalin complexation by **2** [7] indicated that the problem of the reliability of CyDs modeling deserved the detailed analysis [25, 26, 27] that will be presented later.

According to a literature search, more than 140 papers on CyD modeling have been published in 2004 alone, in which QC calculations, [28, 29, 30], MM [31, 32, 33] and MD [34] have been mostly applied. Other theoretical methods, seldom used in CyD studies [35], include MC [36, 37, 38]. It should be stressed, however, that (1) the MC method has been declared to be inefficient for (macro)molecular systems [34], (2) to our knowledge, no comparative check of validity of the MC method using different parameterizations and/or simulation lengths like those described below for MM [25] and MD [26, 27] methods has been reported, and (3) all objections concerning the calculations of small energy differences, e.g. those characterizing chiral recognition, and the CyD complex mobility raised by Pirkle and Pochapsky [39] are also valid for these calculations. The opinion of Pirkle and Pochapsky on the rationalization of chiral recognition is presented in Section 1.4. Here we would like to cite once more their main conclusion: "There is justifiable skepticism concerning the validity of any mechanism purporting to explain such small energy differences, despite a strong tendency among workers in the field to advance chiral recognition rationales, even when comparatively few data are available upon which to base such a rationale."

Modeling on the basis of QSAR [40, 41] chemometric analysis [42], and hybrid methods combining QC with MM [43, 44] or with MD [45] will not be covered in this review. It should be stressed that to the best of our knowledge such hybrid methods have never been applied to CyD studies and they must suffer from most of the deficiencies of semiempirical QC discussed in Section 11.2. As concerns the choice of a specific method to solve the case under study, this problem will be briefly discussed in Section 11.5 after we have presented the methods used in this field.

One more remark should be made here. As will be shown below, some conclusions gained on the basis of sometimes elaborate molecular modeling and used to support experimental results can be drawn from a simple examination of molecular models without any calculations at all. We believe that in such cases one should refrain from using any computations.

11.2 Quantum Chemical Calculations

Quantum mechanics is the most general and well founded theory of molecular structure. However, it can be rigorously applied only to a few very small systems and, as briefly discussed below, its validity in the reliable study of CyDs must be carefully proven in every specific case. Contrary to Lipkowitz's general statement of 1998 cited above [3] and to the opinion expressed by Liu [5], we believe that reliable QC for such large systems as CyDs and their complexes are not feasible even in 2005 for at least two reasons. The first one consists in the inadequate description of nonbonding dispersive interactions by the semiempirical, SCF, and DFT methods with small basis sets. For instance, semiempirical QC have been shown to produce completely unreliable results [46] for the complex formation ability of endohedral fullerene complexes with hydrogen molecules, which are much simpler supramolecular systems, with a rigid host, than CyD complexes. The second reason lies in the fact that only the enthalpic term is calculated even in the most advanced QC for large molecules. Thus, the entropic portion of the free energy is totally neglected in spite of the CyD nonrigidity and the importance of entropy in the stabilization of the relevant complexes discussed in Chapter 8. The neglect of the dynamical aspects associated with CyD nonrigidity in QC can be seen in both reviews on the application of QC [3, 5] in which considerable attention has been focused on the problem of finding the absolute energy minimum in view of the complicated shape of the CyD's energy hypersurface and on the elaborate docking procedures, but the problem of the neglect of the influence of the system's dynamics on its QC calculated energy has practically not been mentioned. Another serious limitation of QC is that they are usually carried out for an isolated molecule or complex. It should be stressed that the inclusion of the solvent effect cannot fully overcome all the deficiencies of such calculations. And, last but not least, contrary to some attempts, e.g. [47, 48, 49], the accuracy of QC does not allow one to apply this method to study chiral recognition where very small energy differences are involved [39].

To our knowledge, with two exceptions [50, 51] only semiempirical CNDO [52, 53], MNDO [54], AM1 [49, 55], PM3 [41], or DFT [56, 57] methods have been applied in cyclodextrin QC. A few representative papers showing the applications of QC in CyDs studies will now be discussed briefly.

In addition to the statistical analysis of literature data to find structure/property relations for a wide variety of drug complexes with α-, β- and γ-CyDs, Bodor and Buchwald [41] carried out AM1 calculations for the complex of the brain-targeted steroid **3**, erroneously called estradiol, with β-CyD **2**. One of the basic findings of their modeling is that "1:1 complexes are first formed with inclusion of the A ring, and at higher CD-concentrations, 1:2 complexes with inclusion at both the A and T ring are formed." This conclusion is not supported by any experimental evidence while carefully proven 1- and 2-dimensional NMR experiments for another steroid [58] indicate that, even at low concentrations, the formation of 1:1 complexes involving inclusion of either A or T rings is possible. Thus, exclusion of T ring involvement for low concentrations in the complex **3@2** could be justified only on the basis of the experimental evidence. The deformation index defined on

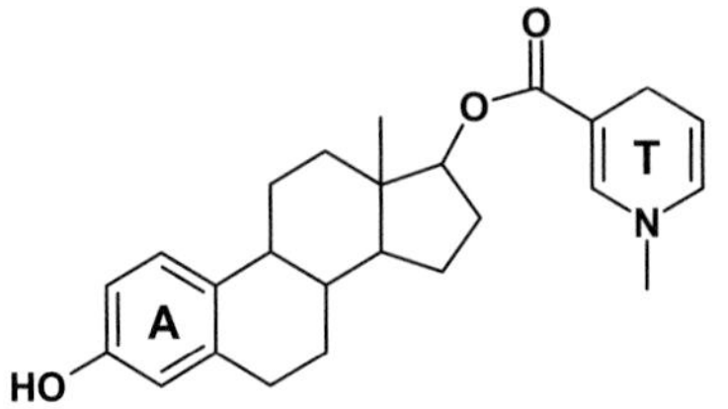

3

the basis of the same AM1 treatment [41] has been calculated for the minimum structure obtained by semiempirical QC subjected to all objections, i.e. the neglect of entropy and the presence of other low-energy structures, presented above.

Botsi et al. [55] reported an AM1 study of β-CyD complexes with 1,7-dioxaspiro[5.5]undecane **4** and nonanal **5** carrying out the calculations for the 1:1 stoichiometries determined in their earlier paper [59] although their experimental data for the latter guest could be also interpreted in terms of 2:2 stoichiometry in line with MM calculations for similar systems [60]. Botsi and coworkers claim that their calculated structure of β-CyD is in a very good agreement with the crystallographic structure. However, what they compare are the average structural parameters calculated for the isolated complexes **4@2** and **5@2** with those in the crystal (see discussion of the X-ray average values above and in Section 1.4) which, in addition to the host under study, contains water molecules either included into the cavity and/or located at interstices. Thus, the "excellent agreement" between the average calculated and X-ray bond lengths and angles seems not to be a sufficient criterion of the reliability of the method.

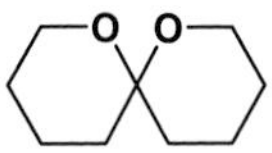

4

5

Similarly, comparing the solution structures of highly mobile complexes in a polar solvent, especially that involving **5**, with those calculated for the isolated complexes in vacuum seems inconclusive. Finally, carrying out the calculations under a C_7 symmetry constraint for the β-CyD host does not allow the authors to fully take into account induced fit adaptation of the host geometry in the complex [61]. Therefore, the authors' conclusion on the reliability of AM1 QC for CyD study seems unfounded. Also their result that the direction of the primary hydroxyl groups determine the dipole moment of β-CyD is neither novel nor precise. Of course, the direction of the hydroxyl groups (the most polar fragments in this host) plays an important role in determining the CyD dipole moment. However, the secondary OH groups (twice as numerous as the primary ones) can also be of importance in polar solvents, which could break the ring of flip-flopping hydrogen bonds present in the solid state [62, 63].

The head-to-head, head-to-tail, and tail-to-tail structures of the α-CyD dimer shown in Fig. 1.9 with a dodecahydrate cluster have been analyzed by Nascimento's group using "BLYP/6-31G(d,p)//PM3 Gibbs free energy" calculations [57]. Adding water molecules known to be highly mobile even in the solid state [64] has made the problem of CyD structure even more complicated. A real system of CyDs in water with hydrogen bonds rapidly reorganizing cannot be mimicked by the CyD dimer with the specific dodecahydrate water cluster; thus the results obtained are really no more reliable than is the case of the isolated cyclodextrin itself. The calculated structures could not and have not been checked against experiment and the authors suggestion of the possibility of an α-CyD nanotube with water tetramers as spacers, instead of the known one with the macrocycles threaded on the PEG chain [65, 66], seems unfounded in view of the rapid reorganization of hydrogen bonds involving water molecules.

Casadesus and coworkers [44] reported an ambitious attempt to test the performance of various theoretical methods by analyzing how they reproduce intermolecular H $\cdots$ H distances and stabilization energies in the inclusion complex of 2-(2′-hydroxyphenyl)-4-methyloxazole **6** with **2**. Molecular Mechanics (discussed below), semiempirical (MNDO, AM1, PM3, PDDG/MNDO, PM5), PDDG/PM3, and four versions of the ONIOM hybrid methods and *ab initio* RHF/STO-3G have been applied. The authors analyzed only two possible relative host–guest orientations and neglected the dynamic character of the complex under study. It should be stressed that the only experimental information the authors used was the inclusion of the oxazole ring of the guest into the host cavity determined in solution by NMR [67].

We can agree with the conclusion that the semiempirical and low-level layer ONIOM methods are unsuitable for the studies of CyD complexes although the authors' results, insufficiently checked against experiment, do not support it. For the same reason, we do not share the authors' cautious optimism concerning the use of the PM5 method while their statement that high-level *ab initio* and ONIOM calculations are not feasible today seems obvious.

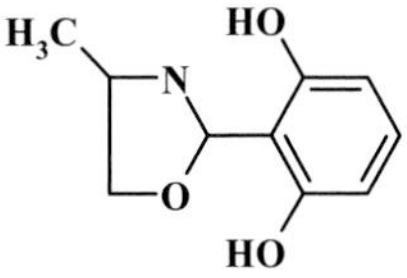

6

The role of charge-transfer interactions in the inclusion complexation of anionic guests by α-CyD for a large series of aromatic guests was studied using the DFT method [56]. In this paper, the CyD was claimed to have a truncated cone structure with all OH groups pointing outside the molecule, thus conferring a highly polar character on the molecular exterior. This picture does not agree with the commonly accepted CyD structure, which has only polar bases (see, for instance, Chapters 1 and 7) and relatively nonpolar sidewalls. Calculations for the host, guests, and complexes with neutral and anionic carboxylic acids have been carried out in vacuum and yielded correlations with the solution association constants obtained for two types of guest in solvents of considerably differing polarity. Fairly good correlations have been obtained for observed association constants and those predicted on the basis of a specific solvation model [68] and $S_{rel} = S_{guest}/S_{host}$ parameter calculated using the DFT method. This parameter is thought to reflect the ease of distortion of the electron cloud of the molecule [69]. However, the physical basis of the correlations reported seems unclear in view of the neglect of entropy factors, which are included in the experimental values of association constants.

Similarly, in spite of extensive experimental study of the complexes of styrene and methylstyrene with β-CyD, the claim by Cao and coworkers [70] on the basis of PM3 calculations that only van der Waals attractions and dipole–dipole interactions influence the complexes' geometry seems ill-founded.

The application of QC in the studies of reaction mechanisms involving CyDs is, probably, the only justifiable use of such methods, if it is carried out in a purely qualitative way and carefully checked against experiment, since MM and MD methods are unsuitable in this case. (Nevertheless, as mentioned in Section 1.3, two MM modelings of the reactions involving the ferrocene unit strongly accelerated by β-CyD have been published [71, 72]). However, in our opinion, QC on the reaction mechanism involving CyDs cannot serve as a decisive argument when one has to choose between conflicting experimental data. For instance, AM1 calculations of phenyl acetate hydrolysis of β-CyD carried out for the assumed macrocycle structure with the unreliable homodromic, i.e. unidirectional, orientations of all primary and secondary OH groups [73] seem ill-founded since at room temperature the hydroxyl groups of the oligosaccharide under scrutiny are known to undergo a rapid flip-flop reorientation even in the solid state (as discussed in Section 7.2.3) while in water, where the reaction takes place, the reorientation of hydroxyl groups should be even more pronounced. Moreover, the disparate experimental evidence cited by the authors concerning the reaction taking place at the 2′- or 3′-hydroxyl groups does not lend support to the results of AM1 modeling.

Once more it should be stressed that the use of quantum calculations to analyze chiral recognition by CyDs [47, 48, 49] is strongly discouraged in view of the very small energy differences involved and all the other arguments described earlier.

11.3
Molecular Mechanics Cyclodextrin Studies

In MM calculations, a molecule is considered on the basis of an extended purely mechanical model as a set of point masses connected by springs, with electrostatic interactions added, in which nonbonding interactions generate distortions from the hypothetical ideal bond lengths and angles [31, 32, 33]. Such a simplified model, in which several parameter sets are used, has been shown to reproduce satisfactorily the experimental structures of small and medium organic molecules and, to a lesser extent, the energy differences between rotational isomers, almost independently of the parameter set used [74]. A check of the predictive power of the MM method against numerous experimental data could be carried out since it has been parameterized for carefully selected sets of small molecules to reproduce their crystal structures. However, it should be stressed that the numerical values of the parameters used in the MM calculations have no clear physical meaning and can differ considerably in different force fields (FFs). For instance, permittivity ε (earlier known under the name dielectric constant) has a well-defined meaning in the macroscopic world but cannot be measured inside a molecule, so-called zero bond lengths and bond angles are purely hypothetical constructs, force constants used in the MM calculations are not the same as those calculated on the basis of vibrational spectra, etc. Similarly, assuming softer H···H and stronger C···C repulsions has been shown to yield similar results for most molecules to the calculations using FFs based on the opposite assumption [74]. In such a situation, developing FFs specific for complexes with CyDs [75] seems inappropriate. Our own experience with the version of AMBER FF [10] especially developed for sugars [76] was not promising. As a consequence of the different FF applied in CyD studies, the partitioning of the steric energy (or of the difference between the steric energies of two complexes) into increments (such as those due to bond distortions, electrostatic interactions, or hydrogen-bond parameters, used in certain parameterizations) to elucidate forces responsible for molecular or chiral recognition has no physical meaning. This problem will be illustrated below when discussing decalin complexes with β-CyD **1@2**.

As mentioned before, the situation with MM calculations for large and very flexible molecules and their inclusion complexes, especially those involving CyDs, is different from that of the calculations for small and medium organic molecules. As discussed in Section 1.3, the former systems are highly flexible, and in the majority of cases they are studied either in the solid state (discussed in Chapter 7) crystallized with, even more mobile, solvent molecules or in solution. Therefore, all the objections raised in Sections 11.1 and 11.2 concerning comparison of the calculated MM results with the corresponding experimental data are also valid.

The pitfalls in MM calculations on CyD complexes can be illustrated by the study of the complexes of isomeric decalins **1@2** [25]. As mentioned in Section 11.1, early MM calculations suggested that, contrary to experimental findings [20], the complex with the *trans*-isomer **1a** is more stable than those with the *cis*-isomers [24], resulting in the abandonment of these studies in our group. The unexpected publication of MM calculations soon afterwards [7], which misinterpreted our experimental ^{13}C NMR results [8] prompted us to analyze in more detail the reliability of MM calculations and their dependence on the assumed parameter values [25]. Four different force fields, (AMBER [10, 76], CVFF [77], CFF91 [78] and MMX [79]) and five permittivity ε values of 1, 2, 4, 10, and 20 have been used in the calculations of the steric energy of the **1a**–**1c** isomers, β-CyD **2**, and their complexes. As is usual in MM studies of CyD complexes, the stabilization energy of the complexes ΔE_{mol} or ΔE_{chir} (characterizing molecular or chiral recognition, respectively) was defined as the difference between the steric energy of the complex and the sum of the corresponding energies of its constituent parts. Contrary to experimental results [20] but in agreement with our earlier attempt to model the complexes **1@2**, the stabilization energy was lowest for the **1a** complex with the *trans*-decalin isomer for the CVFF, CFF91, and MMX FF for all ε values examined and for all but two ε values for the AMBER FF. Moreover, in several cases the energy difference $\Delta\Delta E_{molec}$, defined as the difference between the energy of the complex with the *trans* isomer **1a** and that of the closer in energy *cis* one, characterizing molecular recognition, and the corresponding energy difference $\Delta\Delta E_{chir}$ between the enantiomeric *cis* isomers **1b** and **1c**, characterizing chiral recognition, displayed an unreliable trend with $\Delta\Delta E_{chir}$ values larger than or comparable with that of $\Delta\Delta E_{molec}$. For the complex **1a@2** analyzed in this study [25], the energy partitioning into the increments, which are sometimes interpreted in terms of forces driving the complexation or responsible for chiral recognition, will be briefly discussed later in this section.

The structure of native CyDs was studied using MM calculations by at least three groups [80, 2, 81]. Lipkowitz analyzed several possible conformations of isolated native CyD structures by applying MM2 [82] and AMBER [10] parameterizations. Analysis of the data in this work indicated that the minimum energy conformation found under symmetry constraints depended on the method applied. For instance, for α-CyD with the MM2 FF the most stable of those studied was a conformer of C_3 symmetry while with AMBER a conformer of C_2 symmetry was of the lowest energy (conformers 10 and 19 in Ref. [80]). Moreover, Lipkowitz's comment on the differences, suggesting an examination of energy partitioning to decide which of the FF used, MM2 or AMBER, is more appropriate, seems totally wrong in view of the arguments given below. In addition, it is not clear why conformations of C_1 symmetry were not considered for α-CyD. On the basis of the analysis of only a few minimum conformations of all three native oligosaccharides, two important conclusions were drawn: CyDs were flexible systems of symmetry lower than C_n ($n = 6, 7$, and 8 for α-, β-, and γ-CyDs, respectively) in which the interglycosidic oxygen atoms did not lie in one plane. Of several oligosaccharides studied by Lichtenthaler and Immel [2] of interest here are the α-CyD structure calculated using a specific force field (which to the best of our knowledge has not been used by other research groups) and the beautiful lipophilicity patterns displaying the calculated polarity of the molecular surfaces. We believe that the planarity of the ring of interglycosidic oxygen atoms obtained for α-CyD is an artefact of the calculations since it contradicts the results of both Lipkowitz [80] and Dodziuk and Nowinski presented below [81], in which three different, commonly used FFs were applied. The colorful lipophilicity surfaces determined on the basis of the above MM calculations [2] aroused admiration and were awarded the Science Award of Sugar Processing Research Inc., New Orleans, LA, although the only reliable information they conveyed, namely that the narrower α-CyD base is less polar than the broader one, results from the presence of twelve OH groups on the broader base as compared to six on the narrow base, information which is available without carrying out any computations.

As mentioned before, the partition of steric energy or its differences into increments due to bond and angle distortions, electrostatic terms, van der Waals interactions, etc. was used to propose models of forces driving complex formation and recognition [83]. Since specific parameter values used in the calculations bear no physical significance, it follows that interpreting the large values of energy increments to the steric energy as arising from, for example, electrostatic or van der Waals interactions, is not acceptable. The latter statement can be illustrated by the appropriate data for the complex with *trans*-decalin **1b@2** calculated using AMBER [10, 76], CVFF [77], CFF91 [78], and MMX [79] FFs for $\varepsilon = 4$ [25]. The data collected in Table 3 of the last reference indicate that with the AMBER FF nonbonded repulsions and attractions practically cancel and the Coulomb electrostatic term is the most important. On the other hand, nonbonded energy, i.e. the algebraic sum of the repulsive and attractive nonbonding interactions, is more than twice as large as the Coulomb term in the CVFF parameterization. Calculations using the CFF91 parameters are characterized by a very strong negative, meaning attractive, torsional increment, which in contrast is found positive and large for the MMX FF. Such a dependence of the energy partitioning into increments arising from bond and angle distortions, nonbonded, Coulomb, and some other interactions on the force field used does not allow one to propose driving forces for complexations or separations involving CyDs on the basis of the MM calculations. However, Ivanov and Jaime [84] proposed the interpretation of forces driving the 1-bromoadamantane **7** guest into the cavities of all three native CyDs on the basis of MM calculations including solvent without any check against experiment. Similarly, assigning the role of electrostatic and van der Waals nonbonding interactions on the basis of MM calculations [85, 60] seems unfounded.

Several groups simultaneously applied MM calculations and NOE (or ROE) experiments to determine the mode of guest insertion into the CyD cavity. For instance, Jaime's group [86] studied the inclusion of benzoic acid **8** into the cavity of **2** using these techniques. After calculating the energy minima for two modes of the guest's entrance into the host cavity, the authors correctly put the question "... are energy minima good representatives for the average guest geometry tumbling inside the host?" However, their results seem to indicate a much larger preference for one mode of entrance than that experimentally found. Bekkers et al. [87] correctly determined that anthracycline antibiotics **9** did not form complexes with α- and β-CyD. For γ-CyD the 1:1 inclusion complex was found experimentally and the calculated structure of the complex was in qualitative agreement with proton chemical shifts and NOE results. However, the interaction energy they obtained (the difference between the energy of the complex minus those of its constituent parts) was higher for another isomer than that studied experimentally and this finding was not checked against experiment.

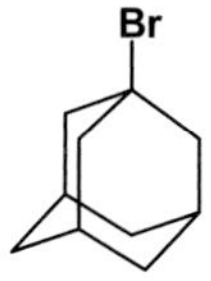

7

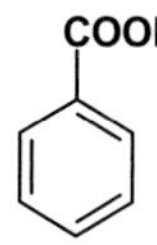

8

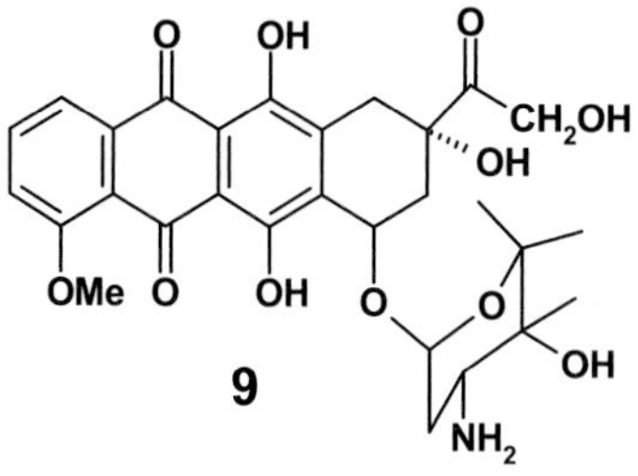

9

Chen and coworkers [88] developed a rather involved extended MM model to analyze CyD binding affinities in terms of free energy, enthalpy, and entropy terms for benzene, resorcinol, and three drugs. They compared their calculated results for the native CyDs with the average solid state geometry and their calculated ΔG°

values with the results obtained for solution, or vapor in some cases. For these five guests their calculated free energy values agree well with the experimental results, subject to the above limitation. However, the $T\Delta S$ values they have calculated for the complexes of all five guests with three native CyDs bear the same sign while the corresponding experimental values [89] do not. Chen et al. also analyzed the physically unfounded partition of the averaged energies into the Coulomb, van der Waals, etc. increments in an effort to provide a physical basis for the observed phenomena.

Two MM studies from 1988 of the mechanism of reactions involving CyDs deserve mention [71, 72] although they are based on many additional assumptions (for instance, the introduction of different special parameters to describe the ferrocenyl group) and lead to partly different conclusions.

Having expressed such deep skepticism towards MM calculations of CyD complexes, we would like to complete this section with a few examples of cases in which they can provide useful information.

Pop and coworkers used the MM method to choose between the possible structures of complexes suggested by powder-diffraction data [90]. Qualitative agreement was found between the relative stabilities of the complexes of the first generation dendrimer with adamantyl end groups **10**@β-CyD studied by chromatography and NMR and the results of MM calculations by Dodziuk et al. [91].

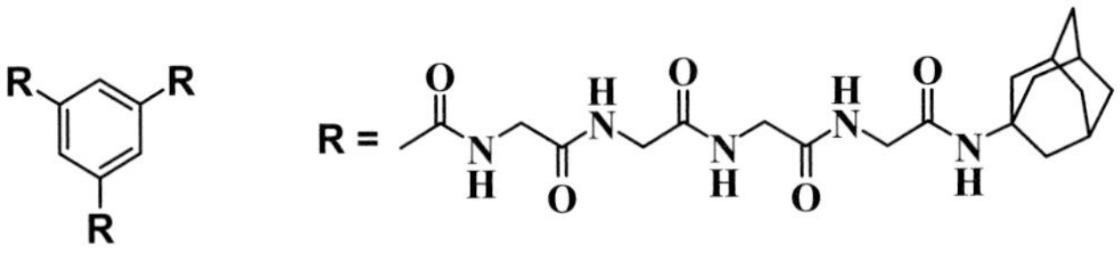

10

Experimental studies on cyanine dyes like **11** [60, 92] indicated that adding β- or γ-CyD in aqueous solution resulted in inclusion of two guest molecules into the CyD cavity if the chain linking was too short and there were no bulky end groups or bulky atoms on the linking chain. These experimental findings were consistent with the rotaxane type of structure (such as those discussed in Chapter 12) with two dye axles threaded through the CyD ring.

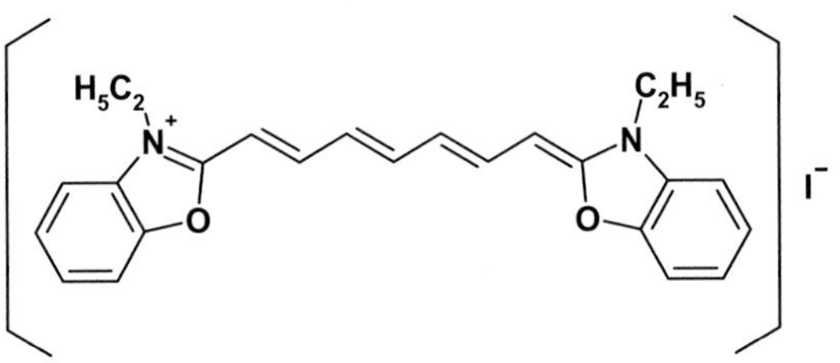

11

The subsequent MM study by Ohashi et al. [60], allowed the authors to rationalize the experimental results. However, in line with the above discussion we believe that their partitioning of the calculated steric energies into components to interpret them in terms of forces responsible for the formation of complexes is not justified.

To conclude this section on MM calculations for CyDs and their complexes, we would like to cite another statement from the review on CyD computations: "Hence molecular mechanics is much faster than quantum mechanics and can be used to reliably predict structures of small- to medium-sized molecules quickly or to evaluate very large molecular assemblies, e.g., cyclodextrins in a water bath." [3]. Let us repeat once more that we do not share this opinion.

11.4 Dynamic Simulations CyD Studies

In dynamic simulations, DS, sometimes called molecular dynamics, MD, Newtonian equations of motion are numerically integrated using a potential simplified in comparison to that used in molecular mechanics [34, 93]. Today, DSs are probably applied to CyD studies as often as MM calculations. However, caution should be exercised when applying these methods since in the opinion of one of the acknowledged authorities in the field, van Gunsteren, "... it is relatively easy, when modeling high-dimensional systems with many parameters, to choose or fit parameters such that good agreement is obtained for a limited number of observable quantities." [34]. We believe that the considerable number of unjustified applications of the simulations, going beyond the field of CyDs, prompted him to meticulously discuss the factors influencing the validity of the calculations, i.e. their reliability [94]. From the point of view of a theoretician, van Gunsteren gave five reasons that could be responsible for a disagreement between the calculated and experimental data (other than a successful theoretical finding of a novel phenomenon). The model applied could be inappropriate for the application, the force field could be inadequate, the sampling insufficient (i.e. too short a simulation time) leading to nonconvergence of the results, the software could contain bugs, or the software could be incorrectly used. According to van Gunsteren, even agreement between simulation and experiment did not guarantee the correctness of the model used (and such an agreement cannot be used, as is frequently done, to justify subsequent uses of the model, HD) since the property examined could be insensitive to the detail of the simulation and/or in the particular case under investigation there could be a compensation of errors. The examples given include, for example, agreement of the NMR spin–spin coupling constant of a β-peptide simulated at two different temperatures with the corresponding experimental values when a large structural difference between the helical and extended structures was not reflected in the J values [95]. Van Gunsteren collected several examples of the documented compensation of errors in simulation works [96] and gave some practical advice concerning carrying out the simulations [94]. For instance, a suggested 1.5-nm cutoff distance could be a reasonably accurate approximation for nonpolar or nonionic systems but has been shown to be totally inadequate for ionic species [97]. Using statistical mechanics van Gunsteren showed [94] that the decomposition of free energy change into different types of interaction, e.g. covalent, van der Waals, electrostatic (corresponding to partitioning of the steric energy discussed in Section 1.2), was, owing to averaging procedures, incorrect from a theoretical point of view even if the results obtained did not depend significantly on the force field applied. The latter dependence is discussed in some detail below.

Numerous structural (average atom positions or atom–atom distances, gyration radius, solvent accessible surface area, NMR order parameters (S^2), crystallographic temperature factors, dipole moment fluctuations (M^2) leading to an estimate of the dielectric permittivity ε, radial distribution functions ($g(r)$) along with density and energetic (heat of vaporization, free energy of solvation, heat capacity, isothermal compressibility, thermal expansion coefficient, and surface tension) parameters as well as dynamic properties (diffusion constants, rotational correlation times, dielectric correlation times, and viscosity) can be obtained on the basis of simulations [94]. In analogy with the free energy of solvation, the free energy of complex formation involving CyDs can be added to this list. Owing to the large sizes of the latter complexes, to the best of our knowledge only a few experimental values of geometrical parameters and energies (free energies, enthalpies, and entropies) of complex formation have been reported and compared with experimental data for CyDs and their complexes.

Van Gunsteren's paper [94] was devoted to an analysis of the problems rooted in the theory that could be encountered when comparing the calculated results of DSs with the experimental data. In this review, the presentation is mainly focused on the other side, i.e. an evaluation of the reliability of the comparison between experimental and computational findings applied to validate the simulations, since, as amply discussed earlier in this chapter, misinterpreted experimental results are often invoked to validate the CyD modeling.

The first important dynamic CyD studies published in 1987 and 1988 were the result of cooperation by the theoreticians van Gunsteren and Koehler with the X-ray specialist Saenger [98, 99, 100, 101, 102]. The aim of these studies was two-fold: on the one hand, they served the development of the GROMOS force field [103] while on the other they had to show that DSs for such complicated systems as CyDs were feasible. Starting from the experimental structures, these simulations of 15 or 20 ps, very short by today's standards, are of historical interest only. Similarly, the work by Mark et al. [104] of 1994 on free perturbation calculations was mainly devoted to the development of the method.

Most of the computer time- and memory-demanding DSs have been carried out for a single parameters set. In continuation of our studies on the validation of CyD modeling [25], MD calculations for decalin **1** complexes with β-CyD **2** [26] in vacuum have been carried out for the CVFF [77] and CFF91 [78] force fields and $\varepsilon = 1, 2, 4, 10$, and 20. (Interestingly, the complexes were not stable with the AMBER FF [10, 76] but no recombination, such as those detected by Varady et al. [105] discussed below, occurred.) As usual in MD simulations, $\Delta\Delta E_{\text{molec}}$ and $\Delta\Delta E_{\text{chir}}$ calculated from the average energies as described in Section 11.3 were taken as measures of molecular and chiral recognition, respectively. The results obtained depended on the assumed parameters. The use of CVFF yielded qualitative agreement relating to molecular recognition, with weak dependence on the assumed ε value within broad limits. For chiral recognition, the $\Delta\Delta E_{chir}$ values calculated with the latter FF were reliably small for $\varepsilon = 1, 2$, and 10. However, its sign for $\varepsilon = 2$ was opposite to that obtained for $\varepsilon = 1$ and 10 denoting the opposite preferences for the assumed different permittivity values. There was no reasonable way to choose between the ε values, and so, in agreement with the Pirkle and Pochapsky warning [39] (cited in Sections 1.4 and 11.1), the modeling of chiral recognition was deemed to be unreliable. Much smaller $\Delta\Delta E_{molec}$ values, comparable to or even larger than the calculated $\Delta\Delta E_{chir}$ ones were obtained using CFF91, disqualifying the results obtained for this FF.

As mentioned before, decalin isomers present an interesting case for analysis of the reliability of modeling of CyD complexation since they enable us to study molecular and chiral recognition for one set of isomeric guests. However, the lack of experimental data on the energy difference between the complexes involving the *cis*-decalin enantiomers prompted us to use another example to model chiral recognition, namely, the 1:2 complexes of enantiomers of α-pinenes **12** with α-CyD.

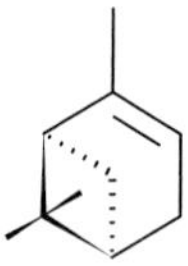

12a

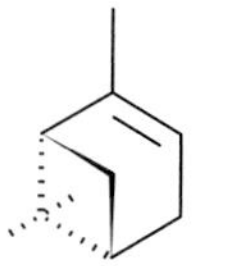

12b

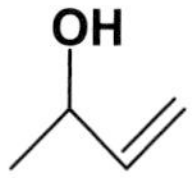

13

In addition to experimental study of these complexes [106], MD simulations in water have been carried out but, similarly to the theoretically unjustified [94] common approach in DSs [3, 4], the difference between the averaged energies of the complexes with **12a** and **12b** was taken as the measure of chiral recognition. The latter value obtained in a 3.5-ns simulation was in excellent, too good to be true [106], agreement with the experimental result by Moeder's group [107]. However, extension of the simulation lengths to 12 ns resulted in peculiar behavior [27]. The energy difference for the complexes involving enantiomers of **12** dropped when the simulation was extended, and between 4.6 and 8 ns the energy difference changed sign yielding, in contrast to experimental results [107], a larger stability of the complex with the (1*R*, 5*R*)-enantiomer and the correct preference was restored only after ca. 10 ns. It should be stressed that even today CyD simulations longer than 10 ns are quite uncommon. The claim by Lipkowitz [4] that 50-ns long simulations have been carried out in his group [108] seems unfounded since what was really done were five long 5-ns long simulations for each enantiomer with different starting positions of but-1-en-3-ol **13** guest with respect to α-CyD cavity.

Lipkowitz and coworkers [83] carried out an extensive NMR study of the complexes of tryptophan enantiomers **14** with α-CyD (in which, however, only one of the protons involved in the coupling described by the spin–spin coupling constants is given) in combination with the calculation of differential binding energies, defined as the energy difference between the average energies of the complexes involving the *R*- and *S*-enantiomers, and analyzed their partitioning into bond, angle, van der Waals, and Coulomb increments looking for the main factors responsible for chiral recognition. As follows from the analysis of the partitioning in the MM calculations discussed in Section 11.3 [25], such a partitioning has no physical meaning since it strongly depends on the force field applied in the calculations.

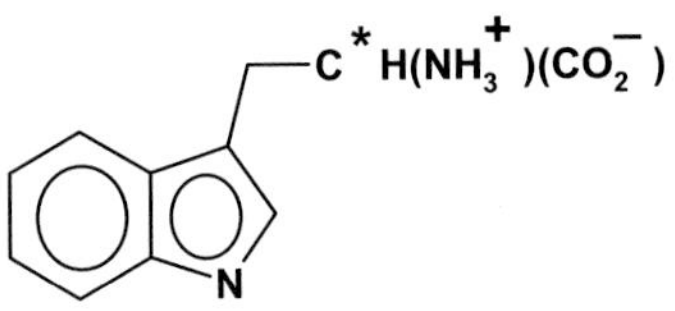

14

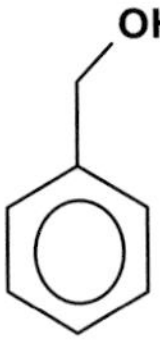

15

Of numerous published simulation calculations, one of the most interesting is the study by Varady and coworkers [105] on the complexation of benzyl alcohol **15** by *β*-CyD applying the long (10 and 12.5 ns) standard MD and their own self-guiding molecular dynamics (SGMD) methods allowing them to better search the conformational space. Notably, the authors studied the system consisting of six benzyl alcohol **15** molecules with one *β*-CyD **2** in a cubic box of water molecules with a 35-Å edge. By starting the simulations with different alcohol positions with respect to the host molecule and by carrying out sufficiently long simulation runs, they were able to observe several times the process that is known to take place in solutions, namely, the guest molecule entering into the host cavity and leaving it. (Many years ago I saw such a wandering of the guest in and out the CyD cavity in a much simpler setup. It was exciting and gave the feeling of observing a real physical phenomenon.)

An MD study of large-ring CyDs (discussed in Chapter 13) [109] encompassed the macrocyclic rings with 26, 30, 55, 70, 85, and 100 pyranose rings in water with simulation lengths of 10.0 ns. We believe that so far simulations of such large systems, for which, except CyD26 [110], no experimental data for comparison have been reported, do not bring any reliable information not available from general reasoning. No calculations are necessary to reach the conclusion that large CyDs are more flexible than native ones. The specific pictures of macroring evolution during the simulations with the portions of double helix and small loops certainly depend on the starting geometry, FF applied, etc. The findings that the results of the simulation in water do not reproduce X-ray-derived geometry for CyD26 is also self-evident, especially in view of the work of Kitamura and coworkers [111]. One could also expect that there are several cavities in the macroring of large CyDs which cannot be stable in view of their flexibility.

Bonnet and coworkers [112] studied the 2:1 complex of *γ*-CyD with C_{60}. The authors developed a novel procedure to account for the nonuniformity of electron density in the fullerene cage, positioning 90 dummy atoms in the bond centers.

This caused severe repulsions in the system and, in consequence, the reparameterization of the AMBER FF [10]. An energy partition was given to interpret the results of the simulations. The full procedure applied in this case casts doubts on the significance of the results since they depend on the parameterization applied and were not checked against experiment.

We also believe that the analysis of the inclusion complex geometry on the basis of the fluorescence spectra and MM modeling without invoking NOE (or ROE), as is done in [113], is not warranted, and also that the authors' interpretation that more than 90% of the stabilization of the 1:1 complexes comes from van der Waals interactions is unfounded since it is based on the partitioning of the difference between the average energies of the complex and the sum of its constituent fragments.

In the MD study of α-, β- and γ-CyD dimers in water [114] several additional assumptions have been made to the AMBER FF, including the calculations of partial charges and solvent effects, with the aim of establishing which of the head-to-head, head-to-tail, and tail-to-tail arrangements (Fig. 1.9) is more stable in solution. The native CyDs are proved to strongly associate in water: see, for instance, [115], (low CyD concentrations are used in the NMR studies to avoid their aggregation). However, to our knowledge, no experimental data on the system's aggregation are available in the literature for comparison with the results of modeling.

Contrary to the average energy values frequently used in DSs, elaborate and computer resources-consuming Free Energy Perturbation, FEP, calculations give theoretically well-based values [116]. The latter laborious method has been rarely used in CyD research. As shown by Mark et al. [104] for four complexes of *para*-substituted phenols with α-CyD, ΔG values calculated using the FEP method are considerably closer to the respective experimental values than the average energies obtained from simulations. Molecular dynamics simulations and FEP methods have been applied [117] to analyze the differences in the complexation of *p-tert*-butylphenyl, *p-tert*-butylbenzoate **16** and *N*-(*p-tert*-butylphenyl)-*p-tert*-butylbenzamide **17** with dimeric β-CyDs **18** studied earlier both experimentally [118] and theoretically [119].

Nevertheless, we believe that in this case also the results obtained depend on the assumed parameter values and their interpretation in terms of energy increments (i.e. energy partitioning in the terms used in this review) is unreliable. It is a pity that such a computer-demanding procedure has not been used to analyze a case studied experimentally in more detail.

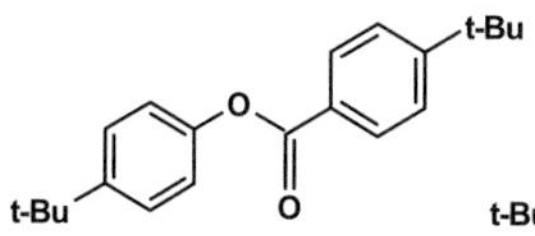

16

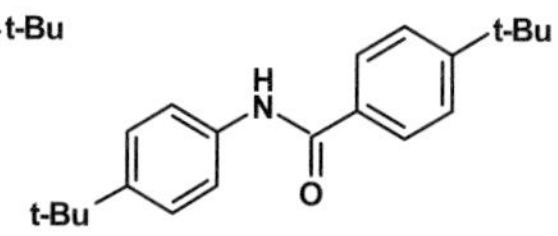

17

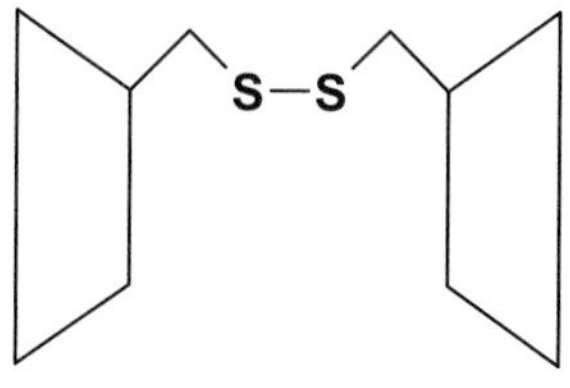

18

To summarize, again as stressed earlier for other computational methods, DSs should not be used to analyze chiral recognition by CyDs [120, 108] in view of the very small energy differences involved and all the other arguments described earlier together with Pirkle's and Pochapsky's reasoning [39]. This opinion is in sharp contrast with the one expressed in Section 6.5 in this book. Similarly, like the energy partitioning applied in numerous papers to rationalize the findings [3, 4, 108, 83, 88, 113, 112], Bea et al's FEP should not be used since it was proved to be strongly dependent on the assumed parameter values [25].

11.5 Conclusions

The examples of molecular modeling for CyDs and their complexes indicate that applications of computational methods are often ill-founded and the claims on the agreement with experimental results are in numerous cases unwarranted. Regrettably, the computations have not been carried out by teams of theoreticians and experimentalists but were rather performed using computer programs as black boxes. Unfortunately, in the majority of CyD physicochemical studies only one method, eventually combined with the modeling, is applied. We believe that, for such flexible systems as CyDs, studies should include simultaneously several experimental methods with a very careful check of computed values against the experimental ones. In particular, the results of NOE/ROE are a must when the geometry of the complexes is analyzed computationally.

Several warnings should be expressed here.

1. For the reasons presented in Section 11.2, quantum chemical calculations for such large systems as those involving CyDs cannot at present yield reliable well-founded information. Some reliable qualitative information based on the proximity of the reacting groups in the complex can be obtained without any QC.
2. As discussed in Sections 11.3 and 11.4, elucidation of the complexation mechanism and that of recognition on the basis of CyD MM and MD is unreasonable in view of the dependence of the partition of the calculated energy on the force fields and other parameters used.
3. In view of Pirkle's and Pochapsky's argument [39] (discussed in detail in Section 1.4) and, in particular, the very small energy differences involved in chiral recognition, the numerous attempts to rationalize the relevant experimental results using modeling seem unscientific.

As concerns the choice of the method applied, QC should not, in our opinion, be applied in CyD studies at all. In studies of reaction mechanisms, analysis of the proximity of the reacting groups on the basis of simple MM calculations of the complexes involved seems to be the most rational approach.

Assuming that the computer power available today is sufficient to carry out a comprehensive evaluation of CyD modeling, a consortium of theoreticians and experimentalists should in our opinion first define a set of benchmarks. That is, in addition to CyDs themselves, there should be a set of guest molecules for which, in addition to the existing literature data, experiments are proposed to create as comprehensive a data set as possible for comparison with the results of calculations. In particular, not only bond lengths and angles but also other geometry- and energy-dependent and parameters listed by van Gunsteren [94], such as NMR order parameters (S^2), crystallographic temperature factors, and dipole moment fluctuations (M^2), should be included in the testing procedure.

Finally, highly time-demanding MD simulations should be carried out for various FF, permittivity, and other parameters, such as cutoff distances, partial charges, simulation lengths, size of solvent boxes, etc., to study the reliability of CyD modeling and, probably, to formulate guidelines for less experienced users. Although, in connection with the last point we believe that the opposition to the effect of the deep compartmentalization of the whole science should lead to common studies led by researchers with various specializations.

To conclude, we believe that until such a comprehensive evaluation of theoretical approaches is carried out CyD modeling can, at best, provide purely qualitative results.

References

1 Dodziuk, H. In *Introduction to supramolecular chemistry*; Kluwer Academic Publishers: Dordrecht, 2002, p 216.

2 Lichtenthaler, F. W.; Immel, S. *Tetrahedr. Asymm.* **1994**, *5*, 2045.

3 Lipkowitz, K. B. *Chem. Rev.* **1998**, *98*, 1829.

4 Lipkowitz, K. B. *Acc. Chem. Res.* **2000**, *33*, 555.

5 Liu, L.; Guo, Q. X. *J. Inclus. Phenom.* **2004**, *50*, 95.

6 Lipkowitz, K. B. *J. Org. Chem.* **1991**, *56*, 6357.

7 Fotiadu, F.; Fathallah, M.; Jaime, C. *J. Inclus. Phenom.* **1993**, *16*, 55.

8 Dodziuk, H.; Sitkowski, J.; Stefaniak, L.; Jurczak, J.; Sybilska, D. *Chem. Commun.* **1992**, 207.

9 Dewar, M. J. S.; Zoebisch, E. G.; Healy, E. F. *J. Am. Chem. Soc.* **1985**, *107*, 3902.

10 Weiner, S. J.; Kollman, P. A.; Case, D. A.; Singh, U. C.; Ghio, C.; Alagona, G.; Profeta, S.; Weiner, P. *J. Am. Chem. Soc.* **1984**, *106*, 765.

11 Becke, A. D. *Phys. Rev. A* **1988**, *38*, 3098.

12 Lee, C.; Yang, W.; Parr, R. G. *Phys. Rev. B* **1988**, *37*, 785.

13 Dewar, M. J. S.; Thiel, W. *J. Am. Chem. Soc.* **1977**, *99*, 4899.

14 Repasky, M. P.; Chandrasekhar, J.; Jorgensen, W. L. *J. Comput. Chem.* **2002**, *23*, 1601.

15 Stewart, J. J. *J. Comput. Chem.* **1989**, *10*, 209, 221.

16 Sundararajan, P. R.; Rao, V. S. R. *Carbohydr. Res.* **1970**, *13*, 351.

17 Nakagawa, T.; Koi, U.; Kashiwa, M.; Watanabe, J. *Tetrah. Lett.* **1994**, *35*, 1921.

18 Tabushi, I.; Kiyosuke, Y.-I.; Sugimoto, T.; Yamamura, K. *J. Am. Chem. Soc.* **1978**, *100*, 916.

19 Tabushi, I.; Mizutani, T. *Tetrahedron* **1987**, *43*, 1447.

20 Koscielski, T.; Sybilska, D.; Lipkowski, J.; Mediokritskaja, A. *J. Chromatogr.* **1986**, *351*, 512.

21 Dodziuk, H.; Sitkowski, J.; Stefaniak, L.; Jurczak, J.; Sybilska, D. *Supramol. Chem.* **1993**, *3*, 79.

22 Dodziuk, H. *Top. Stereochem.* **1994**, *21*, 351.

23 Dodziuk, H. In *Modern Conformational Analysis. Elucidating Novel Exciting Molecular Structures*, 1995, p 157.

24 Dodziuk, H.; Nowinski, K. S. *unpublished results* **1992.**

25 Dodziuk, H.; Lukin, O.; Nowinski, K. S. *J. Mol. Struct. (THEOCHEM)* **2000**, *503*, 221.

26 Dodziuk, H.; Lukin, O. *Pol. J. Chem.* **2000**, *74*, 997.

27 Dodziuk, H.; Lukin, O. *Chem. Phys. Lett.* **2000**, *327*, 18.

28 Piela, L. *Ideas of Quantum Chemistry*; Elsevier: New York, 2006.

29 Szabo, A.; Ostlund, N. S. *Modern Quantum Theory. Introduction to Advanced Electronic Structure Theory*; McGraw-Hill: New York, 1989.

30 Atkins, P. W. *Molecular Quantum Mechanics*; Clarendon Press: Oxford, 1970.

31 Allinger, N. L. *Adv. Phys. Org. Chem.* **1976**, *13*, 1.

32 Osawa, E.; Musso, H. *Angew. Chem, Int. Ed. Engl.* **1983**, *22*, 1.

33 Dodziuk, H. In *Modern Conformational Analysis. Elucidating Novel Exciting Molecular Structures*; VCH-Publisher: New York, 1995, p 90.

34 van Gunsteren, W. F.; Berendsen, H. J. C. *Angew. Chem, Int. Ed. Engl.* **1990**, *29*, 992.

35 Henchman, R. H.; Kilburn, J. A.; Turner, D. L.; Essex, J. W. *J. Phys. Chem. B* **2004**, *108*, 17571.

36 Binder, K., Ed. *Monte Carlo Methods in Statistical Physics*; Springer: Berlin, 1979.

37 Szabelski, P. *Appl. Surface Sci.* **2004**, *227*, 94.

38 Lee, S.; Choi, Y.; Lee, S.; Jeong, K.; Jung, S. *Chirality* **2004**, *16*, 204.

39 Pirkle, W. H.; Pochapsky, T. C. *Chem. Rev.* **1989**, *89*, 347.

40 Park, H. J.; Choi, Y.; Lee, W.; Kim, K. R. *Electrophoresis* **2004**, *25*, 2755.

41 Bodor, N.; Buchwald, P. *J. Inclus. Phenom.* **2002**, *44*, 9.

42 Busch, K. W.; Swamidoss, I. M.; Fakayode, S. O.; Busch, M. A. *Anal. Chim. Acta* **2004**, *525*, 53.

43 Aqvist, J.; Warshel, A. *Chem. Rev.* **1993**, *93*, 2523.

44 Casadesus, R.; Moreno, M.; Gonzales-Lafont, A.; Lluch, M. M.; Repasky, M. P. *J. Comput. Chem.* **2003**, *25*, 99.

45 Car, R.; Parrinello, M. *Phys. Rev. Lett.* **1985**, *55*, 2471.

46 Dodziuk, H. *Chem. Phys. Lett.* **2005**, *410*, 39.

47 da Silva, A. M.; Empis, J.; Teixeira-Dias, J. J. C. *J. Inclus. Phenom.* **1999**, *33*, 81.

48 Cao, Y.; Xiao, X.; Lu, R.; Guo, Q. X. *J. Inclus. Phenom.* **2003**, *46*, 195.

49 Zborowski, K.; Zuchowski, G. *Chirality* **2002**, *14*, 632.

50 Li, X.-S.; Liu, L.; Mu, T. W.; Guo, Q. X. *Monatsh. Chem.* **2000**, *131*, 849.

51 Liu, L.; Li, X.-S.; Guo, Q. X.; Liu, Y. C. *Chin. Chem. Lett.* **1999**, *10*, 1053.

52 Kitagawa, M.; Hoshi, H.; Sakurai, M.; Inoue, Y.; Chujo, R. *Carbohydr. Res.* **1987**, *163*, C1.

53 Sakurai, M.; Kitagawa, M.; Hoshi, H.; Inoue, Y.; Chujo, R. *Carbohydr. Res.* **1990**, *198*, 181.

54 Furuki, T.; Hosokawa, F.; Sakurai, M.; Inoue, Y.; Chujo, R. *J. Am. Chem. Soc.* **1993**, *115*, 2903.

55 Botsi, A.; Yannakopoulou, K.; Hadjoudis, E.; Waite, J. *Carbohydr. Res.* **1996**, *283*, 1.

56 Jimenez, V.; Alderete, J. B. *Tetrahedron* **2005**, *61*, 5449.

57 Nascimento, C. S., Jr.; Anconi, C. P. A.; Dos Santos, H. F.; De Almeida, W. B. *J. Phys. Chem. A* **2005**, *109*, 3209.

58 Jover, A.; Budal, R. M.; Meijide, F.; Soto, V. H.; Tato, J. V. *J. Phys. Chem. B* **2004**, *108*, 18850.

59 Botsi, A.; Yannakopoulou, K.; Perly, B.; Hadjoudis, E. *J. Org. Chem.* **1995**, *60*, 4017.

60 Ohashi, M.; Kasatani, K.;

Shinohara, H.; Sato, H. *J. Am. Chem. Soc.* **1990**, *112*, 5824.
61 Koshland, D. E. *Angew. Chem. Int. Ed. Engl.* **1994**, *33*, 2475.
62 Betzel, C.; Saenger, W.; Hingerty, B. E.; Brown, G. M. *J. Am. Chem. Soc.* **1984**, *106*, 7545.
63 Zabel, W.; Saenger, W.; Mason, S. A. *J. Am. Chem. Soc.* **1986**, *108*, 3664–3673.
64 Dodziuk, H. In *Introduction to Supramolecular chemistry*; Kluwer Academic Publishers: Dordrecht, 2002, p 294.
65 Harada, A.; Kamachi, J. *Nature* **1992**, *356*, 325.
66 Stoddart, J. F. *Angew. Chem, Int. Ed. Engl.* **1992**, *31*, 846.
67 Garcia-Ochoa, L.; Diez Lopez, M.-A.; Vinas, M. H.; Santos, L.; Martinez Ataz, E.; Amat-Guerri, F.; Douhal, A. *Chem. Eur. J.* **1999**, *5*, 897.
68 Miertus, S.; Scrocco, E.; Tomasi, J. *Chem. Phys.* **1981**, *55*, 117.
69 Parr, R. G.; Yang, W. *Density Functional Theory of Atoms and Molecules*; Oxford University Press: Oxford, 1989.
70 Cao, Y.; Xiao, X.; Ji, S.; Lu, R.; Guo, Q. *Spectrochim. Acta A* **2004**, *60*, 815.
71 Menger, F. M.; Sherrod, M. J. *J. Am. Chem. Soc.* **1988**, *110*, 8606.
72 Thiem, H.-J.; Brandl, M.; Breslow, R. *J. Am. Chem. Soc.* **1988**, *110*, 8612.
73 Luzhkov, V. B.; Venanzi, C. A. *J. Phys. Chem.* **1995**, *99*, 2312.
74 Engler, E. M.; Andose, J. D.; von R. Schleyer, P. *J. Am. Chem. Soc.* **1973**, *95*, 8005.
75 Melani, F.; Mura, P.; Adamo, M.; Maestrelli, F.; Gratteri, P.; Bonaccini, C. *Chem. Phys. Lett.* **2003**, *370*, 280.
76 Homans, S. V. *Biochem.* **1990**, *29*, 9110.
77 Dauber-Ogathorpe, P.; Roberts, V. A.; Wolff, J.; Genest, M.; Hagler, A. T. *Proteins: Struct. Funct. Genetics* **1988**, *4*, 31.
78 Maple, J. R.; Dinur, U.; Hagler, A. T. *Proc. Natl. Acad. Sci. USA* **1988**, *85*, 5350.
79 Dr. Gilbert, K. E.; Serena Software: Bloomington, IN 47402.
80 Lipkowitz, K. B. *J. Org. Chem.* **1991**, *56*, 6357.
81 Dodziuk, H.; Nowinski, K. S. *J. Mol. Struct. (THEOCHEM)* **1994**, *304*, 61.
82 Allinger, N. L.; Yuh, Y. H.; University of Georgia: Athens, GA, 1980.
83 Lipkowitz, K. B.; Ragothama, S.; Yang, J.-A. *J. Am. Chem. Soc.* **1992**, *114*, 1554.
84 Ivanov, P. M.; Jaime, C. *J. Mol. Struct.* **1996**, *377*, 137.
85 Lo Meo, P.; D'Anna, F.; Riela, S.; Gruttadauria, M.; Noto, R. *Tetrahedron* **2002**, *58*, 6039.
86 Salvatierra, D.; Jaime, C.; Virgili, A.; Sanchez-Ferrando, F. *J. Org. Chem.* **1996**, *61*, 9597.
87 Bekkers, O.; Kettenes-van den Bosch, J. J.; van Helden, S. P.; Seijkens, D.; Beijnen, J. H.; Bult, A.; Underberg, W. J. M. *J. Inclus. Phenom.* **1991**, *11*, 185.
88 Chen, W.; Chang, C.-E.; Gilson, M. K. *Biophys. J.* **2004**, *87*, 3035.
89 Rekharsky, M. V.; Inoue, Y. *Chem. Rev.* **1998**, *98*, 1875.
90 Pop, M. M.; Goubitz, K.; Borodi, G.; Bogdan, M.; De Ridder, D. J. A.; Peschar, R.; Schenk, H. *Acta Cryst. B* **2002**, *58*, 1036.
91 Dodziuk, H.; Demchuk, O. M.; Bielejewska, A.; Kozminski, W.; Dolgonos, G. *Supramol. Chem.* **2004**, *16*, 287.
92 Kasatani, K.; Ohashi, M.; Sato, H. *Carbohydr. Res.* **1989**, *192*, 197.
93 Dodziuk, H. In *Modern Conformationa Analysis. Elucidating Novel Exciting Molecular Structures*; Wiley-VCH: New York, 1995, p 94.
94 van Gunsteren, W. F.; Mark, A. E. *J. Chem. Phys.* **1998**, *108*, 1609.
95 Daura, X.; van Gunsteren, W. F.; Rigo, D.; Jaun, B.; Seebach, D. *Chem. Eur. J.* **1997**, *3*, 1410.
96 van Gunsteren, W. F. In *Studies in Physical and Theoretical Chemistry*; Rivail, J.-L., Ed.; Elsevier: Amsterdam, 1990, p 463.
97 Smith, P. E.; van Gunsteren, W. F. In *Computer Simulations of*

Biomolecular Systems, Theoretical and Experimental Applications; ESCOM: Leiden, 1993; Vol. 2, p 182.
98 Koehler, J.; Saenger, W.; van Gunsteren, W. F. *Eur. Biophys. J.* **1987**, *15*, 197.
99 Koehler, J.; Saenger, W.; van Gunsteren, W. F. *Eur. Biophys. J.* **1987**, *15*, 211.
100 Koehler, J.; Saenger, W.; van Gunsteren, W. F. *J. Biomol. Struct. Dyn.* **1987**, *6*, 181.
101 Koehler, J.; Saenger, W.; van Gunsteren, W. F. *Eur. Biophys. J.* **1988**, *16*, 153.
102 Koehler, J.; Saenger, W.; van Gunsteren, W. F. *J. Mol. Biol.* **1988**, *203*, 241.
103 van Gunsteren, W. F.; http://www.igc.ethz.ch/gromos/.
104 Mark, A. E.; van Helden, S. P.; Smith, P. E.; Janssen, L. H. M.; van Gunsteren, W. F. *J. Am. Chem. Soc.* **1994**, *116*, 6293.
105 Varady, J.; Wu, X.; Wang, S. *J. Phys. Chem. B* **2002**, *106*, 4863.
106 Dodziuk, H.; Kozminski, W.; Lukin, O.; Sybilska, D. *J. Mol. Struct.* **2000**, *523*, 205.
107 Moeder, C.; O'Brien, T.; Thompson, R.; Bicker, G. *J. Chromatogr. A* **1996**, *735*, 1.
108 Lipkowitz, K. B.; Pearl, G.; Coner, B.; Peterson, M. A. *J. Am. Chem. Soc.* **1997**, *119*, 600.
109 Ivanov, P. M.; Jaime, C. *J. Phys. Chem. B* **2004**, *108*, 6261.
110 Gessler, K.; Uson, I.; Takaha, T.; Krauss, N.; Smith. S. M.; Okada, S.; Sheldrick, G. M.; Saenger, W. *Proc. Natl. Acad. Sci. USA* **1999**, *96*, 4246.
111 Kitamura, S.; Isuda, H.; Shimada, J.; Takada, T.; Okada, S.; Mimura, M.; Kajiwara, K. *Carbohydr. Res.* **1997**, *304*, 303.
112 Bonnet, P.; Bea, I.; Jaime, C.; Morin-Allory, L. *Supramol. Chem.* **2003**, *15*, 251.
113 Pastor, I.; Di Marono, A.; Mendicuti, F. *J. Photochem. Photobiol. A* **2005**, *173*, 238.
114 Bonnet, P.; Jaime, C.; Morin-Allory, L. *MD study of a-, b- and g-CyD dimers in water* **2002**, *67*, 8602.
115 Coleman, A. W.; Nicolis, I.; Keller, N.; Dalbiez, J. P. *J. Incl. Phenom.* **1992**, *13*, 139.
116 Berendsen, H. J. C. *Comp. Phys. Commun.* **1987**, *44*, 233.
117 Bea, I.; Jaime, C.; Kollman, P. *Theor. Chem. Acc.* **2002**, *108*, 286.
118 Breslow, R.; Greenspoon, N.; Guuo, T.; Zarzycki, R. *J. Am. Chem. Soc.* **1989**, *111*, 8296.
119 Bea, I.; Cervello, E.; Kollman, P. A.; Jaime, C. *Comb. Chem. Throughput Screening* **2001**, *4*, 605.
120 Choi, Y.; Jung, S. *Carbohydr. Res.* **2004**, *339*, 1961.
121 Morokuma, K. *Philos. T. Roy. Soc. A.* **2002**, *360*, 1149.
122 Maseras, F.; Morokuma, K.; *J. Comput. Chem.* **1995**, *16*, 1170.

12
Rotaxane and Catenane Structures Involving Cyclodextrins

Konstantina Yannakopoulou and Irene M. Mavridis

12.1
Introduction

According to J.-M. Lehn [1], supramolecular chemistry creates organized, functional molecular-scale devices, which are able to interpret, store, process, and dispatch information just like the sophisticated machines found in natural systems. Representative systems are those involving interlocking molecules such as rotaxanes, catenanes, knots, and other molecular systems with nontrivial topological properties [2]. These systems are appealing to chemists both for their aesthetics and for the challenge of designing their synthesis. Further, owing to their unusual structures and mechanical bonds, they may be endowed with novel properties potentially useful for interesting applications.

Native or synthetically modified cyclodextrins (CyDs) can be used as molecular components for such self-assembled device-like systems. This is due to their characteristic property of forming inclusion complexes in aqueous solutions with a wide variety of lipophilic guests. CyDs have become increasingly important [3] owing to the declining cost of production and their aqueous solubility and environmental compatibility, properties also shared by most of their synthetic derivatives. Their impact in academic research is also high. The number of relevant publications has increased over the years; an entire volume of the *Comprehensive Supramolecular Chemistry Series* [4] is devoted to CyDs as is an entire issue of *Chemical Reviews* [5]. Thus many opportunities emerge in CyD research ranging from new synthetic derivatives, tailor made for certain groups of guests, to novel functional supramolecular systems and "machines at the molecular level" [6, 7] briefly presented in Section 16.4, the latter usually comprising rotaxanes and catenanes (Fig. 12.1). Rotaxanes **1** from *rota* (Latin, meaning wheel) and *axis* (Greek *άξων [axon]*) are composed of a macrocycle, through which a dumbbell-shaped molecule (a rode with bulky stopper groups at both ends) has been threaded. Catenanes **2**, from the *catena* (Latin, meaning chain) are composed of two or more interlocked macrocycles.

These noncovalent supramolecular systems are stabilized via a mechanical interlocking that prevents the components from disassembling. A common precursor

Cyclodextrins and Their Complexes. Edited by Helena Dodziuk

ISBN: 3-527-31280-3

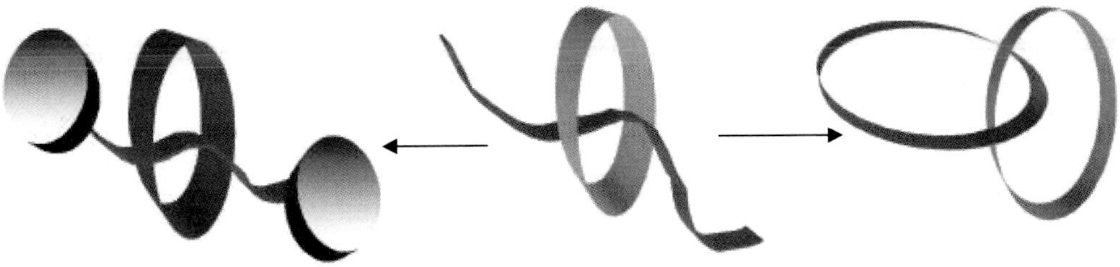

1 3 2

Fig. 12.1. Schematic representation of a [2]rotaxane **1** and a [2]catenane **2** conceptually resulting from the pseudorotaxane **3**.

to both rotaxanes and catenanes is the pseudorotaxane **3** (Greek *pseudo-*, not genuine), which has no stopper groups. Cyclodextrin pseudorotaxanes are stabilized by the cooperative effect of several weak host–guest interactions and H-bonding, depending on the nature of the guest entrapped in the CyD cavity. Since the inclusion takes place mainly in aqueous solution, there is a very subtle balance between the hydrophobicity of the cavity and the guest on the one hand, and the nature, charge, H-bonding ability, or the parts of the guest extending beyond the cavity and interacting with the CyD hydroxyl groups and/or surrounding water molecules, on the other. Several comprehensive reviews on cyclodextrin rotaxanes and catenanes in general [8–10] and of cyclodextrin pseudorotaxanes with long end-functionalized molecules [11] in particular, exist in the literature. This chapter will focus on literature and developments published in the years since 1997.

12.2 Pseudorotaxanes

Pseudorotaxanes **3** are formed via threading of adequately long hydrophobic guest molecules (12–20 Å in length), inside one or more CyD rings, and are designated as [*n*]pseudorotaxanes, *n* being the total number of molecules participating in the assembly. Pseudorotaxanes have the potential to change from the complexed to the noncomplexed state upon an appropriate stimulus and thus they can constitute parts of molecular machines. Therefore, it is important to know the structure and stability of single pseudorotaxanes of CyDs with model compounds in order to be able to judge the suitability of various guest groups in designing novel and/or complex architectures or in seeking novel applications (sensors, molecular machines, or new materials, such as drug carriers, mentioned in Table 14.3). Polymers threaded into many CyD rings, polypseudorotaxanes, are treated in Chapter 3.1.9.

Being in essence CyD complexes, [2]pseudorotaxanes are characterized by medium stability, unless a particular structural host–guest fit has been achieved. Consequently, pseudorotaxanes in aqueous solution behave as complexes, i.e. they are in fast or intermediate exchange with the free components, and therefore only in exceptional cases of a high association constant or slow exchange can signals of the new species **3** be observed separately from those of free constituents. Characterization of pseudorotaxanes by NMR spectroscopy and assessment of the solution structure can therefore be complicated. Low-temperature measurements are usually helpful, while the crystal structure, when available, provides the details of the mutual disposition of host and guest.

Various structural and conformational preferences have been documented by NMR spectroscopy via detailed mapping of NOE effects (discussed in Chapter 9) and by X-ray crystallography (presented in Chapter 7). In solution, a mixture of several threading combinations may be encountered. The host and the guest might associate in stoichiometries 1:1, 1:2, 2:1, or 2:2 (i.e. [2]- or [3]- or higher pseudorotaxanes). Orientational isomerism [9, 12, 13], which refers to the formation of isomeric [2]pseudorotaxanes by threading of an asymmetric guest either from the secondary or from the primary side of a cyclodextrin, is an additional feature that may be revealed. Another possibility is the head-to-head or head-to-tail association of the oligosaccharides upon insertion of more than two CyDs to form [3]- or higher pseudorotaxanes. A last feature that may occur is guest self-association, providing a 1:2 or even a 2:2 host:guest structure [14, 15].

For the narrowest cyclodextrin, α-CyD, the most suitable guests are aliphatic chains. Indeed, this preference has been established in several aliphatic or aliphatic–aromatic systems. Many such systems have been demonstrated in detail by several authors and have been reviewed previously [8, 9]. The presence of charged end-groups on the guest generally slows down the threading process, and thus it can be followed by ^{1}H NMR, e.g. in dodecamethylene end-substituted by alkyl pyridinium groups [16], the rate depending on the nature (hydrophobicity, unsaturation) and size (the alkyl chain) of the latter. 11-Aminoundecanoic acid and 12-aminododecanoic acid [12] form both [2]- and [3]pseudorotaxanes in solution, bearing one and two α-CyD rings, respectively, and they also exhibit orientational isomerism, as ROESY spectra indicated. Of the two stoichiometric species, 1:1 and 2:1, observed in solution, only the 2:1 ([3]pseudorotaxane) is found in the crystal [12]. The same observations were made for 1,12-diaminododecane [17] and 1,13-tridecanedioic acid/β-CyD [18]. Crystallization as [3]pseudorotaxanes of α-CyD is observed also with hexaethyleneglycol or tetraethyleneglycol dibromide [19] and α-CyD/4,4′-biphenyldicarboxylic acid [20]. As the aliphatic chain shortens, protonation effects on the end-groups of aliphatic diamines were found to affect negatively the stability of the pseudorotaxane [21]. The length of the methylene chain (C8–C12) in aliphatic threads bearing ammonium and phosphonium end-groups plays a minor role in the rate of the threading process, but is important in the dethreading process, with the longer, more hydrophobic chains, inducing lower rates of the latter as demonstrated with ^{1}H and ^{31}P NMR [22]. Differential scanning calorimetry (DSC) experiments have shown unusually slow kinetics in the dethreading of one α-CyD ring from the [3]pseudorotaxane of 12-aminododecanoic acid/α-CyD, attributed to the breaking of the H-bonds between the α-CyDs in the dimer cradling the guest [23]. 2,6-Di-*O*-methyl- and 2,3,6-tri-*O*-methyl-α-CyD (TMα-Cyd) are also suitable for aliphatic guests, especially for diynes, and are actually better than cyclophanes [24].

In the next larger cyclodextrin, β-CyD, aliphatic inserts are less popular, although the crystal structures of [3]pseudorotaxanes with aliphatic mono- [25] and diacids [18, 26, 27], diols [28], and bis-imidazolyl hydrocarbons [29] have been reported. β-CyD/monocarboxylic acid complexes pack in channel structures and create supramolecular polymers, whereas β-CyD/diacid complexes break the channel structure, contrary to what could be anticipated based on the propensity of acids to dimerize. With the wide γ-CyD and aromatic guests, such as Congo Red [14] and (*E*)-stilbene [15] dimerization of the guest inside γ-CyD is observed to form [4] and [3]pseudorotaxanes. Irradiation of crystalline complexes of the latter yielded *syn*-tetraphenylcyclobutane as a single isomer.

In the α-CyD cavity long aliphatic guests adopt a regular zig-zag shape [12, 17] and are very well localized, in contrast to the β-CyD complexes, where the aliphatic chain has a bend in the middle of the β-CyD dimer [18, 26–28, 30]. The bending offers stabilization by increasing the host–guest interactions and makes possible inclusion of aliphatic molecules longer than the strict length of the dimer. When the skeleton of the guest includes one or more phenyl groups, the existing X-ray structures indicate that both α-CyD and β-CyD are able to form inclusion complexes. However, the fact that the α-CyD ring is reported to deviate from hexagonal geometry acquiring an ellipsoid cross-section probably indicates that the size of the phenyl ring just fits in the α-CyD cavity (α-CyD/4,4′-biphenylbicarboxylic acid, **4** [20] and α-CyD/methyl orange [31]). The distortion is particularly apparent in the case of the methyl orange complex, where the azo moiety forms a step, thus in-

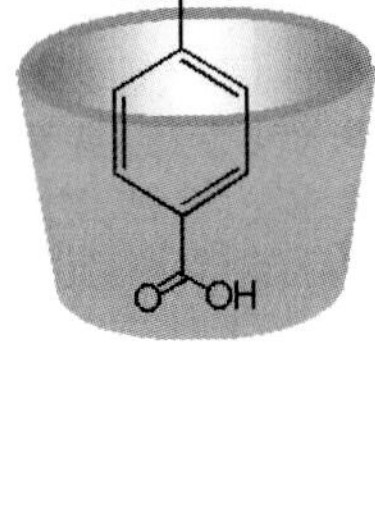

4

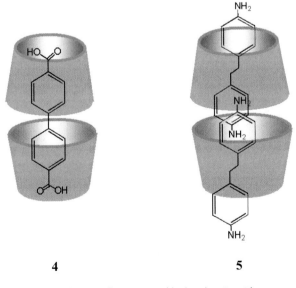

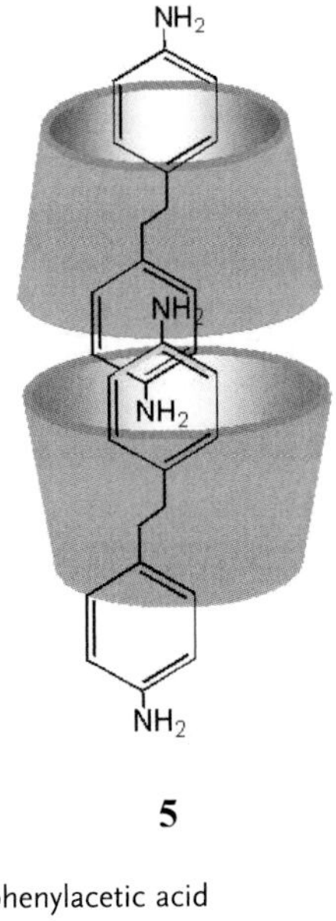

5

Fig. 12.2. In the crystalline state 4,4′-biphenylacetic acid forms the [3]pseudorotaxane **4** with α-CyD, whereas bis(4-aminophenyl)ethane forms a 2:2 complex **5** with the wider β-CyD.

creasing its overall width. The latter might be the reason that the guest cannot fit in the cavity of an α-CyD dimer and forms a [2]pseudorotaxane instead. In contrast, 1,2-bis(4-aminophenyl)ethane with the same 11-bond backbone length and similar width has the freedom to tilt in the wider β-CyD cavity and make space for two guest molecules (Fig. 12.2) in one β-CyD dimer, **5** [32].

12.3 Rotaxanes

The first CyD rotaxanes were prepared in 1981 using metal complexes of cobalt as stoppers [33]. Several other metal complexes of bipyridines, bulky ammonium salts, and bulky metal anions have been synthesized, as described in the detailed review by Nepogodiev and Stoddart [9]. Covalently linked stoppers are more demanding, since the aqueous solutions required for pre-assembly of the pseudorotaxanes allowed use of only a few types of reactions for efficient stoppering and were introduced later (1991) [34]. An outstanding example is the stoppering with 2,4-dinitrobenzene groups of polyethylene glycol (PEG) threaded through 20–23 CyD rings [35]. Stoppers that are widely used are di- and trinitrobenzenes, 2,6-dimethylphenol (for α-CyD), naphthalene sulfonic acids and carbazole (for α- and β-CyDs), covalently linked to the amino-groups of the axles. In one report, the preference of certain stoppers for the secondary side of the cyclodextrin has been confirmed based on NMR spectra [36]. Rotaxanes are kinetically stable, and can be purified by column chromatography, especially if they are soluble in organic solvents. For this reason, methylated CyDs that are both aqueous and organic soluble have been widely employed to construct rotaxanes. Only two detailed structures of rotaxanes as determined by X-ray crystallography are known [37, 38] and both are [2]rotaxanes.

The 1:1 adduct of α-CyD with 4,4′-diaminostilbene has been isolated using trinitrobenzene stoppers ([2]rotaxane), and the double bond was found well localized inside the cavity as shown by NMR spectroscopy, whereas the aromatic rings have been shown to rotate freely [39]. Restricted rotation of α-CyD around a trinitrophenyl-capped stilbene has been imposed by linking covalently the CyD ring to the axle and this proved to be an example of a functional structure [40]. Azobenzene and styrene derivatives have been rotaxanated into α-CyD (or TMα-CyD) [8] and shown [41] to photoisomerize by changing the location of the α-CyD on the backbone of a fully conjugated styrene, as shown by NOE interactions, leading to a photodriven molecular machine. Aromatic conjugated molecules have also been rotaxanated with β-CyD [8, 41, 42] and undergone reversible photoisomerization reactions [41]. In a very similar system α-CyD moves back and forth along the stilbene, as found by NMR experiments, whereas in the solid state it forms cleanly separated and aligned molecular fibers [43]. An azo-coupling reaction of an azo-dye threaded into α-CyD with 2,6-dimethylphenol yielded a [2]rotaxane (12%) and a very stable [3]rotaxane (9%). Thorough NMR investigation has revealed two orientation modes: in the [2]rotaxane **6** the primary side of α-CyD is facing the stop-

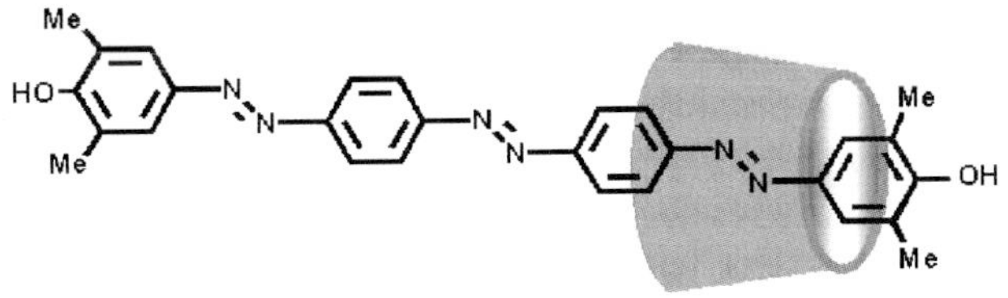

6

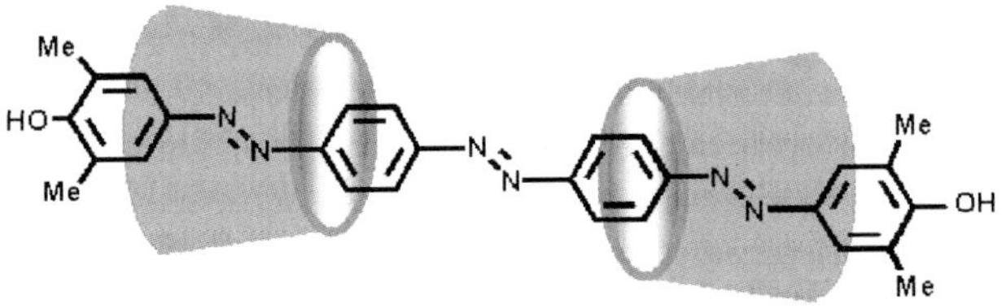

7

Fig. 12.3. Different orientation modes in the [2]rotaxane **6** and the [3]rotaxane **7**.

per, whereas in the [3]rotaxane **7** the secondary sides of both α-CyD rings are facing the stopper (Fig. 12.3) [44]. The above examples are the prototypes of CyD rotaxanes involving conjugated threads. These systems present the case of molecular insulation, which either stabilizes sensitizer dyes toward charge recombination [45] or forms molecular wires, which could be promising components of organic-based electronic devices. Investigation of the latter by Anderson's group has led to the preparation of polyrotaxanes with a conjugated backbone that display enhanced luminescence and aqueous solubility as well as good processability [46].

In an effort to generate novel materials, unconventional ways of stoppering have appeared in the literature. Gold nanoparticles attached via aliphatic thiols to semi-rotaxanes (rotaxanes with one stopper) have subsequently generated α-cyclodextrin-based rotaxanes supported on gold nanospheres [47]. Moreover, cyclodextrins has been threaded by a guest immobilized on an electrode, which serves as the stopper [48]. Fullerenes have been attached as stoppers to a β-CyD pseudorotaxane, via reaction with bis(*p*-aminophenyl)ether [49], resulting in a water-soluble, fullerene-containing polymer. Interestingly, while reaction with the plain diamine results in fullerene polysubstitution, its rotaxanation results only in fullerene disubstitution (every fullerene attached to two CyD/diamine complexes).

An atypical rotaxane structure is that involving a thread covalently attached to the CyD ring. Such a system acts as both host and guest and it has to be specially designed so that the thread is included spontaneously in a cavity other than the one bearing it (*intermolecular* inclusion), whereas *intramolecular* inclusion is prevented. The resulting assembly, named a "daisy chain" by J.F. Stoddart, is either an acyclic supramolecular polymer **8** or a cyclic supramolecular oligomer **9** [50] (Fig. 12.4). Acyclic cyclodextrin polymers were first observed in the solid state in

8

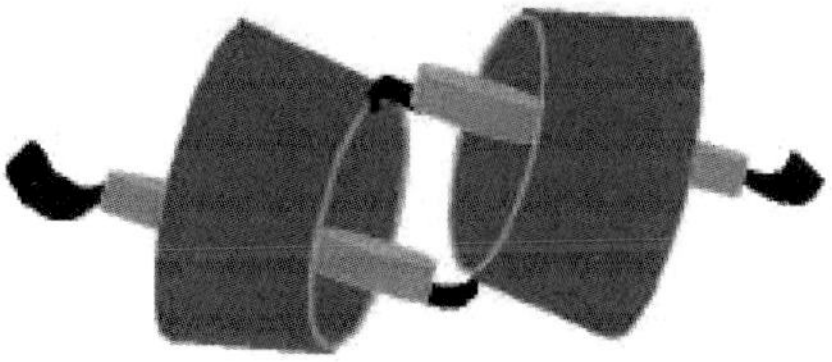

9

Fig. 12.4. Daisy-chain supramolecular polymer **8** and cyclic supramolecular dimer **9**.

the mono-[6-(6-aminohexylamino)-6-deoxy]-β-CyD [51]. Since then threads that assemble spontaneously in aqueous solutions to give cyclic daisy chains have been designed. Thus the mono(4,4′-bipyridinium)-substituted-β-CyD can self-assemble in aqueous solutions and disassemble in a controlled way depending on its oxidation state or other guests present in solution [52]. In systems analogous to **9** the free side of the axle has been capped [53, 54] thus forming *hermaphrodite* or *Janus* [2]rotaxanes. In the same spirit and taking advantage of the molecular recognition of CyD cavities, a host CyD bearing a covalently bound axle threaded into a CyD of different size than the one bearing it, has been prepared [55].

12.4 Applications

It has already been mentioned that isolation of a conjugated axle inside the CyD cavity offers stabilization and is promising for molecular electronics applications [46]. Additional advantages are improved charge- and electron-transfer properties. Rotaxanated cyanine dyes exhibit enhanced fluorescence, photostability, and chemical reversibility of redox forms of the chromophore [56, 57]. Enhancement of fluorescence is also observed in rotaxanated stilbene and tolan [37], while a dramatic increase of stability against reductive-, oxidative-, and photo-bleaching is reported for a rotaxane-encapsulated azo dye [58]. In the case of a β-CyD rotaxane of a non-conjugated molecule **10** with transition metal stopper units of different oxidation

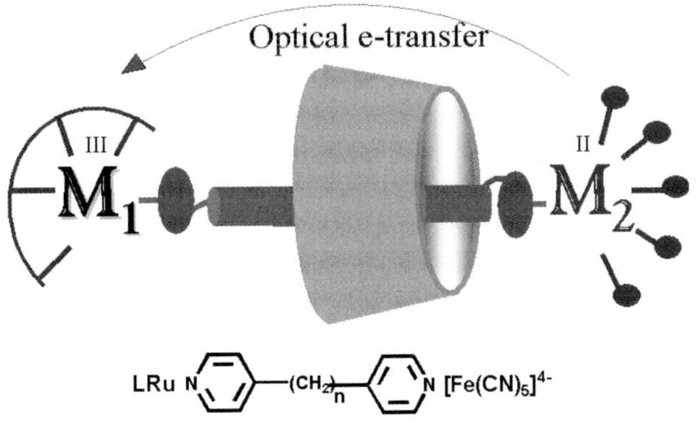

10

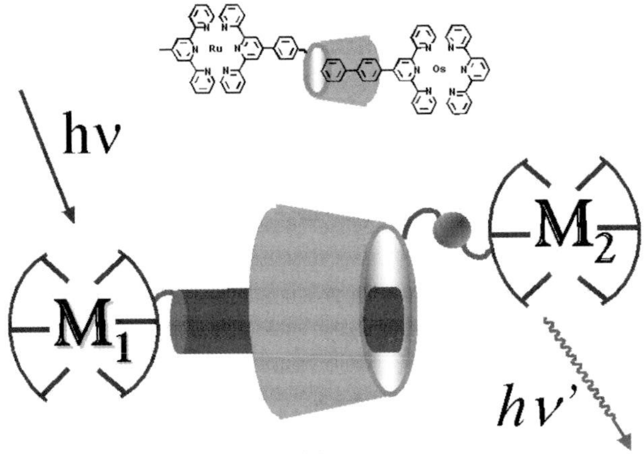

11

Fig. 12.5. Communication between metal centers in **10** [59] and **11** [60] manifested by both electron and energy transfer.

states, complexes of Ru^{III} and Fe^{II} (Fig. 12.5), optical electron transfer was observed from the Fe^{II} to the Ru^{III} center, which was not observed in the absence of inclusion [59]. On the other hand, a modified β-CyD and an appropriate guest molecule form a semirotaxane **11** with two different transition metal complexes on each side, offering a good charge separation architecture that cleanly enabled observation of both energy and electron transfer [60] (Fig. 12.5). The complexes described in Refs. [59, 60] constitute examples of molecular wires, in which, following application of an external stimulus, the ends communicate through space and

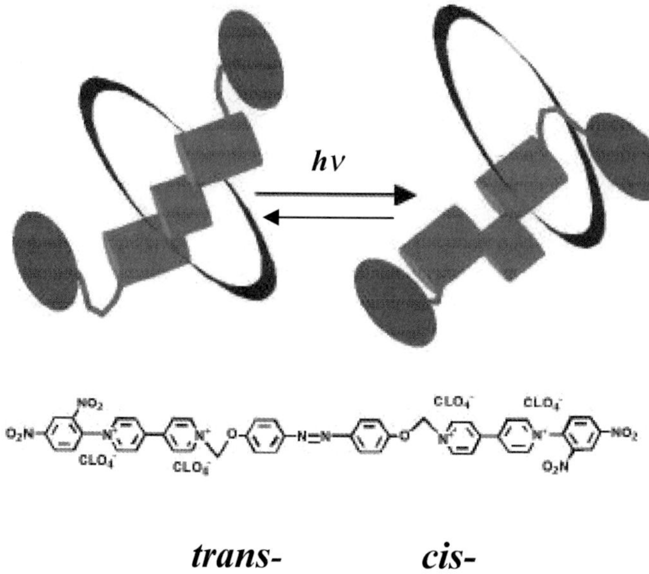

12

Fig. 12.6. A cyclodextrin-based photodriven molecular shuttle [61].

not by a conjugated axle. This approach can be extended to a wide range of self-assembled CyD systems with metal stoppers.

Rotaxanes can work as simple molecular machines. (The application of CyDs in electrochemical molecular devices and molecular machines is briefly presented in Section 16.4.) The rotor CyD can shuttle along a properly designed axle (preferably in a controlled manner), thereby providing a molecular system, which in principle can store information. This is demonstrated in *trans*- to *cis*- isomerizable axles (one example has already been mentioned [41]). Triggered by UV irradiation azobenzene in **12** isomerizes from *trans*- to *cis*- (Fig. 12.6) and *β*-CyD is forced to move. The shuttling back takes place under visible light irradiation [61]. Molecular shuttling of *α*-CyD between two dodecamethylene chains linked on either side by a viologen group has been suggested to take place thermally in a suitably rotaxanated system [62]. A derivatized *β*-CyD bound on an Au electrode has been used to encapsulate an azobenzene derivative, whose *trans*- to *cis*- isomerization transduced optical signals to electrochemical ones [63]. Moreover, a cyclodextrin polyrotaxane applied on a MoS_2 surface has been externally manipulated by the tip of an STM (discussed in Section 10.6) clearly demonstrating the possibility of changing the molecular position of the CyD in various directions reversibly, without destroying the molecule, an example of a bead counting device [64] also discussed in Section 10.6.

12.5
Catenanes

Catenanes **2**, another category of interlocked structures, result from the mechanical linking of two molecular circles, exactly like two rings in a chain. The two components are in principle able to rotate around one another, forming a miniature molecular machine. From the synthetic point of view, the preparation of catenanes is quite challenging, in that it requires a molecular thread penetrating a hollow macrocycle to be able to clip its ends together. This process is unlikely, unless there is a strong templating effect, expressed by a high association constant of the pseudorotaxane precursors [8, 9] that will provide a long enough lifetime of the threaded precursor to further react and clip the ends covalently together. Early attempts [65] to prepare cyclodextrin catenanes **2** using *p*-phenylene and biphenyl guests bearing thiol ends, despite the novelty of the concept, were not successful because the guest molecules were too short.

Cyclodextrin catenanes were synthesized for the first time by Stoddart's group in 1993 [66, 67] (Fig. 12.7). The molecular thread consisted conceptually of three parts, a *hydrophobic* aromatic core bearing on either side *hydrophilic* arms ending with *amino groups* that were clipped together by reaction with terephthaloyl chloride as in **13**.

The architectural exquisiteness of the products was high but the isolated yields of the pure compounds were very low. Not deterred by harvesting only small amounts, the group studied in detail the structural and conformational properties of these compounds and proved that [2]- and [3]catenanes were produced (with one and two CyD rings, respectively), exhibited orientational isomerism ([2]catenanes) and head-to-tail/head-to-tail and head-to-tail/head-to-head isomerization

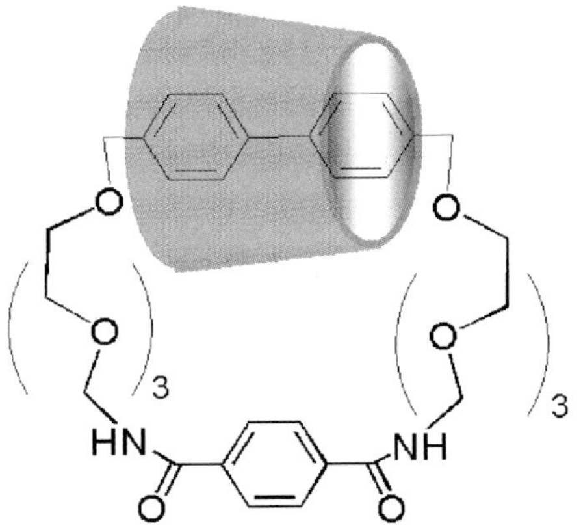

13

Fig. 12.7. Stoddart's cyclodextrin [2]catenane **13**, for which the X-ray structure is also available [66, 67].

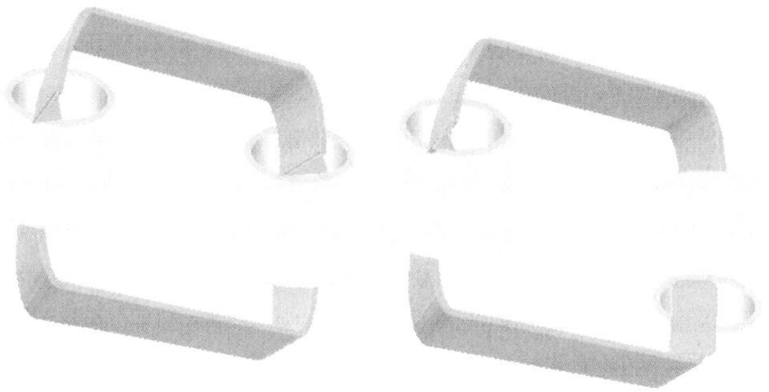

Fig. 12.8. Orientational isomerism in CyD [3]catenanes.

([3]catenanes) (Fig. 12.8). They even managed to crystallize and solve the X-ray structure of a [2]-catenane. However, the templating effect of the cyclodextrin ring is low, since free macrocycles and polymers were also produced. These results are summarized by Stoddart and Nepogodiev in their delightful review [9].

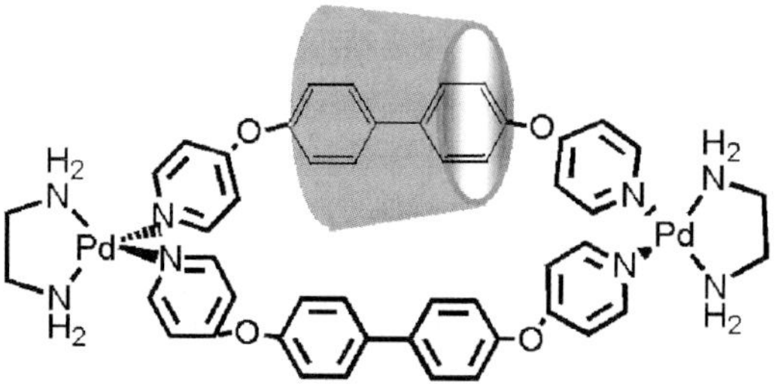

βCD

14

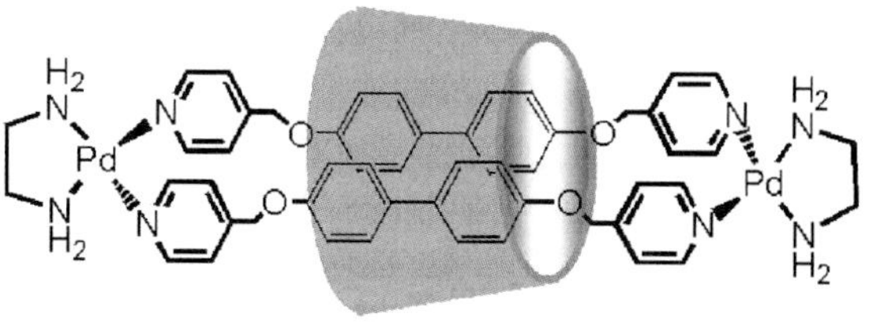

γCD

15

Fig. 12.9. Self-assembly of the biphenyl components through Pd(II) complexation yields the [2]catenane **14** with *β*-CyD but the [3]rotaxane **15** with *γ*-CyD.

Although preparation of cyclodextrin catenanes of polymethylene derivatives and polycatenanes has been claimed, no data are furnished [68]. Since 1993, the only catenane example essentially reported has been the *in situ* formation of a β-cyclodextrin [2]catenane that was characterized in solution by NMR experiments (Fig. 12.9) [69]. A self-assembled macrocyle, through Pd(II)(en) complexation of two threads, forms the [2]catenane **14** with β-CyD at appropriate concentrations but the [3]rotaxane **15** with the wider γ-CyD. Mass spectral characterization has also been provided for both species but none has been isolated.

Extensive search in the literature has not produced any other cyclodextrin catenane. Apparently the difficulty of isolation of such structures in meaningful yields, especially when involving native CyDs, has discouraged further attempts.

References

1 J.-M. Lehn, *Chem. Eur. J.* **2000**, *6*, 2097–2102 and references cited.
2 H. Dodziuk, *Introduction to Supramolecular Chemistry*, Kluwer, Dordrecht, 2002, Sects. 2.3 and 8.2.
3 J. Szejtli, *Chem. Rev.* **1998**, *98*, 1743–1753.
4 J. Szejtli, T. Osa (Eds.) in *Comprehensive Supramolecular Chemistry Series*; Vol. 3, J. L. Atwood, J. E. D. Davies, D. D. MacNicol, F. Vogtle (Eds.), Elsevier, Oxford, **1996**.
5 The entire issue, *Chem. Rev.* **1998**, *98*.
6 V. Balzani, A. Credi, F. M. Raymo, J. F. Stoddart, *Angew. Chem. Int. Ed.* **2000**, *39*, 3349–3391.
7 A. Harada, *Acc. Chem. Res.* **2001**, *34*, 456–464.
8 D. B. Amabilino, J. F. Stoddart, *Chem. Rev.* **1995**, *95*, 2725–2828.
9 S. A. Nepogodiev, J. F. Stoddart, *Chem. Rev.* **1998**, *98*, 1959–1976 and references cited.
10 N. Nakashima, A. Kawabuchi, H. Murakami, *J. Incl. Phenom. Molec. Recogn. Chem.* **1998**, *32*, 363–373.
11 K. Yannakopoulou, I. M. Mavridis, *Current Org. Chem.* **2004**, *8*, 25–34 and references cited.
12 K. Eliadou, K. Yannakopoulou, A. Rontoyianni, I. M. Mavridis, *J. Org. Chem.* **1999**, *64*, 6217–6226.
13 R. Isnin, A. E. Kaifer, *Pure Appl. Chem.* **1993**, *65*, 495–497.
14 N. Mourtzis, G. Cordoyiannis, G. Nounesis, K. Yannakopoulou, *Supramolec. Chem.* **2003**, *15*, 639–649.
15 K. S. S. P. Rao, S. M. Hubig, J. N. Moorthy, J. K. Kochi, *J. Org. Chem.* **1999**, *64*, 8098–8104.
16 A. C. Smith, D. H. Macartney, *J. Org. Chem.* **1998**, *63*, 9243–9251.
17 A. Rontoyianni, I. M. Mavridis, *Supramolec. Chem.* **1999**, *10*, 213–218.
18 S. Makedonopoulou, A. Tulinsky, I. M. Mavridis, *Supramolec. Chem.* **1999**, *11*, 73–81.
19 A. Harada, J. Li, M. Kamachi, Y. Kitagawa, Y. Katsube, *Carbohydr. Res.* **1998**, *305*, 127–129.
20 S. Kamitori, S. Muraoka, S. Kondo, K. Okuyama, *Carbohydr. Res.* **1998**, *312*, 177–181.
21 L. Avram, Y. Cohen, *J. Org. Chem.* **2002**, *67*, 2639–2644.
22 A. P. Lyon, N. J. Banton, D. H. Macartney, *Can. J. Chem.* **1998**, *76*, 843–850.
23 A. Tsortos, K. Yannakopoulou, K. Eliadou, I. M. Mavridis, G. Nounesis, *J. Phys. Chem. B*, **2001**, *105*, 2664–2671.
24 J. Huuskonen, J. E. H. Buston, N. D. Scotchmer, H. L. Anderson, *New J. Chem.* **1999**, *23*, 1245–1252.
25 S. Makedonopoulou, I. M. Mavridis, K. Yannakopoulou,

J. Papaioannou, *Chem. Commun.* **1998**, 2133–2134.

26 S. Makedonopoulou, I. M. Mavridis, *Acta Crystallogr.* **2000**, *B56*, 322–331.

27 S. Makedonopoulou, I. M. Mavridis, *Carbohydr. Res.* **2001**, *335*, 213–220.

28 T. Bojinova, H. Gornitzka, N. Lauth-de Viguerie, I. Rico-Lattes, *Carbohydr. Res.* **2003**, *338*, 781–785.

29 X.-J. Shen, H.-L. Chen, F. Yu, Y.-C. Zhang, X.-H. Yang, Y. Z. Li, *Tetrahedron Lett*, **2004**, *45*, 6813–6817.

30 R. B. Hannak, G. Färber, R. Konrat, B. Kräutler, *J. Am. Chem. Soc.* **1997**, *119*, 2313–2314.

31 K. Harata, *Bull. Chem. Soc. Japan*, **1976**, *49*, 1493–1501.

32 P. Giastas, K. Yannakopoulou, I. M. Mavridis, *Acta Crystallogr.* **2003**, *B59*, 287–299.

33 H. Ogino, *J. Am. Chem. Soc.* **1981**, *103*, 1303–1304.

34 R. Isnin, A. E. Kaifer, *J. Am. Chem. Soc.* **1991**, *113*, 8188–8190.

35 A. Harada, J. Li, M. Kamachi, *Nature*, **1992**, *356*, 325–327.

36 J. W. Park, H.-J. Song, *Org. Lett.* **2004**, *6*, 4869–4872.

37 C. A. Stanier, M. J. O'Connell, W. Clegg, H. L. Anderson, *Chem. Commun.* **2001**, 493–494.

38 J. Terao, A. Tang, J. J. Michels, A. Krivokapic, H. L. Anderson, *Chem. Commun.* **2004**, 56–57.

39 C. J. Easton, S. F. Lincoln, A. G. Meyer, H. Onagi, *J. Chem. Soc., Perkin Trans 1* **1999**, 2501–2506.

40 H. Onagi, C. J. Blake, C. J. Easton, S. F. Lincoln, *Chem. Eur. J.* **2003**, *9*, 5978–5988.

41 C. A. Stanier, S. J. Alderman, T. D. W. Claridge, H. L. Anderson, *Angew. Chem. Int. Ed. Engl.* **2002**, *41*, 1769–1772.

42 S. Anderson, T. D. W. Claridge, H. L. Anderson, *Angew. Chem. Int. Ed.* **1997**, *36*, 1310–1313.

43 H. Onagi, B. Carrozzini, G. L. Cascarano, C. J. Easton, A. J. Edwards, S. F. Lincoln, A. D. Rae, *Chem. Eur. J.* **2003**, *9*, 5971–5977.

44 M. R. Craig, T. D. W. Claridge, M. G. Hutchings, H. L. Anderson, *Chem. Commun.* **1999**, 1537–1538.

45 S. A. Haque, J. S. Park, M. Srinivasarao, J. R. Durrant, *Adv. Mater.* **2004**, *16*, 1177–1181.

46 F. Cacialli, J. S. Wilson, J. J. Michels, C. Daniel, C. Silva, R. H. Friend, N. Severin, P. Samori, J. P. Rabe, M. J. O'Connell, P. N. Taylor, H. L. Anderson, *Nature Mat*, **2002**, *1*, 160–164.

47 J. Liu, R. Xu, A. E. Kaifer, *Langmuir* **1998**, *14*, 7337–7339.

48 I. Willner, V. Pardo-Yissar, E. Katz, K. T. Ranjit, *J. Electroanalyt. Chem.* **2001**, *497*, 72–177.

49 S. Samal, B.-J. Choi, K. E. Geckeler, *Chem. Commun.* **2000**, 1373–1374.

50 P. R. Ashton, I. Baxter, S. J. Cantrill, M. C. T. Fyfe, P. T. Glink, J. F. Stoddart, A. J. P. White, D. J. Williams, *Angew. Chem. Int. Ed.* **1998**, *37*, 1294–1297.

51 D. Mentzafos, A. Terzis, A. W. Coleman, C. de Rango, *Carbohydr. Res.* **1996**, *282*, 125–135.

52 A. Mirzoian, A. E. Kaifer, *Chem. Commun.* **1999**, 1603–1604.

53 T. Fujimoto, Y. Sakata, T. Kaneda, *Chem. Commun.* **2000**, 2143–2144.

54 H. Onagi, C. J. Easton, S. F. Lincoln, *Org. Lett.* **2001**, *3*, 1041–1044.

55 M. Miyauchi, T. Hoshino, H. Yamaguchi, S. Kamitory, A. Harada, *J. Am. Chem. Soc.* **2005**, *127*, 2034–2035.

56 J. E. H. Buston, F. Marken, H. L. Anderson, *Chem. Commun.* **2001**, 1046–1047.

57 J. E. H. Buston, J. R. Young, H. L. Anderson, *Chem. Commun.* **2000**, 905–906.

58 M. R. Craig, M. G. Hutchings, T. D. W. Claridge, H. L. Anderson, *Angew. Chem. Int. Ed.* **2001**, *40*, 1071–1074.

59 A. D. Shukla, H. C. Bajaj, A. Das, *Angew. Chem. Int. Ed.* **2001**, *40*, 446–448.

60 J. M. Haider, M. Chavarot, S. Weidner, I. Sadler, R. M. Williams, L. De Cola, Z. Pikramenou, *Inorg. Chem*, **2001**, *40*, 3912–3921.

61 H. Murakami, A. Kawabuchi, K. Kotoo, M. Kunitake, N. Nakashima, *J. Am. Chem. Soc.* **1997**, *119*, 7605–7606.

62 Y. Kawaguchi, A. Harada, *Org. Lett.* **2000**, *2*, 1353–1356.

63 M. Lahav, K. T. Ranjit, E. Katz, I. Willner, *Chem. Commun.* **1997**, 259–260.

64 H. Shigekawa, K. Miyake, J. Sumaoka, A. Harada, M. Komiyama, *J. Am. Chem. Soc.* **2000**, *122*, 5411–5412.

65 A. Lüttringhaus, F. Cramer, H. Prinzbach, F. M. Hengelein, *Liebigs Ann. Chem.* **1958**, *613*, 185–198.

66 D. Armspach, P. R. Ashton, C. P. Moore, N. Spencer, J. F. Stoddart, T. J. Wear, D. J. Williams, *Angew. Chem. Int. Ed. Engl.* **1993**, *32*, 854–858.

67 D. Armspach, P. R. Ashton, R. Ballardini, A. Godi, C. P. Moore, L. Prodi, N. Spencer, J. F. Stoddart, M. S. Tolley, T. J. Wear, D. J. Williams, *Chem. Eur. J.* **1995**, *1*, 33–55.

68 A. Harada, *Acc. Chem. Res.* **2001**, *34*, 456–464.

69 C. W. Lim, S. Sakamoto, K. Yamaguchi, J.-I. Hong, *Org. Lett.* **2004**, *6*, 1079–1082.

13
Large-ring Cyclodextrins

Haruhisa Ueda and Tomohiro Endo

13.1
Introduction

Cyclodextrins (CyDs) is a common name for cyclic α-1,4-glucans. α-CyD (CyD_6), β-CyD (CyD_7), γ-CyD (CyD_8), and their derivatives have been thoroughly studied and used in many fields. On the other hand, it is very difficult to find reports on Large-ring Cyclodextrin (LR-CyD) with a polymerization degree n greater than 9 glucopyranose units prior to 1985. In 1965, French et al. reported the first definitive evidence for the existence of LR-CyDs with $n = 9$–13 [1]. However, this early LR-CyD study did not reveal anything that attracted attention, and it was also forgotten because of the difficulties in the purification and preparation in reasonable yields of these compounds. In 1986, Kobayashi et al. developed a preparation method for LR-CyD mixtures and succeeded in isolating δ-CyD ($n = 9$) [2]. The crystal structure of δ-CyD was characterized by Fujiwara et al. in 1990 [3]. After this, LR-CyDs were again recognized as worth studying.

Since the mid-1990s, the existence of LR-CyDs has been fully demonstrated [4–11]. So far, LR-CyDs containing up to 35 glucopyranose units have been purified and characterized [11] and the existence of LR-CyDs with a polymerization degree up to several hundreds of glucopyranose units has been revealed [12–18]. The increasing availability of LR-CyDs either as pure substances or as mixtures, at least on the laboratory scale, has stimulated an increasing number of studies of their properties, particularly with regard to inclusion complex formation [19–33]. In 2000, Machida et al. reported that LR-CyD mixtures with a polymerization degree from 22 to 45, and greater than 50, could act as an efficient artificial chaperone for protein refolding [34]. In the autumn of 2001, the new product, a protein refolding kit using LR-CyDs came onto the Japanese market. Other CyD applications are presented in Chapters 14–16. Building on the above situation, LR-CyD studies began to attract considerable attention. Throughout this chapter, the generic names will be used for α-, β- and γ-CyD, whereas the semisystematic names, which include the polymerization degree in the macrocycle, will be used for LR-CyDs with a polymerization degree greater than 9 (abbreviated, CyD_n, where n designates the polymerization degree). LR-CyDs have often been designated "Cycloamylose" (abbrevi-

Cyclodextrins and Their Complexes. Edited by Helena Dodziuk

ISBN: 3-527-31280-3

ated CA_n, where n designates the number of glucose molecules in the macrocycle). However, this is a nonsystematic name whose use has been discouraged [35]. Larsen expounded on the nomenclature of CyDs and LR-CyDs in his reviews [36].

This chapter deals with LR-CyD from basic matters to recent advances.

13.2 Production of LR-CyDs

A mixture of LR-CyDs is produced by the incubation with cyclodextrin glucanotransferases (CGTases, EC 2.4.1.19) originating from bacteria and starches [2, 4–9, 13, 18]. However, it has already been reported that several 4-α-glucanotransferases (EC 2.4.1.25) [16], especially, D-enzyme and amylomaltase, could produce LR-CyDs in a better yield than those from CGTases, when synthetic amylose with a high polymerization degree was used as substrate [12, 15]. However, it has also been known that elongation of the incubation time with CGTase, D-enzyme, and amylomaltase leads to a decrease in the amount of LR-CyDs with a high polymerization degree owing to a transglycosidic linearization (coupling reaction). In 2002, it was found that the glycogen debranching enzyme (GDE, EC 2.4.1.25/EC 3.2.1.33) was also able to produce a mixture of LR-CyDs [37]. This enzyme has 4-α-glucanotransferase activity and amylo-1,6-glucosidase activity in the single polypeptide chain, [38, 39] and had some interesting effects on LR-CyD production, as follows: (a) The GDE from *Saccharomyces cerevisiae* could produce a mixture of LR-CyDs with a polymerization degree range of 11 to around 40, not only from synthetic amyloses but also from starches such as amylopectin and debranched starch (short-chain amylose). (b) The yield of LR-CyDs using the GDE and soluble starch reached about 60% after an incubation time of about 6 h without any other enzymes; however, it was 40% in the case of D-enzyme combined with isoamylase and pullulanase to hydrolyze α-1,6-glucoside bonds. (c) Although the presence of D-glucose enhanced the coupling reaction activity of CGTase, D-enzyme, and amylomaltase, it had no effect on the production of LR-CyDs using the GDE at 0.01 and 0.1% D-glucose concentrations, because the GDE could not use D-glucose and maltose as acceptors of the coupling reaction. These results suggested that the GDE was a useful enzyme for production of a mixture of LR-CyDs , because commercially available starches could be used as a substrate, and inactivation of GDE was not required to simplify the production process. Unfortunately, D-enzyme, amylomaltase, and GDE cannot produce LR-CyDs with smaller polymerization degrees than 17, 22 and 11, respectively, although these enzymes are able to produce mixtures of LR-CyDs in a reasonable yield. If LR-CyDs with a low polymerization degree such as CyD_9 and CyD_{12} are needed for studies of inclusion complex formation, even today CGTases are the only enzyme for their production.

Enzymatic preparations provide mixtures of CyDs with different polymerization degrees. So far, an effective purification method for each LR-CyD has not been developed, which prevents the advance of LR-CyD chemistry. It is expected that chemical syntheses will produce LR-CyD with a single polymerization degree. The first chemical synthesis of LR-CyD was reported in 2002 [40]. In this report, LR-CyD with a polymerization degree of nine (CyD_9) was chemically synthesized from maltotriose as the starting material by the molecular clamp method using a phthaloyl bridge. Unfortunately, there are no other reports of chemical syntheses for LR-CyDs. If chemical syntheses of LR-CyDs with high selectivity and yield were developed, it would be a breakthrough for LR-CyD chemistry.

13.3
Isolation and Purification Procedures

Isolation and purification procedures for LR-CyDs were reported by Ueda's group [4–8, 10, 11] and Koizumi et al. [9]. These procedures included common initial purification steps and several chromatographic methods. In the early stages of LR-CyD study, the isolation of relative large amounts of LR-CyDs required tedious pretreatment and a number of chromatographic separations [4–8, 10]. Since then, chromatographic separations have gradually changed for the better with the use of LR-CyD mixtures of high purity as starting materials. An example of the process is outlined in Fig. 13.1. However, this is still a tedious procedure with a relatively high cost.

The identification of each LR-CyD was carried out by the following routine methods: the purity of each LR-CyD was confirmed by HPLC (presented in Chapter 5) using at least, two different modes; the molecular weight of each LR-CyD was determined by FAB/MS for $n = 9–26$ and MALDI/TOF/MS for $n > 26$ (the MS

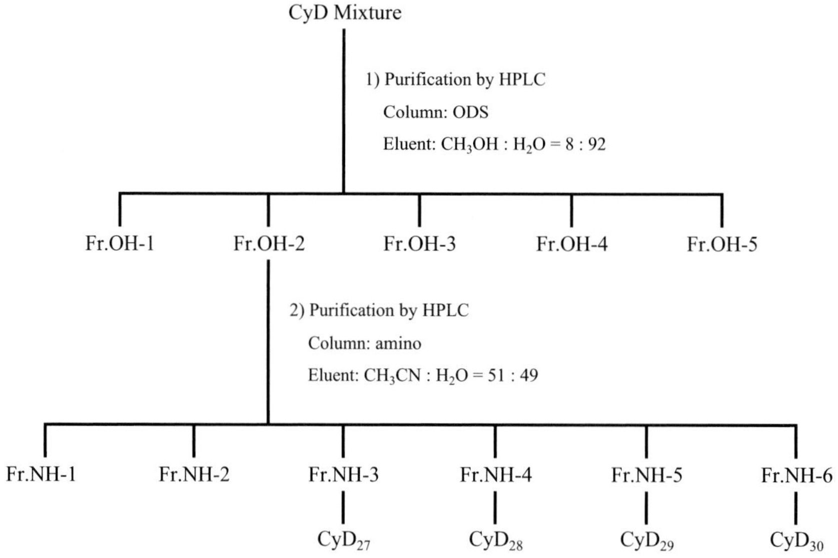

Fig. 13.1. Purification method for CyD_{27}, CyD_{28}, CyD_{29}, and CyD_{30}.

methods are discussed in Section 10.2); the cyclic structure of each LR-CyD was identified by ^{1}H NMR and ^{13}C NMR spectroscopy (reviewed in Chapter 9) [4, 8, 11].

13.4 Physicochemical Properties and Structures of LR-CyDs

Table 13.1 lists some physicochemical properties of regular CyDs and LR-CyDs. The aqueous solubilities of LR-CyDs except CyD_9, CyD_{10}, CyD_{14}, and CyD_{26} are greater than those of α- and γ-CyD. This may be a consequence of their high structural flexibility, on the basis of the formation of intramolecular and intermolecular hydrogen bonds, similar to crystallinity described later. (The problem of CyDs nonrigidity is discussed in detail in Section 1.3.) There are no marked differences in specific rotation among LR-CyDs (CyD_{10}–CyD_{35}). However, the molecular rotation

$$[\Phi]_\lambda^t = M \cdot [\alpha]_\lambda^t/100 \qquad (1)$$

where M is molecular weight, $[\alpha]$ is specific rotation, t is temperature and λ is wavelength reflects the structural changes between regular CyDs and LR-CyDs. If there are no structural differences affecting optical rotatory power, molecular rotation must increase linearly with the increase in number of glucopyranose units. The variation in molecular rotation of CyDs with the number of glucopyranose units is expressed as two or three straight lines. One is plotted against regular CyDs and CyD_9, and the other was plotted against CyD_{10}–CyD_{35}. Therefore, it was suggested that there are specific structural changes such as band flip in LR-CyD molecules, that did not occur in regular CyDs and CyD_9 [41, 11]. There are no marked differences in acid-catalyzed hydrolysis rates among LR-CyDs (CyD_{10}–CyD_{35}). This suggested that the increases in decomposition points (glucosidic linkages) in the macrocyclic rings accompanied by an increasing number of glucopyranose units were not the major reason for the macrocyclic ring-opening reaction catalyzed by protons. On the other hand, it was reported that the half-lives of macrocyclic ring openings (CyD_9–CyD_{21}) paralleled ^{13}C-NMR chemical shifts of C-1 and C-4 in glucopyranose units [10]. This relationship may show that steric strains in macrocyclic rings contribute to the rate of macrocyclic ring-opening under acidic conditions.

The crystal structures of four kinds of LR-CyDs, which are composed of 9 (CyD_9) [3], 10 (CyD_{10}) [42–46], 14 (CyD_{14}) [44, 45, 47], and 26 (CyD_{26}) [48, 49] glucopyranose units, have already been described using X-ray crystallographic analysis presented in Chapter 7. The average structure of regular CyDs, a truncated cone with a round cavity, is well known. The structure of CyD_9 exhibits a distorted elliptical boat-like shape, but it retains a similar structure to regular CyDs. CyD_{10} and CyD_{14} also exhibit a more elliptical macrocyclic ring folded in a saddle-like shape. Furthermore, the arrangement between two adjacent glucopyranose units showed *anti*-type conformation (Band flip) at two sites in their macrocyclic rings, although

Tab. 13.1. Physicochemical properties of regular CyDs and LR-CyDs

	Number of glucopyranose units	Aqueous[a] solubility (g dL^{-1})	Surface[a] tension (mN m^{-1})	Specific rotation $[\alpha]_D^{25}$	Half-life of[b] ring opening (h)
α-CyD	6	14.5	72	+147.8	33
β-CyD	7	1.85	73	+161.1	29
γ-CyD	8	23.2	73	+175.9	15
CyD_9	9	8.19	72	+187.5	4.2
CyD_{10}	10	2.82	72	+204.9	3.2
CyD_{11}	11	>150	72	+200.8	3.4
CyD_{12}	12	>150	72	+197.3	3.7
CyD_{13}	13	>150	72	+198.1	3.7
CyD_{14}	14	2.30	73	+199.7	3.6
CyD_{15}	15	>120	73	+203.9	2.9
CyD_{16}	16	>120	73	+204.2	2.5
CyD_{17}	17	>120	72	+201.0	2.5
CyD_{18}	18	>100	73	+204.0	3.0
CyD_{19}	19	>100	73	+201.0	3.4
CyD_{20}	20	>100	73	+199.7	3.4
CyD_{21}	21	>100	73	+205.3	3.2
CyD_{22}	22	>100	73	+197.7	2.6
CyD_{23}	23	>100	73	+196.6	2.7
CyD_{24}	24	>100	73	+196.0	2.6
CyD_{25}	25	>100	73	+190.8	2.8
CyD_{26}	26	22.4	73	+201.4	2.9
CyD_{27}	27	>125	72	+189.4	2.8
CyD_{28}	28	>125	72	+191.2	2.6
CyD_{29}	29	>125	72	+190.2	2.5
CyD_{30}	30	>125	72	+189.1	2.3
CyD_{31}	31	>125	71	+189.0	2.4
CyD_{32}	32	>125	71	+192.7	2.4
CyD_{33}	33	>125	71	+192.1	2.2
CyD_{34}	34	>125	72	+189.6	2.2
CyD_{35}	35	>125	71	+193.7	2.1

[a] Observed at 25 °C.
[b] In 1-M HCl at 50 °C.

the normal arrangement between two adjacent glucopyranose units is *syn*-type conformation, so that the primary and secondary hydroxyl groups exist on the same side of the macrocyclic ring. CyD_9 had an intermediate structure between that of regular CyDs, and CyD_{10} and CyD_{14}, a distorted elliptical macrocyclic ring without the Band flip.

The structure of CyD_{26}, the largest whose X-ray structure has been determined, no longer has the ring shape but consists of two left-handed amylose-like helixes. The molecule has two-fold symmetry and the two asymmetric parts are bound through the Band flips. To understand these structural features of LR-CyDs in depth, readers should consult some reviews [36, 50–52]. Other LR-CyDs structures have not been reported, because their single crystals could not be prepared. However, several LR-CyD structures have been deduced from molecular dynamics simulations and small-angle X-ray scattering analysis [53–56]. (See, however, the discussion of molecular modeling in Chapter 11.) The low crystallinity of most LR-CyDs might be caused by the difficulties in forming intramolecular and intermolecular hydrogen bonds, owing to the high flexibility of their macrocyclic rings (see also the discussion in Section 1.3). In addition, it has been reported that the rate of nucleation of sugars, which has to precede crystal growth, is generally very low, so that sugar solutions often form supersaturated solutions of syrup-like liquid [57, 58]. The molecular behavior of LR-CyDs with high aqueous solubility in water may resemble that of sugars.

13.5
Inclusion Complex Formation

The effect of complex formation with CyD_9 on the solubility of 22 drugs that are poorly soluble or insoluble in water has been studied, and the first evidence of inclusion complex formation of CyD_9 with spironolactone and digitoxin has been reported [4, 24]. However, CyD_9 did not show any significant solubilization effect on those drugs in comparison with regular CyDs. The inclusion complex forming abilities between LR-CyDs up to CyD_{17} and various anions have been studied by capillary electrophoresis [19, 21, 22, 25]. (Applications of the latter technique to enantioseparations are presented in Chapter 6.) The data obtained showed that 4-*tert*-butyl benzoate and ibuprofen anions formed inclusion complexes with LR-CyDs, except CyD_{10}. The stability constants of complexes of two anions with the LR-CyDs increased from CyD_{11} to CyD_{14} and decreased from CyD_{15} to CyD_{17}. The effect of CyD_9 on the solubilization of buckminster fullerene (C_{60}, C_{70}) into water has been elucidated; its effect was superior to that of regular CyD (γ-CyD) [20, 27]. Dodziuk et al. have reported that not only fullerenes but also single-wall carbon nanotubes (SWNTs), were dissolved in water using LR-CyD (CyD_{12}) [31].

As summarized in Table 13.2, there have been several papers relating to inclusion complex formation between pure LR-CyD or an LR-CyD mixture and guest compounds. In general, the results given in Tab. 13.2 suggest that LR-CyDs may be good host molecules for relative large guest compounds in comparison with regular CyDs. However, there are still only a few reports on the complexation of pure LR-CyDs (especially, for those with $n > 18$ glucopyranose units) with guest molecules [23, 59]. To reveal the relationship between the complex forming ability of LR-CyDs and guest molecule structure, it is necessary to prepare large amounts of each isolated pure LR-CyD efficiently and investigate their complexation with many different guest molecules.

Tab. 13.2. Studies of inclusion complex formation between pure LR-CyD or mixture of LR-CyDs and compounds.

CyD	Indicator or Method	Compound	Ref.
(Pure LR-CyD)			
CyD_9	Enhancement of solubility (UV/VIS absorption)	Anthracene Amphotericin B Ajmalicine Ajmaline Carbamazepine Digitoxin Spironolactone 9,10-Dibromoanthracene Perylene-3,4,9,10-tetracarboxylic dianhydrate	4
	Solubility method	Spironolactone	
CyD_9	Enhancement of solubility (Spectrophotometry)	Fullerene C_{70}	20
CyD_9	Enhancement of solubility (Spectrophotometry)	Reserpine [2,2]-Paracyclophane Perylene Triphenilene 1,8-Naphthalic anhydride Naphthalene-1,4,5,8-tetracarboxylic dianhydride Digitoxin Gitoxin Digoxin Methyldigoxin Lanatoside C G-Strophanthin Proscillaridin A	24
	Solubility method and NMR	Digitoxin	
CyD_9	Simple precipitation	1,5-Cyclooctadiene Cyclononanone Cyclodecanone Cycloundecanone Cyclododecanone Cyclotridecanone Cyclopentadecanone	26
	Powder X-ray diffraction DSC	Cycloundecanone Cyclododecanone	

Tab. 13.2 *(continued)*

CyD	Indicator or Method	Compound	Ref.
CyD_9–CyD_{13}	Capillary electrophoresis	Benzoate 2-Methyl benzoate 3-Methyl benzoate 4-Methyl benzoate 2,4-Dimethyl benzoate 2,5-Dimethyl benzoate 3,5-Dimethyl benzoate 3,5-Dimethoxy benzoate Salicylate 3-Phenyl propionate 4-*tert*-Butyl benzoate Ibuprofen anion 1-Adamantane carboxylate	19, 21, 22
CyD_{14}–CyD_{17}	Capillary electrophoresis	Salicylate 4-*tert*-Butyl benzoate Ibuprofen anion	25
CyD_{21}–CyD_{32}	Isothermal titration calorimetry	Iodine	23
CyD_9	X-ray crystallography	Cycloundecanone	29
CyD_{12}	NMR	Single wall carbon nanotube (SWNT)	31
CyD_{26}	X-ray crystallography	NH_4I_3 $Ba(I_3)_2$ Undecanoic acid Dodecanol	59, 63
(Mixture of LR-CyDs) CA(S)[a] and CA(L)[b]	Spectrofluorometry	8-Anilino-1-naphthalene sulfonic acid	64
CA(S)[a] and CA(L)[b]	Simple precipitation	1-Octanol 1-Butanol Oleic acid	16, 64
CA (with oligomerization degree of 22 to around 60)	Enhancement of solubility (Spectrophotometry)	Fullerene C_{60}	33

[a] CA(S): Mixture of LR-CyDs with polymerization degree around 20 to 55, mainly polymerization degree of 25 to 50.
[b] CA(L): Mixture of LR-CyDs with average polymerization degree of ca. 150, without CyDs with polymerization degree < 50.

13.6
Applications of LR-CyDs

In 2003, the first chemically modified LR-CyD (CyD_9) was synthesized to enhance chemiluminescence efficiency by forming a CyD-bound luciferin analogue, but CyD_9 did not show a high chemiluminescence efficiency in comparison with regular CyDs (especially, γ-CyD) [32].

A mixture of LR-CyDs has also become of interest since 2000, when, as mentioned earlier, the effect of an LR-CyD mixture as an artificial chaperone for protein refolding was reported by Machida et al. [34]. At present, a protein refolding kit containing a mixture of LR-CyDs as one of the active components is on the market [60]. Furthermore, it has also been reported that an LR-CyD mixture provided an efficient method for refolding denatured antibody and biotinylated receptor protein expressed by recombination to correct active structures [61, 62]. In the pharmaceutical applications discussed in Chapters 14 and 15, interactions between drugs and mixture of LR-CyDs have been studied. Tomono et al. evaluated the interactions between a mixture of LR-CyDs with a polymerization degree of 20 to 50 (average molecular weight, 7720) and drugs such as prednisolone, cholesterol, digoxin, digitoxin, and nitroglycerin, using the solubility method [30]. Although nitroglycerin did not interact with the mixture of LR-CyDs, the solubilities of prednisolone, cholesterol, digoxin, and digitoxin were enhanced by the presence of the mixture of LR-CyDs, and the phase solubility diagrams showed complex formations between the mixture of LR-CyDs and the drugs. In particular, the mixture of LR-CyDs showed a higher stabilization effect for cholesterol than regular CyDs. Recently, Tomono's group also reported the improvement of the solubility of fullerene C_{60} by cogrinding with LR-CyDs in the solid state [33]. The molecular state of C_{60} in the complex obtained was evaluated using powder X-ray diffraction, UV-Vis spectroscopy, and dynamic light scattering measurement. The results suggested that C_{60} molecules were dispersed into an LR-CyD micellar system and the UV-Vis spectral change was due to an intermolecular interaction between C_{60} and LR-CyDs. An earlier experiment for the micellar system formation of CyD solutions reveals that α- and γ-CyD can form molecular associates in aqueous solution [65, 66]. Recently, a large scale chain structure of CyD is observed by using cold-spray ionization mass spectrometry. These are first evidence that CyD exists as a cluster in aqueous solutions [67]. Therefore, there is a strong probability that LR-CyDs also can form molecular associates in aqueous solution.

Investigations on further application of LR-CyDs will depend on their availability in larger amounts than has previously been attainable and novel purification methods.

References

1 D. French, A. O. Pulley, J. A. Effenberger, M. A. Rougvie, M. Abdullah, *Arch. Biochem. Biophy.* **1965**, 111, 153–160.

2 S. Kobayashi, M. Fukuda, M. Monma, T. Harumi, M. Kubo, *Abstracts of papers*, the Annual Meetings of the Agriculture chemical

Society of Japan, Kyoto, April, p 649, **1986**.

3 T. Fujiwara, N. Tanaka, S. Kobayashi, *Chem. Lett.* **1990**, 739–742.

4 I. Miyazawa, H. Ueda, H. Nagase, T. Endo, S. Kobayashi, T. Nagai, *Eur. J. Pharm. Sci.* **1995**, 3, 153–162.

5 T. Endo, H. Ueda, S. Kobayashi, T. Nagai, *Carbohydr. Res.* **1995**, 269, 369–373.

6 T. Endo, H. Nagase, H. Ueda, S. Kobayashi, T. Nagai, *Chem. Pharm. Bull.* **1997**, 45, 532–536.

7 T. Endo, H. Nagase, H. Ueda, A. Shigihara, S. Kobayashi, T. Nagai, *Chem. Pharm. Bull.* **1997**, 45, 1856–1859.

8 T. Endo, H. Nagase, H. Ueda, A. Shigihara, S. Kobayashi, T. Nagai, *Chem. Pharm. Bull.* **1998**, 46, 1840–1843.

9 K. Koizumi, H. Sanbe, Y. Kubota, Y. Terada, T. Takaha, *J. Chromatogr. A* **1999**, 852, 407–416.

10 H. Ueda, M. Wakisaka, H. Nagase, T. Takaha, S. Okada, *J. Inclusion Phenom. Macrocyclic Chem.* **2002**, 44, 403–405.

11 S. Nakadate, T. Endo, H. Nagase, H. Ueda, *Abstracts of papers*, the Annual Meeting of the 22nd Cyclodextrin Symposium of Japan, Kumamoto, pp 21–22, **2004**.

12 T. Takaha, M. Yanase, H. Takata, S. Okada, S. M. Smith, *J. Biol. Chem.* **1996**, 271, 2902–2908.

13 Y. Terada, M. Yanase, H. Takata, T. Takaha, S. Okada, *J. Biol. Chem.* **1997**, 272, 15729–15733.

14 T. Takaha, M. Yanase, H. Takata, S. Okada, S. M. Smith, *Biochem. Biophys. Res. Commun.* **1998**, 247, 493–497.

15 Y. Terada, K. Fujii, T. Takaha, S. Okada, *Appl. Environ. Microbiol.* **1999**, 65, 910–915.

16 T. Takaha, S. M. Smith, *Biotechnol. Genet. Eng. Rev.* **1999**, 16, 257–280.

17 Y. Tachibana, T. Takaha, S. Fujiwara, M. Takagi, T. Imanaka, *J. Biosci. Bioeng.* **2000**, 90, 406–409.

18 Y. Terada, H. Sanbe, T. Takaha, S. Kitahata, K. Koizumi, S. Okada, *Appl. Environ. Microbiol.* **2001**, 67, 1453–1460.

19 K. L. Larsen, T. Endo, H. Ueda, W. Zimmermann, *Carbohydr. Res.* **1998**, 309, 153–159.

20 T. Furuishi, T. Endo, H. Nagase, H. Ueda, T. Nagai, *Chem. Pharm. Bull.* **1998**, 46, 1658–1659.

21 K. L. Larsen, T. Endo, H. Ueda, W. Zimmermann, *Proceedings of the Ninth International Symposium on Cyclodextrins*, Kluwer Academic, Dordrecht, pp 93–96, **1999**.

22 K. L. Larsen, W. Zimmermann, *J. Chromatogr. A* **1999**, 836, 3–14.

23 S. Kitamura, K. Nakatani, T. Takaha, S. Okada, *Macromol. Rapid Commun.* **1999**, 20, 612–615.

24 H. Ueda, A. Wakamiya, T. Endo, H. Nagase, K. Tomono, T. Nagai, *Drug Dev. Ind. Pharm.* **1999**, 25, 951–954.

25 B. Mogensen, T. Endo, H. Ueda, W. Zimmermann, K. L. Larsen, *Proceedings of the 10th International Cyclodextrin Symposium*, Ann Arbor, Michigan, USA, Mira Digital, pp 157–162, **2000**.

26 H. Akasaka, T. Endo, H. Nagase, H. Ueda, S. Kobayashi, *Chem. Pharm. Bull.* **2000**, 48, 1986–1989.

27 K. Süvegh, K. Fujiwara, K. Komatsu, T. Marek, H. Ueda, A. Vértes, T. Braun, *Chem. Phys. Lett.* **2001**, 344, 263–269.

28 K. Teranishi, T. Nishiguchi, H. Ueda, *ITE Lett. Batteries, New Technol. Med.* **2002**, 3, 26–29.

29 K. Harata, H. Akasaka, T. Endo, H. Nagase, H. Ueda, *Chem. Commun.* **2002**, 1968–1969.

30 K. Tomono, A. Mugishima, T. Suzuki, H. Goto, H. Ueda, T. Nagai, J. Watanabe, *J. Inclusion Phenom. Macrocyclic Chem.* **2002**, 44, 267–270.

31 H. Dodzuik, A. Ejchart, W. Anczewski, H. Ueda, E. Krinichnaya, G. Dolgonos, W. Kutner, *Chem. Commun.* **2003**, 986–987.

32 K. Teranishi, T. Nishiguchi, H. Ueda, *Carbohydr. Res.* **2003**, 338, 987–993.

33 T. Fukami, A. Mugishima, T. Suzuki, S. Hidaka, T. Endo, H. Ueda, K. Tomono, *Chem. Pharm. Bull.* **2004**, 52, 961–964.

34 S. Machida, S. Ogawa, S. Xiaohua, T. Takaha, K. Fujii, K. Hayashi, *FEBS Lett.* **2000**, 486, 131–135.

35 A. D. McNaught, *Carbohydr. Res.* **1997**, 297, 1–92.
36 (a) K. L. Larsen, *Biol. J. Arm.* **2001**, 53, 9–26. (Special issue: cyclodextrin.); (b) K. L. Larsen, *J. Inclusion Phenom. Macrocyclic Chem.*, **2002**, 43, 1–13.
37 M. Yanase, H. Takaha, T. Takaha, T. Kuriki, S. M. Smith, S. Okada, *Appl. Environ. Microbiol.* **2002**, 68, 4233–4239.
38 B. K. Gillard, T. E. Nelson, *Biochemistry* **1977**, 16, 3978–3987.
39 A. Nakayama, K. Yamamoto, S. Tabata, *J. Biol. Chem.* **2001**, 276, 28824–28828.
40 M. Wakao, K. Fukase, S. Kusumoto, *J. Org. Chem.* **2002**, 67, 8182–8190.
41 S. Motohama, E. Ishii, T. Endo, H. Nagase, H. Ueda, T. Takaha, S. Okada, *Biol. J. Arm.* **2001**, 53, 27–33. (Special issue: cyclodextrin.)
42 H. Ueda, T. Endo, H. Nagase, S. Kobayashi, T. Nagai, *J. Inclusion Phenom. Mol. Recognit. Chem.* **1996**, 25, 17–20.
43 T. Endo, H. Nagase, H. Ueda, S. Kobayashi, M. Shiro, *Anal. Sci.* **1999**, 15, 613–614.
44 J. Jacob, K. Geßler, D. Hoffmann, H. Sanbe, K. Koizumi, S. M. Smith, T. Takaha, W. Saenger, *Angew. Chem. Int. Ed. Engl.* **1998**, 37, 606–609.
45 J. Jacob, K. Geßler, D. Hoffmann, H. Sanbe, K. Koizumi, S. M. Smith, T. Takaha, W. Saenger, *Carbohydr. Res.* **1999**, 322, 228–246.
46 K. Imamura, O. Nimz, J. Jacob, D. Myles, S. A. Mason, S. Kitamura, T. Aree, W. Saenger, *Acta Crystallogr., Sect. B: Struct. Sci.* **2001**, 57, 833–841.
47 K. Harata, T. Endo, H. Ueda, T. Nagai, *Supramol. Chem.* **1998**, 9, 143–150.
48 K. Gessler, I. Usón, T. Takaha, N. Krauss, S. M. Smith, S. Okada, G. M. Sheldrick, W. Saenger, *Proc. Natl. Acad. Sci. USA* **1999**, 96, 4246–4251.
49 O. Nimz, K. Geßler, I. Usón, W. Saenger, *Carbohydr. Res.* **2001**, 336, 141–153.
50 W. Saenger, J. Jacob, K. Gessler, T. Steiner, D. Hoffmann, H. Sanbe, K. Koizumi, S. M. Smith, T. Takaha, *Chem. Rev.* **1998**, 98, 1787–1802.
51 T. Endo, M. Zheng, W. Zimmermann, *Aust. J. Chem.* **2002**, 55, 39–48.
52 K. L. Larsen, *J. Inclusion Phenom. Macrocyclic Chem.* **2002**, 43, 1–13.
53 J. Shimada, S. Handa, H. Kaneko, T. Takada, *Macromolecules* **1996**, 29, 6408–6421.
54 S. Kitamura, H. Isuda, J. Shimada, T. Takada, T. Takaha, S. Okada, M. Mimura, K. Kajiwara, *Carbohydr. Res.* **1997**, 304, 303–314.
55 J. A. Semlyen (Ed.), *Cyclic Polymers*, 2nd edn, Kluwer Academic, Dordrecht, 125–160, **2000**.
56 J. Shimada, H. Kaneko, T. Takada, S. Kitamura, K. Kajiwara, *J. Phys. Chem. B* **2000**, 104, 2136–2147.
57 F. Framks, R. H. M. Hatley, S. F. Mathias, *BioPharm* **1991**, 4, 38, 40–42, 55.
58 B. J. Aldous, A. D. Auffret, F. Franks, *Cryo-Lett.* **1995**, 16, 181–186.
59 O. Nimz, K. Geßler, I. Usón, S. Laettig, H. Welfle, G. M. Sheldrick, W. Saenger, *Carbohydr. Res.* **2003**, 338, 977–986.
60 S. Machida, K. Hayashi, Japanese Patent 2001, Publication number: 2001-261697.
61 S. Machida, K. Hayashi, T. Tokuyasu, T. Takaba, Japanese Patent 2003, Publication number: 2003-128699.
62 S. Machida, K. Hayashi, T. Tokuyasu, Y. Sakakibara, S. Matsunaga, Japanese Patent 2004, Publication number: 2004-125785.
63 O. Nimz, K. Gessler, I. Uson, G. M. Sheldrick, W. Saenger, *Carbohydr. Res.* **2004**, 339, 1427–1437.
64 H. Nakamura, T. Takaha, S. Okada, *Shokuhin Kogyo* **1996**, 39, 52–59. (Japanese)
65 C. LeBas, N. Rysanek, *Cyclodextrins and Their Industrial Uses*; D. Duchêne (Ed.), Editions de Sante, Paris, pp 116–124, **1987**.
66 L. Szente, J. Szejtli, G. L. Kis, *J. Pharm. Sci.* **1998**, 87, 778–781.
67 Y. Sei, T. Koizumi, K. Takahashi, K. Yamaguchi, *Abstracts of papers*, the Annual Meeting of the 22nd Cyclodextrin Symposium of Japan, Kumamoto, pp 159–160, **2004**.

14
Pharmaceutical Applications of Cyclodextrins and Their Derivatives

Kaneto Uekama, Fumitoshi Hirayama, and Hidetoshi Arima

14.1
Introduction

Cyclodextrins, CyDs, are widely applied in pharmaceutical formulations to enhance the solubility, stability, and bioavailability of drug molecules [3–5]. However, native CyDs have relatively low solubility, in both water and organic solvents, which thus limits their uses in pharmaceutical formulations. Various kinds of CyD derivatives such as hydrophilic, hydrophobic, and ionic derivatives have been developed to extend the physicochemical properties and the inclusion capacity of native CyDs [6–12]. They are applicable as functional drug carriers to control the rate and/or time profiles of drug release [13–15]. Hydrophilic CyDs can modify the release rate of poorly water-soluble drugs, which can be used for the enhancement of drug absorption across biological barriers, serving as potent drug carriers in immediate-release formulations [11–15]. Amorphous CyDs such as hydroxyalkylated-β-CyDs are useful for inhibiting polymorphic transitions and crystallization rates of poorly water-soluble drugs during storage, which can consequently maintain higher dissolution characteristics and oral bioavailability of the drugs. On the other hand, hydrophobic CyDs may serve as sustained-release carriers for water-soluble drugs including peptide and protein drugs [14]. Delayed-release formulations can be obtained by the use of enteric-type CyDs such as *O*-carboxymethyl-*O*-ethyl-β-CyD (CME-β-CyD, **18**), which is soluble at the intestinal pH of around 6–7. (The chemical structures of CyDs and their abbreviations used in this chapter are shown in Table 14.1.) A combined use of different CyDs and/or pharmaceutical additives will provide more balanced oral bioavailability with prolonged therapeutic effects. The most desirable attribute for a drug carrier is its ability to deliver a drug to a targeted site [16, 17]. The CyD/drug conjugate can survive passage through the stomach and small intestine, but the drug release will be triggered by enzymatic degradation of the CyD ring in the colon [18]. Such a CyD conjugate can be a versatile means of constructing a new class of colon-targeting prodrug. Moreover, CyD/cationic polymer conjugates may be novel candidates for nonviral vectors to enhance the gene transfer of plasmid DNA [17]. On the basis of the above-

Cyclodextrins and Their Complexes. Edited by Helena Dodziuk

ISBN: 3-527-31280-3

Table 14.1. Pharmaceutically Useful CyDs

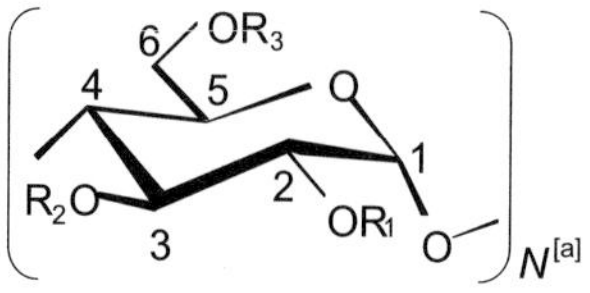

Derivative	Characteristic	Possible use (dosage form)
Hydrophilic derivatives		
Methylated β-CyD		
Me-β-CyD (**1**)	Soluble in cold water and	Oral, dermal,
DM-β-CyD (**2**)	in organic solvents	mucosal[b]
TM-β-CyD (**3**)	Surface active, Hemolytic	
DMA-β-CyD (**4**)	Soluble in water, Low hemolytic	Parenteral, oral, mucosal
Hydroxyalkylated β-CyD		
2-HE-β-CyD (**5**)	Amorphous mixture with	Parenteral, oral, mucosal
2-HP-β-CyD (**6**)	different d.s. (Encapsin^{RT})	Parenteral, oral, mucosal
3-HP-β-CyD (**7**)	Highly water-soluble (>50%)	Parenteral, oral, mucosal
2,3-DHP-β-CyD (**8**)	Low toxicity	Parenteral, oral, mucosal
Branched β-CyD		
G_1-β-CyD (**9**)	Highly water-soluble (>50%)	Parenteral, oral, mucosal
G_2-β-CyD (**10**)	Low toxicity	Parenteral, oral, mucosal
GUG-β-CD (**11**)		Parenteral, oral, mucosal
Hydrophobic derivatives		
Alkylated β-CyD		
DE-β-CyD (**12**)	Water-insoluble, soluble in	Oral, subcutaneous
TE-β-CyD (**13**)	organic solvents, surface-active	(slow-release)
Acylated β-CyD		
TA-β-CyD (**14**)	Water-insoluble, soluble in	Oral, parenteral
	organic solvents	(slow-release)
TB-β-CyD (**15**)	Mucoadhesive	(slow-release)
TV-β-CyD (**16**)	Film formation	(slow-release)
TO-β-CyD (**17**)		
Ionizable derivatives		
Anionic β-CyD		
CME-β-CyD (**18**)	$pK_a = 3$ to 4, Soluble	Oral, dermal, mucosal
	at pH > 4	(delayed-release, enteric[c])
β-CyD·sulfate (**19**)	$pK_a > 1$, Water-soluble	Oral, mucosal
SBE4-β-CyD (**20**)	Water-soluble	Parenteral, oral
SBE7-β-CyD (**21**)	Water-soluble (Captisol^{RT})	Parenteral, oral
Al-β-CyD·sulfate (**22**)	Water-insoluble	Parenteral (slow-release)
Org 25969 (**23**)	Water-soluble	Parenteral

Table 14.1 *(continued)*

Abbreviations:
Me: randomly-methylated; DM: 2,6-di-*O*-methyl; TM: per-2,3,6-tri-*O*-methyl; DMA: peracetylated DM-β-CyD; 2-HE: 2-hydroxyethyl; 2-HP: 2-hydroxypropyl; 3-HP, 3-hydroxypropyl; 2,3-DHP, 2,3-dihydroxypropyl; G_1, glycosyl; G_2, maltosyl; GUG, Glucuronyl-glucosyl; DE: 2,6-di-*O*-ethyl; TE: per-2,3,6-tri-*O*-ethyl; CME: *O*-carboxymethyl-*O*-ethyl; TA: per-2,3,6-tri-*O*-acyl ($C_2 \sim C_{18}$); TB: per-2,3,6-tri-*O*-butanoyl: TV: per-2,3,6-tri-*O*-valeryl; TO: per-2,3,6-tri-*O*-octyl; SBE4: d.s.4 of sulfobutyl ether group; SBE7: d.s.7 of sulfobutyl ether group. Org 25969: octasodium salt of octakis-*S*-(2-carboxyethyl)-octathio-γ-CyD.

[a] Number of glucose units.

[b] Mucosal: nasal, sublingual, ophthalmic, pulmonary, rectal, vaginal, etc.

[c] Enteric: soluble in intestinal fluid (pH 6–7).

mentioned knowledge, the advantages and limitations of CyDs in the design of advanced dosage forms will be discussed [12, 19–23]. This chapter is devoted to the pharmaceutical applications of CyDs in the form of simple inclusion complexes while applications of higher aggregates are presented in Chapter 15.

14.2 Some Characteristics of CyDs as Drug Carriers

The desirable attributes of drug carriers in drug delivery systems are the multifunctional properties such as controlled-release, targeting, and absorption enhancing abilities [22]. From the safety aspect, bioadaptability is a necessity, and quality, cost-performance, etc. are additional requirements for drug carriers. CyDs have such characteristics e.g. they are fairly bioadaptable and hardly absorbable from the gastrointestinal (GI) tract, they interact with specific components of biomembranes such as cholesterol and lipids, their macrocyclic ring survives in the stomach and small intestine, but they are biodegradable in the colon and large intestine, and more-functional CyD derivatives are available to modify the physicochemical and inclusion properties of the host molecules. Table 14.1 contains the pharmaceutically useful β-CyD derivatives, classified into hydrophilic, hydrophobic, and ionic derivatives [22]. Some hydrophilic CyDs have been applied in practice in pharmaceutical preparations (Table 14.2). α- and β-CyDs were first applied in practical dosage forms of E-type prostaglandins, improving the chemical stability and aqueous solubility of drug molecules through the formation of inclusion complexes (Fig. 14.1) (see, however, the stoichiometry of another prostaglandin complex with α-CyD determined by chromatography presented in Fig. 5.1) [23].

Table 14.2. Examples of Marketed Pharmaceutical Preparations Using CyDs

Component	Trade name	Formulation	Country
Prostaglandin E_1-α-CyD	Prostandin	Intraarterial infusion	Japan
Prostaglandin E_2-β-CyD	Prostarmon E	Sublingual tablet	Japan
OP-1206-α-CyD	Opalmon	Tablet	Japan
Piroxicam-β-CyD	Brexin	Tablet	Italy
Garlic oil-β-CyD	Xund, Tegra	Dragees	Germany
Benexate-β-CyD	Ulgut/Lonmiel	Capsules	Japan
Iodine-β-CyD	Mena-Gargle	Gargling	Japan
Dexamethasone, Glyteer-β-CyD	Glymesason	Ointment	Japan
Nitroglycerin-β-CyD	Nitropen	Sublingual tablet	Japan
Cefotiam hexetil hydrochloride-α-CyD	Pansporin T	Tablet	Japan
New oral Cephalosporin (ME1207)-α-CyD	Meiact	Tablet	Japan
Tiaprofenic acid-β-CyD	Suramyl	Tablet	Italy
Chlordiazepoxide-β-CyD	Transillium	Tablet	Argentina
Itraconazole-HP-β-CyD	Spranox	Liquid	Belgium
Hydrocortisone-HP-β-CyD	Dexacort	Liquid	Iceland
Diclofenac-HP-γ-CyD	Voltalen Ophta	Eye drop	Switzerland
Ziprasidone-SBE-β-CyD	Geodone	Injection	USA
Voliconazole-SBE-β-CyD	Vfend	Injection	USA
Estradiol-Methyl β-CyD	Aerodiol	Nasal spray	France

14.2.1
Hydrophilic CyDs

Monographs on the native CyDs can be found in several pharmacopoeias. For example, α- and β-CyDs are listed in the US, European, and Japanese Pharmacopoeias. On the other hand, a monograph for hydroxypropylated β-CyD, HP-β-CyD (Encapsin^{RT}, **6**), has recently appeared in the European Pharmacopoeia [24]. Hydrophilic CyDs such as methylated β-CyDs [1, 5, 25], hydroxypropylated β-CyDs [26–28], and branched β-CyDs [29, 30] have received special attention (see Chapter

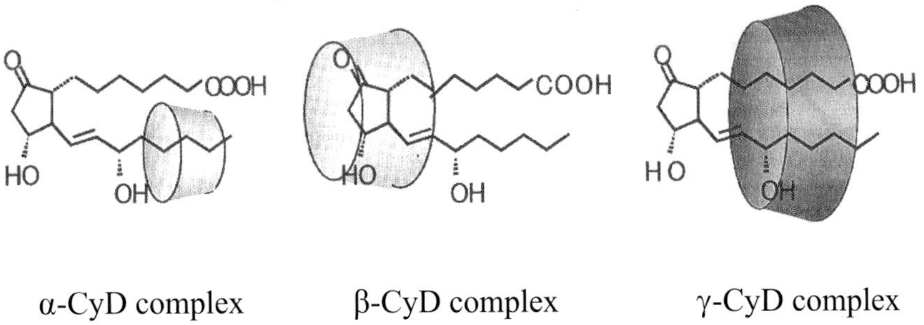

Fig. 14.1. Proposed inclusion modes of prostaglandin E_1 with three native CyDs [23].

2 for the discussion of these and numerous other CyD derivatives), because their solubility in water is extremely high compared to native β-CyD. Methylated CyDs are the most powerful solubilizers presently known and have three main types: so-called randomly-methylated β-CyD (Me-β-CyD, **1**), 2,6-di-*O*-methyl-β-CyD (DM-β-CyD, **2**), and 2,3,6-per-*O*-methyl-β-CyD (TM-β-CyD, **3**), depending on the number of methyl groups attached to each glucose unit (average degree of substitution, d.s.) [24]. HP-β-CyD (**6**) is the first CyD derivative to have been developed as a parenteral drug carrier. The commercially available **6** is a mixture of isomers and homologues, depending on the reaction conditions even at identical d.s. of the samples. The important quality requirements for use as a parenteral drug carrier are the lowest possible content of nonsubstituted β-CyD and of the endotoxin (a toxin produced by certain bacteria and released upon destruction of the bacterial cell).

6-*O*-Maltosyl-β-CyD (G_2-β-CyD, **10**), a typical branched β-CyD, has three different types of glycosidic bond: an α-1,4 bond in the branched sugar, an α-1,4 bond in the CyD ring, and an α-1,6 bond at the junction between the CyD ring and substituents. When **10** was administered intravenously to rats, it disappeared rapidly from the plasma and was recovered almost completely as a form of water-soluble 6-*O*-glycosyl-β-CyD (G_1-β-CyD, **9**) [30]. Since the nephrotoxicity of native β-CyD at higher doses was ascribed to the crystallization of less-soluble β-CyD or its cholesterol complex in renal tissues, the metabolic fate of **10** is suggestive of lower renal toxicity compared with the native β-CyD [31]. Since the solubility of **10** shows little temperature dependence, this kind of property is useful for the design of an aqueous injectable CyD solution that is to be heat-sterilized.

Hemolysis data are known to provide a simple and reliable measure for the estimation of CyD-induced membrane damage or cytotoxicity [10, 14]. Figure 14.2 shows the hemolysis curves of hydrophilic CyDs on rabbit erythrocytes. The hemolytic effects of methylated CyDs are much higher than those of other CyDs [32, 33]. In a series of CyD derivatives, there is a positive correlation between the hemolytic activity and their capacity to solubilize the lipophilic components of the cell membranes (cholesterol and phospholipids). Since the methylated CyDs remove cholesterol significantly from human intestinal epithelial cell monolayers [34], this will

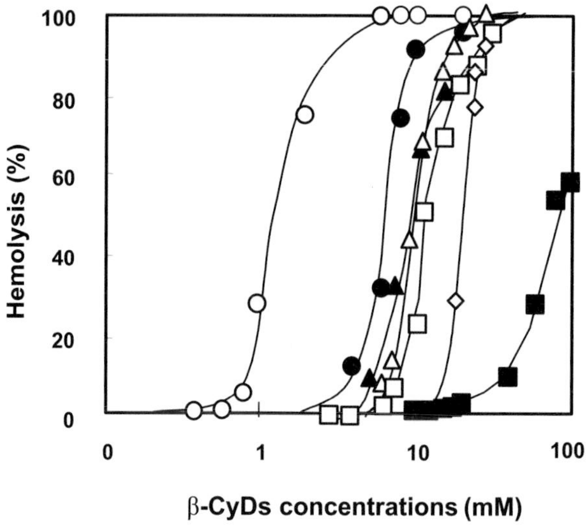

Fig. 14.2. Hemolysis curves of the effects of hydrophilic *β*-CyDs on rabbit erythrocytes in 0.1-M phosphate buffer solution (pH 7.4) at 37 °C. ○: DM-*β*-CyD, **2**; ●: *β*-CyD; △: HP-*β*-CyD, **6** (d.s. 4.8); ▲: G_2-*β*-CyD, **10**; ◇: SBE7-*β*-CyD, **21** (d.s. 6.2); □: GUG-*β*-CyD, **11**; ■: DMA-*β*-CyD, **4**; (d.s. 6.3).

induce membrane invagination through a loss of bending resistance, and consequently lead to death of the cells. The effects of *β*-CyDs on the viability of Caco-2 cells were evaluated by measuring intracellular dehydrogenase activity [33]. For a tentative evaluation of cytotoxicity, Tween 20, a typical nonionic surfactant was used as a positive control, and eventually provided almost complete cytotoxicity at even 20 mM. In contrast to **2**, other CyDs were fairly bioadaptable even in higher concentrations.

Heptakis(2,6-di-*O*-methyl-3-*O*-acetyl)-*β*-CyDs (DMA-*β*-CyDs) have unique pharmaceutical properties such as solubilizing power and hemolytic activity [35]. The hemolytic activity of **2** is drastically decreased by introducing acetyl groups into the secondary hydroxyl groups at the 3-position, while increasing the average d.s. of acetyl groups. DMA-*β*-CyDs are highly water-soluble and maintain a certain inclusion ability comparable to **3**. Interestingly, DMA-*β*-CyDs attenuated nitric oxide (NO) production in macrophages stimulated by lipopolysaccharide and lipoteichoic acid, probably due to the suppression of lipopolysaccharide and lipoteichoic acid binding to their receptors in the cells [36]. In particular, DMA7-*β*-CyD with d.s. 7.0 significantly suppressed the septic shock induced by lipopolysaccharide and D-galactosamine in mice [37].

Figure 14.3 (A) shows the temperature dependence of the solubility of three *β*-CyDs in water [22]. With increasing temperature, the solubility of native *β*-CyD gradually increases, while that of **2** decreases remarkably, similar to a nonionic sur-

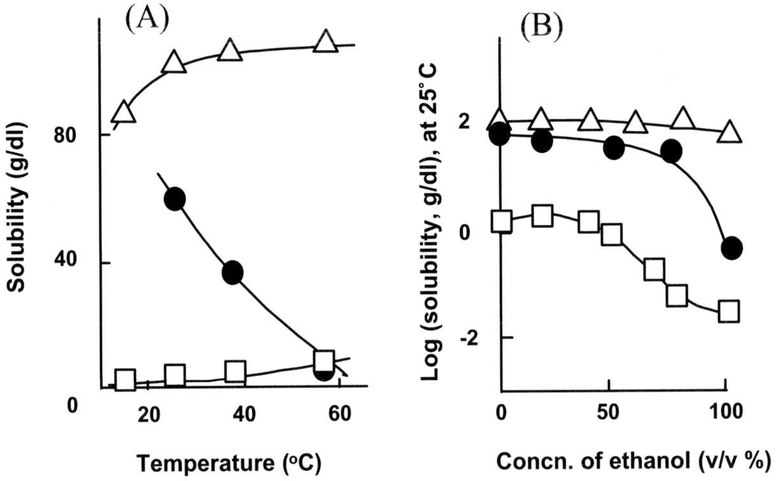

Fig. 14.3. Effects of temperature (A) and solvent polarity (B) on solubilities of three β-CyDs in water. □: β-CyD; ●: DM-β-CyD **2**; △: HP-β-CyD **6** (d.s. 5.8).

factant. In the case of **6**, however, very high solubility was maintained in a wide range of temperatures. In Fig. 14.3 (B), **6** shows the highest solubility in all the ethanol/water solvent systems, compared to other CyDs. This kind of property is advantageous for many purposes in pharmaceutical formulations.

14.2.2
Hydrophobic CyDs

When the hydroxyl groups of CyDs were substituted by ethyl, acetyl, or longer acyl groups (C1–C12), the solubility of these CyDs in water decreased proportionally to their d.s. and/or the length of the alkyl chains [38, 39]. Concentrated solutions of peracylated β-CyDs in organic solvents were highly viscous and sticky, and gelation took place upon evaporation of the solvents. These properties would be useful for a slow-release carrier of water-soluble drugs. The longer acyl side chains, however, severely interfere with the ability of peracylated β-CyDs to form inclusion complexes. Per-*O*-butanoyl-β-CyD (TB-β-CyD, **15**) has a prominent retarding effect for water-soluble drugs, owing to its mucoadhesive property and appropriate hydrophobicity, which differ from those of other derivatives with shorter or longer chains. An X-ray crystallographic study (the method discussed in Chapter 7) was conducted to gain an insight into the interaction of guest molecules with peracylated β-CyDs, and revealed that self-association of acyl groups occurs with increasing chain length; in particular, one butanoyl chain intramolecularly penetrates the cavity [40]. Hence, peracylated β-CyDs will hinder the inclusion of any other guest

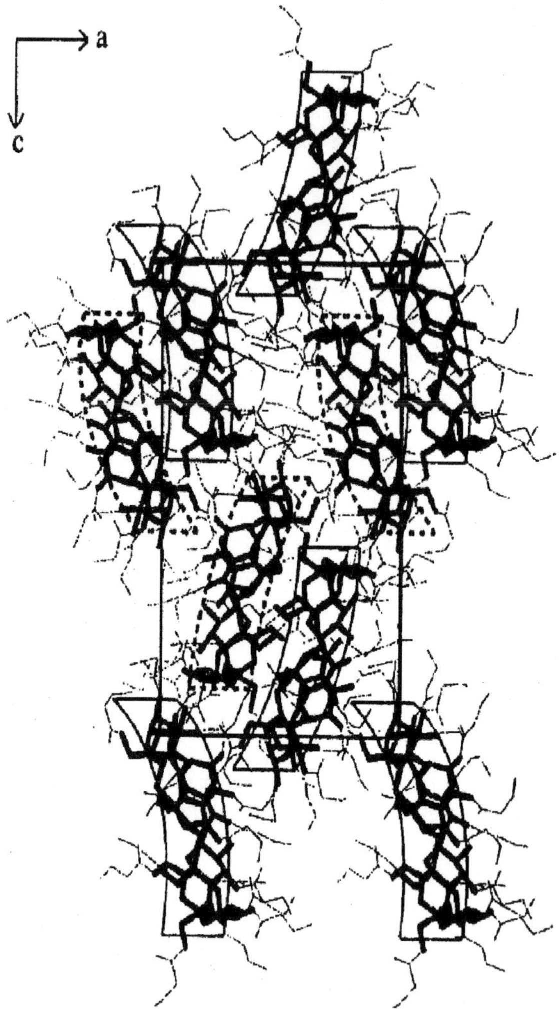

Fig. 14.4. Crystal packing of per-O-butanoyl-β-CyD (**15**), viewed along the *b* axis (thick lines denote the β-CyD backbone and thin ones denote the acyl chains).

molecules into the central cavity, and the guest molecules may be accommodated in the matrix between the acyl chains, forming monomolecularly dispersed system (Fig. 14.4) (see the discussion of dispersed systems involving CyDs in Chapter 15) [40]. In the case of per-O-valeryl-β-CyD (TV-β-CyD, **16**), an adhesive film was obtainable by means of the ordinary casting method, when an ethanol solution containing **16** and drug was spread on a polyethylene backing sheet [41].

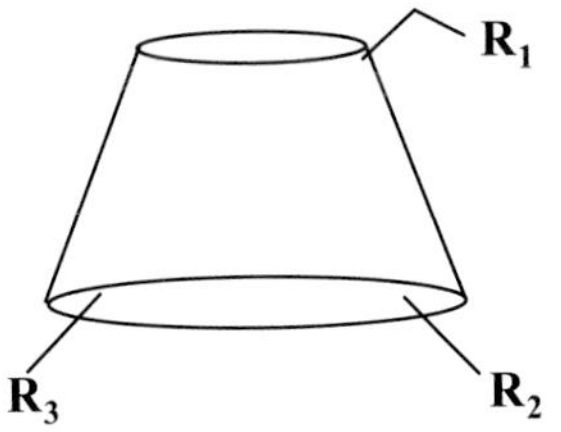

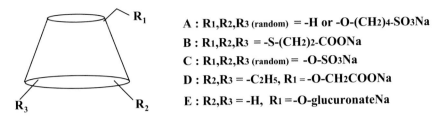

A : R1,R2,R3 (random) = -H or -O-$(CH_2)_4$-SO_3Na
B : R1,R2,R3 = -S-$(CH_2)_2$-COONa
C : R1,R2,R3 (random) = -O-SO_3Na
D : R2,R3 = -C_2H_5, R1 = -O-CH_2COONa
E : R2,R3 = -H, R1 = -O-glucuronateNa

Fig. 14.5. Structures of ionizable CyDs. A: SBE-β-CyD, **20**, **21**; B: Org25969, **23**; C: β-CyD-sul, **19**; D: CME-β-CyD, **18**; E: GUG-β-CyD, **11**.

14.2.3
Ionizable CyDs

SBE7-β-CyD (Captisol[RT], **21**) containing an average of seven sulfobutyl ether groups per CyD molecule, is currently used in parenteral preparations because it is very soluble in water and less toxic than other sulfobutyl ether β-CyD derivatives [11]. Owing to the very low pK_a of the sulfonic acid groups, **21** carries multiple negative charges at physiologically compatible pH values (see Fig. 14.5). The four-carbon butyl chain coupled with repulsion of the end-group negative charges allows an extension of the CyD cavity. This often results in stronger binding to drug candidates than is the case with other modified CyDs. The use of **21** enables the development of parenteral and oral products of active compounds too insoluble to formulate by other solubilization methods. For example, amorphous **21** is effective in improving both the solubility and the chemical stability of ONO-4819, prostaglandin E_1 analogue [42], in contrast to the hydrophilic CyDs. **21** also serves as both a solubility modulating and an osmotic pumping agent for porosity-controlled osmotic pump tablets (see below, Section 14.3.3.3), from which the release rate of both highly and poorly water-soluble drugs can be controlled precisely [43, 44].

A γ-CyD-based synthetic receptor of rocuronium bromide with high stability constant (about 10^7 M^{-1}) has been developed [45]. To increase the total area for hydrophobic interaction inside the cavity, the γ-CyD cavity is extended from the primary side by three carbon atoms through a thioether linkage. In addition, anionic functional groups are introduced at the rim of the cavity. The negative charge of carboxylate groups provides the electrostatic interaction with the positively charged nitrogen atom of the guest molecule, and also maintains the high water solubility of the resulting host molecule. In the crystalline form, the rocuronium is encapsulated in the hydrophobic cavity of the synthetic γ-CyD, octakis-*S*-(2-carboxyethyl)-octathio-γ-CyD octasodium salt (Org25969, **23**, see Fig. 14.5) [45], and the quaternary ammonium group on ring D of the drug is loosely surrounded by the carboxyl groups of **23**. This host molecule reverses the neuromuscular block effect of rocuronium bromide *in vitro* and *in vivo*. The reversal of the biological activity seems to be mediated by encapsulation of the blocker. The supramolecular mechanism of action is superior to current clinically used reversal agents in terms of speed and side effects. This approach appears to be a good example of using a small synthetic host molecule as an antidote (or antagonist) to a biologically active drug.

Complex **18** exhibits pH-dependent solubility for use in selective dissolution of the drug/CyD complex as well as acting as a stabilizer for acid-susceptible drugs [46, 47]. It displays limited solubility under acidic conditions such as those in stomach, with the solubility increasing as the pH passes through the pK_a (ca 3.7) of **18**. The enteric-type **18** complex, which is soluble in the intestinal fluid, has been used in *in vitro* and *in vivo* studies with diltiazem, a calcium channel antagonist, and molsidomine, a peripheral vasodilator [48]. As in the diltiazem studies, a high degree of correlation was noted between the *in vivo* absorption and *in vitro* release measured with the variable-pH dissolution apparatus.

Glucuronyl-glucosyl-β-CyD (GUG-β-CyD, **11**) is a new type of branched CyD, which contains a carboxyl group in the branched maltosyl residue [49, 50]. The hemolytic effect of **11** on rabbit erythrocytes is lower than that of β-CyD and **10** (see Fig. 14.2). This compound shows greater affinity for the basic guest molecules, owing to the electrostatic interaction of carboxyl group with positively charged drug molecule.

14.3 Improvements in the Pharmaceutical Properties of Drugs

An important characteristic of CyDs is the formation of inclusion complexes in both the solution and the solid states, in which each guest molecule is surrounded by the hydrophobic environment of the CyD cavity. This can lead to the alteration of physicochemical properties of guest molecules such as solubility, chemical stability, dispersibility and so on, which can eventually have considerable pharmaceutical potential (Table 14.2) [23]. The stability constant and stoichiometry of the inclusion complexes, depending on the guest/host concentration employed, are useful indices for estimating the binding strength of the complex and changes in the physicochemical properties of the guest molecule in the complex [7, 19]. Moreover, attention should be directed towards various environmental factors, such as dilution, temperature, pH, and additives, in the design of practical formulations and routes of administration [22].

14.3.1 Bioavailability

Since oral administration is the most common and useful, there are many opportunities in developing oral dosage forms. The oral route faces several obstacles to drug absorption, such as the first pass effect, individual variability of bioavailability, and drug-induced local irritation to gastric and intestinal mucosa. According to the US Food and Drug Administration (FDA) Biopharmaceutics Classification System

[51], various drugs are classified as Class I to Class IV: Class I includes highly soluble and highly permeable drugs, Class II includes poorly soluble and highly permeable drugs, Class III includes highly soluble and poorly permeable drugs, and Class IV includes poorly soluble and poorly permeable drugs.

Recently, Loftsson et al. proposed that CyDs can enhance absorption of Class II as well as Class IV drugs, whereas CyD can modify oral absorption and/or reduce local irritation by drugs of Classes I and III [52]. In fact, we have recently reported that hydrophilic CyDs can enhance the oral bioavailability of itraconazol [53] and tacrolimus [33], typical Class II drugs. Additionally, β-CyD is capable of shortening the onset time of action of piroxicam and to reduce the risk of direct-contact gastric irritation [54]. Furthermore, DM-α-CyD and **2** also enhance the bioavailability of cyclosporine A, a Class IV drug [55] (see below, Section 14.4.1).

There are few reports regarding the enhancing effect of CyDs on the oral bioavailability of Class III drugs. For example, Nakanishi et al. reported that β-CyD increased significantly the absorption of sulfanilic acid, a Class III drug, compared to the absorption without pretreatment [56]. Shao et al. reported that **2** increased the bioavailability of insulin in rats under an *in situ* closed-loop condition with no appreciable tissue damage [57]. In addition, **6** was found to suppress bitterness of antihistaminic drugs such as diphenhydramine hydrochloride, hydroxyzine dihydrochloride, cetirizine dihydrochloride, chlorphenylamine maleate and epinastin hydrochloride, which are classified as Class III, although this CyD is unable to further enhance oral bioavailability [23]. Hence, CyDs are promising excipients for a variety of drugs.

Another example of the stabilizing effect of CyDs on labile drugs resulting in enhanced oral bioavailability was reported by Seo et al. [58]. In this case, complexation of γ-CyD with digoxin, a cardiac glycoside and Class II drug, suppressed acid hydrolysis of the drug, and eventually improved the oral absorption of digoxin in dogs by its solubilizing effect. Likewise, Kikuchi et al. demonstrated that β-CyD enhances the oral bioavailability of carmofur, a masked compound of 5-fluorouracil, in rats through its solubilizing and stabilizing effects in the GI tract [59].

Active secretion and intestinal phase I metabolism of absorbed drug have been recognized as major determinants of oral bioavailability. The multidrug efflux pump, P-glycoprotein (P-gp) and multidrug resistant-associated protein 2 (MRP2) as well as cytochrome P450 (CYP) 3A, the major phase I drug metabolizing enzyme, are present at high levels in the villus tip of epithelial cells in the GI tract. In the epithelial cells there is also a variety of transporters for helping drug absorption. Some papers regarding the effects of CyDs on the viability and integrity of human colon adenocarcinoma cell line Caco-2 cell monolayers have been published [60, 61]. Against this background, Ono et al. investigated whether hydrophilic CyDs affect efflux and influx transporters and CYP3A in intestinal epithelial cells *in vitro* [62]. Recently, we have revealed the new enhancing mechanism of **2** with respect to P-gp and MRP2 for the oral bioavailability of lipophilic drugs in Caco-2 cell and vinblastine-resistant Caco-2 (Caco-2R) cell monolayers, i.e. **2** enhances the oral bioavailability of hydrophobic drugs not only by its solubilizing effect but also by its inhibitory effect on the efflux pump activity of P-gp and/or

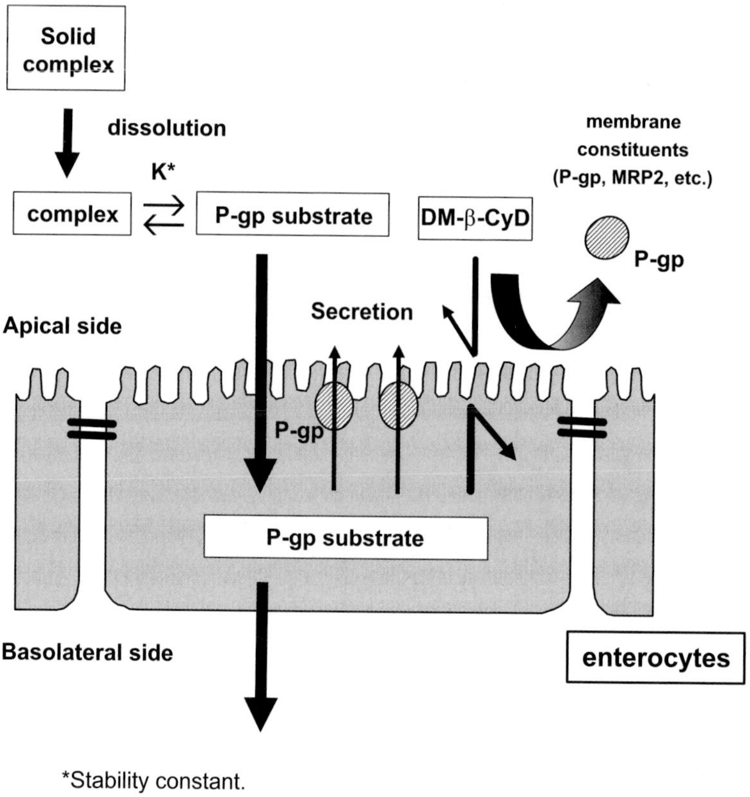

Fig. 14.6. Possible enhancing mechanism of DM-β-CyD **2** on the oral bioavailability of tacrolimus (P-gp substrate).

MRP2 (Fig. 14.6) [34, 63]. In addition, **2** suppresses L-type amino acid transporter 1 and Na^+/glucose cotransporter 1 transporter's function through the release of their transporters from the apical membranes of Caco-2R cell monolayers. Interestingly, DM-α-CyD augments H^+-coupled peptide transporter PepT1-mediated uptake of glycylsarcosine in Caco-2 and Caco-2R cell monolayers.

Hydrophilic CyDs might affect drug metabolism by CYP enzymes, which exist in both intestinal epithelial cells and hepatocytes. However, it has generally been believed that CyDs do not to interact with microsomal proteins, because CyDs cannot enter the cytoplasm. However, Jadot et al. reported that extracellularly added **1** could modify the properties of the lysosomal membrane in human hepatocellular carcinoma HepG2 cells, suggesting that **1** might affect CYP enzymes in microsomes of enterocytes [64]. In our preliminary study using an inverted sac made from rat small intestine, we found that **2** suppresses the bioconversion of testosterone to 6-hydroxy-β-testosterone mediated by CYP3A in the small intestine. Thus, CyDs can enhance the oral bioavailability of drugs in different ways, and the enhancing mechanism of CyDs on the oral bioavailability of drugs may be more complicated than we have so far believed. For further information about the effects of CyDs on drug absorption in other routes such as nasal, rectal, dermal, ophthalmic, and pulmonary, readers are referred to several excellent reviews [19, 65, 66].

14.3.2 Chemical and Physical Stability

The influence of CyDs on the chemical and physical stabilities of drugs in solution and in the solid state is one of the best investigated effects on CyD inclusion complexation [1–6]. The drug must be stable not only during storage but also in biological fluids, since reactions which result in a product that is pharmacologically less active will reduce the therapeutic effectiveness. CyDs are known to accelerate and decelerate various kinds of reactions depending on the nature of the complex formed (see remarks on CyD catalysis in Section 1.1 and the discussion in depth in Chapter 4). Generally, when a drug's active center is included in the CyD cavity, there is a deceleration effect, and the reaction rate is dependent on the free drug concentration arising from dissociation of the complex. On the other hand, if the drug does not totally fit into the cavity or is only partially included, leaving the active center sterically fixed close to the catalysts, it experiences an acceleration effect [19].

14.3.2.1 In Solution

CyDs accelerate or decelerate various reactions, exhibiting many kinetic features shown by enzyme reactions, i.e. catalyst–substrate complex formation, competitive inhibition, saturation, and stereospecific catalysis [67]. CyD-catalyzed reactions can generally be classified in the following three categories according to the type of stimulation: (a) participation of the hydroxyl groups of CyDs; (b) the microsolvent effect of the hydrophobic CyD cavity; and (c) the conformational or steric effect of CyDs [67].

(a) Participation of Hydroxyl Groups of CyDs Alkaline hydrolyses of phenol esters are accelerated by the addition of CyDs, where the secondary hydroxyl groups (pK_a about 12.2) of CyDs work as a nucleophilic catalyst, attacking the carbonyl group of esters to form a covalent intermediate, acylated CyD, that is gradually hydrolyzed to a final product and CyD. Many phenol esters, organophosphonates, carbonates, and β-lactam antibiotics are hydrolyzed according to this covalent catalysis mechanism of CyDs. In contrast to phenol esters, some alkyl benzoates are hydrolyzed through noncovalent catalysis by CyDs, in which the secondary hydroxyl groups work as a general base catalyst, assisting the ionization of a water molecule to form a reactive hydroxide ion that attacks the carbonyl group of the benzoates, producing a final product. These covalent and noncovalent catalyses of CyDs generally accelerate the reactions, i.e. positive catalysis, particularly when the secondary hydroxyl groups of the host molecules are in close contact with a reactive site of the guest molecules. However, negative catalysis by CyDs (deceleration effect) is of greater importance from the drug stabilization perspective. One method of suppressing positive CyD catalysis is to introduce substituents onto the secondary hydroxyl groups of CyDs and to block their catalytic function. A typical example is the effects of methylated β-CyDs on the degradation of E-type prostaglandins [12]. The β-hydroxyketo moiety of E-type prostaglandins is extremely susceptible to dehydration under acidic and alkaline conditions to give A-type prostaglandins, which are isomerized to B-type prostaglandins under alkaline conditions. Native β-CyD accelerates the dehydration and isomerization of E- and A-type prostaglandins, respectively, under alkaline conditions, because the secondary hydroxyl groups of β-CyD act as a general base catalysis. In sharp contrast, methylated β-CyDs significantly decelerated the dehydration and isomerization in neutral and alkaline conditions, where the negative catalytic effect of **2** was larger than that of **3**, because the former includes the reactive moiety of the prostaglandins more tightly than does the latter [68].

(b) The Microsolvent Effect of the Hydrophobic CyD Cavity The CyD cavity has an apolar character similar to an 80% dioxane/water solution and provides a slightly alkaline environment, because it is surrounded by glycosidic ethers [67]. The chemical reactivity of guest molecules is altered by the change in polarity when they are embedded in such a hydrophobic and alkaline environment. For example, the photodecarboxylation of the anti-inflammatory drug benoxaprofen was decelerated by inclusion complexation with β-CyD [69]. The photodecarboxylation of benoxaprofen is known to be slower in less-polar solvents such as ethanol, and is slower in the case of the neutral drug than for its anionic species. Therefore, this deceleration is attributable to the microsolvent effect of CyD, i.e. the drug is included in the hydrophobic environment of CyD cavity and the acid dissociation of the drug is suppressed there. In contrast to benoxaprofen, the decarboxylation of activated α-cyano- and β-keto-acetic acids is faster in solvents with lower dielectric constants, so this reaction is accelerated by the addition of β-CyD [67].

In general, CyD complexation decelerates chemical degradations, in which a highly charged intermediate is evolved in the transition state, because such charged species are energetically unstable in hydrophobic environments such as the CyD cavity. For example, the acid-catalyzed hydrolysis of prostacyclin is decelerated in the β-CyD cavity, because the hydrolysis of its vinyl ether moiety proceeds through a cationic intermediate [70]. The antitumor drug O^6-benzylguanine is susceptible to acid-catalyzed hydrolysis to form guanine and benzyl alcohol according to the S_N1 mechanism with significant charge separation. At pH 4.8 this hydrolysis was decelerated by a factor of 220 by complexation with **21**, because of the unfavorable charge separation in the hydrophobic CyD cavity [71].

As described above, the CyD cavity provides a slightly basic environment, which decelerates acid-catalyzed degradations of drugs. For example, the acid-catalyzed hydrolysis of digoxin (see above, Section 14.3.1), a potent cardiac glycoside, is decelerated by the addition of CyD, particularly β-CyD in which it has a larger stability constant ($K_c = 11200\ M^{-1}$) than in α- and γ-CyD complexes [12]. In particular, the hydrolysis of the glycosidic linkage connecting the A-ring of digitoxin and sugar is completely inhibited by β-CyD: the hydrolysis rate constant in the presence of α-, β- and γ-CyDs (1.0×10^{-2} M) at pH 1.66, 37 °C were 0.108, 0.002, and 0.025 h^{-1}, respectively, whereas that in the absence of CyDs was 0.17 h^{-1}. In a similar manner, the acid-catalyzed dehydration of E-type prostaglandins to A-type prostaglandins was decelerated by the addition of β-CyD. E-type prostaglandins are known to be most stable in weakly acidic solutions (pH about 4). Therefore **18**, which gives a weakly acidic microscopic environment, can decelerate the dehydration of E-type prostaglandins in neutral conditions. The acidic property of **18** is effective in preventing the base- and water-catalyzed hydrolysis of carmofur [12, 59].

(c) Conformational or Steric Effect of CyDs When two reactive sites of molecules are forced to be in close contact by CyD complexation, intra- and inter-molecular reactions are accelerated. On the other hand, the reactions are decelerated when the two sites are forced to be far from each other [67]. For example, the intramolecular transesterification of 2-hydroxymethyl-4-nitrophenol trimethylacetate is accelerated by α-CyD, because the 2-hydroxymethyl group is forcibly oriented in the vicinity of the acyl group in the α-CyD cavity. On the other hand, intramolecular carboxylate ion attack in mono-*p*-carboxyphenyl esters of 3-substituted glutaric acid is depressed by β-CyD, because β-CyD preferentially includes the less-reactive extended conformer rather than the reactive bent conformer. The antiallergic agent tranilast, *N*-(3′,4′-dimethoxycinnamoyl)anthranilic acid, forms inclusion complexes with γ-CyD of different stoichiometries [72]. At lower concentrations of γ-CyD or higher concentrations of tranilast, two drug molecules are included in one γ-CyD cavity, i.e. a 2:1 (guest:host) complex is formed accelerating the photodimerization of the drug while with increasing γ-CyD concentration, the 1:1 and 1:2 complexes are formed as shown in Fig. 14.7, and the dimerization rates decreases. In the 1:2 complex, the dimerization rate is reduced by a factor of 19 300 in comparison to the dimerization rate of the free drug. Chlorpromazine, a tricyclic tranquilizing agent, degrades upon photoirradiation to various oxidized and polymerized products which cause cutaneous photosensitivity, resulting in skin irritation. The attack

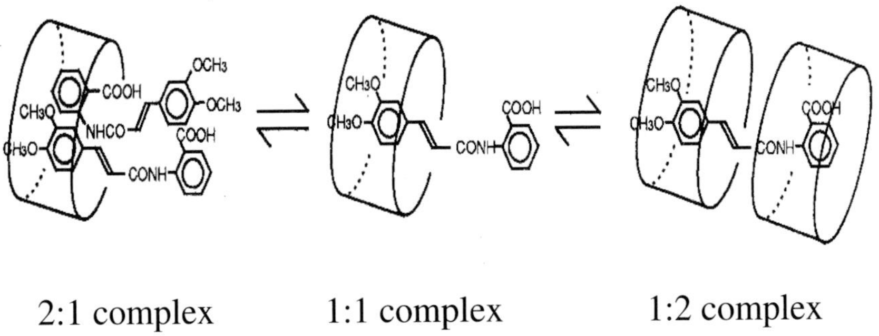

Fig. 14.7. Inclusion structures of tranilast/γ-CyD complexes with 2:1, 1:1, and 1:2 (guest:host) stoichiometries.

of oxygen on chlorpromazine is suppressed by binding to β-CyD, thus depressing the oxidation and polymerization [73]. This steric protection changes the reaction pathway of chlorpromazine, i.e. the dechlorination preferentially proceeds in the hydrophobic CyD cavity, reducing the toxicity of the drug. It is apparent that the steric protection and microsolvent effects of CyDs are involved in the photodegradation of chlorpromazine. The photodimerization of the tricyclic antidepressant protriptyline is also suppressed by the steric protection provided by binding to β-CyD [12].

14.3.2.2 **In the Solid State**

Native CyDs are hygroscopic in nature, and α-, β- and γ-CyDs have, respectively, about 6, 12, and 17 water molecules of hydration in the crystalline state [74]. The CyD complexes are also hygroscopic, although their degree of water uptake is dependent on the hydrophobicity of the included guest molecules. Therefore, native CyDs sometimes accelerate water-sensitive degradation because of their high water-sorption property, which is disadvantageous for water-labile drugs. On the other hand, less-hygroscopic hydroxypropylated and methylated CyDs suppress the degradation of drugs in the solid state. For example, the degradation of carmofur, i.e. hydrolysis of the amide moiety, is accelerated by complexation with native β-CyD in the solid state, whereas it is inhibited by methylated β-CyDs [19]. The thermal degradation of drugs, particu1arly oily substances, is significantly inhibited by complexation with native CyDs and their derivatives. CyDs can convert oily substances to crystalline or powdered forms of the complexes which have low heat conductivity compared with oils, showing thermal resistance. For instance, crystalline E-type prostaglandins are thermally degraded to oily products such as A-type prostaglandins at higher temperatures, and the resulting oils accelerate the degradation. However, such oily products are only produced in small amounts in solid CyD complexes and therefore the thermal degradation of the drug is markedly suppressed by complexation with CyDs [19].

Many solid compounds exist in different crystalline modifications such as amorphous, crystalline, or solvated forms, affecting solubility, dissolution rate, stability, and bioavailability. Among these solid compounds, amorphous forms are pharmaceutically important, because they give a significant increase in the dissolution rate and bioavailability of drugs. However, amorphous forms easily transform to a stable crystalline form during handling and storage. It is therefore necessary to control the crystallization, polymorphic transitions, and whisker generation of solid drugs. CyDs are able to increase the physical stability of drugs even when crystallization and polymorphic transition are involved.

Tolbutamide and chlorpropamide are oral hypoglycemic agents with low solubilities and dissolution rates in water. Both drugs have several polymorphic forms (Forms I–IV etc. for tolbutamide and Forms A–C etc. for chlorpropamide). The dissolution rates of the metastable forms of both drugs are higher than those of the stable forms (Form I and Form A for tolbutamide and chlorpropamide, respectively). The metastable Form IV and Form C are obtained by the spray-drying method [75–77]. However, Form IV of tolbutamide is quickly transformed into the stable Form I via the metastable Form II within about 1 day at 60 °C, 75% relative humidity (R.H.). Form C of chlorpropamide is also quickly transformed into its stable Form A within 1 day under the same conditions. Amorphous complexes of tolbutamide and chlorpropamide with **6** can be obtained by the spray-drying method. Both complexes significantly improve the solubility, dissolution rate, and oral bioavailability of the drugs in dogs. When these amorphous complexes are stored at higher temperatures and humidity (e.g. 60 °C, 75% R.H.), the included drugs crystallize gradually into their metastable forms (Form II and Form C), probably after dissociation of the complexes to each free component. It is interesting to note that the conversion of the resulting metastable Form II and Form C intermediates to the stable forms is markedly suppressed in the **6** matrix. For example, complete conversion of Form IV to the stable Form I takes more than 1 month at 60 °C, 75% R.H. Similarly, the conversion rate of the metastable Form C of chlorpropamide to its stable Form is very slow in the **6** matrix (only 50% conversion after 14 days). These facts indicate that the metastable forms are significantly stabilized in the **6** matrix. Inhibition of the polymorphic transition of a metastable form is also observed for nifedipine [19].

Crystallization of tolbutamide from aqueous solution was significantly affected by CyD complexation (Fig. 14.8) [76]. When the supersaturated transparent tolbutamide solution was stored at 4 °C, the stable Form I crystallized exclusively. On the other hand, a solution containing **2** gave only the metastable Form IV of tolbutamide. DM-α-CyD with the smaller cavity gave the stable Form I crystals. Detailed investigation indicated that crystallization of tolbutamide proceeds via the metastable Form IV, which quickly converts to the stable Form I within 15 h at 4 °C. However, conversion of Form IV to Form I in **2** solutions was markedly suppressed, and continued for at least three more days. This suppression is attributable to inclusion complexation of tolbutamide with **2**, because the addition of competitive inhibitors gave the stable Form I crystals. **2** seems to be effective in suppressing conversion of the metastable form to the stable form in the solution.

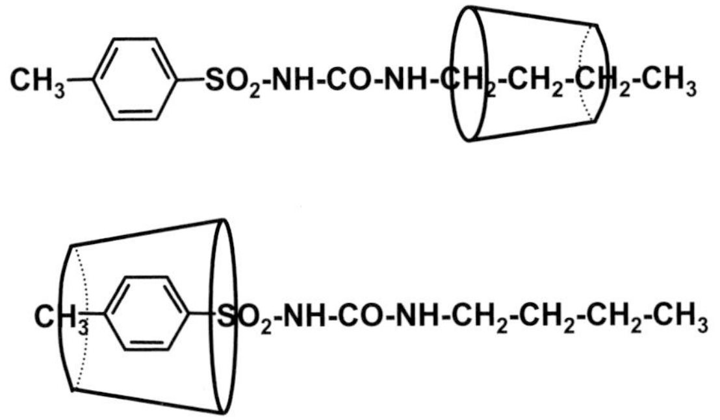

Fig. 14.8. Proposed inclusion structures of tolbutamide/α-CyD (upper) and tolbutamide/β-CyD (lower) complexes.

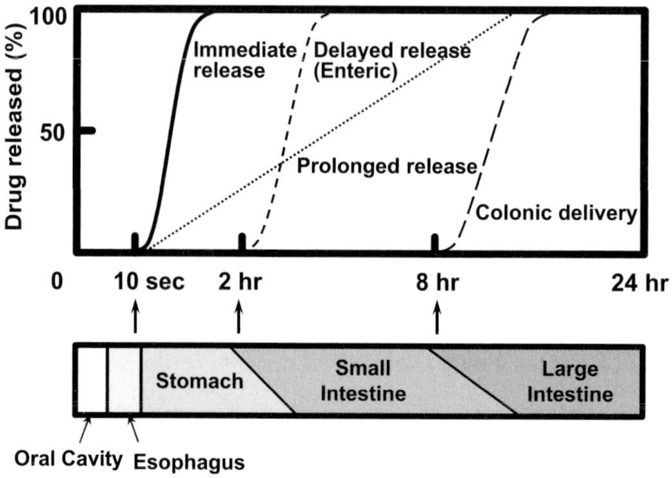

Fig. 14.9. Release profiles of drugs from various dosage forms following oral administration.

14.3.3 Release Control

Drug release should be controlled in accordance with the therapeutic purposes and pharmacological properties of the active substances. The plasma drug levels–time profiles after oral administration can be classified as rate-controlled release and time-controlled release. Rate-controlled release is further classified into three types, i.e. immediate-release, prolonged-release, and modified-release. One typical time-controlled release is delayed-release. More advanced releases can be achieved by combinations of these different release types (Fig. 14.9). Various CyD derivatives have been used in order to modify drug releases in oral preparations [13, 15].

14.3.3.1 Immediate-release

A number of biologically active compounds have been produced by employing advanced combinatorial chemistry and high-throughput screening techniques. However, most of these drug candidates are hydrophobic in nature and thus not very soluble in water. It is necessary to improve the dissolution properties of hydrophobic candidates not only for pharmacological screenings and preclinical tests but also for the design and development of formulations or advanced dosage forms. Hydrophilic CyDs such as native CyDs, **6**, **21**, and branched CyDs are demonstrably useful in improving the dissolving property of poorly water-soluble drugs, because of their high solubilizing ability and safety profiles. The immediate-release of poorly water-soluble drugs through complexation with hydrophilic CyDs are mainly ascribable to increases in the aqueous solubility, wettability, and dispersibility of the drugs and the reduction in particle size of the solid complexes. There are numerous reports showing that the aqueous solubility and dissolution rate of poorly water-soluble drugs are enhanced by complexation with hydrophilic CyDs. Examples of such drugs are prostaglandins, steroids, nonsteroidal anti-inflammatories, benzodiazepines, antidiabetics, fat-soluble vitamins, nifedipine, itraconazole, cyclosporin, and tacrolimus [3, 4, 15, 19].

14.3.3.2 **Prolonged-release**

Prolonged-release (or slow-release) preparations are useful for water-soluble drugs with short biological half-lives. Hydrophobic CyDs, such as ethylated CyDs, with low aqueous solubilities have been demonstrated to work as slow-release carriers of water-soluble drugs such as isosorbide dinitrate, diltiazem hydrochloride, and 5-fluorouracil [19, 38, 39]. For example, **15** has the most prominent retarding effect for water-soluble molsidomine and diltiazem hydrochloride, after oral administration in dogs. The release of isosorbide dinitrate from **16** film was retarded and the plasma drug level after topical application of the film to rat abdominal skin was maintained at 100 ng mL^{-1} for about 10 h [41].

HP-β-CyDs are widely used as solubilizers and a fast-dissolving carriers for many poorly water-soluble drugs. However, they can work as a retarding agent in the release of water-soluble drugs from hydrophobic carriers under certain conditions. We have recently investigated the release behavior of water-soluble drugs such as captopril and metoprolol tartrate from tablets of hydrophobic carriers such as ethylcellulose and acylated β-CyDs [23]. The release of these drugs was significantly decreased owing to the formation of a binary solid dispersion with hydrophobic ethylcellulose or **15**. The release rate was further slowed down by the addition of **6** to the hydrophobic carriers. For example, the release rate of metoprolol decreased with the increase in content of **6** in the ethylcellulose matrix up to (30/10)/60% w/v (metoprolol/**6**)/ethylcellulose, but a further increase in **6** content led to higher release rates. Detailed studies on water penetration, scanning electron microscopic observation, and the physicochemical properties of **6** indicated that the retarding effect of **6** is due to its viscous gel formation in small pores of the ethylcellulose tablet [78]. The **6** gel works as a barrier to water penetration into the tablet and as a rate-decelerating agent for metoprolol when it is formulated in appropriate amounts in a hydrophobic ethylcellulose matrix [79]. The rate-accelerating effect of **6**, when it is present in large amounts, may be attributable to its hydrophilic character overwhelming its gel-forming property. The release rate of metoprolol from the ternary (metoprolol/**6**)/ethylcellulose (30/10)/60% w/v) tablet was barely influenced by the pH of the medium, the paddle rotation rate, the viscosity of the solution, or the storage condition of the tablet. Consequently, a combination of **6** and hydrophobic carriers such as ethylcellulose is useful for the controlled release of water-soluble drugs, and the release control can be tuned by adjusting the composition of components. We have recently demonstrated that metoprolol forms inclusion complexes with three native CyDs in different stoichiometries and that the inclusion modes depend on the cavity size of the CyDs (Fig. 14.10) [79].

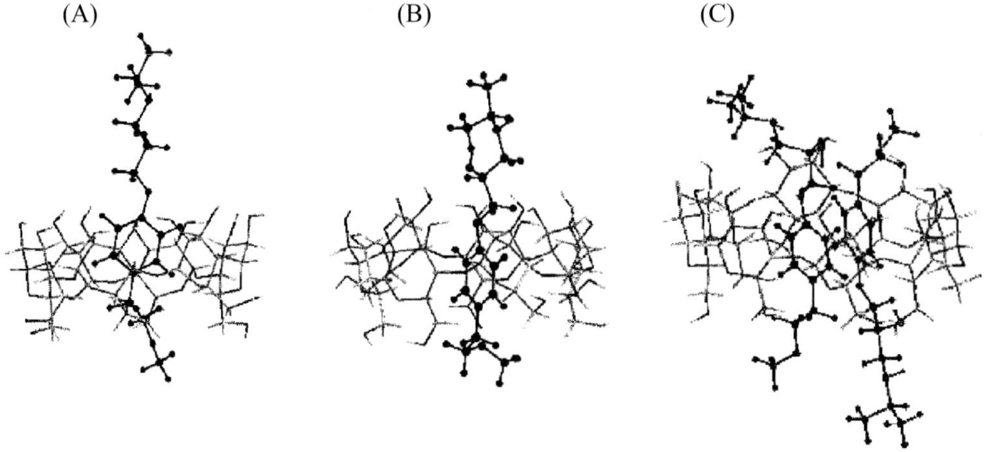

Fig. 14.10. Proposed inclusion modes of metoprolol (Met) with three native CyDs [23] (A): α-CyD complex with the methoxyethylbenzene moiety of Met inserted from the secondary hydroxyl side of α-CyD; (B): β-CyD complex with the methoxyethylbenzene moiety of Met inserted from the secondary hydroxyl side of β-CyD; (C): γ-CyD complex with the methoxyethylbenzene moiety of Met inserted from the secondary hydroxyl side of γ-CyD in an anti-parallel orientation of Met.

14.3.3.3 **Modified-release**

Various modified-release formulations can be obtained by combinations of CyD complexes with different release carriers [15]. For example, a combined use of TA-β-CyD **14** and TO-β-CyD **17** gave a constant plasma level (20–40 ng mL^{-1}) of diltiazem for more than 48 h after oral administration to dogs. A double-layer tablet consisting of a fast-releasing fraction of nifedipine/**6** complex and a slow-releasing fraction of nifedipine/hydroxypropylcellulose (HPC) dispersion gave prolonged plasma drug levels without decrease of bioavailability [80].

Advanced pharmaceutical applications of **21** have been reported in the porosity-controlled osmotic pump tablet (OPT) [43, 44]. The OPT is a tablet coated with a semipermeable membrane containing drugs and leachable pore-forming materials (Fig. 14.11). In this system, after dissolution inside the core the drug is released from the OPT by hydrostatic pressure through pores created by the dissolution of pore formers incorporated into the membrane. The hydrostatic pressure is created by an osmotic agent, the drug itself or tablet components, after water is imbibed across the semipermeable membrane. This system is generally applicable only for water-soluble drugs, because the dissolved drug can pass through the small pore of the membrane, but solids and dispersions with large sizes can not. Poorly water-soluble drugs cannot dissolve adequately in the small volume of water drawn into the OPT, making release from the OPT incomplete. Further, osmotic agents are necessary to create a higher pressure within the membrane, by which the dissolved drug is pushed out from the OPT. Sodium salts of **21** serve both as a solubilizer

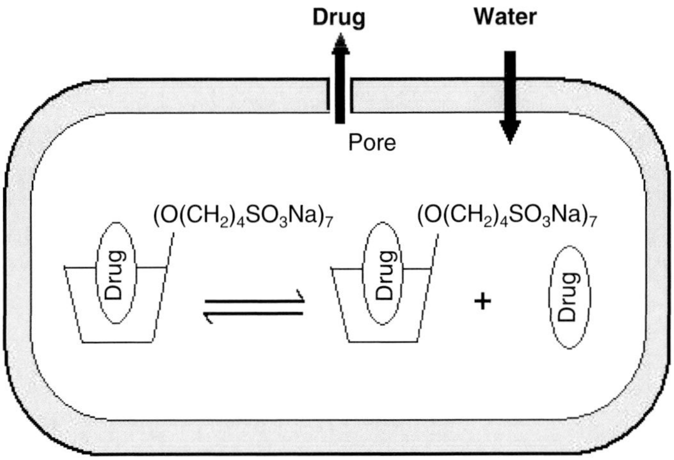

Fig. 14.11. A porosity-controlled osmotic pump tablet utilizing SBE7-β-CyD (**21**) complexation.

and as an osmotic pumping agent. The multiple function of **21** in the OPT system has been demonstrated in the release control of testosterone, prednisolone, chlorpromazine, etc. Drug releases from the OPT system with **21** displayed zero-order release characteristics and were complete, whereas nonionizable CyDs such as **6** showed a first-order release and the release was incomplete. The *in vivo* absorption profiles of prednisolone from the OPT system with **21** correlated well with *in vitro* release profiles.

14.3.3.4 **Delayed-release**

Enteric-type preparations can be classified as exhibiting time-controlled release, since the drug is preferentially released in the small intestinal tract (pH 6.8) after passing the acidic stomach (pH 1.2). Therefore, it takes 1–2 h from the time of oral administration for the drug to appear in the blood, although the lag time changes depending on the physiological conditions. For these purposes, synthetic polymers bearing carboxylic acid groups, such as carboxymethylethylcellulose (CMEC), cellulose acetate phthalate (CAP), and methacrylic acid polymers (Eudragit L 100^{TM}) etc., are used as film-coating agents and additives in tablets. **18**, which has ethyl and carboxymethyl groups in the molecule, is effective in modifying the release rate of water-soluble drugs such as diltiazem hydrochloride and molsidomine sulfate, i.e. the **18** complexes release the drug very slowly in the stomach but rapidly in the intestine [46, 48]. The main advantage of **18** as an enteric-type carrier over the enteric polymers is its solubilizing and stabilizing effect on guest drugs. The delayed-release system with much longer lag times (>6–8 hours) can be obtained by utilizing CyD conjugates in which a drug is covalently bound to the hydroxyl groups of CyDs through an ester linkage. This is described later in Section, 14.5.1 (colon specific delivery).

14.3.4
Enhancement of CyD Action by Combined Use with Additives

In general, practical pharmaceutical preparations include considerable amounts of pharmaceutical additives to maintain the efficacy and safety of the drug molecules. Obviously, CyDs are capable of modifying the physicochemical properties of drugs as well as of the additives and vice versa. Therefore, suitable combinations of CyDs and additives can markedly extend their functions. In this section, therefore, a new strategy of CyD applications will be described for the design of advanced dosage forms. Optimization of practical formulations should be achieved by taking advantage of competitive inclusion, complex dissociation, and pharmacokinetics [22].

Some additives can suppress dissociation of CyD complexes. Morphine occurs naturally in the opium poppy. It is a potent narcotic analgesic, and its primary clinical use is in the suppression of moderately severe pain. In patients with vomiting, a rectal dosage form such as a suppository should be useful, especially a prolonged-release-type suppository. We have demonstrated that a combination of α-CyD and xanthan gum, a polysaccharide-type polymer with high swelling capacity, enhanced rectal absorption of morphine in rabbits, when administered to rabbit's rectum in the form of Witepsol H-15 [81] hollow-type suppositories. The combination of CyDs and viscous polymers may be useful for optimizing the transmucosal delivery of morphine [81].

During the dissolution process, the complex of γ-CyD with diazepam showed an immediate-release profile, but the amount of the drug dissolved gradually decreases through the dissociation of the complex owing to its low stability constant [82]. An attempt to improve the defect has been made by the addition of HPC or polyvinylpyrrolidone (PVP) which removed the defect of diazepam (Fig. 14.12). Likewise, Loftsson et al. have reported the enhancement of oral drug bioavailability as a result of the combined action of CyD complexation and water-soluble polymers such as HPC or PVP. The enhancement is caused by an increase in the apparent stability constant of the drug/CyD complexes via the formation of ternary drug/CyD/polymer complexes [83].

The competitive inclusion phenomenon [84] markedly influences bioavailability after oral administration of a CyD-containing solution. Itraconazole, an orally active triazole antifungal agent, is practically insoluble ($<10^{-10}$ M) in water at physiological pH and only slightly soluble under extremely acidic conditions. Recently, an oral syrup solution (Sporanox^{TM}) of itraconazole containing 40% **6** has been developed to improve its bioavailability [85]. In this formulation, **6** complex (Fig. 14.13) increased bioavailability approximately four times in rats without cannulation of the bile ducts, compared to that of the drug alone [53]. Unexpectedly, the enhancing effect of **6** on the oral bioavailability of itraconazole was significantly decreased by cannulation of the bile ducts of rats. This can be ascribed to complicated factors such as dissociation of the complex via dilution and competitive inclusion, precipitation of the drug due to the pH-change, micellar solubilization of the drug in GI tracts, and so on [53]. Thus, bile acids may play a multi-functional role not only as a competitor, but also as a solubilizer for the improvement of oral bio-

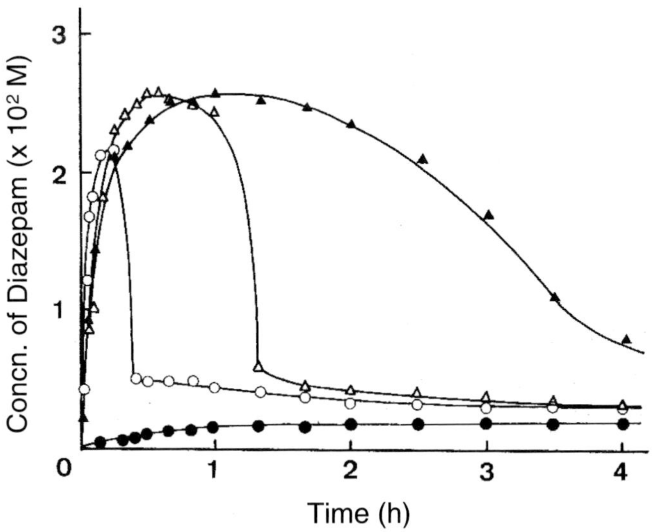

Fig. 14.12. The effects of HPC and PVP on the dissolution of diazepam and its γ-CyD complex in water at 2 °C. ●: diazepam alone; ○: complex ($K = 120\ M^{-1}$); ▲: complex in 10% w/v PVP solution; △: complex in 10% w/v HPC solution.

availability of itraconazole. A similar role of bile acids on the oral bioavailability of cyclosporine A after administration of suspensions containing the drug and **2** was observed [55].

A combination of CyDs with absorption enhancers can enhance bioavailability in various routes. A number of attempts have been made to enhance the absorption of Class III drugs, including not only prostaglandin but also peptides and proteins. One of the methods of improving their bioavailability is to use absorption enhancers. Prostaglandins are chemically unstable and poorly penetrative in epidermis and dermis, and some means of solving these deficiencies of prostaglandins are required.

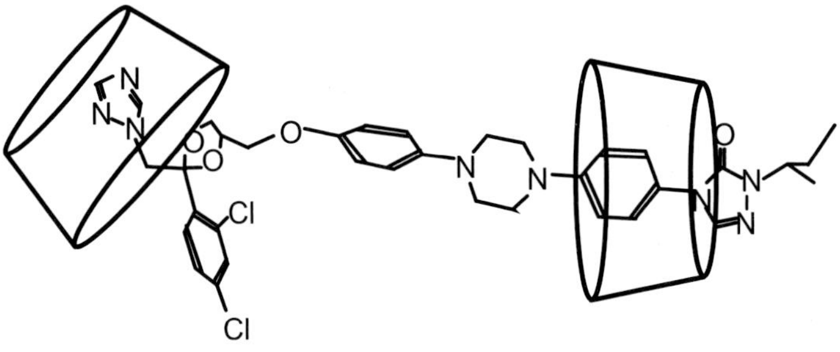

Fig. 14.13. Proposed inclusion structure of 1:2 itraconazole/HP-β-CyD (**6**) complex.

The combination of **18** and HPE-101, an oily absorption enhancer, not only enhanced the percutaneous penetration of prostaglandin E_1 but also prevented its chemical decomposition [86, 87]. Indeed, Fatty Alcohol Propylene Glycol (FAPG) ointments including prostaglandin E_1/**18**/(HPE-101, 1-[2-(decylthio)-ethyl]azacyclopentane-2-one) improved the peripheral vascular occlusive sequelae induced by sodium laurate in the ears of rabbits and increased cutaneous blood flow rate [28]. Another example of the combination of CyDs and absorption enhancers is a nasal formulation including **6** and HPE-101 or oleic acid [88, 89]. The combinatorial formulation was capable of enhancing nasal absorption of buserelin acetate, a luteinizing hormone-releasing hormone analogue, and simultaneously reducing topical irritation caused by HPE-101 alone. Thus, suitable combinations of CyDs and absorption enhancers can offer promising topical formulations.

The combination of CyDs and liposomes (discussed in more detail in Chapter 15) shows great promise as a formulation in cancer therapy. Liposomes are particulate drug carriers and are now considered to be an intriguing drug delivery technology. A major advance in the field of liposomes came with the development of Stealth liposomes, which utilize a surface coating of a hydrophilic carbohydrate or polymer, usually a lipid derivative of polyethylene glycol (PEG), to help evade the mononuclear phagocyte system recognition [90]. We have demonstrated that, when PEG liposomes encapsulating doxorubicin/γ-CyD complexes were intravenously administered in mice inoculated with mouse cancer colon-26 cells, slow disposition from blood and retention of the drug in tumors were observed, leading to an increase in the survival rate of mice [22]. In addition, we have prepared two CyD conjugates with methacrylic acid/methacrylate copolymer, in which 6-deoxy-6-ethylendiamino-β-CyD or 6-deoxy-6-monoamino-β-CyD is covalently bound to carboxylic groups of the methacrylic acid/methyl methacrylate copolymer (MAA, Eudragit^{RT}) through an amide linkage. The solubilizing effects of β-CyD/MAA conjugates were superior to those of native β-CyD, probably because of the cooperative effect of inter/intramolecular interaction of the conjugate with the guest molecule (Fig. 14.14). Since the hemolytic effects of CyD conjugates are much lower than

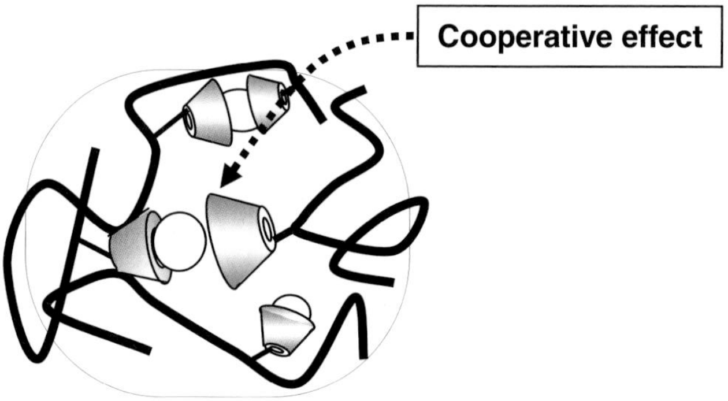

Fig. 14.14. Proposed interaction of CyD/polymer conjugate with a guest molecule.

those of β-CyD and MAA, these new conjugates seem to be useful as multifunctional and bioadaptable drug carriers in controlled-release formulations [22].

14.4 Potential Use of CyDs in Peptide and Protein Formulation

Many attempts have been made to eliminate undesirable properties of peptide and protein drugs such as chemical and enzymatic instability, poor transport through biological membranes, rapid plasma clearance, immunogenicity, etc. [91]. CyDs seem to be an attractive alternative to these approaches [14, 92]. For example: (a) α-CyDs are preferable to solubilize cyclosporine A, a poorly water-soluble peptide; (b) hydrophilic β-CyDs inhibit the adsorption of insulin on the hydrophobic surface of the containers; (c) hydrophilic CyDs improve the nasal and subcutaneous bioavailability of peptide and polypeptide drugs, owing to the enhanced membrane permeation and enzymatic stabilization; (d) branched β-CyDs are particularly effective in inhibiting the aggregation of polypeptide and protein drugs such as insulin and recombinant human growth hormone (rhGH); (e) hydrophobic CyDs are useful in designing sustained-release-type peptide preparations; (f) the aluminum salt of β-CyD-sulfate **22** is effective for sustained-release of basic fibroblast factor (bFGF). This section deals with more recent aspects of utilization of CyDs in peptide and protein drug formulations [22].

14.4.1 Absorption Enhancement by Hydrophilic CyDs

Cyclosporin A, an immunosuppresive drug, is a poorly water-soluble cyclic undecapeptide, exhibiting low oral bioavailability and a wide range of variability in absorption. Hydrophilic CyDs increase the solubility of cyclosporin A in water with a positive deviation from linearity, forming higher-order complexes. The solubilizing effect of native CyDs increases in the order of γ-CyD < β-CyD < α-CyD. Of the hydrophilic CyDs, DM-CyDs showed the greatest solubilizing effect. The dissolution rate of the drug was markedly augmented by complexation with DM-CyDs. In closed loop *in situ* experiments using rat small intestine, **2** considerably augmented the cumulative amounts of the drug in the blood, while decreasing the amount ratio of one of its major metabolites M1 [55]. The inhibiting ability of **6** in the bioconversion of cyclosporin A in the small intestinal microsome of rat was greater than that of DM-α-CyD. An *in vivo* study revealed that DM-CyDs enhance the transfer of cyclosporin A to blood not lymph, with low variability in the absorption after oral administration of the drug suspension to rats. The variability of bioavailability of DM-CyD complexes was lower than that of Sandimmune^{TM}, although the extent of bioavailability of the drug or its **2** complex was appreciably decreased by cannulation of the bile duct of rats, probably due to the competitive inclusion of bile acids, but the extent of the lowering of bioavailability in the presence of **2** was much less serious than that of drug alone.

Internasal delivery of peptide and protein drugs is severely restricted by presystemic elimination by enzymatic degradation or mucocillary clearance and by the limited extent of mucosal membrane permeability. α-CyD has been shown to remove some fatty acids from nasal mucosa and to enhance the nasal absorption of leuprolide acetate in rats and dogs. The utility of chemically modified CyDs as absorption enhancers for peptide drugs in rats has been demonstrated [93]. For example, **2** was shown to be a potent enhancer of insulin absorption in rats, and a minimal effective concentration of **2** for absorption enhancement exerted only a mild effect on *in vitro* ciliary movement [94]. The scope of interaction of insulin with CyDs is limited, because CyDs can only partially include the hydrophobic amino acid residues in peptides, with small stability constants [14]. Under *in vivo* conditions, these complexes will readily dissociate into separate components, and hence the displacement by membrane lipids may further destabilize the complexes. The direct interaction of peptides with CyDs is, therefore, of minor importance in the enhancement of nasal absorption. Of the hydrophilic CyDs tested, **2** had the most prominent inhibitory effect on the enzymatic degradation of both buserelin acetate, an LHRH agonist, and insulin in rat nasal-tissue homogenates [95]. Because of the limited interaction between peptides and CyDs, they may reduce the proteolytic activities of enzyme–substrate complexes. This view is supported by the following observations. Leucine aminopeptidase in the nasal mucosa is known to cleave the B-chain of insulin from the N-terminal end. **2** and **6** reduce the activity of leucine aminopeptidase in a concentration-dependent manner. The inhibition of proteolysis by these CyDs may participate in the absorption enhancement of peptides [96]. Another potential barrier to the nasal absorption of peptide and protein drugs is the limitation in the size of hydrophilic pores through which the drugs are thought to pass. Methylated CyDs significantly extracted membrane lipids, depending on the size and hydrophobicity of the CyD cavity in which the lipids were included [34]. Therefore, lipid solubilization mediated by CyDs may result in transcellular passages, and these changes could be transmitted to the paracellular region, which is the most likely route for the transport of polypeptides. Scanning electron microscopic observations (see Section 10.6) revealed that **2** induced no remarkable changes in surface morphology of the nasal mucosa at the minimum concentration necessary to achieve substantial absorption enhancement. These facts suggest that **2** could improve the nasal bioavailability of buserelin and is well tolerated by the nasal mucosa.

Branched CyDs [29, 30, 97], β-CyD sulfate **19** [98, 99], and sulfoalkyl ethers of CyDs [11, 100] may be a new class of parenteral drug carriers in peptide and protein formulations because they are highly hydrophilic and less hemolytic than native and other hydrophilic CyDs. When insulin solutions containing SBE4-β-CyD **20** were injected into the dorsal subcutaneous tissues of rats, the plasma immunoreactive insulin (IRI) level rapidly increased and higher IRI levels were maintained for at least 8 h (Fig. 14.15) [101]. The bioavailability of the insulin/**20** system was about twice that of insulin alone and approached 96%. The enhancing effects of **20** may be in part due to the inhibitory effect of **20** on the enzymatic degradation

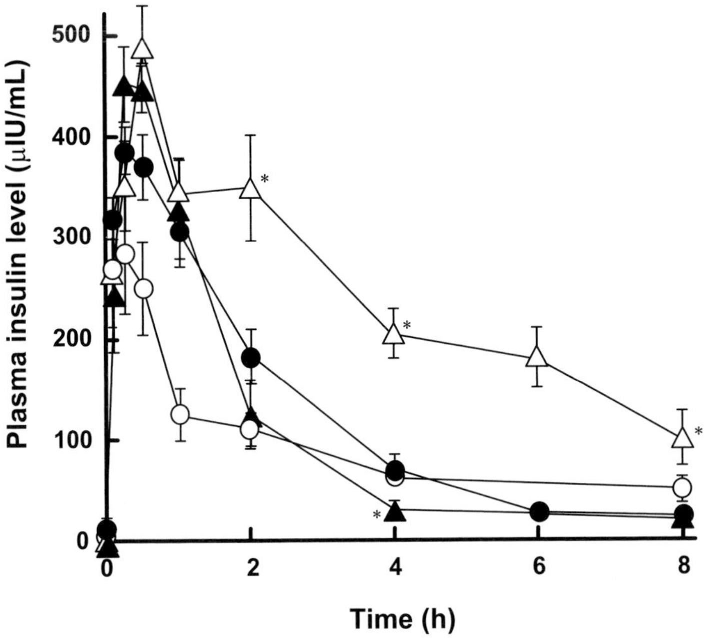

Fig. 14.15. The effects of CyDs on plasma insulin levels after subcutaneous administration of insulin solution (2 IU kg^{-1}) in rats. ●: insulin alone, ○: with G_2-β-CyD (**10**) (100 mM); ▲: with DM-β-CyD (**2**) (25 mM), △: with SBE4-β-CyD (**20**) (100 mM). Each point represents the mean ± S.E. of at least four rats. $^*p < 0.05$ versus insulin alone.

and/or the adsorption of insulin into the subcutaneous tissue at the injection site, although this does not apparently facilitate capillary permeability. Moreover, the prolonged-release profile observed for the **20** system may be due to a decrease in solubility of insulin at the injection site, where the isoelectric point of insulin may shift to neutral pH upon binding to **20**. These results suggest that hydrophilic CyDs are useful for modifying the pharmaceutical properties of insulin injections.

14.4.2
Sustained-release of Peptide and Protein Drugs

Buserelin acetate reduces the plasma levels of endogenous sex hormones to a castrate level through down-regulation, when administered continuously or daily. These paradoxical and pharmacological effects are utilized in the treatment of endocrine-dependent diseases such as endometriosis and precocious and uterine leiomyoma. Oily injection of buserelin with a sustained-release action can be achieved with ethylated β-CyDs [22]. The possible use of bioadaptable per-O-acetylated CyDs (TA-CyDs) as sustained release carriers was studied, since the release rate of buserelin from a peanut oil suspension into the aqueous phase was significantly retarded by the addition of TA-CyDs [102]. A single subcutaneous injection of an oily suspension of buserelin containing **14** or TA-γ-CYD in rats enhanced the retardation of plasma buserelin levels, giving 25 and 30 times longer mean residence times, respectively, than that with buserelin alone. Simultaneously, the suppression of plasma testosterone levels to induce castration, the pharmacological effect of buserelin, continued for one to two weeks and significant weight reduction in genital organs was observed due to the antigonadal effect. Both TA-CyDs were degraded enzymatically in rat skin homogenates. For example, the residual amounts of **14** and TA-γ-CyDs were 72.4 and 59.9% after 8-h incubation, respectively. Thus, both TA-CyDs have potential for use as bioabsorbable sustained-release carriers for water-soluble peptides following subcutaneous injection of oily suspension.

TA-CyDs are also useful to prolong both plasma and lymph levels of cyclosporin A, the drug discussed several times in this chapter [22]. When cyclosporin A was orally administered to rats as complexes with hydrophobic acylated β-CyD derivatives, both plasma and lymph levels of the drug were prolonged up to at least 36 h, although the bioavailability decreased particularly in the case of the **15** and **17** complexes. The administration of a cyclosporin A/olive oil solution in combination with HP-CyDs, especially HP-γ-CyD, further increased the plasma and lymph levels of the drug. Therefore, the less hydrophobic TA-CyDs will be useful for design of the CyD-based sustained-release formulations of poorly water-soluble peptide drugs.

bFGF is a potent mitogen that stimulates the proliferation of a wide variety of cells and could play a crucial role in wound-healing processes. The therapeutic potential of bFGF, however, has not been fully realized because of its susceptibility to proteolytic inactivation and short duration of retention at the site of action. A water-insoluble aluminum salt of sulfated β-CyD, **22**, was prepared, and its possible utility as a stabilizer and sustained-release carrier for bFGF was evaluated [103]. An adsorbate of bFGF with **22** was prepared by incubating the protein with a suspension of **22** in water. The mitogenic activity of bFGF released from the adsorbate, as indicated by the proliferation of kidney cells of baby hamsters (BHK-21), was very close to that of the intact protein. **22** significantly protected bFGF from proteolytic degradation by pepsin of α-chymotrypsin compared with their sodium salts and other oligosaccharides. The *in vitro* release of bFGF from the adsorbate was sustained in proportion to the increase in the **22**/protein ratio in the adsorbate. Of the bFGF preparations tested, the adsorbate of bFGF with **22**, when given subcutaneously to rats, showed the most prominent increase in the formation of granulation tissues, probably because of the stabilization and sustained delivery of the mitogen. These data suggest that the adsorbate of bFGF with **22** has a potent therapeutic efficacy for wound healing, and may be applicable to oral protein formulations for the treatment of intestinal mucosa erosions [22].

14.4.3
Inhibitory Effects of CyDs on Aggregate Formation of Polypeptides

The propensity of polypeptide and protein drugs to form reversible and irreversible aggregates in solution is of great concern as it may lead to the loss of desired pharmacological activity, immunogenic reactions, and unacceptable physical appearance in long-term therapeutic system. To overcome these drawbacks, several approaches have been proposed, including the use of amphiphatic excipients, chemical modification, and site-directed mutation [104].

Self-association of the insulin molecule into oligomers and macromolecular aggregates leads to complications in the development of long-term insulin therapeutic systems and limits the rate of subcutaneous absorptions, a process which is too slow to mimic the physiological plasma insulin profile at the time of meal consumption. These problems are further complicated by the tendency for insulin to adsorb onto the surfaces of containers and devices, perhaps by mechanisms similar to those inducing aggregation. Thus, many attempts have been made to prevent aggregation and surface adsorption in parenteral formulation. We have reported the effects of hydrophilic β-CyDs on the aggregation of bovine insulin in aqueous solution and its adsorption onto hydrophobic surfaces of glass and polypropylene tubes by interaction with hydrophobic regions of the peptide in both concentration- and time-dependent manners. Among the CyDs tested, **6** and **10** significantly inhibited adsorption to containers and self-association of insulin at neutral pH, whereas **2** had only a moderate effect on aggregation [105, 106]. On the other hand, SBE-β-CyDs showed different effects on insulin aggregation, depending on the d.s. of sulfobutyl group i.e. inhibition at relatively low substitution (**20**) and acceleration at higher substitution (**21**). In fact, sulfated β-CyD significantly accelerated insulin aggregation. The sulfate group in **19** and sulfonate groups in **21** would remove the hydration layer from the insulin molecule in a manner similar to lyotropic anions, a situation which makes the intermolecular interaction of the peptide stronger, eventually leading to accelerated aggregation of the peptide.

rhGH is commercially available as a lyophilized powder for treatment of hypopituitary dwarfism, but it exhibits aggregation in industrial production. We have reported the effects of hydrophilic CyDs on the thermally and chemically induced aggregation of rhGH in aqueous solution [49, 50]. Among the CyDs tested, **10** also significantly inhibited the aggregation of rhGH after refolding, where only dimer formation was observed, compared with other CyDs and linear saccharides (Fig. 14.16). This could mean that the β-CyD cavity with branched sugar moieties can prevent the aggregation of rhGH. In contrast, **6** was effective in reduction of the aggregation induced by interfacial denaturation compared with the effect of branched β-CyDs due to their surface activities. On the other hand, hydrophilic CyDs showed no noticeable inhibitory effect on the oxidation and deamidation of rhGH. These results suggest that hydrophilic CyDs may interact with exposed hydrophobic side chains rather than aliphatic side chains of rhGH, resulting in inhibition of aggregation.

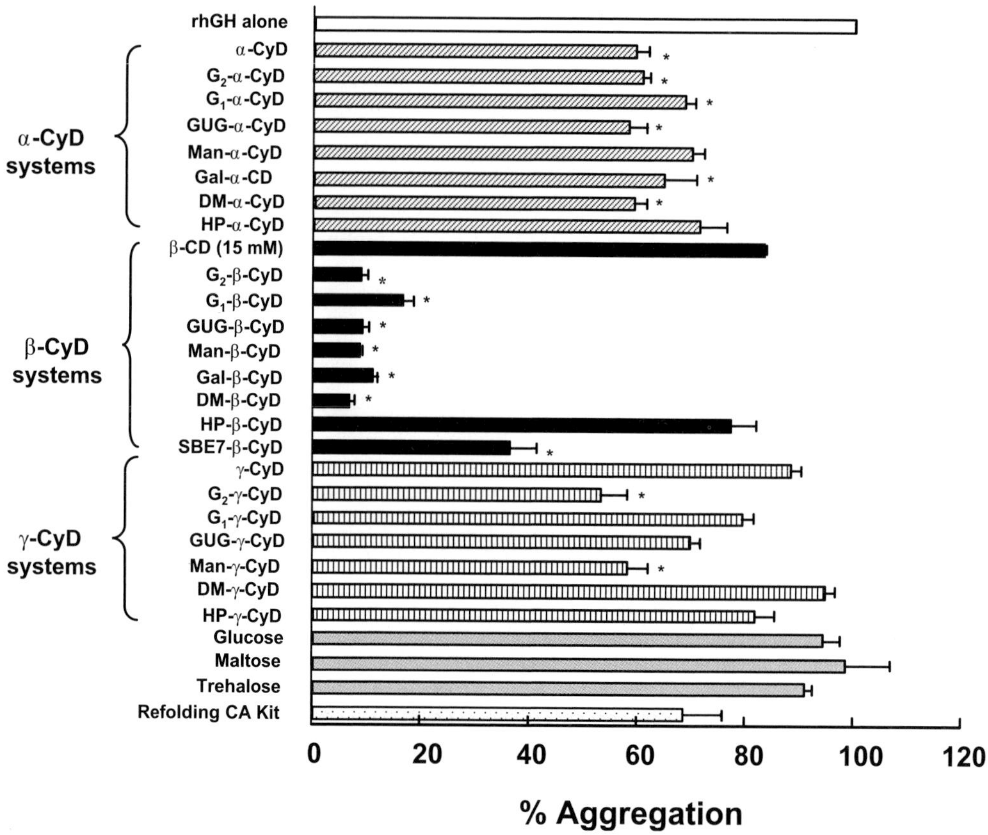

Fig. 14.16. The effects of additives (50 mM) on the aggregation of rhGH during refolding from molten globule-like intermediates in phosphate buffer (pH 5.0) containing 4.5-M GuHCl at 25 °C. Initial concentration of rhGH = 4.73 mg mL^{-1}. The degree of aggregation was determined by measuring the absorbance at 350 nm. Each value represents the mean ± S.E. of 3–4 experiments. $^{*}p < 0.05$ versus rhGH alone.

14.5
Site-specific Drug Delivery

A CyD complex is in equilibrium with guest and host molecules in aqueous solution, with the degree of dissociation determined by the magnitude of the stability constant of the complex. This is desirable because in the appropriate conditions the complex can dissociate to give free CyD and drug at the absorption site, and thus only the drug in a free form enters into the systemic circulation. However, the inclusion equilibrium is sometimes disadvantageous when drug targeting is attempted, because the complex dissociates before it reaches the organ or tissues to which it is to be delivered. One of the methods of preventing the dissociation is to bind a drug covalently to CyD. This type of drug release is essentially classified as delayed-release with a fairly long lag time (see Fig. 14.8). The CyD prodrug approach can provide a versatile means for construction of not only colon-specific delivery systems but also site-specific drug release systems, including gene delivery [19].

14.5.1
Colonic Delivery

The delivery of a drug to the desired site of action is the ultimate purpose of pharmaceutical research, because it maximizes the therapeutic effects and minimizes its adverse effects by augmenting the amount and persistence of a drug in the vicinity of a target site, while reducing the drug exposure of nontargeted sites. Colonic delivery has gainied much attention over past years, because there are a number of colonic diseases, such as ulcerative colitis, colorectal cancer, and Crohn's disease, which could be treated more effectively using colon-specific delivery system. Many kinds of colon-specific drug delivery systems have been developed, some examples of which are film-coating of a drug with pH- or pressure-sensitive polymers, coating of a drug with bacterial degradable polymers, delivery of a drug from time-dependent formulations or biodegradable matrices, and delivery of a drug from prodrugs or polymeric conjugates [107].

Native CyDs are barely capable of being hydrolyzed and are only slightly absorbed in passage through the stomach and small intestine, because they have a stable round average structure (see Section 1.3) and a high hydrophilicity. However, they are fermented by colonic microflora into small saccharides and thus absorbed as maltose or glucose in the large intestine. This biodegradation property of CyDs is useful in a colon-targeting carrier, and thus CyD prodrugs serve as a source of site-specific delivery of drugs to the colon. Therefore, we have prepared CyD conjugates of the nonsteroidal anti-inflammatory drugs ketoprofen and biphenylylacetic acid, the steroidal drug prednisolone, the short-chain fatty acid *n*-butyric acid, and the anticancer agent, 5-fluorouracil (Fig. 14.17), in which the drugs are covalently bound by the primary or secondary hydroxyl groups of CyDs through a spacer and an ester linkage [108–111]. The release behavior of ketoprofen from its α-CyD ester conjugate was investigated in rat gastrointestinal tract contents, intestine and liver homogenates, and blood in isotonic buffer solutions. The ketoprofen/α-CyD conjugate released the drug quantitatively only in cecal and colonic contents, whereas it released no appreciable drug in the other fluids. Detailed studies suggested that two types of enzymes in colon microflora are involved in the drug release from the conjugates, firstly sugar-degrading enzymes that hydrolyze α-1,4 glycosidic bonds of the CyD ring to give drug/linear oligosaccharide conjugates and secondly ester-hydrolyzing enzymes that hydrolyze the ester bond between the oligosaccharide and the drug to give ketoprofen (Fig. 14.18). This consecutive hydrolysis mechanism was demonstrated in hydrolyses of the *n*-butyric acid/CyD conjugate and the 5-fluorouracil/β-CyD conjugate catalyzed by α-amylase (*Aspergillus oryzae*) and carboxylic esterase (porcine liver). The *in vitro* release of the conju-

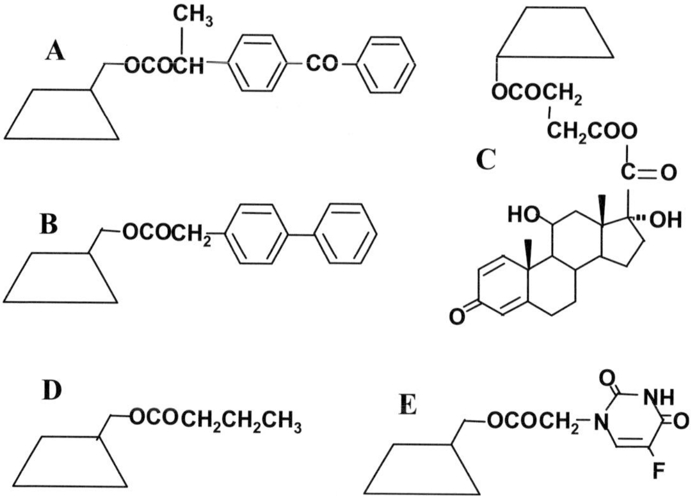

Fig. 14.17. Drug/CyD conjugates for colonic delivery. (A) Ketoprofen; (B) biphenylyl acetic acid; (C) prednisolone; (D) *n*-butyric acid; (E) 5-fluorouracil.

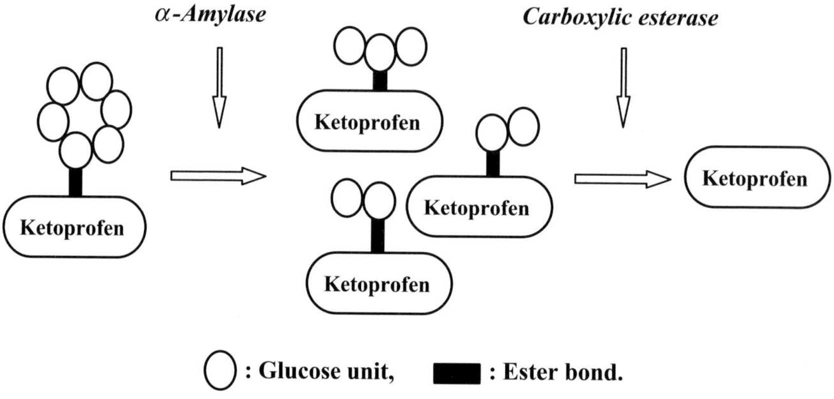

Fig. 14.18. Hydrolysis of ketoprofen/α-CyD conjugate catalyzed by α-amylase and carboxylic esterase.

gate was reflected in the *in vivo* release in rats. The plasma levels of ketoprofen after oral administration of the conjugate increased after a lag time of about 3 h and reached maximum levels at about 7 h, indicating that the conjugate released the drug site-specifically in the cecum or colon. The anti-inflammatory effect of the ketoprofen/α-CyD conjugate after oral administration was much higher than that of the drug alone and was highest 6–12 h after the administration. Similar results were obtained from the biphenylylacetic acid/CyD conjugates [112].

The release of drug from a drug/CyD conjugate after oral administration shows a typical delayed-release behavior. Consequently, when CyD conjugates are combined with other release preparations, we can obtain more advanced and optimized drug release systems, securing balanced oral bioavailability and prominent therapeutic efficacy [22]. For example, a repeated-release preparation may be designed by combining the CyD conjugate with a fast-releasing fraction, while a combined preparation of the conjugate with a slow-releasing fraction may provide a prolonged-release preparation [111]. Modified-releases by means of such combinations were demonstrated using the ketoprofen/α-CyD conjugate (Fig. 14.19). The co-administration of the CyD conjugate and the fast-dissolving ketoprofen/**6** non-covalent complex gave a typical repeated-release profile, i.e. double peaks were observed at about 1–2 hours and 8–12 hours in plasma drug levels after oral administration to rats. On the other hand, the co-administration of the conjugate and the slow-releasing ketoprofen/ethylcellulose solid dispersion gave a typical sustained-release profile, i.e. a constant plasma level was maintained for at least 24 hours. These repeated and long circulating release patterns in plasma drug levels after oral administration were clearly reflected in the anti-inflammatory effect using rat with carageenan-induced acute edema in rat paw.

Glucocorticoids are effective in the treatment of various inflammatory and allergic disorders, and have been used for ulcerative colitis patients. However, oral and intravenous administrations of glucocorticoids to patients with severe ulcerative colitis are restricted, because of the undesirable systemic side effects such as adreno-suppression, immunosuppression, hypertension, and osteoporosis. In order to reduce such side effects, rectal application has been used. However, the chronic use of glucocorticoids in high doses often causes systemic side effects even in topical application. One of the methods to overcome this problem is to develop a prodrug which is poorly absorbed from intestinal tracts and slowly releases the active drug at the site of action. A colon-specific drug delivery system may be particularly useful for the treatment of ulcerative colitis. Therefore, we prepared the prednisolone-appended α-CyD conjugate in which the drug is covalently bound to the secondary hydroxyl groups of α-CyD by an ester linkage through a spacer of succinic acid (see Fig. 14.17). The prednisolone-appended α-CyD ester conjugate was orally administered to rats with 2,4,6-trinitrobenzenesulfonic acid (TNBS)-induced colitis, and its anti-inflammatory and systemic adverse effects were compared with those of prednisolone alone and α-CyD alone [113]. The colonic damage score (CDS), the ratio of distal colon wet weight to body weight (C/B) and the myeloperoxidase (MPO) activity were evaluated as measures of the therapeutic effect of prednisolone, whereas the ratio of thymus wet weight to body weight (T/B) was evaluated as a

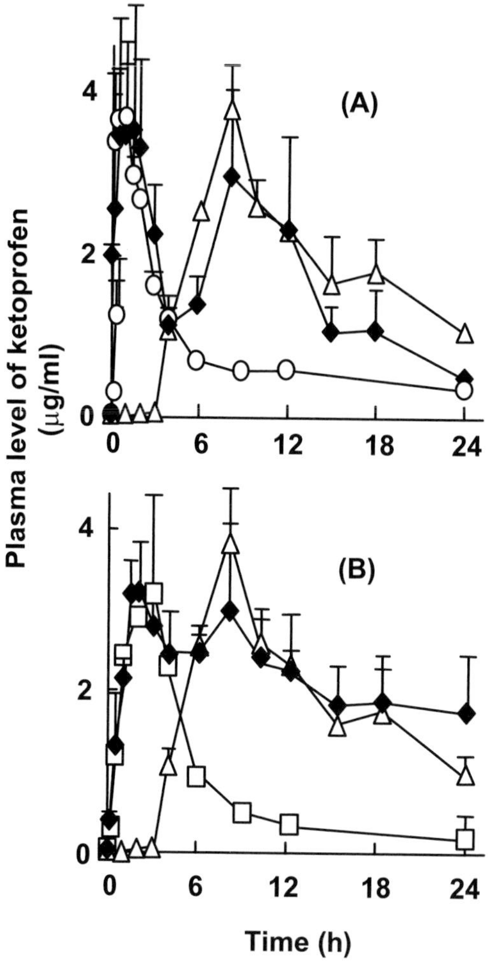

Fig. 14.19. Plasma levels of ketoprofen after oral administration of combined preparations to rats. (A) (Ketoprofen/α-CyD conjugate)/**6** complex system: ○, **6** complex equivalent to 2 mg kg^{-1} of the drug, △, α-CyD conjugate equivalent to 5 mg kg^{-1} of the drug, ◆, combined system of conjugate and complex containing equivalent amounts of the drug; (B) (Ketoprofen/α-CyD conjugate)/ethylcellulose solid dispersion system: □, dispersion equivalent to 6 mg kg^{-1} of the drug, △, α-CyD conjugate equivalent to 5 mg kg^{-1} of the drug, ◆, combined system of conjugate and dispersion containing equivalent amounts of the drug. The drugs were administered as powder-filled capsules. Each point represents the mean ± S.E. of 3–4 experiments.

measure of the side effect of the drug. Healthy rats gave a CDS of 0, a C/B ratio of 0.002, an MPO activity of 0.0001 units mg^{-1} tissues, and a T/B ratio of 0.0014. In the case of the control experiment where the saline solution without drugs was administered after the TNBS treatment, the CDS value, the C/B ratio, and the MPO activity were about 8, 0.007 and 0.012 units mg^{-1} tissue, respectively, indicating severe colitis. The α-CyD alone solution (2.2% w/v) without drugs gave similar therapeutic indices to those of the control experiments, indicating no anti-inflammatory effect of CyD on TNBS-induced colitis under the experimental conditions. On the other hand, the oral administrations of the prednisolone alone and the prednisolone/α-CyD conjugate (equivalent dose of 10 mg kg^{-1} of prednisolone) significantly decreased the CDS (4–5), the C/B ratio (0.004–0.005), and the MPO activity (0.007), indicating higher anti-inflammatory activity. Thymus atrophy is known to be a typical systemic adverse effect in steroid therapy. The administration of the prednisolone alone gave the smallest T/B ratio ($4–5 \times 10^{-4}$), indicating a significantly higher adverse effect. On the other hand, the α-CyD conjugate gave the same T/B ratio of 1.2×10^{-3} as that of the control experiment, indicating a small systemic adverse effect. These results suggest that the conjugate can alleviate the adverse effect of prednisolone without reducing its therapeutic effect. CyDs can serve as pro-moieties for colon-specific targeting prodrugs, and the CyD prodrug approach can provide a versatile means for construction of colon specific delivery systems of certain drugs.

14.5.2
Application in Gene Therapies

An increase in knowledge of the molecular and genetic basis of many diseases has created a unique opportunity for their treatment. Gene, antisense oligonucleotide (ODN), and small interfering RNA (siRNA) therapies attempt to correct pathobiological process by either correcting or inhibiting aberrant cellular functions at the level of gene expression. Since DNA, antisense ODNs, and siRNA are water-soluble and nuclease-sensitive, various strategies to achieve efficient delivery of these functional nucleic acids have been attempted. In this section, we will address the potential use of CyDs and CyD conjugates for this efficient delivery.

The development of nonviral vectors, which have much higher transfection ability, has been progressing. To date, several strategies to enhance the gene expression of nonviral vectors have been developed, e.g. the application of helper and pH-sensitive lipids, endosome-disruptive peptides, nuclear proteins, and nuclear localization signals [114].

Recently, CyDs have been applied to gene transfer. Croyle et al. reported that the cationic β-CyDs acted as viral dispersants, resulting in increasing adenoviral transduction in Caco-2 cells [115]. Roessler et al. demonstrated that amphoteric β-CyD and sulfonated β-CyD as excipients could lead to as much as 200-fold improvement in the efficiency of polyamidoamine (PAMAM) dendrimer-mediated gene transfer to monkey SV40-transformed kidney COS-1 cells [116]. Lawrencia et al. reported that the addition of **1** to DOTAP™, a cationic transfection reagent, and SuperFect™, an activated PAMAM dendrimer, further improved their transduction efficiency in the murine urothelial cell line MB49 by 3.8-fold and 2.6-fold, respectively [117]. Hence CyD can improve the gene transfer activity of viral and nonviral vectors.

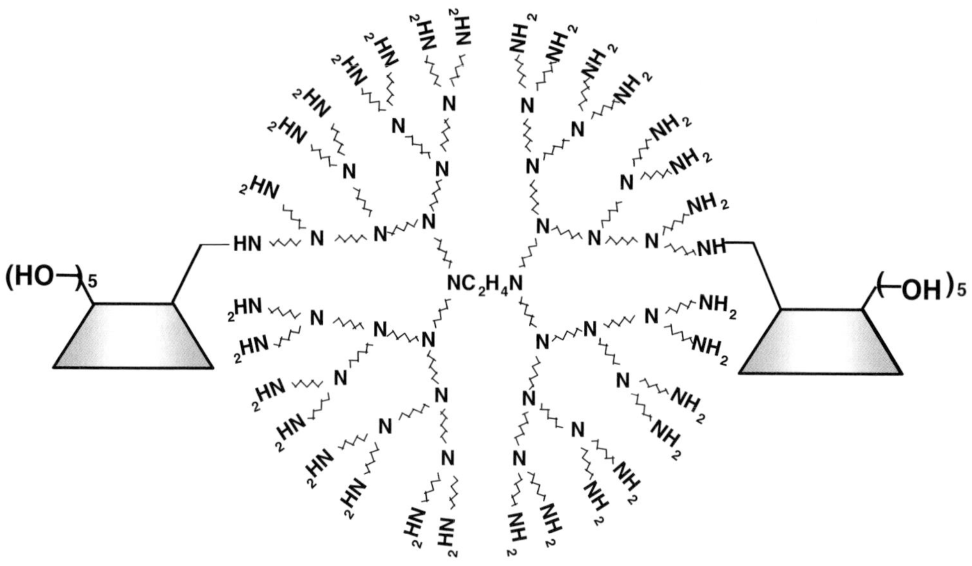

Fig. 14.20. α-CyD/polyamidoamine dendrimer conjugate as a nonviral gene transfer vector.

We have reported the use of Starburst PAMAM dendrimer (generation 2, G2) conjugates with α-, β- or γ-CyDs (CDE conjugates) in a molar ratio of 1:1 [118]. The conjugate of the dendrimer with α-CyD (α-CDE conjugate) revealed a 100-fold increase in the activity, compared to the activity of plasmid DNA complexes with dendrimer. Next, we have reported that of the three generations (G2, G3, and G4) of dendrimers in α-CDE conjugates, α-CDE conjugate (G3) had the most potent gene transfer activity [119]. Furthermore, the α-CDE conjugate (G3) with d.s. value of 2.4 (Fig. 14.20) was found to demonstrate the greatest gene transfer activity, compared with TransFast™ and Lipofection™ [120]. In addition, we are designing a cell-specific gene delivery system using the α-CDE conjugates bearing various sugar moieties, such as the dendrimer (G2) conjugate with α-CyD (α-CDE conjugate (G2)) bearing mannose (Man-α-CDE conjugates) with various degrees of substitution of the mannose moiety (DSM) [121]. As a result, Man-α-CDE conjugates (DSM 3.3 and 4.9) were found to have much higher gene transfer activity than the parent dendrimer, α-CDE conjugate, and other Man-α-CDE conjugates (DSM 1.1 and 8.3) in various cells, which are independent of the expression of cell-surface mannose receptors. Furthermore, the α-CDE conjugates bearing galactose with a spacer between the dendrimer and the galactose residues were found to have hepatocyte-specific gene transfer ability *in vitro*, suggesting the potential use of sugar-bearing α-CDE conjugates as a nonviral vector. Davis et al. reported a number of uses of β-CyD-containing polymers with adamantane-PEG or adamantane-PEG-transferrin for gene transfer [122–126] as well as DNAzyme transfer [127].

Antisense ODNs, small ODNs (15–30 mers) that are complementary to the sequence of pre- or mature mRNA sequences, have been shown to inhibit specifically the expression of numerous genes [128]. Abdou et al. reported the use of β-CyD bearing galactose moiety for antisense phosphodiester-ODN delivery into cells [129]. In addition, Zhao et al. found inhibitory effects of **6** and **5** on immune stimulation by immunostimulatory phosphorothioate-ODN containing the CpG motif in splenocytes isolated from CD1 mice and *in vivo* [130].

CyD conjugates have also been used as novel carriers of antisense ODNs. Habus et al. linked the adamantane molecule (adamantane-ODN conjugate) or amino derivatives of β-CyD (β-CyD-ODN conjugate) to the 3′-end of the ODN and found that stability to nuclease, hybridization properties with their complementary RNA, and their cellular uptake all increased [131]. Thus, the combination of the adamantane-ODN conjugate and **6** may be of value for antisense ODN delivery. In addition, Epa et al. prepared a 5′-adamantyl (linked to the ODN by a carboxylaminohexyl group) antisense phosphorothioate-ODN against the p75 neurotrophin receptor, and proved its efficacy in the presence of **6** both *in vitro* and *in vivo* [132]. Accordingly, CyDs or their conjugates may be useful for antisense therapy.

Gene silencing induced by siRNA is becoming a powerful tool of gene analysis and gene therapy. It has been intensively used as a novel technique instead of traditional gene therapy. However, the clinical application of siRNA is still far away because of the limitation in cell penetration by siRNA. We evaluated α-CDE conjugate as a siRNA carrier for RNA interference. In NIH3T3 cells, the ternary complex of pGL2 control vector encoding luciferase/siRNA/α-CDE conjugate showed higher pGL2 siRNA specific gene silencing effects without off-target effects than those of commercial transfection reagents such as Lipofectamine^{TM}2000 and TransFast^{TM}. Therefore, α-CDE conjugate may be utilized as a novel carrier for siRNA. In conclusion, these *in vitro* and *in vivo* results highlight the potential use of CyDs, CyD conjugates, and CyD polymers for gene, antisense, and siRNA therapies.

14.6 Perspectives

A number of CyD derivatives, CyD polymers, and CyD conjugates have been designed and evaluated for practical uses in pharmaceutical fields. These CyDs are classified into hydrophilic, hydrophobic, and ionic derivatives. Because of their multi-functional characteristics and bioadaptability, CyDs are capable of alleviating the undesirable properties of drug molecules through the formation of inclusion complexes or in the form of CyD/drug conjugates. This chapter has outlined the current application of CyDs in the design and evaluation of CyD-based drug formulations (Table 14.3), focusing on their ability to enhance drug absorption across biological barriers, the ability to control the rate and time profiles of drug release, and the ability to deliver a drug to a targeted site. Since CyDs are able to extend the function of pharmaceutical additives, the combination of molecular encapsulation with other carrier materials will become effective and valuable tools in the

Table 14.3. CyD Based Pharmaceutical Formulations and Their Advantages

- **CyD Complex**
  Hydrophilic CyDs
  - Improvement of solubility and oral bioavailability of poorly water-soluble drugs
  - Control of polymorphic transition and crystallization of drugs

  Hydrophobic CyDs
  - Sustained-release of water-soluble drugs

  Amphiphatic CyDs
  - Improvement of dermal/mucosal absorption of drugs
  - Stabilization of peptide/protein drugs
- **CyD Conjugate**
  - Drug/CyD conjugate: colonic delivery, time-controlled release
  - Cationic polymer/CyD conjugate: gene delivery
- **Combined Use of CyD with Pharmaceutical Excipients**
  - CyD/drug complex: control of equilibrium, competitive inclusion
  - CyD/drug conjugate with liposome: long circulation at specific organ
  - Polyrotaxane: drug targeting, control of cell function, biosensor

improvement of drug formulation. A combined use of different CyDs and various pharmaceutical additives will provide more-balanced oral bioavailability with prolonged therapeutic effects. Equally, drug/CyD conjugates may be useful for colon-specific delivery, time controlled release, gene delivery, and prolonged drug release, with the assistance of other carrier materials such as liposomes, discussed in more detail in Chapter 15. In particular, cationic polymer/CyD conjugates can be a novel candidates for nonviral vectors to enhance gene transfer. Owing to the increasingly globalized nature of CyD-related science and technology and the nontoxicity of CyDs, development of CyD-based pharmaceutical formulation is progressing rapidly. The future should see a number of commercial drugs involving various CyD derivatives.

References

1 Szejtli J. (Ed.), *Cyclodextrins and Their Inclusion Complexes*, Akadémiai Kiadó, Budapest, 1982.

2 Saenger W., *Angew. Chem. Int. Ed. Engl.*, **19**, 344–362 (1980).

3 Szejtli J. (Ed.), *Cyclodextrin Technology*, Kluwer Academic, Dordrecht, 1988.

4 Uekama K., Otagiri M., *CRC Crit. Rev. Ther. Drug Carrier Syst.*, **3**, 1–40 (1987).

5 Duchêne D. (Ed.), *New Trends in Cyclodextrins and Derivatives*, Editions de Santé, Paris, 1991.

6 Frömming K.-H., Szejtli J. (Eds.), *Cyclodextrins in Pharmacy*, Kluwer Academic, Dordrecht, 1994.

7 Uekama K., Irie T., *Comprehensive*

Supramolecular Chemistry, Vol. 3, Szejtli J., Osa T. (Eds.), Pergamon Press, Oxford, UK, 1996, pp. 451–481.
8 Loftsson T., Brewster M.E., *J. Pharm. Sci.*, **85**, 1017–1025 (1996).
9 Rajewski R.A., Stella V.A., *J. Pharm. Sci.*, **85**, 1142–1169 (1996).
10 Irie T., Uekama K., *J. Pharm. Sci.*, **86**, 147–162 (1997).
11 Thompson D.O., *CRC Crit. Rev. Ther. Drug Carrier Syst.*, **14**, 1–104 (1997).
12 Uekama K., Hirayama F., Irie T., *Chem. Rev.*, **98**, 2045–2076 (1998).
13 Uekama K., *Adv. Drug Delivery Rev.*, **36**, 1–2 (1999).
14 Irie T., Uekama K., *Adv. Drug Delivery Rev.*, **36**, 101–123 (1999).
15 Hirayama F., Uekama K., *Adv. Drug Delivery Rev.*, **36**, 125–146 (1999).
16 Duchêne D. (Ed.), *Cyclodextrins and Their Industrial Uses*, Edition de Santé, Paris, France, 1987.
17 Davis M.E., Brewster M.E., *Nature Rev.*, **3**, 1023–1035 (2004).
18 Friend D.R., Chang G.W., *J. Med. Chem.*, **27**, 51–57 (1985).
19 Uekama K., Hirayama F., *The Practice of Medicinal Chemistry*, Chap. 38, Wermuth C.G. (Ed.), Academic Press, London, 2003 (2nd edn.), pp. 649–673.
20 Zhang M.-Q., Rees D.C., *Exp. Opin. Ther. Patents*, **9**, 1697–1717 (1999).
21 Arima H., Hirayama F., Okamoto C.T., Uekama K., *Recent Res. Develop. Chem. Pharm. Sci.*, **2**, 155–193 (2002).
22 Uekama K., *Chem. Pharm. Bull.*, **52**, 900–915 (2004).
23 Uekama K., *Yakugaku Zasshi*, **124**, 909–935 (2004).
24 Szejtli J. (Ed.), *Cyclodextrin News*, **18** **(1)**, 1–4 (2004).
25 Uekama K., *Pharm. Int.*, **6**, 61–65 (1985).
26 Pitha J., Pitha J., *J. Pharm. Sci.*, **74**, 987–990 (1985).
27 Müller B.W., Brauns U., *Int. J. Pharmaceut.*, **26**, 77–88 (1985).
28 Yoshida A., Yamamoto M., Irie T., Hirayama F., Uekama K., *Chem. Pharm. Bull.*, **37**, 1059–1063 (1989).
29 Yamamoto M., Aritomi H., Irie T., Hirayama F., Uekama K., *S.T.P. Pharma Sci.*, **1**, 397–402 (1991).
30 Yamamoto M., Yoshida A., Hirayama F., Uekama K., *Int. J. Pharmaceut.*, **49**, 163–171 (1989).
31 Frank D.W., Gray J.E., Weaver R.N., *Am. J. Pathol.*, **83**, 367–382 (1976).
32 Arima H., Yunomae K., Morikawa T., Hirayama F., Uekama K., *Pharm. Res.*, **21**, 625–634 (2004).
33 Arima H., Yunomae K., Hirayama F., Uekama K., *J. Pharmacol. Exp. Ther.*, **297**, 547–555 (2001).
34 Yunomae K., Arima H., Hirayama F., Uekama K., *FEBS Lett.*, **536**, 225–231 (2003).
35 Hirayama F., Mieda S., Miyamoto Y., Arima H., Uekama K., *J. Pharm. Sci.*, **88**, 970–975 (1999).
36 Arima H., Nishimoto Y., Motoyama K., Hirayama F., Uekama K., *Pharm. Res.*, **18**, 1167–1173 (2001).
37 Arima H., Motoyama K., Nishimoto Y., Matsukawa A., Hirayama F., Uekama K., *Biochem. Pharmacol.*, **70**, 1506–1517 (2005).
38 Hirayama F., Hirashima N., Abe K., Uekama K., Ijitsu T., Ueno M., *J. Pharm. Sci.*, **77**, 233–236 (1988).
39 Hirayama F., Yamanaka M., Horikawa T., Uekama K., *Chem. Pharm. Bull.*, **43**, 130–136 (1995).
40 Anibarro M., Gessler K., Uson I., Sheldrick G.M., Harata K., Uekama K., Hirayama F., Abe Y., Saenger W., *J. Am. Chem. Soc.*, **123**, 11854–11862 (2001).
41 Hirayama F., Zaoh K., Harata K., Saenger W., Uekama K., *Chem. Lett.*, **2001**, 636–637.
42 Uekama K., Hieda Y., Hirayama F., Arima H., Sudoh M., Yagi Y., Terashima H., *Pharm. Res.*, **18**, 1578–1585 (2001).
43 Okimoto K., Rajewski R.A., Uekama K., Jona J.A., Stella V.J., *Pharm. Res.*, **13**, 256–264 (1996).
44 Okimoto K., Tokunaga Y., Ibuki R., Irie T., Uekama K., Rajewsky R.A., Stella V.J., *Int. J. Pharmaceut.*, **286**, 81–88 (2004).
45 Cameron K., Feilden H., *Magn. Res. Chem.*, **40**, S106–S109 Sp. Iss. Dec. 2002.
46 Uekama K., Horiuchi Y., Irie T., Hirayama F., *Carbohydr. Res.*, **192**, 323–330 (1989).

47 Adachi H., Irie T., Uekama K., Manako T., Yano T., Saita M., *Eur. J. Pharm. Sci.*, **1**, 117–123 (1993).

48 Horikawa T., Hirayama F., Uekama K., *J. Pharm. Pharmacol.*, **47**, 124–127 (1995).

49 Tavornvipas S., Hirayama F., Arima H., Uekama K., Ishiguro T., Oka M., Hamayasu K., Hashimoto H., *Int. J. Pharmaceut.*, **249**, 199–209 (2002).

50 Tavornvipas S., Tajiri S., Hirayama F., Arima H., Uekama K., *Pharm. Res.*, **21**, 2370–2377 (2004).

51 Amidon G.L., Lennernas H., Shah V.P., Crison J.R., *Pharm. Res.*, **12**, 413–420 (1995).

52 Loftsson T., Brewster M.E., Masson M., *Am. J. Drug. Deliv.*, **2**, 261–275 (2004).

53 Miyake K., Irie T., Arima H., Hirayama F., Uekama K., Hirano M., Okamoto Y., *Int. J. Pharmaceut.*, **179**, 237–245 (1999).

54 Lister R.E., Acerbi D., Cadel S., *Eur. J. Rheumatol. Inflamm.*, **12**, 6–11 (1993).

55 Miyake K., Arima H., Irie T., Hirayama F., Uekama K., *Biol. Pharm. Bull.*, **22**, 66–72 (1999).

56 Nakanishi K., Nadai T., Masada M., Miyajima K., *Chem. Pharm. Bull.*, **38**, 1684–1687 (1990).

57 Shao Z., Li Y., Chermak T., Mitra A.K., *Pharm. Res.*, **11**, 1174–1179 (1994).

58 Uekama K., Fujinaga T., Hirayama F., Otagiri M., Yamasaki M., Seo H., Hashimoto T., Tsuruoka M., *J. Pharm. Sci.*, **72**, 1338–1341 (1983).

59 Kikuchi M., Hirayama F., Uekama K., *Chem. Pharm. Bull.*, **35**, 315–319 (1987).

60 Hovgaard L., Brondsted H., *Pharm. Res.*, **12**, 1328–1332 (1995).

61 Sattler S., Schaefer U., Schneider W., Hoelzl J., Lehr C.M., *J. Pharm. Sci.*, **86**, 1120–1126 (1997).

62 Ono N., Arima H., Hirayama F., Uekama K., *Biol. Pharm. Bull.*, **24**, 395–402 (2001).

63 Pandalai S. G., *Recent Research Developments in Chemical & Pharmaceutical Sciences*, Vol. 2, Transworld Research Network, India, 2002, pp. 155–193.

64 Jadot M., Andrianaivo F., Dubois F., Wattiaux R., *Eur. J. Biochem.*, **268**, 1392–1399 (2001).

65 Matsuda H., Arima H., *Adv. Drug. Deliv. Rev.*, **36**, 81–99 (1999).

66 Loftsson T., Masson M., *Int. J. Pharmaceut.*, **225**, 15–30 (2001).

67 Bender M.L., Komiyama M. (Eds.), *Cyclodextrin Chemistry*, Springer-Verlag, Berlin, 1978.

68 Hirayama F., Kurihara M., Uekama K., *Chem. Pharm. Bull.*, **34**, 5093–5101 (1986).

69 Hoshino T., Ishida K., Irie T., Hirayama F., Uekama K., *J. Incl. Phenom.*, **6**, 415–423 (1988).

70 Hirayama F., Kurihara M., Uekama K., *Int. J. Pharmaceut.*, **35**, 193–199 (1987).

71 Gorecka B.A., Sanzgiri Y.D., Bindra D.S., Stella V.J., *Int. J. Pharmaceut.*, **125**, 55–61 (1995).

72 Utsuki T., Hirayama F., Uekama K., *J. Chem. Soc. Perkin Trans. 2*, **1993**, 109–114.

73 Uekama K., Irie T., Hirayama F., *Chem. Lett.*, **1978**, 1109–1112.

74 Tanada S., Nakamura T., Kawasaki N., Kurihara T., Umemoto Y., *J. Colloid Interface Sci.*, **181**, 326–330 (1996).

75 Kimura K., Hirayama F., Uekama K., *J. Pharm. Sci.*, **88**, 385–391 (1999).

76 Kimura K., Hirayama F., Arima H., Uekama K., *Chem. Pharm. Bull.*, **48**, 646–650 (2000).

77 Sonoda Y., Hirayama F., Arima H., Uekama K., *J. Incl. Phenom. Macrocycl. Chem.*, **50**, 73–77 (2004).

78 Ikeda Y., Motoune S., Ono M., Arima H., Hirayama F., Uekama K., *J. Drug Del. Sci. Tech.*, **14**, 69–76 (2004).

79 Ikeda Y., Hirayama F., Arima H., Uekama K., Yoshitake K., Harano K., *J. Pharm. Sci.*, **93**, 1659–1671 (2004).

80 Wang Z., Hirayama F., Uekama K., *J. Pharm. Pharmacol.*, **46**, 505–507 (1994).

81 Uekama K., Kondo T., Nakamura K., Irie T., Arakawa K., Shibuya M., Tanaka J., *J. Pharm. Sci.*, **84**, 15–20 (1995).

82 Uekama K., Narisawa S., Irie T., Otagiri M., *J. Incl. Phenom.*, **1**, 309–312 (1984).
83 Loftsson T., *Pharmazie*, **53**, 733–740 (1998).
84 Ono N., Hirayama F., Arima H., Uekama K., *Eur. J. Pharm. Sci.*, **8**, 133–139 (1999).
85 Willems L., van der Geest R., de Beule K., *J. Clin. Pharm. Ther.*, **26**, 159–169 (2001).
86 Uekama K., Adachi H., Irie T., Yano T., Saita M., Noda K., *J. Pharm. Pharmacol.*, **44**, 119–121 (1992).
87 Adachi H., Irie T., Uekama K., Manako T., Yano T., Saita M., *J. Pharm. Pharmacol.*, **44**, 1033–1035 (1992).
88 Abe K., Irie T., Uekama K., *Chem. Pharm. Bull.*, **43**, 2232–2237 (1995).
89 Abe K., Irie T., Uekama K., *Pharm. Sci.*, **1**, 563–567 (1995).
90 Harris J.M., Martin N.E., Modi M., *Clin. Pharmacokinet.*, **40**, 539–551 (2001).
91 Pearman R., Wanf Y.J. (Eds.), *Formulation, Characterization, and Stability of Protein Drugs*, Plenum Press, New York and London, 1996.
92 Brewster M.E., Hora M.S., Simpkins J.W., Stern W.C., Bodor N., *Pharm. Res.*, **8**, 792–795 (1991).
93 Irie T., Wakamatsu K., Arima H., Aritomi H., Uekama K., *Int. J. Phramaceut.*, **84**, 129–139 (1992).
94 Schipper N.G.M., Verhoef J., Romeijin S.G., Merkus F.W.H.M., *J. Control. Rel.*, **21**, 173–186 (1992).
95 Matsubara K., Irie T., Uekama K., *Chem. Pharm. Bull.*, **45**, 378–383 (1997).
96 Matsubara K., Abe K., Irie T., Uekama K., *J. Pharm. Sci.*, **84**, 1295–1300 (1995).
97 Koizumi K., Utamura T., Sato M., Yagi Y., *Carbohydr. Res.*, **153**, 55–67 (1986).
98 Shiotani K., Irie T., Uekama K., Ishimaru Y., *Eur. J. Pharm. Sci.*, **3**, 139–151 (1995).
99 Shiotani K., Uehata K., Irie T., Hirayama F., Uekama K., *Chem. Pharm. Bull.*, **42**, 2332–2337 (1994).
100 Shiotani K., Uehata K., Irie T., Uekama K., Thompson D.O., Stella V.J., *Pharm. Res.*, **12**, 77–83 (1995).
101 Tokihiro K., Arima H., Tajiri S., Irie T., Hirayama F., Uekama K., *J. Pharm. Pharmacol.*, **52**, 911–917 (2000).
102 Matsubara K., Irie T., Uekama K., *J. Control. Rel.*, **31**, 173–180 (1994).
103 Fukunaga K., Hijikata S., Ishimura K., Sonoda R., Irie T., Uekama K., *J. Pharm. Pharmacol.*, **46**, 168–171 (1994).
104 Rozeman D., Gellman S.H., *J. Am. Chem. Soc.*, **117**, 2373–2374 (1995).
105 Tokihiro K., Irie T., Uekama K., Pitha J., *Pharm. Sci.*, **1**, 49–53 (1995).
106 Tokihiro K., Irie T., Hirayama F., Uekama K., *Pharm. Sci.*, **2**, 519–522 (1996).
107 Fiend D.R. (Ed.), *Oral Colon-Specific Drug Delivery*, CRC Press, Boca Raton, FL, 1992.
108 Uekama K., Minami K., Hirayama F., *J. Med. Chem.*, **40**, 2755–2761 (1997).
109 Hirayama F., Ogata T., Yano H., Udo K., Takano M., Uekama K., *J. Pharm. Sci.*, **89**, 1486–1495 (2000).
110 Yano H., Hirayama F., Arima H., Uekama K., *J. Pharm. Sci.*, **90**, 493–503 (2001).
111 Kamada M., Hirayama F., Udo K., Yano H., Arima H., Uekama K., *J. Control. Rel.*, **82**, 402–416 (2002).
112 Minami K., Hirayama F., Uekama K., *J. Pharm. Sci.*, **87**, 715–720 (1998).
113 Yano H., Hirayama F., Kamada M., Arima H., Uekama K., *J. Control. Rel.*, **79**, 103–112 (2002).
114 Pouton C.W., Seymour L.W., *Adv. Drug. Deliv. Rev.*, **46**, 187–203 (2001).
115 Croyle M.A., Roessler B.J., Hsu C.P., Sun R., Amidon G.L., *Pharm. Res.*, **15**, 1348–1355 (1998).
116 Roessler B.J., Bielinska A.U., Janczak K., Lee I., Baker J.R., Jr., *Biochem. Biophys. Res. Commun.*, **283**, 124–129 (2001).
117 Lawrencia C., Mahendran R., Esuvaranathan K., *Gene Ther.*, **8**, 760–768 (2001).
118 Arima H., Kihara F., Hirayama F., Uekama K., *Bioconjug. Chem.*, **12**, 476–484 (2001).
119 Kihara F., Arima H., Tsutsumi T.,

Hirayama F., Uekama K., *Bioconjug. Chem.*, **13**, 1211–1219 (2002).

120 Kihara F., Arima H., Tsutsumi T., Hirayama F., Uekama K., *Bioconjug. Chem.*, **14**, 342–350 (2003).

121 Arima H., Wada K., Tsutsumi T., Hirayama F., Uekama K., *J. Incl. Phenom. Macrocycl. Chem.*, **44**, 361–364 (2002).

122 Gonzalez H., Hwang S.J., Davis M.E., *Bioconjug. Chem.*, **10**, 1068–1074 (1999).

123 Hwang S.J., Bellocq N.C., Davis M.E., *Bioconjug. Chem.*, **12**, 280–290 (2001).

124 Reineke T.M., Davis M.E., *Bioconjug. Chem.*, **14**, 247–254 (2003).

125 Reineke T.M., Davis M.E., *Bioconjug. Chem.*, **14**, 255–261 (2003).

126 Davis M.E., Pun S.H., Bellocq N.C., Reineke T.M., Popielarski S.R., Mishra S., Heidel J.D., *Curr. Med. Chem.*, **11**, 179–197 (2004).

127 Pun S.H., Bellocq N.C., Liu A., Jensen G., Machemer T., Quijano E., Schluep T., Wen S., Engler H., Heidel J., Davis M.E., *Bioconjug. Chem.*, **15**, 831–840 (2004).

128 Pun S.H., Tack F., Bellocq N.C., Cheng J., Grubbs B.H., Jensen G.S., Davis M.E., Brewster M., Janicot M., Janssens B., Floren W., Bakker A., *Cancer. Biol. Ther.*, **3**, 641–650 (2004).

129 Abdou S., Collomb J., Sallas F., Marsura A., Finance C., *Arch. Virol.*, **142**, 1585–1602 (1997).

130 Zhao Q., Temsamani J., Iadarola P.L., Agrawal S., *Biochem. Pharmacol.*, **52**, 1537–1544 (1996).

131 Habus I., Zhao Q., Agrawal S., *Bioconjug. Chem.*, **6**, 327–331 (1995).

132 Epa W.R., Greferath U., Shafton A., Rong P., Delbridge L.M., Bennie A., Barrett G.L., *Antisense Nucleic Acid Drug Dev.*, **10**, 469–478 (2000).

15
Cyclodextrins in Dispersed Systems

Laury Trichard, Dominique Duchêne, and Amélie Bochot

15.1
Introduction

Conventional drug administration does not usually provide rate-controlled release or target specificity. In many cases, conventional drug delivery induces sharp increases of drug concentration to potentially toxic levels. Following a relatively short period at the therapeutic level, drug concentration eventually drops off until re-administration. Several drug delivery technologies have been developed including polymeric (microparticles, nanoparticles) and lipidic systems (emulsions and liposomes) in an attempt to achieve modified drug release. Moreover, colloidal drug carriers such as liposomes and nanoparticles are able to modify the distribution of an associated substance. They can therefore be used to improve the therapeutic index of drugs by increasing their efficacy and/or reducing their toxicity. If these delivery systems are carefully designed with respect to the target and route of administration, they may provide a solution to some of the delivery problems posed by new classes of active molecules such as peptides, proteins, genes, and oligonucleotides.

One of the major problems encountered with dispersed systems appears during their preparation, and often results from poor water solubility of drugs leading to either a low yield in drug loading or a slow or incomplete release of the drug. In order to overcome these drawbacks, several authors have proposed the use of cyclodextrins (CyDs) in the formulation of these systems. Alternatively, CyDs themselves can be employed as raw materials for particle preparation, instead of polymers and lipids.

The first part of this chapter presents dispersed systems involving CyDs, and the methods employed in the preparation of these carriers. The second part details the advantages and potentialities of CyDs associated with dispersed systems, focusing on the different roles of CyDs in such systems. The advantages of simple inclusion complexes involving CyDs in drug administration have been presented in Chapter 14.

Throughout the chapter, reference should be made to Table 15.1, which is placed before the references.

Cyclodextrins and Their Complexes. Edited by Helena Dodziuk

ISBN: 3-527-31280-3

Table 15.1. Uses of Cyclodextrins in dispersed systems:

Dispersed system	Raw materials	Preparation methods	CyD	CyD roles	Encapsulated drugs	Comments	Ref.
Emulsions (>μm)	Soybean oils CyDs +/− Candellila wax	O/W and O/W/O	α-, β- and γ-CyDs	Emulsifying agent	Camphor Benzophenone	Stabilization of O/W or O/W/O α- and β-CyDs are more efficient than γ-CyD	[5]
Microcapsules (>μm)	β-CyD	Interfacial cross-linking	β-CyD	Raw material + Inclusion complex	*p*-Nitrophenol Propranolol	Rapid complexation–decomplexation Application in chiral chromatography	[9–11]
	Ethylcellulose	Emulsion-solvent evaporation	HP-β-CyD	Inclusion complex	Hydrocortisone	Basic technique (no precomplexation) Prolonged drug release	[12]
			HP-α-CyD		Carboplatin	+ Improved drug stability, drug efficiency using cell culture	[13]
	Poly (ε-caprolactone)	W/O/W solvent evaporation technique	Me-β-CyD		Melarsoprol	Increased drug encapsulation; Immediate drug release	[14]

Microspheres (>µm)	Starch/α-, β- or γ-CyD	Interfacial cross-linking	α-, β- or γ-CyD	Raw material + Inclusion complex	Gabexate mesylate	Improved drug stability Immediate drug release due to drug adsorption	[15]
	CyD/Hydroxy propylmethyl cellulose (HPMC)	Spray-drying	β-CyD HP-β-CyD		Carbamazepine Piroxicam	Increased drug dissolution rate	[16, 17] [18]
	HP-β-CyD	Cryogenic spray-freezing into liquid	HP-β-CyD		Danazol	Process adapted to fragile molecules Increased drug dissolution rate	[19]
	Polyacrylic acid (PAA)	Emulsion-solvent evaporation	β-CyD	Inclusion complex	Phenolphthalein Rhodamine-β	Immediate molecule release, reticulation between β-CD and PAA	[20–22]
	Poly (anhydride) acid		HP-β-CyD		Rhodium(II) citrate	Increased hydrophilic drug encapsulation Delayed drug release	[23]
	Poly (lactic-co-glycolic) acid	Spray-drying	HP-β-CyD	Inclusion complex	Insulin	Active insulin form encapsulated Controlled drug release	[24]
		Emulsion-solvent evaporation	HP-β-CyD Me-β-CyD			Regulation of plasma glucose level *in vivo*	[25, 26]

Table 15.1 *(continued)*

Dispersed system	Raw materials	Preparation methods	CyD	CyD roles	Encapsulated drugs	Comments	Ref.
	Poly (lactic-co-glycolic) acid	Emulsion-solvent evaporation	HP-β-CyD Me-β-CyD	Inclusion complex	Chlorhexidine Chlorhexidine digluconate	Increased drug encapsulation, release modified by CyD	[27]
			HP-β-CyD		Zolpidem	Release modified by CyD	[28]
		+ supercritical fluid process			Deslorelin	*In vivo* prolonged drug release	[29]
	Silicone	Emulsion vulcanization	HP-β-CyD		Indomethacin Timolol maleate	Faster drug release by CyD	[30]
	Chitosan	Interfacial cross-linking	HP-β-CyD		Nifedipine	Delayed drug release by CyD	[31]
		Spray-drying			Hydrocortisone		[32]
	Poly (methyl methacrylate)	Radical polymerization	Me-β-CyD	Inclusion complex	Not described	More homogeneous size of microsphere using CyD	[33]
			β-CyD			Faster polymerization Size decreased using CyD	[34]

Nanospheres (<μm)	Poly (alkylcyano-acrylate)	Anionic polymerization	β-CyD γ-CyD HP-β-CyD HP-γ-CyD Sulfobutyl-ether-β-CyD	Inclusion complex	[Progesterone Hydrocortisone Prednisolone Spironolactone Testosterone Danazol Megestrol Saquinavir acetate Metoclopramide]	Increased encapsulation of lipophilic drug by CyD Immediate drug release	[39–43]
	Gelatin	Desolvation	HP-β-CyD Cationic CyD		Hydrocortisone	Modification of particle size using CyD	[44]
	Dextran and β-CyD polymer	Nanoprecipitation	β-CyD polymer	Raw material + Inclusion complex	Not described	Spontaneous formation of nanospheres	[46]
	Amphiphilic CyD	Nanoprecipitation	2,3-di-*O*-acyl-β-CyD	Raw material	Indomethacin Doxorubicin Metronidazole Progesterone Soudan III	High drug encapsulation Decreased undesirable side effects	[47–51]
	Amphiphilic CyD	Nanoprecipitation	2,3-di-*O*-acyl-γ-CyD	Raw material	Progesterone	High drug encapsulation Drug encapsulated and adsorbed onto nanosphere surface Immediate drug release	[52, 53]

Table 15.1 *(continued)*

Dispersed system	Raw materials	Preparation methods	CyD	CyD roles	Encapsulated drugs	Comments	Ref.
Nanospheres (<μm)	Amphiphilic CyD	Nanoprecipitation	2,3-di-*O*-acyl-β-CyD sulfated		Not described	Negatively-charged surface Potential in gene therapy	[54, 55]
			6-*O*-acyl-β-CyD or 6-deoxy-6-*N*-acyl-β-CyD	Raw material + Inclusion complex	Bifonazole Clotrimazole Progesterone	High drug encapsulation Immediate drug release	[56, 57]
		Emulsion-solvent evaporation	2,3-di-*O*-acyl-γ-CyD	Raw material	Progesterone	High drug encapsulation, drug mainly adsorbed onto nanosphere surface Immediate drug release	[53]
			6-*O*-acyl-β-CyD or 6-deoxy-6-*N*-acyl-β-CyD		Caffeine Vitamin E	High drug encapsulation	[58]
		Detergent removal technique	2,3-di-*O*-acyl-γ-CyD		Progesterone	Avoiding organic solvents and high temperatures	[59]

Nanocapsules (<μm)	Amphiphilic CyD	Nanoprecipitation	2,3-di-*O*-acyl-β-CyD (C6 to C14)	Raw material	Indomethacin Amphotericin B Progesterone Doxorubicin Soudan III	High drug encapsulation Decreased undesirable side effects	[60–62]
	Amphiphilic CyD	Nanoprecipitation	6-*O*-acyl-β-CyD or 6-deoxy-6-*N*-acyl-β-CyD (C6)	Raw material	Not described	Influence of physicochemical properties of amphiphilic CyDs on nanocapsule characteristics	[61, 62]
			(2,3-di-*O*-acetyl)-6-deoxy-6-*N*-cholesteryl-βCyD		Vitamin A propionate	Increased drug stability Low drug encapsulation	[63]
			2,3,6-tri-*O*-perfluoroalkyl-β-CyD		ATP Indomethacin Doxorubicin Progesterone	High drug encapsulation Decreased undesirable side effects	[64]
Liposomes (<μm)	DPPC MLV	Hydration film	β-CyD	Inclusion complex	Naproxen Ketoprofen 4-Biphenylacetic acid	Decreased undesirable side effects	[78]
	PC:CH MLV	Dehydration/rehydration	HP-β-CyD β-CyD γ-CyD		Prednisolone	Increased drug encapsulation into the aqueous core	[79]

Table 15.1 *(continued)*

Dispersed system	Raw materials	Preparation methods	CyD	CyD roles	Encapsulated drugs	Comments	Ref.
Liposomes (<μm)	PC or DPPC:CH MLV	α-CyD β-CyD γ-CyD		Riboflavin Ascorbic acid	Indomethacin Dexamethasone DHEA	Increased drug stability Competition between	[80–83]
	DSPC:CH MLV	HP-β-CyD HP-β-CyD			Retinol Retinoic acid	drug and lipids inducing complex dissociation and fast drug release	[84–86]
	DPPC MLV or lecithin MLV	Proliposome dilution	γ-CyD	Inclusion complex	Hydrocortisone	High drug encapsulation	[87]
	Lecithin SUV 150 nm	Ethanol injection	HP-β-CyD		Nifedipine	Increased drug stability	[88]
	Lecithin LUV 300 to 500 nm	Spray-drying (one step)			Metronidazole Verapamil	Competition between drug and lipids inducing complex dissociation and fast drug release	[89]
	DPPC SUV and DPPC LUV + cholesteryl-CyD	Hydration film	Cholesteryl-CyD	Raw material + Inclusion complex	Not described	Physicochemical characterization	[90, 91]

Liposomes	DPPC SUV and DPPC LUV + amphiphilic CyD	Liposomes (sonication or injection) into amphiphilic CyD solution	2,6-di-*O*-polyethoxy-β-CyD		Not described	Physicochemical characterization	[92]
	DPMC SUV + amphiphilic CyD	Liposomes (sonication or injection) into amphiphilic CyD solution	6-deoxy-6-*N*-alkyl-β-CyD		Sodium anthraquinone-2-sulfonate	Molecule inclusion into CyD	[93]
	DPPC/CH SUV + amphiphilic CyD	Spray-drying	2,3-di-*O* alkyl-β-CyD sulfated	Raw material	Not described	Physicochemical characterization	[94]
Niosomes (<μm)	Monostearate sorbitan (= span 60)/CH (12–36 μm)	Hydration film	β-CyD	Inclusion complex	Ciprofloxacin Norfloxacin	Delayed drug Increased drug stability	[95]
					Methotrexate	Delayed drug Increased drug stability	[96]
	CH/Monostearate sorbitan/dicetyl-phosphate (75:75:10 mg) (10–15 μm)				Pilocarpine HCl	Increased drug encapsulation	[97]
					Plumbagin	Delayed drug release Increased drug stability Decreased drug encapsulation	[98]
Vesicles with an aqueous core surrounded by an amphiphilic CyD bilayer		Hydration film + sonication	6-deoxy-6-*S*-alkyl-β-CyD	Raw material	Not described	Physicochemical characterization	[99, 100]
		Not described	Aminated-CyD		Not described		[101]

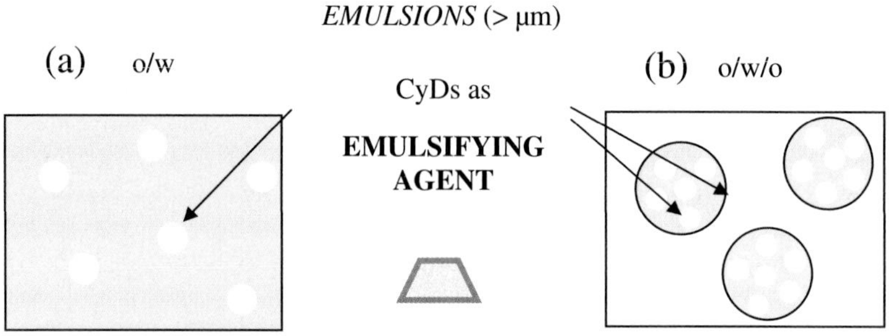

Fig. 15.1. CyDs as stabilizers in (a) o/w emulsions, (b) o/w/o emulsions.

15.2
Dispersed Systems Involving CyDs

15.2.1
Emulsions

A great number of patents involve the use of CyDs in the formulation of emulsions intended for pharmaceutical, cosmetic, or food products. (Pharmaceutical applications of CyDs are discussed in Chapter 14, while those in cosmetic and food products are presented in Section 16.2.) Unfortunately, most often there is no clear explanation of their role. Of course, sometimes they are used for their ability to include molecules, conferring on them new improved physicochemical properties. In some cases they are presented as stabilizers, without any explanation of the mechanism involved.

It is only with the work of Shimada and Yu on oil–water (o/w) [1, 2] or oil–water–oil (o/w/o) emulsions [3–5] that explanations were given of the stabilization process (Fig. 15.1 and Table 15.1).

The main interest in using CyDs to stabilize emulsions is that their irritant potential is very weak in comparison with traditional surfactants, especially hydrophilic surface-active agents [6]. The stabilizing potential of native CyDs is in fact an emulsifier potential that was first studied by Shimada et al. in a simple emulsion composed of soybean oil/aqueous CyD solution [1].

The emulsion stabilization is probably a consequence of the formation of a surface-active CyD-triglyceride inclusion complex at the oil/water interface, the CyD being oriented toward the aqueous phase and the triglyceride fatty acid residues, protruding from the CyD cavity toward the oily phase [7].

In the case of multiple o/w/o emulsions, as prepared by Yu [3–5], the process is that described by Jager-Lezer et al. [8]: a primary o/w emulsion is dispersed in an oily external phase. CyDs are used to stabilize the primary emulsion, while a thickening agent dissolved in oil constitutes the stabilizer of the secondary external emulsion. Of course, some CyDs from the primary phase can also participate in the stabilization of the secondary emulsion.

The best emulsifying effect is observed with α- and β-CyDs whereas the γ-CyD cavity is too wide to lead to an optimal interaction with the fatty acid chains [1, 3].

Active ingredients can be added to emulsions formulated with CyD as the emulsifier, but they must not interact significantly with the CyD cavity, otherwise they displace the fatty acid chains and destabilize the emulsion. From this standpoint, high-molecular-weight active ingredients will probably not destabilize emulsions prepared with α-CyD [2, 4].

15.2.2
Microparticles

Microparticles are defined as particles with a size ranging from 1 μm to 1 mm, and whose main component is a polymer, natural or synthetic.

Most often they are used to stabilize the active ingredient, minimize its toxicity, modify its release, and improve its efficacy. In pharmacy, they can be administered by the oral route, but also by the parenteral route where they constitute a depot system allowing a prolonged release of the drug.

Two types of microparticles are described: microspheres, with a matrix structure in which the active ingredient is dispersed or dissolved, and microcapsules, which are polymeric vesicles with an aqueous or oily core containing the active ingredient.

Microparticles prepared in the presence of CyDs are shown in Fig. 15.2. In microcapsules and microspheres CyDs can either constitute a raw material for the shell of the capsules or the matrix of the spheres, or form with the active ingredient an inclusion complex dissolved or dispersed in the shell or the matrix [9–34].

15.2.2.1 Microcapsules

CyD-based microcapsules

The membrane of the microcapsules is made of β-CyD. They are obtained by interfacial reticulation at room temperature [9]. Briefly, a basic aqueous CyD solution is dispersed in a nonmiscible hydrophobic phase (w/o emulsion) and a cross-linking agent (terephthaloyle chloride) is introduced in the medium to link the free hydroxyls of CyDs, and form a rigid CyD membrane on the droplet surface. Capsules prepared in this way have a diameter from 10 to 30 μm [9]. From an o/w emulsion, it is also possible to obtain microcapsules with an oily core [9]. The loading of a drug, such as *p*-nitrophenol [10] or propranolol [11], is carried out in a second step by incubating microcapsules in drug solution. The final amount of drug associated to the particles depends on the incubation time, capsule size, drug/microcapsule ratio, and rate of CyD reticulation.

Only superficial complexation occurs as a result of this loading method, and the loading cannot be properly called "encapsulation". The main disadvantage associated with this technique, is that the superficial loading leads to a fast release in aqueous medium.

However, a true encapsulation within the capsule cavity seems possible by introducing the drug into the internal phase of the emulsion. That may be feasible for

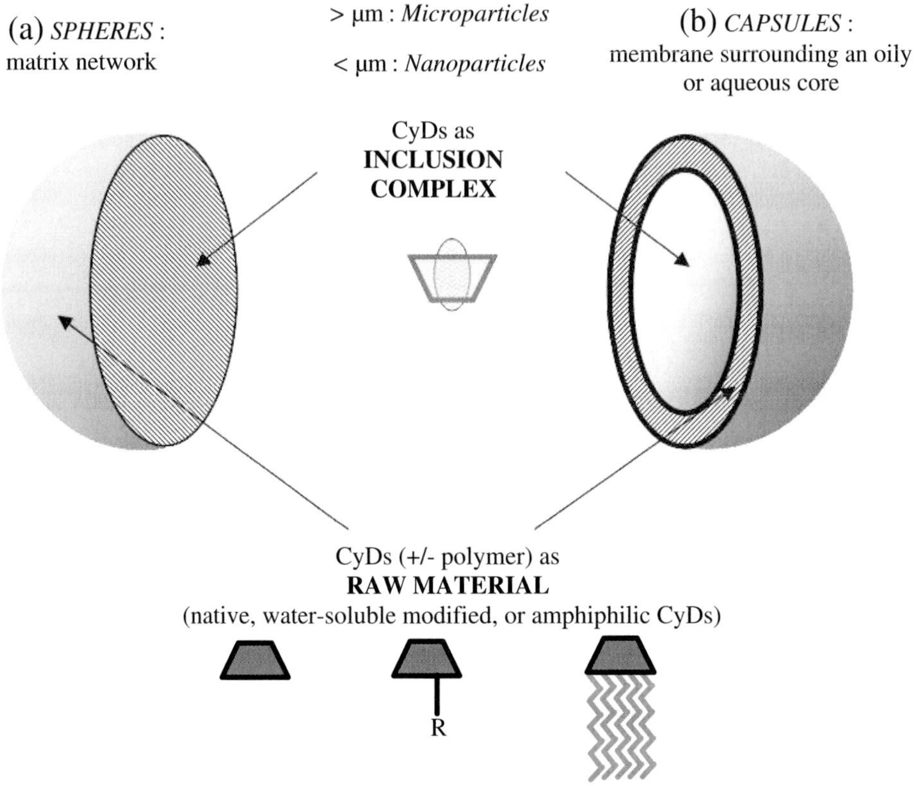

Fig. 15.2. Micro- or nanoparticulate systems involving CyDs: (a) spheres, (b) capsules.

both water-soluble drugs and liposoluble ones. On the other hand, the drug has to be stable at basic pH (cross-linking condition) and must not possess any free hydroxyl function able to interfere with β-CyD cross-linking. (Polymers involving CyDs are discussed in Chapter 3).

Microcapsules of synthetic polymer and cyclodextrin Three examples in the literature mention an association between synthetic polymer microcapsules and CyDs [12–14] as a means of encapsulating drugs complexed with CyDs.

Ethylcellulose capsules can be prepared by a method adapted from that described previously by Huang and Ghebre-Sellassie [35]. An ethanolic phase, containing ethylcellulose and hydroxypropylated CyD (HP-α- or HP-β-CyD), is dispersed in a heptane–paraffin mixture. Ethanol diffuses and dissolves in heptane before being evaporated at room temperature. The microcapsules obtained have a thick membrane (ratio of cavity:membrane radii = 1:2) [12]. Interestingly, this method allows the encapsulation of relatively poorly water-soluble molecules, which have to be soluble in ethanol and insoluble in heptane–paraffin.

In the case of capsules of poly(ε-caprolactone), the method used by Gibaud et al.

is slightly more complicated [14]. A multiple w/o/w emulsion is prepared with the inclusion complex in the internal aqueous phase and the polymer dissolved in the organic phase. The organic solvent is then completely evaporated resulting in polymer precipitation and the formation of capsules with an aqueous core.

The main advantages of CyD addition in these microcapsules are firstly an enhancement of encapsulated drug amount [12, 14], secondly an improvement of its stability and, finally, its efficacy *in vivo* [13]. Moreover, the introduction of CyDs seems to influence *in vitro* release profiles.

15.2.2.2 Microspheres

Cyclodextrin Microspheres

To obtain a drug delivery device adapted to the nasal administration of gabexate mesylate (anti-inflammatory drug), Fundueanu et al. have prepared microparticles from a starch/CyD mixture [15]. Starch (a biodegradable and biocompatible polymer) can confer bioadhesive properties to the particles owing to its strong capacity to be hydrated. CyD is used to complex gabexate mesylate in its hydrophobic cavity, improving its encapsulation rate and its protection from hydrolytic degradation.

The preparation method is based on a w/o emulsion (the basic aqueous phase contains both saccharides, at pH 10.5–14), followed by cross-linking with epichlorhydrin at high temperature (50 °C). The microspheres, which do not contain any drug at this stage, are then filtered, washed, and dried. The microspheres are relatively porous and spherical, ranging in size from 50 to 160 μm. Similarly to CyD microcapsules, drug loading is obtained by incubating the particles in a drug solution followed by water removal (evaporation or freeze-drying). Under these conditions, drug is mainly localized at the particle surface and an immediate drug release can be observed.

Microspheres of Synthetic Polymer and CyD Microspheres combining polymer and CyD as raw materials can be formulated with different polymers, such as silicone, chitosan, poly(lactic-co-glycolic) acid) or poly(acrylic acid). Interestingly, with the last of these polymers, β-CyD is able to form ester bonds (between hydroxyl groups of CyD and the carboxylic acid of the polymer), and acts as a cross-linking agent [21].

To prepare microspheres of synthetic polymer and CyD, conventional techniques are employed, such as solvent emulsion–evaporation, emulsion–vulcanization, reticular interfacial polymerization, or spray-drying [36]. Most of the time, the drug is dissolved according to its solubility in one or other of the phases.

The role of CyDs, more specifically β-CyD and hydroxypropyl-β-CyD, is:

- to improve drug loading by using a preformed inclusion complex to modify the aqueous drug solubility [37]
- to modulate drug release, by conferring an immediate or controlled release profile [24, 28, 38]
- to improve drug efficiency [25, 28]

The physical presence of polymers that are capable of being partly included in the CyD cavity, can hinder the ability of CyD to complex otherwise suitable drugs, and, in such cases, can constitute a drawback.

15.2.3
Nanoparticles

The structure of nanoparticles is similar to that of microparticles but their diameter is less than 1 μm. Similarly, two types of nanoparticle can be distinguished: nanospheres and nanocapsules.

In the pharmaceutical field, nanoparticles can be administered by any route, for instance specifically by the parenteral route. They seem to be very interesting carrier systems because they present higher stability than liposomes which are approximately the same size. Furthermore, they present a larger surface area than microparticles, allowing better contact with biological membranes, leading to higher drug bioavailability.

Nanoparticulate systems can be prepared either by dispersion of preformed polymers or polymerization of monomers. The most frequently employed preparation methods, solvent emulsion evaporation or nanoprecipitation, allow the entrapment of highly hydrophobic drugs but require the use of organic solvents. The latter condition requires an efficient elimination process to avoid any toxicological risk.

Nanoparticles containing CyDs can be classified in three groups described below (see Fig. 15.2).

15.2.3.1 Polymeric Nanoparticles

Poly(alkylcyanoacrylate) nanospheres are prepared by polymerization of alkylcyanoacrylate monomers (isobutyl- or isohexylcyanoacrylate) in water [45]. High concentrations of very lipophilic drugs cannot be dissolved in the aqueous polymerization medium, and, despite a strong affinity for the polymer, this leads to poor drug loading. This limit is easily overcome by using a drug-CyD complex in the aqueous solution. Large amounts of CyD remain associated with the particles, resulting in an increase in drug loading compared with nanoparticles prepared in the absence of CyDs.

More surprisingly, CyDs can in addition directly influence nanosphere characteristics by stabilizing interfaces during the preparation process, as described for emulsions [39]. Thus, it appears that the type of CyD used influences the particle size, their surface potential, and the total amount of CyD associated into the nanospheres [39]. Among CyDs, HP-β-CyD leads to spheres with the smallest size (interesting in parenteral administration) and a neutral surface.

15.2.3.2 Self-assembled Polymeric Nanoparticles

These nanoparticles are not classified as simple polymeric particles because of their ingredients and formation process. They have been developed very recently and can be considered as a new class of submicronic particles [46].

Particles are obtained by combining two polymers: a polymer based on β-CyD units (polyβ-CyD) and a modified polysaccharide. In detail, polyβ-CyD is prepared by a controlled cross-linking of β-CyD by epichlorhydrin at basic pH. The modified polysaccharide used to prepare these nanoparticles is generally dextran grafted with C_{12} chains, which can be included in polyβ-CyD cavities. Complexation between the two polymers results in spontaneous nanoparticle formation.

The extremely rapid fabrication method involves the mixing of two polymeric aqueous solutions. This direct method does not require any further purification. Neither organic solvents nor surfactants are required. The fabrication yield is above 95% and nanoparticle size ranges between 100 and 400 nm. Particles contain empty CyD moieties, which can entrap a variety of compounds depending on the stability constant of the resulting complex. Encapsulation is based on the same, simple, one-step procedure.

These nanoparticles could be used for any therapeutic administration route (ocular, oral, pulmonary, dermal, or intravenous). Another interesting possible application could be the selective extraction of specific compounds from an aqueous solution.

15.2.3.3 **Amphiphilic Cyclodextrin Nanoparticles**

Amphiphilic CyDs (Lollipop, Cup-and-ball, Medusa-like, Skirt-shaped, and Bouquet) have been synthesized by the grafting of various substituents to different faces of native CyDs [65]. (Syntheses of CyD derivatives are presented in Chapter 2). Among them, *Medusa-like cyclodextrins* and *Skirt-shaped cyclodextrins* have demonstrated their ability to form nanocapsules or nanospheres.

Medusa-like CyDs are obtained by grafting hydrophobic groups, aliphatic chains with length between C_4 and C_{16}, to all the primary hydroxyls of the CyD molecule [66–68]. *Skirt-shaped* CyDs consist of β- and γ-CyDs grafted with aliphatic esters or amides (C_2 to C_{14}) on the secondary face [69–72].

The amphiphilic CyD nanoparticle characteristics depend on the degree of substitution, the length of grafted chains, the nature of the grafted chain, and the nature of the chemical bond modification [56, 61, 62]. Most studies focus on the use of amphiphilic CyD with C_6 chains grafted by either ester (biodegradable) or amide bonds. Similarly to parent CyDs, amphiphilic CyDs offer the promising property of including drug molecules in their cavity and preserving this ability during the formation of nanoparticles [56, 57, 68, 73, 74]. In fact, access to the cavity depends on the grafted chain length and the substituted face (primary or secondary) [56, 57]. Hydrophilic active ingredients, such as doxorubicin, have been encapsulated in nanospheres [75, 76]. However, it is much more interesting to use these nanospheres for lipophilic drugs. In this case, specific preparation and loading methods have to be adopted.

Nanoparticles can be prepared by different methods:

(a) Nanospheres can be obtained by the nanoprecipitation method introduced by Fessi et al. [77]. An organic phase consisting of amphiphilic β-CyD dissolved in a water-miscible solvent is added under constant stirring to the aqueous phase. The organic solvent is evaporated under vacuum and the nanosphere dispersion is concentrated to the desired volume. The presence of a surfactant may be necessary to prepare loaded nanospheres [47]. Using this process, nanospheres can be loaded with drugs using different approaches.

(b) Conventionally Loaded Nanospheres (CLNs) are obtained after drug addition to the organic phase during the preparation [47]. The drug is associated mostly by adsorption on the particle surface. Its inclusion in the CyD cavity is a very low possibility.

(c) Preloaded Nanospheres (PNs) are prepared using a preformed drug–CyD complex added to the organic phase during preparation [62]. However, this technique does not significantly increase entrapment efficiencies or associated drug percentages of progesterone, bifonazole, or clotrimazole compared with conventional loading [56, 57]. A variation of this procedure, High Loaded Nanospheres (HLNs), combines the conventional with the preloading method. In this case, PNs are overloaded during the preparation by dissolving an additional amount of drug in the organic phase. The amount of drug encapsulated within the nanospheres can be increased significantly for progesterone, bifonazole, or clotrimazole by this method [56, 57].

The nanoprecipitation method also allows the production of nanocapsules: oil (Miglyols® or benzyl benzoate with or without the drug) is introduced together with the amphiphilic CyD in the organic phase (ethanol or acetone) [47, 61].

(d) The emulsion/solvent evaporation technique can also be employed to prepare nanospheres. The amphiphilic CyD is dissolved in a water-immiscible organic solvent with or without the active ingredient. The organic phase is poured into water under constant stirring and finally the organic solvent is evaporated under vacuum [53]. The encapsulation of progesterone, testosterone, and hydrocortisone within nanospheres of amphiphilic γ-CyD substituted on the secondary face by C_6 chains was performed by Lemos-Senna et al. [52, 53, 59].

(e) Finally, it is possible to reconstitute nanospheres by dilution or dialysis of *n*-octylglucoside, a nonionic detergent, from mixed micelles based on the detergent and amphiphilic CyD. This method named the "detergent removal technique" operates under mild conditions and avoids organic solvents, and high temperatures [59].

15.2.4 Lipidic Vesicles

In this section, lipidic vesicles are classified as follows: liposomes, niosomes, and amphiphilic CyD vesicles (Fig. 15.3).

15.2.4.1 Liposomes and Niosomes Prepared With Native or Water-soluble Modified CyDs

Liposomes are vesicles with size lower than 1 μm, and are composed of one or more phospholipid bilayers enclosing an aqueous phase. They allow the encapsula-

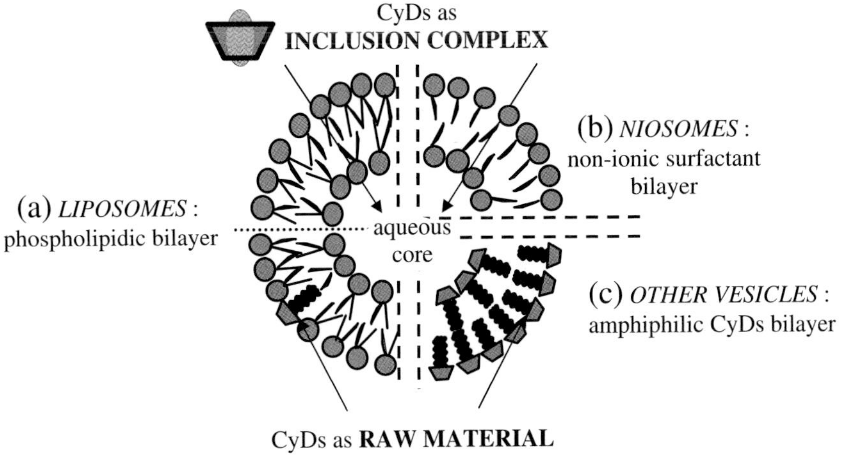

Fig. 15.3. Lipidic vesicles involving CyDs.

tion of either water-soluble drugs in the aqueous cavity or lipophilic drugs in the membrane [102]. Liposomes have a structure which is very similar to that of the central part of the cell membrane and can easily enter the cells to deliver their active content. The more recent niosomes are similar to liposomes but are built of bilayers of nonionic surfactants and lipids.

According to the preparation method chosen, several types of vesicle can be obtained, which differ in their size, their structure, and their capacity to encapsulate. Multilamellar vesicles (MLVs) have an average diameter varying from 400 nm to several micrometers, while unilamellar vesicles can have a large (80 nm to 1 μm: LUV) or small (20–80 nm: SUV) size [102].

Using liposomes can provide a solution to problems related to active ingredient formulation. They can increase the vascular circulation time of active ingredients, target specific sites of action of these molecules, and increase their bioavailability. Unfortunately, they generally show a relatively low encapsulation yield. A novel concept in drug delivery, presented here, takes advantage of certain properties of CyDs and liposomes to combine them into a single system, thus circumventing problems associated with both systems. The concept, entailing entrapment of water-soluble CyD–drug inclusion complexes in liposomes, allows accommodation of insoluble drugs in the aqueous phase of vesicles [85].

In most cases, liposomes are prepared by the film hydration method: lipid mixtures are dissolved in organic solvent and then evaporated to dryness. Liposomes are then formed by rehydration of the dried film with the aqueous phase. To encapsulate CyDs within the aqueous core of liposomes, preformed complexes are generally dissolved in the aqueous rehydration phase. Another technique is to introduce the inclusion complex in solution to rehydrate freeze-dried preformed liposomes. It is also possible to prepare liposomes in only one step, with simultaneous formation of complexes by spray-drying [89].

In these systems, CyDs can improve the drug stability (against light or hydrolysis), and enhance the encapsulation rate of poorly water-soluble drugs.

15.2.4.2 Lipidic Vesicles and Amphiphilic CyDs

Cholesteryl-cyclodextrins Cholesteryl-CyDs have been designed, and are obtained by grafting a cholesterol group onto a CyD. Indeed, cholesterol is described as a phospholipid-bilayer stabilizer and the cholesterol part of cholesteryl-CyDs can enter this bilayer whereas CyD, the hydrophilic part, is located in the internal and external aqueous media between bilayers [90, 91].

Other Amphiphilic Cyclodextrins Other amphiphilic CyDs as well as cholesteryl-CyDs can form lipidic vesicles. These vesicles can be liposomes or vesicles delimited by the bilayer of "amphiphilic" molecules, sometimes incorrectly named "liposomes".

The preparation methods of these vesicles are various. In some cases, the liposomal suspension is firstly prepared and then amphiphilic CyDs are inserted inside the phospholipid-bilayer, for example by sonication [93]. Another method mixes phospholipid (+/− cholesterol) and amphiphilic CyD solutions, to form liposomes by conventional methods [92, 94]. Lastly, vesicles consisting of bilayers of purely amphiphilic CyDs as raw material can be prepared applying a technique used to prepare liposomes, i.e. hydration of "lipidic" film followed by sonication [99–101].

The introduction of amphiphilic CyDs within vesicles presents the following advantages:

- amphiphilic CyDs themselves can form vesicles with the aqueous core [99].
- "bouquet" amphiphilic CyDs (substituted on both faces) can be used to create a "channel" structure in a phospholipid bilayer [92].
- amphiphilic cationic CyDs can be employed to prepare positively charged vesicles, which could be potentially used for gene delivery (DNA is negatively charged) as well as cationic liposomes or polymers [101].

15.3 The Potentialities of CyDs in Dispersed Systems

This section reviews the advantages and potentialities of CyD associated with dispersed systems, focusing on the different roles of CyDs in such systems. CyDs can enhance drug solubility, drug stability, and drug loading. In addition, CyDs themselves can be employed as a raw material for particle preparation, instead of polymers and lipids. Associated to dispersed systems, they can also influence the drug release profile. Finally, studies reporting *in vitro* and *in vivo* evaluations of these systems are very promising.

15.3.1
Conventional Uses of CyDs in Dispersed Systems

15.3.1.1 Enhancement of Drug Solubility and Loading

As summarized in Table 15.1, CyDs are often formulated as inclusion complexes to significantly increase encapsulation of lipophilic drugs within microspheres, nanospheres, and lipidic vesicles.

Most methods of nanosphere preparation described in the literature involve potentially toxic organic solvents, which are difficult to remove from the preparation media. Thus, methods using only aqueous media represent promising alternatives. Unfortunately, these techniques are not suitable for encapsulating large quantities of lipophilic molecules because of their poor solubility in water.

The apparent solubility of a lipophilic drug can be significantly increased in the presence of CyD owing to the formation of a drug–CyD complex. If a CyD, and more specifically HP-β-CyD, is used, a significant increase in the loading capacity of polyalkylcyanoacrylate nanoparticles is observed, varying from seven times for spironolactone to 130 times for prednisolone. The loading depends on both the drug dissolved in the polymerization medium, thanks to the presence of HP-β-CyD, and the partition coefficient of the drug between the polymer and the medium [40].

The use of an inclusion complex instead of the free drug can also theoretically modify the drug localization within liposomes or niosomes. As a result, the included drug is predominantly incorporated within the aqueous core rather than in the lipid bilayer, changing its release profile.

15.3.1.2 Enhancement of Drug Stability

CyDs allow molecular encapsulation inside their cavity. Consequently, labile molecules are protected from environment. This application is employed to significantly improve the stability of a labile molecule in polymeric particles, liposomes (e.q. riboflavin [81] and retinol [85]) and niosomes (e.g. ciprofloxacin and norfloxacin [95]).

15.3.2
Cyclodextrins as New Materials for Particle Preparation

Recently, the ability of β-CyD to form particles without polymers or phospholipids has been studied. Indeed, by employing a reticulation of β-CyD during the preparation process, it is possible to directly obtain microparticles (spheres and capsules) [10, 15]. On the other hand, reticulation of β-CyD can also be performed to synthesize a β-CD polymer, capable of forming nanospheres mixed with modified polysaccharide spontaneously [46].

Amphiphilic CyDs exhibit self-assembly properties and are able to form nanoparticles (spheres or capsules) involving classical preparation methods (nanoprecipitation or emulsion solvent evaporation). Interestingly, amphiphilic CyDs with ester bonds are potentially biodegradable in organisms. Today, up-scaling a well-controlled synthesis of large amount of amphiphilic CyDs constitutes the main limit for their use in drug delivery. In addition, the toxicity of these new derivatives is insufficiently documented.

15.3.3 CyDs to Control Drug Release Profile

15.3.3.1 Microparticles

The three most common dominant, though not exclusive, microparticle drug release mechanisms are: dissolution, diffusion, and erosion. Introduction of CyDs into polymeric network can modify mechanisms by which a drug is released [38], inducing immediate [19, 28], delayed, or prolonged liberation [23, 24].

To summarize, CyDs can induce immediate drug release by:

- improving the aqueous solubility of drugs
- increasing porosity and promoting erosion of the matrix
- enhancing polymer hydration, or
- increasing the concentration of diffusible species (provided that the solid drug exists within the matrix and a diffusion of both free and complexed drug is possible)

Adding CyDs to polymeric matrices can also delay and prolong drug release by:

- complexing the drug, effectively increasing its molecular weight and hence reducing its diffusivity (provided no excess drug is present)
- reducing the concentration of diffusible species by forming poorly soluble complexes
- reducing the concentration of diffusible species by forming drug–CyD complexes in which the host is covalently bound to the polymer backbone, or
- acting as cross-linking agents and decreasing the mesh size of the polymer network

15.3.3.2 Nanoparticles

Polymeric Nanoparticles The mechanisms involved in drug release from polymeric nanoparticles have been studied by Monza da Silveira et al. [39]. The *in vitro* release from poly(alkylcyanoacrylate)/HP-β-CyD nanospheres of a model drug (progesterone) shows a very fast initial release followed by a much slower release phase reaching a plateau. The level of the plateau depends on the nature of the dissolution medium (pH, presence of polyethylene glycol (PEG)) and, consequently, on its ability to solubilize the model drug. In the presence of esterases, which degrade the polymer, progesterone release is almost total after 1 or 2 h depending on the particle size [39]. For its part, HP-β-CyD is totally released in less than 1 h.

These results suggest that a fraction of the drug is simply adsorbed at the nanoparticle surface, either free or included in HP-β-CyD, and can be released rapidly. Another fraction of the drug is molecularly dispersed in the polymeric matrix of the nanoparticles and can be released by diffusion or by penetration of the dissolution medium into the nanoparticles followed by dissolution and diffusion of the progesterone solution. The magnitude of the fraction that can be released by such a process depends on the solubilizing efficacy of the solvent (PEG concentration). Finally, the third fraction of progesterone (plateau region) can be released only after complete hydrolysis of the ester bonds of the side chains of poly(isobutyl cyanoacrylate) by esterases.

Amphiphilic CyD Nanoparticles Only a few studies have evaluated the release kinetics from nanoparticles prepared with amphiphilic CyDs [52, 56, 57, 62]. The parameters involved in *in vitro* release are the nanosphere loading technique, the drug CyD association constant, and the effect of the substitution side.

The release profile of a model drug, bifonazole, was assessed in solution in water:PEG400 (60:40). Conventionally Loaded Nanospheres liberate bifonazole immediately within the first 15 min owing to a burst effect resulting from drug adsorption on the particle surface [56]. On the other hand, Preloaded Nanospheres prolonged the release up to 1 h owing to drug entrapment in the amphiphilic CyDs. Consequently, the intermediate release profiles observed with Highly Loaded Nanospheres arose from a combination of these two phenomena (drug adsorption and drug entrapment). A similar result was obtained with progesterone [57].

The affinity of parent β-CyD for drugs plays a major role in drug release from amphiphilic β-CyD nanospheres [56]. Indeed, bifonazole, which has a very high association constant with β-CyD ($K_{1:1} = 11\,000$ M^{-1}), is released significantly more slowly than clotrimazole, which has a lower association constant ($K_{1:1} = 500$ M^{-1}).

Amphiphilic β-CyD substituted on the primary face displayed slower release profiles for bifonazole than amphiphilic CyDs substituted on the secondary face. This difference is observed for Preloaded and Highly Loaded Nanospheres, both techniques using preformed inclusion complexes. This result is explained by the easier formation of strong inclusions when the wider secondary face remains free of substitution [56].

15.3.3.3 **Liposomes**

The release of active compounds from liposomes is directly related to liposomal bilayer stability. Therefore, a controlled-release profile from CyD-containing liposomes is difficult to obtain because of interaction between inclusion complexes and the lipidic vesicle (cholesterol and phospholipids) [103]. Indeed, lipid/CyD complex formation can occur within the liposomes, thus destabilizing the liposomal bilayer [85]. Liposome destabilization is also closely related to the stability of the encapsulated complex: the higher the affinity between drug and CyD, the slower the drug release [84].

15.3.4 Toxicity and Efficacy Evaluation

In vitro and *in vivo* efficacy of dispersed systems containing CyD is largely undocumented.

15.3.4.1 Microparticles

The efficacy of carboplatin encapsulated in ethylcellulose microcapsules has been studied on a rat glioma model [13]. The safety of microcapsules containing or not containing HP-β-CD was demonstrated. Moreover HP-α-CyD/carboplatin-loaded microcapsules are more efficient than those with free carboplatin. Indeed, rats treated with microcapsules containing the HP-α-CyD/carboplatin complex showed a median survival of 51 days instead of 34 days. This result is certainly due to a better protection of this fragile drug and to a more suitable release from microcapsules. In a preliminary study using microspheres, Trapani et al. have observed in rat that HP-β-CyD is useful to modulate the pharmacological effect of zolpidem (a hypnotic drug) microspheres [28]. Otherwise, DM-β-CD/insulin-loaded PLGA (polylactic-co-glycolic acid) microspheres are able to induce significant reduction of plasma glucose level in two rodent models, normal mice and diabetic rats, after intratracheal administration [25]. This effect could result from a controlled insulin release (zero-order kinetics), as shown for very similar microspheres by Quaglia et al. [24].

15.3.4.2 Polymeric Nanoparticles

In polyalkylcyanoacrylate nanoparticles, CyDs are able to slightly decrease the toxic effect induced by polymer degradation on Caco-2 cells [42]. However, saquinavir-loaded CyD nanospheres are not able to promote the transport of this drug through Caco-2 cell monolayers [104]. In contrast, HP-β-CyD is useful in metoclopramide-loaded nanoparticles, where it enhances the absorption of metoclopramide by a factor of 2 after subcutaneous administration in rat [43].

15.3.4.3 Amphiphilic CyD Nanoparticles

A major drawback of β-CyD for parenteral use is its hemolytic effects due to its interaction with red blood cells, as demonstrated *in vitro* on erythrocyte suspensions. Studies performed on whole blood and erythrocyte suspension samples show that in all concentrations studied amphiphilic β-CyD nanospheres are less hemolytic than native β-CyD, because of their hydrophobic substituents [56]. Besides the hydrophobicity, the self-assembly of amphiphilic β-CyDs in the form of nanospheres is also believed to reduce the interaction and direct contact of CyD with red blood cells.

Antimicrobial activity on *Candida albicans* ATCC 90028 has been reported for amphiphilic β-CyD nanospheres loaded with bifonazole and clotrimazole [56]. When drug-loaded amphiphilic β-CyD nanospheres are incubated with *C. albicans*, a synergistic effect is observed. Indeed, MIC (Minimal Inhibitory Concentration) values are lowered two-fold and ten-fold for bifonazole and clotrimazole, respectively.

Finally, Skiba et al. report the *in vivo* application of indomethacin-loaded nanocapsules, prepared from amphiphilic β-CyD substituted on the secondary face by C_6 chains, β-CyDC_6, in a rat model [48]. They compared the rat gastric ulcerative effect of encapsulated indomethacin with that of the indomethacin solution (Indocid®). Independently of the administered dose, the ulcerative effect was significantly decreased while the bioavailability of indomethacin was maintained. Gastrointestinal ulcer protection by encapsulation of indomethacin in β-CyDC_6 nanocapsules was 82% for 5 mg kg^{-1} and 53% for 10 mg kg^{-1} administered dose.

15.3.4.4 Lipidic Vesicles

CyDs associated with liposomes or niosomes do not play a major part in their *in vitro* and *in vivo* effectiveness [88, 89, 95–98].

15.4 Conclusions

This chapter clearly demonstrates the unique role of CyDs in dispersed systems.

It is possible to use either commercial native CyDs (or classical derivatives) or new synthesized derivatives. In the first case, commercial products are most often used as substitutes for adjuvants (for example, surface-active agents in emulsion preparation) or to improve/control drug loading, release, stability, etc., in carrier systems. In the second case, the synthesized derivatives can represent one of the main inert constituents of completely new types of carrier systems. Whatever their nature or their use, no other compound, or family of compounds, presents equivalent useful properties. Owing to their very special structure, their role in dispersed systems is unique and cannot be compared with that of other compounds. However, for a proper use, it is necessary to have a good knowledge of the CyD(s) one wants or can employ, and of their potential behavior in the presence of chemical or biological entities. With respect to the new synthesized derivatives, there are so many (potential) new compounds that they represent a huge storehouse of new carrier systems.

Of course, there are also drawbacks. Not all commercial CyDs are in either the European or the American Pharmacopoeia, and this can limit their use whatever their input and advantages in carrier formulation. For new synthesized derivatives, there are many problems to overcome, such as the up-scaling of more-or-less complicated syntheses, the toxicological safety of the new products, and recognition and approval by health authorities. But this is the normal lot of new compounds.

As for other chemical entities or carrier systems, we can imagine that the use of CyDs in dispersed systems will develop more rapidly in other industries than the pharmaceutical one (discussed in detail in the Chapter 14), such as cosmetics (the

CyD applications in which are discussed in Section 16.2) or agriculture (the use of CyDs in this domain is briefly presented in Section 16.3.

References

1 Shimada, K.; Ohe, Y.; Ohguni, T.; Kawano, K.; Ishii, J.; Nakamura, T. *Nippon Shokuhin Kogyo Gakkai-Shi*, **1991**, *38*, 16–20.

2 Yu, S.C.; Bochot, A.; Le Bas, G.; Chéron, M.; Grossiord, J.L.; Seiller, M.; Duchêne, D. *STP Pharma Sci.*, **2001**, *11*, 385–391.

3 Yu, S.C.; Bochot, A.; Chéron, M.; Grossiord, J.L.; Le Bas, G.; Duchêne, D. *STP Pharma Sci.*, **1999**, *9*, 273–277.

4 Yu, S.C.; Bochot, A.; Le Bas, G.; Chéron, M.; Mahuteau, J.; Grossiord, J.L.; Seiller, M.; Duchêne, D. *Int. J. Pharm.*, **2003**, *261*, 1–8.

5 Duchêne, D.; Bochot, A.; Yu, S.C.; Pépin, C.; Seiller, M. *Int. J. Pharm.*, **2003**, *266*, 85–90.

6 Duchêne, D.; Wouessijewe, D.; Poelman, M.C. In: *New Trends in Cyclodextrins and Derivatives*; Duchêne, D. (Ed.), Editions Santé: Paris, **1991**, 448–481.

7 Shimada, K.; Kawano, K.; Ihsii, J.; Nakamura, T. *J. Food Sci.*, **1992**, *57*, 655–656.

8 Jager-Lezer, N.; Terrisse, I.; Bruneau, F.; Tokgoz, S.; Ferreira, L.; Clausse, D.; Seiller, M.; Grossiord, J.L. *J. Control. Release*, **1997**, *45*, 1–13.

9 Lévy, M.C.; Pariot, N.; Edwards, F.; Andry, M.C.; Rey Goutenoire, S.; Buffevant, C.; Perrier, E. French Patent 9808809, **1998**.

10 Pariot, N.; Edwards-Levy, F.; Andry, M.C.; Levy, M.C. *Int. J. Pharm.*, **2000**, *211*, 19–27.

11 Pariot, N.; Edwards-Levy, F.; Andry, M.C.; Levy, M.C. *Int. J. Pharm.*, **2002**, *232*, 175–181.

12 Loftsson, T.; Kristmundsdottir, T.; Ingvarsdottir, K.; Olafsdottir, B.J.; Baldvinsdottir, J. *J. Microencapsul.*, **1992**, *9*, 375–382.

13 Utsuki, T.; Brem, H.; Pitha, J.; Loftsson, T.; Kristmundsdottir, T.; Tyler, B.M.; Olivi, A. *J. Control. Release*, **1996**, *40*, 251–260.

14 Gibaud, S.; Gaia, A.; Astier, A. *Int. J. Pharm*, **2002**, *243*, 161–166.

15 Fundueanu, G.; Constantin, M.; Dalpiaz, A.; Bortolotti, F.; Cortesi, R.; Ascenzi, P.; Menegatti E. *Biomaterials*, **2004**, *25*, 159–170.

16 Koester, L.S.; Myaorga, P.; Bassani, V.L. *Drug Dev. Ind. Pharm.*, **2003**, *29*, 139–144.

17 Koester, L.S.; Myaorga, P.; Pereira, V.P.; Petzhold, C.L.; Bassani, V.L. *Drug Dev. Ind. Pharm.*, **2003**, *29*, 145–154.

18 Jug, M.; Becirevic-Lacan, M.; Cetina-Cizmek, B.; Horvat, M. *Pharmazie*, **2004**, *59*, 686–691.

19 Rogers, T.L.; Nelsen, A.C.; Hu, J.; Brown, J.N.; Sarkari, M.; Young, T.J.; Johnston, K.P.; Williams III, R.O. *Eur. J. Pharm. Biopharm.*, **2002**, *54*, 271–280.

20 Bibby, D.C.; Davies, N.M.; Tucker, I.G. *J. Microencapsul.*, **1998**, *15*, 629–637.

21 Bibby, D.C.; Davies, N.M.; Tucker, I.G. *Int. J. Pharm.*, **1999**, *180*, 161–168.

22 Bibby, D.C.; Davies, N.M.; Tucker, I.G. *Int. J. Pharm.*, **1999**, *187*, 243–250.

23 Sinisterra, R.D.; Shastri, V.P.; Najjar, R.; Langer, R. *J. Pharm. Sci.*, **1999**, *88*, 574–576.

24 Quaglia, F.; De Rosa, G.; Granata, E.; Ungaro, F.; Fattal, E.; La Rotonda, M.I. *J. Control. Release*, **2003**, *86*, 267–278.

25 Rodrigues Jr., J.M.; De Melo Lima, K.; De Matos Jensen, C.E.; De Aguiar, M.M.G.; Da Silva Cunha Jr., A. *Artif. Organs*, **2003**, *27*, 492–497.

26 Aguiar, M.M.G.; Rodrigues, J.R.; Silva Cunha, A. *J. Microencapsul.*, **2004**, *21*, 553–564.

27 Yue, I.C.; Poff, J.; Cortes, M.E.; Sinisterra, R.D.; Faris, C.B.; Hildgen, P.; Langer, R.; Shastri, V.P. *Biomaterials*, **2004**, *25*, 3743–3750.

28 Trapani, G.; Lopedota, A.; Boghetich, G.; Latrofa, A.; Franco, M.; Sanna, E.; Liso, G. *Int. J. Pharm.*, **2003**, *268*, 47–57.

29 Koushik, K.; Dhanda, D.S.; Cheruvu, N.P.S.; Kompella, U.B. *Pharm. Res.*, **2004**, *21*, 1119–1126.

30 Sutinen, R.; Laasanen, V.; Paronen, P.; Urtii, A. *J. Control. Release*, **1995**, *33*, 163–171.

31 Filipovic-Grcic, J.; Becirevic-Lacan, M.; Skalko, N.; Jaisenjak, I. *Int. J. Pharm.*, **1996**, *135*, 183–190.

32 Filipovic-Grcic, J.; Voinovich, D.; Moneghini, M.; Becirevic-Lacan, M.; Magarotto, L.; Jalsenjak, I. *Eur. J. Pharm. Sci.*, **2000**, *9*, 373–379.

33 Storsberg, J.; Van Aert, H.; Van Roost, C.; Ritter, H. *Macromolecules*, **2003**, *36*, 50–53.

34 Li, S.; Hu, J.; Liu, B.; Li, H.; Wang, D.; Liao, X. *Polymer*, **2004**, *45*, 1511–1516.

35 Huang, H.P.; Ghebre-Sellassie, I. *J. Microencapsul.*, **1989**, *6*, 219–225.

36 Freiberg, S.; Zhu, X.X. *Int. J. Pharm*, **2004**, *282*, 1–18.

37 Loftsson, T.; Brewster, M.E. *J. Pharma. Sci.*, **1996**, *85*, 1017–1025.

38 Bibby, D.C.; Davies, N.M.; Tucker, I.G. *Int. J. Pharm.*, **2000**, *197*, 1–11.

39 Monza da Silveira, A.; Ponchel, G.; Puisieux, F.; Duchêne, D. *Pharm. Res.*, **1998**, *15*, 1051–1055.

40 Monza da Silveira, A.; Ponchel, G.; Duchêne, D.; Couvreur, P.; Puisieux, F. European Patent 1056477, **2000**.

41 Boudad, H.; Legrand, P.; Le Bas, G.; Chéron, M.; Duchêne, D.; Ponchel, G. *Int. J. Pharm.*, **2001**, *218*, 113–124.

42 Boudad, H.; Legrand, P.; Appel, M.; Coconnier, M.H.; Ponchel, G. *STP Pharma Sci.*, **2001**, *11*, 369–375.

43 Radwan, M.A. *J. Microencapsul.*, **2001**, *18*, 467–477.

44 Vandervoort, J.; Ludwig, A. *Eur. J. Pharm. Biopharm.*, **2004**, *57*, 251–261.

45 Vauthier, C.; Dubernet, C.; Fattal, E.; Pinto-Alphandary, H.; Couvreur, P. *Adv. Drug Deliv. Rev.*, **2003**, *55*, 519–548.

46 Gref, R.; Amiel, C.; Sébille, B.; Couvreur, P. International Patent WO 2004/006897, **2004**.

47 Skiba, M.; Wouessidjewe, D.; Coleman, A.; Devissaguet, J.P.; Duchêne, D.; Puisieux, F. French Patent 9207287, **1992**.

48 Skiba, M.; Morvan, C.; Duchêne, D.; Puisieux, F.; Wouessidjewe, D. *Int. J. Pharm.*, **1995**, *126*, 275–279.

49 Skiba, M.; Duchêne, D.; Puisieux, F.; Wouessidjewe, D. *Int. J. Pharm.*, **1996**, *129*, 113–121.

50 Skiba, M.; Nemati, F.; Puisieux, F.; Duchêne, D.; Wouessidjewe, D. *Int. J. Pharm.*, **1996**, *145*, 241–245.

51 Gèze, A.; Aous, S.; Baussanne, I.; Putaux, J.; Defaye, J.; Wouessidjewe, D. *Int. J. Pharm*, **2002**, *242*, 301–305.

52 Lemos-Senna, E.; Wouessidjewe, D.; Lesieur, S.; Puisieux, F.; Couarraze, G.; Duchêne, D. *Pharm. Dev. Technol.*, **1998**, *3*, 85–94.

53 Lemos-Senna, E.; Wouessidjewe, D.; Lesieur, S.; Duchêne, D. *Int. J. Pharm*, **1998**, *170*, 119–128.

54 Dubes, A.; Parrot-Lopez, H.; Shahgaldian, P.; Coleman, A.W. *J. Colloid Interface Sci.*, **2003**, *259*, 103–111.

55 Dubes, A.; Parrot-Lopez, H.; Abdelwahed, W.; Degobert, G.; Fessi, H.; Shahgaldian, P.; Coleman, A.W. *Eur. J. Pharm. Biopharm.*, **2003**, *55*, 279–282.

56 Memişoğlu, E.; Bochot, A.; Ozalp, M.; Sen, M.; Duchêne, D.; Hincal, A.A. *Pharm. Res.*, **2003**, *20*, 117–125.

57 Memişoğlu, E.; Bochot, A.; Sen, M.; Duchêne, D.; Hincal, A.A. *Int. J. Pharm.*, **2003**, *251*, 143–153.

58 Terry, N.; Rival, D.; Coleman, A.; Perreir, E. French Patent 0006102, **2000**.

59 Lemos-Senna, E.; Wouessidjewe, D.; Duchêne, D.; Lesieur, S. *Colloids Surf.*, **1998**, *10*, 291–301.

60 Skiba, M.; Wouessidjewe, D.; Fessi,

H.; Devissaguet, J.P.; Duchêne, D.; Puisieux, F. French Patent 9207285, **1992**.

61 Ringard-Lefebvre, C.; Bochot, A.; Memisoglu, E.; Charon, D.; Duchêne, D.; Baszkin, A. *Colloids Surf.*, **2002**, *25*, 109–117.

62 Memişoğlu, E.; Bochot, A.; Sen, M.; Charon, D.; Duchêne, D.; Hincal, A.A. *J. Pharm. Sci.*, **2002**, *91*, 1214–1224.

63 Weisse, S. Ph.D. thesis, Université Paris XI, Châtenay-Malabry, France, **2002**.

64 Skiba, M.; Skiba, M.; Duclos, R.; Combret, J.C.; Arnaud, P. French Patent 9909863, **1999**.

65 Duchêne, D.; Ponchel, G.; Wouessidjewe, D. *Adv. Drug Deliv. Rev.*, **1999**, *36*, 29–40.

66 Kawabata, Y.; Matsumoto, M.; Tanaka, M.; Takahashi, H.; Irinatsu, Y.; Tagaki, W.; Nakahara, H.; Fukuda, K. *Chem. Lett.*, **1983**; *1933–1934*.

67 Djedaini, F.; Coleman, A.W.; Perly, B. In: *Minutes of the 5th International Symposium on Cyclodextrins*; Duchêne, D. (Ed.), Editions Santé: Paris, **1990**, 328–331.

68 Liu, F.Y.; Kildsig, D.O.; Mitra, A.K. *Drug. Dev. Ind. Pharm.*, **1992**; *18*, 1599–1612.

69 Zhang, P.; Ling, C.C.; Coleman, A.W.; Parrot-Lopez, H.; Galons, H. *Tetrahedron. Lett.*, **1991**; *32*, 2769–2770.

70 Zhang, P.; Parrot-Lopez, H. *J. Phys. Org. Chem.*, **1992**; 518–528.

71 Memişoğlu, E.; Charon, D.; Duchêne, D.; Hincal, A.A. In: *Proceedings of the 9th International Symposium on Cyclodextrins*; Torres-Labandeira, J.; Vila-Jato, J. (Eds.), Kluwer Academic: Dordrecht, **1999**: 622–624.

72 Lesieur, S.; Charon, D.; Lesieur, P.; Ringard-Lefebvre, C.; Muguet, V.; Duchêne, D.; Wouessidjewe, D. *Chem. Phys. Lip.*, **2000**; *106*, 127–144.

73 Kasselouri, A.; Munoz, M.; Parrot-Lopez, H.; Coleman, A.W. *Polish. J. Chem.*, **1993**, *67*, 1981–1985.

74 Takahashi, H.; Irinatsu, Y.; Kazuka, S.; Tagaki, W. *Mem. Fac. Eng., Osaka City Univ.*, **1985**; *26*, 93–99.

75 Leroy-Lechat, F.; Wouessidjewe, D.; Duchêne, D.; Puisieux, F. In: *Proceedings of the 1st World Meeting on Pharmaceutics, Biopharmaceutics and Pharmaceutical Technology*, Budapest, **1995**; 499.

76 Wouessidjewe, D.; Skiba, M.; Leroy-Lechat, F.; Lemos-Senna, E.; Puisieux, F.; Duchêne, D. *STP Pharma Sci.*, **1996**, *6*, 21–26.

77 Fessi, H.; Devissaguet, J.P.; Puisieux, F.; Thies, C. European Patent 0274961, **1987**.

78 Castelli, F.; Puglisi, G.; Giammona, G.; Ventura, C.A. *Int. J. Pharm.*, **1992**, *88*, 1–8.

79 Fatouros, D.G.; Hatzidimitriou, K.; Antimisiriasis, S.G. *Eur. J. Pharm. Sci.*, **2001**, *13*, 287–296.

80 Loukas, Y.L.; Jayasekera, P.; Gregoriadis, G. *J. Phys. Chem.*, **1995**, *99*, 11035–11040.

81 Loukas, Y.L.; Jayasekera, P.; Gregoriadis, G. *Int. J. Pharm.*, **1995**, *117*, 85–94.

82 Loukas, Y.L.; Vraka, V.; Gregoriadis, G. *Pharm. Sci.*, **1996**, *2*, 523–527.

83 Loukas, Y.L.; Vraka, V.; Gregoriadis, G. *Int. J. Pharm.*, **1998**, *162*, 137–142.

84 McCormack, B.; Gregoriadis, G. *Int. J. Pharm.*, **1994**, *112*, 249–258.

85 McCormack, B.; Gregoriadis, G. *J. Drug Target.*, **1994**, *2*, 449–454.

86 McCormack, B.; Gregoriadis, G. *Biochim. Biophys. Acta*, **1996**, 237–244.

87 Becirevic-Lacan, M.; Skalko, N. *STP Pharma Sci.*, **1997**, *7*, 343–347.

88 Skalko, N.; Brandl, M.; Becirevic-Lacan, M.; Filipovic-Grcic, J.; Jalsenjak, I. *Eur. J. Pharm. Sci.*, **1996**, *4*, 359–366.

89 Skalko-Basnet, N.; Pavelic, Z.; Becirevic-Lacan, M. *Drug Dev. Ind. Pharm.*, **2000**, *26*, 1279–1284.

90 Auzély-Velty, R.; Perly, B.; Taché, O.; Zemb, T.; Jéhan, P.; Guénot, P.; Dalbiez, J.P.; Djedaïni-Pilard, F. *Carbohydr. Res.*, **1999**, *318*, 82–90.

91 Roux, M.; Auzély-Velty, R.; Djedaini-Pilard, F.; Perly, B. *Biophys. J.*, **2002**, *82*, 813–822.

92 Jullien, L.; Lazrak, T.; Canceill, J.; Lacombe, L.; Lehn, J.M. *J. Chem. Soc., Perkin Trans. 2*, **1993**, 1011–1020.

93 Lin, J.; Creminon, C.; Perly, B.; Djedaini-Pilard, F. *J. Chem. Soc., Perkin Trans. 2*, **1998**, 2639–2646.

94 Sukegawa, T.; Furuike, T.; Niikura, K.; Yamagishi, A.; Monde, K.; Nishimura, S.I. *Chem. Comm.*, **2002**, 430–431.

95 D'Souza, S.A.; Ray, J.; Pandey, S.; Udupa, N. *J. Pharm. Pharmacol.*, **1997**, *49*, 145–149.

96 Oomen, E.; Shenoy, B.D.; Udupa, N.; Kamath, R.; Devi, P.U. *Pharm. Pharmacol. Comm.*, **1999**, *5*, 281–285.

97 Sheena, I.P.; Singh, U.V.; Aithal, K.S.; Udupa, N. *Pharm. Sci.*, **1997**, *3*, 383–386.

98 Oomen, E.; Tiwari, S.B.; Udupa, N.; Kamath, R.; Devi, P.U. *Indian J. Pharmacol.*, **1999**, *31*, 279–284.

99 Ravoo, B.J.; Darcy, R. *Angew. Chem., Int. Ed.*, **2000**, *39*, 4324–4326.

100 Ravoo, B.J.; Jacquier, J.C.; Wenz, G. *Angew. Chem., Int. Ed.*, **2003**, *42*, 2066–2070.

101 Donohue, R.; Mazzaglia, A.; Ravoo, B.J.; Darcy, R. *Chem. Commun.*, **2002**, 2864–2865.

102 Dodziuk, H. *Introduction to Supramolecular Chemistry*, Kluwer Academic, Dordrecht, **2003**, Ch. 4.

103 Thompson, D.O. *Crit. Rev. Ther. Drug Carr. Syst.*, **1997**, *14*, 1–104.

104 Boudad, H. Ph.D. thesis, Université Paris XI, Châtenay-Malabry, France, **2002**.

16
Applications Other Than in the Pharmaceutical Industry

16.1
Introductory Remarks

Helena Dodziuk

Cyclodextrin complexes are the only group of supramolecular entities that have reached the stage of significant commercialization [1]. In the author's opinion, they will only be surpassed in the future by carbon nanotubes when the latter's usage in molecular electronics is realized, allowing the "bottom-up approach" in computer architecture to be realized.

Large-scale applications of native CyDs have been made possible by their ease of manufacture and low cost, not only of β-CyD but also of α- and γ-CyDs. Their lack of toxicity has formed the basis for their applications in the pharmaceutical, agrochemical, and food industries as well as in toiletries and cosmetics [2–4]. They are also used, or have been proposed to be applied for impregnating paper, as sensitizers and stabilizers of dyes [5], as corrosion inhibitors and rust proofing materials, as UV stabilizers, and as antioxidants (see below).

CyD applications in the pharmaceutical industry are discussed in Chapters 14 and 15 while the analytical ones and those in separation science are presented in Chapters 5 and 6. In particular, enantioselective chromatography using these oligosaccharides as chiral stationary or mobile phases allows us to carry out stereochemical analyses of natural compounds, determine the enantiomeric excess in asymmetric syntheses, and prove the enantiomeric purity of pharmaceuticals, fragrances, and chiral reagents. In addition, CDs are of great value in studies of the metabolism of chiral compounds, not only in clinical tests but also in environmental research. They are also of importance in investigations of the mechanism of reactions involving inversion or retention of configuration of chiral molecules and racemization and chirality transfer in rearrangement reactions. In spite of great achievements in analytical applications of CyDs, their application in large-scale industrial separations still remains a challenge. Similarly, there have been numerous exciting results concerning CyD catalysis, and these macrocycles have been successfully used as enzyme models (sometimes, as discussed in Section 1.1, leading

Cyclodextrins and Their Complexes. Edited by Helena Dodziuk

ISBN: 3-527-31280-3

to the abandonment of the generally accepted models of their action). However, to the best of our knowledge, no industrial application of CyDs in catalysis have been reported.

Numerous proposals of other applications are scattered throughout this book. To name but a few, CyD-based optical sensors, nanowires, and biosensors are presented in Sections 10.3 and 10.6 and in Table 14.3, respectively. In this chapter a few applications in food and drinks, in cosmetics and toiletry, in the textile and wrapping industries, and in agrochemistry are shown, while the applications of rotaxanes (discussed in Chapter 12) in molecular devices are briefly discussed at the end.

Limitations of space necessarily mean that the presentation of CyD applications in this chapter is very limited and does not reflect the richness of this domain, which also covers their use in ecology for waste water [6] and contaminated soil [7, 8] recovery as well as for the production of biodegradable polymers [8] for packaging and conferring flame retardant, antibacterial, and other properties to polymers [10]. The use of CyDs in removing cholesterol [11, 12] and other health-hazardous substances from food are also of considerable significance. A citation from a website could be mentioned here: "With increasing in the health consciousness of consumers, the future for products containing oligosaccharides seems to be greatly promising. These oligosaccharides may play an important role especially for the reduction of lifestyle-related diseases in the near future as well as the improvement of human health." (http://www.mindbranch.com/listing/product/R328-0005.html)

Without knowing it, we probably use products involving CyDs in everyday life, for instance when drinking grapefruit juice with the bitter-tasting naringin removed by complexation with β-CyD or Earl Grey tea in which the flavoring agents are stabilized in the form of CyD complexes. The future is bright for CyD research and their abundant applications.

References

1 Dodziuk, H. In *Introduction to Supramolecular Chemistry*; Kluwer: Dordrecht, 2002, pp 207–219.
2 *Comprehensive Supramolecular Chemistry*, Szejtli, J., Ed., Pergamon: Oxford, 1996; Vol. 3.
3 Szejtli, J. *Cyclodextrin Technology*; Kluwer: Dordrecht, 1988.
4 Duchene, D. *Cyclodextrins and Their Industrial Uses*; Edition Sante: Paris, 1991.
5 Craig, M. R.; Hutchings, M. G.; Claridge, T. D.; Anderson, H. L. *Angew. Chem. Int. Ed. Engl.* 2001, *40*, 1071.
6 Crini, G.; *Progr. Polym. Sci.*. 2005, *30*, 38.
7 Reddy, K. R.; Ala, P. R. *Separ. Sci. Technol.* 2005, *40*, 1701.
8 Villaverde, J.; Maqueda, C.; Morillo E. *J. Agr. Food Chem.* 2005, *55*, 5366.
9 Huang, L.; Taylor, H.; Gerber, M.; Orndorff, P.; Horton, J.; Tonelli, A. E. *J. Appl. Polym. Sci.* 1990, *74*, 937.
10 Huang, L.; Gerber, M.; Lu, J.; Tonelli, A. E. *Polym. Degrad. Stab.* 2001, *71*, 279.
11 Hwang, J. H.; Lee, S. J.; Kwak, H. S. *Asian Austr. J. Anim.* 2005, *18*, 1041.
12 Kim, S. H.; H. E.; Ahn, J.; Kwak, H. S. *Asian Austr. J. Anim.* 2005, *18*, 584.

16.2
CyD Applications in Food, Cosmetic, Toiletry, Textile and Wrapping Material Fields

Hitoshi Hashimoto

Taking advantage of their inclusion capacity, CyDs are being used in various fields to improve stability of guest substances, e.g. to protect any components that are unstable when exposed to light or oxygen, to prevent volatile components evaporating, to modify physicochemical properties, or to eliminate unpleasant tastes or odors. In addition, CyDs are also utilized to control release of a fragrance or drug [1].

This section will focus on industrial applications in the fields of foods, cosmetics, textiles, wrapping materials, etc. to show that numerous CyD-containing commodities have been put to practical use.

16.2.1
In Foods

16.2.1.1
Stabilization by Powdering

- Flavors or Spices
  Flavor components such as apple, citrus fruits, and plums, and spices such as allspice, cinnamon, garlic, or ginger, and herbs such as peppermint and basil, included into CyDs are available on the market and have a good reputation for the high stability they exhibit when they are heated during industrial food processing.
  For example, in the production of candies, cookies, and chewing gum using these flavors it is normally necessary to use a larger quantity of flavor than that which remains in the finished product. However, when CyD inclusion complexes of these flavors are used for processing, a product can be obtained using far smaller amounts of flavor than would be required conventionally. Furthermore, CyD inclusion complexes of these components are stable and last longer than liquid essences or the components themselves [2–12].
- Fish Oil
  Docosahexaenoic acid (DHA) is one of the ω-polyunsaturated fatty acids (PUFAs) derived from fish oil, which are believed to have physiological functions such as antithrombotic and cholesterol-lowering effects. But DHA smells bad and is unstable when exposed to heat, light, or oxygen. By powdering it with CyDs, however, the unpleasant smell and taste can be reduced [13], and the stability can

also be improved [14]. Powdered DHA in the form of a CyD complex has been processed into tablets and soft capsules, which are available on the market as dietary supplements.

- Coffee
  A stable instant coffee powder can be produced which retains the volatile coffee-flavored components [15–18].
- Green Tea
  CyDs are used to maintain flavor and color. This makes it possible for us to enjoy green tea flavor and color in ice cream, mousse, and other Japanese confectioneries [19–21].
- Dietary Supplements
  Many health-care products containing amino acids and vitamins are available. The amino acids and vitamins can be stabilized and protected by CyDs.

16.2.1.2
Taste Modification

CyDs are mainly used for improving the tastes and flavors of beverages and herbal medicine.

- Reduction of Bitter/Astringent Tastes of Tea-leaf Extracts or Tea Drinks
  The catechins contained in tea leaves are called also "tannin" and are the source of the astringent taste specific to green tea. The physiological effects of catechins include cholesterol/lipid-lowering, anti-oxidant, antibacterial, anti-hypertensive, antitumor, and anti-hyperglycemic effects. But the strong bitter/astringent taste of catechins is a big problem in the marketing of teas as daily drinks. By adding β-CyD to catechins, the bitter taste can be reduced allowing less-bitter tea drinks to be prepared. Currently, several tea drinks containing catechins at high concentrations are marketed as popular health commodities.
- Improvement of Taste and Other Properties
  Various physiologically active substances need modification before they are palatable. For example, ginseng (*Araliaceae*) roots contain many physiologically active ingredients and are utilized as a raw material for drugs and foods. When γ-CyD is added to ginseng extract powder, the hygroscopic nature is reduced and the bitter taste specific to ginseng can be masked, resulting in improvement of the taste. In addition, the solubility and dispersibility in water can be improved. Such ginseng extract with added γ-CyD can be consumed either as the powder or after it has been dissolved in water [22].

16.2.1.3
Anti-oxidation, Stabilization, and Improvement of Bioavailability

Many vitamins, fatty acids, pigments, and various other physiologically active substances are unstable and easily destroyed by heat, light (UV radiation), or acids, or are susceptible to oxidation, and they are not efficiently absorbed into the body. By adding such physiologically active supplements, CyD inclusion technology can be utilized to solve various problems such as unfavorable taste, disagreeable odor, instability, low absorption rate, and low bioavailability.

- Stabilization of Vitamins
  Since vitamins are unstable and easily deteriorate, it is necessary to devise a formulation that will stabilize them. One effective method of achieving this is the use of various CyD inclusion complexes. In particular, γ-CyD is becoming popular for the purposes of making liposoluble vitamins soluble in water, powdering them, or stabilizing them and increasing their bioavailability.
  - Vitamin E
    Even the liposoluble vitamin E (tocopherol), which is unstable under sunlight, UV light, or heat, can be stabilized with γ-CyD. When a mixture of tocopherol and starch was kept at 45 °C, 60% of its activity was lost in 17 weeks. But under the same conditions, γ-CyD-included tocopherol lost only 20% of its activity.
  - Vitamin C and Anthocyanin
    It is known that when vitamin C is mixed with anthocyanin as an antioxidant, the two substances form a condensation product. If vitamin C is present in various functional health foods, this quality deterioration is a problem. Furthermore, since vitamin C is highly soluble in water, it is impossible to stabilize it by CyD inclusion. However, a method of stabilizing vitamin C has been developed in which the antioxidant is included in a CyD to keep it remote from the vitamin C. In such cases, depending on the type of anthocyanin, either β- or γ-CyD is more effective.
- Unsaturated Fatty-acid Triglycerides
  Unsaturated fatty acids such as DHA (mentioned earlier), eicosapentaenoic (EPA), arachidonic, γ-linolenic and linoleic acids, and their alkali metal salts or esters are unstable, since air oxidation easily occurs at the double-bond site. Hence, plant and fish oils, which contain many triglycerides with a high proportion of unsaturated fatty acids, are easily oxidized, resulting in loss of stability in storage. It was considered difficult to stabilize triglycerides with CyDs since triglycerides are large molecules, but the use of γ-CyD has allowed plant and fish oils to be stabilized [22].
- Natural Colorants
  Researchers have begun to pay attention to the physiological functions of natural colorants and have tried to apply them for use in food products. Many papers and patents have been published relating to the effects of CyDs on natural colorants. CyDs are used for the stabilization and solubilization of natural colorants [24–27]. There is a strange common characteristic among the natural red colorants such as anthocyanin, lycopene, and astaxanthine, namely that of the native α-, β-, and γ-CyDs it is γ-CyD that is most effective in stabilizing all three natural red colorants.

16.2.1.4
Solubilization of Insoluble Substances

As discussed in Chapter 14, one of the main advantages of CyDs complexation is an increase in the solubility of drugs. Similarly, by forming CyD inclusion complexes it is possible to solubilize, as stable solutions, some substances that are hardly soluble in water or that have been regarded as unsuitable for the preparation of drinks.

- Flavonoids
  In general, many flavonoids are solubilized, as stable solutions giving no precipitates, by adding both β- and γ-CyDs in appropriate amounts tailored to each flavonoid instead of using either β- or γ-CyDs alone.
- Isoflavones
  Isoflavone derivatives are flavonoid glycosides contained in *Leguminosae*. Their physiological activities include estrogen-like, antioxidant, and antibacterial effects. They are thought to exert a prophylactic effect against climacteric disturbance, breast cancer, or prostate cancer. When isoflavone derivatives are converted into inclusion complexes with β-CyD or γ-CyD, the bitter/astringent taste of the derivatives is suppressed and the water-solubility is improved. Furthermore, it has also been reported that such inclusion complexes of isoflavone derivatives are absorbed more efficiently into human bodies and exhibit more pronounced physiological effects than uncomplexed ones [28].
- Raspberry Ketones
  Raspberry ketones prevent obesity or counteract a constitutional tendency toward obesity by promoting the decomposition of lipids accumulated in the adipose tissues. But these ketones are poorly soluble in water and have a characteristic aroma even at an extremely low concentration, so it was impossible to add an amount capable of exerting a lipolytic or anti-obese effect to cosmetics, drinks, or pet foods [29]. However, by including raspberry ketones into CyDs the solubility is increased and the specific odor is reduced, resulting in a marked improvement of the taste and stability. Importantly, the lipolytic and anti-obese effects are not affected by the complexation.
- Caffeine and Catechin
  The curdling of beverages such as tea, caused by formation and precipitation of caffeine/catechin complexes is prevented by adding small amount of β-CyD to the beverage. β-CyD forms more stable soluble complexes with catechins than with caffeine [30].

16.2.2
In Cosmetics and Toiletries

CyDs are used in the cosmetics field to solubilize fragrances, to suppress their volatility, and to allow perfume-containing products to be sprayed as micropowders.

CyDs are also used as stabilizers, emulsifiers, and deodorants [30, 31]. In Europe, about 400 tonnes of CyDs are used every year, of which 75% is used as deodorants.

- Long-lasting Perfumes and Creams
  These products contain fragrance@CyD complexes that permit slow release of the fragrance.
- Ceramide-containing Cosmetics
  Ceramide is a complex lipid that is found in the cytoplasm and plasma membrane. It has a moisture-maintaining capacity, keeping the skin moist and soft and increasing its elasticity. Stable ceramide powder formulations involving CyDs are dispersible and suitable for cosmetics.
- Deodorant Spray
  A spray for fabrics and clothes given the brand name "Febreze Fabric Refresher" has been marketed by Procter & Gamble Inc. It contains CyDs to neutralize smoke, mold, cooking, stale, pet, and other odors sticking to fabrics.
- Skin-care Products
  For products such as soaps and shampoos containing cypress oil extract, Hinokitiol has been developed. Hinokitiol has an antibacterial or bactericidal effect, but it is volatile and rapidly decomposes by oxidation. CyDs can stabilize Hinokitiol.

16.2.3
In Textiles

Textile products with specific functions have been developed. CyDs are essential ingredients for the production of such new materials in the textile field. CyDs are mainly used for keeping moisture in squalane fibers, reducing odors in fibers added with plant extracts, and retaining the anti-bacterial activity in fibers containing an added antibacterial agent. Examples of products utilizing such functions include underwear suitable for dry skin, underwear beneficial for people with allergies, and socks treated to reduce unpleasant smells.

Underwear containing the squalane@CyD inclusion complex as an active ingredient are already on the market in Japan. Powdered squalane is attached to fibers with a special binder. Squalane is extracted from the liver oil of deep-sea sharks. It can prevent human skin from drying and keep skin velvety. CyDs are used to squalane in powder form.

Underwear for patients with atopic dermatitis can effectively prevent skin itchness. CyDs are used to assist the production of γ-linolenic acid in powder form. The powdered γ-linolenic acid is fixed on fibers of the underwear. γ-Linolenic acid can be absorbed directly by the skin. This substance is believed to effectively maintain the water-retention ability of the skin. Special processing is adopted to ensure that it remains effective even after repeated washing.

Underwears and socks with antibacterial properties, made from fabrics containing the Hinokitiol@CyD inclusion complex mentioned earlier, are marketed in Japan.

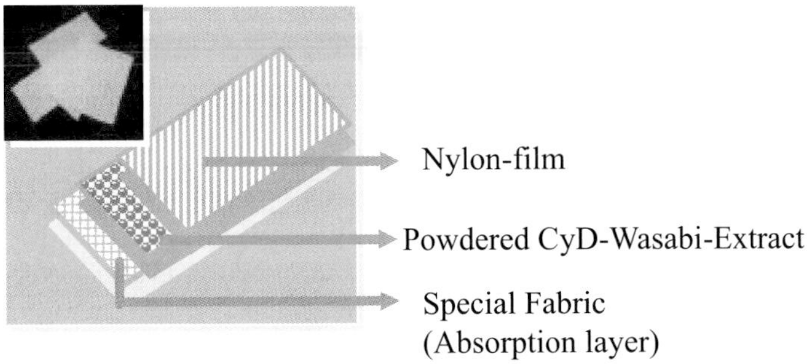

Fig. 16.2.1. Structure of antibacterial sheet.

16.2.4
In Wrapping Materials

Packaging specialists has developed some unique applications of CyDs. The containers and wrapping materials for foods are produced utilizing foodstuff materials that have preservative and antibacterial effects. For instance, wasabi is a plant with a hot-tasting spicy root. It is a traditionally widely used as a spice in Japan. Allylisothiocyanate (AITC), the main active ingredient of wasabi, is known to exhibit antibacterial action. However, it is volatile and decomposes rapidly by oxidation. CyDs are used to stabilize it in wrapping materials. Antibacterial cooking sheets and trays containing the wasabi@CyD inclusion complex have been introduced to the Japanese market. The outer layer of the sheet is nylon film coated with poly(vinylidene chloride), the middle layer is the wasabi@CyD inclusion complex, and the inner layer is a special fabric (Fig. 16.2.1). When moisture or food juice makes contact with this sheet, AITC vaporises, and the packed foods are surrounded by AITC gas. As a result, the growth of bacteria or mold is suppressed. CyDs permit favorable release control and stabilize the active ingredients. These products are used to pack meats and raw fish, such as Sashimi, to keep them fresh.

16.2.5
Future Prospects

Potential markets for CyDs in future are expected to be huge, although the current market is not so large and development efforts, exploring various end uses, are currently going on worldwide. The future huge market is easily foreseeable from the numerous and intriguing CyD-related scientific results presented at the International Cyclodextrin Symposiums held every other year. The recent trend in development of new CyD applications is characterized by accelerated progress in developing practical uses in the fine-chemical areas related to electricity or electronics, the prevention of environmental pollution, and technology for sophisticated separation and fractionation, in addition to the conventional chemical areas related to petrochemistry, such as the liquid–liquid extraction of xylene isomers [32, 33].

When CyD applications are realized in these nonfood areas, where much larger consumption is expected, the total market size may exceed 20 000 tonnes per year in the near future.

It is believed that CyDs have many potential applications, some of which we already know while others as yet remain unknown. Future research efforts will further advance the CyD application frontier, and I believe that we can make a significant contribution to overcoming some important problems that we currently face.

References

1 J. Szejtli, T. Osa (Eds.), *Comprehensive Supramolecular Chemistry*, Vol. 3, Pergamon, Oxford (1996), pp 483–502.
2 H. Hashimoto, K. Hara, N. Kuwahara, K. Mikuni, K. Kainuma, S. Kobayashi, Jpn. Kokai, JP85 234 564(1985).
3 Y. Kominato, S. Nishimura, Jpn. Kokai, JP8 321 620(1983).
4 I. Kijima, Jpn. Kokai, JP81 137867(1981).
5 M. Kobayashi, K. Yamashita, K. Matukura, J. Okumura, Jpn. Kokai, JP856 176(1985).
6 T. Miyake, Jpn. Kokai, JP78 124657(1978).
7 D. Marmo, F. L. Rocco (IFF), US. Patent 4 001 438 (1977).
8 H. Hashimoto, K. Hara, N. Kuwahara, I. Yashiki, Jpn. Kokai, JP8 755 100(1987).
9 H. Hashimoto, K. Hara, N. Kuwahara, K. Mikuni, K. Kainuma, S. Kobayashi, Jpn. Kokai, JP85 232070(1985).
10 N. Kitamura, Jpn. Kokai, JP81 113273(1981).
11 K. Lindner, L. Szente, J. Szejtli, *Acta Alimentaria*, 1981, 10, 175, (*Chem. Abstr.*, 1981, 96, 5067).
12 K. Lindner, *Nahrung*, 1982, 26, 675, (*Chem. Abstr.*, 1982, 97, 214428).
13 A. Tamagawa, Jpn. Kokai, JP 8 5204 739(1985).
14 M. Masui, Ger. Pat. Offen., DE3419796(1984), (*Chem. Abstr.*, 1984, 102, 80747).
15 L. Szente, J. Szejtli, *J. Food Sci.*, 51, 1024–1027 (1986).
16 O. Hirai, A. Kawaide, K. Suzumura, Jpn. Kokai Tokyo Koho JP 2004049200 A2 19 Feb 2004 (Chem. Abstr., 140:180661).
17 G. Schmid, M. Regiert, G. Antlsperger, *Applicability of γ-cyclodextrin in food*, 209th ACS National Meeting, American Chemical Society, Anaheim, CF, April 2–6 1995.
18 R. Taniguchi, R. Senoo, Y. Ooishi, Y. Fukui, S. Yokoyama, Jpn. Kokai, JP89196257 (1989).
19 H. Hashimoto, K. Hara, N. Kuwahara, I. Yashiki, Jpn. Kokai, JP8 719 051(1987).
20 J. Sato, T. Kurisu, Y. Takahama, H. Hirai, Jpn. Kokai, JP87241(1987).
21 K. Yamamoto, M. Sya, Jpn. Kokai, JP85 137249(1985).
22 Y. Akiyama, Jpn. Kukai, JP7980463 (*Chem. Abstr.*, 1979, 92, 28556).
23 K. Terao, *Cyclodextrin "Entrance Book for R&D technologists of food industries"*, Nihon Shokuryo Sinbunsha Japan (2004) p. 67.
24 H. Hashimoto, *1st International Symposium on Natural Colors for Foods Nutraceuticals, Beverages and Confectionery*, Amherst, MA, Nov. 7–10, (1993).
25 H. Hashimoto, K. Hara, N. Kuwahara, I. Yashiki, Jpn. Kokai, JP8 719051(1987).
26 J. Sato, T. Kurisu, Y. Takahama, H. Hirai, Jpn. Kokai, JP87 241(1987).
27 (a) K. Yamamoto, M. Sya, Jpn. Kokai,

JP 85 137249(1985); (b) K. Terao, *Cyclodextrin "Entrance Book for R&D technologists of food industries"*, Nihon Shokuryo Sinbunsha Japan (2004) p. 79.
28 K. Mikuni, E. Shibata, N. Kagei, K. Hara, H. Hashimoto, Seito Gijutsu Kenkyu Kaishi 41,71-5(1993)(C.A.: 121:203856).
29 Gunze, Jpn Tokukai JP 2004–27372 (2004).
30 K. Miyao, S. Nishimara, Jpn. Kokai, JP 870150467(1987).
31 M. Yajima (Asama Chem. Co. Ltd.), Ger. Offen., 3008663(1980), (*Chem. Abstr.*, 1980, 94, 36133).
32 I. Uemasu, H. Takahasi, *J. Jpn. Petroleum Inst.*, 36, 210–214 (1993).
33 I. Uemasu, H. Takahashi, K. Hara, *Proceedings of the 12th International Cyclodextrin Symposium*, Montpellier, 2004, pp 545–548.

16.3 Application of Cyclodextrins in Agrochemistry

Esmeralda Morillo

The ability of cyclodextrins, CyDs, to form inclusion complexes with a wide variety of hydrophobic guest molecules has been used in agriculture. Their ability to alter the physical, chemical, and biological properties of guest molecules has been used for the preparation of new formulations of pesticides. CyDs form complexes with a wide variety of agricultural chemicals including herbicides, insecticides, fungicides, repellents, pheromones, and growth regulators [1, 2].

Each CyD has its own ability to form inclusion complexes with specific pesticides, depending on a proper fit of the pesticide molecule into the hydrophobic CyD cavity. The principal advantage is that the binding of pesticide molecules within the host molecule is not fixed or permanent but rather is a dynamic equilibrium. Dissociation of the inclusion complex is a relatively rapid process usually driven by a large increase in the number of water molecules in the surrounding environment [3].

Similarly to CyD applications in the pharmaceutical (Chapters 14 and 15) and other industries, the most common benefits of CyD applications in agriculture include, among others, alterations of the solubility of the pesticide, stabilization against the effects of light or biochemical degradation, and a reduction of volatility. In some applications, more than one benefit is obtained by complexation with CyDs. However, the application of these hosts in pesticide formulations is very modest, and represents less than 1% of the CyD literature [4]. Moreover, most of these publications are not directly practice oriented. Instead they deal with the preparation of pesticides-CyD complexes using different oligosaccharides and processing methods and their characterization using a wide variety of techniques.

Among these papers, some are more or less theoretical papers dealing with the correlation between the properties of the new complexes obtained and the properties of the CyDs, pesticides, or external parameters used. Viernstein et al.

[5] compared the abilities of β- and dimethyl-β-CyDs to increase the solubility of the systemic fungicide triflumizole. Pospisil et al. [6, 7] studied the influence of difenzoquat-CyD complexation on the electron-transfer reaction of the herbicide molecule and on its solution conductivity and fluorescence intensity. Ishiwata and Kamiya [8] estimated the depth of guest insertion in CyD inclusion complexes of the organophosphorus pesticides parathion and paraoxon, using rotational strength analysis. Pérez-Martínez et al. [9] employed ^{1}H NMR spectroscopy (see Chapter 9) to confirm the complexation and study structural aspects of the inclusion complexes formed between the herbicide 2,4-D (2,4-dichlorophenoxyacetic acid) and α- and β-CyDs. Also Consonni et al. [10] have used ^{1}H and ^{13}C NMR spectroscopic techniques to establish the three-dimensional structure of the fungicide imazalil-β-CyD complex. A study of the interaction of 18 pesticides with a water-soluble β-CyD polymer (BCDP) demonstrated that the lipophilicity of the majority of them decreased in relation to the strength of the BCDP–pesticide inclusion complex [11].

In addition to the techniques previously mentioned, a wide variety of methods has been used to characterize the new inclusion compounds in solution and in the solid state, in both directly practice-oriented and theoretical papers, to elucidate the relationship between the relative strength of interaction and some surface parameters of the guest molecules. Complexes obtained in solution are frequently studied by phase-solubility, to obtain the stoichiometric ratio for the complex and an apparent stability constant [12–14], but spectral studies including UV, infrared, fluorescence, and NMR spectroscopy (see Section 10.3 and Chapter 9) can also be used for characterization [6, 15–17]. Inclusion compounds obtained in the solid state are frequently characterized using infrared spectroscopy, X-ray diffraction (Chapter 7), scanning electron microscopy techniques [18, 19] (Section 10.6), differential scanning calorimetry (DSC) (Chapter 8) [20, 21], and/or fluorescence (Section 10.3) and voltammetric measurements (Section 10.5) [16, 22].

However, a large number of papers present research in which complexation with CyD is studied using pesticides that really present problems, from both an agricultural and/or an environmental point of view with the aim of obtaining certain benefits, although the majority of these papers do not address the experimental application of the complexes obtained. Most pesticide–CyD complexes have been prepared to improve their solubility in water. Owing to the importance of solubility in foliar translocation and the penetration of systemic herbicides and fungicides, Manolikar and Sawant [13] prepared and characterized β-CyD complexes of the herbicide isoproturon. Saikosin et al. [19] prepared inclusion complexes with the insecticide carbaryl by kneading and freeze-drying processing methods to obtain formulations with lower toxicological effects. They obtained an 18.4-fold solubility increase when the insecticide was complexed to methyl-β-CyD, showing a lower toxicity than commercial carbaryl. Owing to the wide use of the herbicides diuron and isoproturon, considered as important environmental pollutants, Dupuy et al. [23] have prepared and characterized their inclusion complexes with β-CyD in solution and in the solid state.

2,4-D is also one of the most widely used herbicides and causes contamination of soils and waters owing to its physicochemical properties and persistence. For this reason, some 2,4-D-CyD complexes has been prepared to provide different properties. Although only a slight increase of the herbicide's solubility was obtained after its complexation to β-CyD [12], the removal of herbicide previously adsorbed on the soil was improved by the use of β-CyD solutions [24]. Morillo et al. [25] observed the complete leaching of 2,4-D by the application of β-CyD solution to soil columns in which this herbicide was previously adsorbed. However, preliminary application of CyD to the soil retarded leaching of the herbicide through the soil column, probably because the 2,4-D was adsorbed on the soil though β-CyD previously adsorbed. The complexation of 2,4-D with α-, HP-β-, and PM-β-CyDs has been also studied [18, 20, 26], and 3-, 9- and 17-fold increases of herbicide solubility were found, respectively, relative to the solubility of the uncomplexed compound.

To achieve good biological activity, fungicides must be used dissolved in water, which, in most cases, involves the preparation of ionized derivatives, and, if this is not feasible, the use of organic solvents. For this reason, Lezcano et al. [16, 17] selected eight fungicides with quite low water solubilities to increase their solubility by complexation with CyDs, and to study the feasibility of preparing solid fungicide-CyD complexes. However, the only two that showed a marked increase of water solubility when complexed with β-CyD were prochloraz (800%) and benalaxyl (440%). The isolation of solid complexes from these fungicides was also possible.

Some hydrophobic organic pesticides have proven to be easily sorbed by soil, decreasing their effectiveness and making it difficult to remove their residues from soils. Agents such as organic co-solvents and surfactants have been considered for improving the solubility of such pesticides, but both present disadvantages from an environmental point of view. As an alternative, CyDs may have potential for use as solubility-enhancement agents. Luo et al. [27] evaluated the ability of β-, HP-β-, and HE-β-CyD to increase the aqueous solubility of methyl parathion, carbofuran, and pentachlorophenol, observing no significant toxicity effect to nontarget organisms, such as tadpoles.

Norflurazon is another herbicide that presents some agricultural and environmental problems, such as prolonged persistence in some kinds of soils, and significant dissipation in the field due to photodegradation and volatilization. Villaverde et al. [14, 28] have studied the possibility of obtaining inclusion complexes with natural CyDs to obtain formulations that improve the behavior of this herbicide in soils. An increase in norflurazon solubility up to five-fold was obtained with α- and β-CyDs and up to four-fold with γ-CyD. Solid complexes were obtained using different processing methods (kneading, spray drying, and vacuum evaporation). Desorption studies of norflurazon from soils in the presence of α- and γ-CyDs showed that both CyDs greatly increased the removal of norflurazon previously adsorbed, proving their potential use for *in situ* remediation of pesticide-contaminated soils [28].

Concern over the contamination of soils and water has led to the development of formulations that prevent entry of the pesticides into the groundwater while maintaining effective pest control. The release characteristics of guest compounds can be modified by encapsulation of the compounds in CyDs. When complexes are exposed to water, dissolution and release of pesticides occurs. Dailey et al. [29, 30] prepared complexes with selected herbicides (metribuzin, atrazine, alachlor, simazine, and metolachlor) in an attempt to develop formulations that prevent leaching while maintaining effective pest control. β-CyD complexes of atrazine and simazine were prepared only after forcing reaction conditions, but they are impervious to dissociation, owing to their high stability. On the contrary, a metribuzin-β-CyD formulation controlled selected weed species in a greenhouse efficacy study.

Dailey et al. [15] also prepared and characterized β-CyD complexes of the insecticides aldicarb and sulprofos. Groundwater contamination by aldicarb is of particular concern owing to its acute toxicity to mammals; longer residual effects of sulprofos and a reduction in its phytotoxicity and operator hazards were the benefits expected from its complexation with CyDs. Formulations of the β-CyD complex of sulprofos were tested for toxicity against the tobacco budworm on cotton [31]. Although none of the new formulations was as efficacious as the commercial formulation, the addition of additives such as Airvol 205 increased their toxicity against budworm.

Pesticides can be complexed with CyD to reduce their volatility. The interaction of the guest with the CyD produces a higher energy barrier to volatilization. Szente and Szejtli [32] prepared the inclusion complex of the volatile insecticide DDVP with β-CyD obtaining a crystalline substance with a much more persistent contact effect than free DDVP.

Controlled-release solid formulations of selected volatile organophosphorus pesticides (malathion, DDVP, sumithion, chlorpyriphos, and sulprofos) were studied by Szente [33]. These solid formulations exhibited negligible vapor pressure and preserved their entrapped pesticide content even at elevated temperature. Malathion and chlorpyriphos formulations showed increased physical stability, and resulted in an effective masking of the unpleasant smell while the complex formulations existed as dry solid. Sulfluramid is an expensive insecticide that is lost by volatilization, but complexation to β-CyD reduced the loss [21].

When a pesticide is complexed with CyD, the interaction of the side groups on the guest pesticide molecule with the hydroxyl or substituted hydroxyl groups of the host can have an effect on the reactivity of the guest molecule. Depending upon the group on the pesticide molecule, this can result in catalysis of the reaction or stabilization of the guest molecule by prevention of chemical reactions [34]. Studies of the catalytic effects of CyDs on the degradation of pesticides are important for an understanding of their persistence and fate in natural environments, in order to promote the degradation of such hazardous pollutants. Kamiya et al. [35–37] studied the effect of natural and methylated CyDs on the hydrolysis rate of some pesticides. They observed the double merit of β-CyD, which both stabilized the labile parathion, methyl parathion, and fenitrothion and at the same time accelerated the alkaline hydrolysis of paraoxon, which is much more toxic and is produced by the oxidation of parathion in natural environments. This behavior was explained in terms of the geometry of the inclusion complexes which determine the degree of proximity between the pesticide reaction site and the CyD catalytic site. This work also revealed that methylation of the catalytically active secondary hydroxyl groups of the CyD hosts altered the stabilizing effect.

Ishiwata and Kamiya [38] observed the promotive inclusion-catalytic effect of α-, β-, and γ-CyDs on the degradation of eight organophosphorus pesticides in neutral aqueous media. Pesticide degradations were particularly accelerated in diazinon-α-CyD and chloropyrifos-β-CyD.

Pospisil et al. studied the reduction mechanism of the pesticide vinclozoline in a nonaqueous environment in the presence of β-CyD [39], which changed it from the predominant formation of 3,5-dichloroaniline to the formation of dechlorinated products. These authors had previously demonstrated that the redox properties of the fungicide difenzoquat were also affected by stronger complexation of its reduction product in the presence of β-CyD [7]. In the case of atrazine, which forms complexes with all three natural CyDs, its inclusion into the host cavity enables its reduction even in the non-protonized form [40].

Similarly to CyD complexes with drugs, these hosts can also be used to stabilize unstable pesticide molecules to prevent their interaction and reaction with other molecules, ions, or radicals, since the pesticide is isolated from them. Association with the cavity, or with the hydroxyl groups surrounding the cavity, can also stabilize the guest in less reactive forms. Some pesticides exhibit an increase or decrease in the intensity of light absorption when included in the CyD cavity. Kamiya et al. [35, 41, 42] studied the inclusion effects of α-, β-, γ-, hexa-2,6-dimethyl α- and hepta-2,6-dimethyl-β-CyDs on the photodegradation rates of the pesticide parathion and its oxidation product paraoxon. Some of the CyDs promoted the photodegradation of the pesticides, while the others inhibited it. These different effects were related to the inclusion depth parameters of the pesticides into their cavity. The inhibition effect may be caused by too deep inclusion of the phosphorus atom, the reaction centre of the pesticides, which prevent it from interacting with the catalytic sites of the host cavity (the secondary hydroxyl groups) and the water or oxygen molecules from the solvent.

On the other hand, Kamiya et al. [43] observed an increased photodegradation of eleven organophosphorus pesticides in humic water after complexation to α-, β-, and γ-CyDs. They concluded that the promotion effects of CyDs towards photodegradation were attributable to their inclusion-trapping abilities, which are active not only to the pesticides, but also to photoradical species, such as superoxide and hydroxyl radicals, discharged by the photosensitization action of humic acids.

An enhancement of biological activity of some pesticides can also take place as a consequence of the synergistic activity of CyDs, as shown in the case of benomyl [44]. The effectiveness of isoxaben, a benzamide herbicide, is enhanced by CyDs, the growth of *Amarantus retroflexus* and *Solanum nigrum* being inhibited more than by isoxaben alone [45]. Glyphosate formulations with these oligosaccharides enhance plant penetration of the herbicide [46], and aqueous formulations are storage stable and easily diluted with water.

Another advantage of pesticides-CyD complexes is the masking of undesirable effects of the guest molecule. The irritating or toxic effects of some pesticides can be reduced or eliminated. When a guest is included in a molecule of CyD, it is isolated and prevented from coming into contact with body surfaces where it could cause unwanted side effects such as irritation. Loukas et al. [47] prepared γ-CyD complexes of the insecticide DCPE and observed a decrease of its acute oral toxicity, indicating that a safer product for the applicator had been prepared.

The application of CyDs to obtain new formulations of pesticides leads to the presence of CyDs in soils, but little is so far known about their effects on soil physicochemical properties. Jozefaciuk et al. [48] studied the effect of applications of a randomly methylated β-CyD (RAMEB) on three typical clay minerals present in soils, bentonite, illite, and kaolinite, selected as a rational base for understanding soil processes. Their findings suggested that the surface properties and pore structure of minerals changed dramatically when CyDs were introduced, leading to a decrease in surface area, an increase in adsorption energy, and decreases in the volumes of micro and mesopores. These authors also observed a great modification of physical properties in some selected soils after RAMEB addition, such as increasing water adsorption and surface area in sandy soils and their decreasing in clay soils, and an increasing proportion of coarse-size soil fractions owing to aggregation of smaller particles [49]. However, it has not been determined if certain surface and pore properties return to the original state after the removal of the CyD, either by leaching or by biodegradation, or if the phenomena observed for RAMEB are generally true for other CyDs. Probably due in part to these changes in soil properties, crop yield increased when β-CyD was used, although the germination process and development of shoots and roots was initially retarded strongly, allowing protection against the influence of herbicides administrated during the sowing [50].

On the other hand, the inclusion effects of CyDs on certain pesticides could also be influenced by the presence of different soil components. Ishiwata and Kamiya [51] studied the concentration effects of humic acids on the inclusion of organophosphorus pesticides (parathion, methyl parathion, paraoxon) by CyDs. They found that humic acids exert characteristic inhibition effects on the pesticide inclusion to CyDs, which correlate with the inclusion-depth parameters of the pesticide-CyD complexes. These effects can be explained in terms of the complexation and thus solubilization functions associated with humic materials, and may be considered as one of the potential factors to affect the inclusion functions of CyDs applied to control hazardous pollutants in water–soil environments rich in humic substances.

In spite of the relatively large number of papers published on the preparation, characterization, and properties of a wide variety of pesticide-CyD complexes, very few patents exist relating to their real application in the development of pesticide formulations. Most of them are Japanese, although German and Chinese patents can be also found. Some of them are related to the increasing stability of the new pesticide formulations [52–55]. The acaricide amitraz was stabilized by conversion into an inclusion complex with β-CyD [56], and microcapsules of the pesticide fenitrothion, among others, were prepared using a water-soluble coating material (α-CyD) to obtain stabilized microcapsules that could be easily handled [57].

Some patents refer to the controlled release of certain pesticides [58, 59]. Matolcsi et al. [60] used β-CyD to prepare inclusion compounds with benzenesulfonylurea derivatives with herbicidal or plant growth regulator properties [61], obtaining a prolonged controlled release of the active ingredient. Ikeuchi et al. [62] prepared inclusion complexes of triazole derivatives which presented potent insecticidal activities at low concentrations for an extended period.

Pesticidal compositions containing CyDs to increase their solubility have also been patented [63]. Azadirachtin, a biopesticide of plant origin, highly unstable in aqueous media and extremely sensitive to sunlight, has been formulated using β- and β-methyl-CyD to enhance its solubility [64], avoiding the use of surfactants and organic solvents, which do not sufficiently enhance the shelf life of the formulations.

There are some other patents in which the use of CyD gives an increased efficacy of the pesticide [46, 65, 66]. Xiao and Wang [67] use CyD as a synergist to prepare floating-type agrochemicals for rice fields. Enhancement of the activity of benzamide herbicides has been obtained with CyDs [45]. The association of isoxaben with at least one CyD improved its mobility in the ground and its biological efficacy against dicotyledons in maize and winter cereals and in ornamental trees, especially conifers. Aven et al. [68] have observed a greater degree of enhancement of the herbicidal efficacy of certain herbicides by addition of a larger content of α-, β-, and/or γ-CyDs in the formulation's solid carrier. The addition of CyDs can reduce the recommended amount of active ingredient per hectare, so that additional weeds can be controlled.

Conclusions

This section amply demonstrates the possibility of preparing cyclodextrin inclusion complexes with a wide variety of pesticides, that possess particularly advantageous properties for some specific applications. These complexes would improve the behavior of the pesticides in comparison to the current commercial formulations, but, according to Szejtli [69], the pesticide industry is very raw-material price sensitive, and it seems that up to now the price of even the cheapest technical-quality β-CyD is too high to allow the commercialization of these pesticide formulations.

Although most of the pesticides-CyD inclusion complexes studied have used β-CyD because of its lower price, there are also advantages to other CyDs, depending on a proper fit of the pesticide molecule into the hydrophobic CyD cavity. Although most of the published papers related to pesticides are not directly practice-oriented, a wide variety of CyD derivatives have been studied as hosts for pesticides at the laboratory scale. For the same reason, many different processing methods, some of them highly sophisticated, have been used to prepare pesticides-CyD inclusion complexes, even though there was no possibility of using them to prepare pesticide formulations on an industrial scale.

References

1 J. Szejtli, *Starch/Staerke* 1985, 37, 382–386.

2 L. Szente, J. Szejtli, In *Comprehensive Supramolecular Chemistry*, Vol. 3, Szejtli, J., Osa, T., (Eds.); Elsevier Science: Oxford, 1996; pp 503–514.

3 E.M. Martin del Valle, *Process Biochem.* 2004, 39, 1033–1046.

4 J. Szejtli, *J. Mater. Chem.* 1997, 7, 575–587.

5 H. Viernstein, P. Weiss-Greiler, P. Wolschann, *J. Incl. Phenom. Macro. Chem.* 2002, 44, 235–239.

6 L. Pospisil, M.P. Colombini, *J. Incl. Phenom. Mol.* 1993, 16, 255–266.

7 L. Pospisil, J. Hanzlik, R. Fuoco, M.P. Colombini, *J. Electroanal. Chem.* 1994, 368, 149–154.

8 S. Ishiwata, M. Kamiya, *Chemosphere* 2000, 41, 701–704.

9 J.I. Pérez-Martínez, J.M. Ginés, E. Morillo, J.R. Moyano, *J. Incl. Phenom. Macro.* 2000, 37, 171–178.

10 R. Consonni, T. Recca, M.A. Dettori, D. Fabri, G. Delogu, *J. Agric. Food Chem.* 2004, 52, 1590–1593.

11 T. Cserhati, E. Forgacs, Y. Darwish, G. Oros, Z. Illes, *J. Incl. Phenom. Macrocycl. Chem.* 2002, 42, 235–240.

12 J.M. Gines, J.I. Perez-Martinez, M.J. Arias, J.R. Moyano, E. Morillo, A. Ruiz-Conde, P.J. Sanchez-Soto, *Chemosphere* 1996, 33, 321–334.

13 M.K. Manolikar, M.R. Sawant, *Chemosphere* 2003, 51, 811–816.

14 J. Villaverde, E. Morillo, J.I. Perez-Martinez, J.M. Gines, C. Maqueda, *J. Agric. Food Chem.* 2004, 52, 864–869.

15 O.D. Dailey, J.M. Bland, B.J. Trask-Morrell, *J. Agric. Food Chem.* 1993, 41, 1767–1771.

16 M. Lezcano, W. Al-Soufi, M. Novo, E. Rodriguez-Nuñez, J. Vazquez Tato, *J. Agric. Food Chem.* 2002, 50, 108–112.

17 M. Lezcano, M. Novo, W. Al-Soufi, E. Rodriguez-Nuñez, J. Vazquez Tato, *J. Agric. Food Chem.* 2003, 51, 5036–5040.

18 J.I. Pérez-Martínez, J.M. Ginés, E. Morillo, M.L. Rodríguez, J.R. Moyano, *Environ. Technol.* 2000b, 21, 209–216.

19 R. Saikosin, T. Limpaseni, P. Pongsawasdi, *J. Incl. Phenom. Macro. Chem.* 2002, 44, 191–196.

20 J.I. Perez-Martinez, M.J. Arias, J.M. Gines, J.R. Moyano, E. Morillo, P.J. Sanchez-Soto, C. Novak, *J. Thermal Anal.* 1998, 51, 965–972.

21 R.C. Bergamasco, G.M. Zanin, F.F. De Moraes, *J. Agric. Food Chem.* 53, 1139–1143.

22 M. Hromadová, L. Pospisil, S. Zalis, N. Fanelli, *J. Incl. Phenom. Macro. Chem.* 2002, 373–380.

23 N. Dupuy, S. Marquis, G. Vanhove, M. Bria, J. Kister, L. Vrielynck, *Appl. Spectrosc.* 2004, 58, 711–718.

24 J.I. Pérez-Martínez, E. Morillo, J.M. Ginés, *Chemosphere* 1999, 39, 2047–2056.

25 E. Morillo, J.I. Pérez-Martínez, J.M. Ginés, *Chemosphere* 2001, 44, 1065–1069.

26 J.I. Pérez-Martínez, J.M. Ginés, E. Morillo, M.L. Rodríguez, J.R. Moyano, *Pest Manag. Sci.* 2000, 56, 425–430.

27 Y.C. Luo, Q.R. Zeng, G. Wu, Z.K. Luan, R.B. Yan, B.H. Liao, *Bull. Environ. Contam. Toxicol.* 2003, 70, 998–1005.

28 J. Villaverde, J.I. Pérez-Martínez, C. Maqueda, J.M. Ginés, E. Morillo, *Chemosphere* 2005, 60, 656–664.

29 O.D. Dailey, C.C. Dowler, N.C. Glaze, In: *Pesticide Formulations and Application Systems*; Bode, L.E.,

HAZEN, J.L., CHASIN, D.G., (Eds.); American Society for Testing and Materials, Philadelphia, PA, 1990.
30 O.D. DAILEY, In *Biotechnology of Amylodextrin Oligosaccharides*; FRIEDMAN, R.B., (Ed.); ACS Symposium Series 458. American Chemical Society. Washington DC, 1991.
31 M.A. LATHEEF, O.D. DAILEY, *Southwest Entomol.* 1995, 20, 351–356.
32 L. SZENTE, J. SZEJTLI, *Acta Chim. Acad. Sci. Hung.* 1981, 107, 195–202.
33 L. SZENTE, *J. Thermal Anal.* 1998, 51, 957–963.
34 A.L. HEDGES, *Chem. Rev.* 1998, 98, 2035–2044.
35 M. KAMIYA, K. NAKAMURA, *Pestic. Sci.* 1994, 41, 305–309.
36 M. KAMIYA, S. MITSUHASHI, M. MAKINO, *Chemosphere* 1992, 25, 783–796.
37 M. KAMIYA, K. NAKAMURA, C. SASAKI, *Chemosphere* 1995, 30, 653–660.
38 S. ISHIWATA, M. KAMIYA, *Chemosphere* 1999, 39, 1595–1600.
39 L. POSPISIL, R. SOKOLOVA, M. HROMADOVA, S. GIANNARELI, R. FUOCO, M.P. COLOMBINI, *J. Electroanal. Chem.* 2001, 517, 28–36.
40 L. POSPISIL, R. TRSKOVA, M.P. COLOMBINI, R. FUOCO, *J. Incl. Phenom.* 1998, 31, 57–70.
41 M. KAMIYA, K. NAKAMURA, C. SASAKI, *Chemosphere* 1994, 28, 1961–1966.
42 M. KAMIYA, K. NAKAMURA, *Environ. Intern.* 1995, 21, 299–304.
43 M. KAMIYA, K. KAMEYAMA, S. ISHIWATA, *Chemosphere* 2001, 42, 251–255.
44 J. SZEJTLI, P. TETENYI, M. KINICZKY, J. BERNARTH, M. TETENYI NEE ERDOSI, E. DOBOS, E. BANKY NEE ELOD, Patent No. US 4923853, 1990.
45 S. GOSSET, C. GAUVRIT, Patent No. WO 9222204, 1992.
46 H.W. WOLLENWEBER, A. RATHJENS, H.G. MAINX, Patent No. WO 2002034051, 2002.
47 Y. LOUKAS, E. ANTONIADOU-VYZA, A. PAPADAKI-VALIRAKI, K. MACHERA, *J. Agric. Food Chem.* 1994, 42, 944–948.
48 G. JOZEFACIUK, A. MURANYI, E. FENYVESI, *Environ. Sci. Technol.* 2001, 35, 4947–4952.
49 G. JOZEFACIUK, A. MURANYI, E. FENYVESI, *Environ. Sci. Technol.* 2003, 37, 3012–3017.
50 J. SZEJTLI, *Starch-Staerke* 1983, 35, 433–438.
51 S. ISHIWATA, M. KAMIYA, *Chemosphere* 1999, 38, 2219–2226.
52 Y. GOTO, M. SAWAMURA, T. OKAUCHI, Patent No. JP 5065202, 1993.
53 M. KOIKE, M. SAWAMURA, Patent No. JP 07291803, 1995.
54 M. KOIKE, M. SAWAMURA, K. AKASHI, Patent No. JP 08225404, 1996.
55 M. KAWASHIMA, M. IMAI, Patent No. JP 08113504, 1996.
56 G. KULCSAR, L. SZENTE, A. UJHAZY, J. SZEJTLY, J. SZEMAN, Patent No. DE 3908687, 1989.
57 K. AKASHI, T. TANABAYASHI, K. KITAGAWA, Patent No. JP 5238904, 1993.
58 K. AKASHI, Y. EBISAWA, Patent No. WO 9626719, 1996.
59 J. LI, Patent No. CN 1168761, 1997.
60 G. MATOLCSY, A. GIMESI, K. PELEJTEI, J. SZIAISZ, Patent No. CN 85104674, 1986.
61 F. TSORTEKI, K. BETHANIS, D. MENTZAFOS, *Carbohydr. Res.* 2004, 339, 233.
62 T. IKEUCHI, J. MISUMI, M. GOTO, K. ADACHI, J. NAKANO, Patent No. JP 05331012, 1993.
63 E. NAKAMURA, A. AZUMA, M. FUKADA, Patent No. JP 63079802, 1988.
64 R. SUBBA, V. PILLARISETTI, S.P. KUMBLE, R.S. ANNADURAI, M. SRINIVAS, A.S. RAO, C.S. RAMADOSS, Patent No. WO 2000054596, 2000.
65 G. XIAO, Y. NA, K. FENG, Patent No. CN 1180476, 1998.
66 G. WULFF, A. STEINERT, W. ANDERSCH, K. STENZEL, J. HOELTERS, U. PRIESNITZ, Patent No. DE 19751631, 1999.
67 G. XIAO, R. WANG, Patent No. CN 1252218, 2000.
68 M. AVEN, A. BRANDT, N. NELGEN, Patent No. WO 2001097613, 2001.
69 J. SZEJTLI, *Chem. Rev.* 1998, 98, 1743–1753.

16.4
Cyclodextrins in Molecular Devices

Renata Bilewicz and Kazimierz Chmurski

16.4.1
Introduction

One of the focuses in CyD applications is their use as components of devices in systems with nontrivial topological properties, i.e. catenanes, rotaxanes [1–8], and polyrotaxanes [1, 9, 10] presented in some detail in Chapter 12. For a recent review of interlocked assemblies see Ref. [11]. This is a part of novel "bottom up" approach to the construction of devices on the basis of one molecule or one molecular aggregate [12].

Stoddart et al. [13–14] describe the design, preparation, and testing of electronic devices based on molecular switches and show in the form of a scheme (Fig. 16.4.1) the molecular and device properties that need to be satisfied. The numerous demands articulated in this scheme demonstrate how many fundamental and practical problems would have to be solved before such devices could be commercialized.

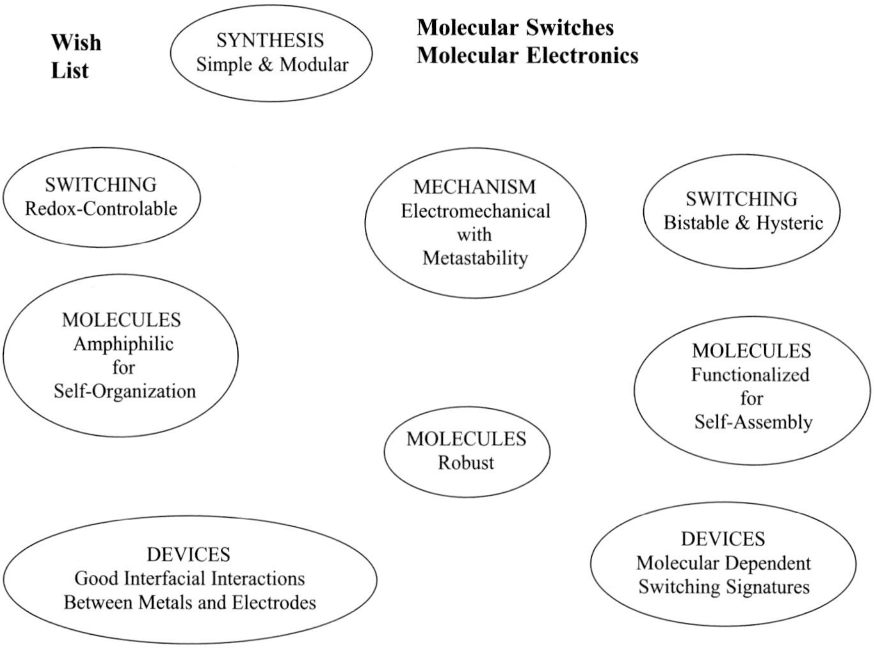

Fig. 16.4.1. Stoddart's scheme for the design, preparation, and testing electronic devices. Adapted from Ref. [13].

Easton et al. have discussed the applications, potential, and limitations of molecular machines and reactors [15]. They define molecular reactors as miniature vessels for the assembly of reactants at the molecular level, in order to change the nature of chemical transformations. Molecular machines consist of interrelated parts with separate functions, and perform some kind of work, at the molecular level. However, practical examples are not based on individual functions of single-molecule devices.

The signal-triggered functions of these molecular assemblies have to be first characterized in bulk solution. Then, extensive efforts have been directed to integrate these photoswitchable chemical assemblies with transducers in order to tailor switchable molecular devices. The redox properties of photoisomerizable monolayers assembled on an electrode surface are employed for controlling interfacial electron transfer [16]. Specifically, electrical transduction of photonic information recorded by photosensitive monolayers on electrode supports can be used in developing monolayer optoelectronic systems [16–19]. Electrodes with receptor sites exhibiting controlled binding of photoisomerizable redox-active substrates from the solution [20] also allow the construction of molecular optoelectronic devices.

16.4.2 Electrochemically Controlled Molecular Machines

The investigation of self-assembled molecular machines based on CyDs with electrochemical control of the switching process is a challenging research area [13, 21–23] because they are foreseen or already applied as working units [15] in electronic devices e.g. logic gates in computer technology, sensing devices, and bioengineering [24–28].

A tailored molecular machine requires integration of the chemical component – a molecular assembly with a transducer, e.g. an electrode, by which it communicates with the environment. The molecular machine, molecular switch, or shuttle can be switched on or off in response to external stimuli, which are often electrical or photonic signals [17, 29–31].

Switchable binding, the formation of a donor–acceptor complex, intramolecular translocation, or a change of redox state are the triggered functions characterized by electroanalytical or spectroscopic techniques [31–34].

Interlocked molecular compounds consisting of a π-electron-rich component and a π-electron-deficient unit can be assembled to form nanoscale molecular switches influenced by external electrochemical stimuli. Materials composed of CyDs include polyrotaxanes [34–36].

Intermolecular interaction between the axle and ring of pseudorotaxane molecules (Fig. 16.4.2) involving $N \cdots H{-}O$ hydrogen bonds with the *cis* but not the *trans* azobenzene group was found to retard photoinduced *cis* to *trans* isomerization [36, 37].

Haider and Pikramenou [38] focus on strategies involving metallocyclodextrins for the construction of supramolecular arrays with light-activated functions. The

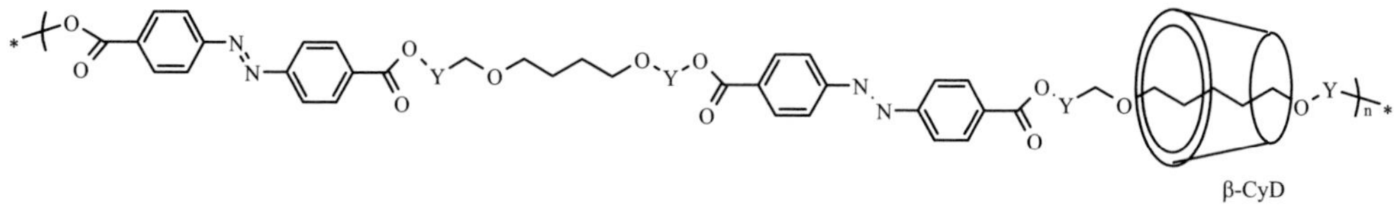

Fig. 16.4.2. Yamaguchi photoisomerizable pseudorotaxane [37].

introduction of photoactive metal centers onto cyclodextrin receptors opens up new possibilities for the design of sensors, wires, and energy conversion systems. The authors present assembly procedures for building such arrays, together with the features required for them to function both as sensors for ion or small-molecule detection and as wires for photoinduced long-range energy or electron transport.

16.4.3 Applications of CyDs in Bioelectrocatalysis

An important application of CyD complexes is in the field of mediated electrocatalysis and bioelectrocatalysis. Detection of a target analyte using a biosensor based on a redox enzyme is well recognized as a more convenient solution than one based on the electrochemistry of reaction products. An ideal mediator for electrocatalysis should be soluble in water or easily anchored on an electrode surface. CyDs have often been employed in solubilizing hydrophobic molecules used either as mediators in catalysis or occurring as the products of catalytic reactions. In the latter case, their role was to avoid fouling of the electrode surface.

Since the first report of Siegel and Breslow [39], the inclusion of ferrocene derivatives into the hydrophobic cavity of β-CyD has been shown repeatedly to result in changes of their electrochemical properties and an increase of their aqueous solubility. Unfortunately, ferrocenes incorporated into β-CyD lead rather to the inhibition of the mediation of, for example, glucose oxidase (GOx) catalytic processes than to their efficient mediation [40–44].

On the other hand, a half-sandwich complex of manganese with β-CyD $[(\eta\text{-}MeC_5H_4)Mn(NO)(S_2CNMe_2)]$ (Fig. 16.4.3) was found to be an efficient mediator in enzyme-based sensors including those useful for blood glucose sensing such as those involving GOx and horseradish oxidase [44].

Tetrathiafulvalene (TTF), for example, is a versatile mediator, but it is not soluble in water. A simple method for its solubilization is to bind it in the CyD cavity. Hydroxypropyl-β-CyD (HP-β-CyD) is even more soluble than the nonsubstituted CyD and was chosen by Schmidt et al. [45] to bind TTF with the aim of providing direct electrical communication between enzyme, GOx, and the electrode. Without it, the redox center of the enzyme is inaccessible and no redox chemistry can be observed. The cyclic voltammograms for TTF-HP-β-CyD in the presence of glucose, show a large anodic peak without a reduction counterpart (Fig. 16.4.4) indicating bioelec-

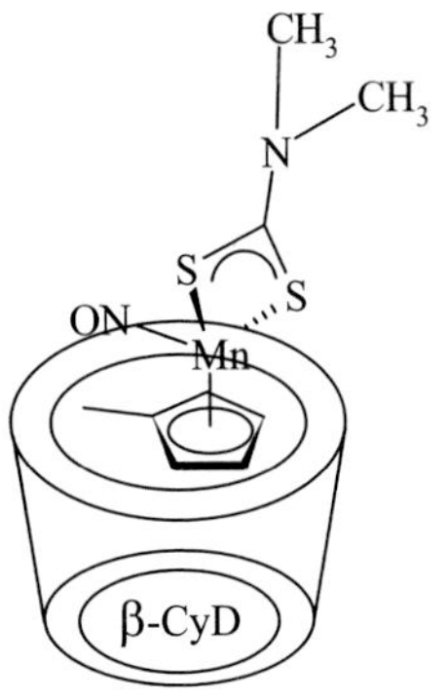

Fig. 16.4.3. Mn half-sandwich complex used as a mediator for glucose sensing [44].

trocatalysis. The ratio of this current to the diffusion-limited current allows determination of the second-order homogeneous rate constant of the process in the presence of the inclusion complex as mediator. Reaction with GOx reflects the efficiency of the inclusion complex as mediator.

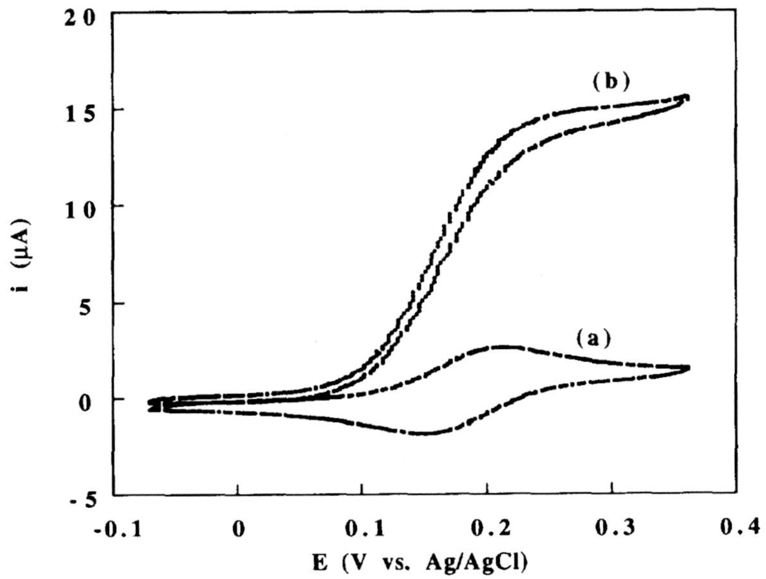

Fig. 16.4.4. Cyclic voltammograms at 15 mV s^{-1} for 1 mol m^{-3} TTF in 4.5 mol m^{-3} HP-β-CyD showing: (a) diffusion-limited current in 50 mol m^{-3} glucose solution; (b) catalytic current in 50 mol m^{-3} glucose solution with 12×10^{-3} mol m^{-3} GOx [45]. Reprinted with permission of Elsevier.

References

1 S. A. Nepogodiev, J. F. Stoddart, *Chem. Rev.*, 1998, *98*, 1959–1976.
2 H. Murakami, A. Kawabuchi, K. Kotoo, M. Kunitake, N. Nakashima, *J. Am. Chem. Soc.*, 1997, *119*, 7605–7606.
3 M. Kunitake, K. Kotoo, O. Manabe, T. Muramatsu, N. Nakashima, *Chem. Lett.*, 1993, 1033–1036.
4 A. Ohira, T. Ishizaki, M. Sakata, I. Taniguchi, C. Hirayama, M. Kunitake, *Colloids Surfaces*, 2000, *169*, 27–33.
5 B. L. Feringa, (Ed.) In *Molecular Switches*, Wiley, Weinheim, 2001. Chs. 5, 7.
6 H. Onagi, C. J. Blake, C. J. Easton, S. F. Lincoln, *Chem. Eur. J.*, 2003, *9*, 5978–5988.
7 N. Kihara, M. Hashimoto, T. Takata, *Org. Lett.*, 2004, *6*, 1693–1696.
8 C. A. Stanier, S. J. Alderman, T. D. W. Claridge, H. L. Anderson, *Angew. Chem. Int. Ed.*, 2002, *41*, 1769–1772.
9 A. Harada, J. Ji, M. Kamachi, *Nature*, (London) 1992, *356*, 325–327.
10 I. G. Panova, I. N. Topchieva, *Usp. Khim.*, 2001, *70*, 28–51.
11 Y. K. Agrawal, C. R. Sharma, *Rev. Anal. Chem.*, 2005, *24*, 35–74.
12 H. Dodziuk, *Introduction to Supramolecular Chemistry*, Kluwer, Dordrecht, 2002, Ch. 6.
13 A. H. Flood, R. J. A. Ramirez, W.-Q. Deng, R. P. Muller, W. A. Goddard III, J. F. Stoddart, *Aust. J. Chem.*, 2004, *57*, 301–322.
14 H. R. Tseng, D. M. Wu, N. X. L. Fang, X. Zhang, J. F. Stoddart, *Chem. Phys. Chem*, 2004, *5*, 111–116.
15 C. J. Easton, S. F. Lincoln, L. Barr, H. Onagi, *Chem. Eur. J.*, 2004, *10*, 3120–3128.
16 A. Doron, E. Katz, G. Tao, I. Willner, *Langmuir*, 1997, *13*, 1783–1790.
17 I. Willner, V. Pardo-Yissar, E. Katz, K. T. Ranjit, *J. Electroanal. Chem.*, 2001, *497*, 172–177.
18 Z. Liu, K. Hashimoto, A. Fujishima, *Nature*, 1990, *347*, 658–660.
19 A. Doron, M. Portnoy, M. Lion-Dagan, E. Katz, I. Willner, *J. Am. Chem. Soc.*, 1996, *118*, 8937–8944.
20 K. T. Ranjit, S. Marx-Tibbon, I. Ben-Dov, I. Willner, *Angew. Chem. Int. Ed. Engl.*, 1997, *36*, 147–150.
21 V. Balzani, M. Gomez-Lopez, J. F. Stoddart, *Acc. Chem. Res.* 1998, *31*, 405–414.
22 P. R. Ashton, V. Balzani, O. Kocian, L. Prodi, N. Spencer, J. F. Stoddart, *J. Am. Chem. Soc.* 1998, *120*, 11190–11191.
23 V. Balzani, A. Credi, F. M. Raymo, J. F. Stoddart, *Angew. Chem.*, 2000, *39*, 3348–3391.
24 K. D. Schierbaum, T. Weiss, E. U. T. van Velzen, E. F. J. Engkersen, D. A. Reinhoudt, W. Göpel, *Science*, 1994, *265*, 1413–1415.
25 F. L. Carter, R. E. Siatkowski, H. Wojten, (Eds.) *Molecular Electronics Devices*, Elsevier, Amsterdam, 1988.
26 P. A. de Silva, N. H. Q. Gunaratne, C. P. McCoy, *Nature*, 1993, *364*, 42–44.
27 K. E. Dexler (Ed.) *Nanosystems, Molecular Machinery, Manufacturing and Computation*, J. Wiley & Sons, New York, 1992.
28 I. Willner, *Acc. Chem. Res.* 1997, *30*, 347–356.
29 S. H. Kawai, S. L. Gilat, R. Ponsinet, J.-M. Lehn, *Chem. Eur. J.*, 1995, *1*, 285–293.
30 R. Blonder, E. Katz, I. Willner, V. Wray, A. F. Buckmann, *J. Am. Chem. Soc.*, 1997, *119*, 11747–11757.
31 I. Willner, S. Marx, Y. Eichen, *Angew. Chem. Int. Ed. Engl.*, 1992, *31*, 1243–1244.
32 F. Würthner, J. Rebek Jr., *Angew. Chem. Int. Ed. Engl.*, 1995, *34*, 446–448.
33 S. Shinkai, T. Minami, Y. Kusano, O. Manabe, *J. Am. Chem. Soc.*, 1982, *104*, 1967–1972.
34 A. Harada, *Acta Polymerica*, 1998, *49*, 3–17.
35 H. Murakami, A. Kawabuchi, K.

Kotoo, M. Kunitake, N. Nakashima, *J. Am. Chem. Soc.*, 1997, *119*, 7605–7606.
36 I. Yamaguchi, K. Osakada, T. Yamamoto, *J. Am. Chem. Soc.*, 1996, *118*, 1811–1812.
37 I. Yamaguchi, K. Osakada, T. Yamamoto, *Chem. Commun.*, 2000, *14*, 1335–1336.
38 J. M. Haider, Z. Pikramenou, *Chem. Soc. Rev.*, 2005, *34*, 120–132.
39 B. Siegel, R. Breslow, *J. Am. Chem. Soc.*, 1975, *97*, 6869–6870.
40 A. D. Ryabow, E. M. Tyapochlein, S. D. Varfolomeev, A. Karyakin, *Bioelectrochem. Bioenergy*, 1990, *24*, 257–262.
41 R. Isnin, C. Salam, A. E. Kaifer, *J. Org. Chem.*, 1991, *56*, 35–41.
42 P. M. Bersier, J. Bersier, B. Klingert, *Electroanalysis*, 1991, *3*, 443–455.
43 C. A. Groom, J. H. T. Luong, R. Thatipalmala, *Anal. Biochem.*, 1995, *231*, 393–399.
44 N. J. Forrow, S. J. Walters, *Biosens. Bioelectron.*, 2004, *19*, 763–770.
45 P. M. Schmidt, R. S. Brown, J. H. T. Luong, *Chem. Eng. Sci.*, 1995, *50*, 1867–1876.

Index

Cyclodextrins and Their Complexes. Edited by Helena Dodziuk

ISBN: 3-527-31280-3

d

e

k

l

m

n

o

p